T0325826

WEAK SCALE SUPERSYMMETRY
From Superfields to Scattering Events

Supersymmetry has been proposed as a new symmetry of nature. Supersymmetric models of particle physics predict new superpartner matter states for each known particle in the Standard Model. The existence of such superpartners will have wide-ranging implications, from the early history of the Universe to what is observed at high energy accelerators such as CERN's LHC.

In this text, the authors develop the concepts of supersymmetry from first principles, and show how it can be incorporated into a theoretical framework for describing unified theories of elementary particles. They develop the technical tools of supersymmetry using four-component spinor notation familiar to high energy experimentalists and phenomenologists. The text takes the reader from an abstract formalism to a straightforward recipe for writing supersymmetric gauge theories of particle physics, and ultimately to the calculation of cross sections and decay rates necessary for practical applications to experiments both at colliders and for cosmology.

This advanced text is a comprehensive, practical, and accessible introduction to supersymmetry for experimental and phenomenological particle physicists and graduates studying supersymmetry. Some familiarity with the Standard Model and tree-level calculations in quantum field theory is required. Exercises and worked examples that supplement and clarify the material are interspersed throughout.

HOWARD BAER is currently the J. D. Kimel Professor of Physics at Florida State University. He received his Ph.D. in theoretical physics from the University of Wisconsin in 1984 and has held postdoctoral appointments at the European Laboratory for Particle Physics (CERN), Argonne National Laboratory, and Florida State University. Dr. Baer is a Fellow of the American Physical Society. He has published over a hundred articles on the physics of elementary particles.

XERXES TATA is a professor in the Department of Physics and Astronomy, University of Hawaii. He received his Ph.D. in theoretical high energy physics from the University of Texas at Austin. He has done postdoctoral work at the University of Oregon, the European Laboratory for Particle Physics (CERN), and the University of Wisconsin. He was also a visiting scientist at the High Energy Accelerator Research Organization (KEK), Japan. He joined the University of Hawaii in 1988. Dr. Tata has published over a hundred research articles in high energy physics, and is a Fellow of the American Physical Society.

WEAK SCALE SUPERSYMMETRY
From Superfields to Scattering Events

HOWARD BAER
Florida State University

XERXES TATA
University of Hawaii

CAMBRIDGE
UNIVERSITY PRESS

CAMBRIDGE
UNIVERSITY PRESS

University Printing House, Cambridge CB2 8BS, United Kingdom

One Liberty Plaza, 20th Floor, New York, NY 10006, USA

477 Williamstown Road, Port Melbourne, VIC 3207, Australia

314-321, 3rd Floor, Plot 3, Splendor Forum, Jasola District Centre, New Delhi - 110025, India

79 Anson Road, #06-04/06, Singapore 079906

Cambridge University Press is part of the University of Cambridge.

It furthers the University's mission by disseminating knowledge in the pursuit of education, learning and research at the highest international levels of excellence.

www.cambridge.org
Information on this title: www.cambridge.org/9780521857864

First published 2006

A catalogue record for this publication is available from the British Library

ISBN 978-0-521-85786-4 Hardback
ISBN 978-0-521-29031-9 Paperback

To Adrienne and Kalpana, Madeleine, Kashmira, and Jake

Contents

Preface		*page* xiv
1	**The Standard Model**	**1**
	1.1 Gauge invariance	1
	1.2 Spontaneous symmetry breaking	3
	1.3 Brief review of the Standard Model	6
	1.3.1 QCD	6
	1.3.2 The electroweak model	6
2	**What lies beyond the Standard Model?**	**11**
	2.1 Scalar fields and quadratic divergences	12
	2.2 Why is the TeV scale special?	14
	2.3 What could the New Physics be?	16
3	**The Wess–Zumino model**	**23**
	3.1 The Wess–Zumino Lagrangian	23
	3.1.1 The field content	23
	3.1.2 SUSY transformations and invariance of the action	25
	3.1.3 The chiral multiplet	28
	3.1.4 Algebra of the SUSY charges	29
	3.2 Quantization of the WZ model	32
	3.3 Interactions in the WZ model	35
	3.4 Cancellation of quadratic divergences	36
	3.5 Soft supersymmetry breaking	39
4	**The supersymmetry algebra**	**41**
	4.1 Rotations	41
	4.2 The Lorentz group	42
	4.3 The Poincaré group	44
	4.4 The supersymmetry algebra	46
5	**Superfield formalism**	**49**
	5.1 Superfields	49

5.2	Representations of symmetry generators: a recap	52
5.3	Representation of SUSY generators as differential operators	54
5.4	Useful θ identities	56
5.5	SUSY transformations of superfields	58
5.6	Irreducible SUSY multiplets	60
	5.6.1 Left-chiral scalar superfields	61
	5.6.2 Right-chiral scalar superfields	63
	5.6.3 The curl superfield	64
5.7	Products of superfields	64
5.8	Supercovariant derivatives	65
5.9	Lagrangians for chiral scalar superfields	68
	5.9.1 Kähler potential contributions to the Lagrangian density	70
	5.9.2 Superpotential contributions to the Lagrangian density	72
	5.9.3 A technical aside	74
	5.9.4 A master Lagrangian for chiral scalar superfields	75
5.10	The action as an integral over superspace	76
6	**Supersymmetric gauge theories**	**79**
6.1	Gauge transformations of superfields	79
6.2	The Wess–Zumino gauge	84
	6.2.1 Abelian gauge transformations	84
	6.2.2 Non-Abelian gauge transformations	86
6.3	The curl superfield in the Wess–Zumino gauge	89
6.4	Construction of gauge kinetic terms	92
6.5	Coupling chiral scalar to gauge superfields	95
	6.5.1 Fayet–Iliopoulos D-term	98
6.6	A master Lagrangian for SUSY gauge theories	98
6.7	The non-renormalization theorem	104
7	**Supersymmetry breaking**	**105**
7.1	SUSY breaking by elementary fields	106
7.2	F-type SUSY breaking: the O'Raifeartaigh model	107
	7.2.1 Mass spectrum: Case A	109
	7.2.2 Mass spectrum: Case B	111
7.3	D-type SUSY breaking	113
	7.3.1 Case A	114
	7.3.2 Case B	114
7.4	Composite goldstinos	115
7.5	Gaugino condensation	116
7.6	Goldstino interactions	117
7.7	A mass sum rule	118
	7.7.1 Scalar contributions	118

	7.7.2 Vector contributions	119
	7.7.3 Fermion contributions	119
7.8	Explicit supersymmetry breaking	121
7.9	A technical aside: γ_5-dependent fermion mass matrices	124
8	**The Minimal Supersymmetric Standard Model**	**127**
8.1	Constructing the MSSM	127
	8.1.1 Parameter space of the MSSM	134
	8.1.2 A simplified parameter space	136
8.2	Electroweak symmetry breaking	138
8.3	Particle masses in the MSSM	141
	8.3.1 Gauge bosons	141
	8.3.2 Matter fermions	142
	8.3.3 Higgs bosons	144
	8.3.4 Gluinos	148
	8.3.5 Charginos and neutralinos	149
	8.3.6 Squarks and sleptons	155
8.4	Interactions in the MSSM	161
	8.4.1 QCD interactions in the MSSM	161
	8.4.2 Electroweak interactions in the MSSM	164
	8.4.3 Interactions of MSSM Higgs bosons	174
8.5	Radiative corrections	184
	8.5.1 Higgs boson masses	184
	8.5.2 Squark mass	187
	8.5.3 Chargino and neutralino masses	187
	8.5.4 Yukawa couplings and SM fermion masses	188
8.6	Should the goldstino be part of the MSSM?	188
9	**Implications of the MSSM**	**190**
9.1	Low energy constraints on the MSSM	191
	9.1.1 The SUSY flavor problem	191
	9.1.2 The SUSY CP violation problem	195
	9.1.3 Large CP-violating parameters in the MSSM?	196
9.2	Renormalization group equations	199
	9.2.1 Gauge couplings and unification	199
	9.2.2 Evolution of soft SUSY breaking parameters	204
	9.2.3 Radiative breaking of electroweak symmetry	209
	9.2.4 Naturalness constraint on superparticle masses	211
9.3	Constraints from $b \to s\gamma$ decay	214
9.4	$B_s \to \mu^+\mu^-$ decay	217
9.5	Muon anomalous magnetic moment	220
9.6	Cosmological implications	221

	9.6.1	Relic density of neutralinos	223
	9.6.2	Direct detection of neutralino dark matter	228
	9.6.3	Indirect detection of neutralinos	230
9.7		Neutrino masses	231
	9.7.1	The MSSM plus right-handed neutrinos	232
10		**Local supersymmetry**	**235**
10.1		Review of General Relativity	236
	10.1.1	General co-ordinate transformations	236
	10.1.2	Covariant differentiation, connection fields, and the Riemann curvature tensor	238
	10.1.3	The metric tensor	240
	10.1.4	Einstein Lagrangian and field equations	242
	10.1.5	Spinor fields in General Relativity	243
10.2		Local supersymmetry implies (super)gravity	245
10.3		The supergravity Lagrangian	251
10.4		Local supersymmetry breaking	257
	10.4.1	Super-Higgs mechanism	258
11		**Realistic supersymmetric models**	**261**
11.1		Gravity-mediated supersymmetry breaking	264
	11.1.1	Hidden sector origin of soft supersymmetry breaking terms	264
	11.1.2	Why is the μ parameter small?	268
	11.1.3	Supergravity Grand Unification (SUGRA GUTs)	269
11.2		Anomaly-mediated SUSY breaking	278
	11.2.1	The minimal AMSB (mAMSB) model	280
	11.2.2	D-term improved AMSB model	284
11.3		Gauge-mediated SUSY breaking	285
	11.3.1	The minimal GMSB model	287
	11.3.2	Non-minimal GMSB models	293
11.4		Gaugino-mediated SUSY breaking	294
11.5		An afterword	296
12		**Sparticle production at colliders**	**298**
12.1		Sparticle production at hadron colliders	299
	12.1.1	Chargino–neutralino production	301
	12.1.2	Chargino pair production	308
	12.1.3	Neutralino pair production	310
	12.1.4	Slepton and sneutrino pair production	312
	12.1.5	Production of gluinos and squarks	314
	12.1.6	Gluino or squark production in association with charginos or neutralinos	319

	12.1.7	Higher order corrections	321
	12.1.8	Sparticle production at the Tevatron and LHC	322
12.2	Sparticle production at e^+e^- colliders		322
	12.2.1	Production of sleptons, sneutrinos, and squarks	325
	12.2.2	Production of charginos and neutralinos	328
	12.2.3	Effect of beam polarization	331
	12.2.4	Bremsstrahlung and beamstrahlung	335
13	**Sparticle decays**		**338**
13.1	Decay of the gluino		342
	13.1.1	$\tilde{g} \to u\bar{d}\widetilde{W}_j$: a worked example	342
	13.1.2	Other gluino decays	346
13.2	Squark decays		350
13.3	Slepton decays		353
13.4	Chargino decays		357
	13.4.1	A chargino degenerate with the LSP	360
13.5	Neutralino decays		361
13.6	Decays of the Higgs bosons		364
	13.6.1	Light scalar h	365
	13.6.2	Heavy scalar H	366
	13.6.3	Pseudoscalar A	366
	13.6.4	Charged scalar $H^\pm$	367
13.7	Top quark decays to SUSY particles		367
13.8	Decays to the gravitino/goldstino		368
	13.8.1	Interactions	368
	13.8.2	NLSP decay to a gravitino within the mGMSB model	371
14	**Supersymmetric event generation**		**374**
14.1	Event generation		377
	14.1.1	Hard scattering	377
	14.1.2	Parton showers	377
	14.1.3	Cascade decays	379
	14.1.4	Models of hadronization	382
	14.1.5	Beam remnants	383
14.2	Event generator programs		383
14.3	Simulating SUSY with ISAJET		384
	14.3.1	Program set-up	384
	14.3.2	Models for SUSY in ISAJET	385
	14.3.3	Generating events with ISAJET	388
15	**The search for supersymmetry at colliders**		**394**
15.1	Early searches for supersymmetry		395

	15.1.1	e^+e^- collisions	395
	15.1.2	Searches at the CERN S$p\bar{p}$S collider	397
	15.1.3	A light gluino window?	397
15.2	Search for SUSY at LEP and LEP2		398
	15.2.1	SUSY searches at the Z pole	398
	15.2.2	SUSY searches at LEP2	399
	15.2.3	SUSY Higgs searches at LEP2	401
15.3	Supersymmetry searches at the Tevatron		402
	15.3.1	Supersymmetry searches at run 1	403
	15.3.2	Prospects for future SUSY searches	407
15.4	Supersymmetry searches at supercolliders		414
	15.4.1	Reach of the CERN LHC	416
	15.4.2	SUSY reach of e^+e^- colliders	423
15.5	Beyond SUSY discovery		427
	15.5.1	Precision SUSY measurements at the LHC	427
	15.5.2	Precision measurements at a LC	437
	15.5.3	Models of sparticle masses: a bottom-up approach	450
15.6	Photon, muon, and very large hadron colliders		452
16	**R-parity violation**		**454**
16.1	Explicit (trilinear) R-parity violation		457
	16.1.1	The TRV Lagrangian	457
	16.1.2	Experimental constraints	459
	16.1.3	s-channel sparticle production	465
	16.1.4	$\not{R}$ decay of the LSP	466
	16.1.5	Collider signatures	468
16.2	Spontaneous (bilinear) R-parity violation		470
17	**Epilogue**		**474**
Appendix A	**Sparticle production cross sections**		**476**
A.1	Sparticle production at hadron colliders		476
	A.1.1	Chargino and neutralino production	476
	A.1.2	Gluino and squark production	478
	A.1.3	Gluino and squark associated production	481
	A.1.4	Slepton and sneutrino production	482
A.2	Sparticle production at e^+e^- colliders		483
Appendix B	**Sparticle decay widths**		**491**
B.1	Gluino decay widths		491
	B.1.1	Two-body decays	491
	B.1.2	Three-body decays to light quarks	492
	B.1.3	$\tilde{g} \to \widetilde{Z}_i t\bar{t}$ and $\tilde{g} \to \widetilde{Z}_i b\bar{b}$	493
	B.1.4	$\tilde{g} \to \widetilde{W}_i t\bar{b}$ decays	499

B.2 Squark decay widths 501

B.3 Slepton decay widths 506

B.4 Neutralino decay widths 509

  B.4.1 Two-body decays 509

  B.4.2 $\widetilde{Z}_i \to \widetilde{Z}_j f \bar{f}$ decays 511

  B.4.3 $\widetilde{Z}_j \to \widetilde{W}_i^+ \tau^- \nu_\tau$ decays 516

B.5 Chargino decay widths 516

  B.5.1 Two-body decays 516

  B.5.2 Three-body decay: $\widetilde{W}_i \to \widetilde{Z}_j \tau \bar{\nu}_\tau$ 519

B.6 Top quark decay to SUSY particles 523

Appendix C Higgs boson decay widths **524**

C.1 Decays to SM fermions 524

C.2 Decays to gauge bosons 525

C.3 Decays to sfermions 526

C.4 Decays to charginos and neutralinos 528

C.5 Decays to Higgs bosons 529

Bibliography 531

Index 533

Preface

Supersymmetry (SUSY) is a lovely theoretical construct, and has captured the imagination of many theoretical physicists. It allows for a new synthesis of particle interactions, and offers a new direction for the incorporation of gravity into particle physics. The supersymmetric extension of the Standard Model also ameliorates a host of phenomenological problems in the physics of elementary particles, if superpartners exist at the TeV scale. These new states may well be discovered in experiments at high energy colliders or in non-accelerator experiments within the next few years!

There are several excellent books that explore the theoretical structure of supersymmetry. These advanced texts are rather formal, and focus more on the theoretical structure rather than on the implications of supersymmetry. This makes them somewhat inaccessible to a large number of our experimental as well as phenomenological particle physics colleagues, working on the search for the new particles predicted by supersymmetry. Our goal in this book is to provide a comprehensive (and comprehensible) introduction to the theoretical structure of supersymmetry, and to work our way towards an exploration of its experimental implications, especially for collider searches. Although we have attempted to orient this book towards experimentalists and phenomenologists interested in supersymmetry searches, we hope that others will also find it interesting. In particular, we hope that it will provide theorists with an understanding and appreciation of some of the experimental issues that one is confronted with in the search for new physics.

We use the language of four-component relativistic spinors throughout this text, rather than the sometimes more convenient approach using two-component spinors. Although this makes some of the manipulations, especially in Chapters 5–6, appear to be somewhat more cumbersome, we felt that the use of four-spinors, which is familiar to most "practical particle physicists," would make up for this. For this reason, and also because we did not want to adopt a schizophrenic approach using two-component spinors for some things, and four-component spinors for others, we have eschewed the use of two-component spinors throughout.

After a review of the Standard Model (mainly to set up notation) and an examination of the motivations for weak scale supersymmetry, the text naturally divides into three parts. The first part (Chapters 3–7) of the book introduces supersymmetry, and details how to construct globally supersymmetric relativistic quantum field theories. We provide a "master formula" for the Lagrangian of a general, globally supersymmetric non-Abelian gauge theory that can serve as the starting point for the construction of supersymmetric models of particle physics. The inclusion of supersymmetry breaking is discussed in Chapter 7.

The second part of the book applies these lessons and develops the so-called Minimal Supersymmetric Standard Model, the MSSM, which is (almost) the direct supersymmetrization of the Standard Model. The physical particles of the MSSM are identified, and their various couplings, which are necessary for exploration of the broad phenomenological implications of the theory, are calculated. An assortment of implications of the MSSM are examined in Chapter 9, including the SUSY flavor and CP problems, renormalization group running, cosmological dark matter, and more. We discuss local supersymmetry (which, we show, includes general relativity) in Chapter 10, and in the following chapter present an overview of some of the specific mechanisms by which Standard Model superpartners may acquire supersymmetry breaking masses and couplings.

The final third of the book is oriented towards collider physics. We detail the calculations of scattering cross sections and decay rates starting from the couplings of supersymmetric particles that were found in Chapter 8. We focus on technical aspects of these calculations, including methods for dealing with Majorana particles, which the reader may not be familiar with. We also outline methods for simulation of collider scattering events in which supersymmetric matter has been produced. We then discuss what has been learned, and what may be learned, about weak scale supersymmetry from past, present, and future experiments at both hadron and e^+e^- colliders. In a final chapter, we go beyond the MSSM, but only insofar as to introduce R-parity violation, which changes the phenomenology considerably. In three appendices, we present formulae for evaluating tree-level scattering cross sections at electron–positron as well as hadron colliders, decay rates of supersymmetric particles, and decay rates of the several Higgs bosons present in all SUSY models. Various exercises are interspersed throughout the text. Some of these are pedagogical in nature, asking the reader to fill in or complete a calculation, while others develop the subject beyond the discussion in the text.

We have not attempted to make a comprehensive list of references to the vast literature on supersymmetry. Where we develop a topic from scratch, we reference only some of the classic papers on the subject. Sometimes, this means that we may not reference earlier pioneering work in favor of more complete studies that may prove more useful to a reader attempting to learn the subject. However, where we content ourselves with stating a particular result rather than deriving it completely – this is more frequently the case in the latter part of the text where we discuss supersymmetry phenomenology – we provide a reference where the reader may find further details. Thus, except for referencing some classic papers, we generally provide references only to papers where necessary details not presented in the text may be found. We apologize to the reader for this shortcoming, and also to the many researchers whose work has not been explicitly referenced.

Although we hope that the interested reader will work through the entire book, those who are interested only in phenomenology and are willing to accept supersymmetric couplings from the MSSM at face value, can skip Chapters 3–7 altogether. Chapter 10 can also be omitted without essential loss of continuity. Alternatively, the reader who is interested in model-building but does not want to work through the machinery of SUSY may use the "master formula" in Chapter 6 as a starting point, focussing on its use for writing down supersymmetric models. We urge all readers to visit Chapter 3, where many of the extraordinary properties of supersymmetric theories are explicitly illustrated.

We assume that the reader is familiar with tree-level calculations in quantum field theory through QED, as presented, for instance, by the first seven chapters of *Introduction to Quantum Field Theory*, by M. Peskin and D. V. Schroeder. We also assume some familiarity with the Standard Model of particle physics, but just in the unitarity gauge, as presented for instance in *Collider Physics*, by V. Barger and R. J. N. Phillips. No prior knowledge of supersymmetry is assumed. Indeed we have done our utmost to develop this subject from scratch, paying attention both to concepts as well as to technical details that will enable the reader to carry out research in the field. However, while we have emphasized pedagogy in our development of topics to do with supersymmetry, it is not possible to be as detailed on every topic that is necessary for describing the implications of supersymmetry for particle physics. Aside from tree-level quantum field theory and the basics of the Standard Model that we have already mentioned, these might include the ideas of the parton model, collider kinematics, Grand Unification, renormalization group methods, Big Bang cosmology etc. that have become part of the repertoire of many working particle physicists. Although we develop these ideas enough for the reader to be able to follow along, the reader who is interested in their detailed development is urged to consult the references in the text, and also the excellent treatment in the many text books listed in the Bibliography.

In writing this book, we are indebted to an enormous number of people, including teachers, students and colleagues from whom we have learned much. One of us (XT) benefited vastly from S. Weinberg's lectures on supersymmetry at the University of Texas at Austin in Spring 1982. Much of what we know is the result of collaborations and discussions over the years with many people, including S. Abdullin, G. Anderson, R. Arnowitt, D. Auto, J. Bagger, C. Balázs, V. Barger, R. M. Barnett, A. Bartl, A. Belyaev, M. Bisset, A. Box, M. Brhlik, C. Burgess, D. Castaño, C. H. Chen, M. Chen, D. Denegri, M. Díaz, D. Dicus, C. Dionisi, M. Drees, D. Dzialo-Karatas-Giudice, J. Ellis, J. Feng, J. Ferrandis, J. Freeman, R. Godbole, J. Gunion, H. Haber, K. Hagiwara, B. Harris, S. Hesselbach, K. Hikasa, C. Kao, T. Krupovnickas, T. Kamon, C. S. Kim, W. Majerrotto, S. Martin, M. Martinez, P. Mercadante, J. K. Mizukoshi, R. Munroe, A. Mustafayev, S. Nandi, D. Nanopoulos, U. Nauenberg, P. Nath, M. Nojiri, J. O'Farrill, F. Paige, M. Paterno, S. Pakvasa, F. Pauss, M. Peskin, S. Profumo, S. Protopopescu, P. Quintana, W. Repko, S. Rindani, D. P. Roy, P. Roy, L. Rurua, J. Schechter, T. Schimert, J. Sender, N. Stepanov, E. C. G. Sudarshan, T. ter Veldhuis, Y. Wang, J. Woodside, and A. White. We are also grateful to S. F. Tuan for his careful reading of the text.

HB would like to thank especially Adrienne, Madeleine, and Jake, and also Iranus M. Baer, for their support and encouragement, and as always, Norman C. Rone for his guidance and support, while completing this text. XT is grateful to Kalpana and Kashmira for their support and patience during the time that he was working on this text, and to his late parents who always encouraged him to further his studies.

Corrections to this book

A list of misprints and corrections to this book is posted on the World Wide Web at the URL `www.cambridge.org/9780521857864/`. Reports of additional errors or misprints in this book would be appreciated by the authors.

We thank those readers who sent us corrections to the hardback edition of this text.

1

The Standard Model

The 1970s witnessed the emergence of what has become the Standard Model (SM) of particle physics. The SM describes the interactions of quarks and leptons that are the constituents of all matter that we know about. The strong interactions are described by quantum chromodynamics (QCD) while the electromagnetic and the weak interactions have been synthesized into a single electroweak framework. This theory has proven to be extremely successful in describing a tremendous variety of experimental data ranging over many decades of energy. The discovery of neutral currents in the 1970s followed by the direct observation of the W and Z bosons at the CERN $Sp\bar{p}S$ collider in the early 1980s spectacularly confirmed the ideas underlying the electroweak framework. Since then, precision measurements of the properties of the W and Z bosons at both e^+e^- and hadron colliders have allowed a test of electroweak theory at the 10^{-3} level. QCD has been tested in the perturbative regime in hard collision processes that result in the breakup of the colliding hadrons. In addition, lattice gauge calculations allow physicists to test non-perturbative QCD via predictions for the observed properties of hadrons for which there is a wealth of experimental information.

1.1 Gauge invariance

One of the most important lessons that we have learned from the SM is that dynamics arises from a symmetry principle. If we require the Lagrangian density to be invariant under *local* gauge transformations, we are *forced* to introduce a set of gauge potentials with couplings to elementary scalar and fermion matter fields that, apart from an overall scale, are completely determined by symmetry principles. The most familiar example of such a field theory is the electrodynamics of Dirac fermions or complex scalars, where the invariance of the Lagrangian under

1

spacetime-dependent phase transformations,

$$\psi(x) \to e^{iq_\psi \alpha(x)} \psi(x),$$

or

$$\phi(x) \to e^{iq_\phi \alpha(x)} \phi(x),$$

requires us to introduce the vector potential A_μ, with a coupling given by,

$$\mathcal{L} = i\bar{\psi} \gamma_\mu D^\mu \psi - m\bar{\psi}\psi - \frac{1}{4} F^{\mu\nu} F_{\mu\nu}, \tag{1.1a}$$

or

$$\mathcal{L} = (D^\mu \phi)^* (D_\mu \phi) - m^2 \phi^* \phi - \frac{1}{4} F^{\mu\nu} F_{\mu\nu}. \tag{1.1b}$$

Here, D_μ is the gauge covariant derivative given by $D_\mu = \partial_\mu + iq_{\psi/\phi} A_\mu(x)$, $F_{\mu\nu} = \partial_\mu A_\nu - \partial_\nu A_\mu$ and $q_{\psi/\phi}$ is any real number identified with the charge of the field. It is easy to check that if, in addition to the local phase transformation of the fields ψ and ϕ, the vector potential transforms inhomogeneously as

$$A_\mu(x) \to A'_\mu(x) = A_\mu(x) - \partial_\mu \alpha(x),$$

the Lagrangians of Eq. (1.1a) and Eq. (1.1b) will be invariant under the set of local gauge transformations. The phase transformations of the fermion or scalar "matter" fields form the group $U(1)$. We will thus regard electrodynamics as a gauge theory based on the group $U(1)$, which is an Abelian group – i.e. its elements commute with one another. We stress two features of these Lagrangians.

- The coupling of the vector potential (identified with the photon field when the theory is quantized) to matter fields is given by the "minimal coupling principle" where the ordinary derivative is replaced by the gauge-covariant derivative. For fermionic matter, this gives the familiar fermion–antifermion–photon "vector" coupling (proportional to the charge q_ψ), while in the case of scalar matter, we have both a three-point derivative coupling to the photon proportional to the charge q_ϕ *and* a four-point non-derivative scalar–scalar–photon–photon coupling proportional to q_ϕ^2. The point to be made is that *the form of the interactions of the photons with matter is completely fixed by the requirement of local gauge invariance.*
- The photon field is massless because a mass term $\frac{1}{2} m_\gamma^2 A_\mu A^\mu$ would not be locally gauge invariant. The matter fields may, however, be massive.

Yang and Mills, and independently Shaw (and later Utiyama), generalized this idea to more complicated transformations of matter fields that form a non-Abelian group rather than the group $U(1)$. The construction of these Yang–Mills gauge theories is given in many texts and will not be repeated here. Instead of a single photon field, we now have several gauge fields (equal to the number of generators) in the adjoint representation of the gauge group. Matter and gauge fields ($V_{A\mu}$) are again "minimally coupled" via the prescription

$$\partial_\mu \to D_\mu = \partial_\mu + igt_A V_{A\mu},$$

where t_A is the matrix representation of the group generator in the representation to which the matter field belongs (for example, if the gauge group is $SU(2)$ with matter forming $SU(2)$ doublets, $t_A = \frac{1}{2}\sigma_A$, where σ_A ($A = 1, 2, 3$) are the Pauli matrices), and g is a universal (gauge) coupling constant. Again, as before, there can be no mass term for the gauge potentials, and the interaction of matter and gauge fields is fixed by the local gauge symmetry. There are some important distinctions from the Abelian case.

- The gauge field strength $F_{A\mu\nu} = \partial_\mu V_{A\nu} - \partial_\nu V_{A\mu} - gf_{ABC}V_{B\mu}V_{C\nu}$, where f_{ABC} are the structure constants of the gauge group, contains a new term in addition to the curl that is present in electromagnetism. This results in self-interactions of the non-Abelian gauge fields, and has important physical consequences such as the well-known asymptotic freedom of QCD.
- The "charge" factor in the minimal coupling principle for the non-Abelian case is replaced by $g \times t_A$. As a result, for simple groups, the coupling of matter to gauge bosons is determined to be the universal coupling g times a determined group theory factor. Thus the gauge boson couplings to matter are considerably more restrictive than in the $U(1)$ case where the charge q was any real number. In particular, the ratio of charges in the $U(1)$ case need not be a rational number.

1.2 Spontaneous symmetry breaking

We are familiar with the fact that the symmetries of the Hamiltonian (or the symmetries of the equations of motion) do not coincide with the symmetries of the solutions of these equations. For instance, Newton's laws governing the gravitational force between the Earth and the Sun are rotationally symmetric, yet the motion of the Earth around the Sun (i.e. a solution to the rotationally invariant equations of motion) is confined to a plane. Moreover, the orbit of the Earth, in general, is elliptical, and not even invariant under rotations about a single axis. This is also true in quantum theory. The $p, d, f \ldots$ orbitals of the hydrogen atom are rotationally variant solutions of the rotationally invariant Schrödinger equation.

What then does it mean for a Hamiltonian to be invariant under some symmetry transformation? These symmetries do not reflect themselves as symmetries of the solutions to the corresponding equations of motion. What is true, however, is that given a solution to the equations of motion, then we can find other solutions with the same energy by acting on the known solution by the symmetry transformation: for the example of the Earth's orbit that we considered, orbits where this ellipse is differently oriented (but with the Sun still at the focus) correspond to allowed motions and have the same total energy as the original orbit. If, however, the solution that we found itself happens to be invariant under the symmetry transformation, new solutions cannot be generated in this way.

In quantum field theory, we are especially interested in the symmetries of the ground state of the system, since it is the excitations of the ground state that are identified as particles. If, however, a ground state is not invariant under a symmetry transformation, we know there must be another solution with the same energy; i.e. the ground state must be degenerate. If the symmetry transformation that leaves the equations of motion invariant is labeled by a continuous parameter, in general there will be a continuous infinity of ground states. Which one should we choose to build the spectrum of excitations upon? The answer is that it does not matter. It is, however, important to note that once we make this choice, and express the Hamiltonian (or the Lagrangian) in terms of fields whose quanta correspond to excitations about any *one particular* ground state, the original symmetry of the action is no longer manifest. The underlying symmetry is hidden, and is (perhaps misleadingly) generally referred to as being spontaneously broken.

Although the symmetry is not really broken, it will not be obvious to an observer doing experiments with particles that are excitations of one of the many ground states of the theory. This is not to say that the underlying symmetry has no experimental implications. For instance, in a renormalizable theory, all coupling constant relationships for dimension four operators implied by the symmetry are preserved even when this symmetry may be spontaneously broken. It is this feature that gives us the universality of gauge interactions even though the gauge symmetry is spontaneously broken. Relationships between lower dimensional operators can, however, be modified by spontaneous symmetry breaking. A familiar example of this is the fact that gauge bosons may acquire mass via the Higgs mechanism even though, as we have seen, the explicit inclusion of such a mass term is forbidden by gauge invariance. Indeed our interest in gauge theories with spontaneous symmetry breaking stems mainly from this single observation, which allows the construction of gauge theories where (some of) the gauge bosons acquire mass, resulting in short-range forces as required by phenomenology.

We assume that the reader is sufficiently familiar with the physics of spontaneous symmetry breaking which is discussed in many excellent text books, so that we will not describe the Goldstone and Higgs phenomena here. Instead, we confine our discussion to some very general features of symmetry and spontaneous symmetry breaking. Our purpose is mainly to illustrate that these familiar considerations also apply to supersymmetry.

We begin by considering the action of a symmetry transformation (which, by definition, leaves the equations of motion invariant) on a state. This can be written as,

$$|\psi\rangle \rightarrow |\psi'\rangle = e^{i\alpha Q}|\psi\rangle, \tag{1.2}$$

where Q is the generator of the transformation and α is a real parameter. The symmetry in question may be a spacetime symmetry in which case Q would be one of the generators of the Poincaré group, or it may be an internal symmetry. In general, we get a new state. As we have mentioned, the action of this transformation on the ground state is especially important: the symmetry is spontaneously broken, unless

$$e^{i\alpha Q}|0\rangle = |0\rangle, \tag{1.3a}$$

or equivalently,

$$Q|0\rangle = 0. \tag{1.3b}$$

The symmetry transformation changes the dynamical variables (which are operators $\mathcal{O}$ acting on the states) as,

$$\mathcal{O}' = e^{i\alpha Q}\mathcal{O}e^{-i\alpha Q} \approx \mathcal{O} + i\alpha[Q, \mathcal{O}], \tag{1.4}$$

where the last equality holds for an infinitesimal transformation. We thus see that in order for a symmetry not to be spontaneously broken, we must have

$$\langle 0|\delta\mathcal{O}|0\rangle \equiv i\alpha\langle 0|[Q, \mathcal{O}]|0\rangle = 0, \tag{1.5}$$

where $\delta\mathcal{O}$ is the change in $\mathcal{O}$ under the (infinitesimal) symmetry transformation. Of course, $\delta\mathcal{O}$ is itself a dynamical variable.

In quantum field theory,[1] the field operators are the dynamical variables $\mathcal{O}$. In this case, as we have just seen, the vacuum expectation value (VEV) of some (possibly composite) field operator acts as the order parameter for symmetry breaking. In order that Poincaré invariance not be spontaneously broken, only spin zero field operators may acquire a VEV. If this is to result in the spontaneous breaking of a symmetry generated by Q, then the field operator in question must transform non-trivially under this symmetry. In the SM one is led to introduce a weak isodoublet of spin zero fields that acquires a VEV and results in the spontaneous breaking of

---

[1] In this case, the operator Q is obtained as the space integral of the time component of a Noether current. We will not enter into discussions as to whether the integral is defined or whether we necessarily have to discuss these issues in terms of densities. We will merely state that as long as we refer only to commutator brackets of Q with some dynamical variable, we appear to be safe.

electroweak gauge symmetry. We will see in Chapter 7 that these general considerations apply to supersymmetry, so that at least in this sense supersymmetry is not
different from other familiar symmetries.

1.3 Brief review of the Standard Model

The SM is a non-Abelian Yang–Mills type gauge theory based on the group
$SU(3)_C \times SU(2)_L \times U(1)_Y$, with $SU(2)_L \times U(1)_Y$ spontaneously broken to
$U(1)_{em}$. Color $SU(3)_C$ is assumed to be unbroken.

1.3.1 QCD

The $SU(3)_C$ gauge bosons are the gluons and the resulting gauge theory is QCD.
Quarks are assigned to the fundamental **3** representation. Thus antiquarks are assigned to the conjugate **3*** representation. All other particles are $SU(3)_C$ singlets,
and do not directly couple to the gluons. The QCD Lagrangian is given by

$$\mathcal{L}_{QCD} = -\frac{1}{4}G_{A\mu\nu}G_A^{\mu\nu} + \sum_{i=\text{flavors}} \bar{q}_i(i\not{D} - m_i)q_i \qquad (1.6)$$

where $G_{\mu\nu A} = \partial_\mu G_{A\nu} - \partial_\nu G_{A\mu} - gf_{ABC}G_{B\mu}G_{C\nu}$, $D_\mu = \partial_\mu + ig_s\frac{\lambda_A}{2}G_{A\mu}$ and q_i
contains a color triplet of quarks of flavor i. Quantization of the theory is possible
if appropriate gauge fixing terms are added to the QCD Lagrangian.

The QCD couplings of matter fermions with the gluons can now be extracted by
expanding the QCD Lagrangian. The self-interactions of the gluons are completely
fixed by gauge invariance. The interaction Lagrangian reads,

$$\mathcal{L}_{QCD} \ni -g_s \sum_i \bar{q}_i\gamma^\mu\frac{\lambda_A}{2}G_{A\mu}q_i + \tfrac{1}{2}g_s f_{ABC}(\partial_\mu G_{A\nu} - \partial_\nu G_{A\mu})G_B^\mu G_C^\nu$$
$$-\tfrac{1}{4}g_s^2 f_{ABC}f_{AB'C'}G_{B\mu}G_{C\nu}G_{B'}^\mu G_{C'}^\nu. \qquad (1.7)$$

A summation over the repeated color indices $A, B \ldots$ is implied, and the sum in the
first term is again over all quark flavors.

1.3.2 The electroweak model

In order to allow a chiral structure for the weak interactions,[2] the left- and right-
handed components of quark and lepton fields are assigned to different representations of the electroweak gauge group $SU(2)_L \times U(1)_Y$. The $SU(3)_C \times SU(2)_L \times$
$U(1)_Y$ assignment for the matter fields of the first generation of quarks and leptons

---

[2] The QCD and QED couplings of fermions are vectorial because their left- and right-chiral components are
assumed to have the same charge.

Table 1.1 *The matter and Higgs boson field content of the Standard Model along with the gauge quantum numbers.*

Field	$SU(3)_C$	$SU(2)_L$	$U(1)_Y$
$L = \begin{pmatrix} \nu_L \\ e_L \end{pmatrix}$	1	2	-1
e_R	1	1	-2
$Q = \begin{pmatrix} u_L \\ d_L \end{pmatrix}$	3	2	$\frac{1}{3}$
u_R	3	1	$\frac{4}{3}$
d_R	3	1	$-\frac{2}{3}$
$\phi = \begin{pmatrix} \phi^+ \\ \phi^0 \end{pmatrix}$	1	2	1

is shown in Table 1.1. The other generations are copies of this in that they have the same pattern of quantum numbers.

We should mention that we could equally well have written the SM field content solely in terms of left-handed fermion fields. In that case, instead of the right-handed e_R, u_R and d_R, we can work with their charge conjugates, $(e_R)^c$, $(u_R)^c$, and $(d_R)^c$, which are left-handed fields that have opposite hypercharge assignments from those shown in Table 1.1. Needless to say, these charge-conjugated quark fields would transform according to the **3*** representation of $SU(3)_C$. This way of writing the field content of the SM will be useful when we consider the supersymmetrization of the SM.

The electroweak Lagrangian is given by

$$\mathcal{L}_{EW} = \mathcal{L}_{gauge} + \mathcal{L}_{matter} + \mathcal{L}_{Higgs} + \mathcal{L}_{Yukawa}, \tag{1.8}$$

where

$$\mathcal{L}_{gauge} = -\frac{1}{4} W_{A\mu\nu} W_A^{\mu\nu} - \frac{1}{4} B_{\mu\nu} B^{\mu\nu}, \tag{1.9}$$

$$\mathcal{L}_{matter} = \sum_{generations} \left[i\bar{L}\, \not{D}L + i\bar{Q}\, \not{D}Q + i\bar{u}_R\, \not{D}u_R + i\bar{d}_R\, \not{D}d_R + i\bar{e}_R\, \not{D}e_R \right], \tag{1.10}$$

$$\mathcal{L}_{Higgs} = |D_\mu\phi|^2 + \mu^2\phi^\dagger\phi - \lambda(\phi^\dagger\phi)^2, \tag{1.11}$$

and

$$\mathcal{L}_{\text{Yukawa}} = \sum_{\text{generations}} \left[-\lambda_e \bar{L} \cdot \phi e_R - \lambda_d \bar{Q} \cdot \phi d_R - \lambda_u \epsilon^{ab} \bar{Q}_a \phi_b^\dagger u_R + \quad \text{h.c.} \right], \quad (1.12)$$

where the Ds are appropriate covariant derivatives for each matter multiplet, and ϵ^{ab} is the completely antisymmetric $SU(2)$ tensor with $\epsilon^{12} = 1$.

The interaction Lagrangian for the electroweak theory is more complicated since the $SU(2)_L \times U(1)_Y$ symmetry is assumed to be spontaneously broken to $U(1)_{\text{em}}$. The electroweak symmetry breaking sector of the SM is particularly simple, and consists of a single complex $SU(2)_L$ doublet ϕ of spin zero fields with gauge quantum numbers shown in Table 1.1. The field ϕ acquires a VEV signaling the spontaneous breakdown of electroweak symmetry. This VEV is left invariant by one combination of $SU(2)_L$ and $U(1)_Y$ generators which generates a different $U(1)$ group identified as $U(1)_{\text{em}}$. The corresponding linear combination of gauge fields remains massless and is identified as the photon,

$$A_\mu = \sin\theta_W W_{3\mu} + \cos\theta_W B_\mu \qquad (1.13)$$

with $\sin\theta_W = g'/\sqrt{g^2 + g'^2}$ and $\cos\theta_W = g/\sqrt{g^2 + g'^2}$, while all other gauge fields acquire mass via the Higgs mechanism. The physical particles in the bosonic sector of the SM are: the photon, a pair of charged massive spin 1 bosons $W^\pm$

$$W_\mu^\pm = (W_{1\mu} \mp iW_{2\mu})/\sqrt{2}, \qquad (1.14)$$

a massive spin 1 neutral boson Z^0

$$Z_\mu^0 = -\cos\theta_W W_{3\mu} + \sin\theta_W B_\mu, \qquad (1.15)$$

and finally, one neutral scalar boson H_{SM}, the Higgs boson, which is left over as the relic of spontaneous symmetry breaking. In order to establish our notation, and also for the convenience of the reader, we list the interactions of the physical particles of the SM that we will use later when we discuss phenomenological issues.

The interactions of quarks and leptons with gauge bosons can, as usual, be worked out from the minimal coupling prescription discussed above, and simply rewriting the $SU(2)_L \times U(1)_Y$ gauge fields in terms of the mass eigenstate photon, $W^\pm$, and Z^0 fields. For the electroweak gauge couplings of matter we find,

$$\mathcal{L}_{\text{neutral}} = -e \sum_f q_f \bar{f} \gamma^\mu f A_\mu + e \sum_f \bar{f} \gamma^\mu (\alpha_f + \beta_f \gamma_5) f Z_\mu, \qquad (1.16a)$$

Table 1.2 *The constants α_f and β_f that appear in Eq. (1.16a). The couplings are independent of the fermion generation. Here $t \equiv \tan \theta_W$ and $c \equiv \cot \theta_W$.*

f	q_f	α_f	β_f
ℓ	-1	$\frac{1}{4}(3t - c)$	$\frac{1}{4}(t + c)$
ν_ℓ	0	$\frac{1}{4}(t + c)$	$-\frac{1}{4}(t + c)$
u	$\frac{2}{3}$	$-\frac{5}{12}t + \frac{1}{4}c$	$-\frac{1}{4}(t + c)$
d	$-\frac{1}{3}$	$\frac{1}{12}t - \frac{1}{4}c$	$\frac{1}{4}(t + c)$

and

$$\mathcal{L}_{\text{charged}} = -\frac{g}{\sqrt{2}} \left(\bar{u} \gamma^\mu \frac{1 - \gamma_5}{2} V_{\text{KM}} d W_\mu^+ + \bar{\nu} \gamma^\mu \frac{1 - \gamma_5}{2} \ell W_\mu^+ + \quad \text{h.c.} \right). \quad (1.16b)$$

Here g is the $SU(2)_L$ gauge coupling, and $e \equiv g \sin \theta_W$ is the electromagnetic coupling. The weak mixing angle θ_W is given in terms of g and the weak hypercharge coupling g' by $g' \equiv g \tan \theta_W$. The constants α_f and β_f that appear in Eq. (1.16a) are listed in Table 1.2. In Eq. (1.16b), V_{KM} is the Kobayashi–Maskawa matrix that arises because the weak interaction quark eigenstates and the corresponding mass eigenstates do not coincide. It should also be understood that u and d in Eq. (1.16b) contain all three generations of up- and down-type quarks, respectively, with matrix multiplication implied over the generation indices.

Exercise *Verify that the gauge interactions in Eq. (1.16a) are reproduced when we replace the right-handed fermions with $(E^c)_L$, $(U^c)_L$, and $(D^c)_L$ in our assignment of quantum numbers for the fundamental fields. We use capital letters to denote these fields only to match the notation that we will use later, but here $(E^c)_L$ is just left-handed $SU(2)$ singlet positron field, $(e_R)^c$, and likewise for $(U^c)_L$ and $(D^c)_L$.*

The couplings of the Higgs boson to the gauge bosons are given by,[3]

$$\mathcal{L}_{HVV} = g M_W H_{\text{SM}} (W_\mu^+ W^{\mu-} + \tfrac{1}{2} \sec^2 \theta_W Z_\mu Z^\mu) \quad (1.17a)$$

and

$$\mathcal{L}_{HHVV} = \frac{g^2}{4} (W_\mu^+ W^{\mu-} + \tfrac{1}{2} \sec^2 \theta_W Z_\mu Z^\mu) H_{\text{SM}}^2, \quad (1.17b)$$

---

[3] We write all interactions in the unitary gauge where there are no unphysical fields.

while the self-interactions of the SM Higgs boson are given in terms of its mass $m_{H_{SM}} = \sqrt{-2\mu^2}$ by,

$$\mathcal{L}_H = -\frac{g m_{H_{SM}}^2}{4 M_W} H_{SM}^3 - \frac{g^2 m_{H_{SM}}^2}{32 M_W^2} H_{SM}^4. \tag{1.18}$$

The electroweak vector bosons also have self-interactions with the couplings given by,

$$\mathcal{L}_{WWV} = -ig \left[W_{\mu\nu}^+ W^{\mu-} - W_{\mu\nu}^- W^{\mu+} \right] (A^\nu \sin\theta_W - Z^\nu \cos\theta_W)$$
$$- ig\, W_\nu^- W_\mu^+ (A^{\mu\nu} \sin\theta_W - Z^{\mu\nu} \cos\theta_W), \tag{1.19a}$$

and

$$\mathcal{L}_{WWVV} = -\frac{g^2}{4} \left\{ \left[2W_\mu^+ W^{\mu-} + (A_\mu \sin\theta_W - Z_\mu \cos\theta_W)^2 \right]^2 \right.$$
$$- \left[W_\mu^+ W_\nu^- + W_\nu^+ W_\mu^- + (A_\mu \sin\theta_W - Z_\mu \cos\theta_W) \right.$$
$$\left. \times (A_\nu \sin\theta_W - Z_\nu \cos\theta_W) \right]^2 \right\}. \tag{1.19b}$$

In Eq. (1.19a), $A_{\mu\nu} = \partial_\mu A_\nu - \partial_\nu A_\mu$, and likewise for $Z_{\mu\nu}$ and $W_{\mu\nu}$.

Since the two chiralities of matter fermions belong to different representations of $SU(2)_L \times U(1)_Y$, it is not possible to include fermion mass terms without explicitly breaking gauge invariance. As for electroweak gauge bosons, these masses are also generated when electroweak symmetry is spontaneously broken. Fortunately, one does not have to introduce additional fields for this purpose. The scalar doublet ϕ in Table 1.1 (or its charge conjugate) has gauge invariant, renormalizable Yukawa interactions of the form $\bar{Q}\phi d_R$ or $\bar{L}\phi \ell_R$ ($\bar{Q}\phi^c u_R$) to down-type (up-type) fermions, which acquire mass when the field acquires a VEV. These Yukawa interactions result in a scalar coupling of the Higgs boson H_{SM} to SM fermions that is proportional to the corresponding fermion mass, and is given by,

$$\mathcal{L}_{Yukawa} = -\sum_i \frac{\lambda_{f_i}}{\sqrt{2}} \bar{f}_i f_i H_{SM}, \tag{1.20}$$

where $\lambda_{f_i} = \frac{g m_{f_i}}{\sqrt{2} M_W}$. The sum extends over all flavors of quarks and leptons. Notice that the Yukawa couplings of all but the top quark in Eq. (1.20) are much smaller than all the gauge couplings.

2
What lies beyond the Standard Model?

The Standard Model is a consistent, renormalizable quantum field theory that accounts for a wide variety of experimental data over an energy range that encompasses a fraction of an electron volt to about 100 GeV, a range of over twelve orders of magnitude.[1] Initially, the SM was tested at the tree level, but the remarkable agreement between SM predictions and the precision measurements at the CERN LEP collider have tested the SM to at least a part per mille and, more importantly, have established that radiative corrections as given by the SM are essential for agreement with these data. Quite aside from this, the SM also qualitatively explains why baryon and lepton numbers appear to be approximately conserved: with the particle content of the SM, it is not possible to write renormalizable interactions that do not conserve baryon and lepton numbers, so that these interactions (if they exist) must be suppressed by (powers of) some new physics scale.

The SM is nevertheless incomplete. Experimental arguments in support of this are:

- **E1** The solar and atmospheric neutrino data, interpreted as neutrino oscillations, strongly suggest neutrinos have mass.
- **E2** Observations, starting with Zwicky in 1933 and continuing to this day with studies of the fluctuations in the spectrum of the relic microwave background from the Big Bang, have established the existence of cold dark matter in the Universe for which there is no candidate in the SM.[2]
- **E3** Observations of type Ia supernovae at large red shifts as well as the cosmic microwave background radiation both suggest that the bulk of the energy of

---

[1] This is a rather conservative estimate. For instance, it may be argued from the dipole nature of planetary magnetic fields that (at least) electrodynamics has been tested out to Solar System distance scales.

[2] Gravitational microlensing data disfavor black holes (at least in our galactic halo) as dark matter, while dark matter in the form of ordinary baryons condensed into brown dwarfs is excluded, both because it would lead to conflicts with Big Bang nucleosynthesis as well as with the baryon density as determined from the acoustic peaks in the cosmic microwave background spectrum.

the Universe resides in a novel form dubbed "dark energy". This could be the cosmological constant first introduced by Einstein, or something more bizarre.
- **E4** Gravity exists.

There are also theoretical or aesthetic considerations that suggest that the SM cannot be the complete story.

- **T1** We lack any understanding of particle masses and mixing patterns, which results in the large number of underlying parameters in the SM.
- **T2** The choice of gauge group and particle representations is completely ad hoc.
- **T3** Although we can incorporate the spontaneous breakdown of electroweak symmetry by introducing new scalar fields, we have to do so "by hand" via an arbitrary scalar potential; i.e. there is no understanding of why the squared mass parameter for the Higgs field is negative. Indeed, it remains to be seen whether VEVs of elementary scalar fields are the origin of electroweak symmetry breaking.

While these arguments (especially **E1–E3**) all point to new physics, without further assumptions, they do not point decisively to the scale for this new physics. Fortunately, there is a somewhat different argument that not only suggests that there should be new physics, but also that the scale of the new physics is close to the electroweak scale. To understand this, we must first examine the divergence structure of the radiative corrections in the SM.

2.1 Scalar fields and quadratic divergences

Let us begin by considering radiative corrections to quantum electrodynamics described by the Lagrangian (1.1a). The theory describes the interactions of a fermion with a photon. As discussed in the exercise below, these interactions conserve *chirality*: a left-handed fermion, when it emits (or absorbs) a photon, remains left handed, while a right-handed fermion remains right handed. The kinetic energy term also conserves chirality. Indeed the only term in (1.1a) that does not conserve chirality is the mass term. This observation has an immediate consequence. Because the emission or absorption of a photon cannot change the chirality of the fermion, any radiative correction to the fermion mass (which by the exercise below is an operator that connects ψ_L with ψ_R) must vanish to all orders in perturbation theory if the fermion mass is zero! In other words,

$$\delta m \propto m.$$

It is well known that the loop integrals that enter these calculations are divergent. If we regularize these using a Lorentz invariant cut-off Λ, by dimensional analysis the cut-off dependence has to be given by,

$$\delta m \propto m \ln \frac{\Lambda}{m}.$$

Naive dimensional analysis would have suggested that $\delta m \propto \Lambda$. However, because of chiral symmetry, the actual divergence is much milder. We say that chiral symmetry protects fermion masses from large radiative corrections. This terminology may seem strange since the correction diverges when we take the cut-off to infinity – what we mean by this will be explained below.

We could similarly ask how interactions modify the photon mass. If one naively introduces the same cut-off to regularize the integrals for vacuum polarization, it is well known that the corrections to the squared mass of the photon are quadratically divergent. This would imply that electromagnetic gauge invariance is broken! The cut-off is, however, not a gauge invariant regulator.[3] If instead we use a gauge-invariant regulator (such as dimensional regularization) the radiative correction to the photon mass vanishes. We say that gauge invariance protects the photon from acquiring a mass.[4] Indeed in quantum electrodynamics of fermions, the leading divergence in *any* quantity is logarithmic, as the reader may readily argue from dimensional analysis. There are no quadratic or linear divergences.

The divergence structure of field theories with elementary scalars is, however, quite different. To see this, we will examine the radiative corrections to the scalar boson mass in the SM. We could equally well have performed the same analysis for scalar electrodynamics defined by (1.1b) and arrived at the same conclusion. Examples of one-loop corrections to m_H are shown in Fig. 2.1. The first of these graphs gives a momentum independent energy shift to the single scalar boson state. Using standard quantum mechanics perturbation theory, this may be evaluated by computing the diagonal element of the interaction Hamiltonian given by the second term in (1.18). Since the answer is independent of the momentum of the state, it corresponds to a mass correction. We find,

$$\delta m_{H_{SM}}^2 = \langle H_{SM} | \frac{g^2 m_{H_{SM}}^2}{32 M_W^2} H_{SM}^4 | H_{SM} \rangle = 12 \frac{g^2 m_{H_{SM}}^2}{32 M_W^2} \int \frac{d^4 k}{(2\pi)^4} \frac{i}{k^2 - m_{H_{SM}}^2},$$

$$= \frac{12 g^2 m_{H_{SM}}^2}{32 M_W^2} \frac{1}{16\pi^2} \left(\Lambda^2 - m_{H_{SM}}^2 \ln \frac{\Lambda^2}{m_{H_{SM}}^2} + \mathcal{O}(\frac{1}{\Lambda^2}) \right), \quad (2.1)$$

---

[3] This is simple to see. Under gauge transformations, $A_\mu(x) \to A_\mu(x) - \partial_\mu \alpha(x)$, or in k-space, $A_\mu(k) \to A_\mu(k) - ik_\mu \alpha(k)$, for all values of k. We see that even if we cut off the modes with $k > \Lambda$ in one gauge, these do not vanish in a different gauge – i.e. a cut-off on the momentum integrals is a gauge-dependent notion.

[4] We are tacitly ignoring the possibility of dynamical gauge symmetry breaking first pointed out by J. Schwinger, *Phys. Rev.* **125**, 397 (1962).

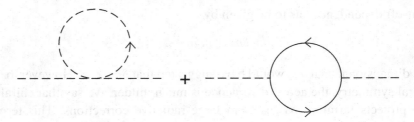

Figure 2.1 Examples of quadratically divergent Feynman graphs contributing to the corrections to the Higgs boson mass in the Standard Model.

where 12 comes from 12 possible field contractions. We see that this correction is quadratically divergent. Mass corrections are more conventionally computed by evaluating the correction to the scalar propagator. We will do so in Chapter 3 for scalar fields in a simple supersymmetric model introduced therein, and see that, once again, there are quadratically divergent contributions to the scalar boson mass.

Within the SM, there are other quadratically divergent contributions to the scalar mass from gauge boson loops, as well as from fermion loops. The computation of the gauge boson loop contribution is very similar to the one that we have just performed, and will be left as an exercise for the reader. We will not evaluate the fermion loop contribution at this point since we perform a very similar calculation in the next chapter. We only note that we expect the top loop contribution to dominate, and further, that this correction is also quadratically divergent, but with opposite sign for the Λ^2 term to the Higgs loop contribution. This difference in signs between contributions from boson and fermion loops is a general feature.

Exercise *Define the chiral projections $\psi_{L/R} \equiv P_{L/R}\psi$, where $P_L = \frac{1-\gamma_5}{2}$ and $P_R = \frac{1+\gamma_5}{2}$. For any two Dirac spinors ψ and χ, verify that,*

$$\bar{\psi}\gamma_\mu\chi = \overline{\psi_L}\gamma_\mu\chi_L + \overline{\psi_R}\gamma_\mu\chi_R,$$
$$\bar{\psi}\chi = \overline{\psi_L}\chi_R + \overline{\psi_R}\chi_L.$$

Convince yourself that these identities imply that the kinetic energy term and the interaction term in (1.1a) preserve chirality, while the mass term does not. Would chirality be conserved had the interaction been axial vector instead of vector?

2.2 Why is the TeV scale special?

Having established the difference between the divergence structure of quantum field theories with and without elementary scalar fields, let us examine the sense in which this difference may be of relevance. We note at the outset that quadratic divergences do not imply any logical problem. The SM is a renormalizable quantum field theory, and we can use it to evaluate radiative corrections to any precision that we may choose. Indeed, we have already observed that unless these corrections are included, SM predictions are at variance with precision measurements from LEP and other colliders. What then is the problem?

To understand this, we write the one-loop corrected physical Higgs boson mass as,

$$m_{H_{SM}}^2(\text{phys}) \simeq m_{H_{SM}}^2 + \frac{c}{16\pi^2}\Lambda^2, \qquad (2.2)$$

where $m_{H_{SM}}^2$ is the Higgs mass squared parameter in the Lagrangian and the second term denotes the quadratically divergent correction in (2.1); we have dropped the $\ln\Lambda$ terms in writing this formula. The coefficient c depends on the various coupling constants of the SM. In writing (2.2), we only integrate over the energy–momentum range for which we expect the SM to provide a reasonable description. In other words, we interpret the cut-off Λ as the scale at which the SM ceases to be valid. This may be because new degrees of freedom that are not included in the SM begin to become important. These new degrees of freedom may be unknown heavy particles whose effects are negligible at low energy, for instance, new particles associated with grand unification. It is also possible that the SM breaks down because of new form factors (whose origin may be some unknown strongly coupled dynamics) that develop at the scale Λ. The scale Λ might be as low as several TeV, but certainly no higher than the reduced Planck scale $M_P \simeq 2.4 \times 10^{18}$ GeV, the scale at which quantum gravity corrections are expected to become important.

How do we judge what values of Λ are reasonable? *Perturbative unitarity* arguments imply that the physical Higgs boson mass $m_{H_{SM}}(\text{phys})$ that appears on the left-hand side of (2.2) has to be smaller than a few hundred GeV.[5] If we now require (2.2) to be satisfied without excessive fine tuning between the two terms on its right-hand side, we would have to deduce that $\Lambda \le \mathcal{O}(\text{TeV})$. Again, we stress that this conclusion stems not from a logical inconsistency of the SM but from the additional "no fine-tuning requirement" that we impose on our theory. In the SM the fine tuning that is required can be truly incredible: if we assume the validity of the SM as a low energy effective theory below the GUT scale, and take $\Lambda = M_{GUT} \sim 10^{16}$ GeV, then the Lagrangian mass parameter $m_{H_{SM}}^2$ will have to be fine-tuned to 1 part in 10^{26} to provide the needed cancellation that will maintain a physical Higgs mass below its unitarity limit. In contrast, the logarithmic term in (2.1) contributes a correction which is $\sim m_{H_{SM}}^2$ even for $\Lambda \sim M_P$. This is what

[5] D. Dicus and V. Mathur, *Phys. Rev.* **D7**, 3111 (1973); B. Lee, C. Quigg and H. Thacker, *Phys. Rev.* **D16**, 1519 (1977).

we meant when we said above that the ln Λ corrections, which are also present for fermions, are not large. Put somewhat differently, the large Λ^2 corrections imply that if we use the high energy theory (from which the SM originates as the effective low energy theory) to make predictions at TeV energies, these predictions would be extremely sensitive to the parameters of the high energy theory if $\Lambda \gg 1$ TeV. While this is not logically impossible, this is usually thought to be symptomatic of a deeper problem. We refer to this as the *fine-tuning problem* of the SM. If we take this seriously, we are led to the conclusion that *there must be new degrees of freedom that manifest themselves in high energy collisions at the TeV energy scale.* This is especially exciting because we expect this scale to be directly probed by experiments at the Large Hadron Collider (LHC) which is expected to begin operation at CERN in 2007.

2.3 What could the New Physics be?

Although the arguments that we have made point to the existence of new degrees of freedom at the TeV scale, they do not by themselves provide clues as to what this new physics might be. Before proceeding further, however, let us carefully examine if it is possible to evade the conclusion about the existence of new physics at the TeV energy scale.

1. An obvious out is to accept that nature is fine-tuned, and proceed. We would then have to give up on deducing the parameters of the low energy theory from high scale physics, in the manner that we were able to deduce the Fermi coupling G_F in terms of the parameters of electroweak gauge theory. Some authors have recently suggested such a philosophy. They argue that while there have been several proposals that allow us to solve the fine-tuning problem of the SM that we have discussed, it is still necessary to fine-tune the cosmological constant, and to much greater extent than the scalar mass. They argue that accepting the greater fine tuning but not the lesser one seems artificial. We will not pursue this line of reasoning any further here.
2. There are no elementary scalar fields in nature (and hence no associated fine-tuning issues), but composite states of tightly bound fermions play the role of the Higgs boson. This is the idea behind the technicolor model, which posits new *technifermions* that interact and bind together via an asymptotically free QCD-like technicolor interaction, that becomes confining at the TeV scale.[6] Technicolor dynamics causes technifermions to condense, leading to a breakdown of electroweak gauge invariance. This attractive picture runs into problems when we attempt to use the same idea to give masses to fermions.

---

[6] For reviews, see e.g. E. Fahri and L. Susskind, *Phys. Rep.* **74**, 277 (1981) and K. Lane, hep-ph/0202255.

Construction of realistic models with massive fermions requires the introduction of yet other *extended technicolor* interactions. The simplest models predict flavor-changing neutral current processes at unacceptably large rates. More complicated technicolor theories can be constructed to avoid these problems, but the models become cumbersome and are often in conflict with data involving precision electroweak measurements. We remark that the technicolor approach predicts several pseudo-Goldstone bosons with masses in the several hundred GeV to TeV range as their signature.

3. The arguments leading to (2.2) are inherently perturbative, and would break down if interactions of Higgs bosons with themselves or with the gauge or fermion sector became strong at the scale Λ. One might then expect that these strong interactions would result in new resonances, especially in the Higgs boson and electroweak gauge boson scattering amplitudes, that would reveal themselves in high energy collisions. Even if there is no resonance, just an increase in the scattering amplitude may be experimentally observable in some channels, for instance in $W^{\pm}W^{\pm}$ scattering, where the background is small.[7]

4. A very radical alternative that has received some attention in the last several years is that gravitational effects become strong at an energy scale close to the weak scale. The motivation is to explain the evident difference in the size of gravitational and gauge interactions. Specifically, it is envisioned that there are additional compact spatial dimensions with a size $M_c \ll M_P$, and that (unlike SM gauge interactions) gravitational interactions permeate all these dimensions. Gravity appears weak at distances $\gg M_c^{-1}$ because most of the flux is "lost" in these additional dimensions. These extra dimensions are directly probed in particle collisions at energy scales $\gtrsim M_c$ where effects of gravitation become important. In such a scenario, the cut-off Λ in (2.2) would be $\sim M_c$ and the fine-tuning problem disappears if $M_c \sim \mathcal{O}(\text{TeV})$. In this case, we would expect to see exotic effects from Kaluza–Klein resonances of ordinary particles, production of black holes of masses a few times M_c, and strong gravity at high energy colliders.[8]

5. We have implicitly assumed that if a theory has quadratic divergences, these will necessarily show up at the lowest order. It is logically possible that the quadratic divergence only appears at the multi-loop level, but cancels at the one-loop level. In this case, the second term on the right-hand side of (2.2) will have additional powers of $16\pi^2$ in the denominator, and the scale Λ will be correspondingly pushed up. Scenarios where quadratic divergences appear only at the two-loop level have recently been constructed, and go under the rubric

---

[7] See e.g. J. Bagger *et al.*, *Phys. Rev.* **D49**, 1246 (1994) and *Phys. Rev.* **D52**, 3878 (1995).
[8] For principles and overview, see C. Csaki, hep-ph/0404096; see also J. Hewett and M. Spiropulu, *Ann. Rev. Nucl. Part. Sci.* **52**, 397 (2002) for a review of the phenomenological implications of extra dimensional models.

of *Little Higgs* models.[9] They require new particles to ensure this cancellation. It appears that the parameter space of the simplest of these models is already severely constrained by experimental data. Also, it is clear that this approach only postpones the problem by a couple of orders of magnitude in energy, so that we will need yet more new physics at the 100 TeV scale.

We see that all but the first of these alternatives to attempt to evade the perturbative argument that led us to conclude that there would be new degrees of freedom that could be explored at the TeV scale invoke new strong interactions at this scale, and so lead to potentially observable signatures in particle scattering at $\sqrt{s} \sim \mathcal{O}(\text{TeV})$. Naively, we would also expect that these strong interactions might not decouple, and potentially lead to observable effects in precision measurements at LEP.

The alternative is to assume that the arguments that lead to (2.2) are valid, and that there are new degrees of freedom that are *perturbatively coupled* to the particles of the SM. These new degrees of freedom must then serve to cancel the quadratic divergence that appears because of the presence of elementary scalars. Moreover, it is desirable to have this cancellation occur to all orders, not just at the one-loop level. We have already had a glimmer of how such a cancellation may be arranged when we remarked that the boson and fermion loops led to opposite signs for the coefficient of the Λ^2 term in (2.2). However, in general, one would not expect this cancellation to be complete except by accident, and further, even if we somehow got fermion loops to cancel boson loops at the one-loop level, we would not expect the cancellation to continue at higher orders, *unless the couplings of fermions and bosons are somehow related*. Such relations occur only due to symmetries.[10] However, all symmetries that we know (Lorentz symmetry, baryon or lepton $U(1)$ symmetry, isospin, ...) relate the properties of bosons (fermions) to those of other bosons (fermions), but never those of bosons to those of fermions. A symmetry that relates properties of bosons and fermions is a truly novel symmetry referred to as a *supersymmetry*.

We will see that supersymmetry requires that for every boson, a fermion partner should exist, and vice versa. In other words, every SM boson (fermion) has an as yet unseen fermionic (bosonic) supersymmetric partner. Moreover, supersymmetry relates the interactions of SM particles and their supersymmetric partners in the same way that isospin relates the interactions of protons and neutrons. It is these

---

[9] For a review, see M. Schmaltz, hep-ph/0210415.

[10] Relativistic quantum electrodynamics is a well-known precedent for a symmetry changing the divergence structure of the theory. This theory has charge conjugation invariance, and requires the presence of antiparticles. In classical theory, it is well known that the self energy of the electron diverges inversely as its size (or equivalently, linearly as the cut-off Λ). The existence of the positron is irrelevant for the classical calculation. The quantum fluctuations include the effects of all particles and thus know about the positron whose existence serves to cancel the linear divergence, leaving the electron self energy that depends logarithmically on the size of the electron.

supersymmetric particles that serve as the new perturbatively coupled degrees of freedom that act to cancel the quadratic divergences of the SM.

We should make a clear distinction between the fine-tuning problem, and the issue of *the origin* of the very small ratio between the weak and GUT or Planck scales. Supersymmetry, by itself, does not explain *why* we have such a small dimensionless ratio. However, once we introduce this ratio by choosing the parameters of the Lagrangian accordingly, this ratio is not destabilized by radiative corrections if the theory is supersymmetric, or if (as we will see in Chapter 3 and again in Chapter 7) supersymmetry is broken appropriately. In the literature, this is also stated by saying that supersymmetry solves the *technical aspect* of the gauge hierarchy problem that is endemic to theories with elementary scalar fields.

Supersymmetry was discovered in the late 1960s and early 1970s under quite different motivations.[11] The first four-dimensional globally supersymmetric quantum field theory was written down in 1974 by Wess and Zumino, and supergravity was discovered shortly thereafter (though attempts to construct locally supersymmetric theories had been made much earlier). While Fayet had already pioneered many phenomenological studies of supersymmetry, it was only after it was recognized that supersymmetry provided a solution to the fine-tuning problem of the SM that there was an explosion of interest in the particle physics community.[12] The simplest viable supersymmetric version of the SM – the Minimal Supersymmetric Standard Model, or MSSM – was developed in the early 1980s, and has since served as the starting point for many phenomenological analyses.

Although it is possible that the fine-tuning problem is solved by something no one has yet thought of, from our vantage point today, the motivations for examining supersymmetry are numerous, and remain as strong as ever.

• *Aesthetics*

Two pillars upon which the fundamental laws of physics are formulated are the theories of relativity and quantum mechanics. These two are successfully merged within the highly constrained (and hence predictive) framework of relativistic quantum field theory. Haag, Lopuszanski, and Sohnius generalized earlier work by O'Raifeartaigh and by Coleman and Mandula, and showed that the most general symmetries of the *S*-matrix are a direct product of the super-Poincaré group, which includes supersymmetry transformations linking bosons with fermions in addition to translations, rotations and boosts, with the internal symmetry group. It would be a pity if nature did not make use of this additional mathematical structure at some level.

[11] For an early history, see G. L. Kane and M. Shifman, *The Supersymmetric World: The Beginning of the Theory*, World Scientific (2000).
[12] E. Witten, *Nucl. Phys.* **B188**, 513 (1982); R. Kaul, *Phys. Lett.* **B109**, 19 (1982).

- *Ultra-violet behavior and fine tuning*
  We have already seen that, unlike the case of the Standard Model, the scalar potential of supersymmetric models is stable under radiative corrections, provided that supersymmetric particle masses are comparable to the weak scale. Supersymmetric grand unified models are thus technically natural.
- *Connection to gravity*
  By elevating global supersymmetry transformations to local ones, one is forced into introducing a spin 2 massless gauge field, the graviton, which mediates gravitational interactions (together with its superpartner, the gravitino) in much the same way as local gauge invariance requires us to introduce gauge bosons. Just as gauge invariance is sufficient to fix the dynamics, local (super)symmetry dictates the dynamics of supergravity, which includes Einstein's general relativity. Like any four-dimensional theory of gravity, this supergravity theory is not renormalizable. Nonetheless, this connection to gravity is very tantalizing.
- *Ultra-violet completeness*
  Except for supersymmetry, all the extensions of the SM that we considered above to ameliorate the fine-tuning problem are effective theories and require even more new physics at a scale that is only a couple of orders of magnitude above the TeV scale.[13] While this is not necessarily an argument against such scenarios, the fact that supersymmetric theories can in principle be extrapolated all the way to the GUT or Planck scales is especially attractive.
- *Connection to superstrings*
  Supersymmetry is an essential ingredient of superstring theories, thought to be candidates for a consistent, finite quantum theory of gravitation. In this framework, the problem of non-renormalizability of gravitational theory is bypassed by moving away from point-like particles to intrinsically finite theories of extended objects, open or closed loops of a fundamental string. Superstring theories have enjoyed an important success in the counting of microscopic states of a string black hole.

In addition to these aesthetic considerations, there are several experimental arguments that also highlight the promise of supersymmetric models.

- *Unification of gauge couplings*
  The values of running gauge couplings measured at LEP do not unify if we evolve these to high energies using the renormalization group equations of the Standard Model. If instead, we extrapolate the gauge couplings from $Q = M_Z$ to $Q = M_{GUT}$ using supersymmetric evolution equations, they unify remarkably well provided superpartner masses are in the range 100 GeV–10 TeV. It is extremely

---

[13] Models with warped extra dimensions may be a possible exception to this.

suggestive that this scale agrees so well with the scale that we inferred from fine-tuning arguments. Unless this is a perverse accident, this strongly points toward supersymmetric Grand Unification. Moreover, the value of M_{GUT} that is obtained is somewhat higher than in non-supersymmetric GUTs; this reduces the amplitude for proton decay by GUT boson exchange, bringing this contribution in accord with lower limits on the proton lifetime. Potentially larger superpotential contributions to the proton decay amplitude have to be controlled, however.

- *Cold dark matter*
All supersymmetric models with a conserved R-parity quantum number include a stable massive particle which is usually electrically and color neutral, and so makes an excellent candidate for the observed cold dark matter in the Universe.

- *Radiative breakdown of electroweak symmetry*
In the SM, electroweak symmetry breaking (EWSB) can be accommodated by appropriate choice of the parameters of the scalar potential, without any explanation for this choice. We will see in Chapter 11 that in many supersymmetric models, scalars with the same gauge quantum numbers have the same mass parameters, renormalized at some high scale. Over a large part of the model parameter space, renormalization effects drive the Higgs boson squared mass parameters to negative values, while those for scalars with non-trivial $SU(3)_C \times U(1)_{em}$ are left positive, resulting in the observed electroweak symmetry breaking pattern. Radiative electroweak symmetry breaking, as this mechanism is known, occurs naturally if $m_t \sim 100$–200 GeV. While this mechanism is very attractive, it cannot be regarded as a complete explanation since there are parameter ranges for which color and electromagnetic gauge invariance may be broken. We should add that it also requires that the soft SUSY breaking parameters are $\sim$TeV.

- *Decoupling in SUSY theories*
Radiative corrections from SUSY particles in loops to electroweak observables in LEP experiments rapidly decouple with SUSY particle masses, so that SUSY models can readily replicate the apparent successes of the SM in explaining the LEP data. This is not to say that *all* SUSY loops decouple. For instance, SUSY loop corrections to the couplings of quarks and leptons to Higgs scalars (and hence to the relation between fermion masses and Yukawa couplings) do not necessarily decouple.

- *Mass of the Higgs boson*
The Higgs boson of the SM can have mass of any value between the lower limits set by LEP and LEP2 experiments ($m_{H_{SM}} > 114.4$ GeV), up to ~ 800 GeV. The much more constrained Higgs sector of the MSSM requires the lightest Higgs scalar h to have mass $m_h \lesssim 135$ GeV. Meanwhile, precision measurements of electroweak parameters which are sensitive to the mass of the Higgs boson point towards $m_h \sim 120$ GeV, with $m_h \lesssim 200$ GeV at the 95% CL.

   The wide array of issues that are addressed by the inclusion of supersymmetry in particle physics has led many physicists to suspect that supersymmetry is realized in nature, perhaps at or just above the TeV energy scale. The goal of this book is to show how weak scale supersymmetry can be incorporated into the basic laws of physics, and detail how to extract the observable consequences of the supersymmetry hypothesis. The important and exciting conclusion is that the idea of weak scale supersymmetry can be tested at various collider and non-accelerator experiments.

3

The Wess–Zumino model

The simplest four-dimensional quantum field theory with supersymmetry realized linearly, i.e. where the transformed field is a linear function of the original fields, was written down in 1974 by Julius Wess and Bruno Zumino.[1] The Wess–Zumino (WZ) model is interesting not only because it illustrates many of the characteristics of more complicated supersymmetric models within a toy framework (this forms the subject of this chapter), but also because Yukawa interactions of the supersymmetric SM can be written as a straightforward extension of this model.

3.1 The Wess–Zumino Lagrangian

3.1.1 The field content

Let us consider a field theory with the Lagrangian given by

$$\mathcal{L} = \mathcal{L}_{\text{kin}} + \mathcal{L}_{\text{mass}}. \tag{3.1a}$$

---

[1] J. Wess and B. Zumino, *Nucl. Phys.* **B70**, 39 (1974). This is not, however, the first paper on (relativistic) spacetime supersymmetry. This distinction belongs to Y. Golfand and E. Likhtman, *JETP Lett.* **13**, 323 (1971) who introduced the supersymmetric extension of the Poincaré algebra. Motivated by the possibility that the neutrino could be the Goldstone fermion (see Chapter 7) associated with the spontaneous breakdown of a fermionic symmetry, D. Volkov and V. Akulov, *JETP Lett.* **16**, 621 (1972) and *Phys. Lett.* **B46**, 109 (1973) *independently* constructed a model with non-linearly realized supersymmetry. Local supersymmetry was first considered by D. Volkov and V. Soroka, *JETP*, **18**, 312 (1973). In this remarkable paper, they noticed the need for dynamical spin 2 and spin $\frac{3}{2}$ fields, noted the connection with gravity, and also what we now refer to as the super-Higgs mechanism; see Chapter 10. Wess and Zumino wrote their seminal paper quite unaware of any of these developments in what was formerly the Soviet Union. Two-dimensional world sheet supersymmetry (which is conceptually distinct from the spacetime supersymmetry that is the subject of this book) was discovered in 1971 in string models by A. Neveu and J. Schwarz, *Nucl. Phys.* **B31**, 86, (1971), and by P. Ramond, *Phys. Rev.* **D3**, 2415 (1971), and recognized as such by J. Gervais and B. Sakita, *Nucl. Phys.* **B34**, 632 (1971). We refer the interested reader to SUSY 30, *Proc. of the International Symposium Celebrating 30 Years of Supersymmetry*, K. Olive, S. Rudaz and M. Shifman, Editors, *Nucl. Phys.* **B 101** (Proc. Suppl.) (2001), and to *The Supersymmetric World*, G. Kane and M. Shifman, Editors (World Scientific, 2000) for a view of these developments through the eyes of the pioneers of supersymmetry.

$$\mathcal{L}_{\text{kin}} = \frac{1}{2}(\partial_\mu A)^2 + \frac{1}{2}(\partial_\mu B)^2 + \frac{i}{2}\bar{\psi}\partial\!\!\!/\psi + \frac{1}{2}(F^2 + G^2). \qquad (3.1b)$$

$$\mathcal{L}_{\text{mass}} = -m[\frac{1}{2}\bar{\psi}\psi - GA - FB]. \qquad (3.1c)$$

Here, A and B are real scalar fields with mass dimension $[A] = [B] = 1$, while ψ is a 4-component *Majorana* spinor field with mass dimension $[\psi] = 3/2$. A Majorana spinor is its own charge conjugate, so that

$$\psi = \psi^c = C\bar{\psi}^T, \qquad (3.2a)$$

where the charge conjugation matrix C satisfies

$$C\gamma_\mu^T C^{-1} = -\gamma_\mu \qquad (3.2b)$$
$$C^T = C^{-1} = -C \qquad (3.2c)$$

and

$$[C, \gamma_5] = 0. \qquad (3.2d)$$

Notice that (3.2a) is a constraint equation that says only two of the four components of ψ are independent. For instance, projecting out the right chiral component of (3.2a) yields

$$\psi_{\text{R}} \equiv \frac{1 + \gamma_5}{2}\psi = C\gamma^0\frac{1 - \gamma_5}{2}\psi^* = C\gamma^0\psi_{\text{L}}^* \qquad (3.3)$$

which shows that ψ_{R} is completely determined by ψ_{L}.[2]

The fields F and G in (3.1b) and (3.1c) are also real scalar fields with mass dimension $[F] = [G] = 2$. Since they have no kinetic energy term, these fields do not propagate, and their Euler–Lagrange equations are purely algebraic. It is, therefore, simple to write F and G in terms of the propagating fields, and eliminate them from the Lagrangian altogether. For this reason, these fields are customarily referred to as *auxiliary* fields. Explicitly, the Euler–Lagrange equations,

$$\frac{\partial \mathcal{L}}{\partial \phi_i} - \partial_\mu \frac{\partial \mathcal{L}}{\partial(\partial_\mu \phi_i)} = 0, \qquad (3.4)$$

for the Lagrangian (3.1a), with $\phi_i = F$ and G, give

$$F = -mB, \qquad G = -mA. \qquad (3.5)$$

We thus see that F and G are not dynamically independent. The reason for introducing the auxiliary fields F and G, as we will soon see, is that it allows us to

---

[2] Throughout this book we use the convention that γ_5 is a real, symmetric matrix with $\gamma_5^2 = 1$.

write supersymmetric variations as *linear* transformations on the fields, even in an interacting theory. It is interesting to see that the number of bosonic and fermionic degrees of freedom in the Lagrangian (3.1a) exactly balance, regardless of whether the Euler–Lagrange equations are satisfied: without equations of motion, there are four real components for the Majorana spinor field which are balanced by the four real scalars, A, B, F, and G. We can, however, eliminate the auxiliary fields using (3.5) to obtain the Lagrangian for the dynamically independent fields which then takes the form,

$$\mathcal{L} = \frac{1}{2}(\partial_\mu A)^2 + \frac{1}{2}(\partial_\mu B)^2 + \frac{i}{2}\bar{\psi}\partial\psi - \frac{1}{2}m^2(A^2 + B^2) - \frac{1}{2}m\bar{\psi}\psi. \tag{3.6}$$

This is the Lagrangian for free fields A, B, and ψ. When these fields obey their respective equations of motion, their quanta correspond to two spin zero particles A and B, and a self-conjugate, spin $\frac{1}{2}$ particle, all with the same mass. Once again, we see that there is an exact match between the bosonic and fermionic degrees of freedom.

3.1.2 SUSY transformations and invariance of the action

In quantum field theory, a symmetry transformation is a transformation which leaves the equations of motion for the fields of the theory invariant. This is guaranteed if the action $S = \int d^4x \mathcal{L}$ is left invariant under the transformation. In particular, if the Lagrangian $\mathcal{L}$ is invariant, or if it changes by a total derivative $\mathcal{L} \to \mathcal{L}' = \mathcal{L} + \partial_\mu \Lambda^\mu$, the action remains invariant. This can be seen by applying the four-dimensional version of Gauss' theorem, $\int_V d^4x \partial_\mu \Lambda^\mu = \int_{\partial V} d\sigma \Lambda^\mu n_\mu$, to the transformed Lagrangian. The quantity Λ^μ vanishes on the boundary ∂V as long as it is assumed that the fields vanish at spatial infinity, and field variations are equal to zero at the end points of the time integration.

Wess and Zumino noted that under the following set of *infinitesimal* field transformations, where $A \to A + \delta A$, etc., with

$$\delta A = i\bar{\alpha}\gamma_5\psi, \tag{3.7a}$$

$$\delta B = -\bar{\alpha}\psi, \tag{3.7b}$$

$$\delta\psi = -F\alpha + iG\gamma_5\alpha + \partial\gamma_5 A\alpha + i\partial B\alpha, \tag{3.7c}$$

$$\delta F = i\bar{\alpha}\partial\psi, \tag{3.7d}$$

$$\delta G = \bar{\alpha}\gamma_5\partial\psi, \tag{3.7e}$$

the Lagrangian (3.1a) changes by a *total derivative*. Here, α is a spacetime-independent anticommuting Majorana spinor parameter with dimension $[\alpha] = -1/2$. The linear transformations (3.7a–3.7e), which clearly mix boson fields with fermion fields, are known as *supersymmetry* transformations.

To verify the invariance of the action under the above transformations, we first note that bilinears of Majorana spinors have special symmetry properties. For example, for Majorana spinors ψ and χ,

$$\bar{\psi}\chi = \psi^T C \chi = \psi_a C_{ab} \chi_b = -\chi_b(-C_{ba})\psi_a = \chi^T C \psi = \bar{\chi}\psi, \qquad (3.8a)$$

where the first minus sign in step three is due to the anticommutativity of spinor fields and the second is due to the antisymmetry of C.[3] In a similar fashion, using the properties $\gamma_5^T = \gamma_5$ and $C^{-1}\gamma_\mu^T C = -\gamma_\mu$, it is straightforward to show that

$$\bar{\psi}\gamma_5\chi = \bar{\chi}\gamma_5\psi, \qquad (3.8b)$$

$$\bar{\psi}\gamma_\mu\chi = -\bar{\chi}\gamma_\mu\psi, \qquad (3.8c)$$

$$\bar{\psi}\gamma_\mu\gamma_5\chi = \bar{\chi}\gamma_\mu\gamma_5\psi, \qquad (3.8d)$$

$$\bar{\psi}\sigma_{\mu\nu}\chi = -\bar{\chi}\sigma_{\mu\nu}\psi. \qquad (3.8e)$$

Exercise *As discussed in the previous footnote, when $\chi = \psi$, we have to worry that χ and $\bar{\psi}$ do not perfectly anticommute. Except for the case $\mu = 0$, in (3.8c), the unwanted delta function term disappears because $Tr(\gamma^0\Gamma) = 0$, for $\Gamma = \gamma_5, \gamma_k, \gamma_5\gamma_\mu$ and $\sigma_{\mu\nu}$. This trace does not, however, vanish for $\Gamma = \gamma_0$. Show that (3.8c) still holds if we understand the field product to be normal ordered.*

Now we apply the supersymmetry transformations to each term of $\mathcal{L}_{\text{kin}}$, and make use of the product rule $\partial_\mu(f \cdot g) = \partial_\mu f \cdot g + f \cdot \partial_\mu g$ and the relations (3.8a–3.8e):

$$\frac{1}{2}\delta[(\partial_\mu A)^2] = (\partial^\mu A)\partial_\mu \delta A = i\partial^\mu A\bar{\alpha}\gamma_5\partial_\mu\psi,$$

$$= \partial^\mu(i\partial_\mu A\bar{\alpha}\gamma_5\psi) - i\Box A\bar{\alpha}\gamma_5\psi, \qquad (3.9a)$$

$$\frac{1}{2}\delta[(\partial_\mu B)^2] = -\partial^\mu B\bar{\alpha}\partial_\mu\psi,$$

$$= \partial^\mu(-\partial_\mu B\bar{\alpha}\psi) + \Box B\bar{\alpha}\psi, \qquad (3.9b)$$

$$\frac{i}{2}\delta[\bar{\psi}\partial\!\!\!/\psi] = \frac{i}{2}[\delta\bar{\psi}\partial\!\!\!/\psi + \bar{\psi}\partial\!\!\!/\delta\psi]$$

$$= \partial^\mu(-\frac{i}{2}F\bar{\alpha}\gamma_\mu\psi) + i\bar{\alpha}\partial\!\!\!/F\psi + \partial^\mu(-\frac{1}{2}G\bar{\alpha}\gamma_5\gamma_\mu\psi)$$

$$- \bar{\alpha}\partial\!\!\!/G\gamma_5\psi + \partial^\mu(\frac{-i}{2}\bar{\alpha}\gamma_5\partial\!\!\!/A\gamma_\mu\psi) + i\bar{\alpha}\gamma_5\Box A\psi$$

$$+ \partial^\mu(\frac{1}{2}\bar{\alpha}\partial\!\!\!/B\gamma_\mu\psi) - \bar{\alpha}\Box B\psi, \qquad (3.9c)$$

$$\frac{1}{2}\delta(F^2) = iF\bar{\alpha}\partial\!\!\!/\psi$$

---

[3] If $\chi = \psi$, then the fields (at equal times) do not anticommute to zero but to a multiple of γ_0 times a delta function. Since γ_0 is traceless, the result in (3.8a) still holds.

$$= \partial^\mu(iF\bar{\alpha}\gamma_\mu\psi) - i\bar{\alpha}\partial\!\!\!/F\psi, \qquad (3.9d)$$

$$\frac{1}{2}\delta(G^2) = G\bar{\alpha}\gamma_5\partial\!\!\!/\psi$$

$$= \partial^\mu(G\bar{\alpha}\gamma_5\gamma_\mu\psi) + \bar{\alpha}\partial\!\!\!/G\gamma_5\psi, \qquad (3.9e)$$

where $\Box = \partial_\mu\partial^\mu = \partial^2/\partial t^2 - \partial^2/\partial x^2 - \partial^2/\partial y^2 - \partial^2/\partial z^2 = \partial\!\!\!/\partial\!\!\!/$. By combining the terms contributing to $\mathcal{L}_{\text{kin}}$ in Eq. (3.9a–3.9e), we see that

$$\delta\mathcal{L}_{\text{kin}} = \partial^\mu(-\frac{1}{2}\bar{\alpha}\gamma_\mu\partial\!\!\!/B\psi + \frac{i}{2}\bar{\alpha}\gamma_5\gamma_\mu\partial\!\!\!/A\psi + \frac{i}{2}F\bar{\alpha}\gamma_\mu\psi + \frac{1}{2}G\bar{\alpha}\gamma_5\gamma_\mu\psi), \quad (3.10a)$$

so that $\mathcal{L}_{\text{kin}}$ changes by a total derivative under a SUSY transformation. The reader can similarly check that $\delta\mathcal{L}_{\text{mass}}$ is a total derivative.

Exercise *Show that*

$$\delta\mathcal{L}_{\text{mass}} = \partial^\mu(mA\bar{\alpha}\gamma_5\gamma_\mu\psi + imB\bar{\alpha}\gamma_\mu\psi) \qquad (3.10b)$$

under the supersymmetry transformations (3.7a–3.7e).

We now recall Noether's theorem which states that for every continuous symmetry transformation in a field theory, there is a corresponding current which is conserved, as long as the field equations are satisfied. For the case at hand, where $\delta\mathcal{L} = \partial^\mu\Lambda_\mu$, the current is given by

$$\bar{\alpha}j^\mu(x) = \sum_{\text{fields }\phi_i}\frac{\partial\mathcal{L}}{\partial(\partial_\mu\phi_i)}\delta\phi_i - \Lambda^\mu, \qquad (3.11)$$

with $\phi_i = A, B$, and ψ. The variations $\delta\phi_i$ as well as the quantity Λ^μ depend linearly on the transformation parameter $\bar{\alpha}$. The contributions to j^μ from the A, B, and ψ fields are

$$\frac{\partial\mathcal{L}}{\partial(\partial_\mu A)}\delta A = \partial^\mu A i\bar{\alpha}\gamma_5\psi, \qquad (3.12a)$$

$$\frac{\partial\mathcal{L}}{\partial(\partial_\mu B)}\delta B = \partial^\mu B(-\bar{\alpha}\psi), \quad \text{and} \qquad (3.12b)$$

$$\frac{\partial\mathcal{L}}{\partial(\partial_\mu\psi)}\delta\psi = \bar{\alpha}[\frac{1}{2}(iF + G\gamma_5)\gamma^\mu + \frac{1}{2}\partial\!\!\!/(-iA\gamma_5 - B)\gamma^\mu]\psi. \qquad (3.12c)$$

Combining the above with (3.10a) and (3.10b), we can explicitly construct the current (known in this case as the *supercurrent*). Notice that the supercurrent itself carries a spinorial index since its time component has to integrate to the spinor generator of supersymmetry transformations. We find

$$j^\mu = \partial\!\!\!/(-iA\gamma_5 - B)\gamma^\mu\psi + im(iA\gamma_5 - B)\gamma^\mu\psi. \qquad (3.13a)$$

For later use, notice that the supercurrent may also be written as,

$$j^\mu = \partial(-iA\gamma_5 - B)\gamma^\mu\psi + (G\gamma_5 + iF)\gamma^\mu\psi. \tag{3.13b}$$

Exercise *Show that $\partial_\mu j_a^{\prime\mu} = 0$ if the fields A and B satisfy the Klein–Gordon equation, and ψ satisfies the Dirac equation.*

The conserved charges associated with the current $j_a^\mu(x)$ are then given by

$$Q_a = \int j_a^0(x)\mathrm{d}^3 x. \tag{3.14}$$

In the next section, we will explicitly compute the super-charge Q_a for the WZ Model, and show that it indeed generates the SUSY transformations (3.7a–3.7e) as long as the field equations hold.

Exercise *Verify that if we substitute the solutions (3.5) to the Euler–Lagrange equations for F and G into the SUSY transformation laws (3.7d–3.7e), the resulting "on-shell" transformations are consistent with (3.7a) and (3.7b) as long as A, B, and ψ satisfy their equations of motion.*

3.1.3 The chiral multiplet

For the purposes of the development of superfield calculus, we remark that the fields of the WZ model can be conveniently written in terms of complex fields,

$$\mathcal{S} = \frac{1}{\sqrt{2}}(A + iB)$$

$$\psi_{\mathrm{L}} = \frac{1 - \gamma_5}{2}\psi \tag{3.15}$$

$$\mathcal{F} = \frac{1}{\sqrt{2}}(F + iG)$$

where $\mathcal{S}$, ψ_{L}, and $\mathcal{F}$ transform into one another under the SUSY transformations (3.7a–3.7e). It is straightforward to check that these transformations can be written as,

$$\delta\mathcal{S} = -i\sqrt{2}\bar{\alpha}\psi_{\mathrm{L}}, \tag{3.16a}$$

$$\delta\psi_{\mathrm{L}} = -\sqrt{2}\mathcal{F}\alpha_{\mathrm{L}} + \sqrt{2}\partial\mathcal{S}\alpha_{\mathrm{R}}, \tag{3.16b}$$

$$\delta\mathcal{F} = i\sqrt{2}\bar{\alpha}\partial\psi_{\mathrm{L}}. \tag{3.16c}$$

Since ψ_R is not independent of ψ_L, we only have to specify how ψ_L transforms. Thus, S, ψ_L, and $\mathcal{F}$ together constitute an irreducible supermultiplet in much the same way that the proton and neutron form a doublet of isospin. Further, analogous to the isospin formalism that treats the nucleon doublet as a single entity, there is a formalism known as the superfield formalism that combines all three components of the supermultiplet into a superfield $\hat{S}$. Since only one chiral component of the Majorana spinor ψ enters the transformations, such superfields are referred to as (left) chiral superfields. Because the lowest spin component of the multiplet has spin zero, this superfield is known as a left-chiral scalar superfield. We will defer detailed discussion of the superfield formalism until Chapter 5.

3.1.4 Algebra of the SUSY charges

We have already seen in Chapter 1 that a continuous symmetry transformation can be written in terms of the corresponding generator. This is also true of supersymmetry transformations, the difference being that the parameter of the transformation α is a Majorana spinor whose components anticommute with themselves and also with fermionic operators. Just as in (1.4), we may write the change of the field S under an infinitesimal SUSY transformation as,

$$S \rightarrow S' = e^{i\bar{\alpha}Q} S e^{-i\bar{\alpha}Q} \approx S + [i\bar{\alpha}Q, S] = S + \delta S \equiv (1 - i\bar{\alpha}Q)S . \qquad (3.17)$$

Here, Q is the (Majorana) spinor generator of the SUSY transformation except in the last equality, where we have abused notation in that Q there denotes the representation of the super-charge generator (explicitly worked out in Chapter 5), in the same way that the translation generator P_μ is represented by $i\partial_\mu$ when we write $[P_\mu, S] = -i\partial_\mu S$. We thus write the change of the field S as $\delta S = -i\bar{\alpha}QS$. We can now work out the algebra for the Q's and their conjugates $\bar{Q}$ by considering the commutator of two successive SUSY transformations – the first by parameter α_1, and the second by parameter α_2. For the case of the scalar field S, since $\delta_1 S = -\sqrt{2}i\bar{\alpha}_1\psi_L$, then

$$(\delta_2\delta_1 - \delta_1\delta_2)S = -2i\left\{ -\mathcal{F}\bar{\alpha}_1\frac{1-\gamma_5}{2}\alpha_2 + \bar{\alpha}_1\gamma^\mu\frac{1+\gamma_5}{2}\alpha_2\partial_\mu S \right\}$$

$$+2i\left\{ -\mathcal{F}\bar{\alpha}_2\frac{1-\gamma_5}{2}\alpha_1 + \bar{\alpha}_2\gamma^\mu\frac{1+\gamma_5}{2}\alpha_1\partial_\mu S \right\}$$

$$= 2i\bar{\alpha}_2\gamma^\mu\alpha_1\partial_\mu S$$

$$= -2\bar{\alpha}_2\gamma^\mu\alpha_1[P_\mu, S]. \qquad (3.18)$$

We can work out the same commutator in terms of the SUSY generator Q using $\delta S = [i\bar{\alpha}Q, S]$ to obtain,

$$\delta_2\delta_1 S = [i\bar{\alpha}_2 Q, \delta_1 S] = \left[i\bar{\alpha}_2 Q, [i\bar{Q}\alpha_1, S]\right]$$
$$= -\left[i\bar{Q}\alpha_1, [S, i\bar{\alpha}_2 Q]\right] - \left[S, [i\bar{\alpha}_2 Q, i\bar{Q}\alpha_1]\right], \qquad (3.19)$$

where in the last step we have used the Jacobi identity,

$$[[A, B], C] + [[B, C], A] + [[C, A], B] = 0,$$

that holds for any three bosonic operators, A, B, and C, as the reader may easily verify. Applying (3.19) twice, we readily obtain

$$(\delta_2\delta_1 - \delta_1\delta_2)S = -\bar{\alpha}_{2a}\alpha_{1b}[\{Q_a, \bar{Q}_b\}, S]. \qquad (3.20)$$

Finally, by equating the right-hand sides of (3.18) and (3.20) we can write the algebra obeyed by the SUSY generators as,

$$\{Q_a, \bar{Q}_b\} = 2(\gamma^\mu)_{ab} P_\mu \qquad (3.21)$$

where P_μ is the Poincaré group generator of spacetime translations.

A similar calculation can be performed for the commutator of SUSY transformations on the field ψ_L. It is straightforward to show

$$(\delta_2\delta_1 - \delta_1\delta_2)\psi_L = -2i[(\bar{\alpha}_2 \partial\!\!\!/ \psi_L)\alpha_{1L} - (\bar{\alpha}_1 \partial\!\!\!/ \psi_L)\alpha_{2L}]$$
$$- 2i[(\bar{\alpha}_2 \partial_\mu \psi_L)\gamma^\mu \alpha_{1R} - (\bar{\alpha}_1 \partial_\mu \psi_L)\gamma^\mu \alpha_{2R}]. \qquad (3.22)$$

To proceed further, we need to apply a Fierz re-arrangement to the spinors, and combine the two α's into a bilinear.

Exercise *The set of 16 matrices $\Gamma_i = \{\mathbf{1}, \gamma_5, \gamma_\mu, i\gamma_\mu\gamma_5, \sigma_{\mu\nu}\}$ (with $\sigma_{\mu\nu} = \frac{i}{2}[\gamma_\mu, \gamma_\nu]$ for $\mu > \nu$), with Γ^i defined the same way except with all the indices upstairs, have the properties $\mathrm{Tr}\Gamma^i = 0$ (for $\Gamma^i \neq 1$) and $\mathrm{Tr}\Gamma^i\Gamma_j = 4\delta^i_j$. These matrices can be used as a basis of expansion for any other 4×4 matrix. In particular, for the combination of spinors*

$$\bar{\psi}(1)\psi(2)\psi_b(3) \equiv \bar{\psi}_a(1)\psi_a(2)\psi_b(3) = \psi_b(3)\bar{\psi}_a(1)\psi_a(2),$$

the quantity can be written as $\psi_b(3)\bar{\psi}_a(1) = \sum_i c_i\Gamma^i_{ba}$. Multiplying both sides of this expansion by Γ_{jab} and summing over a and b, show that $c_j = -\frac{1}{4}\bar{\psi}(1)\Gamma_j\psi(3)$, so that

$$\bar{\psi}(1)\psi(2)\psi_b(3) = -\frac{1}{4}\sum_j \bar{\psi}(1)\Gamma^j\psi(3)(\Gamma_j\psi(2))_b. \qquad (3.23)$$

Applying the Fierz re-arrangement to (3.22) yields

$$(\delta_2\delta_1 - \delta_1\delta_2)\psi_L = \frac{2i}{4}\sum_i(\bar{\alpha}_2\Gamma^i\alpha_1 - \bar{\alpha}_1\Gamma^i\alpha_2)P_L\Gamma_i\gamma^\mu\partial_\mu\psi_L$$

$$+ \frac{2i}{4}\sum_i(\bar{\alpha}_2\Gamma^i\alpha_1 - \bar{\alpha}_1\Gamma^i\alpha_2)P_L\gamma^\mu\Gamma_i\partial_\mu\psi_L, \quad (3.24)$$

where the chiral projection operators P_L allow only the vector and axial-vector forms of Γ_i to contribute. Using relations (3.8c) and (3.8d) on the Γ_A and Γ_V terms, we find

$$(\delta_2\delta_1 - \delta_1\delta_2)\psi_L = i\bar{\alpha}_2\gamma_\mu\alpha_1[\gamma^\mu\gamma^\nu + \gamma^\nu\gamma^\mu]\partial_\nu\psi_L$$

$$= 2i\bar{\alpha}_2\gamma^\mu\alpha_1\partial_\mu\psi_L. \quad (3.25)$$

Comparison of this expression with the corresponding expression involving the Q and $\bar{Q}$ operators again verifies the relation (3.21).

Exercise *Show that the commutator of two SUSY transformations applied to the auxiliary field $\mathcal{F}$ again leads to the anticommutator (3.21).*

We thus see that (3.21) is valid acting on each component of an arbitrary field, so that it may be regarded as an operator relation.

The appearance of the translation generator in (3.21) shows that supersymmetry is a spacetime symmetry. Conservation of supersymmetry implies

$$[Q_a, P^0] = 0, \quad (3.26a)$$

or, from Lorentz covariance,

$$[Q_a, P^\mu] = 0. \quad (3.26b)$$

The commutators of Q with the Lorentz group generators $J_{\mu\nu}$ are fixed because we have already declared Q to be a spin $\frac{1}{2}$ Majorana spinor.

The supersymmetry algebra described above is not a Lie algebra since it includes anticommutators. Such algebras are referred to as graded Lie algebras. Haag, Lopuszanski, and Sohnius have shown that (except for the possibility of neutral elements and of more than one spinorial charge Q) the algebra that we have obtained above is the most general graded Lie algebra consistent with rather reasonable physical assumptions. Models with more than one SUSY charge in the low energy theory do not lead to chiral fermions and so are excluded for phenomenological reasons. We will henceforth assume that there is just a single super-charge.

3.2 Quantization of the WZ model

The main purpose of this section is to review the implementation of the WZ model as a quantum field theory. This provides us with an opportunity to set up our conventions for the field expansions as well as for the (anti)commutators of the creation and annihilation operators. In the process we will also see how the quantization of the Majorana field differs from the more familiar quantization of the Dirac field. We always use the four-component spinor notation that many particle physicists are most familiar with.

We adopt the canonical quantization procedure, wherein the fields are regarded as quantum operators acting upon a Fock space of particle states. For the scalar fields A and B, the conjugate field momenta are $\Pi_A = \partial \mathcal{L}/\partial(\frac{\partial A}{\partial t}) = \partial A/\partial t \equiv \dot{A}$ and $\Pi_B = \dot{B}$. The equal time commutators for the A and B fields are stipulated to be

$$[A(\mathbf{x}), \dot{A}(\mathbf{y})] = i\delta^3(\mathbf{x} - \mathbf{y}), \quad [A(\mathbf{x}), A(\mathbf{y})] = [\dot{A}(\mathbf{x}), \dot{A}(\mathbf{y})] = 0, \tag{3.27a}$$

$$[B(\mathbf{x}), \dot{B}(\mathbf{y})] = i\delta^3(\mathbf{x} - \mathbf{y}), \quad [B(\mathbf{x}), B(\mathbf{y})] = [\dot{B}(\mathbf{x}), \dot{B}(\mathbf{y})] = 0. \tag{3.27b}$$

The Hermitian field operators A and B can be Fourier expanded such that

$$A(x) = \int \frac{d^3k}{(2\pi)^3} \frac{1}{2E_\mathbf{k}} \left(a_\mathbf{k} e^{-ikx} + a_\mathbf{k}^\dagger e^{ikx} \right), \tag{3.28a}$$

$$B(x) = \int \frac{d^3k}{(2\pi)^3} \frac{1}{2E_\mathbf{k}} \left(b_\mathbf{k} e^{-ikx} + b_\mathbf{k}^\dagger e^{ikx} \right), \tag{3.28b}$$

where the a ($a^\dagger$) and b ($b^\dagger$) operators are annihilation (creation) operators satisfying

$$[a_\mathbf{k}, a_\mathbf{l}^\dagger] = (2\pi)^3 2E_\mathbf{k} \delta^3(\mathbf{k} - \mathbf{l}), \quad [a_\mathbf{k}, a_\mathbf{l}] = [a_\mathbf{k}^\dagger, a_\mathbf{l}^\dagger] = 0, \tag{3.29a}$$

$$[b_\mathbf{k}, b_\mathbf{l}^\dagger] = (2\pi)^3 2E_\mathbf{k} \delta^3(\mathbf{k} - \mathbf{l}), \quad [b_\mathbf{k}, b_\mathbf{l}] = [b_\mathbf{k}^\dagger, b_\mathbf{l}^\dagger] = 0. \tag{3.29b}$$

The usual four-component Dirac spinor field ψ_D is quantized by stipulating the equal-time anticommutators,

$$\{\psi_{Da}(\mathbf{x}), \psi_{Db}^\dagger(\mathbf{y})\} = \delta_{ab}\delta^3(\mathbf{x} - \mathbf{y}),$$
$$\{\psi_{Da}(\mathbf{x}), \psi_{Db}(\mathbf{y})\} = \{\psi_{Da}^\dagger(\mathbf{x}), \psi_{Db}^\dagger(\mathbf{y})\} = 0. \tag{3.30}$$

The field is expanded using distinct creation and annihilation operators for particles and antiparticles as,

$$\psi_D(x) = \int \frac{d^3k}{(2\pi)^3} \frac{1}{2E_\mathbf{k}} \sum_s [c_{\mathbf{k},s} u_{\mathbf{k},s} e^{-ikx} + d_{\mathbf{k},s}^\dagger v_{\mathbf{k},s} e^{ikx}]. \tag{3.31}$$

These creation and annihilation operators satisfy the well-known *anticommutation* relations which we will not write out here. Note that for a Dirac spinor,

$$\langle 0|T\psi_{Da}(x)\bar{\psi}_{Db}(y)|0\rangle = S_{Fab}(x-y), \quad \text{and} \tag{3.32a}$$

$$\langle 0|T\psi_{Da}(x)\psi_{Db}(y)|0\rangle = \langle 0|T\bar{\psi}_{Da}(x)\bar{\psi}_{Db}(y)|0\rangle = 0. \tag{3.32b}$$

A similar procedure for quantizing a four-component Majorana field ψ *cannot* be followed because the Majorana spinor is constrained by the Majorana condition $\psi = \psi^c = C\bar{\psi}^T$, i.e. only two of the four components of the Majorana spinor are independent. To proceed further, we evaluate the field expansion for the conjugate Dirac field ψ_D^c. Using the spinor relations $u^c \equiv C\bar{u}^T = v$ and $v^c \equiv C\bar{v}^T = u$, we find

$$\psi_D^c(x) = \int \frac{d^3k}{(2\pi)^3} \frac{1}{2E_\mathbf{k}} \sum_s [c_{\mathbf{k},s}^\dagger v_{\mathbf{k},s} e^{ikx} + d_{\mathbf{k},s} u_{\mathbf{k},s} e^{-ikx}]. \tag{3.33}$$

Next, impose the constraint $\psi = \psi^c$. The constraint is respected if we require $c = d$ and $c^\dagger = d^\dagger$, so that the Majorana spinor field expansion is just

$$\psi(x) = \int \frac{d^3k}{(2\pi)^3} \frac{1}{2E_\mathbf{k}} \sum_s [c_{\mathbf{k},s} u_{\mathbf{k},s} e^{-ikx} + c_{\mathbf{k},s}^\dagger v_{\mathbf{k},s} e^{ikx}], \tag{3.34}$$

with

$$\{c_{\mathbf{k},r}, c_{\mathbf{l},s}^\dagger\} = (2\pi)^3 2E_k \delta_{rs} \delta^3(\mathbf{k}-\mathbf{l}), \quad \{c_{\mathbf{k},r}, c_{\mathbf{l},s}\} = \{c_{\mathbf{k},r}^\dagger, c_{\mathbf{l},s}^\dagger\} = 0. \tag{3.35}$$

The condition $\psi = \psi^c$ is the analogue of the reality condition for the scalar fields A and B; the condition $c_\mathbf{k} = d_\mathbf{k}$ implies the identity of the particle and antiparticle quanta of this field. For a Majorana spinor field, we still have

$$\langle 0|T\psi_a(x)\bar{\psi}_b(y)|0\rangle = S_{Fab}(x-y), \tag{3.36}$$

but now, because $\psi = C\bar{\psi}^T$ and $\bar{\psi} = \psi^T C$, $\langle 0|T\psi_a(x)\psi_b(y)|0\rangle$ and $\langle 0|T\bar{\psi}_a(x)\bar{\psi}_b(y)|0\rangle$ do not vanish as in the case of a Dirac field. It is easy to show that

$$\langle 0|T\psi_a(x)\psi_b(y)|0\rangle = S_{Fac}(x-y)C_{cb}^T \quad \text{and} \tag{3.37a}$$

$$\langle 0|T\bar{\psi}_a(x)\bar{\psi}_b(y)|0\rangle = C_{ac}^T S_{Fcb}(x-y). \tag{3.37b}$$

We must not forget to include these contractions when computing matrix elements of operators involving products of Majorana spinor fields.

The four-momentum operator P^μ for the WZ model can now be explicitly constructed from the energy–momentum tensor $T^{\mu\nu}$. Recall

$$T^{\mu\nu} = \sum_{\text{fields } \phi_i} \frac{\partial \mathcal{L}}{\partial(\partial_\mu \phi_i)} \partial^\nu \phi_i - g^{\mu\nu} \mathcal{L}, \tag{3.38}$$

where

$$P^\mu = \int T^{0\mu} d^3x. \tag{3.39}$$

Substituting the field expansions (3.28a), (3.28b), and (3.34) into (3.39) and performing a rather lengthy calculation leads to

$$P^\mu = \int \frac{d^3k}{(2\pi)^3 2E_\mathbf{k}} k^\mu [a_\mathbf{k}^\dagger a_\mathbf{k} + \frac{1}{2}\delta^3(\mathbf{0}) + b_\mathbf{k}^\dagger b_\mathbf{k} + \frac{1}{2}\delta^3(\mathbf{0}) \tag{3.40}$$
$$+ \sum_s (c_{\mathbf{k},s}^\dagger c_{\mathbf{k},s} - \frac{1}{2}\delta^3(\mathbf{0}))]$$

$$= \int \frac{d^3k}{(2\pi)^3 2E_\mathbf{k}} k^\mu [a_\mathbf{k}^\dagger a_\mathbf{k} + b_\mathbf{k}^\dagger b_\mathbf{k} + \sum_s c_{\mathbf{k},s}^\dagger c_{\mathbf{k},s}]. \tag{3.40}$$

Thus, in the WZ model, we see that for the field four-momentum operator, the zero-point energy terms exactly cancel due to equal and opposite bosonic and fermionic contributions. This is the first of several examples of the cancellation of infinities in supersymmetric models. Expressions for the rotation and boost generators of the Poincaré group can be similarly constructed, but we will not do so here.

It is, however, instructive to explicitly construct the super-charge from the supercurrent (3.13a) for the WZ model. We find,

$$Q = \int j^0 d^3x \tag{3.41}$$

$$= \sum_s \int \frac{d^3k}{(2\pi)^3 2E_\mathbf{k}} \left\{ (a_\mathbf{k}\gamma_5 + ib_\mathbf{k})c_{\mathbf{k},s}^\dagger v_{\mathbf{k},s} - (a_\mathbf{k}^\dagger\gamma_5 + ib_\mathbf{k}^\dagger)c_{\mathbf{k},s}u_{\mathbf{k},s} \right\}.$$

It should be apparent from this expression that the action of Q on a bosonic (fermionic) state results in an admixture with a fermionic (bosonic) state.

Exercise *Verify Eq. (3.41).*

It is now possible to explicitly show that the generators P^μ and Q obtained above commute with each other as indeed they should.

We can now use the expression (3.41) for the super-charge to work out the effect on the dynamically independent field operators of the WZ model.

Exercise *Using the expression (3.41) for the super-charge in the WZ model, verify that for an infinitesimal SUSY transformation,*

$$\delta A = [i\bar{\alpha} Q, A] = i\bar{\alpha}\gamma_5\psi,$$
$$\delta B = [i\bar{\alpha} Q, B] = -\bar{\alpha}\psi,$$
$$\delta\psi = [i\bar{\alpha} Q, \psi] = \partial\gamma_5 A\alpha + i\partial B\alpha - im A\gamma_5\alpha + m B\alpha.$$

The first two of these expressions are just the transformations of the A and B fields in (3.7a) and (3.7b), whereas the last of these corresponds to the transformation (3.7c) for $\delta\psi$ where the auxiliary fields are eliminated via their Euler–Lagrange equations. The fact that F and G do not appear on the right-hand side could have been anticipated since these do not appear in the form of the supercurrent.

3.3 Interactions in the WZ model

Up to this point we have been discussing free field theory which, despite being supersymmetric, would not be of interest if interactions could not be incorporated. Following Wess and Zumino, we add interaction terms given by

$$\mathcal{L}_{\text{int}} = -\frac{g}{\sqrt{2}}A\bar{\psi}\psi + \frac{ig}{\sqrt{2}}B\bar{\psi}\gamma_5\psi + \frac{g}{\sqrt{2}}(A^2 - B^2)G + g\sqrt{2}ABF, \quad (3.43)$$

to the Lagrangian (3.1a). It can be verified by brute force that $\mathcal{L}_{\text{int}}$ is separately supersymmetric. The calculation is rather messy. We will demonstrate the super-symmetry of this Lagrangian more elegantly in Chapter 5 using the superfield formalism.

Once again we can eliminate the auxiliary fields F and G via their Euler–Lagrange equations which get modified to,

$$F = -mB - g\sqrt{2}AB \quad (3.44a)$$
$$G = -mA - \frac{g}{\sqrt{2}}(A^2 - B^2), \quad (3.44b)$$

and obtain the total Lagrangian in terms of the dynamical fields as,

$$\mathcal{L} = \frac{1}{2}(\partial_\mu A)^2 + \frac{1}{2}(\partial_\mu B)^2 + \frac{i}{2}\bar{\psi}\partial\psi - \frac{1}{2}m^2(A^2 + B^2) - \frac{1}{2}m\bar{\psi}\psi$$
$$- \frac{g}{\sqrt{2}}A\bar{\psi}\psi + \frac{ig}{\sqrt{2}}B\bar{\psi}\gamma_5\psi - gm\sqrt{2}AB^2 - \frac{gm}{\sqrt{2}}A(A^2 - B^2)$$
$$- g^2A^2B^2 - \frac{1}{4}g^2(A^2 - B^2)^2. \quad (3.45)$$

Several features of the Lagrangian in (3.45) are worth stressing.

1. It describes the interaction of two real spin zero fields and a Majorana field with spin half. As before, the number of bosonic and fermionic degrees of freedom match.
2. There is a single mass parameter m common to all the fields.
3. Although the interaction structure of the model is very rich and includes parity-conserving scalar and pseudoscalar interactions of the scalar A and pseudoscalar B with the fermion, as well as all possible (renormalizable) parity conserving trilinear and quartic scalar interactions, there is just *one* single coupling constant g. We thus see that supersymmetry is like other familiar symmetries in that it relates the various interactions as well as masses. The mass and coupling constant relationships inherent in (3.45) are completely analogous to the familiar (approximate) equality of neutron and proton masses or the relationships between their interactions with the various pions implied by (approximate) isospin invariance.

Before closing we remark that Iliopoulos and Zumino observed that unlike (3.13a), the expression (3.13b) for the supercurrent holds also in the presence of interactions, provided of course that the auxiliary fields satisfy (3.44a) and (3.44b).[4] We will use this observation in Chapter 7 when we discuss the interactions of the massless Goldstone fermion that appears as a result of spontaneous supersymmetry breaking.

3.4 Cancellation of quadratic divergences

We have already mentioned that the existence of supersymmetric partners serves to remove the quadratic divergences that destabilize the scalar sector of a generic field theory. We will illustrate this cancellation of quadratic divergences for the simple case of the WZ model. Consider the corrections to the "one-point function"

$$\langle\Omega|A(x)|\Omega\rangle = \text{ sum of all connected diagrams with one external point}$$

of the field A to first order in the coupling constant g in (3.45). Here $|\Omega\rangle$ is the ground state of the interacting theory. The relevant interaction Hamiltonian from (3.45) is

$$\mathcal{H}_{\text{int}} = -\mathcal{L}_{\text{int}} \ni \frac{g}{\sqrt{2}} A\bar{\psi}\psi + \frac{g}{\sqrt{2}} m A B^2 + \frac{g}{\sqrt{2}} m A^3. \tag{3.46}$$

The loop corrections to the one-point function are represented by the tadpole diagrams shown in Fig. 3.1. Expanding the matrix element $\langle\Omega|T A(x)|\Omega\rangle$

---

[4] J. Iliopoulos and B. Zumino, *Nucl. Phys.* **B76**, 310 (1974).

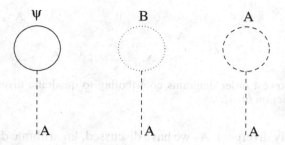

Figure 3.1 Lowest order diagrams contributing to quadratic divergences in the one-point function of A.

perturbatively to order g gives,[5]

$$-i\frac{g}{\sqrt{2}} \int d^4y\, D_F^A(x-y)\left[(-1)\mathrm{Tr}S_F(y-y) + m D_F^B(y-y) + 3m D_F^A(y-y)\right],$$

where the factor 3 in the last term arises from three possible contractions involving the A^3 interaction term. The factor in the square brackets is proportional to

$$\mathrm{Tr}\int \frac{d^4p}{\not{p}-m_\psi} - m\int \frac{d^4p}{p^2-m_B^2} - 3m\int \frac{d^4p}{p^2-m_A^2}$$

$$= \int \frac{d^4p}{p^2-m_\psi^2} 4m_\psi - m\int \frac{d^4p}{p^2-m_B^2} - 3m\int \frac{d^4p}{p^2-m_A^2}. \qquad (3.47)$$

Here, we have deliberately denoted the masses that enter via the propagators by m_A, m_B, and m_ψ, although these are exactly the same as the mass parameter m that enters via the trilinear scalar couplings in (3.45). We first see that because all these masses are exactly equal in a supersymmetric theory, the three contributions in (3.47) add to zero. Thus although each diagram is separately quadratically divergent, the divergence from the fermion loop exactly cancels the sum of divergences from the boson loops. Two remarks are in order.

1. In order for this cancellation to occur, it is crucial that the A^3, AB^2, and $A\bar{\psi}\psi$ couplings be exactly as given in (3.45).
2. The *quadratic* divergence in the expression (3.47) is independent of the scalar masses, m_A and m_B. It is, however, crucial that the fermion mass m_ψ is exactly equal to the mass m that enters via the trilinear scalar interactions in order for the cancellation of the quadratic divergence to be maintained. If the boson masses differ from the fermion mass m_ψ, the expression in (3.47) is at most

---

[5] For a review, see *Introduction to Quantum Field Theory*, M. Peskin and D. Schroeder, Perseus Press (1995), Chapter 4, where $D_F(x-y)$ is defined.

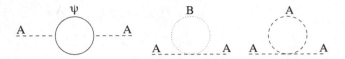

Figure 3.2 Lowest order diagrams contributing to quadratic divergences in the two-point function of A.

logarithmically divergent. As we have discussed, logarithmic divergences do not severely destabilize scalar masses.

It is also instructive to inspect the lowest order quadratic divergences in the two-point function of A defined as $\langle \Omega | T A(x) A(y) | \Omega \rangle$. The one-loop contributions to the quadratic divergences are shown in Fig. 3.2. [6] The first diagram of Fig. 3.2 gives a contribution

$$-\frac{g^2}{2} \int d^4z d^4z' D_F^A(x - z) D_F^A(z' - y)$$
$$\times \left[(-1) \mathrm{Tr} S_F(z - z') S_F(z' - z) + \mathrm{Tr} C^T C^T S_F(z - z') S_F(z - z') \right],$$

where the second term in the square parenthesis arises because contractions of the Majorana ψ (and $\bar{\psi}$) field with itself do not vanish as noted in (3.37a) and (3.37b). The integration over the intermediate points z and z' can be performed by writing the Fourier expansions of each of the propagators. One then finds that the correction from the fermion loop in Fig. 3.2 is given by,

$$g^2 \int \frac{d^4p}{(2\pi)^4} \frac{1}{p^2 - m_A^2} e^{-ip(x-y)} \int \frac{d^4q}{(2\pi)^4} \mathrm{Tr} \left[\frac{1}{(\slashed{q} - m_\psi)} \frac{1}{(-\slashed{p} + \slashed{q} - m_\psi)} \right] \frac{1}{p^2 - m_A^2}.$$

From the second diagram where the fields $A(x)$ and $A(y)$ can be contracted in two ways we get,

$$-i\frac{g^2}{2} 2 \int d^4z D_F^A(x - z) D_F^A(z - y) D_F^B(z - z),$$

while the third diagram for which we have twelve possible contractions yields,

$$-i\frac{g^2}{4} 12 \int d^4z D_F^A(x - z) D_F^A(z - y) D_F^A(z - z).$$

Once again, we can do the integration over z using the momentum expansion of the propagators. By combining the contributions from the diagrams in Fig. 3.2, we see that including the lowest order correction to the two-point function of A changes

---

[6] There are additional quadratic divergences in the two-point function from the tadpoles of Fig. 3.1 which, as we have just seen, separately cancel.

the momentum space propagator as

$$\frac{i}{p^2 - m_A^2} \rightarrow \frac{i}{p^2 - m_A^2} + \frac{i}{p^2 - m_A^2}(-i\Pi(p))\frac{i}{p^2 - m_A^2},$$

with the divergences all being contained in the function $\Pi(p)$ given by

$$i\Pi(p) = g^2 \int \frac{d^4q}{(2\pi)^4} \left[\text{Tr}\left(\frac{(\slashed{q} + m_\psi)}{q^2 - m_\psi^2} \cdot \frac{(-\slashed{p} + \slashed{q} + m_\psi)}{(q-p)^2 - m_\psi^2}\right) \right.$$

$$\left. - \frac{1}{q^2 - m_B^2} - 3\frac{1}{q^2 - m_A^2} \right]. \tag{3.48}$$

It is now straightforward to see that once again the quadratic divergences cancel between fermionic and bosonic loops. Moreover, this cancellation occurs for *all* values of particle masses. This is because trilinear scalar interactions do not contribute to the quadratic divergence that we have just computed. It is, however, crucial that the fermion Yukawa coupling $(g/\sqrt{2})$ is related to the quartic scalar couplings on the last line of (3.45).

Exercise *Verify Eq. (3.48) and check that the quadratic divergence cancels.*

Exercise *Verify that the quadratic divergence cancels in the one-loop tadpole and mass corrections to the B field.*

3.5 Soft supersymmetry breaking

The fact that the quadratic divergences continue to cancel even if the scalar boson masses are not exactly equal to fermion masses (as implied by SUSY) is absolutely critical for the construction of phenomenologically viable models. We know from observation that SUSY cannot be an exact symmetry of nature. Otherwise, there would have to exist a spin zero or spin one particle *with exactly the mass and charge of an electron*. Such a particle could not have evaded experimental detection. The only way out of this conundrum is to admit that supersymmetric partners cannot be degenerate with the usual particles. Thus, supersymmetry must be a broken symmetry.

Would the breaking of supersymmetry destroy the delicate cancellation of quadratic divergences in field theoretic models? Fortunately, it does not. We have just seen (by the two examples above) that if SUSY is explicitly broken because scalar masses differ from their fermion counterparts, no new quadratic divergences occur. We state here (without proof) that this is true for all processes, and to all orders in perturbation theory. It is, therefore, possible to introduce new terms such as independent additional masses for the scalars which break SUSY without the reappearance of quadratic divergences. Such terms are said to break SUSY *softly*. Not all SUSY breaking terms are soft. We have already seen that if $m_\psi \neq m$, the expression in (3.47) is quadratically divergent. Thus additional contributions to the fermion mass in the Wess–Zumino model results in a *hard* breaking of supersymmetry. Similarly, any additional contribution to just the quartic scalar interactions will result in the reappearance of a quadratic divergence in the correction to m_A^2 since these contributions only affect the last two diagrams in Fig. 3.2.

Are there other soft SUSY breaking terms possible for the WZ model? Recall the combinatorial factor 3 in the last term in (3.47). This tells us that the contribution of the A loop from the trilinear A^3 interaction is exactly three times bigger than the contribution from the B loop from the AB^2 interaction (the coupling constants for these interactions are exactly equal). Thus, there will be no net quadratic divergence in the expression (3.47) even if we add a term of the form,

$$\mathcal{L}_{\text{soft}} = k(A^3 - 3AB^2) \tag{3.49}$$

to our model, where k is a dimensional coupling constant. Obviously, this interaction does not give a quadratically divergent correction to the one-loop, contribution to m_A^2. It is an example of a soft supersymmetry breaking interaction term. We remark that this term can be written in terms of $\mathcal{S} = \frac{A+iB}{\sqrt{2}}$ as

$$\mathcal{L}_{\text{soft}} = \sqrt{2}k(\mathcal{S}^3 + \quad \text{h.c.}) \tag{3.50a}$$

while an arbitrary splitting in the masses of A and B can be incorporated by including a term,

$$\mathcal{L}_{\text{soft}} = m'^2(\mathcal{S}^2 + \quad \text{h.c.}) \tag{3.50b}$$

into the Lagrangian. It will turn out that super-renormalizable terms that are analytic in $\mathcal{S}$ are soft, while terms that involve products of $\mathcal{S}$ and $\mathcal{S}^*$ (except supersymmetric terms such as $\mathcal{S}^*\mathcal{S}$ already present in (3.45)) result in a hard breaking of SUSY.

Exercise *Check that an interaction proportional to $(\mathcal{S}^2\mathcal{S}^* + \text{h.c.}) \sim (A^2 + B^2)A$ leads to a quadratically divergent contribution to the expression in (3.47).*

Although we have illustrated the cancellation of quadratic divergences with just a few examples, it is important to stress that this is a general feature of supersymmetric theories. As we will elaborate upon in Section 6.7, this cancellation of quadratic divergences occurs to all orders in perturbation theory.

4

The supersymmetry algebra

4.1 Rotations

In classical mechanics, rotations of three-vectors can be represented by a rotation matrix $\mathbf{R}$ acting upon vectors such as $\mathbf{x} = (x, y, z)$ as

$$x_i \rightarrow x_i' = \mathbf{R}_{ij} x_j. \tag{4.1}$$

In quantum mechanics, rotation transformations are represented by unitary operators $U(\boldsymbol{\theta})$ acting upon state vectors $|\psi\rangle$ such that

$$|\psi\rangle \rightarrow |\psi'\rangle = U(\boldsymbol{\theta})|\psi\rangle, \tag{4.2}$$

where the direction of $\boldsymbol{\theta}$ is along the axis about which the rotation occurs, and its magnitude is the rotation angle. For infinitesimal rotations, the operator $U(\boldsymbol{\theta})$ can be written as $U(\boldsymbol{\theta}) \simeq 1 + i\boldsymbol{\theta} \cdot \mathbf{J}$, where the Hermitian operators $\mathbf{J}$ are the rotation generators. For spinless states, J_i can be represented as differential operators ($J_k = \frac{1}{2}\epsilon_{ijk} J_{ij}$, with $J_{ij} = -i(x_i \partial_j - x_j \partial_i)$), it is easy to check that the commutation relations

$$[J_i, J_j] = i\epsilon_{ijk} J_k \tag{4.3}$$

are satisfied for $i, j = 1, 2, 3 \leftrightarrow x, y, z$. Finite rotations can be built up from an infinite product of infinitesimal ones, so that the operator $U(\boldsymbol{\theta}) = \exp(i\boldsymbol{\theta} \cdot \mathbf{J})$. The operators $U(\boldsymbol{\theta})$ form a representation of the Lie group $SU(2)$, for which the J_i are the group generators, and where Eq. (4.3) defines the Lie algebra associated with the group $SU(2)$. Since the parameters θ_i each run over a compact domain 0 to 2π, we say that $SU(2)$ is a compact Lie group.

A Casimir operator is an operator that commutes with all of the group generators. The eigenvalues of a Casimir operator are unchanged under group transformations, so they serve as a useful tool to classify group representations. The representations of $SU(2)$ can be labelled according to the eigenvalues of the quadratic Casimir operator

41

$J^2 = \mathbf{J} \cdot \mathbf{J}$, for which $J^2|jm\rangle = j(j+1)|jm\rangle$, with $j = 0, 1/2, 1, 3/2, 2, \ldots$ For the $j = 1/2$ representation, the operators J_i can be represented by the Pauli spin matrices $J_i = \sigma_i/2$, and the state vectors can be represented by 2-component spinors. Higher j representations can be constructed by taking direct products of lower j representations. For higher j representations of $SU(2)$, the J_i's can be represented by $(2j+1) \times (2j+1)$ matrices, and the corresponding state vectors by $2j+1$ component column matrices.

4.2 The Lorentz group

We want to build a quantum theory that is invariant under Lorentz transformations. We restrict our discussion to proper, orthochronous Lorentz transformations, i.e. boosts and rotations, and neglect parity and time reversal. In addition to rotations which mix the spatial coordinates amongst themselves, we now have boost transformations which mix the time co-ordinate x^0 with the spatial co-ordinates; e.g. a boost along the x^1 direction can be written as $x'^0 = x^0 \cosh\phi + x^1 \sinh\phi$, $x'^1 = x^0 \sinh\phi + x^1 \cosh\phi$, $x'^2 = x^2$ and $x'^3 = x^3$. The usual velocity parameter β that characterizes the boost is given in terms of the rapidity ϕ by $\beta = \tanh\phi$. The infinitesimal transformation matrix U that transforms quantum mechanical states can then be augmented to,

$$U(\boldsymbol{\theta}, \boldsymbol{\phi}) \simeq 1 + i\boldsymbol{\theta} \cdot \mathbf{J} + i\boldsymbol{\phi} \cdot \mathbf{K}, \tag{4.4}$$

where K_i are the boost generators and $\boldsymbol{\phi}$ points along the direction of the boost.[1] A Lorentz transformation is thus characterized by the six parameters (θ_i, ϕ_j). Since the parameters ϕ_j are not restricted to a compact interval, the Lorentz group, unlike the rotation group, is not compact.

The Lorentz group generators satisfy

$$[J_i, J_j] = i\epsilon_{ijk} J_k, \quad [K_i, J_j] = i\epsilon_{ijk} K_k, \quad [K_i, K_j] = -i\epsilon_{ijk} J_k. \tag{4.5}$$

The first of these relations shows that rotation generators form a closed sub-algebra, so that the rotation group forms a subgroup of the Lorentz group. The commutator of two boost generators is a rotation generator (this is the origin of Thomas precession) so that the boosts, by themselves, do not form a sub-algebra.

The Lorentz algebra that we have introduced above can be written in a manifestly covariant form by writing the generators as the six components of an antisymmetric second rank tensor generator $M_{\mu\nu}$, with $M_{ij} = \epsilon_{ijk} J_k$ and $M_{0i} = -M_{i0} = -K_i$. The commutators for the Lorentz group generators can then be recast into covariant

---

[1] For infinitesimal boosts, notice that $|\boldsymbol{\phi}| = |\boldsymbol{\beta}|$.

form

$$[M_{\mu\nu}, M_{\rho\sigma}] = -i(g_{\mu\rho}M_{\nu\sigma} - g_{\mu\sigma}M_{\nu\rho} - g_{\nu\rho}M_{\mu\sigma} + g_{\nu\sigma}M_{\mu\rho}). \qquad (4.6)$$

To find the finite-dimensional unitary representations of the Lorentz group, the generators can be alternatively written by defining $S_i = \frac{1}{2}(J_i + iK_i)$ and $T_i = \frac{1}{2}(J_i - iK_i)$. In this case, it is easy to check that the commutators of the generators become

$$[S_i, S_j] = i\epsilon_{ijk}S_k, \quad [T_i, T_j] = i\epsilon_{ijk}T_k, \quad [S_i, T_j] = 0, \qquad (4.7)$$

i.e. the algebra decomposes into the product of two independent $SU(2)$ groups, for which we know the representations. For the Lorentz group, there are thus *two* Casimir operators, S^2 and T^2, with eigenvalues $s(s+1)$ and $t(t+1)$, again with $s, t = 0, 1/2, 1, \ldots$ (Note that J^2 is no longer a Casimir operator since it no longer commutes with all the group generators, e.g. $[J^2, K_1] \neq 0$.) The irreducible representations can be categorized according to values of (s, t). A Lorentz scalar transforms as the $(0, 0)$ representation while a four-vector transforms as $(1/2, 1/2)$ representation. There are *two* distinct fundamental representations $(1/2, 0)$ and $(0, 1/2)$, each of which corresponds to two-spinors. The $(1/2, 0)$ object, as we will soon see, transforms as a left-handed Weyl two-spinor whose components are usually denoted by ψ_{LA}, with $A = 1, 2$. The $(0, 1/2)$ object transforms as a right-handed two-spinor with components $\psi_R^{\dot{A}}$, where the dot on the index calls attention to the fact that the spinor transforms under the second of the two $SU(2)$ groups.

Exercise *Verify that the boost transformation for a state transforming as the $(1/2, 0)$ and $(0, 1/2)$ representations of the Lorentz group are respectively given by $\psi'_{L,R} = (\cosh\frac{\phi}{2} \mp \sigma \cdot \hat{\mathbf{p}} \sinh\frac{\phi}{2})\psi_{L,R}$, where $\hat{\mathbf{p}}$ is a unit vector along the direction of $\mathbf{p}$. Recalling that $\tanh\phi = \beta$, show that the spinors for states with momentum $\mathbf{p}$ can be obtained from the corresponding rest frame states as,*

$$\psi_{L,R}(\mathbf{p}) = \frac{E + m \mp \sigma \cdot \mathbf{p}}{\sqrt{2m(E+m)}}\psi_{L,R}(0).$$

Noting that $\psi_L(0) = \psi_R(0)$ because there is no preferred direction in the rest frame to define the particle's handedness, show that

$$(E \pm \sigma \cdot \mathbf{p})\psi_{L,R}(\mathbf{p}) = m\psi_{R,L}(\mathbf{p}).$$

This is just the Dirac equation in two-component notation. Notice that for $m = 0$, we have $(\sigma \cdot \hat{\mathbf{p}})\psi_{L,R}(\mathbf{p}) = \mp\psi_{L,R}(\mathbf{p})$, which justifies the use of our labels left and right for the states transforming as the $(1/2, 0)$ and $(0, 1/2)$ representations of the Lorentz group.

A four-component Dirac spinor can be built out of the direct sum of two-component spinors $(1/2, 0) \oplus (0, 1/2)$ so that

$$\psi_a^D = \begin{pmatrix} \psi_{LA} \\ \chi_R^{\dot{A}} \end{pmatrix}. \qquad (4.8)$$

The two spinors ψ_L and χ_R are independent. It is simple to check that the two-component spinor $-i\sigma_2\psi_L^*$ transforms as a $(0, 1/2)$ representation of the Lorentz group, i.e. it transforms as χ_R. We can thus construct a different four-component spinor whose right-handed piece is completely determined by its left-handed pieces via $\chi_R = -i\sigma_2\psi_L^*$. This four-spinor would transform as the $(1/2, 0) \oplus (0, 1/2)$ representation of the Lorentz group, but would have just half as many independent components as ψ^D above. It can be expressed as

$$\psi_a = \begin{pmatrix} \psi_{LA} \\ (-i\sigma_2\psi_L^*)^{\dot{A}} \end{pmatrix}. \qquad (4.9)$$

This object is the Majorana spinor that we have already encountered in Chapter 3, and the relation $\chi_R = -i\sigma_2\psi_L^*$ is simply (3.3).[2] Since the Dirac spinor contains twice as many independent components as a Majorana spinor, it can be thought of as a combination of two Majorana spinors, in much the same way that we can think of a complex number as a combination of two real numbers.

Many textbooks and review articles use the more fundamental two-component spinor notation. Here, we formulate everything in terms of four-component spinors, which are perhaps more familiar to particle physicists interested in performing phenomenological calculations.

4.3 The Poincaré group

In addition to rotations and boosts, the other spacetime transformations include translations in space and time. Translations in space and time are generated by the energy–momentum operator P_μ, which can be represented by the differential operator $P_\mu = i\partial_\mu$. The Poincaré group is formed by combining rotations, boosts, and translations. We then have *ten* independent generators: the six $M_{\mu\nu}$ plus the four P_μ. It is then straightforward to work out the commutation relations for the

---

[2] In Chapter 3 and elsewhere, $\psi_{L,R}$ is also used to denote the four-component spinor $P_{L,R}\psi$ which has only two non-vanishing components in the representation where the matrix γ_5 is diagonal. These non-vanishing components are just the components of the two-spinor $\psi_{L,R}$ discussed in this chapter. Although this is an abuse of notation, it should be clear from the context whether we are using $\psi_{L,R}$ to denote four- or two-component spinors.

generators, using their representation as differential operators. One finds:

$$[P_\mu, P_\nu] = 0, \tag{4.10a}$$

$$[M_{\mu\nu}, P_\lambda] = i(g_{\nu\lambda} P_\mu - g_{\mu\lambda} P_\nu), \tag{4.10b}$$

$$[M_{\mu\nu}, M_{\rho\sigma}] = -i(g_{\mu\rho} M_{\nu\sigma} - g_{\mu\sigma} M_{\nu\rho} - g_{\nu\rho} M_{\mu\sigma} + g_{\nu\sigma} M_{\mu\rho}). \tag{4.10c}$$

To classify the representations of the Poincaré group, we again look for Casimir operators. One of these is the operator P^2, which certainly commutes with all the group generators. Its eigenvalue operating on particle state vectors is just the squared mass $P^2|\psi\rangle = m^2|\psi\rangle$. The other Casimir operator is obtained from the Pauli–Lubanski four-vector $W_\mu = \frac{1}{2}\epsilon_{\mu\nu\rho\sigma} P^\nu M^{\rho\sigma}$, with $W^\mu P_\mu = 0$. The square of the Pauli–Lubanski vector, W^2, can be shown to commute with all the generators of the Poincaré group. Notice also that in the rest frame (of a massive state) W^i is proportional to the rotation generator J^i. The various representations of the Poincaré group were first worked out by Wigner. The physically realized unitary representations which are of interest to us are:

- $P^2 \equiv m^2 > 0$, with $W^2 = -m^2 s(s+1)$, where s denotes the spin quantum number $s = 0, \frac{1}{2}, 1, \ldots$ Thus, these states correspond to particles of definite mass and discrete spin values.
- $P^2 = 0$, $W^2 = 0$ so that $W_\mu = \lambda P_\mu$. Here λ is the state helicity value, and $\lambda = \pm s$, for $s = 0, \frac{1}{2}, 1, \ldots$ These correspond e.g. to single particle states of massless particles such as photons with $\lambda = \pm 1$ or the graviton with $\lambda = \pm 2$.
- Finally, $P^\mu \equiv 0$, corresponding to the vacuum state which is invariant under Poincaré transformations.

In the 1960s, a number of papers were written about the possibility of embedding the spacetime symmetries (i.e. the Poincaré group) into some larger *master* group such as $SU(6)$ that would serve as a more general framework for the symmetry of the laws of physics. These efforts culminated in several no-go theorems, the most general of which was the Coleman–Mandula theorem.[3] It states the following.

Theorem Let G be a connected symmetry group of the S matrix (its generators commute with the S matrix), and assume the following.

- G contains a subgroup which is locally isomorphic to the Poincaré group (Poincaré invariance).
- All particle types correspond to positive energy representations of the Poincaré group. For any finite mass m, there are only a finite number of types of particles with mass less than m.

[3] S. Coleman and J. Mandula, *Phys. Rev.* **159**, 1251 (1967).

- Elastic scattering amplitudes are analytic functions of the Mandelstam variables s and t in some neighborhood of the physical region, except at normal thresholds, and the S matrix is non-trivial in the sense that essentially any two one-particle momentum states scatter, except perhaps for isolated values of s.
- Finally, a technical assumption: the generators of G, considered as integral operators in momentum space, have distributions as their kernels.

Coleman and Mandula asserted that if these conditions hold, *G is locally isomorphic to the direct product of a compact symmetry group and the Poincaré group.*

Stated more simply, under a number of physically reasonable assumptions, it is not possible to form a non-trivial merger of the Poincaré symmetry with other symmetries of the S matrix into a bigger group. It is not possible to have a larger spacetime symmetry and, further, internal symmetries such as local gauge symmetries or additional global symmetries (e.g. isospin) can only be realized as a direct product of these symmetry groups with the Poincaré group. It is intriguing that all the Poincaré group symmetries of the S matrix are in fact realized in nature. It is important to recognize that Coleman and Mandula did not envisage the possibility of anticommuting spinorial charges in their analysis. It is precisely the inclusion of these that allows us to enlarge the spacetime symmetry group to include supersymmetry, as we have already seen in Chapter 3.

4.4 The supersymmetry algebra

Our investigation of the Wess–Zumino model in Chapter 3 shows that it is possible to construct a relativistic quantum field theory that is invariant under supersymmetry transformations, for which the generators are anticommuting spinorial charges Q_a. We saw that the algebra of the Q_a's (this involved anticommutators, which is how the Coleman–Mandula theorem is circumvented) closes to yield P_μ, so that the supersymmetry is, in effect, a spacetime symmetry. In this sense, supersymmetry can be looked upon as a generalization of the special theory of relativity.

We had already worked out the algebra of the spinorial generators Q_a amongst themselves and with the translation generators in Chapter 3. Since this involves anticommutators of the super-charges, it is called a graded Lie algebra. The commutator of the Lorentz generators with the super-charges Q_a is simply given by the fact that these are spin $\frac{1}{2}$ objects. We can thus write the supersymmetric extension of the Poincaré algebra, known as the super-Poincaré algebra, as

$$[P_\mu, P_\nu] = 0, \tag{4.11a}$$

$$[M_{\mu\nu}, P_\lambda] = i(g_{\nu\lambda} P_\mu - g_{\mu\lambda} P_\nu), \tag{4.11b}$$

$$[M_{\mu\nu}, M_{\rho\sigma}] = -i(g_{\mu\rho} M_{\nu\sigma} - g_{\mu\sigma} M_{\nu\rho} - g_{\nu\rho} M_{\mu\sigma} + g_{\nu\sigma} M_{\mu\rho}), \tag{4.11c}$$

$$[P_\mu, Q_a] = 0, \tag{4.11d}$$

$$[M_{\mu\nu}, Q_a] = -(\tfrac{1}{2}\sigma_{\mu\nu})_{ab} Q_b, \tag{4.11e}$$

$$\{Q_a, \bar{Q}_b\} = 2(\gamma^{\mu})_{ab} P_{\mu}. \tag{4.11f}$$

Since Q is a Majorana spinor charge, we can use the last of these relations to work out the anticommutators between two Q's or $\bar{Q}$'s.

Exercise: *Verify that*

$$\{Q_a, Q_b\} = -2(\gamma^{\mu}C)_{ab} P_{\mu},$$

$$\{\bar{Q}_a, \bar{Q}_b\} = 2(C^{-1}\gamma^{\mu})_{ab} P_{\mu}.$$

An extension of the Coleman–Mandula type analysis that allows for spinorial charges was worked out by Haag, Lopuszanski, and Sohnius who showed that the super-Poincaré algebra above is indeed the most general extension of the Poincaré algebra, provided we have just a single spinorial charge Q.[4] These authors also showed that theories with more than one spinorial generator are possible. These are referred to as extended supersymmetry theories. Such theories do not allow chiral representations which, as we know, are crucial for phenomenology. Only theories with a single spinorial generator Q_a, known as $N = 1$ supersymmetry theories, allow chiral representations. For this reason, we restrict our attention only to $N = 1$ supersymmetry.[5] To sum up, the Haag–Lopuszanski–Sohnius theorem tells us that the most general symmetry of the S matrix is the direct product of some internal symmetry with super-Poincaré invariance.

The irreducible representations of the super-algebra can be worked out as usual by finding the relevant Casimir operators. For the SUSY algebra above, the operator P^2 again commutes with all generators, so that all particles occurring in a super-multiplet will have the same mass. However, the square of the Pauli–Lubanski pseudovector W^2 is no longer a Casimir invariant, so that supermultiplets can now contain particles of differing spins. We will not discuss the construction of a new Casimir operator for this case, but instead focus on the particle supermultiplets that furnish representations of the super-Poincaré algebra.

For the massive case $P^2 \equiv m^2 > 0$, the representations are labeled by (m, j) with $j = 0, 1/2, 1, \ldots$ For fixed m, the complete supermultiplet contains a state each corresponding to spin $s = j \pm 1/2$, and two states with spin $s = j$ (except for the case $j = 0$ where the state with $s = j - 1/2$ is absent), all of which have the same mass. Notice that the number of helicity states for the two objects with spin j $[2(2j + 1)]$ is exactly balanced by the corresponding number of helicity

---

[4] R. Haag, J. Lopuszanski and M. Sohnius, *Nucl. Phys.* **B88**, 257 (1975).

[5] The underlying fundamental theory could be an extended supersymmetric theory, but all but one (or none!) of the supersymmetries must somehow be broken at much higher scales, leaving an $N = 1$ supersymmetric theory as the extension of the Standard Model that could have phenomenological relevance.

states for the two states with spins $j \pm \frac{1}{2}$ $[2(j + 1/2) + 1 + 2(j - 1/2) + 1]$. This is just the statement that the number of bosonic and fermionic helicity states are the same. If $j = 0$, we have the multiplet of the Wess–Zumino model – two spin zero states and two spin half states – as we discussed in Chapter 3. If $j = 1/2$, there are four bosonic degrees of freedom (three spin 1 and one spin zero) balanced by four fermionic degrees of freedom corresponding to two Majorana spin half fermions as we will see when we study spontaneously broken gauge theories.

For the massless case, one can show that if j is the state with largest helicity in a supermultiplet, it is always accompanied by another state with helicity $j - 1/2$. Furthermore, a Lorentz invariant field theory always contains these states together with their CPT conjugates which have opposite helicities, $-j$ and $-j + 1/2$ and which are also massless. These states constitute a complete massless supermultiplet. If $j = 1/2$, this multiplet consists of two fermionic states with helicities $\pm 1/2$, and a pair of spin zero bosonic states. This multiplet occurs in the massless Wess–Zumino model. Such a multiplet would also describe a massless neutrino and antineutrino together with its supersymmetric partners which would be two spin zero states (which can be regarded as quanta of one massless *complex* scalar field). For $j = 1$, the bosonic states would correspond to a massless gauge boson (helicities ± 1); the fermionic partner states, which have helicity $\pm 1/2$, then correspond to a Majorana fermion referred to as a *gaugino*. This "gauge multiplet" is a crucial ingredient of supersymmetric gauge theories, which form the basis of all supersymmetry phenomenology. Finally, if $j = 2$, we see that the two bosonic states have helicities ± 2. These are thought to correspond to a graviton, the massless spin two quantum that mediates gravity. The fermionic partners of these $j = 2$ states have helicities $\pm 3/2$ and describe a spin $\frac{3}{2}$ massless Majorana fermion referred to as a *gravitino*. This "gravity multiplet" is essential for theories in which the parameter α of SUSY transformations depends on x^μ. These locally supersymmetric theories necessarily involve Einsteinian gravity, and are referred to as *supergravity* theories.

5

Superfield formalism

We saw in Chapter 3 how the Wess–Zumino model could be formulated in terms of the fields S, ψ_L, and the auxiliary field $\mathcal{F}$, which transform into each other under a supersymmetry transformation. Here, we simply "pulled a Lagrangian out of a hat", and verified by brute force that (at least the free part of) this Lagrangian led to a supersymmetric action. While this example was instructive, it provided no guidance as to how to write down other more complicated supersymmetric theories. We alluded, however, to the fact that we could think of the fields, S, ψ_L, and $\mathcal{F}$ as the components of a single entity, a chiral superfield.[1]

The superfield formalism provides a convenient way to formulate general rules for the construction of supersymmetric Lagrangians, even for theories with non-Abelian gauge symmetry that are the foundation of modern particle physics. The superfield calculus that we develop in this and succeeding chapters will provide us with a constructive procedure for writing down theories that are guaranteed to be supersymmetric. This procedure will ultimately be used to write down the simplest supersymmetric extension of the Standard Model. This theory, augmented with suitable soft supersymmetry breaking terms, is known as the Minimal Supersymmetric Standard Model, or MSSM.

5.1 Superfields

To begin, we would like to somehow combine the fields S, ψ_L, and $\mathcal{F}$ into a single "superfield", in much the same way that the neutron and proton fields are combined into a single "nucleon" field in the isospin formalism. The fields S and ψ transform differently under Lorentz transformations so that it is by no means obvious how to combine these fields into a single entity called a superfield in which the component fields all enter on the same footing, i.e. we do not combine the scalar bilinear in ψ

---

[1] A. Salam and J. Strathdee, *Nucl. Phys.* **B76**, 477 (1974).

49

with the scalars S and $\mathcal{F}$. We are thus led to introduce a new Majorana spinor θ, with components θ_1, θ_2, θ_3, and θ_4, which can be combined with ψ to make a Lorentz scalar that can then be "added" to S. Furthermore, since the components of ψ obey anticommutation relations, the components of θ will be taken to be anticommuting *Grassmann* numbers, so that

$$\{\theta_a, \theta_b\} = 0. \tag{5.1}$$

We will further assume that

$$\{\theta_a, \psi_b\} = 0. \tag{5.2}$$

Note that Eq. (5.1) implies that $\theta_a \theta_a = 0$ (no sum on a).

The spinor θ is determined by the four independent quantities θ_a that we have introduced. We emphasize that these are not complex numbers, but are a new type of object, a Grassmann number. Although these do not commute, we should be clear that they are *not* operators, but anticommuting numbers, in the same sense that usual complex numbers are commuting numbers. These Grassmann numbers (sometimes also referred to as a-numbers in analogy with commuting c-numbers) also anticommute with fermionic operators, but commute with bosonic operators.

The Majorana condition, $\bar{\theta} = \theta^T C$ means that the components of the conjugate spinor $\bar{\theta}$ are completely determined in terms of the four independent θ_as. Thus a product of any chain of larger than a total of four θ or $\bar{\theta}$s is identically zero. Alternatively, it will sometimes be convenient to think of two components of θ and two components of $\bar{\theta}$ as independent, or that each of the two components of θ_L and θ_R are independent.

A superfield is a function of x^μ and θ. The spinor θ (together with the coordinate vector x^μ) is a superfield label in exactly the same way that the coordinate vector x^μ is a label in the conventional formulation of field theory. We will denote superfields by carets and let $\hat{\Phi}(x, \theta)$ stand for a general superfield. The field $\hat{\Phi}$ thus depends on four (commuting) spacetime co-ordinates x^μ and on four anticommuting co-ordinates, θ_a. The extension of four-dimensional spacetime to include the four anticommuting dimensions is usually referred to as superspace. Whether the anticommuting variables have a physical significance, or whether they serve only as bookkeeping devices is something we will not dwell upon.

An important property of functions of Grassmann variables follows from the fact that any power series expansion in terms of the anticommuting co-ordinates always terminates because the square of any Grassmann variable is zero. For instance, if η is a Grassmann variable, and we have a function $f(\eta)$, then $f(\eta) = A + B\eta$, where A and B are just ordinary (real or complex) numbers. The power series expansion terminates with the first term in η. A function $g(x, \eta)$ would have a similar expansion, except that the coefficients A and B would now be (real or complex) functions of x. We can similarly write the superfield $\hat{\Phi}$ in terms of independent products of the four θ_a variables, with coefficients that are functions of spacetime co-ordinates x^μ.

Exercise *Verify that from the four Grassmann variables θ_a, $a = 1, 2, 3, 4$, one can make exactly 16 independent products of $0, 1 \ldots 4$ θs. The most obvious choice is 1, θ_a (4 terms), $\theta_a \theta_b$ (6 terms, because of the anticommutativity of the θ_as), $\theta_a \theta_b \theta_c$ (4 terms, as any one θ_a from the unique quartic term in the θs can be omitted) and finally one quartic term, $\theta_1 \theta_2 \theta_3 \theta_4$.*

We could thus expand $\hat{\Phi}(x, \theta) = A + B\theta_1 + \cdots + P\theta_1\theta_2\theta_3\theta_4$. It is, however, more convenient to expand the superfield in terms of

$$1 \text{ term independent of } \theta \;\; ; \;\; \mathbf{1}, \tag{5.3a}$$

$$4 \text{ terms linear in } \theta \;\; ; \;\; \text{choose } \bar{\theta}\gamma_5, \tag{5.3b}$$

$$6 \text{ terms bilinear in } \theta \;\; ; \;\; \text{choose } \bar{\theta}\theta, \; \bar{\theta}\gamma_5\theta, \; \bar{\theta}\gamma_\mu\gamma_5\theta, \tag{5.3c}$$

$$4 \text{ terms trilinear in } \theta \;\; ; \;\; \text{choose } \bar{\theta}\gamma_5\theta \cdot \bar{\theta}, \tag{5.3d}$$

$$1 \text{ term quartic in } \theta \;\; ; \;\; \text{choose } (\bar{\theta}\gamma_5\theta)^2, \tag{5.3e}$$

since this manifestly displays the Lorentz properties of the "expansion coefficients" which will ultimately be the usual fields in the theory. Terms such as $\bar{\theta}\gamma_\mu\theta$ and $\bar{\theta}\sigma_{\mu\nu}\theta$ are identically zero due to Eqs. (3.8c) and (3.8e). We can thus write a general superfield as,[2]

$$\hat{\Phi}(x, \theta) = \mathcal{S} - i\sqrt{2}\bar{\theta}\gamma_5\psi - \frac{i}{2}(\bar{\theta}\gamma_5\theta)\mathcal{M} + \frac{1}{2}(\bar{\theta}\theta)\mathcal{N} + \frac{1}{2}(\bar{\theta}\gamma_5\gamma_\mu\theta)V^\mu$$

$$+ i(\bar{\theta}\gamma_5\theta)[\bar{\theta}(\lambda + \frac{i}{\sqrt{2}}\partial\!\!\!/\psi)] - \frac{1}{4}(\bar{\theta}\gamma_5\theta)^2[\mathcal{D} - \frac{1}{2}\Box\mathcal{S}]. \tag{5.4}$$

Thus, the coefficients in the above expansion are the sixteen component fields

$$\mathcal{S}, \; \psi, \; \mathcal{M}, \; \mathcal{N}, \; V^\mu, \; \lambda, \; \text{and } \mathcal{D}. \tag{5.5}$$

Here, V^μ is a vector field and ψ and λ are spinor fields. In general, the bosonic fields are complex, while ψ and λ are Dirac fields. The peculiar form of the coefficients of trilinear and quartic terms in θ in this expansion as well as the factors of half and $\sqrt{2}$ is chosen for future convenience. It should be obvious to the reader that although any scalar superfield can be written as in Eq. (5.4), this form is not unique. We will

---

[2] Actually, this is not the most general superfield since we have assumed that the θ independent term in the expansion is a Lorentz scalar. It is possible, and indeed necessary as we will see when we consider supersymmetric gauge theories, to introduce superfields where this is not the case. Such superfields will carry an additional index which specifies the Lorentz transformation property of their leading, i.e. θ-independent component. We will refer to the superfield in Eq. (5.4) as a *scalar* superfield.

regard (5.4) as the canonical form. Any other expansion can be straightforwardly reduced to this canonical form using identities amongst the Grassmann variables introduced later in this chapter.

Let us compute the Hermitian conjugate superfield $\hat{\Phi}^\dagger$. We will need the identities,

$$(\bar{\psi}\chi)^\dagger = \bar{\chi}\psi = \overline{\psi^c}\chi^c, \tag{5.6a}$$

$$(\bar{\psi}\gamma_5\chi)^\dagger = -\bar{\chi}\gamma_5\psi = -\overline{\psi^c}\gamma_5\chi^c, \quad \text{and} \tag{5.6b}$$

$$(\bar{\psi}\partial\!\!\!/\chi)^\dagger = \partial_\mu\bar{\chi}\gamma_\mu\psi = -\overline{\psi^c}\partial\!\!\!/\chi^c, \tag{5.6c}$$

so that

$$(\bar{\theta}\theta)^\dagger = \bar{\theta}\theta, \tag{5.7a}$$

$$(i\bar{\theta}\gamma_5\theta)^\dagger = i\bar{\theta}\gamma_5\theta, \quad \text{and} \tag{5.7b}$$

$$(\bar{\theta}\gamma_5\gamma_\mu\theta)^\dagger = \bar{\theta}\gamma_5\gamma_\mu\theta. \tag{5.7c}$$

Then,

$$\hat{\Phi}^\dagger(x,\theta) = \mathcal{S}^\dagger - i\sqrt{2}\bar{\theta}\gamma_5\psi^c - \frac{i}{2}(\bar{\theta}\gamma_5\theta)\mathcal{M}^\dagger + \frac{1}{2}(\bar{\theta}\theta)\mathcal{N}^\dagger + \frac{1}{2}(\bar{\theta}\gamma_5\gamma_\mu\theta)V^{\mu\dagger}$$

$$+ i(\bar{\theta}\gamma_5\theta)[\bar{\theta}(\lambda^c + \frac{i}{\sqrt{2}}\partial\!\!\!/\psi^c)] - \frac{1}{4}(\bar{\theta}\gamma_5\theta)^2[\mathcal{D}^\dagger - \frac{1}{2}\Box\mathcal{S}^\dagger]. \tag{5.8}$$

We define the superfield $\hat{\Phi}$ to be real if $\hat{\Phi} = \hat{\Phi}^\dagger$. In this case, we see that the bosonic fields are real and the fermionic fields are Majorana ($\psi = \psi^c$ and $\lambda = \lambda^c$). It was for this reason that we inserted the factors of i in our superfield expansion in Eq. (5.4). In general, however, $\hat{\Phi}$ need not be real.

Exercise *Verify the relations in (5.6a), (5.6b), and (5.6c). Notice that these hold regardless of whether the spinors are Dirac or Majorana.*

5.2 Representations of symmetry generators: a recap

In quantum field theory, symmetry transformations act on field operators which are the dynamical variables. We focus on symmetries which are linear transformations of the field operators. A symmetry operation, with a set of parameters α_a, due to the action of the set of generators Q_a can thus be written as,

$$e^{i\alpha_a Q_a}\phi_m e^{-i\alpha_b Q_b} = \left(e^{-i\alpha_a t_a}\right)_{mn}\phi_n. \tag{5.9a}$$

It is important to understand that $\left(e^{-i\alpha_a t_a}\right)_{mn}$ are simply numerical coefficients. There are, of course, as many parameters α_a as there are generators Q_a, and for each Q_a we have a matrix coefficient (t_a). For an infinitesimal transformation, this becomes,

$$\delta\phi_m = i\alpha_a[Q_a, \phi_m] = -i(\alpha_a t_a)_{mn}\phi_n. \tag{5.9b}$$

By considering the action of successive symmetry transformations and using the Jacobi identity,

$$[[a, b], c] + [[b, c], a] + [[c, a], b] = 0,$$

in the last step, it is straightforward to show that

$$\delta_2\delta_1\phi_m = [i\alpha_{2b}Q_b, \delta_1\phi_m] = -[i\alpha_{1a}Q_a, [\phi_m, i\alpha_{2b}Q_b]] - [\phi_m, [i\alpha_{2b}Q_b, i\alpha_{1a}Q_a]],$$

which then yields,

$$(\delta_2\delta_1 - \delta_1\delta_2)\phi_m = [[i\alpha_{2b}Q_b, i\alpha_{1a}Q_a], \phi_m]. \tag{5.10a}$$

The result of the successive transformations can also be written in terms of the numerical coefficients t_{mn} introduced above as,

$$\delta_2\delta_1\phi_m = -\alpha_{1a}(t_a)_{mn}\alpha_{2b}(t_b)_{np}\phi_p,$$

so that the right-hand side of Eq. (5.10a) can also be written as,

$$(\delta_2\delta_1 - \delta_1\delta_2)\phi_m = -\alpha_{1a}\alpha_{2b}[t_a, t_b]_{mp}\phi_p, \tag{5.10b}$$

with the usual matrix multiplication rule for the product of the matrices t_a and t_b appearing on the right-hand side.

The set of generators Q_a satisfies algebraic commutation relations that are determined by the symmetry in question. If these are the generators of spacetime symmetries, these are the commutation relations of the Poincaré algebra. If these are generators of internal symmetry transformations, they satisfy the commutation relations of the corresponding symmetry algebra. In both these cases (and many others that we encounter), the algebra is a Lie algebra, so that the commutation rules can be written as,

$$[Q_a, Q_b] = if_{abc}Q_c,$$

where the coefficients f_{abc} are the structure constants of the algebra. Requiring the right-hand sides of (5.10a) and (5.10b) to be the same, we see that the set of coefficient matrices t_a must satisfy,

$$[t_a, t_b] = -if_{bac}t_c = if_{abc}t_c.$$

In other words, these coefficient matrices obey the same commutation relations as the abstract generators Q. We say that these furnish a representation of the symmetry algebra.

Exercise *We implicitly assumed that the parameters α_a are commuting numbers when we showed that the matrices t_a obey the same commutation relations as the generators Q_a. If instead the parameters are anticommuting numbers, and the generators Q_a and Q_b obey an anticommutation relation, show that corresponding matrices t_a and t_b obey these same relations, and so, furnish a representation of this graded algebra. In this case, of course, the exponential in Eq. (5.9a) becomes a polynomial.*

The familiar Pauli matrices or the Gell-Mann matrices are examples of matrix representations of the generators of internal symmetry groups $SU(2)$ and $SU(3)$. But what does all this have to do with the representation of spacetime symmetry generators by differential operators that we have seen in Chapter 4? The underlying idea is the same. For instance, the momentum, defined as the generator of translations, satisfies

$$\phi \to \phi' = \phi(x+a) = e^{ia^\mu P_\mu}\phi e^{-ia^\mu P_\mu} \simeq \phi(x) + a^\mu \frac{\partial\phi}{\partial x^\mu} + \cdots \qquad (5.11)$$

For an infinitesimal translation we find,

$$\phi' = \phi + \delta\phi = (1 + ia^\mu P_\mu)\phi(1 - ia^\mu P_\mu) = \phi + a^\mu \frac{\partial\phi}{\partial x^\mu}, \qquad (5.12)$$

or

$$[P_\mu, \phi] = -i\partial_\mu\phi. \qquad (5.13)$$

Using Eq. (5.9b), we see that the translation generator P_μ can be represented by $\delta(x - x') \times i\partial_\mu$ (where the indices m and n are the continuous spacetime indices x and x'). It is customary to omit the "identity matrix" $\delta(x - x')$ when writing this, and we frequently say that P_μ is represented by the differential operator $i\partial_\mu$. The other generators of the Poincaré algebra can be similarly represented by differential operators. It is then straightforward to check that the differential operators furnish a representation of the Poincaré algebra, i.e. they obey the same commutation relations as the generators.

5.3 Representation of SUSY generators as differential operators

We have just seen that the generators of the Poincaré algebra can be represented by differential operators, where the derivative is with respect to the spacetime co-ordinate. We now want to realize the spinorial generator of supersymmetry transformations Q as a differential operator in superspace acting on the superfield $\hat{\Phi}(x, \theta)$.

This requires us to first explain what is meant by derivatives with respect to Grassmann numbers θ_a. First, since the four θ_as (or, the four $\bar{\theta}_a$s) are independent we *define*,

$$\frac{\partial \theta_a}{\partial \theta_b} = \delta_{ab} \quad \text{and} \quad \frac{\partial \bar{\theta}_a}{\partial \bar{\theta}_b} = \delta_{ab}. \tag{5.14}$$

Then, since $\theta_a = C_{ab}\bar{\theta}_b$, we have

$$\frac{\partial \theta_a}{\partial \bar{\theta}_b} = C_{ab}. \tag{5.15}$$

If we have a product of θs, we must bring $\partial/\partial\theta_a$ next to the θ we wish to differentiate, e.g.

$$\frac{\partial}{\partial \theta_c}(\theta_a \theta_b) = \frac{\partial \theta_a}{\partial \theta_c}\theta_b - \theta_a \frac{\partial \theta_b}{\partial \theta_c} = \delta_{ac}\theta_b - \theta_a \delta_{bc}, \tag{5.16}$$

where the "$-$" sign arises because the θs anticommute. Differentiation of a product of $\bar{\theta}$s or a combination of θs and $\bar{\theta}$s is analogously defined.

Since Q is a spinor operator, its action on a superfield $\hat{\Phi}$ correspondingly changes its Lorentz transformation properties by either taking away or adding a θ to each term. Since, as we have just seen, differentiation with respect to θ_a removes a θ, we are led to try,

$$[Q_m, \hat{\Phi}] = \left(M_{mn}\frac{\partial}{\partial \bar{\theta}_n} + N_{mn}\theta_n \right) \hat{\Phi}(x, \theta), \tag{5.17}$$

where the matrices M_{mn} and N_{mn} (which may depend on x) have to be determined. The reader may wonder why we wrote the derivative with respect to $\bar{\theta}$ rather than θ. By Eq. (5.15), these are the same up to the numerical matrix C_{ab}. We will see shortly that by writing it as in Eq. (5.17), the matrix M becomes a multiple of the identity matrix, and we can write the representation of a SUSY transformation with the Majorana spinor parameter α as,

$$[\bar{\alpha}Q, \hat{\Phi}] = \left(\bar{\alpha}\frac{\partial}{\partial \bar{\theta}} + \bar{\alpha}N\theta \right) \hat{\Phi}. \tag{5.18}$$

We can work out what N must be by applying the Jacobi identity to two successive SUSY transformations by amounts α_1 and α_2. This gives,

$$[[\bar{\alpha}_1 Q, \bar{\alpha}_2 Q], \hat{\Phi}] = [\bar{\alpha}_1 Q, [\bar{\alpha}_2 Q, \hat{\Phi}]] - [\bar{\alpha}_2 Q, [\bar{\alpha}_1 Q, \hat{\Phi}]]. \tag{5.19}$$

We then write each term on the RHS as an action of successive SUSY transformations using (5.18) to obtain,

$$\left(\bar{\alpha}_1 \frac{\partial}{\partial \bar{\theta}} + \bar{\alpha}_1 N\theta \right) \left(\bar{\alpha}_2 \frac{\partial}{\partial \bar{\theta}} + \bar{\alpha}_2 N\theta \right) \hat{\Phi} - (2 \leftrightarrow 1).$$

A little manipulation of indices (and remembering that both θs and αs are anticommuting variables) shows that the terms involving no derivatives with respect to $\bar{\theta}$ give zero, as do the terms involving two such derivatives. We are then left only with two terms, each involving a θ derivative from one factor multiplying $N\theta$ from the other factor. We leave the following as an exercise for the reader.

Exercise *Verify that the RHS of (5.19) reduces to*
$$\left[-\bar{\alpha}_{1a}\bar{\alpha}_{2b}\,(NC)_{ba} + \bar{\alpha}_{2b}\bar{\alpha}_{1a}\,(NC)_{ab}\right]\hat{\Phi}\,.$$

On the other hand, the inner commutator of the LHS of (5.19) becomes

$$\bar{\alpha}_{2b}\bar{\alpha}_{1a}\{Q_a, Q_b\} = -2\bar{\alpha}_{2b}\bar{\alpha}_{1a}(\gamma_\mu C)_{ab} P_\mu$$

so that

$$\left[[\bar{\alpha}_1 Q, \bar{\alpha}_2 Q], \hat{\Phi}\right] = 2i\bar{\alpha}_{2b}\bar{\alpha}_{1a}(\gamma_\mu C)_{ab}\partial_\mu \hat{\Phi}.$$

We are thus led to require that the matrix N must satisfy,

$$(NC)_{ba} + (NC)_{ab} = 2i\,(\partial C)_{ba}\,,$$

whose solution may be written as $N = i\partial$. Of course, because each term in the Jacobi identity is quadratic in Q, we cannot fix the overall factor in front of (5.18) from this. The choice of this factor is a convention. We will choose it to be i, which as we will see later is consistent with the SUSY transformations of chiral scalar superfields that we have already introduced in Chapter 3. We thus obtain the desired realization of the SUSY generator,

$$[\bar{\alpha} Q, \hat{\Phi}] = i\left(\bar{\alpha}\frac{\partial}{\partial\bar{\theta}} + i\bar{\alpha}\partial\theta\right)\hat{\Phi}. \tag{5.20}$$

This expression for the supersymmetry generator is the analogue of (5.13) for the translation generator.

5.4 Useful θ identities

Before proceeding further, we have a short digression to establish a number of useful identities for Grassmann numbers θ that we have introduced into our formalism. These identities are especially useful when we do superfield manipulations. For instance, we may need to take a product of two (or more) superfields which, since it is just a function of x and θ coordinates, is itself a superfield, but not in the canonical form of Eq. (5.4). Indeed most manipulations will leave us with a superfield which is not in this canonical form. However, in order to read off the components of the resulting superfield, or simply to add superfields, we will need to be able to recast any superfield into canonical form. We have found the following identities to be very useful for this purpose, and we will use them repeatedly in our subsequent manipulations.

One manipulation that we need repeatedly is regrouping θs and $\bar{\theta}$s into a common set of bilinears. For this purpose, it is very useful to note that,

$$\theta_a\bar{\theta}_b = -\frac{1}{4}\left\{\bar{\theta}\gamma_5\theta(\gamma_5)_{ab} + \bar{\theta}\theta\delta_{ab} - (\bar{\theta}\gamma^\mu\gamma_5\theta)(\gamma_\mu\gamma_5)_{ab}\right\}. \tag{5.21}$$

This is the basic formula that underlies the Fierz re-arrangement discussed in Chapter 3.

We list below various relations that we have found very useful for superfield manipulation. We outline how to establish these, and leave it to the reader to verify these in detail.

Bilinear Identities

$$\bar{\theta}\gamma_\mu\theta = 0, \tag{5.22a}$$

$$\bar{\theta}\sigma_{\mu\nu}\theta = 0, \tag{5.22b}$$

$$\bar{\theta}\gamma_\mu\gamma_\nu\theta = g_{\mu\nu}\bar{\theta}\theta, \tag{5.22c}$$

$$\bar{\theta}\gamma_5\gamma_\mu\gamma_\nu\theta = g_{\mu\nu}\bar{\theta}\gamma_5\theta, \tag{5.22d}$$

$$\bar{\theta}\gamma_\mu\theta_{L/R} = -\bar{\theta}\gamma_\mu\theta_{R/L}, \tag{5.22e}$$

$$\bar{\theta}\gamma_\mu\gamma_5\theta_{L/R} = \bar{\theta}\gamma_\mu\gamma_5\theta_{R/L}. \tag{5.22f}$$

The first two are the result of the Majorana character of θ and follow immediately from (3.8b) and (3.8c) of Chapter 3. To establish the next two, decompose $\gamma_\mu\gamma_\nu$ into its symmetric and antisymmetric parts, and use (3.8e) to see that the latter gives zero. Finally, the last two follow from the fact the vector bilinear identically vanishes.

Trilinear Identities

$$\bar{\theta}\theta \cdot \theta = -\bar{\theta}\gamma_5\theta \cdot (\gamma_5\theta), \tag{5.23a}$$

$$\bar{\theta}\theta \cdot \bar{\theta} = -\bar{\theta}\gamma_5\theta \cdot (\bar{\theta}\gamma_5), \tag{5.23b}$$

$$\bar{\theta}\gamma_5\gamma_\mu\theta \cdot \theta = -\bar{\theta}\gamma_5\theta \cdot (\gamma_\mu\theta), \tag{5.23c}$$

$$\bar{\theta}\gamma_5\gamma_\mu\theta \cdot \bar{\theta} = \bar{\theta}\gamma_5\theta \cdot (\bar{\theta}\gamma_\mu). \tag{5.23d}$$

To prove the first, we note that we can write the left-hand side in terms of θ alone (using $\bar{\theta} = \theta^T C$) as $\theta_L^T C\theta_L\theta_R + \theta_R^T C\theta_R\theta_L$. Here, we have used the fact that any product of three θ_Ls or three θ_Rs identically vanishes as only two of these are independent (and θs anticommute). The reader can similarly check that the right-hand side of (5.23a) reduces to this same quantity. Eq. (5.23b) can be proven in the same manner, or alternatively, by taking the Dirac conjugate of (5.23a).

To establish (5.23c) we first show using Eq. (5.21) that

$$(\overline{\theta_L}\gamma_5\gamma_\mu\theta_L)\theta_R = (\overline{\theta_R}\gamma_5\gamma_\mu\theta_R)\theta_R = -\frac{1}{2}(\bar{\theta}\gamma_5\theta)P_R\gamma_\mu\theta,$$

$$(\overline{\theta_L}\gamma_5\gamma_\mu\theta_L)\theta_L = (\overline{\theta_R}\gamma_5\gamma_\mu\theta_R)\theta_L = -\frac{1}{2}(\bar{\theta}\gamma_5\theta)P_L\gamma_\mu\theta.$$

Combining these appropriately immediately leads to (5.23c). Eq. (5.23d) may be obtained by taking the Dirac conjugate of (5.23c).

Quartic Identities

$$\bar{\theta}\gamma_5\theta \cdot \bar{\theta}\theta = 0, \tag{5.24a}$$

$$\bar{\theta}\gamma_5\theta \cdot \bar{\theta}\gamma_\mu\gamma_5\theta = 0, \tag{5.24b}$$

$$\bar{\theta}\theta \cdot \bar{\theta}\gamma_\mu\gamma_5\theta = 0, \tag{5.24c}$$

$$(\bar{\theta}\theta)^2 = -(\bar{\theta}\gamma_5\theta)^2, \tag{5.24d}$$

$$\bar{\theta}\gamma_5\gamma_\mu\theta \cdot \bar{\theta}\gamma_5\gamma_\nu\theta = -g_{\mu\nu}(\bar{\theta}\gamma_5\theta)^2. \tag{5.24e}$$

The first of these follows if we recognize that $\bar{\theta}\Gamma\theta = \theta_R^T C\theta_R \pm \theta_L^T C\theta_L$ where the upper (lower) sign corresponds to $\Gamma = I(\gamma_5)$, and use the fact that a product of three or more θ_Ls or θ_Rs identically vanishes. Writing the left-hand side of (5.24b) or (5.24c) in terms of its chiral components immediately shows that it vanishes. Multiplying Eq. (5.23a) on the left by $\bar{\theta}$ immediately leads to (5.24d). Finally, the last of these identities may be obtained from $\bar{\theta}\gamma_5\gamma_\mu\theta \cdot \bar{\theta}\gamma_5\gamma_\nu\theta = -\bar{\theta}\gamma_5\theta \cdot \bar{\theta}\gamma_5\gamma_\nu\gamma_\mu\theta$ which follows from Eq. (5.23c); then using (5.22d) immediately yields (5.24e).

Exercise *Convince yourself that the θ identities that we have listed are valid.*

The trilinear [quartic] identities show how various trilinear [quartic] terms in θ can be recast as $\bar{\theta}\gamma_5\theta \cdot \bar{\theta}\ [(\bar{\theta}\gamma_5\theta)^2]$ that appear in our canonical form of the superfield in (5.4). Quadratic terms can be similarly cast into the forms appearing there. We expect that it is now clear to the reader how any other form for the expansion of the superfield may be reduced to this canonical form.

5.5 SUSY transformations of superfields

We are now in a position to compute how a general superfield $\hat{\Phi}(x, \theta)$ changes under an infinitesimal SUSY transformation. Our starting point is the relation

$$\delta\hat{\Phi} = i\left[\bar{\alpha}Q, \hat{\Phi}\right] = \left(-\bar{\alpha}\frac{\partial}{\partial\bar{\theta}} - i\bar{\alpha}\partial\!\!\!/\theta\right)\hat{\Phi}. \tag{5.25}$$

To proceed, we must work out the action of $\partial/\partial\bar{\theta}$ on various terms in $\hat{\Phi}$. For instance, to work out $\frac{\partial}{\partial\bar{\theta}}(\bar{\theta}\theta)$, it helps again to keep track of spinor indices:

$$\frac{\partial}{\partial\bar{\theta}_a}(\bar{\theta}_b\theta_b) = \theta_a - \bar{\theta}_b C_{ba}.$$

But

$$\bar{\theta}_b C_{ba} = C_{ab}^T \bar{\theta}_b^T = -(C\bar{\theta}^T)_a = -\theta_a$$

so that

$$\frac{\partial}{\partial\bar{\theta}}(\bar{\theta}\theta) = 2\theta. \tag{5.26a}$$

In a similar fashion, we can show that,

$$\frac{\partial}{\partial\bar{\theta}}(\bar{\theta}\gamma_5\theta) = 2\gamma_5\theta, \tag{5.26b}$$

$$\frac{\partial}{\partial\bar{\theta}}(\bar{\theta}\gamma_\mu\gamma_5\theta) = 2\gamma_\mu\gamma_5\theta, \tag{5.26c}$$

$$\frac{\partial}{\partial\bar{\theta}_a}(\bar{\theta}\gamma_5\theta) \cdot \bar{\theta}_b = 2(\gamma_5\theta)_a\bar{\theta}_b + \bar{\theta}\gamma_5\theta\delta_{ab}, \tag{5.26d}$$

$$\frac{\partial}{\partial\bar{\theta}}(\bar{\theta}\gamma_5\theta)^2 = 4(\bar{\theta}\gamma_5\theta) \cdot (\gamma_5\theta). \tag{5.26e}$$

We can now evaluate the RHS of Eq. (5.25). First, using $\chi \equiv \lambda + \frac{i}{\sqrt{2}}\partial\psi$, we find

$$-\bar{\alpha}\frac{\partial}{\partial\bar{\theta}}\hat{\Phi} = i\sqrt{2}\bar{\alpha}\gamma_5\psi + i\bar{\theta}\gamma_5\alpha\mathcal{M} - \bar{\theta}\alpha\mathcal{N} - \bar{\theta}\gamma_5\gamma_\mu\alpha V^\mu + \frac{i}{2}\theta\theta\bar{\alpha}\gamma_5\chi$$

$$-\frac{i}{2}\bar{\theta}\gamma_5\theta\bar{\alpha}\chi + \frac{i}{2}\bar{\theta}\gamma_\mu\gamma_5\theta\bar{\alpha}\gamma^\mu\chi + \bar{\theta}\gamma_5\theta\bar{\theta}\gamma_5\alpha\left[\mathcal{D} - \frac{1}{2}\Box\mathcal{S}\right]. \tag{5.27}$$

Next,

$$-i\bar{\alpha}\partial[\theta\hat{\Phi}] = -i\bar{\alpha}\partial\mathcal{S}\theta - \sqrt{2}\bar{\alpha}\partial\theta\bar{\theta}\gamma_5\psi - \frac{1}{2}\bar{\theta}\gamma_5\theta\bar{\alpha}\partial\theta\mathcal{M}$$

$$-\frac{i}{2}\theta\theta\bar{\alpha}\partial\theta\mathcal{N} - \frac{i}{2}\bar{\theta}\gamma_5\gamma_\mu\theta\bar{\alpha}\partial\theta V^\mu + \bar{\theta}\gamma_5\theta\bar{\alpha}\partial\theta\bar{\theta}\chi. \tag{5.28}$$

The superfield in Eq. (5.27) is already in the canonical form. We have used the (anti)symmetry properties of Majorana spinor bilinears as well as (5.21) to write it this way. We must similarly re-arrange the last expression so that we can combine it with (5.27) to obtain $\delta\hat{\Phi}$ (with components $\delta\mathcal{S}, \delta\psi, \ldots$) in the canonical form. By comparing "coefficients", we obtain the transformation laws for the components of a general scalar superfield:

$$\delta S = i\sqrt{2}\bar{\alpha}\gamma_5\psi, \tag{5.29a}$$

$$\delta\psi = -\frac{\alpha\mathcal{M}}{\sqrt{2}} - i\frac{\gamma_5\alpha\mathcal{N}}{\sqrt{2}} - i\frac{\gamma_\mu\alpha V^\mu}{\sqrt{2}} - \frac{\gamma_5\partial S\alpha}{\sqrt{2}}, \tag{5.29b}$$

$$\delta\mathcal{M} = \bar{\alpha}\left(\lambda + i\sqrt{2}\partial\psi\right), \tag{5.29c}$$

$$\delta\mathcal{N} = i\bar{\alpha}\gamma_5\left(\lambda + i\sqrt{2}\partial\psi\right), \tag{5.29d}$$

$$\delta V^\mu = -i\bar{\alpha}\gamma^\mu\lambda + \sqrt{2}\bar{\alpha}\partial^\mu\psi, \tag{5.29e}$$

$$\delta\lambda = -i\gamma_5\alpha\mathcal{D} - \frac{1}{2}[\partial, \gamma_\mu]V^\mu\alpha, \tag{5.29f}$$

$$\delta\mathcal{D} = \bar{\alpha}\partial\gamma_5\lambda. \tag{5.29g}$$

Exercise *Perform the required algebra to obtain the transformation laws for the components of the scalar superfield.*

Equations (5.29a)–(5.29g) define a linear transformation of the component fields, and, as expected for a SUSY transformation, the variation of a bosonic (fermionic) field is proportional to a fermionic (bosonic) field.

5.6 Irreducible SUSY multiplets

We have just seen that the components of a general scalar superfield transform into one another under supersymmetry. This does not, however, mean that we require *all* the components to be simultaneously present. Of course, we cannot arbitrarily leave out any component since these would "be generated" by the transformation. For instance, if we said S was absent, we would see that it would be generated by the transformation as long as $\psi \neq 0$. It is, however, possible that there might be a smaller set of component fields which transform into just one another under SUSY. If we find such a set, we say the representation furnished by the original (larger) set is *reducible*. If this set cannot be reduced any further, we say that it furnishes an *irreducible* representation of supersymmetry.

Exercise *A familiar example of the concept of irreducibility is the representation (of the 3-D rotation transformation) furnished by the tensor $T^{ij} = x^i y^j$, where x^i and y^j are the components of two co-ordinate vectors. Under rotations, the nine components of T^{ij} clearly transform into one another. Show that the six components of $S^{ij} = x^i y^j + x^j y^i$ as well as the three components of $A^{ij} = x^i y^j - x^j y^i$*

separately transform into one another. Show further that while A^{ij} furnishes an irreducible representation, the representation furnished by S^{ij} can be reduced further into a traceless symmetric tensor $\bar{S}^{ij} = S^{ij} - \frac{1}{3}\text{Trace}(S)\delta^{ij}$ whose five components transform among themselves, and the unit tensor δ^{ij}, which is inert under the transformations. This is, of course, the familiar statement that the combination of two angular momentum **1** states gives states with angular momenta **0**, **1**, and **2**.

5.6.1 Left-chiral scalar superfields

Our examination of the Wess–Zumino model in Chapter 3 showed us that there is a consistent supersymmetric model that can be written down in terms of just the $\mathcal{S}$, ψ_L, and $\mathcal{F}$ fields. Furthermore, Eq. (3.16a)–(3.16c) show that these three fields (which are contained in our general superfield) form a multiplet under SUSY transformations. It should, therefore, be possible to find a representation where several of the components of the general superfield $\hat{\Phi}$ are zero or unphysical. In other words, the representation furnished by the components of $\hat{\Phi}$ should be reducible. Since the Wess–Zumino multiplet (3.15) does not include any vector field, we naturally look for a representation where the field strength $(\partial_\mu V_\nu - \partial_\nu V_\mu)$ vanishes, i.e. $V_\mu = \partial_\mu \zeta$. We must, of course, require that this is not altered by the SUSY transformations (5.29a)–(5.29g). In order that δV^μ is also a pure gradient, we infer from (5.29e) $\lambda = 0$. Then, requiring $\delta\lambda = 0$, gives us

$$\delta\lambda = -i\gamma_5\alpha\mathcal{D} - \frac{1}{2}[\gamma_\rho, \gamma_\mu]\partial^\rho\partial^\mu\zeta\alpha = 0,$$

which, in turn, implies $\mathcal{D} = 0$. We can thus consistently choose

$$\lambda = \mathcal{D} = 0, \quad V_\mu = \partial_\mu\zeta.$$

The set of SUSY transformations then reduces to,

$$\delta\mathcal{S} = i\sqrt{2}\bar{\alpha}\gamma_5\psi, \tag{5.30a}$$

$$\delta\psi = -\frac{\alpha\mathcal{M}}{\sqrt{2}} - i\frac{\gamma_5\alpha\mathcal{N}}{\sqrt{2}} - i\frac{\gamma_\mu\alpha V^\mu}{\sqrt{2}} - \frac{\gamma_5\partial\mathcal{S}\alpha}{\sqrt{2}}, \tag{5.30b}$$

$$\delta\mathcal{M} = i\sqrt{2}\bar{\alpha}\partial\psi, \tag{5.30c}$$

$$\delta\mathcal{N} = \sqrt{2}\bar{\alpha}\partial\gamma_5\psi, \tag{5.30d}$$

$$\delta V^\mu = \sqrt{2}\bar{\alpha}\partial^\mu\psi. \tag{5.30e}$$

These can then be written as,

$$\delta\left[\frac{\partial^\mu S \mp iV^\mu}{\sqrt{2}}\right] = \mp 2i\bar{\alpha}\,\partial^\mu\psi_{\substack{L\\R}}, \tag{5.31a}$$

$$\delta\psi_{\substack{L\\R}} = -\frac{M \mp i\mathcal{N}}{\sqrt{2}}\alpha_{\substack{L\\R}} \pm \frac{\partial^\mu S \mp iV^\mu}{\sqrt{2}}\gamma_\mu\alpha_{\substack{R\\L}}, \tag{5.31b}$$

$$\delta\left[\frac{M \mp i\mathcal{N}}{\sqrt{2}}\right] = 2i\bar{\alpha}\,\partial\!\!\!/\,\psi_{\substack{L\\R}}. \tag{5.31c}$$

We then see that the fields

$$\frac{(\partial^\mu S - iV^\mu)}{\sqrt{2}}, \quad \psi_L, \quad \frac{M - i\mathcal{N}}{\sqrt{2}} \tag{5.32}$$

transform into one another, as does the set

$$\frac{(\partial^\mu S + iV^\mu)}{\sqrt{2}}, \quad \psi_R, \quad \frac{M + i\mathcal{N}}{\sqrt{2}}. \tag{5.33}$$

Let us recapitulate what we have accomplished. Starting with a scalar multiplet, by choosing $\lambda = \mathcal{D} = 0$ and $V_\mu = \partial_\mu\zeta$, we have *reduced* the original multiplet into two multiplets such that the component fields of each multiplet transform only among themselves. If the superfield $\hat{\Phi}$ that we started with was real, then these two reduced multiplets are conjugates of one another. If, however, we had started with a complex field $\hat{\Phi}$, the two multiplets are unrelated. A superfield transforming as the set (5.32) is called a left-chiral superfield, while one transforming as the set (5.33) is called a right-chiral superfield. We trust that it is clear that our reduction procedure is conceptually identical to the example of reducing the second rank co-ordinate tensor into its scalar and the traceless symmetric and antisymmetric parts, discussed in the last exercise.

Finally, let us recover the field content of the Wess–Zumino model. We can reduce a complex superfield $\hat{\Phi}$ as described above, and set all the components in the set (5.33) to zero, consistent with SUSY transformations. In other words, we can choose $\psi_R = 0$, $V^\mu = i\partial^\mu S$ and let $\mathcal{N} = i\mathcal{M} \equiv i\mathcal{F}$. Then, the field content of our model will be a complex spin zero field S, ψ_L (or equivalently, a four-component Majorana spinor ψ whose right-handed components are chosen to make it Majorana) and a complex field $\mathcal{F}$. Making the appropriate substitutions in (5.4), we obtain the expansion of a left-chiral scalar superfield,

$$\hat{S}_L = S + i\sqrt{2}\bar{\theta}\psi_L + i\bar{\theta}\theta_L\mathcal{F} + \frac{i}{2}(\bar{\theta}\gamma_5\gamma_\mu\theta)\partial^\mu S$$

$$- \frac{1}{\sqrt{2}}\bar{\theta}\gamma_5\theta \cdot \bar{\theta}\partial\!\!\!/\,\psi_L + \frac{1}{8}(\bar{\theta}\gamma_5\theta)^2\Box S. \tag{5.34}$$

The transformation laws for the component fields are then

$$\delta S = -i\sqrt{2}\bar{\alpha}\psi_L, \tag{5.35a}$$

$$\delta\psi_L = -\sqrt{2}\mathcal{F}\alpha_L + \sqrt{2}\partial S\alpha_R, \tag{5.35b}$$

$$\delta\mathcal{F} = i\sqrt{2}\bar{\alpha}\partial\psi_L, \tag{5.35c}$$

which is exactly the same as in Eq. (3.16a)–(3.16c). Throughout the remainder of this book, we will reserve S, ψ, and $\mathcal{F}$ to denote components of chiral superfields.

Exercise *Convince yourself that the components of the left-chiral superfield form an irreducible multiplet. In other words, show that it is not possible to set any of the components (or combinations thereof) to zero.*

Exercise *In our reduction of the general superfield to the left-chiral scalar superfield, we took $V_\mu = \partial_\mu\zeta$, and $\lambda = \mathcal{D} = 0$. Show that any attempt to reduce the system by setting $V^\mu = 0$ with $i\gamma_\mu\lambda + \sqrt{2}\partial^\mu\psi = 0$, etc. collapses the system of equations.*

5.6.2 Right-chiral scalar superfields

In order to obtain a right-chiral scalar superfield, we set $\psi_L = 0$, $V^\mu = -i\partial^\mu S$ and $\mathcal{N} = -i\mathcal{M} \equiv \mathcal{F}$ in (5.4) so that

$$\hat{S}_R = S - i\sqrt{2}\bar{\theta}\psi_R - i\bar{\theta}\theta_R\mathcal{F} - \frac{i}{2}(\bar{\theta}\gamma_5\gamma_\mu\theta)\partial^\mu S$$

$$-\frac{1}{\sqrt{2}}\bar{\theta}\gamma_5\theta \cdot \bar{\theta}\partial\psi_R + \frac{1}{8}(\bar{\theta}\gamma_5\theta)^2\Box S. \tag{5.36}$$

We note that the field

$$\hat{S}_L^\dagger = S^\dagger - i\sqrt{2}\bar{\psi}\theta_R - i\bar{\theta}\theta_R\mathcal{F}^\dagger - \frac{i}{2}(\bar{\theta}\gamma_5\gamma_\mu\theta)\partial^\mu S^\dagger$$

$$-\frac{1}{\sqrt{2}}\bar{\theta}\gamma_5\theta \cdot \bar{\theta}\partial\psi_R + \frac{1}{8}(\bar{\theta}\gamma_5\theta)^2\Box S^\dagger \tag{5.37}$$

has the form of a right-chiral scalar superfield.

5.6.3 The curl superfield

Let us define the field strength tensor field $F^{\mu\nu} \equiv \partial^\mu V^\nu - \partial^\nu V^\mu$. It is then straightforward to check that,

$$\delta F^{\mu\nu} = -i\bar{\alpha}[\gamma^\nu \partial^\mu - \gamma^\mu \partial^\nu]\lambda, \tag{5.38a}$$

$$\delta\lambda = -i\gamma_5\alpha\mathcal{D} + \frac{1}{4}[\gamma_\nu, \gamma_\mu]F^{\mu\nu}\alpha, \quad \text{and} \tag{5.38b}$$

$$\delta\mathcal{D} = \bar{\alpha}\partial\gamma_5\lambda, \tag{5.38c}$$

so that the components $F^{\mu\nu}$, λ, and $\mathcal{D}$ transform into each other. Nevertheless, it is not possible to choose $\mathcal{S}$, ψ, $\mathcal{M}$, and $\mathcal{N}$ all equal to zero, since (because of (5.29b) this choice is not invariant under a SUSY transformation. In a gauge theory, however, which is where we will have need for the curl superfield, there is more freedom because of gauge invariance. We will see in the next chapter how it is possible to work with a multiplet containing only the $F^{\mu\nu}$, λ, and $\mathcal{D}$ fields. Such a gauge multiplet will be derived from a *real* superfield that contains the gauge potential V^μ.

5.7 Products of superfields

We begin by noting that the expansion,

$$\hat{\mathcal{S}}_L(x,\theta) = \mathcal{S}(x) + i\sqrt{2}\bar{\theta}\psi_L(x) - \frac{i}{2}(\bar{\theta}\gamma_5\theta)\mathcal{F} + \frac{i}{2}(\bar{\theta}\theta)\mathcal{F} + \frac{i}{2}(\bar{\theta}\gamma_5\gamma_\mu\theta)\partial^\mu\mathcal{S}(x)$$

$$-\frac{1}{\sqrt{2}}\bar{\theta}\gamma_5\theta \cdot \bar{\theta}\partial\psi_L(x) + \frac{1}{8}(\bar{\theta}\gamma_5\theta)^2\Box\mathcal{S}(x), \tag{5.39}$$

for a left-chiral scalar superfield can be succinctly written in terms of a new variable $\hat{x}_\mu = x_\mu + \frac{1}{2}\bar{\theta}\gamma_5\gamma_\mu\theta$ as,

$$\hat{\mathcal{S}}_L(x,\theta) = \mathcal{S}(\hat{x}) + i\sqrt{2}\bar{\theta}\psi_L(\hat{x}) + i\bar{\theta}\theta_L\mathcal{F}(\hat{x}). \tag{5.40}$$

To see this, we can expand each of the fields in (5.40) as power series around $\hat{x} \simeq x$. Since any term can contain at most two θs and two $\bar{\theta}$s, this expansion must terminate. We can thus write $\mathcal{S}(\hat{x})$ as

$$\mathcal{S}(\hat{x}) = \mathcal{S}(x) + \frac{i}{2}(\bar{\theta}\gamma_5\gamma_\mu\theta)\partial^\mu\mathcal{S}(x) + \frac{1}{2!}(\frac{i}{2})^2(\bar{\theta}\gamma_5\gamma_\mu\theta)(\bar{\theta}\gamma_5\gamma_\nu\theta)\partial^\mu\partial^\nu\mathcal{S}(x)$$

$$= \mathcal{S}(x) + \frac{i}{2}(\bar{\theta}\gamma_5\gamma_\mu\theta)\partial^\mu\mathcal{S}(x) + \frac{1}{8}(\bar{\theta}\gamma_5\theta)^2\Box\mathcal{S} \tag{5.41}$$

where we have used identity (5.24e) to obtain the last term. Likewise, using (5.23d) we have

$$\bar{\theta}\psi_{\mathrm{L}}(\hat{x}) = \bar{\theta}\left[\psi_{\mathrm{L}}(x) + \frac{i}{2}(\bar{\theta}\gamma_5\gamma_\mu\theta)\partial^\mu\psi_{\mathrm{L}}(x)\right]$$

$$= \bar{\theta}\psi_{\mathrm{L}}(x) + \frac{i}{2}(\bar{\theta}\gamma_5\theta)\bar{\theta}\,\partial\!\!\!/\,\psi_{\mathrm{L}}. \tag{5.42}$$

Finally, from (5.24b) and (5.24c) we have

$$\bar{\theta}\theta\mathcal{F}(\hat{x}) = \bar{\theta}\theta\mathcal{F}(x) \quad \text{and} \quad \bar{\theta}\gamma_5\theta\mathcal{F}(\hat{x}) = \bar{\theta}\gamma_5\theta\mathcal{F}(x). \tag{5.43}$$

Combining these results, we arrive at Eq. (5.40).

The important point about Eq. (5.40) is that it shows that a left-chiral scalar superfield is a function of just $\hat{x}$ and θ_{L} (recall that $\bar{\theta}_{\mathrm{R}} = \theta_{\mathrm{L}}^T C$). The θ_{R} dependence of $\hat{\mathcal{S}}_{\mathrm{L}}$ enters only via $\hat{x}$. If we take the product of two (or more) left-chiral scalar superfields, it will again be a function of just $\hat{x}$ and θ_{L}, and can be written in the form of Eq. (5.40). We thus conclude that *a product of any number of left-chiral scalar superfields is itself a left-chiral scalar superfield.*

In a similar fashion, a right-chiral scalar superfield $\hat{\mathcal{S}}_{\mathrm{R}}$ can be written as just a function of $\hat{x}^\dagger$ and θ_{R}:

$$\hat{\mathcal{S}}_{\mathrm{R}}(x, \theta) = \mathcal{S}(\hat{x}^\dagger) - i\sqrt{2}\bar{\theta}\psi_{\mathrm{R}}(\hat{x}^\dagger) - i\bar{\theta}\theta_{\mathrm{R}}\mathcal{F}(\hat{x}^\dagger). \tag{5.44}$$

This then establishes that the product of two (or more) right-chiral scalar superfields is a right-chiral scalar superfield.

Exercise *By explicit multiplication, or otherwise, convince yourself that the product of a left-chiral superfield with a right-chiral superfield is a general superfield.*

5.8 Supercovariant derivatives

Covariant derivatives are defined so that when these act on any object, they yield a new object with the same transformation properties as the original one. For instance, in gauge theories, unlike the ordinary derivative, the gauge covariant derivative acting on a field whose components transform according to a representation **R** of the gauge group, is a new field with components that transform in the same way.

Since the representation (5.25) of the generator for supersymmetry includes the second term with a θ in it, it is clear that, under SUSY, the components $\partial\hat{\Phi}/\partial\bar{\theta}$ transform differently from those of $\hat{\Phi}$. This is in contrast to spatial derivatives where, because P_μ commutes with the super-charge, the components of $\partial_\mu\hat{\Phi}$ transform the same way as those of $\hat{\Phi}$. Thus ordinary spacetime derivatives are automatically covariant with respect to SUSY transformations.

To facilitate the construction of invariant functions of superfields and their derivatives with respect to θ, we want to define a supersymmetric covariant derivative D so that the components of $D\hat{\Phi}$ transform the same way as the components of $\hat{\Phi}$ under a supersymmetry transformation. We thus require,

$$\left[-\bar{\alpha}\frac{\partial}{\partial\bar{\theta}} - i\bar{\alpha}\partial\theta \right] D\hat{\Phi} = D\left[-\bar{\alpha}\frac{\partial}{\partial\bar{\theta}} - i\bar{\alpha}\partial\theta \right]\hat{\Phi}. \tag{5.45}$$

We will leave it to the reader to verify that the fermionic derivative operator,

$$D = \frac{\partial}{\partial\bar{\theta}} - i\partial\theta, \tag{5.46}$$

anticommutes with $-\frac{\partial}{\partial\bar{\theta}} - i\partial\theta$ and satisfies (5.45) because the fermionic parameter α anticommutes with θ.

Exercise *Verify that the expression for D in (5.46) satisfies (5.45).*

For later use, we will *define* a related derivative $\bar{D}$ so that $D = C\bar{D}^T$ so that D satisfies the "Majorana condition". We can readily find the explicit form for $\bar{D}$. Starting with $\bar{D} = D^T C$, we find,

$$\bar{D}_b = \left[\frac{\partial}{\partial\bar{\theta}_a} - i(\partial\theta)_a \right] C_{ab}$$

$$= \frac{\partial}{\partial\theta_c}\frac{\partial\theta_c}{\partial\bar{\theta}_a}C_{ab} - i(\partial C\bar{\theta}^T)_a C_{ba}^T$$

$$= -\frac{\partial}{\partial\theta_b} + i(\bar{\theta}\partial)_b$$

so that

$$\bar{D} = -\frac{\partial}{\partial\theta} + i\bar{\theta}\partial. \tag{5.47}$$

We can also *define* left and right SUSY covariant derivatives by acting on D with the projectors P_L or P_R. To do so, we note that $\bar{\theta}_a = \bar{\theta}_{Lb}P_{Rba} + \bar{\theta}_{Rb}P_{Lba}$ immediately gives us,

$$\frac{\partial\bar{\theta}_a}{\partial\bar{\theta}_{Lb}} = P_{Rba} \quad \text{and} \quad \frac{\partial\bar{\theta}_a}{\partial\bar{\theta}_{Rb}} = P_{Lba}, \tag{5.48}$$

which implies,

$$\frac{\partial}{\partial\bar{\theta}_{La}} = \frac{\partial}{\partial\bar{\theta}_b}\frac{\partial\bar{\theta}_b}{\partial\bar{\theta}_{La}} = \frac{\partial}{\partial\bar{\theta}_b}P_{Rab}. \tag{5.49}$$

Although it should be clear from the context, we clarify that we are taking the derivative with respect to the conjugates of the spinors θ_L or θ_R. A similar relation holds for $\partial/\partial\bar{\theta}_R$, so that

$$\frac{\partial}{\partial\bar{\theta}_L} = P_R \frac{\partial}{\partial\bar{\theta}} \quad \text{and} \quad \frac{\partial}{\partial\bar{\theta}_R} = P_L \frac{\partial}{\partial\bar{\theta}}. \tag{5.50}$$

We then have

$$D_L \equiv P_L D = \frac{\partial}{\partial\bar{\theta}_R} - i\partial\theta_R \tag{5.51a}$$

$$D_R \equiv P_R D = \frac{\partial}{\partial\bar{\theta}_L} - i\partial\theta_L, \tag{5.51b}$$

where $D_L + D_R = D$. Finally, let us also *define*,

$$\bar{D}_R \equiv D_L^T C \quad \text{and} \quad \bar{D}_L \equiv D_R^T C. \tag{5.52a}$$

Clearly, $\bar{D}_R + \bar{D}_L = (D_L^T + D_R^T)C = D^T C = \bar{D}$. Note that once again our definition is consistent with the "Majorana condition" for the spinorial operator D. Notice also that,

$$\bar{D}_L = D_R^T C = D^T P_R^T C = \bar{D} P_R, \tag{5.52b}$$

and

$$\bar{D}_R = D_L^T C = D^T P_L^T C = \bar{D} P_L. \tag{5.52c}$$

We will leave it to the reader to verify that by steps very similar to those that led us to (5.51a) and (5.51b) we obtain,

$$\bar{D}_L = -\frac{\partial}{\partial\theta_R} + i\bar{\theta}_R\partial, \tag{5.53a}$$

and

$$\bar{D}_R = -\frac{\partial}{\partial\theta_L} + i\bar{\theta}_L\partial. \tag{5.53b}$$

As one more exercise in the manipulation of the supercovariant derivative, we establish an identity involving D_L and D_R to be used in the next chapter. We compute the anticommutation relation

$$\{D_{La}, D_{Rb}\} = \left\{ \frac{\partial}{\partial\bar{\theta}_{Ra}} - i(\partial\theta_R)_a, \frac{\partial}{\partial\bar{\theta}_{Lb}} - i(\partial\theta_L)_b \right\}$$

$$= -i\left\{ (\partial\theta_R)_a, \frac{\partial}{\partial\bar{\theta}_{Lb}} \right\} - i\left\{ \frac{\partial}{\partial\bar{\theta}_{Ra}}, (\partial\theta_L)_b \right\}$$

$$= -i\frac{\partial}{\partial\theta_{Lb}}(\partial\theta_R)_a - i\frac{\partial}{\partial\theta_{Ra}}(\partial\theta_L)_b.$$

To obtain the last step we can explicitly act on a superfield and, since the θs anticommute, see that just the terms shown survive. Finally, since $\theta_R = C\bar{\theta}_L^T$ and $\theta_L = C\bar{\theta}_R^T$, we have $\partial\theta_{Rb}/\partial\bar{\theta}_{La} = C_{ba}$ and $\partial\theta_{Lb}/\partial\bar{\theta}_{Ra} = C_{ba}$, so that

$$\{D_{La}, D_{Rb}\} = -i(\partial C)_{ab} - i(\partial C)_{ba}$$
$$= -2i(\partial C)_{ab}. \tag{5.54}$$

Exercise *Similarly show that*

$$\{D_{La}, D_{Lb}\} = \{D_{Ra}, D_{Rb}\} = 0.$$

Thus any term with a product of three D_Ls or three D_Rs vanishes.

To conclude this section, we show that the action of D_R on a left-chiral superfield $\hat{S}_L(\theta_L, \hat{x})$ gives zero. To evaluate the first term, we note that the $\bar{\theta}_L$ dependence enters only via $\hat{x}$, and we can write,

$$\frac{\partial\hat{S}_L}{\partial\bar{\theta}_L} = \frac{\partial\hat{S}_L}{\partial\hat{x}^\mu}\frac{\partial\hat{x}^\mu}{\partial\bar{\theta}_L} = \frac{\partial\hat{S}_L}{\partial\hat{x}^\mu}\frac{i}{2}\frac{\partial(\bar{\theta}\gamma_5\gamma^\mu\theta)}{\partial\bar{\theta}_L}$$
$$= \frac{\partial\hat{S}_L}{\partial x^\mu}\cdot i\gamma^\mu\theta_L,$$

where in the last step we used $\bar{\theta}\gamma_5\gamma_\mu\theta = 2\bar{\theta}_L\gamma_\mu\theta_L$. We thus establish the important property,

$$D_R\hat{S}_L = \left(\frac{\partial}{\partial\bar{\theta}_L} - i\partial\theta_L\right)\hat{S}_L = 0. \tag{5.55}$$

Working the steps backwards, we see that this is also a sufficient condition for any field to be a left-chiral superfield. The reader can similarly show,

$$D_L\hat{S}_R = 0. \tag{5.56}$$

We remark that the result of the last exercise in the previous section follows immediately from Eq. (5.55) and (5.56).

5.9 Lagrangians for chiral scalar superfields

Our goal in this section is to present a systematic strategy to construct actions that are invariant under supersymmetric transformations. This means that the variation of the Lagrangian density can at most be a total derivative. In fact, the Lagrangian density can *never* be a SUSY invariant. This follows simply from the SUSY algebra.

By manipulations similar to (3.20) of Chapter 3 we get,

$$(\delta_2\delta_1 - \delta_1\delta_2)\mathcal{L} = -2\bar{\alpha}_2\gamma_\mu\alpha_1\partial_\mu\mathcal{L}.$$

This would, of course, have to vanish if $\mathcal{L}$ were truly a SUSY invariant. We would then be led to conclude that $\mathcal{L}$ is a constant and that the theory has no dynamics. Thus, SUSY transformations always change the Lagrangian density by a (non-vanishing) total derivative.

The first observations toward our goal stem from Eq. (5.29g) and Eq. (5.35c) which show that the D-component (the coefficient of $(\bar{\theta}\gamma_5\theta)^2$) of *any* superfield and the F-component of *chiral* superfields (the coefficient of $\bar{\theta}\theta_L$ of a left-chiral superfield, or the coefficient of $\bar{\theta}\theta_R$ of a right-chiral superfield) transform as a total derivative under a SUSY transformation. This leads us to two important conclusions:

- if we take the product of any number of chiral superfields and their Hermitian conjugates, the D-term of the product superfield will change only by a total derivative under SUSY transformations, and
- if we take the product of only left- (or only right-) chiral superfields, the F-term of the product will also change by just a total derivative. The would-be D-term (i.e. the coefficient of $(\bar{\theta}\gamma_5\theta)^2$) of this product is already a total derivative.

These D- or F-components of the composite (product) superfield are themselves products of the ordinary fields that were the components of the individual superfields in the product. Thus, *these D- and F-terms are candidates for a SUSY Lagrangian.* With just chiral scalar multiplets, we can only obtain a theory with spin 0 and spin $\frac{1}{2}$ fields.

The recipe for obtaining SUSY invariant actions (given a set of N chiral superfields) is now in hand. We start with two functions $K(\hat{\mathcal{S}}_{Li}^\dagger, \hat{\mathcal{S}}_{Lj})$ and $\hat{f}(\hat{\mathcal{S}}_{Li})$ of a set of left-chiral superfields $\hat{\mathcal{S}}_{Li}$, where $i = 1, \ldots, N$. Since $\hat{\mathcal{S}}_{Li}^\dagger$ is a right-chiral superfield, K is a general superfield, while $\hat{f}$ is a left-chiral superfield. Then, the D-term of K and the F-term of $\hat{f}$ are candidates for a SUSY Lagrangian density. The function K is called the Kähler potential and the function $\hat{f}$ is known as the superpotential. We make two clarifying remarks.

- There is no loss of generality in writing the superpotential as a function of just left-chiral superfields because every right-chiral superfield $\hat{\mathcal{S}}_{Rj}$ can, by the analogue of Eq. (5.37), be written as a left-chiral superfield $\left(\hat{\mathcal{S}}_{Rj}\right)^\dagger$.
- The reader may wonder why we do not include the D-term of the superpotential in the Lagrangian. We see from Eq. (5.34) that the coefficient of the $(\bar{\theta}\gamma_5\theta)^2$ term of any left-chiral superfield is itself a total derivative, and so does not contribute to

the action. For the same reason, terms in the Kähler potential that do not depend on *both* $\hat{S}_L$ and $\hat{S}_L^\dagger$ would also be irrelevant.

Just as the scalar potential specifies any theory of spin zero and spin half fields in usual field theory, a supersymmetric field theory with chiral superfields is specified by the Kähler potential together with the superpotential. We now compute the Lagrangian density for any SUSY theory with just spin zero and spin half fields in terms of these functions. For simplicity, we will restrict our discussion to theories that are power counting renormalizable.

5.9.1 Kähler potential contributions to the Lagrangian density

We begin with the computation of the Kähler potential contribution to the action. This requires us to compute the coefficient of the $(\bar{\theta}\gamma_5\theta)^2$ term (or the D-term) of the function K. For this reason, this contribution is frequently known as the "D-term contribution" to the Lagrangian density.

Renormalizability imposes stringent restrictions on the form of K, and also as we will see below, on the form of the superpotential. To see this, we have to do some dimensional analysis. We will denote the mass dimension of any quantity X as $[X]$. Since $[P] = 1$, from the SUSY algebra $\{Q, \bar{Q}\} = 2\gamma^\mu P_\mu$, we must have $[Q] = [\bar{Q}] = 1/2$. (Remember that $Q = C\bar{Q}^T$ implies $[Q] = [\bar{Q}]$.) Then from Eq. (5.25), we obtain $[\theta] = [\bar{\theta}] = -1/2$.

If, in our expansion (5.34) of the chiral superfield, we now choose the scalar field S to have the canonical dimension $[S] = 1$, then $[\psi] = 3/2$ and $[\mathcal{F}] = 2$, just as for the Wess–Zumino model. Indeed the left-chiral superfield $\hat{S}_L$ can be assigned $[\hat{S}_L] = 1$. Since $[(\bar{\theta}\gamma_5\theta)^2] = -2$, then

$$[K_{D\text{-term}}] = [K] + 2. \tag{5.57}$$

If this D-term is to represent a *renormalizable* Lagrangian, then $[K_{D\text{-term}}] \leq 4$, so that

$$[K] \leq 2 \quad \text{(renormalizable theory)} \tag{5.58}$$

and the Kähler potential is at most a quadratic polynomial of $\hat{S}$ and $\hat{S}^\dagger$. In non-renormalizable theories (such as supergravity), higher powers may be present. In fact, then K need not even be a polynomial.

As already noted, chiral superfields have only gradient D-terms, so there is no point writing linear terms (or for that matter terms involving just $\hat{S}$ or just $\hat{S}^\dagger$) in the Kähler potential. In a renormalizable theory, since cubic and higher terms are not allowed in K, the most general form of K is a real function (to ensure the

Hermiticity of the Lagrangian density)

$$K = \sum_{i,j=1}^{N} A_{ij}\hat{S}_i^\dagger \hat{S}_j. \tag{5.59}$$

Without loss of generality, we can choose a basis so that A_{ij} is diagonal, and the fields $\hat{S}_i$ can be normalized so that the $A_{ii} = 1$. Then

$$K[\hat{S}^\dagger, \hat{S}] = \sum_{i=1}^{N} \hat{S}_i^\dagger \hat{S}_i \tag{5.60}$$

is a general choice for K in a renormalizable theory.

Exercise *For the curl superfield, show that if we choose $[V^\mu] = 1$, we would have $[\lambda] = 3/2$ and $[D] = 2$. Notice that unlike the chiral supermultiplet, the mass dimension of the curl superfield vanishes, so that renormalizability considerations do not restrict the power of this multiplet in the Kähler potential. It cannot, of course, enter the superpotential since it is not a chiral superfield. We will exploit this in the next chapter when we discuss supersymmetric gauge theories.*

We now have only to compute the coefficient of the $(\bar{\theta}\gamma_5\theta)^2$ term in the product $\hat{S}_i^\dagger \hat{S}_i$. In so doing, we need only keep terms with four θ's or $\bar{\theta}$'s in any combination. Multiplying the expansion (5.37) by (5.34), four sets of terms will arise.

The first set of terms is

$$\bullet \; \frac{1}{8}S^\dagger \Box S(\bar{\theta}\gamma_5\theta)^2 + \frac{1}{8}\Box S^\dagger S(\bar{\theta}\gamma_5\theta)^2 + \frac{1}{4}(\bar{\theta}\gamma_5\gamma_\mu\theta)(\bar{\theta}\gamma_5\gamma_\nu\theta)\partial^\mu S^\dagger \partial^\nu S.$$

We integrate by parts on each of the first two terms above, and discard the surface terms. For the third term, apply identity (5.24e). The result is that

$$\hat{S}_L^\dagger \hat{S}_L \ni -\frac{1}{2}(\bar{\theta}\gamma_5\theta)^2\partial_\mu S^\dagger \partial^\mu S. \tag{5.61}$$

The second set of terms is

$$\bullet \; i\bar{\psi}\theta_R(\bar{\theta}\gamma_5\theta)\bar{\theta}\partial\psi_L - i(\bar{\theta}\gamma_5\theta)\bar{\theta}\partial\psi_R\bar{\theta}\psi_L.$$

In the first term, re-write $\bar{\psi}\theta_R \rightarrow \bar{\psi}_L\theta$ and apply the Fierz re-arrangement identity (5.21) to the $\theta\bar{\theta}$ product. The second and third terms of (5.21) lead to vanishing contributions via the identities (5.24a) and (5.24b), respectively, while the second term of (5.21) leads to a contribution $-\frac{i}{4}(\bar{\theta}\gamma_5\theta)^2\bar{\psi}_L\partial\psi_L$. Similarly, the second term of $\bullet$ above leads to a contribution $-\frac{i}{4}(\bar{\theta}\gamma_5\theta)^2\bar{\psi}_R\partial\psi_R$. Since $\psi = \psi_L + \psi_R$ is Majorana, we find

$$\hat{S}_L^\dagger \hat{S}_L \ni -\frac{1}{2}(\bar{\theta}\gamma_5\theta)^2\frac{i}{2}\bar{\psi}\partial\psi. \tag{5.62}$$

The third set of terms consists of

$$\bullet \; \bar{\theta}\theta_R\bar{\theta}\theta_L\mathcal{F}^\dagger\mathcal{F}.$$

By expanding the P_L and P_R projection operators and using (5.24a) and (5.24d), we find

$$\hat{\mathcal{S}}_L^\dagger\hat{\mathcal{S}}_L \ni -\frac{1}{2}(\bar{\theta}\gamma_5\theta)^2\mathcal{F}^\dagger\mathcal{F}. \qquad (5.63)$$

A fourth set of terms

$$\bullet \; \frac{1}{2}\bar{\theta}\theta_R(\bar{\theta}\gamma_5\gamma_\mu\theta)\mathcal{F}^\dagger\partial^\mu\mathcal{S} + \frac{1}{2}(\bar{\theta}\gamma_5\gamma_\mu\theta)\bar{\theta}\theta_L\partial_\mu\mathcal{S}^\dagger\mathcal{F}$$

will identically vanish due to identities (5.24b) and (5.24c).

Putting all the pieces together, we find

$$\hat{\mathcal{S}}_L^\dagger\hat{\mathcal{S}}_L \ni -\frac{1}{2}(\bar{\theta}\gamma_5\theta)^2\{\partial_\mu\mathcal{S}^\dagger\partial^\mu\mathcal{S} + \frac{i}{2}\bar{\psi}\!\!\not{\partial}\psi + \mathcal{F}^\dagger\mathcal{F}\}. \qquad (5.64)$$

We will *define* the D-term to be the coefficient of the $-\frac{1}{2}(\bar{\theta}\gamma_5\theta)^2$ term in the product $\hat{\mathcal{S}}_L^\dagger\hat{\mathcal{S}}_L$ since this gives us the canonically normalized kinetic energy terms for the scalar field $\mathcal{S}$ and the Majorana spinor field ψ. The D-term contribution to the Lagrangian density for a single chiral scalar superfield is thus

$$\mathcal{L}_D = \partial_\mu\mathcal{S}^\dagger\partial^\mu\mathcal{S} + \frac{i}{2}\bar{\psi}\!\!\not{\partial}\psi + \mathcal{F}^\dagger\mathcal{F}. \qquad (5.65)$$

The field $\mathcal{F}$ enters without any derivative. It turns out to be an auxiliary field that satisfies an *algebraic* equation of motion.

5.9.2 *Superpotential contributions to the Lagrangian density*

We now turn to the computation of the superpotential contributions to the Lagrangian density. This is proportional to the coefficient of the $\bar{\theta}\theta_L$, or the F-term, of the superpotential function. These contributions are therefore frequently referred to as F-term contributions. Dimensional analysis tells us that the F-term of the superpotential $\hat{f}$ has dimensions $[\hat{f}] - 1$. In a renormalizable theory, therefore, the superpotential is at most a cubic polynomial in $\hat{\mathcal{S}}_i$.

We can formally write any superpotential as a power series about $\hat{S} = S$ as,

$$\hat{f}(\hat{S}) = \hat{f}(\hat{S} = S) + \sum_i \left.\frac{\partial \hat{f}}{\partial \hat{S}_i}\right|_{\hat{S}=S} (\hat{S} - S)_i$$

$$+ \frac{1}{2} \sum_{ij} \left.\frac{\partial^2 \hat{f}}{\partial \hat{S}_i \partial \hat{S}_j}\right|_{\hat{S}=S} (\hat{S} - S)_i (\hat{S} - S)_j$$

$$+ \frac{1}{3!} \sum_{ijk} \left.\frac{\partial^3 \hat{f}}{\partial \hat{S}_i \partial \hat{S}_j \partial \hat{S}_k}\right|_{\hat{S}=S} (\hat{S} - S)_i (\hat{S} - S)_j (\hat{S} - S)_k$$

$$+ \cdots \tag{5.66}$$

Here, $\hat{S} = S$ means that after the derivative is evaluated, each superfield is set to be the scalar component so that these "derivative coefficients" are functions of just the scalar fields. The terms $(\hat{S} - S)_i$ are at least linear in θ so that there can be at most four factors of this type because any product of five θs and $\bar{\theta}$s vanishes. In fact, because the superpotential is a function of only left-chiral superfields, even the product of four factors vanishes, so that there really are no terms represented by the ellipsis in the expansion above.

Let us now isolate the potential sources of the $\bar{\theta}\theta_L$ terms in $\hat{f}(\hat{S})$ whose coefficient is the item of interest to us. We see that:

1. the first term in the expansion will not contribute since it has no θs,
2. the last terms cannot contribute since they all contain at least three θs,
3. the $\sum_i \partial \hat{f}/\partial \hat{S}_i|_{\hat{S}=S}(\hat{S} - S)_i$ term contributes with the $\bar{\theta}\theta_L$ coefficient from $\hat{S} - S$, and
4. the $\frac{1}{2}\sum_{ij} \partial^2 \hat{f}/\partial \hat{S}_i \partial \hat{S}_j|_{\hat{S}=S}(\hat{S} - S)_i(\hat{S} - S)_j$ term contributes when $(\hat{S} - S)_i$ and $(\hat{S} - S)_j$ each contribute a term linear in θ.

The form of the term from item 3 above is easy to write down; using (5.34) it is just

$$\left.\frac{\partial \hat{f}}{\partial \hat{S}_i}\right|_{\hat{S}=S} (\hat{S} - S)_i = \left.\frac{\partial \hat{f}}{\partial \hat{S}_i}\right|_{\hat{S}=S} \left(i\mathcal{F}_i \bar{\theta}\theta_L\right). \tag{5.67}$$

The term from item 4 can be written as

$$\frac{1}{2}\left.\frac{\partial^2 \hat{f}}{\partial \hat{S}_i \partial \hat{S}_j}\right|_{\hat{S}=S} (\hat{S} - S)_i(\hat{S} - S)_j = \frac{1}{2}\left.\frac{\partial^2 \hat{f}}{\partial \hat{S}_i \partial \hat{S}_j}\right|_{\hat{S}=S} (i\sqrt{2}\bar{\psi}_i P_L\theta)(i\sqrt{2}\bar{\theta}\psi_{jL})$$

$$= \frac{1}{4} \frac{\partial^2 \hat{f}}{\partial \hat{S}_i \partial \hat{S}_j} \Big|_{\hat{S}=S} \bar{\psi}_i P_L \left[\bar{\theta}\theta \mathbf{1} + \bar{\theta}\gamma_5\theta \cdot \gamma_5 - \bar{\theta}\gamma_\mu\gamma_5\theta \cdot \gamma^\mu\gamma_5 \right] P_L \psi_j$$

$$= \frac{1}{2} \frac{\partial^2 \hat{f}}{\partial \hat{S}_i \partial \hat{S}_j} \Big|_{\hat{S}=S} \bar{\theta}\theta_L \bar{\psi}_i P_L \psi_j, \tag{5.68}$$

where we have used identity (5.21).

The coefficient of $\bar{\theta}\theta_L$ in $\hat{f}(\hat{S})$ is thus

$$i \frac{\partial \hat{f}}{\partial \hat{S}_i} \Big|_{\hat{S}=S} \mathcal{F}_i + \frac{1}{2} \frac{\partial^2 \hat{f}}{\partial \hat{S}_i \partial \hat{S}_j} \Big|_{\hat{S}=S} \bar{\psi}_i P_L \psi_j, \tag{5.69}$$

where a sum over the various fields is implied. This term is not Hermitian, since $\hat{f}$ is intrinsically complex. However, we note that the F-term of the right-chiral superfield $[\hat{f}(\hat{S})]^\dagger$ which also leads to a SUSY-invariant action gives just the Hermitian conjugate of the expression (5.69). We will add this to obtain a Hermitian Lagrangian density.

In defining the F-terms, we will actually take the coefficient of $-\bar{\theta}\theta_L$ as the choice for a Lagrangian. This is purely conventional. We will choose the size of the terms in the superpotential to give mass and interaction terms with usual normalizations in the Lagrangian density. Thus,

$$\mathcal{L}_F = -i \sum_i \frac{\partial \hat{f}}{\partial \hat{S}_i} \Big|_{\hat{S}=S} \mathcal{F}_i - \frac{1}{2} \sum_{i,j} \frac{\partial^2 \hat{f}}{\partial \hat{S}_i \partial \hat{S}_j} \Big|_{\hat{S}=S} \bar{\psi}_i P_L \psi_j$$

$$+ i \sum_i \left(\frac{\partial \hat{f}}{\partial \hat{S}_i} \right)^\dagger \Big|_{\hat{S}=S} \mathcal{F}_i^\dagger - \frac{1}{2} \sum_{i,j} \left(\frac{\partial^2 \hat{f}}{\partial \hat{S}_i \partial \hat{S}_j} \right)^\dagger \Big|_{\hat{S}=S} \bar{\psi}_i P_R \psi_j. \tag{5.70}$$

We remark that nowhere in our derivation of (5.70) did we need to assume the dimensionality of the superpotential.

5.9.3 A technical aside

The careful reader may have noticed that we did not allow the superpotential to contain terms involving supercovariant derivatives of the superfield. This is because the supercovariant derivative of a chiral superfield is *not*, in general, a chiral superfield. However, by the exercise immediately following (5.54), we see that the product of any three right (or left) supercovariant derivatives vanishes. Hence, even for a general superfield $\hat{\Phi}$, $\bar{D}_L D_R \hat{\Phi}$ must be a left-chiral superfield (since D_R acting on this vanishes). This raises the question whether such terms (or functions thereof) may be included in the superpotential of a more general theory not involving just chiral superfields.

First, we note that up to total derivatives, this term is just $-\partial^2 \hat{\Phi}/\partial\theta_R \partial\bar{\theta}_L$ so that it just removes one θ_R and one $\bar{\theta}_L$ from the general expansion (5.4) of $\hat{\Phi}$. Aside from total derivatives, this then leaves only terms with $\mathcal{M}$ or $\mathcal{N}$ (with no θ or $\bar{\theta}$), a term with λ (with one $\bar{\theta}$) and a term with $\mathcal{D}$ (with a $\bar{\theta}\gamma_5\theta$). Up to total derivatives, the F-term (which is proportional to the coefficient of $\bar{\theta}\theta_L$) of $\bar{D}_L D_R \hat{\Phi}$ is then just a multiple of the $\mathcal{D}$ component of $\hat{\Phi}$ and would be included in our general list of contributions from the Kähler potential.

Next, the reader may worry about terms like $\hat{S}_L \bar{D}_L D_R \hat{\Phi}$ since this is also a left-chiral superfield. However, since $D_R \hat{S} = \bar{D}_L \hat{S} = 0$, this can be written as $\bar{D}_L D_R(\hat{S}\hat{\Phi})$ which we just argued that we do not need to include. Powers of $\bar{D}_L D_R \hat{\Phi}$ are just a special case of this. We thus see that there is no loss of generality in not including supercovariant derivatives of superfields in the superpotential *as long as we allow for a general Kähler potential* (which can include terms involving these derivatives). In a renormalizable theory, however, the choice of Kähler potential is greatly restricted as we have already noted. Finally, we remark that our analysis above shows that certain F-terms (which lead to non-renormalizable interactions in four dimensions) can be rewritten as D-terms.

5.9.4 A master Lagrangian for chiral scalar superfields

We can now combine the D- and F-term Lagrangian candidates above to arrive at the general Lagrangian for renormalizable theories involving only chiral scalar superfields:

$$
\begin{aligned}
\mathcal{L} = \mathcal{L}_D + \mathcal{L}_F \\
= \sum_i \left[\partial_\mu \hat{S}_i^\dagger \partial^\mu S_i + \frac{i}{2}\bar{\psi}_i \partial\!\!\!/\psi_i + \mathcal{F}_i^\dagger \mathcal{F}_i \right] \\
- i\sum_i \frac{\partial \hat{f}}{\partial \hat{S}_i}\bigg|_{\hat{S}=S} \mathcal{F}_i - \frac{1}{2}\sum_{i,j}\frac{\partial^2 \hat{f}}{\partial \hat{S}_i \partial \hat{S}_j}\bigg|_{\hat{S}=S} \bar{\psi}_i P_L \psi_j \\
+ i\sum_i \left(\frac{\partial \hat{f}}{\partial \hat{S}_i}\right)^\dagger\bigg|_{\hat{S}=S} \mathcal{F}_i^\dagger - \frac{1}{2}\sum_{i,j}\left(\frac{\partial^2 \hat{f}}{\partial \hat{S}_i \partial \hat{S}_j}\right)^\dagger\bigg|_{\hat{S}=S} \bar{\psi}_i P_R \psi_j. \quad (5.71)
\end{aligned}
$$

We see that while the fields S_i and ψ_i have conventional kinetic energy terms, the fields $\mathcal{F}_i$ have no kinetic energy term, and so are not dynamical fields.

At this stage we eliminate these auxiliary fields from the Lagrangian by using their (algebraic) Euler–Lagrange equations:

$$\frac{\partial \mathcal{L}}{\partial \mathcal{F}_i^\dagger} = 0 \Rightarrow \mathcal{F}_i + i \left(\frac{\partial \hat{f}}{\partial \hat{S}_i} \right)^\dagger = 0, \tag{5.72a}$$

$$\frac{\partial \mathcal{L}}{\partial \mathcal{F}_i} = 0 \Rightarrow \mathcal{F}_i^\dagger - i \frac{\partial \hat{f}}{\partial \hat{S}_i} = 0. \tag{5.72b}$$

We thus obtain the general supersymmetric Lagrangian for theories with just scalars and spinors to be

$$\mathcal{L} = \sum_i (\partial_\mu \mathcal{S}_i)^\dagger (\partial^\mu \mathcal{S}_i) + \frac{i}{2} \sum_i \bar{\psi}_i \partial\!\!\!/ \psi_i - \sum_i \left| \frac{\partial \hat{f}}{\partial \hat{S}_i} \right|^2_{\hat{S}=\mathcal{S}}$$

$$- \frac{1}{2} \sum_{i,j} \left[\frac{\partial^2 \hat{f}}{\partial \hat{S}_i \partial \hat{S}_j} \Big|_{\hat{S}=\mathcal{S}} \bar{\psi}_i \frac{1 - \gamma_5}{2} \psi_j + \text{h.c.} \right]. \tag{5.73}$$

The third term yields the scalar potential (which is quartic if the superpotential is cubic). The masses and Yukawa interactions of fermions are all included in the last term. The model dependence of the theory enters via the choice of the superpotential which can be an arbitrary function (at most a cubic polynomial for renormalizable theories) of left-chiral superfields, but not their Hermitian conjugates.

Exercise (Recovering the Wess–Zumino model) *To recover the Wess–Zumino model, complete with interactions, create a theory with a single left-chiral scalar superfield $\hat{S}_L \ni (\mathcal{S}, \psi_L, \mathcal{F})$. Let $\mathcal{S} = \frac{A+iB}{\sqrt{2}}$ and $\mathcal{F} = \frac{F+iG}{\sqrt{2}}$, where A, B, F, and G are real scalar fields. Assume a superpotential of the form $\hat{f} = \frac{1}{2} m \hat{S}^2 + \frac{1}{3} g \hat{S}^3$. Recover the Lagrangian terms given in Eq. (3.1b), (3.1c), and (3.43). This exercise completes the proof that the WZ model interaction terms Eq. (3.43) are, in fact, supersymmetric.*

5.10 The action as an integral over superspace

Supersymmetric actions are commonly expressed as integrals over superspace. To understand how this is accomplished, we must first define integration over Grassmann numbers. Consider the integral over the entire range of η of a function $f(\eta)$ of a single Grassmann variable η:

$$\int f(\eta) d\eta = \int (A + B\eta) d\eta,$$

where we have expanded f as a power series in η. Following Berezin,[3] we *define*

$$\int d\eta = 0 \tag{5.74a}$$

$$\int d\eta \cdot \eta = 1. \tag{5.74b}$$

Notice that (5.74b) implies that the dimension of η is the negative of that of $d\eta$ – hence $d\eta$ should not be thought of as an increment of η. This then gives

$$\int f(\eta)d\eta = \int (A + B\eta)d\eta = B.$$

Exercise *Verify that with this definition, Berezin integration is a linear operation, i.e. that $\int d\eta[af(\eta) + bg(\eta)] = a \int d\eta f(\eta) + b \int d\eta g(\eta)$, where a and b are (commuting) constants and f and g are functions. Show also that the integral $\int d\eta f(\eta)$ over the entire range of η is invariant under finite shifts $\eta \to \eta + \eta'$ of the integration variable by a Grassmann-valued constant.*

For integrals over several Grassmann variables, there is a sign ambiguity. We define the integral over several variables by requiring that the variable to be integrated first be moved to the extreme left: we thus have,

$$\int d\eta_1 d\eta_2 \cdot \eta_1 \eta_2 = -\int d\eta_1 d\eta_2 \cdot \eta_2 \eta_1 = -1. \tag{5.75}$$

We are now ready to see how to write the D- and F-term contributions to the action as integrals over superspace. The D-term contribution to the Lagrangian density was defined as the coefficient of $-\frac{1}{2}(\bar{\theta}\gamma_5\theta)^2$ when the Kähler potential is expanded in the canonical form. Since $(\bar{\theta}\gamma_5\theta)^2$ is quartic in the θs, it must be proportional to $\theta_1\theta_2\theta_3\theta_4$. Plugging in an explicit representation for the Dirac γ matrices shows that $(\bar{\theta}\gamma_5\theta)^2 = 8\,\theta_4\theta_3\theta_2\theta_1$. A look at (5.4) then tells us that

$$\int d\theta_1 d\theta_2 d\theta_3 d\theta_4 K\left(\hat{S}^\dagger, \hat{S}\right) \equiv \int d^4\theta K\left(\hat{S}^\dagger, \hat{S}\right)$$

equals 8 times the coefficient of $(\bar{\theta}\gamma_5\theta)^2$ in the expansion of K. Since we have *defined* the D-term as the coefficient of $-\frac{1}{2}(\bar{\theta}\gamma_5\theta)^2$, we can write the D-term part of the action as,

$$\int d^4x \mathcal{L}_D = -\frac{1}{4}\int d^4x d^4\theta K\left(\hat{S}^\dagger, \hat{S}\right). \tag{5.76}$$

---

[3] See *The Method of Second Quantization*, F. A. Berezin, Academic Press (1966).

To see how to write the F-term action as a superspace integral, it is most straight-forward to work in the chiral representation where, as in (4.9), the upper (lower) components of the spinor correspond to the two left-(right-)chiral components; i.e. $(\theta_1, \theta_2, \theta_3, \theta_4) = (\theta_{L1}, \theta_{L2}, \theta_{R1}, \theta_{R2})$. It is then easy to check that $\bar{\theta}\theta_L = 2\theta_{L2}\theta_{L1}$. From the form (5.34) for the expansion of an (elementary or composite) left-chiral superfield, we see that

$$\int d\theta_{L1} d\theta_{L2} \hat{f}(\hat{S}_L) \equiv \int d^2\theta_L \hat{f}(\hat{S}_L)$$

is exactly twice the coefficient of $\bar{\theta}\theta_L$ in the expansion of the superpotential. Since the F-term contribution to the Lagrangian density was *defined* to be the coefficient of $-\bar{\theta}\theta_L$ in this expansion, we see that the F-term part of the action can be expressed as

$$\int d^4x \mathcal{L}_F = -\frac{1}{2} \left[\int d^4x d^2\theta_L \hat{f}(\hat{S}) + \quad \text{h.c.} \right]. \tag{5.77}$$

While the F-term of left-chiral superfields involves integration over just the two Grassmann co-ordinates θ_{L1} and θ_{L2}, the D-term involves an integration over θ_{R1} and θ_{R2} as well. In the literature, it is instead common to see integrations over $d^2\theta$ and $d^2\bar{\theta}$, where θ and $\bar{\theta}$ are two-component spinors. Eq. (4.9) provides the connection. The two undotted components in (4.9) of the Majorana spinor θ (i.e. the two components of our θ_L) are frequently denoted by θ_i while the two dotted components are denoted by $\bar{\theta}^i$ ($i = 1, 2$). The analogue of our integration over the two θ_L (θ_R) co-ordinates is then integration over the two components of θ ($\bar{\theta}$).

6

Supersymmetric gauge theories

Local gauge invariance is a very powerful requirement. It is a symmetry principle that provides a powerful rationale for (Yang–Mills type) dynamics which, together with the ideas of spontaneous symmetry breaking, forms the basis of the Standard Model. To preserve the spectacular success of the Standard Model, it is reasonable to expect that its supersymmetric extension will incorporate the gauge principle. This then leads us to consider theories that are both supersymmetric and locally gauge invariant. In this chapter, we develop a formula analogous to Eq. (5.73) for a gauge invariant supersymmetric model with an arbitrary gauge group and any number of "matter" chiral superfields in specified representations of this group. This formula will then be our starting point for developing the Minimal Supersymmetric Standard Model or, for that matter, globally supersymmetric grand unified theories.

6.1 Gauge transformations of superfields

We saw in Chapter 4 that internal symmetry transformations must commute with the super-charge. Thus the various components of the superfields must transform in the same way under any internal symmetry transformation and, in particular, under a local gauge transformation. Hence, for a chiral scalar supermultiplet with components $(\mathcal{S}, \psi, \mathcal{F})$, we want the Lagrangian density to be invariant under the local gauge transformations,

$$\mathcal{S}_a(x) \to \left[e^{igt_A\omega_A(x)}\right]_{ab} \mathcal{S}_b(x), \tag{6.1a}$$

$$\psi_a(x) \to \left[e^{igt_A\omega_A(x)}\right]_{ab} \psi_b(x), \tag{6.1b}$$

$$\mathcal{F}_a(x) \to \left[e^{igt_A\omega_A(x)}\right]_{ab} \mathcal{F}_b(x), \tag{6.1c}$$

where we write the local transformation parameters as $\omega_A(x)$ to distinguish them from the SUSY transformation parameter α or the Majorana coordinate θ. The $\omega_A(x)$ are, of course, real functions of x, g is the gauge coupling constant, and the

t_A are matrix representations of the generators of the gauge group that satisfy the Lie algebra $[t_A, t_B] = i f_{ABC} t_C$.

What would be the corresponding transformation property of the superfield $\hat{S}(x)$? The answer would be easy if the ω_A were independent of x: then we would have simply that $\hat{S}_a \to \left[e^{i g t_A \omega_A}\right]_{ab} \hat{S}_b$. However, such a form cannot be correct for a local gauge transformation because of the derivatives that appear in the expansion in Eq. (5.34) or (5.36).

Recall, however, that the expansion (5.40) of the superfield in terms of $\hat{x} = x + \frac{1}{2}\bar{\theta}\gamma_5\gamma_\mu\theta$ has no derivatives. We would then be tempted to consider the transformations,

$$\hat{S}_a(\hat{x}, \theta) \to \left[e^{i g t_A \omega_A(\hat{x})}\right]_{ab} \hat{S}_b(\hat{x}, \theta).$$

The component fields would then transform as

$$S_a(\hat{x}) \to \left[e^{i g t_A \omega_A(\hat{x})}\right]_{ab} S_b(\hat{x}),$$
$$\psi_a(\hat{x}) \to \left[e^{i g t_A \omega_A(\hat{x})}\right]_{ab} \psi_b(\hat{x}),$$
$$\mathcal{F}_a(\hat{x}) \to \left[e^{i g t_A \omega_A(\hat{x})}\right]_{ab} \mathcal{F}_b(\hat{x}).$$

These reduce to (6.1a)–(6.1c) for $\theta = 0$. This cannot, however, be right either because after the gauge transformation, the components of $\hat{S}$, which was a left-chiral superfield, no longer transform as a left-chiral superfield. This is because $e^{i g t_A \omega_A(\hat{x})}$ (which has only one component field ω_A) is *not* a left-chiral superfield.

To ensure that the gauge transformed left-chiral superfield remains a left-chiral superfield, we are forced to introduce a set of left-chiral scalar superfields $\hat{\Omega}_A$ with as many members as the generators of the gauge group. We then consider the superfield transformation,

$$\hat{S}_a(\hat{x}, \theta) \to \left[e^{i g t_A \hat{\Omega}_A(\hat{x})}\right]_{ab} \hat{S}_b(\hat{x}, \theta), \tag{6.2}$$

which at least has the virtue that the transform of a left-chiral superfield remains a left-chiral superfield. The components then transform as

$$S_a(\hat{x}) \to \left[e^{i g t_A \hat{\Omega}_A(\hat{x})}\right]_{ab} S_b(\hat{x}), \tag{6.3a}$$

$$\psi_a(\hat{x}) \to \left[e^{i g t_A \hat{\Omega}_A(\hat{x})}\right]_{ab} \psi_b(\hat{x}), \tag{6.3b}$$

$$\mathcal{F}_a(\hat{x}) \to \left[e^{i g t_A \hat{\Omega}_A(\hat{x})}\right]_{ab} \mathcal{F}_b(\hat{x}). \tag{6.3c}$$

Setting $\theta = 0$, i.e. looking at the scalar components of (6.3a)–(6.3c), we see that we almost recover (6.1a)–(6.1c), except for what looks like a complex gauge transformation parameter (since the scalar components of $\hat{\Omega}_A$ are complex functions of x). We will return to this later.

We stress here that the "parameter superfield" $\hat{\Omega}$ is not a dynamical degree of freedom. Its components are just classical functions (Grassman-valued in the fermion case) of the spacetime co-ordinates. We need to introduce such a field so that after a gauge transformation, chiral superfields transform appropriately under a supersymmetry transformation.

Now that we have been able to sensibly extend the notion of gauge transformations to chiral superfields, we proceed to examine how to couple these in a gauge invariant way. Here, and in the following, by gauge invariance we will mean invariance under this extended gauge transformation. Since all interactions of chiral superfields with one another are given by the superpotential, we begin our study with that. However, because the superpotential is simply a polynomial of chiral superfields and *does not* contain any spacetime or supercovariant derivatives, it is clear that choosing it to be invariant under global gauge transformations ensures it is also invariant under the transformations (6.2). The Lagrangian density derived from this is then also invariant.

We thus have only to worry about the Kähler potential contributions which give rise to the kinetic terms for the component fields. For renormalizable theories, the Kähler potential is given by (5.60). We see immediately that this term is not invariant under the transformation (6.2) for the chiral superfield because the gauge parameter superfield $\hat{\Omega}_A$ is intrinsically complex. We have (with matrix multiplication implied),

$$\hat{S}^\dagger \rightarrow \hat{S}^\dagger e^{-igt_A \hat{\Omega}_A^\dagger}$$

and, as a result, the Kähler potential term,

$$\hat{S}^\dagger \hat{S} \rightarrow \hat{S}^\dagger e^{-igt_A \hat{\Omega}_A^\dagger} e^{igt_B \hat{\Omega}_B} \hat{S}$$

is no longer a gauge invariant. This should not be surprising. In the usual formulation of gauge theories, kinetic terms for the scalar or fermion fields are also not gauge invariant. We have to introduce new fields (the gauge potentials) and couple these to the scalars or fermions via a gauge covariant derivative to obtain a gauge invariant Lagrangian that includes these kinetic terms.[1]

Towards this end, we are led to introduce a set of gauge potential superfields $\hat{\Phi}_A$ in which the vector potentials reside. These are not chiral superfields, but are chosen to satisfy the reality conditions $\hat{\Phi}_A^\dagger = \hat{\Phi}_A$ so that their bosonic components are real while their fermionic components are Majorana. This ensures that the vector potential and the gauge field strength are real. The SUSY transformation rules for

---

[1] In fact, the parallel is exact since global gauge invariance of the Yukawa interactions of fermions as well as the scalar potential ensures these are also locally gauge invariant, just as the global gauge invariance of the superpotential (which leads to Yukawa interactions and the scalar potential in a supersymmetric theory) also guarantees its local gauge invariance.

its components are given by Eq. (5.29a)–(5.29g) of Chapter 5 (with the index A implied). We will see shortly that it is possible to work with the components $\mathcal{S}$, ψ, $\mathcal{M}$, and $\mathcal{N}$ set to zero. The curl supermultiplet that contains the field strengths is then constructed from $\hat{\Phi}_A$, just as the field strengths $F_A^{\mu\nu}$ are constructed from the vector potentials.

In order to maintain local gauge invariance of the Kähler potential, we modify it to,[2]

$$K = \hat{\mathcal{S}}^\dagger e^{-2gt_A\hat{\Phi}_A}\hat{\mathcal{S}} \tag{6.4}$$

where it is, of course, implicit that the dimensionality of the matrix t_A depends on the representation to which the chiral superfield belongs. We then *require* that the Kähler potential remains invariant under a gauge transformation, i.e.

$$\hat{\mathcal{S}}^\dagger e^{-igt_P\hat{\Omega}_P^\dagger}e^{-2gt_A\hat{\Phi}'_A}e^{igt_Q\hat{\Omega}_Q}\hat{\mathcal{S}} = \hat{\mathcal{S}}^\dagger e^{-2gt_A\hat{\Phi}_A}\hat{\mathcal{S}}.$$

This then fixes the gauge transformation rule for the fields $\hat{\Phi}_A$ to be,

$$e^{-igt_P\hat{\Omega}_P^\dagger}e^{-2gt_A\hat{\Phi}'_A}e^{igt_Q\hat{\Omega}_Q} = e^{-2gt_A\hat{\Phi}_A}. \tag{6.5}$$

Notice that the Kähler potential is now not a polynomial in the fields since the field $\hat{\Phi}$ is exponentiated. It still has mass dimension 2, however, because as noted in the exercise following Eq. (5.60) of the last chapter, $[\hat{\Phi}] = 0$, and renormalizability is not affected.

Let us *define* a left-chiral superfield

$$gt_A\hat{W}_A \equiv -\frac{i}{8}\bar{D}D_R\left[e^{2gt_C\hat{\Phi}_C}D_L e^{-2gt_B\hat{\Phi}_B}\right] \tag{6.6}$$

where the $D_{R/L}$ are the right/left supercovariant derivatives defined in Chapter 5. Its chiral nature follows because we have already checked (see the exercise below (5.54)) that the components of D_R anticommute, so that by the "Majorana character" of D, $D_R(\bar{D}D_R) = D_R(D_R^T C D_R) = 0$. The leading component (i.e. the θ-independent term) of the superfield $\hat{W}_A$ is a spinor, and we will call this a left-chiral *spinor* superfield (as opposed to a left-chiral scalar superfield).

Exercise *Convince yourself that none of the properties that we have derived for chiral superfields depended upon the fact that the leading component was a scalar. In other words, these properties of chiral superfields hold for $\hat{W}_A$ also. In particular, powers of $\hat{W}_A$ are left-chiral superfields (though not necessarily Lorentz scalars).*

---

[2] The Kähler potential (5.60) of a renormalizable theory is trivially invariant under global gauge transformations. More generally, if the Kähler potential $K(\hat{\mathcal{S}}_L, \hat{\mathcal{S}}_L^\dagger)$ is chosen to be globally gauge invariant, then $K(\hat{\mathcal{S}}_L, \hat{\mathcal{S}}_L^\dagger e^{-2gt_A\hat{\Phi}_A})$ will also be locally gauge invariant if $\hat{\Phi}_A$ transform as discussed below. This is because the product of any representation times the adjoint contains the original representation.

Since we know the corresponding transformation rule for $\hat{\Phi}_A$, we can now work out how the fields $\hat{W}_A$ transform under a gauge transformation. We have,

$$gt_A \hat{W}_A \rightarrow$$
$$-\tfrac{i}{8}\bar{D}D_R \left[e^{igt_P \hat{\Omega}_P} e^{2gt_C \hat{\Phi}_C} e^{-igt_{P'} \hat{\Omega}^\dagger_{P'}} D_L e^{igt_{Q'} \hat{\Omega}^\dagger_{Q'}} e^{-2gt_B \hat{\Phi}_B} e^{-igt_Q \hat{\Omega}_Q} \right].$$

Now, since $D_L \hat{\Omega}^\dagger_{Q'} = 0$ (because $\hat{\Omega}^\dagger_{Q'}$ is a right-chiral superfield),

$$e^{-igt_{P'} \hat{\Omega}^\dagger_{P'}} D_L e^{igt_{Q'} \hat{\Omega}^\dagger_{Q'}} = D_L,$$

and our gauge transformation simplifies to,

$$gt_A \hat{W}_A \rightarrow -\frac{i}{8}\bar{D}D_R \left[e^{igt_P \hat{\Omega}_P} e^{2gt_C \hat{\Phi}_C} D_L e^{-2gt_B \hat{\Phi}_B} e^{-igt_Q \hat{\Omega}_Q} \right].$$

The same type of argument allows us to move $\bar{D}D_R$ past the left-chiral superfield $e^{igt_P \hat{\Omega}_P}$, and we find that

$$gt_A \hat{W}_A \rightarrow -\frac{i}{8}e^{igt_P \hat{\Omega}_P} \left[\bar{D}D_R e^{2gt_C \hat{\Phi}_C} D_L e^{-2gt_B \hat{\Phi}_B} e^{-igt_Q \hat{\Omega}_Q} \right].$$

The D_L in the square brackets may act on either $e^{-2gt_B \hat{\Phi}_B}$ or $e^{-igt_Q \hat{\Omega}_Q}$. When it acts on the latter, the corresponding contribution to the square bracket becomes,

$$\bar{D}D_R D_L e^{-igt_Q \hat{\Omega}_Q}.$$

Since $D_R e^{-igt_Q \hat{\Omega}_Q} = 0$, we can replace $D_R D_L$ by the anticommutator, and then using (5.54), obtain $D_R D_L e^{-igt_Q \hat{\Omega}_Q} = -2i\partial\!\!\!/ C e^{-igt_Q \hat{\Omega}_Q}$. Then, since spacetime and supersymmetric covariant derivatives commute, we get

$$\bar{D}D_R D_L e^{-igt_Q \hat{\Omega}_Q} = -2i\partial_\mu \bar{D}_L \gamma^\mu C e^{-igt_Q \hat{\Omega}_Q} =$$
$$2i\partial_\mu \bar{D}_L C \gamma^{\mu T} e^{-igt_Q \hat{\Omega}_Q} = -2i\partial_\mu (\gamma^\mu D_R e^{-igt_Q \hat{\Omega}_Q})^T = 0$$

where the expression in the brackets vanishes because $\hat{\Omega}_Q$ is left handed. Thus, we only get a contribution when D_L acts on $e^{-2gt_B \hat{\Phi}_B}$ thereby giving us our final result for the gauge transformation of $\hat{W}_A$,

$$t_A \hat{W}_A \rightarrow e^{igt_P \hat{\Omega}_P} t_B \hat{W}_B e^{-igt_Q \hat{\Omega}_Q}. \tag{6.7}$$

Notice that unlike the transformation law (6.5) for the gauge potential superfields $\hat{\Phi}_A$ which entailed both $\hat{\Omega}_A$ and $\hat{\Omega}^\dagger_A$, the transformation law for $t_A \hat{W}_A$ brings in only the fields $\hat{\Omega}_A$. In fact, $t_A \hat{W}_A$ transforms as a *gauge field strength* $F^A_{\mu\nu}$ (except that the local transformation parameter is a superfield).[3]

---

[3] See, for example, *Introduction to Quantum Field Theory* by M. Peskin and D. Schroeder, Perseus Press (1995), Eq. (15.36), where the field strength transforms as $t_A F_{\mu\nu A} \rightarrow e^{igt_P \alpha_P} t_B F_{\mu\nu B} e^{-igt_Q \alpha_Q}$.

Let us summarize the main results for local gauge transformations of superfields. Chiral superfields transform as

$$\hat{S} \to e^{igt_A \hat{\Omega}_A} \hat{S} \quad \text{and} \quad \hat{S}^\dagger \to \hat{S}^\dagger e^{-igt_A \hat{\Omega}_A^\dagger}; \tag{6.8a}$$

the gauge potential superfield transforms as

$$e^{-2gt_A \hat{\Phi}_A} \to e^{igt_P \hat{\Omega}_P^\dagger} e^{-2gt_B \hat{\Phi}_B} e^{-igt_Q \hat{\Omega}_Q}; \tag{6.8b}$$

finally, the superfields

$$gt_A \hat{W}_A = -\frac{i}{8} \bar{D} D_R \left[e^{2gt_C \hat{\Phi}_C} D_L e^{-2gt_B \hat{\Phi}_B} \right]$$

transform as

$$t_A \hat{W}_A \to e^{igt_P \hat{\Omega}_P} t_B \hat{W}_B e^{-igt_Q \hat{\Omega}_Q}. \tag{6.8c}$$

We will see below that it is the superfield $\hat{W}_A$ that contains the field strength $F_A^{\mu\nu}$, and we will work out its other components. But first, to connect up this rather formal discussion with the usual formulation of gauge theories, let us work out the transformations (6.8b), and later (6.8c), in terms of the component fields.

6.2 The Wess–Zumino gauge

In the last chapter, we showed that under supersymmetry the $F^{\mu\nu}$, λ, and the $\mathcal{D}$ components of the curl superfield transformed into one another, but we did not discuss the other components of this multiplet. The reason for this, as we show next, is that we can work with all but the λ, V^μ, and $\mathcal{D}$ components of the *gauge potential superfield* set to zero. Then, because the curl superfield is derived from the gauge potential, the question of the other components does not arise.

6.2.1 Abelian gauge transformations

We begin by working out the transformations (6.8b) in component form for an Abelian theory. In this case, (6.8b) reads,

$$g\hat{\Phi}' = g\hat{\Phi} + i\frac{g}{2}(\hat{\Omega} - \hat{\Omega}^\dagger). \tag{6.9}$$

Notice that $i\frac{g}{2}(\hat{\Omega} - \hat{\Omega}^\dagger)$ is a real superfield so that $\hat{\Phi}$ remains real under a gauge transformation.

Recall that $\hat{\Omega}$ is a classical left-chiral scalar superfield. We denote its components by ω, ξ_L, and ζ. We are abusing notation here by using the symbol ω both for the (real) parameter of the local gauge transformation in (6.1a)–(6.1c) as well as for the (complex) scalar component of $\hat{\Omega}$, but we trust that this will not cause confusion. We can expand $\hat{\Omega}$ in its canonical form,

$$\hat{\Omega} = \omega(x) + i\sqrt{2}\bar{\theta}\xi_L(x) + i\bar{\theta}\theta_L\zeta(x) + \frac{i}{2}\bar{\theta}\gamma_5\gamma_\mu\theta\partial^\mu\omega(x)$$

$$- \frac{1}{\sqrt{2}}\bar{\theta}\gamma_5\theta\bar{\theta}\partial\!\!\!/\xi_L(x) + \frac{1}{8}(\bar{\theta}\gamma_5\theta)^2\Box\omega(x). \tag{6.10}$$

Then,

$$i\frac{g}{2}(\hat{\Omega} - \hat{\Omega}^\dagger) = i\frac{g}{2}\left\{ i\sqrt{2}\omega_I + i\sqrt{2}\bar{\theta}\xi + \frac{i}{\sqrt{2}}\bar{\theta}\theta\zeta_R + \frac{1}{\sqrt{2}}\bar{\theta}\gamma_5\theta\zeta_I \right.$$

$$\left. + \frac{i}{\sqrt{2}}(\bar{\theta}\gamma_5\gamma_\mu\theta)\partial^\mu\omega_R + \frac{1}{\sqrt{2}}\bar{\theta}\gamma_5\theta\bar{\theta}\partial\!\!\!/\gamma_5\xi + i\frac{\sqrt{2}}{8}(\bar{\theta}\gamma_5\theta)^2\Box\omega_I \right\},$$

where, $\omega = \frac{\omega_R + i\omega_I}{\sqrt{2}}$ and $\zeta = \frac{\zeta_R + i\zeta_I}{\sqrt{2}}$. Reading off the components of (6.9) immediately tells us that under a gauge transformation, the various components of $\hat{\Phi}$ transform as,

$$\mathcal{S}' = \mathcal{S} - \frac{1}{\sqrt{2}}\omega_I, \tag{6.11a}$$

$$\psi' = \psi - \frac{i}{2}\gamma_5\xi, \tag{6.11b}$$

$$\mathcal{M}' = \mathcal{M} - \frac{1}{\sqrt{2}}\zeta_I, \tag{6.11c}$$

$$\mathcal{N}' = \mathcal{N} - \frac{1}{\sqrt{2}}\zeta_R, \tag{6.11d}$$

$$V^{\mu\prime} = V^\mu - \frac{1}{\sqrt{2}}\partial^\mu\omega_R, \tag{6.11e}$$

$$\lambda' = \lambda, \tag{6.11f}$$

$$\mathcal{D}' = \mathcal{D}. \tag{6.11g}$$

This transformation preserves the reality of the Bose fields and the Majorana nature of the Fermi fields.

The important thing to note is that even if we started with a multiplet with non-zero $(\mathcal{S}, \psi, \mathcal{M}, \mathcal{N}, V^\mu, \lambda, \mathcal{D})$, by choosing ω_I, ξ, ζ_I, and ζ_R appropriately, we can set $\mathcal{S}'$, ψ', $\mathcal{M}'$, and $\mathcal{N}'$ to zero! This choice is called the *Wess–Zumino gauge*. Of course, if after setting these to zero we perform another SUSY transformation, we will re-generate these components again. Thus, the Wess–Zumino (WZ) gauge is not supersymmetric.

The local parameter $\omega_R(x)$ is not fixed by our choice of the WZ gauge, and the transformation corresponding to just this parameter reads,

$$V^{\mu\prime} = V^\mu - \frac{1}{\sqrt{2}}\partial^\mu \omega_R, \tag{6.12a}$$

$$\lambda' = \lambda, \tag{6.12b}$$

$$\mathcal{D}' = \mathcal{D}, \tag{6.12c}$$

while the other components are not affected by it. We still have the freedom to perform the gauge transformations (6.12a)–(6.12c). But these are the usual gauge transformations for an Abelian theory. The gauge field changes by a gradient, and the transformation parameter is the real part of the scalar component of the parameter superfield $\hat{\Omega}$, while the other components (which, being partners of a gauge field, must be neutral) remain invariant under the gauge transformation. In other words, the choice of the WZ gauge does not fix the gauge in the usual sense of the term.

6.2.2 Non-Abelian gauge transformations

We will now work out the transformation laws for the gauge potential superfields of a non-Abelian gauge theory. Our starting point will be Eq. 6.8b:

$$e^{-2gt_A\hat{\Phi}'_A} = e^{igt_P\hat{\Omega}^\dagger_P}e^{-2gt_B\hat{\Phi}_B}e^{-igt_Q\hat{\Omega}_Q}. \tag{6.13}$$

In this case, because the matrices t_A do not commute with one another, it is not possible to explicitly display the transformation to the WZ gauge as we did for the Abelian case above. Using the fact that a product of the exponential of three arbitrary matrices $\mathbf{u}$, $\mathbf{v}$, and $\mathbf{w}$ can be written (using the Baker–Campbell–Hausdorff formula) as,

$$e^{\mathbf{u}}e^{\mathbf{v}}e^{\mathbf{w}} = e^{\mathbf{z}},$$

with

$$\mathbf{z} = \mathbf{u} + \mathbf{v} + \mathbf{w} + \frac{1}{2}[\mathbf{u}, \mathbf{v}] + \frac{1}{2}[\mathbf{u}, \mathbf{w}] + \frac{1}{2}[\mathbf{v}, \mathbf{w}] + \cdots,$$

where the ellipsis denotes terms with nested commutators, we see that (6.13) gives us,

$$2gt_A\hat{\Phi}'_A = 2gt_B\hat{\Phi}_B + igt_B(\hat{\Omega}_B - \hat{\Omega}^\dagger_B) + g^2 f_{BCD}t_D(\hat{\Phi}_B\hat{\Omega}_C - \hat{\Phi}_C\hat{\Omega}^\dagger_B)$$

$$+ \frac{ig^2}{2} f_{BCD}t_D\hat{\Omega}_B\hat{\Omega}^\dagger_C + \cdots \tag{6.14}$$

The nested commutators that we have ignored have even higher powers of couplings. We first observe that the first two terms of (6.14) are the same as the corresponding equation for the Abelian case. Moreover all other terms, including the ellipsis, vanish in this case since the structure constants are zero. We now see that there is an iterative procedure for going to the WZ gauge for the non-Abelian case. To zeroth order in g, the gauge transformation that we need is identical to (6.11a)–(6.11g) discussed above. But this must be corrected to the next order in g to include terms on the second line of (6.14), and then again to include the yet higher order terms denoted by the ellipsis. The point of this argument is only to convince the reader that there is a Wess–Zumino gauge even for non-Abelian theories, where the S_A, ψ_A, $\mathcal{M}_A$, and $\mathcal{N}_A$ components of the field $\hat{\Phi}_A$ can be set to zero.

From now on, we will work in the WZ gauge where the gauge potential superfield can be written as,

$$\hat{\Phi}_A = \frac{1}{2}(\bar{\theta}\gamma_5\gamma_\mu\theta)V_A^\mu + i\bar{\theta}\gamma_5\theta\bar{\theta}\lambda_A - \frac{1}{4}(\bar{\theta}\gamma_5\theta)^2 D_A. \qquad (6.15)$$

We must remember that the components ω_{IA}, ξ_A, and ζ_A of the parameter superfield $\hat{\Omega}_A$ are now fixed, and the only gauge freedom corresponds to transformations that depend on the parameter ω_{RA} (which, we will see, is the gauge transformation in the conventional sense). We will thus compute how the components of $\hat{\Phi}_A$ change under the gauge transformation (6.13) taking the parameter superfield with $\omega_{IA} = \xi_A = \zeta_A = 0$, and $\omega_{RA}/\sqrt{2} \equiv \alpha_A$. In other words, we take,

$$\hat{\Omega}_A = \alpha_A(x) + \frac{i}{2}(\bar{\theta}\gamma_5\gamma_\mu\theta)\partial^\mu\alpha_A(x) + \frac{1}{8}(\bar{\theta}\gamma_5\theta)^2\Box\alpha_A(x) \qquad (6.16)$$

and

$$\hat{\Omega}_A^\dagger = \alpha_A(x) - \frac{i}{2}(\bar{\theta}\gamma_5\gamma_\mu\theta)\partial^\mu\alpha_A(x) + \frac{1}{8}(\bar{\theta}\gamma_5\theta)^2\Box\alpha_A(x). \qquad (6.17)$$

This transformation clearly preserves the WZ gauge. For simplicity, we will only compute an infinitesimal gauge transformation.

In evaluating the LHS of (6.13), we need only keep terms in the expansion of the exponential to second order, since each term in (6.15) is at least quadratic in θ. Thus,

$$e^{-2gt_A\hat{\Phi}'_A} = 1 - g\bar{\theta}\gamma_5\gamma_\mu\theta(t \cdot V^{\mu\prime}) - 2ig(\bar{\theta}\gamma_5\theta)\bar{\theta}(t \cdot \lambda')$$
$$+ \frac{1}{2}(\bar{\theta}\gamma_5\theta)^2 \left\{ gt \cdot \mathcal{D}' - (gt \cdot V^{\mu\prime})(gt \cdot V'_\mu) \right\}. \qquad (6.18)$$

We have used (5.24e) to cast the expression in canonical form and introduced the notation $t \cdot X$ as a shorthand for $t_A X_A$.

A straightforward substitution of (6.16), (6.17), and (6.15) into the RHS of Eq. (6.13) yields, to first order in α_A,

$$1 - gt_A \left[\bar\theta\gamma_5\gamma_\mu\theta\, V_A^\mu + 2i(\bar\theta\gamma_5\theta)\bar\theta\lambda_A - \frac{1}{2}(\bar\theta\gamma_5\theta)^2 \mathcal{D}_A \right] - \frac{1}{2}g^2 t_A t_B (\bar\theta\gamma_5\theta)^2 V_A^\mu V_{\mu B}$$
$$+ ig\left[(\alpha\cdot t), e^{-2gt_A\hat\Phi_A} \right] + \frac{g}{2}\bar\theta\gamma_5\gamma_\mu\theta \left\{ \partial^\mu\alpha\cdot t, e^{-2gt_A\hat\Phi_A} \right\}$$
$$+ \frac{i}{8}g(\bar\theta\gamma_5\theta)^2 \left[\Box\alpha\cdot t, e^{-2gt_A\hat\Phi_A} \right].$$

The last term of this expression vanishes because the commutator has at least two θs yielding a term with more than four θs. The second last term involving an anticommutator has two non-vanishing terms:

$$g\bar\theta\gamma_5\gamma_\mu\theta\partial^\mu\alpha\cdot t + \frac{g^2}{2}(\bar\theta\gamma_5\theta)^2\partial^\mu\alpha_A V_{\mu B}\{t_A, t_B\}.$$

The third last term becomes

$$- ig\left[(\alpha\cdot t), gt\cdot V^\mu\bar\theta\gamma_5\gamma_\mu\theta \right] + 2g\left[(\alpha\cdot t), g\bar\theta t\cdot\lambda \right]\bar\theta\gamma_5\theta$$
$$+ i\frac{g}{2}\left[(\alpha\cdot t), gt\cdot\mathcal{D} \right](\bar\theta\gamma_5\theta)^2 - i\frac{g}{2}\left[(\alpha\cdot t), (gt\cdot V^\mu)(gt\cdot V_\mu) \right](\bar\theta\gamma_5\theta)^2.$$

Putting all the pieces together, the RHS of (6.13) becomes

$$1 - \bar\theta\gamma_5\gamma_\mu\theta\, (gt\cdot V^\mu - g\partial^\mu\alpha\cdot t + ig[\alpha\cdot t, gt\cdot V^\mu])$$
$$- 2i(\bar\theta\gamma_5\theta)\bar\theta\,(gt\cdot\lambda + ig[\alpha\cdot t, gt\cdot\lambda])$$
$$+ \frac{1}{2}(\bar\theta\gamma_5\theta)^2\,\left(g\mathcal{D}\cdot t - (gt\cdot V^\mu)(gt\cdot V_\mu) + g^2\partial^\mu\alpha_A V_{\mu B}\{t_A, t_B\} \right.$$
$$\left. + ig^2[\alpha\cdot t, t\cdot\mathcal{D}] - ig[\alpha\cdot t, (gt\cdot V^\mu)(gt\cdot V_\mu)] \right). \tag{6.19}$$

Equating the coefficients of $-g\bar\theta\gamma_5\gamma_\mu\theta$ in (6.19) and (6.18) leads to,

$$(t\cdot V^{\mu\prime}) = t\cdot V^\mu - \partial^\mu\alpha\cdot t + ig\alpha_A V_B^\mu[t_A, t_B]. \tag{6.20}$$

Comparing coefficents of $-2ig(\bar\theta\gamma_5\theta)\bar\theta$ gives,

$$(t\cdot\lambda') = t\cdot\lambda + ig\alpha_A\lambda_B[t_A, t_B]. \tag{6.21}$$

Finally, by equating coefficients of $\frac{g}{2}(\bar\theta\gamma_5\theta)^2$ we get,

$$t\cdot\mathcal{D}' - g(t\cdot V^{\mu\prime})(t\cdot V'_\mu) = t\cdot\mathcal{D} + ig\alpha_A\mathcal{D}_B[t_A, t_B] - g(t\cdot V^\mu)(t\cdot V_\mu)$$
$$+ g\partial^\mu\alpha_A V_{\mu B}\{t_A, t_B\} - ig^2\alpha^A V_\mu^B V^{\mu C}[t_A, t_B t_C].$$

Exercise *Using the relation*

$$i[t_A, t_B t_C] = -f_{ABD} t_D t_C - f_{ACD} t_B t_D,$$

show that Eq. (6.20) leads to

$$-g(t \cdot V^{\mu'})(t \cdot V'_\mu) = - g(t \cdot V^\mu)(t \cdot V_\mu) + g V_{\mu A} \partial^\mu \alpha_B \{t_A, t_B\}$$
$$- ig^2 [t_A, t_B t_C] \alpha_A V_\mu^B V^{\mu C}.$$

Using the result of the exercise above, it is easy to show that

$$\mathcal{D}'_C = \mathcal{D}_C - g f_{ABC} \alpha_A \mathcal{D}_B. \tag{6.22}$$

It is now easy to see from (6.20) and (6.21) that for an infinitesimal gauge transformation by a parameter α_A, the component fields of the gauge potential superfield transform as,

$$V_C^{\mu'} = V_C^\mu - \partial^\mu \alpha_C - g f_{ABC} \alpha_A V_B^\mu, \tag{6.23a}$$
$$\lambda'_C = \lambda_C - g f_{ABC} \alpha_A \lambda_B, \tag{6.23b}$$
$$\mathcal{D}'_C = \mathcal{D}_C - g f_{ABC} \alpha_A \mathcal{D}_B. \tag{6.23c}$$

The first of these is exactly what we expect for the gauge transformation of a non-Abelian gauge potential. The vector field V_C^μ does not transform covariantly in that its transformation includes the inhomogeneous $\partial^\mu \alpha_C$ piece. The fields λ_C and $\mathcal{D}_C$ transform covariantly under the gauge transformation (i.e. the transformation is homogeneous). For the λ_C, for instance, this is just what we expect since it corresponds to fermions in the adjoint representation of the gauge group.

6.3 The curl superfield in the Wess–Zumino gauge

Before we can proceed with the construction of supersymmetric Lagrangians for gauge theories, we need to work out the explicit form of the curl superfield $\hat{W}_A$ introduced in Eq. (6.6):

$$g t_A \hat{W}_A = -\frac{i}{8} \bar{D} D_R \left[e^{2g t_A \hat{\Phi}_A} D_L e^{-2g t_B \hat{\Phi}_B} \right].$$

We will work in the WZ gauge where $\hat{\Phi}_A$ is given by (6.15). The calculation is rather lengthy, so we will break it up into a number of steps, and leave it to the reader to work through the details.

Step 1: Act with $D = \partial/\partial\bar{\theta} - i\partial\theta$ on $e^{-2g t_A \hat{\Phi}_A}$ to obtain

$$De^{-2g t_A \hat{\Phi}_A} = -2gt \cdot V^\mu (\gamma_5 \gamma_\mu \theta) + ig \left[\bar{\theta}\theta \gamma_5 t \cdot \lambda - (\bar{\theta}\gamma_5\theta) t \cdot \lambda - \bar{\theta}\gamma_5\gamma_\alpha\theta\gamma^\alpha t \cdot \lambda \right]$$
$$+ 2(\bar{\theta}\gamma_5\theta)\gamma_5\theta \{gt \cdot \mathcal{D} - g^2(t \cdot V)^2\} - ig(t \cdot \partial \not{V})\theta(\bar{\theta}\gamma_5\theta)$$
$$- \frac{1}{2} g(\bar{\theta}\gamma_5\theta)^2 \gamma_5(t \cdot \partial\lambda), \tag{6.24}$$

where we have, as usual, made use of various θ identities to cast the result in canonical form.

Step 2: Act with P_L on the above expression to obtain

$$D_L e^{-2gt_A \hat{\Phi}_A} = 2gt \cdot \not{V} \theta_R - ig\left[2\bar{\theta}\theta_R t \cdot \lambda_L + \bar{\theta}\gamma_5\gamma_\alpha\theta\gamma^\alpha t \cdot \lambda_R\right]$$
$$- 2(\bar{\theta}\gamma_5\theta)\theta_L\{gt \cdot \mathcal{D} - g^2(t \cdot V)^2\}$$
$$- ig(t \cdot \not{\partial} \not{V})\theta_L(\bar{\theta}\gamma_5\theta) + \frac{1}{2}(\bar{\theta}\gamma_5\theta)^2(gt \cdot \not{\partial}\lambda_R). \quad (6.25)$$

Step 3: Next we multiply this by $e^{2gt \cdot \hat{\Phi}}$ to find after some tedious algebra,

$$e^{2gt \cdot \hat{\Phi}} D_L e^{-2gt \cdot \hat{\Phi}} = 2gt \cdot \not{V}\theta_R - 2ig\bar{\theta}\theta_R t \cdot \lambda_L - ig(\bar{\theta}\gamma_5\gamma_\alpha\theta)\gamma^\alpha t \cdot \lambda_R$$
$$- 2\bar{\theta}\gamma_5\theta\{gt \cdot \mathcal{D} + i\frac{g}{2}(t \cdot \not{\partial} \not{V}) + \frac{1}{2}ig^2 f_{ABC} t_C \not{V}_B \not{V}_A\}\theta_L$$
$$+ \frac{1}{2}(\bar{\theta}\gamma_5\theta)^2 \{(gt \cdot \not{\partial}\lambda_R) + 2g^2 f_{ABC} \not{V}_B t_C \lambda_{AR}\}. \quad (6.26)$$

The structure constants in (6.26) come from writing the product $t_A t_B$ of Lie algebra generators as the sum of a commutator and an anticommutator. The former gives the structure constants, while the symmetry of the latter (under interchange of A and B) helps to reduce the expression to the form given above.

Step 4: Work out $\bar{D}\frac{1+\gamma_5}{2} D$ to find

$$\bar{D}\frac{1 + \gamma_5}{2} D = -\frac{\partial}{\partial\theta_a} P_{Rab}\frac{\partial}{\partial\bar{\theta}_b} + \bar{\theta}\theta_L\Box - 2i(P_R\not{\partial}\theta)_c\frac{\partial}{\partial\theta_c}. \quad (6.27)$$

We can now work out the action of $\frac{\partial}{\partial\theta_a} P_{Rab}\frac{\partial}{\partial\bar{\theta}_b}$ on various terms in (6.26) involving θ. Using (5.26a)–(5.26e), we obtain:

$$\frac{\partial}{\partial\theta_a} P_{Rab}\frac{\partial}{\partial\bar{\theta}_b}\bar{\theta} P_R\theta = 4, \quad (6.28a)$$

$$\frac{\partial}{\partial\theta_a} P_{Rab}\frac{\partial}{\partial\bar{\theta}_b}(\bar{\theta}\gamma_5\gamma_\alpha\theta) = 0, \quad (6.28b)$$

$$\frac{\partial}{\partial\theta_a} P_{Rab}\frac{\partial}{\partial\bar{\theta}_b}(\bar{\theta}\gamma_5\theta)\theta_{Lc} = 4\theta_{Lc}, \quad \text{and} \quad (6.28c)$$

$$\frac{\partial}{\partial\theta_a} P_{Rab}\frac{\partial}{\partial\bar{\theta}_b}(\bar{\theta}\gamma_5\theta)^2 = -8\bar{\theta}\theta_L. \quad (6.28d)$$

We thus obtain the action of $\bar{D}\frac{1+\gamma_5}{2}D$ on (6.26) to find

$$8igt\cdot\lambda_L + 8\{gt\cdot\mathcal{D}+i\frac{g}{2}(t\cdot\partial\!\!\!/\,\slashed{V})+i\frac{g^2}{2}f_{ABC}tc\,\slashed{V}_B\,\slashed{V}_A\}\theta_L$$

$$+4\bar{\theta}\theta_L\{(gt\cdot\partial\lambda_R)+2g^2f_{ABC}\,\slashed{V}_Btc\lambda_{AR}\}+2g\bar{\theta}\theta_L(t\cdot\Box\,\slashed{V})\theta_R$$

$$+ig(\bar{\theta}\gamma_5\theta)^2(t\cdot\Box\lambda_L)-4ig\partial_\mu(t\cdot\slashed{V})\gamma^\mu\theta_L-8g\bar{\theta}\partial\theta_L(t\cdot\lambda_L)+4g\partial t\cdot\lambda_R\bar{\theta}\theta_L$$

$$+8i\bar{\theta}\partial\theta_L\{gt\cdot\mathcal{D}+i\frac{g}{2}(t\cdot\partial\!\!\!/\,\slashed{V})+\frac{i}{2}g^2f_{ABC}tc\,\slashed{V}_B\,\slashed{V}_A\}\theta_L. \tag{6.29}$$

Step 5: To complete our calculation, we will exploit the fact that $\hat{W}_A$ is a left-chiral superfield; hence, its dependence on $\bar{\theta}_L$ and θ_R can arise only through $\hat{x}$. To obtain $\hat{W}_A$, we can thus pick off the terms involving only $\bar{\theta}_R$ and θ_L from Eq. (6.29), including of course the θ independent term, and then simply change the argument in the component fields from x to $\hat{x}$. These terms are,

$$8igt\cdot\lambda_L + 8\{gt\cdot\mathcal{D}+i\frac{g}{2}(t\cdot\partial\!\!\!/\,\slashed{V})+i\frac{g^2}{2}f_{ABC}tc\,\slashed{V}_B\,\slashed{V}_A\}\theta_L$$

$$+4\bar{\theta}\theta_L\{(gt\cdot\partial\lambda_R)+2g^2f_{ABC}\,\slashed{V}_Btc\lambda_{AR}\}-4ig\partial_\mu t\cdot\,\slashed{V}\gamma^\mu\theta_L+4g\bar{\theta}\theta_L\partial t\cdot\lambda_R$$

$$=8igt\cdot\lambda_L+4ig\gamma^\mu\gamma^\nu[(\partial_\mu V_{\nu A}-\partial_\nu V_{\mu A})t_A+gf_{ABC}V_{\mu B}V_{\nu A}tc]\theta_L$$

$$+8g\bar{\theta}\theta_Ltc\,[\partial\delta_{AC}+gf_{ABC}\,\slashed{V}_B]\lambda_{RA}+8gt\cdot\mathcal{D}\theta_L. \tag{6.30}$$

Since $F_{\mu\nu A}=\partial_\mu V_{\nu A}-\partial_\nu V_{\mu A}-gf_{ABC}V_{\mu B}V_{\nu C}$, the term in the first square brackets above is just $t_A F_{\mu\nu A}$. Also, recall that the gauge group structure constants furnish a representation – the adjoint representation of the gauge group: $[t_C^{\text{adj}}]_{AB}=-if_{CAB}$. Using this, the second set of square brackets above yields

$$[\partial\delta_{AC}+gf_{ABC}\,\slashed{V}_B]\lambda_{RA}=[\partial\delta_{CA}+ig(t_B^{\text{adj}})_{CA}\,\slashed{V}_B]\lambda_{RA},$$

which is the gauge covariant derivative acting on the field λ_A that always belongs to the adjoint representation of the gauge group.

Thus, the $\bar{\theta}_L$ and θ_R independent part of $\bar{D}D_R\{e^{2gt\cdot\Phi}D_Le^{-2gt\cdot\Phi}\}$ is:

$$8igt_A\lambda_{AL}+4ig\gamma^\mu\gamma^\nu t_A F_{\mu\nu A}\theta_L$$

$$+8g\bar{\theta}\theta_Lt_A(\slashed{\mathcal{D}}\lambda_R)_A+8gt_A\mathcal{D}_A\theta_L.$$

Comparing this with our definition (6.6) of $gt_A\hat{W}_A$, and then replacing the argument x with $\hat{x}$ we find that, in the WZ gauge,

$$\hat{W}_A(\hat{x},\theta)=\lambda_{LA}(\hat{x})+\frac{1}{2}\gamma^\mu\gamma^\nu F_{\mu\nu A}(\hat{x})\theta_L-i\bar{\theta}\theta_L(\slashed{\mathcal{D}}\lambda_R)_A-iD_A(\hat{x})\theta_L, \tag{6.31}$$

where $\slashed{\mathcal{D}}_{AC}=\partial\delta_{AC}+ig(t_B^{\text{adj}})_{AC}\,\slashed{V}_B$ is the gauge covariant derivative in the adjoint representation. The interested reader can explicitly check that, aside from the factor $8igt_A$, expanding the component fields about $\hat{x} = x$ indeed reproduces the expression in (6.29).

We note the following:

- We have already remarked that $\hat{W}_A$ is a left-chiral *spinor* superfield. We see explicitly that the θ independent term in $\hat{W}_A$ is a spinor. In addition to its gauge index A, it carries a spinor index which we have suppressed.
- $\hat{W}_A$ has components λ_A, $F_{\mu\nu A}$, and $\mathcal{D}_A$; the spinor λ_A and the scalar $\mathcal{D}_A$ are the same components that are in the gauge potential superfield $\hat{\Phi}_A$, but instead of the vector potential, the third component of $\hat{W}_A$ is the field strength.
- The product $\hat{W}_A \hat{W}_A$ is gauge invariant but not Lorentz invariant. Since $\hat{W}_A^c = C \overline{\hat{W}}_A^T$ transforms as the adjoint representation also, but is a right-chiral superfield, the combination

$$\overline{\hat{W}_A^c} \hat{W}_A$$

is a gauge-invariant, Lorentz-invariant bilinear in $\hat{W}_A$, *and is a product of only left-chiral superfields*. Its F-term is, therefore, a candidate for the Lagrangian density.

Exercise *Show that*

$$\overline{\hat{W}_A^c} = \bar{\lambda}_{RA} + \frac{1}{2} F_{\mu\nu A} \bar{\theta}_R \gamma^\nu \gamma^\mu$$

$$- i\bar{\theta}\theta_L \left[-\bar{\lambda}_{LA} \overleftarrow{\partial\!\!\!/} - gf_{CBA}\bar{\lambda}_{LC} \not{V}_B \right] - i\mathcal{D}_A \bar{\theta}_R. \qquad (6.32)$$

6.4 Construction of gauge kinetic terms

We have just seen that the F-term of $\overline{\hat{W}_A^c} \hat{W}_A$ is a candidate for a supersymmetric action. Moreover, inspection of (6.31) and (6.32) shows that this term contains a contribution proportional to $F_A^{\mu\nu} F_{\mu\nu A}$ so that this term potentially contains the gauge kinetic term. Before computing it, however, let us do some dimensional analysis to see the constraints imposed by renormalizability.

The dimensionality of the superfield $\hat{W}_A$ can be worked out as

$$[\hat{W}_A] = [\bar{D}DD] = [(\frac{\partial}{\partial\theta})^3] = \frac{3}{2}.$$

Since renormalizability requires that the (composite) superfield whose F-term is proportional to the Lagrangian density have mass dimension ≤ 3, this function can at most be quadratic in $\hat{W}_A$. We are thus left with just $\overline{\hat{W}_A^c} \hat{W}_A$ as the most general

Lorentz and gauge invariant bilinear.[4] Notice also that since $\hat{W}_A$ carries with it a spinor index, it can only enter via even powers, assuming that the other fields in the theory are just chiral scalar superfields. We do not, therefore, have to worry about products of $\hat{W}_A$s and $\hat{S}_{Li}$ in a renormalizable theory.

We are thus led to compute the $\bar{\theta}\theta_L$ term of $\overline{\hat{W}_A^c}\hat{W}_A$. We can do so by simply using the form (6.31) for $\hat{W}_A$ and setting $\hat{x} = x$ because, since $\hat{x} - x$ is already bilinear in θ, any other contribution to the $\bar{\theta}\theta_L$ term can only come from the θ independent term of this product. But this contribution is proportional to $\bar{\theta}\gamma_\mu\theta$ and not $\bar{\theta}\theta_L$ that we are looking for, and so does not contribute. We thus have,

$$\overline{\hat{W}_A^c}\hat{W}_A\bigg|_{\bar{\theta}\theta_L \text{ term}}$$

$$= \left[\bar{\lambda}_{RA} + \frac{1}{2}F_{\mu\nu A}\bar{\theta}_R\gamma^\nu\gamma^\mu + i\bar{\theta}\theta_L\left(\bar{\lambda}_{LA}\overset{\rightarrow}{\partial} + gf_{CBA}\bar{\lambda}_{CL}\,\rlap{/}{V}_B\right) - iD_A\bar{\theta}_R\right]$$

$$\times \left[\lambda_{LA} + \frac{1}{2}\gamma^{\mu'}\gamma^{\nu'}F_{\mu'\nu'A}\theta_L - i\bar{\theta}\theta_L\left(\overset{\leftarrow}{\partial}\lambda_{RA} + gf_{C'B'A}\,\rlap{/}{V}_B'\lambda_{RC'}\right) - iD_A\theta_L\right]\bigg|_{\bar{\theta}\theta_L \text{ term}}.$$

The sources of $\bar{\theta}\theta_L$ terms are,

1. $\left(-i\bar{\lambda}_{RA}(\overset{\leftarrow}{\partial}\lambda_{RA} + gf_{CBA}\,\rlap{/}{V}_B\lambda_{RC}) + i(\bar{\lambda}_{LA}\overset{\rightarrow}{\partial} + gf_{CBA}\bar{\lambda}_{CL}\,\rlap{/}{V}_B)\lambda_{LA}\right)\bar{\theta}\theta_L$

2. $\frac{1}{4}F_{\mu\nu A}F_{\mu'\nu'A}\bar{\theta}_R\gamma^\nu\gamma^\mu\gamma^{\mu'}\gamma^{\nu'}\theta_L$,

3. $-D_A D_A\bar{\theta}\theta_L$, and

4. $-\frac{i}{2}F_{\mu\nu A}\bar{\theta}_R\gamma^\nu\gamma^\mu D_A\theta_L - \frac{i}{2}F_{\mu\nu A}\bar{\theta}_R\gamma^\mu\gamma^\nu D_A\theta_L$.

In the first term above, we can integrate by parts and shift the derivative acting on $\bar{\lambda}_A$ to one acting on λ_A, at the cost of a sign. Then, up to a surface term that we will not display, the derivative terms together yield the usual kinetic term for the spinor field λ aside from a factor of -2.

The second term can be simplified by noting that,

$$\bar{\theta}_R\gamma^\nu\gamma^\mu\gamma^{\mu'}\gamma^{\nu'}\theta_L = \frac{1}{4}Tr\left[\gamma^\nu\gamma^\mu\gamma^{\mu'}\gamma^{\nu'}P_L\{\bar{\theta}\theta\mathbf{I} + \bar{\theta}\gamma_5\theta\cdot\gamma_5 - \bar{\theta}\gamma_5\gamma_\rho\theta\cdot\gamma_5\gamma^\rho\}\right]$$

$$= \frac{1}{2}(\bar{\theta}\theta_L)Tr\left[\gamma^\nu\gamma^\mu\gamma^{\mu'}\gamma^{\nu'}P_L\right]$$

where we have used (5.21). This then simplifies to

$$\left(\frac{1}{2}F_{\mu\nu A}F_A^{\mu\nu} + \frac{i}{4}\epsilon^{\nu\mu\mu'\nu'}F_{A\mu\nu}F_{A\mu'\nu'}\right)\bar{\theta}\theta_L. \tag{6.33}$$

---

[4] It is clear from (6.31) that $\gamma_5\hat{W}_A = -\hat{W}_A$ so that $\overline{\hat{W}_A^c}\gamma_5\hat{W}_A$ is not an independent term.

The first term in (6.33) is the usual gauge kinetic term apart from a factor -2, while the second term can be re-written as

$$\frac{1}{4}\epsilon^{\nu\mu\mu'\nu'}F_{A\mu\nu}F_{A\mu'\nu'}$$

$$= -\epsilon^{\mu\nu\mu'\nu'}(\partial_\mu A_{A\nu} - \frac{g}{2}f_{ABC}A_{B\mu}A_{C\nu})(\partial_{\mu'}A_{A\nu'} - \frac{g}{2}f_{AB'C'}A_{B\mu'}A_{C\nu'})$$

$$= -\epsilon^{\mu\nu\mu'\nu'}\partial_\mu(A_{A\nu}\partial_{\mu'}A_{A\nu'} - \frac{g}{3}f_{ABC}A_{A\nu}A_{B'\mu'}A_{C'\nu'})$$

$$- \epsilon^{\mu\nu\mu'\nu'}\frac{g^2}{4}f_{ABC}f_{AB'C'}A_{B\mu}A_{C\nu}A_{B'\mu'}A_{C'\nu'}.$$

In the last step the 1/3 enters because it does not matter upon which of the three gauge potentials the derivative acts – they all give the same contribution. We will leave it to the reader (see exercise below) to check that the last line of the expression above vanishes, so that the second term of (6.33) turns out to be a total derivative and makes no contribution to the equations of motion.

Finally, the last term in our list contracts a symmetric and antisymmetric tensor, and so identically vanishes.

Exercise *Verify that*

$$\epsilon_{\mu\nu\mu'\nu'}\frac{g^2}{4}f_{ABC}f_{AB'C'}A_B^\mu A_C^\nu A_{B'}^{\mu'}A_{C'}^{\nu'} = 0.$$

Hints: one way to verify this is to note that we may write,

$$t_P f_{PBC}A_B^\mu A_C^\nu = [t_B, t_C]A_B^\mu A_C^\nu \equiv [\mathbf{A}^\mu, \mathbf{A}^\nu]$$

$$t_Q f_{QB'C'}A_{B'}^{\mu'}A_{C'}^{\nu'} = [t_{B'}, t_{C'}]A_{B'}^{\mu'}A_{C'}^{\nu'} \equiv [\mathbf{A}^{\mu'}, \mathbf{A}^{\nu'}],$$

where we have introduced matrices, $\mathbf{A}^\mu \equiv A_B^\mu t_B$ etc. Then since $Tr(t_P t_Q) \propto \delta_{PQ}$, the term in question becomes $\epsilon_{\mu\nu\mu'\nu'}Tr(\mathbf{A}^\mu\mathbf{A}^\nu\mathbf{A}^{\mu'}\mathbf{A}^{\nu'})$ (aside from a multiplicative constant), and so vanishes because of the cyclic property of the trace.

Collecting all terms from the computation of the coefficient of $\bar\theta\theta_L$ in $\overline{\hat{W}_A^c}\hat{W}_A$ and inserting an additional $-1/2$ to put the gauge kinetic terms in canonical form, we obtain a supersymmetric and gauge-invariant Lagrangian density $\mathcal{L}_{\text{GK}}$ for the gauge field kinetic terms,

$$\mathcal{L}_{\text{GK}} = \frac{i}{2}\bar\lambda_A \slashed{\mathcal{D}}_{AC}\lambda_C - \frac{1}{4}F_{\mu\nu A}F_A^{\mu\nu} + \frac{1}{2}\mathcal{D}_A\mathcal{D}_A, \qquad (6.34)$$

where, as before,

$$F_{\mu\nu A} = \partial_\mu V_{\nu A} - \partial_\nu V_{\mu A} - gf_{ABC}V_{\mu B}V_{\nu C} \quad \text{and}$$

$$(\slashed{\mathcal{D}}\lambda)_A = \slashed{\partial}\lambda_A + ig(t_B^{\text{adj}}\,\slashed{V}_B)_{AC}\lambda_C.$$

We see that we have the usual gauge kinetic term for the Yang–Mills field. We also have a gauge invariant kinetic term for the massless spin $\frac{1}{2}$ fields λ_A in the adjoint representation of the gauge group. The spin zero fields $\mathcal{D}_A$ enter without any derivatives so these will turn out to be auxiliary fields that satisfy *algebraic* equations of motion. The quanta of the theory whose Lagrangian density is $\mathcal{L}_{GK}$ would thus be massless vector bosons together with a set of massless spin $\frac{1}{2}$ fermions, both in the adjoint representation of the gauge group. These fermions are termed *gauginos*.

Exercise *In our derivation of the Lagrangian density (6.34), we dropped surface terms. Show that these can be written as,*

$$\frac{1}{4}\epsilon_{\mu\nu\rho\sigma} F_A^{\mu\nu} F_A^{\rho\sigma} - \frac{i}{2}\partial_\mu \left(\overline{\lambda_A}\gamma^\mu\gamma_5\lambda_A\right).$$

These terms do not contribute to the field equations. Even so, they might be relevant in a non-Abelian gauge theory when instanton effects are important. In Abelian gauge theories, they have no effect.

We remark that the superfield $\overline{\hat{W}_A^c}\hat{W}_A$ is intrinsically complex, and it was only fortuitous that the surface terms that we ignored were anti-Hermitian. This would not be a problem, since, as in the case of the Lagrangian density for chiral superfields that we obtained in the last chapter, we would simply have added the Hermitian conjugate. The point, however, is that (since the Hermitian and anti-Hermitian parts are separately supersymmetric and gauge invariant) a supersymmetric gauge theory may include terms proportional to the surface terms shown in the preceding exercise. In non-Abelian gauge theories, the corresponding constant of proportionality is conventionally written as θ (not to be confused with the Grassmann number θ that we have been using).

Finally we note that it is only in the WZ gauge that we can set the scalar components of $\hat{\Phi}_A$ to zero. This is what gave us only a finite number of terms in the expansion (6.6) of $\hat{W}_A$, and not an infinite series. The latter would have resulted in a non-polynomial Lagrangian where renormalizability would not have been at all clear.

6.5 Coupling chiral scalar to gauge superfields

We have already seen that the gauge interactions of chiral superfields enter via the Kähler potential (6.4),

$$\hat{S}_L^\dagger e^{-2gt_A\hat{\Phi}_A}\hat{S}_L. \tag{6.35}$$

There is one such term for every chiral scalar superfield that we introduce. The $(\bar{\theta}\gamma_5\theta)^2$ component is a candidate Lagrangian density. In the previous chapter we had seen that, for $g = 0$, these gave rise to the kinetic energy terms for the scalar and fermion components of the chiral supermultiplet.

To work out this quartic term in θ, we substitute the expansions for $\hat{S}_L^\dagger$ and $\hat{S}_L$ from Eqs. (5.34) and (5.37) together with the unprimed version of (6.18) for the exponential of $\hat{\Phi}$ in the WZ gauge. There are then four sources of θ^4 terms in (6.35). First, we have the quartic terms from just $\hat{S}_L^\dagger \hat{S}_L$ multiplying the 1 from the exponential, which was exactly what we had worked out in Chapter 5. These terms are,

$$\hat{S}_L^\dagger \hat{S}_L \Big|_{\theta^4} = -\frac{1}{2}(\bar{\theta}\gamma_5\theta)^2 \{\frac{i}{2}\bar{\psi}\partial\!\!\!/\psi + \mathcal{F}^\dagger\mathcal{F} + (\partial_\mu S)^\dagger(\partial^\mu S)\}. \tag{6.36a}$$

Next we have contributions from,

$$-\hat{S}_{La}^\dagger g(t \cdot V^\mu)_{ab}\hat{S}_{Lb}(\bar{\theta}\gamma_5\gamma_\mu\theta), \tag{6.36b}$$

where the chiral superfields contribute two factors of θ. Then we have another contribution from,

$$-2ig(\bar{\theta}\gamma_5\theta)\hat{S}_{La}^\dagger\bar{\theta}(t \cdot \lambda)_{ab}\hat{S}_{Lb} \tag{6.36c}$$

with one θ from the chiral superfields, and finally, we have a contribution from,

$$\frac{1}{2}(\bar{\theta}\gamma_5\theta)^2\hat{S}_{La}^\dagger\{gt \cdot \mathcal{D} - g^2(t \cdot V)^2\}_{ab}\hat{S}_{Lb}. \tag{6.36d}$$

Non-zero terms from (6.36b) can only come when we have either θ_L and a $\bar{\theta}_L$ or θ_R and a $\bar{\theta}_R$ from the chiral superfields. These contributions are:

$$2\bar{\psi}_a\theta_R[-gt \cdot V^\mu]_{ab}\bar{\theta}\psi_{Lb}(\bar{\theta}\gamma_5\gamma_\mu\theta) - \frac{i}{2}(\bar{\theta}\gamma_5\gamma_\mu\theta)\partial_\mu S_a^\dagger[-gt \cdot V^\nu]_{ab}S_b(\bar{\theta}\gamma_5\gamma_\nu\theta)$$

$$+ \frac{i}{2}(\bar{\theta}\gamma_5\gamma_\mu\theta)S_a^\dagger[-gt \cdot V^\nu]_{ab}\partial_\mu S_b(\bar{\theta}\gamma_5\gamma_\nu\theta).$$

Using (5.21) on the first of these terms, and (5.24d) on the remaining terms to cast this in the canonical form, we obtain,

$$\frac{1}{2}g\bar{\psi}(t \cdot V\!\!\!/)\psi_L(\bar{\theta}\gamma_5\theta)^2 - \frac{i}{2}(\bar{\theta}\gamma_5\theta)^2\partial_\mu S^\dagger[gt \cdot V^\mu]S + \frac{i}{2}(\bar{\theta}\gamma_5\theta)^2 S^\dagger[gt \cdot V^\mu]\partial_\mu S.$$

Note that in the first term here we can rewrite (for reasons that will become clear shortly) the fermion bilinear using,

$$-\bar{\psi}_a(t \cdot V\!\!\!/)_{ab}\frac{1 - \gamma_5}{2}\psi_b = \bar{\psi}_b(t \cdot V\!\!\!/)_{ab}\frac{1 + \gamma_5}{2}\psi_a$$

$$= \frac{1}{2}[-\bar{\psi}(t \cdot V\!\!\!/)\psi_L + \bar{\psi}(t^* \cdot V\!\!\!/)\psi_R].$$

The terms from (6.36c) are

$$-2ig(\bar{\theta}\gamma_5\theta)(-i\sqrt{2}\bar{\psi}_a\theta_R)\bar{\theta}(t \cdot \lambda)_{ab}S_b - 2ig(\bar{\theta}\gamma_5\theta)S_a^\dagger\bar{\theta}(t \cdot \lambda)_{ab}(-i\sqrt{2}\bar{\theta}\psi_{Lb}).$$

Again, casting these in canonical form using the relations for bilinears of Majorana spinors together with (5.21), we obtain

$$\frac{g}{\sqrt{2}}(\bar{\theta}\gamma_5\theta)^2(t\cdot\lambda)_{ab}\psi_{Ra}S_b + \frac{g}{\sqrt{2}}(\bar{\theta}\gamma_5\theta)^2 S_a^\dagger(t_A)_{ab}\bar{\lambda}_A\psi_{Lb}.$$

We take the Lagrangian density to be the coefficient of $-\frac{1}{2}(\bar{\theta}\gamma_5\theta)^2$ in (6.35) which, as we saw in the last chapter, gives the correctly normalized kinetic terms for the scalar and fermion components of $\hat{S}$.

Collecting all the terms from Eq. (6.36a)–(6.36d), we find that the contribution to the Lagrangian density from $\hat{S}^\dagger e^{-2gt_A\hat{\Phi}_A}\hat{S}$ is,

$$\begin{aligned}
\mathcal{L}_{\text{gauge}} = &\frac{i}{2}\bar{\psi}\,\partial\!\!\!/\,\psi + (\partial_\mu S)^\dagger(\partial^\mu S) + \mathcal{F}^\dagger\mathcal{F} \\
&+ i(\partial_\mu S)^\dagger g(t\cdot V^\mu)S - iS^\dagger g(t\cdot V^\mu)\partial_\mu S - S^\dagger\left[gt\cdot\mathcal{D} - g^2(t\cdot V)^2\right]S \\
&+ \frac{1}{2}\left[-g\bar{\psi}(t\cdot\slashed{V})\psi_L + g\bar{\psi}(t^*\cdot\slashed{V})\psi_R\right] \\
&- \left(\sqrt{2}gS^\dagger t_A\bar{\lambda}_A\frac{1-\gamma_5}{2}\psi + \text{h.c.}\right).
\end{aligned} \tag{6.37}$$

We can now cast the interactions of the scalar and fermion components of the chiral superfields with gauge bosons in the familiar form using gauge covariant derivatives introduced in Chapter 1. The covariant derivatives on S are,

$$D_\mu S = \partial_\mu S + igt\cdot V_\mu S \tag{6.38a}$$
$$(D_\mu S)^\dagger = (\partial_\mu S)^\dagger - igS^\dagger t\cdot V_\mu. \tag{6.38b}$$

For the action of the covariant derivative on the Majorana spinor ψ, we must be careful because (3.3) shows that its left- and right-handed components are complex conjugates of one another.[5] Thus, if ψ_L transforms according to a representation given by t_A, then ψ_R transforms according to the conjugate representation whose generators are given by $-t_A^*$.

An aside on conjugate representations *Consider a field ϕ that transforms under some representation of a group, and let t_A be a matrix representation of the corresponding generators. Then if $\phi \to e^{i\alpha_A t_A}\phi$, $\phi^* \to e^{-i\alpha_A t_A^*}\phi^* = e^{i\alpha_A(-t_A^*)}\phi^*$. In other words, the conjugate field transforms with generators $(-t_A)^*$.*

It is easy to see that these satisfy the same Lie algebra $[t_A, t_B] = if_{ABC}t_C$ as the generators t_A. Since the structure constants can be chosen to be real, we have

---

[5] Only if we insist that the left- and right-handed components transform the same way, can we conclude that the fermion must belong to a real representation. For the case of the $U(1)$ group, this will mean that the charge of the fermion is zero. But we stress that it is possible to represent each chirality of a charged particle by a Majorana field. Then one of the chiral components of this Majorana field corresponds to the field of the particle, while the other corresponds to the antiparticle field.

$[t_A^*, t_B^*] = -\mathrm{i} f_{ABC} t_C^*$, or $[-t_A^*, -t_B^*] = \mathrm{i} f_{ABC}(-t_C^*)$. *Thus, if* t_A *is a set of representation matrices, then* $-t_A^*$ *is equally good. If the set of* $d \times d$ *matrices* t_A *furnish a representation denoted by* **d**, *the matrices* $-t_A^*$ *provide another equally good representation of the same dimensionality. This is known as the conjugate representation and is denoted by* **d***.

The gauge covariant derivative on a Majorana fermion ψ whose *left-chiral* component transforms via a representation furnished by t_A is thus given by,

$$D_\mu \psi = \partial_\mu \psi + \mathrm{i} g(t \cdot V_\mu)\psi_{\mathrm{L}} - \mathrm{i} g(t^* \cdot V_\mu)\psi_{\mathrm{R}}. \tag{6.38c}$$

We can then write the Lagrangian (6.37) as,

$$\mathcal{L}_{\text{gauge}} = \frac{\mathrm{i}}{2}\bar{\psi}\,\not{D}\psi + (D_\mu \mathcal{S})^\dagger (D^\mu \mathcal{S}) + \mathcal{F}^\dagger \mathcal{F}$$
$$-g\mathcal{S}^\dagger t \cdot \mathcal{D}\mathcal{S} + \left(-\sqrt{2}g\mathcal{S}^\dagger t_A \bar{\lambda}_A \frac{1 - \gamma_5}{2}\psi + \quad \text{h.c.}\right). \tag{6.39}$$

6.5.1 Fayet–Iliopoulos D-term

We have seen that the D-term of *any* superfield is a candidate for a Lagrangian. The D-term of a product of chiral superfields, being a total derivative, is not interesting. However, the D-term of the gauge potential multiplet $\hat{\Phi}_A$ is not a derivative of anything. It is independent of the terms that we have considered so far. As we saw in (6.23c), it is however gauge covariant and not gauge invariant, unless of course $f_{ABC} = 0$, i.e. when the gauge group is Abelian. We can thus include

$$\mathcal{L}_{\text{FI}} = \xi_p \mathcal{D}_p \tag{6.40}$$

in the Lagrangian density, where p runs over each $U(1)$ factor of the gauge group, where ξ_p are coupling constants with mass dimension $[\xi_p] = 2$.

It is easy to see that the D-term of higher powers of $\hat{\Phi}$ is not gauge invariant.

6.6 A master Lagrangian for SUSY gauge theories

We now collect the various contributions to the Lagrangian density of a renormalizable supersymmetric gauge theory that we have obtained into a single master formula which will serve as the starting point for SUSY model building. Our Lagrangian density consists of,

$$\mathcal{L} = \mathcal{L}_{\text{GK}} + \mathcal{L}_{\text{gauge}} + \mathcal{L}_F + \mathcal{L}_{\text{FI}}, \tag{6.41}$$

where $\mathcal{L}_{\text{GK}}$, $\mathcal{L}_{\text{gauge}}$, and $\mathcal{L}_{\text{FI}}$ have been constructed in this chapter, and $\mathcal{L}_F$ is as given in Eq. (5.70) of Chapter 5. $\mathcal{L}_{\text{GK}}$ and $\mathcal{L}_{\text{gauge}}$ have been explicitly constructed to be gauge invariant. Since the superpotential $\hat{f}$ is a sum of products of superfields without any (spacetime or supercovariant) derivatives, $\mathcal{L}_F$ will be just a product of fields with no derivatives. Thus, the condition for $\mathcal{L}_F$ to be locally gauge invariant is that it simply be globally gauge invariant. This is guaranteed if the superpotential is globally gauge invariant.

The complete Lagrangian for renormalizable, supersymmetric gauge theories is

$$
\begin{aligned}
\mathcal{L} = {} & \sum_i (D_\mu \mathcal{S}_i)^\dagger (D^\mu \mathcal{S}_i) + \frac{i}{2} \sum_i \bar{\psi}_i \, \slashed{D} \psi_i + \sum_i \mathcal{F}_i^\dagger \mathcal{F}_i \\
& + \frac{i}{2} \sum_{A,B} \bar{\lambda}_A \, \slashed{D}_{AB} \lambda_B - \frac{1}{4} \sum_A F_{\mu\nu A} F_A^{\mu\nu} + \frac{1}{2} \sum_A \mathcal{D}_A \mathcal{D}_A \\
& + \left(-\sqrt{2} g \sum_i \mathcal{S}_i^\dagger t \cdot \bar{\lambda} \frac{1 - \gamma_5}{2} \psi_i + \text{h.c.} \right) - g \sum_{i,A} \mathcal{S}_i^\dagger (t_A \mathcal{D}_A) \mathcal{S}_i \\
& - \sum_p \xi_p \mathcal{D}_p + \sum_i \left\{ -i \left(\frac{\partial \hat{f}}{\partial \mathcal{S}_i} \right)_{\hat{\mathcal{S}} = \mathcal{S}} \mathcal{F}_i + i \left(\frac{\partial \hat{f}}{\partial \mathcal{S}_i} \right)^\dagger_{\hat{\mathcal{S}} = \mathcal{S}} \mathcal{F}_i^\dagger \right\} \\
& - \frac{1}{2} \sum_{i,j} \bar{\psi}_i \left[\left(\frac{\partial^2 \hat{f}}{\partial \mathcal{S}_i \partial \mathcal{S}_j} \right)_{\hat{\mathcal{S}} = \mathcal{S}} \frac{1 - \gamma_5}{2} + \left(\frac{\partial^2 \hat{f}}{\partial \mathcal{S}_i \partial \mathcal{S}_j} \right)^\dagger_{\hat{\mathcal{S}} = \mathcal{S}} \frac{1 + \gamma_5}{2} \right] \psi_j, \quad (6.42)
\end{aligned}
$$

where i, j denote the matter field types, A is the gauge group index, and p runs over all the $U(1)$ factors of the gauge group.

To obtain our final formula, we may eliminate the auxiliary fields $\mathcal{F}_i$ and $\mathcal{D}_A$ via their equations of motion, which are purely algebraic:

$$
\mathcal{F}_i = -i \left(\frac{\partial \hat{f}}{\partial \mathcal{S}_i} \right)^\dagger_{\hat{\mathcal{S}} = \mathcal{S}} \quad \text{and} \quad \mathcal{F}_i^\dagger = i \left(\frac{\partial \hat{f}}{\partial \mathcal{S}_i} \right)_{\hat{\mathcal{S}} = \mathcal{S}} \quad (6.43a)
$$

$$
\mathcal{D}_A = g \sum_i \mathcal{S}_i^\dagger t_A \mathcal{S}_i + \xi_A. \quad (6.43b)
$$

Substituting into Eq. (6.42), we arrive at the master formula for supersymmetric gauge theories:

$$
\begin{aligned}
\mathcal{L} = {} & \sum_i (D_\mu \mathcal{S}_i)^\dagger (D^\mu \mathcal{S}_i) + \frac{i}{2} \sum_i \bar{\psi}_i \, \slashed{D} \psi_i + \sum_{\alpha,A} \left[\frac{i}{2} \bar{\lambda}_{\alpha A} (\slashed{D}\lambda)_{\alpha A} - \frac{1}{4} F_{\mu\nu\alpha A} F_{\alpha A}^{\mu\nu} \right] \\
& - \sqrt{2} \sum_{i,\alpha,A} \left(\mathcal{S}_i^\dagger g_\alpha t_{\alpha A} \bar{\lambda}_{\alpha A} \frac{1 - \gamma_5}{2} \psi_i + \text{h.c.} \right)
\end{aligned}
$$

$$-\frac{1}{2}\sum_{\alpha,A}\left[\sum_i \mathcal{S}_i^\dagger g_\alpha t_{\alpha A}\mathcal{S}_i + \xi_{\alpha A}\right]^2 - \sum_i \left|\frac{\partial \hat{f}}{\partial \hat{\mathcal{S}}_i}\right|^2_{\hat{\mathcal{S}}=\mathcal{S}} \tag{6.44}$$

$$-\frac{1}{2}\sum_{i,j}\bar{\psi}_i\left[\left(\frac{\partial^2 \hat{f}}{\partial\hat{\mathcal{S}}_i\partial\hat{\mathcal{S}}_j}\right)_{\hat{\mathcal{S}}=\mathcal{S}}\frac{1-\gamma_5}{2}+\left(\frac{\partial^2 \hat{f}}{\partial\hat{\mathcal{S}}_i\partial\hat{\mathcal{S}}_j}\right)^\dagger_{\hat{\mathcal{S}}=\mathcal{S}}\frac{1+\gamma_5}{2}\right]\psi_j,$$

where the covariant derivatives are given by,

$$D_\mu\mathcal{S} = \partial_\mu\mathcal{S} + i\sum_{\alpha,A}g_\alpha t_{\alpha A}V_{\mu\alpha A}\mathcal{S}, \tag{6.45a}$$

$$D_\mu\psi = \partial_\mu\psi + i\sum_{\alpha,A}g_\alpha(t_{\alpha A}V_{\mu\alpha A})\psi_\mathrm{L} - i\sum_{\alpha,A}g_\alpha(t^*_{\alpha A}V_{\mu\alpha A})\psi_\mathrm{R}, \tag{6.45b}$$

$$(\not{D}\lambda)_{\alpha A} = \not{\partial}\lambda_{\alpha A} + ig_\alpha\left(t^\mathrm{adj}_{\alpha B}\not{V}_{\alpha B}\right)_{AC}\lambda_{\alpha C}, \tag{6.45c}$$

$$F_{\mu\nu\alpha A} = \partial_\mu V_{\nu\alpha A} - \partial_\nu V_{\mu\alpha A} - g_\alpha f_{\alpha ABC}V_{\mu\alpha B}V_{\nu\alpha C}. \tag{6.45d}$$

The index α that suddenly appears in (6.44) is simply to allow for several gauge couplings that would be present if the gauge group is not simple.

Exercise *Observe that unlike ordinary derivatives, the covariant derivatives defined above do not commute. Show that their commutator is given by,*

$$\left[D_\mu, D_\nu\right] = i\sum_{\alpha,A}g_\alpha t_{\alpha A}F_{\mu\nu\alpha A}. \tag{6.46}$$

We will return to this result when we consider the covariant derivative in general relativity.

We note the following features of our master Lagrangian density (6.44).

1. The first line is the usual gauge-invariant kinetic energies for the components of the chiral and gauge superfields. The derivatives that appear are gauge-covariant derivatives appropriate to the particular representation in which the field belongs. For example, if we are talking about SUSY QCD, for quark fields in the first line of Eq. (6.44) the covariant derivative contains triplet $SU(3)_\mathrm{C}$ matrices: i.e. $D_\mu = \partial_\mu + ig_s\frac{\lambda_A}{2}V_A^\mu$, whereas the covariant derivative acting on the gauginos in the following line will contain octet matrices. These terms completely determine how *all* particles couple to gauge bosons. As in any gauge theory, this coupling is fixed by the minimal coupling prescription.

2. The next line describes the interactions of gauginos with the scalar and fermion components of chiral superfields. We will see later that matter particles as well as Higgs bosons together with their superpartners belong to chiral scalar supermultiplets. Thus, this term describes how gauginos couple matter fermions

to their superpartners, or Higgs bosons to their superpartners. Notice that these interactions are completely determined by the gauge couplings. Here $t_{\alpha A}$ is the appropriate dimensional matrix represention of the group generators for the αth factor of the gauge group, while g_α is the corresponding gauge coupling constant (one for each factor of the gauge group). Matrix multiplication is implied.

3. The third line describes the scalar potential. This has two distinct contributions. The first term on this line is determined solely by the gauge interactions and has its origin in the auxiliary field $\mathcal{D}_A$. This term is referred to as the D-term contribution to the scalar potential. The second term comes from the superpotential $\hat{f}$. We saw in the last chapter that this term arises when the auxiliary fields $\mathcal{F}_i$ are eliminated from the Lagrangian density. This set of terms is, therefore, referred to as F-term contributions to the scalar potential.

4. Finally, the last line of Eq. (6.44) describes the non-gauge, superpotential interactions of matter and Higgs fields as well as fermion mass terms. Since this line describes the interaction of fermion pairs with scalars, the Yukawa interactions of the SM can arise from this term. In other words, all the Yukawa couplings are contained in the superpotential.

5. We note here that in a supersymmetric theory, the scalar potential contains no new couplings other than the gauge couplings and the "Yukawa couplings" and fermion mass terms already present in the superpotential. This is the result of supersymmetry which relates the masses as well as couplings of fermions and bosons within a supermultiplet. Additional terms in the scalar potential are possible if supersymmetry is softly broken.

We conclude this chapter by presenting a recipe for the construction of renormalizable supersymmetric gauge theories.

(*a*) Choose a gauge group and the representations for the various supermultiplets, taking care to ensure that the theory is free of chiral anomalies. Matter fermions and Higgs bosons form parts of chiral scalar supermultiplets, $\hat{S}_{Li}$, while gauge bosons reside in the real gauge supermultiplet $\hat{\Phi}_A$. Keep in mind that we will need a chiral scalar superfield for every chiral component of matter fermions that we want to introduce.

(*b*) Choose a superpotential function which is a globally gauge-invariant polynomial (of degree ≤ 3 for renormalizable interactions) of the various left-chiral superfields.

(*c*) The interactions of all particles with gauge bosons are given by the usual "minimal coupling" prescription.

(*d*) Couple the gauginos to matter via the gauge interactions given in the second line of (6.44).

(*e*) Write down the additional self-interactions of the scalar matter fields as given by the third line of (6.44).

(*f*) Write down the non-gauge interactions of matter fields coming from the superpotential. The form of these is given by the last two terms of (6.44).

This theory is, of course, exactly supersymmetric. The final step for obtaining realistic models is to incorporate supersymmetry breaking. This forms the subject of the following chapter.

Exercise *Construct the Lagrangian density for supersymmetric quantum electrodynamics using (6.44) and the recipe just mentioned.*

Remember that you will need to introduce **two** *left-chiral scalar supermultiplets in order to obtain a massive Dirac electron. The left-handed part of the Majorana fermion field in the first multiplet will annihilate the left-handed (Dirac) electron, while the corresponding component in the second multiplet will annihilate a left-handed positron. By the Majorana property, the right-handed part of this fermion will annihilate right-handed electrons. The Dirac electron field is then the sum of the left-handed part of the first and the right-handed part of the second. There are, therefore, two scalar partners (one for each chiral component) of the Dirac electron.*

Show that the interaction of the photon with the Dirac electron is exactly as you would expect in QED, while the corresponding couplings to the scalar electrons are as in scalar QED. Work out the couplings of the photino (the SUSY partner of the photon) to the electron and the scalar electron.

Before concluding, we remark that the action for superymmetric gauge theories can also be written as an integral over superspace. We have,

$$S = -\frac{1}{4} \int d^4x d^4\theta \left[\hat{S}^\dagger e^{-2g t_A \hat{\Phi}_A} \hat{S} + 2\xi_p \hat{\Phi}_p \right]$$
$$- \frac{1}{2} \left[\int d^4x d^2\theta_L \hat{f}(\hat{S}) + \text{h.c.} \right] - \frac{1}{4} \int d^4x d^2\theta_L \overline{\hat{W}_A^c} \hat{W}_A, \qquad (6.47)$$

where the ξ_p are dimensionful couplings for Fayet–Iliopoulos terms, one for each $U(1)$ factor of the gauge group.

It is, perhaps, worth emphasizing here that supersymmetry is also restrictive in a sense that we have not yet encountered because we have been dealing with renormalizable theories. In this case, supersymmetry mandated the existence of superpartners with well-defined interactions, but (aside from the holomorphy requirement on the superpotential), did not restrict the spacetime structure of the interactions. However, not all interactions that we might imagine in ordinary field theory can be incorporated in a supersymmetric theory. This is exemplified in the exercise below where we assert that an arbitrary "Pauli magnetic moment" of fermions is forbidden in a $U(1)$ gauge theory.

Exercise *Show that supersymmetry precludes the introduction of the "Pauli term" (even if it is generalized to include transitional magnetic moments), $\overline{\psi}_1 \sigma_{\mu\nu} \psi_2 F^{\mu\nu} +$ h.c., in a globally supersymmetric Abelian gauge theory.*

One way to proceed is as follows. Since the Pauli term is a dimension 5 operator, in a supersymmetric theory it must arise either from a dimension 4 term in the superpotential, or from a dimension 3 term in the Kähler potential. Moreover, since this term is (anti-)linear in ψ_1, ψ_2, and $F_{\mu\nu}$, it must originate in a superfield term that includes (at least) one power of $\hat{S}_1$ and $\hat{S}_2$ (or in the case of the Kähler potential, possibly $\hat{S}_1^\dagger$ or $\hat{S}_2^\dagger$), the left-chiral superfields whose spinor components are ψ_1 and ψ_2, and one power of the left-chiral spinor curl superfield $\hat{W}$ exhibited in (6.31) whose θ component is the gauge field strength $F_{\mu\nu}$. But the mass dimension $[\hat{S}_1] = [\hat{S}_2] = 1$, and $[\hat{W}] = 3/2$, so that $[\hat{S}_1 \hat{S}_2 \hat{W}] = 7/2 > 3$, showing that this term cannot originate in a dimension 3 superfield operator in the Kähler potential.

*Finally, note that though $\hat{S}_1 \hat{S}_2 \hat{W}$ is a (possibly) gauge invariant left-chiral superfield, it is not Lorentz invariant because it is a spinor under Lorentz transformations: in order to be able to include it in the superpotential, we have to contract the spinor index on $\hat{W}$. We do so by letting a supersymmetric covariant derivative (remember that this also has a spinor index) act on any **one** of the superfields in the product: this then results in a dimension 4 superfield product as required. We have, however, already seen that the supercovariant derivative acting on a left-chiral superfield does not leave it as a left-chiral superfield, so that terms that include such a supercovariant derivative are not allowed in the superpotential. We thus conclude that the "Pauli term" is absent if supersymmetry is unbroken.[6]*

Notice that our argument relies only upon dimensional counting and hence applies equally to electric as well as magnetic dipole moments. Also, its validity is independent of whether these dipole moments are diagonal (for Dirac fermions) or transitional.

We thus conclude that in supersymmetric models, anomalous magnetic moments or radiative transitions of elementary fermions (contained in chiral supermultiplets) are possible only if supersymmetry is broken. In other words, contributions from supersymmetric partners in the loops exactly cancel SM contributions if supersymmetry is unbroken. Measurements of anomalous magnetic moments of SM fermions or radiative decays of heavy quarks or leptons potentially provide information about supersymmetry breaking. We will return to this in Chapter 9.

---

[6] This was first noted by S. Ferrara and E. Remiddi, *Phys. Lett.* **B53**, 347 (1974).

6.7 The non-renormalization theorem

Supersymmetric theories have better ultra-violet behavior than their non-supersymmetric counterparts. We have already seen an illustration of this in our examination of the one-loop corrections in the Wess–Zumino model, where it was shown that quadratically divergent loop integrals all cancelled. It is now understood that this apparently miraculous cancellation of quadratic divergences is a general consequence of the *SUSY non-renormalization theorem* which states that to any order in perturbation theory, any loop correction can be written as a *D*-term, i.e. one particle irreducible loop corrections do not generate *F*-terms. In particular, there are no loop corrections to the superpotential.

This was first established by using supergraph methods,[7] a perturbative technique that maintains manifest supersymmetry throughout the calculation in the same way that Feynman diagram techniques keep the Lorentz covariance manifest.[8] A more direct proof of this theorem was given by Seiberg who recognized that the holomorphy of the superpotential (which is a direct consequence of supersymmetry) suffices to establish that there are no perturbative loop corrections to the superpotential, as long as the regularization procedure preserves supersymmetry and gauge invariance.[9]

D-terms in the action of a supersymmetric theory lead to the kinetic energy terms for the components of chiral superfields, so that corrections to these lead to so-called "wave function renormalization". Since loop corrections do not change the superpotential, superpotential masses and couplings are renormalized only because of the wave function renormalization; i.e. *supersymmetry precludes additional renormalization of the mass terms in the superpotential.* The reader familiar with the basics of renormalization in quantum field theory will immediately recognize that the wave function renormalization is at most logarithmically divergent in the cut-off, thereby establishing that supersymmetric theories are free of quadratic divergences to all orders in perturbation theory. This is important because the existence of quadratic divergences played the central role in persuading us that there must be new physics at the TeV scale. It is the non-renormalization theorem that assures us that TeV scale superpartners can stabilize the electroweak symmetry breaking sector of the supersymmetric extension of the SM in the sense discussed in Chapter 2.

[7] M. T. Grisaru, W. Siegel and M. Roček, *Nucl. Phys.* **B159**, 429 (1979).

[8] Supergraph methods were introduced by A. Salam and J. Strathdee, *Phys. Rev.* **D11**, 1521 (1975) and developed by other authors. See e.g. J. Honerkamp *et al.*, *Nucl. Phys.* **B95**, 397 (1975) and S. Ferrara, *Nucl. Phys.* **B93**, 261 (1975).

[9] N. Seiberg, *Phys. Lett.* **B318**, 469 (1993).

7

Supersymmetry breaking

The fundamental relation

$$[Q, P^\mu] = 0$$

of the supersymmetry algebra tells us that if supersymmetry is exact, every bosonic state other than the vacuum state must have a corresponding fermionic partner with exactly the same energy, assuming that we can identify P^0 with the Hamiltonian. To see this, we simply note that if $|\mathcal{B}\rangle$ is a bosonic eigenstate of the Hamiltonian with eigenvalue E_B, we must have,

$$P^0(Q|\mathcal{B}\rangle) = QP^0|\mathcal{B}\rangle = E_B Q|\mathcal{B}\rangle,$$

so that $|\mathcal{F}\rangle \equiv Q|\mathcal{B}\rangle$ is a fermionic eigenstate of this same Hamiltonian, with the same energy E_B. Thus the only bosonic states which are *not* paired with a fermionic state are those that are annihilated by Q. States with non-vanishing four-momenta transform non-trivially under supersymmetry (and so, are not annihilated by Q), and the only candidate for an unpaired bosonic state is the vacuum state. For massive single particle states in the rest frame, this implies that in a supersymmetric theory bosons and fermions must come in mass-degenerate pairs.

This is, of course, experimentally excluded since we know, for instance, that there is no integer spin charged particle with the same mass as that of the electron. Supersymmetry must, therefore, be a broken symmetry. While we cannot exclude the possibility that SUSY is explicitly broken by soft terms, it is much more appealing to consider that, like electroweak gauge symmetry, SUSY is broken spontaneously.

As with bosonic symmetries, if the generator Q of a supersymmetry transformation does not annihilate the vacuum, then supersymmetry is spontaneously broken. In correspondence with Eq. (1.5), we then write the condition for supersymmetry *not* to be spontaneously broken as

$$\langle 0|\delta\mathcal{O}|0\rangle \equiv i\langle 0|[\bar{\alpha}Q, \mathcal{O}]|0\rangle = 0, \tag{7.1}$$

where, in field theory, the dynamical variable $\mathcal{O}$ is a field operator and $\delta\mathcal{O}$ is its variation under a supersymmetry transformation with a Grassmann parameter α. If we find a field operator such that its *variation* is non-zero in the ground state, then supersymmetry will be spontaneously broken. Just as familiar gauge symmetries may be broken by vacuum expectation values (VEVs) of either elementary or composite field operators, $\mathcal{O}$ may be either elementary or composite. In order for Poincaré invariance not to be spontaneously broken, $\delta\mathcal{O}$ must be a spinless operator. Since SUSY connects fields whose spins differ by $1/2$, $\mathcal{O}$ must thus be a spinorial operator.

7.1 SUSY breaking by elementary fields

Up to this point, we have two classes of fields in a SUSY theory: the chiral scalar superfield and the curl superfield (or equivalently the gauge potential superfield). To identify potential order parameters for SUSY breaking, let us look at the transformation of their spinor components.

The variation of the spinor component of the chiral scalar superfield is

$$\delta\psi_{\rm L} = -\sqrt{2}\mathcal{F}\alpha_{\rm L} + \sqrt{2}\partial\!\!\!/ S\alpha_{\rm R}$$

while for that of a gauge superfield we have the variation of the spinor component as

$$\delta\lambda_A = -i\gamma_5\alpha\mathcal{D}_A + \frac{1}{4}[\gamma_\nu, \gamma_\mu]F_A^{\mu\nu}\alpha,$$

(we may equivalently discuss the gauge potential superfield $\hat{\Phi} \ni (V^\mu, \lambda, \mathcal{D})$ with similar results). Since Poincaré invariance requires,

$$\langle 0|\partial_\mu S|0\rangle = \langle 0|F^{\mu\nu}|0\rangle = 0,$$

the condition for SUSY to be spontaneously broken is,

$$\langle 0|\mathcal{F}_i|0\rangle \neq 0 \quad \text{or} \quad \langle 0|\mathcal{D}_A|0\rangle \neq 0 \tag{7.2}$$

for some fields. We will refer to these two possibilities as F-type or D-type SUSY breaking.

Since the auxiliary fields are given by (6.43a) and (6.43b) of the last chapter, we conclude that supersymmetry is spontaneously broken if the system of equations,

$$\left(\frac{\partial\hat{f}}{\partial S_i}\right)_{\hat{S}=S} = 0 \quad (F\text{-type}) \tag{7.3a}$$

or

$$g \sum_i \mathcal{S}_i^\dagger t_A \mathcal{S}_i + \xi_A = 0 \quad (D\text{-type}) \tag{7.3b}$$

does not have any solutions. Otherwise, supersymmetry is unbroken. In this chapter we will examine several toy models to illustrate both the F- and D-type SUSY breaking mechanisms.

Two comments are in order.

- The master formula (6.44) for the Lagrangian for SUSY gauge theories contains the terms,

$$\mathcal{L} \ni -V_{\text{scalar}} \equiv -\frac{1}{2} \sum_A \mathcal{D}_A \mathcal{D}_A - \sum_i |\mathcal{F}_i|^2.$$

Thus, if any of the auxiliary fields develop a VEV, then so will the scalar potential.
- If $Q|0\rangle \neq 0$, then the state has infinite norm. This is because

$$\| \, Q|0\rangle \, \|^2 = \int d^3x \, \langle 0|j^{0\dagger}(x) Q|0\rangle$$

(where $j^\mu(x)$ is the spinorial Noether current corresponding to the super-charge Q) diverges in a translationally invariant theory unless Q annihilates the vacuum. This is exactly as for the case of spontaneous breaking of bosonic symmetries. It is often loosely stated that spontaneous SUSY breaking is signalled by the VEV of the Hamiltonian. This is not the case since if the Hamiltonian density develops a constant VEV, its integral does not exist. In fact, just as in the familiar case of ordinary symmetries where the charges do not exist when the symmetry is spontaneously broken, the generators of the super-algebra do not exist if supersymmetry is spontaneously broken; the charge and current densities are, however, well defined and it is only these that we need for most manipulations in field theory.

7.2 *F*-type SUSY breaking: the O'Raifeartaigh model

A simple supersymmetric model exhibiting spontaneous breaking of supersymmetry was written down by O'Raifeartaigh in 1975.[1] It contains three chiral scalar superfields $\hat{X} \ni (X, \psi_X, \mathcal{F}_X)$, $\hat{Y} \ni (Y, \psi_Y, \mathcal{F}_Y)$, and $\hat{Z} \ni (Z, \psi_Z, \mathcal{F}_Z)$ interacting via the superpotential,

$$\hat{f}(\hat{X}, \hat{Y}, \hat{Z}) = \lambda(\hat{X}^2 - \mu^2)\hat{Y} + m\hat{X}\hat{Z}, \tag{7.4}$$

[1] L. O'Raifeartaigh, *Nucl. Phys.* **B96**, 331 (1975).

with m and λ as real parameters. Since $\mathcal{F}_i = -i(\partial \hat{f}/\partial \hat{\mathcal{S}}_i)^\dagger_{\hat{\mathcal{S}}=\mathcal{S}}$, we have

$$i\mathcal{F}_X = 2\lambda Y^\dagger X^\dagger + mZ^\dagger, \tag{7.5a}$$

$$i\mathcal{F}_Y = \lambda(X^{\dagger 2} - \mu^2), \quad \text{and} \tag{7.5b}$$

$$i\mathcal{F}_Z = mX^\dagger. \tag{7.5c}$$

Note that both $\langle \mathcal{F}_Y \rangle$ and $\langle \mathcal{F}_Z \rangle$ cannot simultaneously be zero. Hence, supersymmetry must be broken.

The scalar potential of this model is,

$$V(X, Y, Z) = \sum_i |\mathcal{F}_i|^2 = |2\lambda XY + mZ|^2 + \lambda^2|X^2 - \mu^2|^2 + m^2|X|^2. \tag{7.6}$$

Notice that the potential is a sum of non-negative terms. This is a general feature of theories with global supersymmetry. Indeed, we see from the master formula (6.44) that the D- and F-term contributions to the scalar potential are separately non-negative.

To find the minimum of this potential, observe that the first term can be made to be zero no matter what $\langle X \rangle$ and $\langle Y \rangle$ are since $\langle Z \rangle$ is chosen to cancel it. The vacuum state is, therefore, infinitely degenerate. The direction (in field space) along which the first term vanishes is referred to as an F-flat direction (since the value of the potential is flat along this direction). Flat directions frequently occur in supersymmetric models. We should add here that the flatness of the (tree-level) potential is generally removed when quantum corrections are taken into account.

Returning to the potential of the O'Raifeartaigh model, the minimum thus depends only on the last two terms of (7.6) that define the self-couplings V_X for the X field. We break the complex field X into real and imaginary parts $X = \frac{X_R + iX_I}{\sqrt{2}}$, so that

$$\begin{aligned} V_X &= \lambda^2|X^2 - \mu^2|^2 + m^2|X|^2 \\ &= \frac{\lambda^2}{4}(X_R^2 + X_I^2)^2 + \frac{1}{2}(m^2 - 2\lambda^2\mu^2)X_R^2 \\ &\quad + \frac{1}{2}(m^2 + 2\lambda^2\mu^2)X_I^2 + \lambda^2\mu^4. \end{aligned} \tag{7.7}$$

We will examine two cases for V_X, illustrated in Fig. 7.1.

Case A: If $m^2 > 2\lambda^2\mu^2$, the minimum of V_X is clearly at $\langle X \rangle = 0$. In this case, $\langle Z \rangle = 0$ but $\langle Y \rangle$ is undetermined, and $V_{\min} = \lambda^2\mu^4$. Y is a flat direction of the scalar potential.

Case B: If $m^2 < 2\lambda^2\mu^2$, $\langle X_R \rangle \neq 0$ but $\langle X_I \rangle = 0$ since the coefficient of X_I^2 is positive. The minimum will occur at $\langle X_R \rangle^2 = 2\mu^2 - m^2/\lambda^2$ and $\langle Z \rangle = -\frac{2\lambda}{m}\sqrt{\mu^2 - \frac{m^2}{2\lambda^2}}\langle Y \rangle$. At the minimum, $V_{\min} = -\frac{\lambda^2}{4}(2\mu^2 - m^2/\lambda^2)^2 + \mu^4\lambda^2$. Note that the minimum does *not* occur at $\langle X \rangle^2 = \mu^2$.

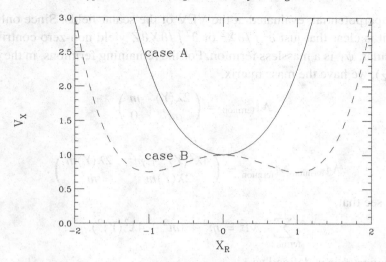

Figure 7.1 Scalar potential in the O'Raifeartaigh model for case A with $m/\mu = 2$ and $\lambda = 1$, and case B with $m/\mu = \lambda = 1$. In both cases, $X_I = 0$.

7.2.1 Mass spectrum: Case A

Next we will construct the mass matrix for the scalar fields of case A in the O'Raifeartaigh model. As usual, we first shift the fields by their VEVs, and then rewrite the scalar potential in terms of the shifted field $Y_S = Y - \langle Y \rangle$ together with X and Z to obtain,

$$V = |2\lambda X(Y_S + \langle Y \rangle) + mZ|^2 + \lambda^2|X^2 - \mu^2|^2 + m^2|X|^2. \qquad (7.8)$$

There are no bilinear terms in the field Y_S which must, therefore, be massless. Let us write $X = \frac{X_R + iX_I}{\sqrt{2}}$ and $Z = \frac{Z_R + iZ_I}{\sqrt{2}}$. We can now work out the scalar mass squared matrix for the four real fields. In the basis (X_R, Z_R, X_I, Z_I) it is given by,

$$\begin{pmatrix} m^2 + 4\lambda^2\langle Y \rangle^2 - 2\lambda^2\mu^2 & 2\lambda m\langle Y \rangle & 0 & 0 \\ 2\lambda m\langle Y \rangle & m^2 & 0 & 0 \\ 0 & 0 & m^2 + 4\lambda^2\langle Y \rangle^2 + 2\lambda^2\mu^2 & 2\lambda m\langle Y \rangle \\ 0 & 0 & 2\lambda m\langle Y \rangle & m^2 \end{pmatrix}$$

$$(7.9)$$

where we have taken $\langle Y \rangle$ to be real. If $\langle Y \rangle \neq 0$, then X and Z mix. Regardless of the mixing, however, the trace of the matrix gives,

$$\sum_{\text{bosons}} \mathcal{M}^2 = 2m^2 + 2(m^2 + 4\lambda^2\langle Y \rangle^2). \qquad (7.10)$$

To find the fermion masses, we must examine fermionic bilinear terms that can be derived from the superpotential. These terms will arise from second derivatives of the superpotential, evaluated at the VEV of the scalar fields. Since only Y has a VEV, it is clear that just $\partial^2 \hat{f}/\partial \hat{X}^2$ or $\partial^2 \hat{f}/\partial \hat{X} \partial \hat{Z}$ yield non-zero contributions. In particular, ψ_Y is a massless fermion. For the remaining fermions, in the basis of (ψ_X, ψ_Z), we have the mass matrix,

$$\mathcal{M}_{\text{fermion}} = \begin{pmatrix} 2\lambda\langle Y\rangle & m \\ m & 0 \end{pmatrix} \tag{7.11}$$

so that

$$\mathcal{M}_{\text{fermion}}\mathcal{M}_{\text{fermion}}^{\dagger} = \begin{pmatrix} 4\lambda^2\langle Y\rangle^2 + m^2 & 2\lambda\langle Y\rangle m \\ 2\lambda\langle Y\rangle m & m^2 \end{pmatrix}. \tag{7.12}$$

We thus see that,

$$\sum_{\text{fermions}} \mathcal{M}^2 = m^2 + (m^2 + 4\lambda^2\langle Y\rangle^2). \tag{7.13}$$

The *supertrace* is defined as

$$STr\mathcal{M}^2 = \sum_{\text{particles}} (-1)^{2J}(2J+1)m_J^2, \tag{7.14}$$

where the sum is over all particles in the theory, J is the spin and m_J is the mass. In any model where supersymmetry is unbroken, the degeneracy of the fermion and boson masses, together with the equality of bosonic and fermionic degrees of freedom obviously means that the supertrace must vanish. In the case under study, summing over all bosons and fermions, we obtain

$$STr\mathcal{M}^2 = 0 \tag{7.15}$$

even though supersymmetry is spontaneously broken. We will shortly see that (at tree level) the supertrace is always zero for theories with only chiral scalar superfields, even if supersymmetry is spontaneously broken.

Exercise *Work out the mass spectrum of the model. Show that aside from the complex massless boson field Y and the massless fermion ψ_Y, the remaining boson squared masses are,*

$$m^2 + 2\lambda^2\langle Y\rangle^2 - \lambda^2\mu^2 \pm \left[(m^2 + 2\lambda^2\langle Y\rangle^2 - \lambda^2\mu^2)^2 - m^2(m^2 - 2\lambda^2\mu^2)\right]^{\frac{1}{2}},$$

$$m^2 + 2\lambda^2\langle Y\rangle^2 + \lambda^2\mu^2 \pm \left[(m^2 + 2\lambda^2\langle Y\rangle^2 + \lambda^2\mu^2)^2 - m^2(m^2 + 2\lambda^2\mu^2)\right]^{\frac{1}{2}},$$

while the remaining fermion masses are,

$$\sqrt{\lambda^2\langle Y\rangle^2 + m^2} \pm \lambda\langle Y\rangle.$$

You can now confirm that the supertrace formula is satisfied. Notice also that if $V_{\min} = \lambda^2\mu^4$ vanishes, supersymmetry is restored in the spectrum.

The fact that there is a massless fermion in the spectrum is a general feature in theories with spontaneous supersymmetry breaking. This fermion is called the goldstino. It is the analogue of the Goldstone boson that arises when global bosonic symmetries are spontaneously broken. We will not present a general argument that shows that spontaneous SUSY breaking always results in a goldstino. The proof parallels that given by Goldstone, Salam, and Weinberg for the Goldstone theorem.[2] The goldstino is a spin $\frac{1}{2}$ fermion because the SUSY generator itself carries spin $\frac{1}{2}$. It is the fermionic partner of the auxiliary field that develops a SUSY breaking VEV.

Notice that aside from the $\pm 2\lambda^2\mu^2$ terms in the diagonal X_R and X_I entries in the mass matrix for scalars, the mass matrices look supersymmetric. In other words, but for these terms, the mass matrices (7.9) and (7.12) would be those of a theory with unbroken SUSY. This should not be surprising because, at tree level, the order parameter $\mathcal{F}_Y$ for SUSY breaking (and hence also the goldstino) couples to only the X field, as is evident from the form of the superpotential.

It is easy to see from the result of the exercise above that the heaviest boson is heavier than the heaviest fermionic state. But because the sums over the squared masses of the bosons and fermions are the same, this also means that the lightest of the massive bosons must be lighter than the lightest of the massive fermions. In other words, the spontaneous breaking of supersymmetry results in fermion masses that are bracketed between the boson masses. The "SUSY breaking" $2\lambda^2\mu^2$ contribution does not enter the fermion masses, but splits the boson masses about their would-be value (equal to the fermion masses) in the absence of SUSY breaking.[3] Since this pattern of mass splittings has its origin in the vanishing of the supertrace – a general feature in models with global supersymmetry broken spontaneously by F-terms – it is very difficult to use this mechanism to get realistic models with global supersymmetry broken at the TeV scale: these models typically give a spin zero superpartner lighter than all the fermions, and so run into conflict with experiment.

7.2.2 Mass spectrum: Case B

We can now similarly work out the mass spectrum for Case B, where the minimum occurs at $\langle X_R \rangle^2 = 2\mu^2 - m^2/\lambda^2$, $\langle X_I \rangle = 0$, with a flat direction along $\langle Z \rangle = -\frac{2\lambda}{m}\sqrt{\mu^2 - \frac{m^2}{2\lambda^2}}\langle Y \rangle$. For the most part, we will leave it to the reader to work out the details for this case. From the scalar potential, it is straightforward to work

---

[2] J. Goldstone, A. Salam and S. Weinberg, *Phys. Rev.* **127**, 965 (1962), Sec. III.
[3] It is reasonable that spontaneous SUSY breaking yields new contributions to boson masses but not to fermions. We saw in Chapter 3 that a SUSY breaking mass for a fermion in a chiral scalar multiplet would have been a hard breaking of supersymmetry.

out the scalar mass matrix. The bosonic contribution to the supertrace is as given in the exercise below.

Exercise *For case B, calculate the bilinear terms in the scalar potential of the shifted fields and show that*

$$\sum_{\text{bosons}} \mathcal{M}^2 = 4\lambda^2(\langle X_R\rangle^2 + \langle Y_R\rangle^2 + \langle Y_I\rangle^2 + 2\mu^2). \qquad (7.16)$$

(Remember that $\langle X_R\rangle \neq 0$, while $\langle X_I\rangle = 0$.)

To identify the goldstino, we recall that the auxiliary fields obtain VEVs

$$i\langle\mathcal{F}_X\rangle = 0, \qquad (7.17a)$$

$$i\langle\mathcal{F}_Y\rangle = \lambda(\frac{1}{2}\langle X_R\rangle^2 - \mu^2), \qquad (7.17b)$$

$$i\langle\mathcal{F}_Z\rangle = m\frac{\langle X_R\rangle}{\sqrt{2}}. \qquad (7.17c)$$

We can then work with orthogonal linear combinations of the superfields $\hat{Y}$ and $\hat{Z}$, so that the auxiliary component of just one of the linear combinations develops a VEV:

$$\hat{P} = \frac{\frac{m\langle X_R\rangle}{\sqrt{2}}\hat{Y} - \lambda(\frac{1}{2}\langle X_R\rangle^2 - \mu^2)\hat{Z}}{\sqrt{\frac{1}{2}m^2\langle X_R\rangle^2 + \lambda^2(\frac{1}{2}\langle X_R\rangle^2 - \mu^2)^2}} \equiv \hat{Y}\cos\theta - \hat{Z}\sin\theta \qquad (7.18a)$$

and

$$\hat{Q} = \frac{\lambda(\frac{1}{2}\langle X_R\rangle^2 - \mu^2)\hat{Y} + \frac{m\langle X_R\rangle}{\sqrt{2}}\hat{Z}}{\sqrt{\frac{1}{2}m^2\langle X_R\rangle^2 + \lambda^2(\frac{1}{2}\langle X_R\rangle^2 - \mu^2)^2}} \equiv \hat{Y}\sin\theta + \hat{Z}\cos\theta. \qquad (7.18b)$$

In this case, we have $\langle\mathcal{F}_P\rangle = 0$ and $\langle\mathcal{F}_Q\rangle \neq 0$. We then expect that ψ_Q, the fermionic component of $\hat{Q}$, will be the massless goldstino field. To establish this, as well as to obtain the fermionic contribution to the supertrace, we write the superpotential in terms of $\hat{X}$, $\hat{P}$, and $\hat{Q}$,

$$\hat{f}(\hat{X}, \hat{P}, \hat{Q}) = \lambda(\hat{X}^2 - \mu^2)(\hat{P}\cos\theta + \hat{Q}\sin\theta) + m\hat{X}(\hat{Q}\cos\theta - \hat{P}\sin\theta). \qquad (7.19)$$

We see from (6.44) that ψ_Q is massless since there is neither a diagonal mass for it (no $\hat{Q}^2$ term in the superpotential) nor a bilinear mixing with either ψ_P or ψ_X.

Exercise *That there is no mixing of ψ_Q with ψ_P is obvious from the superpotential. Verify that the ψ_X–ψ_Q mixing term also vanishes.*

In the (ψ_X, ψ_P) basis, the non-vanishing fermion mass submatrix can be written as,

$$\mathcal{M}_{\text{fermion}} = \mathcal{M}\frac{1 - \gamma_5}{2} + \mathcal{M}^\dagger\frac{1 + \gamma_5}{2}, \tag{7.20a}$$

with,

$$\mathcal{M} = \begin{pmatrix} 2\lambda\frac{\langle Y_R\rangle + i\langle Y_I\rangle}{\sqrt{2}} & \sqrt{2}\lambda\cos\theta\langle X_R\rangle - m\sin\theta \\ \sqrt{2}\lambda\cos\theta\langle X_R\rangle - m\sin\theta & 0 \end{pmatrix}. \tag{7.20b}$$

Except when $\langle Y_I\rangle = 0$, the fermion "mass matrix" is γ_5-dependent. The reader who is not familiar with how to deal with this is referred to the technical note at the end of this chapter.[4] There, we show that the squared masses of the fermions are given by the eigenvalues of the matrix $\mathcal{M}^\dagger\mathcal{M}$.

Exercise *By explicitly computing the sum of the squared masses for the fermions in case B, verify that the supertrace once again vanishes.*

7.3 D-type SUSY breaking

As an illustration of SUSY breaking by D-terms, we consider a simple model with just one chiral superfield coupled to a $U(1)$ gauge field. We include a Fayet–Iliopoulos (FI) D-term. Gauge symmetry precludes any superpotential interactions.

The scalar potential for this model is just

$$V = \frac{1}{2}\mathcal{D}^2 = \frac{1}{2}(g\mathcal{S}^\dagger\mathcal{S} + \xi)^2. \tag{7.21}$$

The minimum of the potential occurs at

(a) $\langle \mathcal{S}^\dagger\mathcal{S}\rangle = 0$ if $\xi > 0$,

(b) $\langle \mathcal{S}^\dagger\mathcal{S}\rangle = -\dfrac{\xi}{g} = \dfrac{|\xi|}{g}$ if $\xi < 0$.

In case (a), SUSY is spontaneously broken because the D-term acquires a vacuum expectation value. The gauge symmetry remains unbroken because $\langle \mathcal{S}^\dagger\mathcal{S}\rangle = 0$. In case (b), the $U(1)$ gauge symmetry is spontaneously broken but SUSY remains intact.

---

[4] The matrices M and N of the note can be identified as $\frac{M + M^\dagger}{2} = M$ and $\frac{M^\dagger - M}{2} = iN$.

7.3.1 Case A

In this case, the auxiliary field $\mathcal{D} = g\mathcal{S}^\dagger\mathcal{S} + \xi$ acquires a VEV $\langle\mathcal{D}\rangle = \xi$. Because there is no superpotential, the chiral fermion ψ remains massless. The FI term causes a mass splitting with the scalar $\mathcal{S}$, which acquires a mass $\sqrt{g\xi}$.

The $U(1)$ gauge boson and gaugino are massless at tree level. The gaugino, which is the partner of the $\mathcal{D}$ field that acquires a VEV, plays the role of the goldstino. In this toy theory, since the complex field $\mathcal{S}$ is the only state to acquire mass, the supertrace is just

$$STr\mathcal{M}^2 = 2g\xi$$

in accord with the general sum rule Eq. (7.35) discussed later in this chapter. Notice that, unlike the O'Raifeartaigh model, this model does not suffer from the problem of "scalars lighter than fermions."

7.3.2 Case B

In this case, gauge symmetry is spontaneously broken because $\mathcal{S}$ acquires a VEV, but SUSY remains intact. Let us work out the spectrum of the model to see explicitly how this works.

Bosons

The relevant piece of the Lagrangian for vector bosons is

$$\mathcal{L} \ni \left[(\partial_\mu\mathcal{S})^\dagger - igV_\mu\mathcal{S}^\dagger\right][\partial^\mu\mathcal{S} + igV^\mu\mathcal{S}] - \frac{1}{4}F_{\mu\nu}F^{\mu\nu}. \qquad (7.22)$$

We note that by redefining the phase of $\mathcal{S}$, $\langle\mathcal{S}\rangle$ can be chosen to be real without loss of generality. As usual, we then shift $\mathcal{S} \to \mathcal{S} + \langle\mathcal{S}\rangle$ and then re-write the Lagrangian in terms of $\mathcal{S} = \frac{\mathcal{S}_R + i\mathcal{S}_I}{\sqrt{2}}$ (here, $\mathcal{S}_R$ and $\mathcal{S}_I$ are fluctuations about the vacuum) to obtain

$$\mathcal{L} \ni \frac{1}{2}(\partial_\mu\mathcal{S}_R)^2 + \frac{1}{2}(\partial_\mu\mathcal{S}_I)^2 + \sqrt{2}g\langle\mathcal{S}\rangle V_\mu\partial^\mu\mathcal{S}_I + g^2\langle\mathcal{S}\rangle^2 V_\mu V^\mu - \frac{1}{4}F_{\mu\nu}F^{\mu\nu}.$$

The field $\mathcal{S}_I$ can be absorbed by a local gauge transformation by an amount $\frac{-\mathcal{S}_I}{\sqrt{2}g\langle\mathcal{S}\rangle}$. This piece of the Lagrangian becomes

$$\mathcal{L} \ni \frac{1}{2}(\partial_\mu\mathcal{S}_R)^2 + g^2\langle\mathcal{S}\rangle^2 V_\mu V^\mu - \frac{1}{4}F_{\mu\nu}F^{\mu\nu}. \qquad (7.23)$$

The vector boson has developed a mass $m_V^2 = 2g^2\langle\mathcal{S}\rangle^2 = 2g|\xi|$, while the $\mathcal{S}_I$ field has disappeared, being eaten by the vector boson field. This is, of course, the familiar Higgs mechanism. A real scalar $\mathcal{S}_R$ remains. Its mass can be obtained from the scalar potential (7.21).

Exercise *Show that the Higgs field S_R of our toy model has the same mass as the vector boson.*

Fermions

Since there is no superpotential, bilinear terms for fermions come only from gaugino chiral fermion mixing induced by a VEV of the scalar field. The relevant term is,

$$\mathcal{L} \ni -\sqrt{2}\langle S \rangle g \bar{\lambda} \frac{1-\gamma_5}{2} \psi + \quad \text{h.c.}$$

We see that there are two degenerate Majorana fermions that can be combined into a single Dirac fermion $\chi_D = \frac{1-\gamma_5}{2}\psi + \frac{1+\gamma_5}{2}\lambda$, with a mass again equal to that of the bosons.

Thus, in case B, the physical particles are one massive spin 1 boson and a spin 0 boson, each with mass $\sqrt{2g|\xi|}$ and one Dirac fermion with the same mass. The spectrum is clearly supersymmetric, but the original gauge symmetry is hidden.[5]

7.4 Composite goldstinos

We have considered examples where the goldstino is an elementary field that occurs in the Lagrangian. This need not always be the case. The goldstino may be a composite fermion just as the composite pion may be regarded as a (pseudo)-Goldstone boson.

This is realized if chiral fermions condense. If $\hat{S}$ and $\hat{S}'$ are two left-chiral superfields,

$$\hat{S} = S(\hat{x}) + i\sqrt{2}\bar{\theta}\psi_L + i\bar{\theta}\theta_L\mathcal{F}(\hat{x}) \quad \text{and}$$
$$\hat{S}' = S'(\hat{x}) + i\sqrt{2}\bar{\theta}\psi_L' + i\bar{\theta}\theta_L\mathcal{F}'(\hat{x}),$$

then the composite field $\hat{S}\hat{S}'$ is given by,

$$\hat{S}\hat{S}' = SS'(\hat{x}) + i\sqrt{2}\bar{\theta}(S'\psi_L + S\psi_L')(\hat{x}) - 2\bar{\theta}\psi_L\bar{\theta}\psi_L'(\hat{x}) + i\bar{\theta}\theta_L(\mathcal{F}S' + S\mathcal{F}')(\hat{x}).$$

Using the fact that $-2\bar{\theta}\psi_L\bar{\theta}\psi_L' = \bar{\theta}\theta_L\bar{\psi}'\psi_L$, we have

$$\hat{S}\hat{S}' = SS' + i\sqrt{2}\bar{\theta}(S\psi_L' + S'\psi_L) + i\bar{\theta}\theta_L(S\mathcal{F}' + \mathcal{F}S' - i\bar{\psi}'\psi_L). \quad (7.24)$$

---

[5] This spectrum corresponds to that of the $j = 1/2$ supermultiplet discussed toward the end of Section 4.4.

Thus, the auxiliary component of the composite superfield contains the product of the fermion components of the elementary superfields.

This led theorists in the 1980s to consider SUSY technicolor-like models where $\hat{S}$ and $\hat{S}'$ were taken to be $\mathbf{n}$ and $\mathbf{n}^*$ representations of a confining group $SU(n)$.[6] It was assumed that the chiral fermions would condense forming an $SU(n)$ singlet condensate. If such a condensate forms, then SUSY is dynamically broken by the F-term of the composite superfield. The goldstino field would then be a composite object, the fermionic component $(\mathcal{S}'\psi_{\mathrm{L}} + \mathcal{S}\psi'_{\mathrm{L}})$ of a composite superfield. Whether or not such a condensate forms is a dynamical question, and is more difficult to address.

7.5 Gaugino condensation

If we allow non-renormalizable interactions (as we must if we want to include gravity in our effective low energy theory), some of our considerations have to be suitably generalized. Of importance to us here is the fact that instead of starting with just $\hat{W}_A^c \hat{W}_A$ whose $\bar{\theta}\theta_{\mathrm{L}}$ component led to the kinetic term for gauge fields and gauginos, we could have started with

$$f_{AB}(\hat{S}_{\mathrm{L}i})\overline{\hat{W}_A^c}\hat{W}_B$$

whose $\bar{\theta}\theta_{\mathrm{L}}$ component also leads to a SUSY invariant action. To maintain gauge invariance, we must require that the dimensionless function f_{AB} transform as a representation contained in the symmetric product of two adjoints. The function f_{AB} is known as the *gauge kinetic function*. For a renormalizable gauge theory, we must have $f_{AB} = \delta_{AB}$, but otherwise more general forms are possible.

We will explore some important implications of a non-trivial gauge kinetic function in later chapters. For the present purposes, we only note that in supergravity models the expression (6.43a) for the auxiliary component of chiral superfields is modified: in particular, it picks up a term proportional to,

$$\left.\frac{\partial f_{AB}}{\partial \hat{S}_{\mathrm{L}i}}\right|_{\hat{S}=S} \bar{\lambda}_A\lambda_B.$$

Thus if there are new strong gauge interactions that result in a non-vanishing condensate $\langle\bar{\lambda}_A\lambda_A\rangle$ of gauginos, supersymmetry may be dynamically broken.[7] Gaugino condensation is considered by many authors as a promising way of breaking super-symmmetry.

[6] M. Dine, W. Fischler and M. Srednicki, *Nucl. Phys.* **B189**, 575 (1981); S. Dimopoulos and S. Raby, *Nucl. Phys.* **B192**, 353 (1981).
[7] S. Ferrara, L. Girardello and H. P. Nilles, *Phys. Lett.* **B125**, 457 (1983).

7.6 Goldstino interactions

It is well known that at low energy, the couplings of Goldstone bosons to other particles are fixed only by symmetry considerations, and do not depend on the details of the model. One might similarly expect that the low energy interactions of the goldstino with other multiplets are similarly model-independent. To understand how this comes about, we begin by recalling that (3.13b) tells us that each chiral multiplet contributes

$$j^\mu = \partial(-iA\gamma_5 - B)\gamma^\mu\psi + (G\gamma_5 + iF)\gamma^\mu\psi,$$

to the supercurrent. We may thus write the supercurrent as,

$$j^\mu = \partial(-iA_g\gamma_5 - B_g)\gamma^\mu\psi_g + (G_g\gamma_5 + iF_g)\gamma^\mu\psi_g + j^{\mu,\text{rest}}, \qquad (7.25a)$$

where the subscript g refers to the fields in the goldstino multiplet, and $j^{\mu,\text{rest}}$ includes contributions to the supercurrent from all other supermultiplets. If SUSY is spontaneously broken by the vacuum expectation value of the complex auxiliary field

$$\langle \mathcal{F} \rangle = \left\langle \frac{F_g + iG_g}{\sqrt{2}} \right\rangle,$$

that we take to be real, the supercurrent acquires a term *linear* in the goldstino field, and can be written as,

$$j^\mu = i\sqrt{2}\langle\mathcal{F}\rangle\gamma^\mu\psi_g + \partial(-iA_g\gamma_5 - B_g)\gamma^\mu\psi_g + (G_g\gamma_5 + iF_g)\gamma^\mu\psi_g + j^{\mu,\text{rest}}, \qquad (7.25b)$$

where F_g and G_g now denote the shifted fields.

Conservation of the supercurrent then implies that

$$0 = \partial_\mu j^\mu = i\sqrt{2}\langle\mathcal{F}\rangle\partial\psi_g + \partial_\mu j^{\mu,\text{rest}} + \cdots \qquad (7.26)$$

where the ellipsis denotes bilinear (or higher, after the auxiliary fields F_g and G_g are eliminated) terms in fields from the goldstino supermultiplet. This is the equation of motion for the goldstino. It may be obtained from the phenomenological Lagrangian density,

$$
\begin{aligned}
\mathcal{L}_{\text{goldstino}} &= \frac{i}{2}\bar\psi_g\partial\psi_g + \left[\frac{1}{2\sqrt{2}\langle\mathcal{F}\rangle}\bar\psi_g\partial_\mu j^{\mu,\text{rest}} + \quad\text{h.c.}\right] + \cdots \\
&= \frac{i}{2}\bar\psi_g\partial\psi_g + \left[\frac{-1}{2\sqrt{2}\langle\mathcal{F}\rangle}(\partial_\mu\bar\psi_g)\,j^{\mu,\text{rest}} + \quad\text{h.c.}\right] + \cdots, \qquad (7.27)
\end{aligned}
$$

where in the last step, we have omitted a term that is a total derivative. Again, the ellipsis denotes couplings of the goldstino to its superpartner which are not relevant to our discussion. By using the explicit form of the supercurrent (3.13b) it is now straightforward to work out the couplings of the goldstino with fields in *other* supermultiplets. The first term in the supercurrent that originates in the kinetic energy piece of the Lagrangian gives rise to a *model-independent* interaction of the goldstino with the scalar and fermion members of the chiral multiplet with components $(\mathcal{S}, \psi)$. We will leave it to the reader to work out that

$$\mathcal{L}_{\text{goldstino}} = \frac{i}{2} \bar{\psi}_g \partial \psi_g - \frac{i}{\langle \mathcal{F} \rangle} (\partial_\mu \bar{\psi}_g) \left[\partial \mathcal{S} \gamma^\mu \frac{1+\gamma_5}{2} \psi - \partial_\mu \mathcal{S}^\dagger \gamma^\mu \frac{1-\gamma_5}{2} \psi \right] + \cdots,$$

(7.28)

where the ellipsis denotes other interactions of the goldstino. There is one such term for each chiral multiplet. In a gauge theory, gauginos and gauge bosons also contribute to the supercurrent, and there will be an analogous gauge boson–gaugino–goldstino interaction. We will return to these goldstino couplings when we consider decays of supersymmetric particles into gravitinos (the superpartners of the graviton) in Chapter 13.

7.7 A mass sum rule

In previous sections, we alluded to the fact that the superparticle spectrum is significantly constrained even when SUSY is spontaneously broken. For instance, for F-type breaking, we saw that $STr\mathcal{M}^2 = 0$, which implied that at least one of the scalar components of chiral scalar superfields must be lighter than any of the fermions. For D-type breaking, we saw that this was not always the case. These features are not particular to the specific model that we considered. To see this, we can compute the squared masses of each particle using the Lagrangian density in (6.44), and hence the respective contributions to the supertrace.

Tree-level masses are defined by the coefficients of bilinear terms in fields expanded about the minimum of the scalar potential. The mass sum rule we obtain holds somewhat more generally, in that it holds for "masses" defined about any scalar field configuration, not just a local extremum. These "masses" are, of course, field-dependent. The immediate payoff is that we can immediately infer that our STr formula is also valid for the special case of spontaneously broken gauge symmetries where we are computing the coefficients of field bilinears about a non-trivial classical minimum.

7.7.1 Scalar contributions

The scalar potential of a supersymmetric theory has the form,

$$V(\mathcal{S}, \mathcal{S}^*) = \sum_i \left| \frac{\partial \hat{f}}{\partial \hat{\mathcal{S}}_i} \right|^2_{\hat{\mathcal{S}}=\mathcal{S}} + \frac{1}{2} \sum_A \left(\sum_i \mathcal{S}_i^\dagger g t_A \mathcal{S}_i + \xi_A \right)^2,$$

(7.29)

where we suppress the index α in (6.44).

Potential mass terms include terms like $S^\dagger S$ as well as terms like $S^2 + \text{h.c.}$ We will let the reader check (see exercise below) that while the latter terms may affect the individual masses, they never contribute to the trace of the scalar mass matrix. Hence the scalar boson contribution to the supertrace can be obtained from,

$$m_{ij}^2 = \frac{\partial^2 V}{\partial S_i^\dagger \partial S_j}.$$

We can then write this as,

$$STr\mathcal{M}_{\text{scalars}}^2 = 2 \sum_{i,j} \left(\frac{\partial^2 \hat{f}}{\partial \hat{S}_i \partial \hat{S}_j} \right)_{\hat{S}=S} \left(\frac{\partial^2 \hat{f}}{\partial \hat{S}_i \partial \hat{S}_j} \right)_{\hat{S}=S}^*$$
$$+ 2 \sum_A \mathcal{D}_A Tr(g t_A) + 2 \sum_{i,A} g^2 S_i^\dagger t_A t_A S_i, \tag{7.30}$$

where the 2 comes from the fact that each complex scalar is really two degrees of freedom, and $\mathcal{D}_A$ is a shorthand for $\sum_i S_i^\dagger g t_A S_i + \xi_A$.

7.7.2 Vector contributions

The vector boson mass matrix arises from the kinetic energy terms for scalars:

$$\mathcal{L} \ni (D_\mu S_i)^\dagger D^\mu S_i = (\partial_\mu S_i + i g t_A V_{\mu A} S_i)^\dagger (\partial^\mu S_i + i g t_B V_B^\mu S_i).$$

Here i labels different chiral scalar multiplets. Every multiplet that transforms non-trivially under the gauge group contributes to the "field-dependent" vector mass matrix. Of course, the tree-level physical masses will get contributions from only those multiplets that develop a gauge symmetry breaking VEV. The vector contribution to the supertrace is,

$$STr\mathcal{M}_{\text{vectors}}^2 = 2 \times 3 \times g^2 \sum_{A,i} (S_i^\dagger t_A)(t_A S_i), \tag{7.31}$$

where the factor of 2 arises because the vector fields are real, and the 3 comes from the three degrees of freedom for each massive spin 1 field (the factor $2J + 1$ in the definition of the STr).

7.7.3 Fermion contributions

The technical note on γ_5-dependent fermion mass matrices at the end of this chapter shows that if the fermion bilinears in the Lagrangian density for Majorana fermions is given by,

$$\mathcal{L} \ni -\frac{1}{2}\bar{\chi}_a \mathcal{M}_{ab}\frac{1-\gamma_5}{2}\chi_b + \quad \text{h.c.}$$

$$= -\frac{1}{2}\bar{\chi}_R \mathcal{M}\chi_L - \frac{1}{2}\bar{\chi}_L \mathcal{M}^\dagger \chi_R, \tag{7.32}$$

then the squared masses of the fermions are given by the eigenvalues of the matrix $\mathcal{M}\mathcal{M}^\dagger$.

In our master formula (6.44), fermion bilinears arise from the superpotential interactions in the last line, or from mixing between gauginos and chiral fermions in line 2. The relevant terms can be written as,

$$\mathcal{L} \ni -\frac{1}{2}\begin{pmatrix}\bar{\lambda}_A & \bar{\psi}_i\end{pmatrix}\begin{pmatrix} 0 & \sqrt{2}g(\mathcal{S}^\dagger t_A)_j \\ \sqrt{2}g(\mathcal{S}^\dagger t_B)_i & \left(\frac{\partial^2 \hat{f}}{\partial \hat{\mathcal{S}}_i \partial \hat{\mathcal{S}}_j}\right)_{\hat{\mathcal{S}}=\mathcal{S}}\end{pmatrix}\frac{1-\gamma_5}{2}\begin{pmatrix}\lambda_B \\ \psi_j\end{pmatrix} + \quad \text{h.c.} \tag{7.33}$$

We can now easily obtain the fermionic contribution to the supertrace,

$$STr\mathcal{M}^2_{\text{fermions}} = (-1) \times 2 \times \left[\sum_{i,A} 4g^2(\mathcal{S}^\dagger t_A)_i(t_A\mathcal{S})_i\right.$$

$$\left. + \sum_{i,j}\left(\frac{\partial^2 \hat{f}}{\partial \hat{\mathcal{S}}_i \partial \hat{\mathcal{S}}_j}\right)_{\hat{\mathcal{S}}=\mathcal{S}}\left(\frac{\partial^2 \hat{f}}{\partial \hat{\mathcal{S}}_i \partial \hat{\mathcal{S}}_j}\right)^*_{\hat{\mathcal{S}}=\mathcal{S}}\right], \tag{7.34}$$

where the 2 comes from the two spin degrees of freedom for the Majorana fermions.

We can now combine the contributions (7.30), (7.31), and (7.34) to obtain the tree-level mass sum rule for globally supersymmetric models with spontaneously broken supersymmetry,

$$STr\mathcal{M}^2 = 2\sum_A \mathcal{D}_A \text{Tr}(gt_A). \tag{7.35}$$

Here, the trace refers to a sum over *complex* fields in the chiral supermultiplets. Nowhere in its derivation did we assume that we are at an extremum of the scalar potential. If we now evaluate this at a non-trivial classical minimum, the masses entering on the left-hand side are simply the tree-level masses in the theory, while the right-hand side is the D term whose VEV is one of the order parameters for SUSY breaking.

We now understand why the supertrace vanished for both cases in the O'Raifeartaigh model but not for the model with D-type SUSY breaking. We also note that the right-hand side of (7.35) vanishes if the gauge group is simple. For a model such as the MSSM, with a $U(1)$ hypercharge symmetry, the right-hand side will vanish since the representations are chosen to be anomaly free, i.e. the sum of all the $U(1)_Y$ charges cancels. The problem of light scalars then re-surfaces even in models with D-type SUSY breaking.[8]

Exercise *By decomposing the complex scalar fields into their real and imaginary parts, show that terms such as $m'^2_{ij} S_i S_j +$ h.c. cannot contribute to the supertrace.*

Hint: Aside from "off-diagonal" terms involving products of real and imaginary parts of fields which cannot contribute to the trace, show that these terms can be written as $\frac{1}{2}(m'^2_{ij} + m'^{2*}_{ij})(S_{Ri} S_{Rj} - S_{Ii} S_{Ij})$, so that the real and imaginary components make equal and opposite contributions to the trace.

7.8 Explicit supersymmetry breaking

There is as yet no compelling theory of SUSY breaking. We have alluded to potential phenomenological problems that arise if global supersymmetry is spontaneously broken at the TeV scale. Indeed, the strategy most common to model-building today is to assume that supersymmetry is broken in a sector of a theory that is essentially decoupled from our world of quarks, leptons, and gauge and Higgs bosons (and their superpartners). The effects of SUSY breaking in this "hidden sector" are then communicated to our world by messenger interactions. The low energy phenomenology that results is qualitatively dependent on what these messenger interactions are, but we will return to this in later chapters.

It is fair to say that we have not yet discovered the dynamics which causes the breaking of supersymmetry. Hopefully, when this dynamics is discovered, we will find that supersymmetry, like gauge symmetry, is spontaneously broken. The spontaneous breaking of supersymmetry does not alter the supersymmetry relationship between various (tree level) dimensionless couplings in the Lagrangian density in Eq. (6.44).[9] For instance, (tree level) chiral fermion–scalar–gaugino interactions are *fixed* by the usual gauge coupling. As we saw in Chapter 3, altering these would be a hard breaking of supersymmetry in that it would result in a re-appearance of quadratic divergences that we have worked so hard to eliminate. Spontaneous breaking of supersymmetry does not lead to new quadratic divergences.

In the absence of knowledge about SUSY breaking dynamics, the best that we can do is to parametrize the effects of SUSY breaking by adding to the Lagrangian all possible SUSY breaking terms, consistent with all desired (unbroken) symmetries at the SUSY breaking scale, that do not lead to the re-appearance of quadratic

---

[8] In realistic models because electric charge is strictly conserved and particles with different charges cannot mix, the supertrace vanishes separately in each charge sector. In other words, there should be an up-type scalar quark lighter than the up quark, a down-type scalar quark lighter than the down quark, and an integrally charged scalar lighter than an electron!

[9] Spontaneous SUSY breaking means $\langle \mathcal{F} \rangle$ or $\langle \mathcal{D} \rangle \neq 0$, but the dimensionless (gauge or superpotential) couplings come from the unshifted parts of $\mathcal{F}$ and $\mathcal{D}$.

divergences. In Chapter 3, we referred to such operators as *soft* SUSY breaking operators, and gave examples of these in the context of the Wess–Zumino model.

Girardello and Grisaru have classified the forms of the soft breaking operators in a general theory.[10] They have shown that to all orders in perturbation theory,

- linear terms in the scalar field S_i (relevant only for singlets of all symmetries),
- scalar masses,
- and bilinear or trilinear operators of the form $S_i S_j$ or $S_i S_j S_k$ (where $\hat{S}_i \hat{S}_j$ and $\hat{S}_i \hat{S}_j \hat{S}_k$ occur in the superpotential),
- and finally, in gauge theories, gaugino masses, one for each factor of the gauge group,

break supersymmetry softly. In general, masses of fermions in chiral supermultiplets, chiral fermion–gaugino mixing masses (these are relevant only if there are chiral supermultiplets in the adjoint representation of the gauge group), and trilinear scalar interactions involving S_i and $S_j^\dagger$ are hard. Finally, all dimension four SUSY breaking couplings are hard.

It is not hard to understand why the dimensionful mass terms and trilinear interactions listed above lead to softly broken supersymmetry. If SUSY is explicitly broken by an operator with a coupling M_{SUSY} with dimension of mass, a quadratic divergence in any operator would have a coefficient proportional to $M_{\mathrm{SUSY}} \Lambda^2$, where Λ is the ultra-violet cut-off. Only dimension one operators (i.e. operators linear in a spin-zero field) can have such a coefficient. Thus, in a theory in which there are no scalars that are singlets of all the symmetries, *all* dimensionful, renormalizable SUSY breaking operators are soft. In theories with singlets, we get further restrictions by studying the quadratic divergences in tadpole graphs, and the further restrictions listed by Girardello and Grisaru apply.

Spontaneous breaking of supersymmetry leads to soft SUSY breaking operators. We illustrate this with examples of scalar and gaugino mass terms, as well as trilinear SUSY breaking scalar interactions. These operators, as we will see, play an important role in realistic model building. If there is a left-chiral superfield $\hat{U}$ whose F-term develops a SUSY breaking VEV, then the terms,

$$\frac{1}{M^2} \int \mathrm{d}^4\theta\, \hat{U}\hat{U}^\dagger \hat{S}_{\mathrm{L}} \hat{S}_{\mathrm{L}}^\dagger = \frac{|\langle F \rangle|^2}{M^2} S^\dagger S + \cdots, \tag{7.36a}$$

$$\frac{1}{M} \int \mathrm{d}^2\theta_{\mathrm{L}}\, \hat{U}\, \overline{\hat{W}_A^c}\, \hat{W}_A = \frac{\langle F \rangle}{M} \bar{\lambda}_A \lambda_A + \cdots, \tag{7.36b}$$

and

$$\frac{1}{M} \int \mathrm{d}^2\theta_{\mathrm{L}}\, \hat{U} \hat{S}_i \hat{S}_j \hat{S}_k = \frac{\langle F \rangle}{M} S_i S_j S_k + \cdots \tag{7.36c}$$

---

[10] L. Girardello and M. Grisaru, *Nucl. Phys.* **B194**, 65 (1982).

are, respectively, mass terms for chiral supermultiplet scalars, gauginos, and trilinear interactions of chiral supermultiplet scalars in an effective theory (below the scale M) with spontaneously broken supersymmetry.

In the literature one sometimes sees soft terms of an explicitly broken SUSY theory written in this way. Then $\hat{U}$, which has only a non-vanishing F-component (equal to the SUSY breaking parameter), is not a dynamical superfield. This is then a technical device, and $\hat{U}$ is referred to as a *spurion*.

We summarize this section by listing all SUSY breaking operators consistent with the absence of quadratic divergences in any renormalizable theory. These are:

1. Linear, bilinear, and trilinear scalar self-interactions analytic in the complex scalar field, consistent with gauge and other symmetries. Linear terms are obviously absent if there are no singlet superfields. It is customary to write the bilinear and trilinear soft breaking interactions as

$$B_{ij}\mu_{ij}\mathcal{S}_i\mathcal{S}_j \quad \text{and} \quad A_{ijk}f_{ijk}\mathcal{S}_i\mathcal{S}_j\mathcal{S}_k$$

   where the terms $\mu_{ij}\hat{\mathcal{S}}_i\hat{\mathcal{S}}_j$ and $f_{ijk}\hat{\mathcal{S}}_i\hat{\mathcal{S}}_j\hat{\mathcal{S}}_k$ occur in the superpotential. It should, however, be kept in mind that such soft SUSY breaking terms are possible even if the corresponding terms have been set to zero in the superpotential.
2. Scalar mass terms, and
3. gaugino mass terms.

The soft SUSY breaking Lagrangian may thus be written as

$$\mathcal{L}_{\text{soft}} = \sum_i C_i \mathcal{S}_i + \sum_{i,j} B_{ij}\mu_{ij}\mathcal{S}_i\mathcal{S}_j + \sum_{i,j,k} A_{ijk}f_{ijk}\mathcal{S}_i\mathcal{S}_j\mathcal{S}_k + \quad \text{h.c.}$$
$$- \sum_{i,j} \mathcal{S}_i^\dagger m_{ij}^2 \mathcal{S}_j - \frac{1}{2}\sum_{A,\alpha} M_{A\alpha}\bar{\lambda}_{A\alpha}\lambda_{A\alpha} - \frac{i}{2}\sum_{A,\alpha} M'_{A\alpha}\bar{\lambda}_{A\alpha}\gamma_5\lambda_{A\alpha}, \quad (7.37)$$

where α runs over the different factors of the gauge group. We note that there are two types of gaugino bilinears that we have introduced above. The first of these is what the reader will recognize as a usual mass term for gauginos. The second term is a CP-odd "mass term" that is not precluded unless we further assume that the SUSY breaking sector does not contain additional sources of CP violation. In models without gauge singlet superfields (of which the MSSM, discussed in the next chapter, is an example), additional terms may be allowed. These include,

4. mixing mass terms between gauginos and fermion members of chiral supermultiplets in an adjoint representation, and
5. trilinear scalar interactions of the form $\mathcal{S}_i\mathcal{S}_j\mathcal{S}_k^*$.

Mass terms for fermions in a chiral supermultiplet are redundant since these can be reabsorbed into the bilinear terms in the superpotential, together with appropriate redefinition of soft SUSY breaking masses and couplings in the scalar sector.

It is important to stress that the introduction of explicit SUSY breaking terms into the Lagrangian is a *parametrization* of our ignorance about the dynamics of SUSY breaking. An understanding of SUSY breaking (which will hopefully be obtained in the future) should lead to $\mathcal{L}_{\text{soft}}$, but with the various soft SUSY breaking parameters being *determined* in terms of the (presumably much fewer) fundamental parameters of a more complete theory.

7.9 A technical aside: γ_5-dependent fermion mass matrices

We see from the last line of the master formula (6.44) that the bilinear terms in (Majorana) fermion fields will, in general, be γ_5-dependent. This is what we encountered in our discussion of Case B of the O'Raifeartaigh model. Similar terms can also arise from mixing between the gauginos and "matter" fermions when the scalar fields on the second line of (6.44) acquire complex VEVs. We thus have to understand how to obtain the fermion mass spectrum from these γ_5-dependent fermion mass matrices.

We write the fermion bilinear terms in the Lagrangian density as,

$$-\mathcal{L} = \frac{1}{2}\bar{\mathcal{N}}_i[M_{ij} + i\gamma_5 N_{ij}]\mathcal{N}_j, \tag{7.38}$$

(summation is implied) where $\mathcal{N}_i$ are Majorana spinors and i is a label that distinguishes different particle types. Hermiticity of $\mathcal{L}$ requires that M and N are Hermitian matrices. Since the $\mathcal{N}_i$ are Majorana spinors, M and N also have to be symmetric (and hence real) matrices since $\bar{\mathcal{N}}_i\Gamma\mathcal{N}_j = \bar{\mathcal{N}}_j\Gamma\mathcal{N}_i$ for $\Gamma = I$ or γ_5. This will be crucial later. The Lagrangian density can be written by separating the left- and right-chiral parts of the spinors as,

$$-2\mathcal{L} = \bar{\mathcal{N}}_{\text{L}i}[M_{ij} + iN_{ij}]\mathcal{N}_{\text{R}j} + \bar{\mathcal{N}}_{\text{R}i}[M_{ij} - iN_{ij}]\mathcal{N}_{\text{L}j}.$$

We can always find unitary matrices U and V such that $V^\dagger[M + iN]U = D$, and $U^\dagger[M - iN]V = D^\dagger$, where D is a diagonal (but not necessarily real) matrix. V is the unitary matrix that diagonalizes the Hermitian matrix $[M + iN][M - iN]$ to give,

$$V^\dagger[M + iN][M - iN]V = DD^\dagger.$$

U is the corresponding matrix for $[M - iN][M + iN]$ which is also Hermitian.

It is important to note that $[M + iN][M - iN]$ and $[M - iN][M + iN]$ have the same (real and positive) eigenvalues. Furthermore, since M and N are real matrices, if the column X is an eigenvector of $[M + iN][M - iN]$, then X^* is an eigenvector of $[M - iN][M + iN]$ with the same eigenvalue. As a result, we can choose $V = U^*$. This guarantees that the spinor ψ' defined by,

$$\mathcal{N}_L = V\psi'_L, \quad \mathcal{N}_R = U\psi'_R = V^*\psi'_R$$

is Majorana when the original spinor $\mathcal{N}$ is Majorana. Writing the Lagrangian density (7.38) in terms of ψ' we obtain (with matrix multiplication implied),

$$-\mathcal{L} = \frac{1}{2}[\bar{\psi}'_L D\psi'_R + \bar{\psi}'_R D^\dagger \psi'_L], \quad (7.39)$$

which (though it still contains a γ_5-dependent mass term) is now diagonal in particle type. We can now get rid of this γ_5 dependence in the fermion bilinears by performing chiral rotations,

$$\psi'_{Lj} = e^{-i\phi_{Lj}}\psi_{Lj}, \quad \psi'_{Rj} = e^{-i\phi_{Rj}}\psi_{Rj},$$

(no summation over j). These transformations leave the kinetic terms invariant. If we write the elements of the diagonal matrix D by

$$D_i = m_i e^{ia_i},$$

with m_i and a_i as real numbers, the γ_5 dependence in (7.39) is removed if we choose,

$$a_i + \phi_{Li} - \phi_{Ri} = 0.$$

This, of course, fixes only the difference $\phi_{Li} - \phi_{Ri}$, but not the two separately. In order to maintain the Majorana character of ψ_i, we should also choose,

$$\phi_{Li} = -\phi_{Ri}.$$

We can now write (7.39) in terms of ψ to obtain,

$$-\mathcal{L} = \sum_i \frac{m_i}{2}[\bar{\psi}_{Li}\psi_{Ri} + \bar{\psi}_{Ri}\psi_{Li}] = \sum_i \frac{m_i}{2}\bar{\psi}_i\psi_i. \quad (7.40)$$

We see that the m_i are then the positive masses for the fermions. A straightforward way to obtain these is to note that the m_i^2 are the eigenvalues of $D^\dagger D$ which, of course, coincide with eigenvalues of $(M - iN)(M + iN)$ (or of $(M + iN)(M - iN)$).

We should remember several things from this discussion.

- "Fermion masses" (by this we mean the coefficient of $\bar{\psi}\psi$ in the Lagrangian density) are not physical objects. It is only the squares of these that give the squared masses of the fermions. A special case of this that we will have frequent occasion to use is when a fermion mass has the "wrong sign". In this case, the reader can easily check that the transformation, $\psi \to \gamma_5\psi$ fixes this sign.

This does not, however, preserve the Majorana nature of ψ. If ψ is Majorana, the appropriate transformation should be $\psi \rightarrow i\gamma_5\psi$. Both these transformations preserve the kinetic terms, but may introduce additional γ_5 matrices as well as i's in interaction terms. These are important as they lead to physically observable changes in amplitudes.

- For a system of fermions, the physical masses are given not by the eigenvalues of the "fermion mass matrix" (which need not even be Hermitian). Instead, the eigenvalues of the mass matrix times its Hermitian adjoint are the squares of the fermion masses.

- We can eliminate γ_5 dependence in fermion bilinears by separately rotating the left- and right-chiral components. Care must be taken, however, if we are dealing with Majorana fermions, to preserve their Majorana character. Such γ_5-dependent mass terms, which are not precluded by Poincáre invariance, frequently signal CP violation.[11] In two-component notation, the analogue of these is phases in masses for spinor fields.

---

[11] Recall that $\bar{\psi}\gamma_5\psi$ is odd under CP.

8

The Minimal Supersymmetric Standard Model

At this point, we have all the ingredients necessary for constructing a supersymmetric version of the Standard Model, complete with explicit soft SUSY breaking terms. The simplest such model, known as the Minimal Supersymmetric Standard Model, or MSSM, is a direct supersymmetrization of the Standard Model (except for the fact that one needs to introduce two Higgs doublet fields). It is minimal in the sense that it contains the smallest number of new particle states and new interactions consistent with phenomenology.

To construct the MSSM, we follow the recipe for the construction of supersymmetric gauge theories at the end of Chapter 6 and proceed as follows:

1. We choose the gauge symmetry group for the theory to be the Standard Model gauge group, $SU(3)_C \times SU(2)_L \times U(1)_Y$.
2. We select the matter content of the theory, to be realized as left-chiral scalar superfields, with gauge quantum numbers exactly as in the Standard Model. The Higgs sector is chosen to consist of two left-chiral scalar superfields with opposite hypercharge.
3. We choose the form of the superpotential.
4. Finally, we compute the supersymmetric Lagrangian using the master formula Eq. (6.44), and augment it by all possible soft SUSY breaking terms consistent with the gauge and Poincaré symmetries as discussed in Chapter 7.

8.1 Constructing the MSSM

As mentioned, we choose the gauge symmetry of the Standard Model: $SU(3)_C \times SU(2)_L \times U(1)_Y$. The gauge bosons of the SM are promoted to gauge superfields. In the Wess–Zumino gauge,

$$B_\mu \to \hat{B} \ni (\lambda_0, B_\mu, \mathcal{D}_B),$$
$$W_{A\mu} \to \hat{W}_A \ni (\lambda_A, W_{A\mu}, \mathcal{D}_{WA}), \quad A = 1, 2, 3, \text{ and}$$
$$g_{A\mu} \to \hat{g}_A \ni (\tilde{g}_A, G_{A\mu}, \mathcal{D}_{gA}), \quad A = 1, \ldots, 8.$$

The second step is to stipulate the matter content of the MSSM. The fermion fields of the SM are promoted to chiral scalar superfields, with one superfield for each chirality of every SM fermion. Since the superpotential must be a function of just *left*-chiral superfields, instead of using the right-handed fermions as the building blocks, we will, as mentioned in Chapter 1, use their left-handed charge conjugates. The matter superfields then consist of,

$$\begin{pmatrix} \nu_{iL} \\ e_{iL} \end{pmatrix} \rightarrow \hat{L}_i \equiv \begin{pmatrix} \hat{\nu}_i \\ \hat{e}_i \end{pmatrix},$$

$$(e_{iR})^c \rightarrow \hat{E}_i^c,$$

$$\begin{pmatrix} u_{iL} \\ d_{iL} \end{pmatrix} \rightarrow \hat{Q}_i \equiv \begin{pmatrix} \hat{u}_i \\ \hat{d}_i \end{pmatrix},$$

$$(u_{iR})^c \rightarrow \hat{U}_i^c,$$

$$(d_{iR})^c \rightarrow \hat{D}_i^c,$$

where $i = 1, 2, 3$ refers to the *generation* of each field, i.e. $\hat{u}_3$ contains the dynamical fields $\tilde{t}_L$ and ψ_{tL} (in addition to the corresponding auxiliary field).[1] To be explicit, we write down the superfield expansions which contain the electron fields:

$$\hat{e} = \tilde{e}_L(\hat{x}) + i\sqrt{2}\bar{\theta}\psi_{eL}(\hat{x}) + i\bar{\theta}\theta_L\mathcal{F}_e(\hat{x}) \qquad (8.1)$$

while

$$\hat{E}^c = \tilde{e}_R^\dagger(\hat{x}) + i\sqrt{2}\bar{\theta}\psi_{E^cL}(\hat{x}) + i\bar{\theta}\theta_L\mathcal{F}_{E^c}(\hat{x}). \qquad (8.2)$$

In Eq. (8.2), the scalar component destroys the superpartner of the $SU(2)$ singlet (left-handed) positron, or creates the superpartner of the $SU(2)$ singlet (right-handed) electron, and so is written as $\tilde{e}_R^\dagger$.

The familiar four-component Dirac spinor for the massive electron is built from the two Majorana spinors ψ_e and ψ_{E^c}. Since ψ_{eL} and ψ_{E^cR} have the same electric charge (see the discussion immediately following Eq. (6.38b) of Chapter 6), we may write this Dirac field as,

$$e = P_L\psi_e + P_R\psi_{E^c}. \qquad (8.3)$$

The other massive matter fermions of the SM are similarly constructed.

---

[1] We do not introduce fields for the right-handed neutrinos. Although such fields are very likely to be present in nature, they will be part of some extension of the MSSM.

Exercise *The construction of the massive Dirac spinor in terms of two Majorana spinors can be most easily seen in the chiral representation for γ matrices, with*

$$\gamma^5 = \begin{pmatrix} -\mathbf{1} & \mathbf{0} \\ \mathbf{0} & \mathbf{1} \end{pmatrix},$$

(the bold-face entries are 2×2 matrices). Check that the Majorana spinors

$$\psi_e = \begin{pmatrix} e_1 \\ e_2 \\ -e_2^* \\ e_1^* \end{pmatrix} \quad \text{and} \quad \psi_{E^c} = \begin{pmatrix} e_4^* \\ -e_3^* \\ e_3 \\ e_4 \end{pmatrix}.$$

can be combined via Eq. (8.3) into an arbitrary Dirac spinor.

Exercise *Check that the kinetic energy terms for the Majorana spinors ψ_e and ψ_{E^c} in our master formula yield the kinetic energy term for the Dirac spinor e; i.e. verify that (up to a total derivative),*

$$\frac{\mathrm{i}}{2}\bar{\psi}_e \partial\!\!\!/ \psi_e + \frac{\mathrm{i}}{2}\bar{\psi}_{E^c} \partial\!\!\!/ \psi_{E^c} = \mathrm{i}\bar{e}\partial\!\!\!/ e. \tag{8.4}$$

The reader will have noticed that in promoting the SM fields to superfields, we have introduced many new particles, in order to complete the multiplets of supersymmetry. The existence of these new states is a prediction of supersymmetry, in exactly the same way the existence of the Ω^- was the prediction of flavor $SU(3)$ way back in the 1960s, or the existence of the Z^0 boson is a prediction of the SM symmetries. The superpartners of matter fermions are spin zero particles, known as sfermions. There is a sfermion pair (the spin zero particle and its antiparticle) for each chiral fermion in the SM, with the same internal quantum numbers as the fermion. The spin zero partners of quarks are the scalar quarks, or *squarks* for short. Likewise, the spin zero partners of the leptons are the scalar leptons or *sleptons*. Other s-words such as *selectron*, *smuon*, and *stau* are analogously defined. The subscripts L and R on the scalar fields in (8.1) and (8.2) refer to the chirality of the corresponding *electron*. These selectrons are referred to as left-(right-)selectrons, and sometimes loosely referred to as left-handed (right-handed) selectrons. It should, of course, be clear that selectrons, being spinless, cannot have handedness or chirality. Left- and right-squarks, sleptons, smuons, staus are similarly defined. The Higgs and gauge fields have fermionic superpartners respectively known as higgsinos and gauginos.

Next, we introduce the Higgs multiplets of the theory. The usual Higgs doublet of the SM is promoted to a doublet of left-chiral superfields:

$$\phi = \begin{pmatrix} \phi^+ \\ \phi^0 \end{pmatrix} \rightarrow \hat{H}_u = \begin{pmatrix} \hat{h}_u^+ \\ \hat{h}_u^0 \end{pmatrix}. \tag{8.5}$$

It transforms as a doublet **2** under $SU(2)_L$ and carries weak hypercharge $Y = 1$. The usual Yukawa interactions of its scalar component with matter fermions must arise via the superpotential, since our list of soft SUSY breaking interactions does not include interactions of chiral fermions. The VEV of the scalar component of $\hat{h}_u^0$ gives mass to up-type quarks but, unlike in the SM, cannot give a mass to the $T_3 = -1/2$ fermions. This is because the $Y = -1$ field needed to give mass to these would have to be the scalar component of the right-chiral superfield $\hat{h}_u^{0\dagger}$, and so, not allowed in the superpotential. We contrast this with the situation in the SM where the charge conjugate field $\phi^c = i\sigma_2\phi^*$ with weak hypercharge $Y = -1$ could be responsible also for the mass of the down-type fermions. We are thus forced to introduce a *second* left-chiral scalar doublet superfield,

$$\hat{H}_d = \begin{pmatrix} \hat{h}_d^- \\ \hat{h}_d^0 \end{pmatrix}, \tag{8.6}$$

which transforms as a **2*** under $SU(2)_L$ and has weak hypercharge $Y = -1$. The VEV of $\hat{h}_d^0$ can give mass to the down-type quarks and the charged leptons.

Remarkably, the introduction of this second doublet also solves another problem that we have unwittingly created. In promoting $\phi \rightarrow \hat{H}_u$, we have introduced new fermions, the hypercharge $Y = 1$ higgsinos $\psi_{h_u^+}$ and $\psi_{h_u^0}$ into the theory, which upsets the successful cancellation of triangle anomalies in the SM. However, the higgsinos in the $Y = -1$ doublet have just the right quantum numbers to restore the anomaly cancellation.

The third step in our construction procedure is to choose a superpotential to describe the interactions between the various chiral superfields. For the MSSM, we take this to be,

$$\hat{f} = \mu \hat{H}_u^a \hat{H}_{da} + \sum_{i,j=1,3} \left[(\mathbf{f}_u)_{ij} \epsilon_{ab} \hat{Q}_i^a \hat{H}_u^b \hat{U}_j^c + (\mathbf{f}_d)_{ij} \hat{Q}_i^a \hat{H}_{da} \hat{D}_j^c + (\mathbf{f}_e)_{ij} \hat{L}_i^a \hat{H}_{da} \hat{E}_j^c \right].$$

$$\tag{8.7}$$

The indices a and b are $SU(2)$ doublet indices, and explicitly exhibit the contractions needed for the invariance of the superpotential under $SU(2)_L$ transformations. In all but the second term, a doublet **2** is contracted with an antidoublet **2***, and this contraction is trivial. In the second term, ϵ_{ab} is the completely antisymmetric $SU(2)$ tensor with $\epsilon_{12} = 1$. Its presence reflects the fact (familiar from elementary quantum mechanics) that it is the antisymmetric combination of two doublets that is an $SU(2)$ singlet. The color indices on the triplet (antitriplet) superfields $\hat{Q}$ ($\hat{U}^c$, $\hat{D}^c$) contract

Table 8.1 *The matter and Higgs superfield content of the MSSM along with gauge transformation properties and weak hypercharge assignments, for a single generation.*

Field	$SU(3)_C$	$SU(2)_L$	$U(1)_Y$
$\hat{L} = \begin{pmatrix} \hat{v}_{eL} \\ \hat{e}_L \end{pmatrix}$	1	2	-1
$\hat{E}^c$	1	1	2
$\hat{Q} = \begin{pmatrix} \hat{u}_L \\ \hat{d}_L \end{pmatrix}$	3	2	$\frac{1}{3}$
$\hat{U}^c$	3^*	1	$-\frac{4}{3}$
$\hat{D}^c$	3^*	1	$\frac{2}{3}$
$\hat{H}_u = \begin{pmatrix} \hat{h}_u^+ \\ \hat{h}_u^0 \end{pmatrix}$	1	2	1
$\hat{H}_d = \begin{pmatrix} \hat{h}_d^- \\ \hat{h}_d^0 \end{pmatrix}$	1	2^*	-1

trivially, and have been suppressed. Also, it is easily checked that the hypercharge of each term sums to zero, so the superpotential is invariant under $U(1)_Y$. The **f** terms are elements of 3×3 Yukawa coupling matrices with indices $i, j = 1$–3 corresponding to the various generations. In general, the $(\mathbf{f})_{ij}$ as well as μ are complex numbers.

The reader can easily check that the superpotential in Eq. (8.7) respects baryon and lepton number conservation, where these are defined in their natural manner: $B = 1/3 \, (-1/3)$ for quark (antiquark) superfields, $L = 1 \, (-1)$ for the lepton (antilepton) superfields, and zero for the Higgs and gauge superfields. The gauge (and gaugino) interactions on the first three lines of our master formula (6.44) obviously conserve B and L also.

Within the SM, the requirement of gauge invariance automatically guarantees baryon and lepton number conservation for all renormalizable interactions. Unfortunately, this is not the case in the MSSM. Because there are scalar fields that carry baryon or lepton number (the scalar components of quark and lepton superfields), it is possible to write down renormalizable operators that do not conserve B or L that are consistent with the SM gauge symmetries as well as supersymmetry. To see this, we simply note that the additional superpotential interactions, the terms

$$\hat{f}_{\cancel{L}} = \sum_{i,j,k} [\lambda_{ijk}\epsilon_{ab}\hat{L}_i^a\hat{L}_j^b\hat{E}_k^c + \lambda'_{ijk}\epsilon_{ab}\hat{L}_i^a\hat{Q}_j^b\hat{D}_k^c] + \sum_i \mu'_i\epsilon_{ab}\hat{L}_i^a\hat{H}_u^b, \qquad (8.8a)$$

and

$$\hat{f}_{B\!\!\!/} = \sum_{i,j,k} \lambda''_{ijk} \hat{U}_i^c \hat{D}_j^c \hat{D}_k^c, \tag{8.8b}$$

are consistent with $SU(3)_C \times SU(2)_L \times U(1)_Y$ symmetry, but violate the conservation of lepton and baryon number, respectively. Since the superpotential terms (8.8a) and (8.8b) are at most cubic in the superfields, they result in renormalizable interactions that do not conserve L or B.

Obviously, the presence of such terms is dangerous since B- or L-violating processes are strongly constrained by experiment. For instance, if conservation of baryon number and lepton number are both violated, protons will decay at extremely rapid rates. In the spirit of minimality of new interactions, we will insist upon B and L conservation, and set these terms to zero. The SUSY non-renormalization theorem then ensures that these will not be radiatively generated.

Before proceeding to construct the Lagrangian for the MSSM, let us digress to discuss alternative symmetries that can be invoked to justify the absence of these terms. After all, the conservation of baryon number and lepton number are broken by non-perturbative effects, and so cannot be exact. The unwanted terms can also be forbidden by requiring that the superpotential be invariant under a new type of parity (often referred to as matter parity), where quark and lepton superfields are odd, while the gauge and Higgs superfields are even. This requirement then allows the superpotential terms in (8.7), while forbidding those in (8.8a) and (8.8b).

Exercise *Convince yourself that all the kinetic terms as well as the non-superpotential interactions in our master formula (6.44) conserve matter parity. It may be simplest to do so by observing that, except for the term involving the superpotential, all terms in (6.47) are manifestly invariant under the matter parity transformation.*

The conservation of matter parity works out to be equivalent to the conservation of R-parity defined (for the component fields) by,

$$R = (-1)^{3(B-L)+2s}, \tag{8.9}$$

where s is the spin of the field. Note that because of the $(-1)^{2s}$ dependence, the scalar and fermion (spinor and vector) components of a chiral scalar (spinor) superfield have opposite R-parities. If we now take the Grassmann co-ordinate θ to be odd under R, we see that R-parity transformation of the superfield is just the matter parity transformation discussed above. A look at (6.47) of Chapter 6 shows that R-parity violation can only come from R-odd B- and L-violating terms in the superpotential.

Exercise *Starting with the definition (8.9) for R-parity, verify that the SM fermions, gauge bosons, and both Higgs doublets are R-even, while their superpartners are R-odd. In models with conserved R-parity, this quantum number therefore provides an unambiguous distinction between "ordinary particles" and superpartners.*

It may appear that the assumption of R-parity conservation is equivalent to the conservation of B and L. This is the case for renormalizable operators in a theory whose field content is that of the MSSM. For higher dimensional operators, this need not be the case as is exemplified by the exercise below.

Exercise *Verify that the low energy superpotential of an effective low energy theory could contain the R-parity invariant operators*

$$\epsilon_{ab} \hat{L}^a \hat{H}_u^b \epsilon_{cd} \hat{L}^c \hat{H}_u^d \text{ or } \hat{U}^c \hat{U}^c \hat{D}^c \hat{E}^c.$$

Observe that the first of these violates the conservation of lepton number while the latter violates both lepton and baryon number conservation. These operators could be responsible for neutrino masses and proton decay, respectively, even if R-parity is conserved.

Can you construct an R-parity and gauge invariant operator that conserves L but not B? Such an operator could, for instance, be responsible for neutron anti-neutron oscillations.

We repeat that our choice of the MSSM superpotential to be that given by (8.7) is dictated only by constraints of minimality of new interactions. The resulting conservation of R-parity has important phenomenological consequences as we will see. We should mention, however, that it is possible to construct phenomenologically viable models in which R-parity is not conserved. Indeed, we will discuss such models in Chapter 16, but now we proceed with the construction of the MSSM.

Up to this point, we have stipulated the symmetries, field content, and superpotential of the MSSM. We can now use the master formula (6.44) to write down the complete globally supersymmetric Lagrangian. The final step is to write down the various soft SUSY breaking terms for the MSSM.

We may use Eq. (7.37) to write all gauge invariant soft SUSY breaking terms. They are

$$
\begin{aligned}
\mathcal{L}_{\text{soft}} = & -\left[\tilde{Q}_i^\dagger \mathbf{m}_{\mathbf{Q}ij}^2 \tilde{Q}_j + \tilde{d}_{\mathrm{R}i}^\dagger \mathbf{m}_{\mathbf{D}ij}^2 \tilde{d}_{\mathrm{R}j} + \tilde{u}_{\mathrm{R}i}^\dagger \mathbf{m}_{\mathbf{U}ij}^2 \tilde{u}_{\mathrm{R}j} \right.\\
& \left. + \tilde{L}_i^\dagger \mathbf{m}_{\mathbf{L}ij}^2 \tilde{L}_j + \tilde{e}_{\mathrm{R}i}^\dagger \mathbf{m}_{\mathbf{E}ij}^2 \tilde{e}_{\mathrm{R}j} + m_{H_u}^2 |H_u|^2 + m_{H_d}^2 |H_d|^2 \right]\\
& -\frac{1}{2}\left[M_1 \bar\lambda_0 \lambda_0 + M_2 \bar\lambda_A \lambda_A + M_3 \bar{\tilde{g}}_B \tilde{g}_B \right]\\
& -\frac{i}{2}\left[M_1' \bar\lambda_0 \gamma_5 \lambda_0 + M_2' \bar\lambda_A \gamma_5 \lambda_A + M_3' \bar{\tilde{g}}_B \gamma_5 \tilde{g}_B \right]\\
& +\left[(\mathbf{a_u})_{ij}\epsilon_{ab} \tilde{Q}_i^a H_u^b \tilde{u}_{\mathrm{R}j}^\dagger + (\mathbf{a_d})_{ij} \tilde{Q}_i^a H_{da} \tilde{d}_{\mathrm{R}j}^\dagger + (\mathbf{a_e})_{ij} \tilde{L}_i^a H_{da} \tilde{e}_{\mathrm{R}j}^\dagger + \quad \text{h.c.} \right]\\
& +\left[(\mathbf{c_u})_{ij}\epsilon_{ab} \tilde{Q}_i^a H_d^{*b} \tilde{u}_{\mathrm{R}j}^\dagger + (\mathbf{c_d})_{ij} \tilde{Q}_i^a H_{ua}^* \tilde{d}_{\mathrm{R}j}^\dagger + (\mathbf{c_e})_{ij} \tilde{L}_i^a H_{ua}^* \tilde{e}_{\mathrm{R}j}^\dagger + \quad \text{h.c.} \right]\\
& +\left[b H_u^a H_{da} + \quad \text{h.c.} \right],
\end{aligned}
\tag{8.10}
$$

where the generation indices i, j, as well as the $SU(2)$ indices a, b, are implicitly summed over. Hermiticity requires that the scalar mass squared matrices are 3×3 Hermitian matrices, each of which can be written in terms of 6 real and 3 imaginary parameters. The six gaugino mass parameters (M_i, M_i') with $i = 1$–3 corresponding to the three factors of the MSSM gauge group, are real. The terms with M''s violate CP invariance. The $\mathbf{a}$ and $\mathbf{c}$ matrices that describe trilinear scalar interactions are general 3×3 complex matrices, just like the Yukawa matrices. The parameters $m_{H_u}^2$ and $m_{H_d}^2$ are real, while the b bilinear term is, in general, complex. The trilinear interactions involving $\mathbf{c}$ matrices are frequently not written down because such terms are strongly suppressed in many models, but there is really no reason to exclude these within the MSSM framework.

At this point, we have the complete Lagrangian for the MSSM. Of course, it is written in terms of fields with definite quantum numbers for the gauge group. Upon spontaneous symmetry breaking, fields with the same color, electric charge, and spin may mix. The spectrum and couplings of the mass eigenstates have to be extracted from this Lagrangian.

8.1.1 Parameter space of the MSSM

It is now worthwhile to count the free parameters that enter the MSSM Lagrangian. Recall that the SM has nineteen free parameters: three gauge couplings g_1, g_2, and g_3, the parameter θ_{QCD}, μ and λ from the Higgs potential, six quark and three lepton masses, plus three mixing angles, and one CP-violating phase in the Kobayashi–Maskawa matrix.

In the MSSM, we have in the gauge sector again g_1, g_2, g_3, and θ_{QCD}, plus we have six gaugino masses M_1, M_2, M_3 and M_1', M_2', and M_3'. As noted in the exercise below, one of the CP-violating gaugino masses can be removed by performing a chirality transformation of the gaugino field. By convention, we choose this to be M_3'. Thus, we have nine parameters in the gauge sector of the model.

In the Higgs sector, we have the real mass terms $m_{H_u}^2$ and $m_{H_d}^2$, together with μ from the superpotential and its corresponding soft SUSY breaking term b. The latter two are complex but one of the phases, usually taken to be the one associated with b, can be absorbed by redefining the overall phase of one of the Higgs fields. Thus, in the Higgs sector of the MSSM, we have five real parameters.

Finally, we turn to the matter fermions and their superpartners. First, there are five soft SUSY breaking Hermitian mass matrices for the scalar partners of the quarks and leptons, with six real parameters plus three phases each, for a total of 45 parameters. Then, we have three 3×3 complex Yukawa coupling matrices ($18 \times 3 = 54$ parameters). There are another 54 terms in three corresponding **a**-parameter matrices and the same number in the **c** matrices. This gives a total of 207 parameters in the flavor sector, but not all of them are physical.

To count the number of unphysical parameters, i.e. those parameters that can be removed by field redefinitions, we first note that the kinetic terms and gauge interactions are invariant under a global $U(3)^5$ transformation, one $U(3)$ corresponding to transformations amongst each of the three $\hat{L}_i$, $\hat{E}_i^c$, $\hat{Q}_i$, $\hat{U}_i^c$, and $\hat{D}_i^c$. It is just the superpotential Yukawa terms, and the SUSY breaking **a** and **c** terms that are not invariant under these global chiral transformations, which can thus be used to remove some of these parameters. Since any $U(3)$ can be parametrized by three angles and six phases, $5 \times (3 + 6) = 45$ parameters of the 207 that we obtained above should be removable. However, two of the phases in $U(3)^5$ correspond to the conservation of the total B and L: since the corresponding transformations leave the Lagrangian invariant, they cannot be used to do any useful field redefinitions. Summing up the gauge, Higgs, and matter sectors, we have a model with a total of $9 + 5 + 207 - 43 = 178$ parameters in the MSSM. As we stated above, the 54 **c** parameters are usually not included in the MSSM which is then said to contain 124 parameters.

Presumably, once we understand the mechanism underlying SUSY breaking (including how it is conveyed to the superpartners of the SM particles), it will be possible to reduce this plethora of parameters to a handful of truly fundamental parameters. But until then, one of the principal goals of model builders is to arrive at phenomenologically viable but economic models, based on well-motivated assumptions of physics at high energy scales, each with just a few model parameters. It is reasonable to expect that once sparticles are discovered, a determination of their properties will serve to discriminate between these models, thereby pointing the way to the underlying theory.

Exercise *Show that, for an appropriate choice of ϕ, the transformations $\tilde{g}_L \to e^{-i\phi}\tilde{g}_L$, $\tilde{g}_R \to e^{i\phi}\tilde{g}_R$ (this maintains the Majorana property of $\tilde{g}$) can be used to eliminate the CP-violating mass parameter M_3' of the gluino field. In fact, we used this procedure in our discussion following Eq. (7.39) of the technical note in Chapter 7. The phase ϕ, which then shows up in the $\tilde{g}q\tilde{q}_{L,R}$ couplings, can be absorbed by redefinition of the squark fields. We should emphasize that this does not mean that the M_3' term is irrelevant, but only that this term does not give rise to observable CP-violation. The physical mass of the gluino is $m_{\tilde{g}} = \sqrt{M_3^2 + M_3'^2}$, and it is this quantity that appears as the coefficient of the "usual gluino mass term" after the CP-violating gluino bilinear is rotated away.*

Note also that we cannot simultaneously remove M_1' or M_2' since the phase that needs to be absorbed into the squark field will now be different. That we choose to remove the CP-violating mass of the gluino rather than the $SU(2)$ or $U(1)$ gaugino is, of course, only a convention.

Exercise *Using arguments similar to the ones for the MSSM, show that the Yukawa sector of the SM with n generations (assuming neutrinos are massless) contains n real parameters in the lepton sector and $n(n + 3)/2$ real parameters and $n(n - 3)/2 + 1$ phases in the quark sector.*

Note that, unlike the MSSM, the SM with massless neutrinos separately conserves the lepton number for each generation.

8.1.2 A simplified parameter space

We have just seen that the MSSM contains an intractably large number of parameters for meaningful phenomenological analyses. While we have no direct knowledge of these parameters, we can nonetheless make reasonable simplifying assumptions to facilitate our discussion.

We begin by recalling that our motivation for weak scale supersymmetry was to stabilize the electroweak symmetry breaking sector of the SM which suffered from the presence of quadratic divergences. We saw, at least by example, that softly broken SUSY theories have the virtue that they do not suffer from these: the masses of the superpartners set the scale for radiative corrections to the Higgs boson mass, and hence the weak scale. This is the *raison d'être* for weak scale supersymmetry. We thus *require* that the SUSY breaking parameters as well as μ are in the range of the weak scale, or at least not larger than a few TeV. This is our most important assumption.

Next, we note that the scalar matter sector of the MSSM generically would have large flavor violation in both the squark and slepton sectors if the off-diagonal terms in the corresponding mass matrices or the **a** or **c** matrices are comparable to the diagonal terms. Moreover, we saw that the scalar sector of the MSSM has many physical phases that serve as novel sources of CP violation. Even at low energies (well below the SUSY threshold) SM particles would "feel" these sources of flavor and CP violation through SUSY particles in loop diagrams. The magnitude of these effects, of course, depends on the sparticle masses. There are experimental bounds on lepton flavor violation and on CP violation that stringently restrict the size of some of the off-diagonal terms as well as phases referred to in Section 8.1.1. For instance, large off-diagonal contributions to slepton mass matrices would lead to large decay rates for $\mu \rightarrow e\gamma$. Large off-diagonal terms in the squark mass matrices are greatly restricted by K^0–$\overline{K}^0$, D–$\overline{D}$, and B–$\overline{B}$ mixing, and by processes such as $b \rightarrow s\gamma, b \rightarrow s\ell\bar{\ell}$ or $K^0 \rightarrow \mu^+\mu^-$ decays. Large off-diagonal terms in the trilinear **a** and **c** matrices are similarly restricted. There are also strong constraints on CP-violating parameters from measurements of the electron and neutron electric dipole moments.

In the following, we will for simplicity set all SUSY sources of CP violation to zero. In addition, we will also assume that squark and slepton matrices as well as the **a** matrices are diagonal in the same basis that the fermion Yukawa couplings are diagonal. Following common practice, we will set the **c** terms to zero. This is because these terms are frequently small in many models. These simplifying assumptions may well prove to be incorrect. It could turn out that experiments may show that nature requires sources of flavor or CP violation beyond those present in the SM. While there is scant evidence for this at the present time, things could be different in the future. We should also stress that our predictions for even the simplest properties (such as mass) of SUSY particles are sensitive to these assumptions. In the interest of pedagogy, however, we will continue to work within the simplified framework, and leave it to the reader to make the appropriate modifications in more complicated frameworks.

Finally, since our main focus is on SUSY particles, we will keep track of only the third generation Yukawa couplings, and neglect Yukawa interactions of the first two generations. This is obviously unrealistic, but has little effect for most things that we will study. In other words, we will approximate the Yukawa matrices in the superpotential by,

$$\mathbf{f}_e \sim \begin{pmatrix} 0 & 0 & 0 \\ 0 & 0 & 0 \\ 0 & 0 & f_\tau \end{pmatrix}, \quad \mathbf{f}_u \sim \begin{pmatrix} 0 & 0 & 0 \\ 0 & 0 & 0 \\ 0 & 0 & f_t \end{pmatrix}, \quad \mathbf{f}_d \sim \begin{pmatrix} 0 & 0 & 0 \\ 0 & 0 & 0 \\ 0 & 0 & f_b \end{pmatrix}.$$

Frequently, the matrices $\mathbf{a}_{ij}$ are written as $A_{ij}\mathbf{f}_{ij}$. Within our approximation, these will then take the form,

$$\mathbf{a}_e \sim \begin{pmatrix} 0 & 0 & 0 \\ 0 & 0 & 0 \\ 0 & 0 & f_\tau A_\tau \end{pmatrix}, \quad \mathbf{a}_u \sim \begin{pmatrix} 0 & 0 & 0 \\ 0 & 0 & 0 \\ 0 & 0 & f_t A_t \end{pmatrix}, \quad \mathbf{a}_d \sim \begin{pmatrix} 0 & 0 & 0 \\ 0 & 0 & 0 \\ 0 & 0 & f_b A_b \end{pmatrix},$$

with A_τ, A_t, and A_b being real parameters (the A-terms). The bilinear b term is, likewise, written as $b = B\mu$, where B is taken to be real. The parametrization of the $\mathbf{a}$ and b terms in terms of the corresponding superpotential interactions is motivated by gravity-mediated models (to be discussed later). Indeed, within the MSSM framework, the soft breaking scalar parameters are completely unrelated to the parameters in the superpotential, i.e. $\mathbf{a}$ may be non-zero even if the Yukawa couplings vanish and, further, $\mathbf{c}$ terms need not be small.

From this point onwards, unless explicitly stated, we will assume that we are working within the simplified parameter space.

8.2 Electroweak symmetry breaking

The theory we have written down so far respects the gauge symmetry $SU(3)_C \times SU(2)_L \times U(1)_Y$. Our next task is to ensure that the gauge symmetry of the MSSM can be successfully broken down to observed $SU(3)_C \times U(1)_{em}$, so that W and Z bosons and fermions may receive mass as they do in the SM.

To investigate electroweak symmetry breaking, we must examine the minima of the scalar potential in the MSSM. The tree-level scalar potential consists of three parts

$$V_{MSSM} = V_F + V_D + V_{soft}, \tag{8.13}$$

where

$$V_F = \sum_i \left| \frac{\partial \hat{f}}{\partial \hat{\mathcal{S}}_i} \right|^2_{\hat{\mathcal{S}}=\mathcal{S}}, \tag{8.14a}$$

$$V_D = \frac{1}{2} \sum_A \left[\sum_i \mathcal{S}_i^\dagger g t_A \mathcal{S}_i \right]^2 \quad \text{and} \tag{8.14b}$$

$$V_{soft} = \sum_i m_{\phi_i}^2 |\phi_i|^2 - B\mu \left(H_d H_u + \quad \text{h.c.} \right) + \text{a-terms}. \tag{8.14c}$$

The sum over i is over all scalar fields in the model. Each real component of each scalar field may be regarded as a separate direction in "field space". Thus, the scalar "field space" of the MSSM, with 14 real matter scalars per generation, plus four complex Higgs scalars, is a 50-dimensional space. We look for parameter regions where this scalar potential develops a minimum along "directions" of the Higgs scalars. If a deeper minimum develops along other scalar field directions, then the ground state of the theory could develop such that electric charge, color or lepton number symmetry is broken. In fact, these considerations can be used to put constraints on the parameters of the theory. We will assume here that such non-standard minima do not develop.

We can then restrict our attention to the scalar potential involving only the Higgs scalar fields. We may use the $SU(2)_L$ gauge symmetry freedom to rotate the VEV of H_u to its lower component which we have *defined* to be neutral. Minimization of the potential with respect to the other component of H_u then requires that $\langle h_d^- \rangle = 0$ as demonstrated in the following exercise. The MSSM Higgs potential, therefore, allows only charge-conserving vacua.[2]

Exercise *Verify that for the Higgs fields, V_D can be written as,*

$$V_D^{\text{Higgs}} = \frac{g^2 + g'^2}{8}(\mathcal{A}_u^2 + \mathcal{A}_d^2) + \frac{g^2 - g'^2}{4}\mathcal{A}_u\mathcal{A}_d - \frac{g^2}{2}|\mathcal{A}_{ud}|^2,$$

where $\mathcal{A}_u = |H_u|^2$, $\mathcal{A}_d = |H_d|^2$, and $\mathcal{A}_{ud} = H_u H_d$. The tree-level Higgs potential to be minimized is,

$$V^{\text{Higgs}} = (m_{H_u}^2 + \mu^2)\mathcal{A}_u + (m_{H_d}^2 + \mu^2)\mathcal{A}_d - B\mu(\mathcal{A}_{ud} + \mathcal{A}_{ud}^\dagger) + V_D^{\text{Higgs}}.$$

We see that for fixed magnitudes of H_u and H_d, i.e. fixed values of $\mathcal{A}_u$ and $\mathcal{A}_d$, the minimum of V^{Higgs} is obtained by making $|\mathcal{A}_{ud}|$ as large as possible. This means, of course, that H_d and H_u are aligned, so that $\langle h_d^- \rangle = 0$. Moreover, for real values of $B\mu$ the second last term in V^{Higgs} is minimized when $\mathcal{A}_{ud}$ is real and positive (negative) if $B\mu$ is positive (negative). Thus as long as the parameters of the Higgs potential are real, no CP-violating phases are induced by the interactions of Higgs bosons.

Notice that there is no loss of generality if we choose the VEVs of both fields to have the same sign as long as the sign of $B\mu$ can always be appropriately chosen.

We then only have to minimize the scalar potential for the "neutral fields" which now reads,

$$V_{\text{scalar}} = (m_{H_u}^2 + \mu^2)|h_u^0|^2 + (m_{H_d}^2 + \mu^2)|h_d^0|^2$$

$$- B\mu(h_u^0 h_d^0 + \quad \text{h.c.}) + \frac{1}{8}(g^2 + g'^2)\left(|h_u^0|^2 - |h_d^0|^2\right)^2. \quad (8.15)$$

---

[2] Of course, we still have to assume that the matter scalars do not develop VEVs.

To find the minimum of the scalar potential, we set the first derivatives of this potential with respect to the fields as well as to their conjugates to zero:

$$\frac{\partial V}{\partial h_u^{0*}} = (m_{H_u}^2 + \mu^2)h_u^0 - B\mu h_d^{0*} + \frac{1}{4}(g^2 + g'^2)h_u^0(|h_u^0|^2 - |h_d^0|^2)$$

$$= 0, \tag{8.16a}$$

$$\frac{\partial V}{\partial h_d^{0*}} = (m_{H_d}^2 + \mu^2)h_d^0 - B\mu h_u^{0*} - \frac{1}{4}(g^2 + g'^2)h_d^0(|h_u^0|^2 - |h_d^0|^2)$$

$$= 0. \tag{8.16b}$$

The point(s) in field space where these equations are satisfied is an extremum of the (tree-level) potential. One possible solution is $\langle h_u^0 \rangle = \langle h_d^0 \rangle = 0$, i.e. no electroweak symmetry breaking. To ensure that this does not occur, the origin must be a local maximum of the potential. In other words, the determinant of the matrix of second derivatives should be negative at the origin. Since we are interested in the evaluation of the second derivatives at the origin of field space just the bilinear terms contribute, and we must have,

$$(B\mu)^2 > (m_{H_u}^2 + \mu^2)(m_{H_d}^2 + \mu^2). \tag{8.17a}$$

We must also check that the scalar potential indeed has a stable minimum, and is not unbounded from below. For most field values this is not an issue because the positive definite quartic term dominates the scalar potential for large field values. However, in the direction of field space where $|h_u^0| = |h_d^0|$, the quartic term vanishes. This is a D-flat direction in field space, and in this direction we must require the scalar potential to be positive. This leads to

$$m_{H_u}^2 + m_{H_d}^2 + 2\mu^2 > 2|B\mu|. \tag{8.17b}$$

If these conditions are met, then the scalar potential should develop a well-defined local minimum in which electroweak symmetry is spontaneously broken. We write $\langle h_u^0 \rangle \equiv v_u$ and $\langle h_d^0 \rangle \equiv v_d$ with the VEVs as real numbers, and define a parameter,

$$\tan \beta \equiv \frac{v_u}{v_d} \tag{8.18}$$

that will play an important role in phenomenological studies of the MSSM. It is simple to see that the potential minimization conditions can be written as:

$$B\mu = \frac{(m_{H_u}^2 + m_{H_d}^2 + 2\mu^2)\sin 2\beta}{2} \quad \text{and} \tag{8.19a}$$

$$\mu^2 = \frac{m_{H_d}^2 - m_{H_u}^2 \tan^2 \beta}{(\tan^2 \beta - 1)} - \frac{M_Z^2}{2}. \tag{8.19b}$$

To obtain (8.19b), we have used the relation (to be derived shortly), $M_Z^2 = \frac{(g^2+g'^2)}{2}(v_u^2 + v_d^2)$. The first of these equations allows one to trade the parameter $B\mu$ for the more commonly used parameter $\tan\beta$. Given the soft SUSY breaking Higgs masses $m_{H_u}^2$ and $m_{H_d}^2$, we use the second to fix the magnitude (but not the sign) of μ to reproduce the observed value of M_Z.

Up to now, we have focussed on the tree-level potential (8.15) for the electroweak symmetry breaking sector of the MSSM. A characteristic feature of this potential is that the quartic self-interactions of the Higgs fields are determined solely by the $SU(2) \times U(1)$ gauge couplings. This implies that the Higgs sector of the MSSM automatically satisfies perturbative unitarity constraints, in sharp contrast to the SM where the Higgs self-coupling constant is an independent parameter. This important feature of the MSSM can be traced to the fact that the μ term is the only possible superpotential term bilinear in the Higgs superfields. Indeed, as we will see, the structure of the self-couplings in the Higgs sector of the MSSM implies an *upper* limit of M_Z on the mass of the SM-like Higgs boson! This is a tree-level result, and radiative corrections modify it in an important way. We will, however, postpone any further discussion about this until we are ready to examine the spectrum of the relics of the electroweak symmetry breaking sector of the MSSM.

8.3 Particle masses in the MSSM

8.3.1 Gauge bosons

Once we are assured of the correct pattern of electroweak symmetry breaking, we can proceed to calculate the masses of the vector bosons. Since the vacuum does not spontaneously break the $U(1)_{em}$ associated with electromagnetic gauge invariance, we expect that the photon will remain massless, while the $W^\pm$ and Z^0 will acquire a mass via the Higgs mechanism. As in the SM, these vector boson mass terms arise from the kinetic energy terms of the Higgs fields:

$$\mathcal{L} \ni |D_\mu H_u|^2 + |D_\mu H_d|^2, \tag{8.20}$$

where

$$D_\mu H_u = (\partial_\mu + ig\frac{\tau_A}{2}W_{A\mu} + i\frac{g'}{2}B_\mu)H_u \quad \text{and}$$

$$D_\mu H_d = (\partial_\mu + ig(-\frac{\tau_A^*}{2})W_{A\mu} - i\frac{g'}{2}B_\mu)H_d.$$

The vector boson masses are obtained by making the replacement,

$$\langle H_u \rangle \to \begin{pmatrix} 0 \\ v_u \end{pmatrix} \quad \text{and} \quad \langle H_d \rangle \to \begin{pmatrix} 0 \\ v_d \end{pmatrix}. \tag{8.21}$$

Identifying the charged fields,

$$W_\mu^\pm = \frac{1}{\sqrt{2}}(W_{1\mu} \mp i W_{2\mu}),$$

we find,

$$M_W^2 = \frac{g^2}{2}(v_u^2 + v_d^2). \tag{8.22a}$$

As in the SM, the neutral fields $W_{3\mu}$ and B_μ mix, and the neutral mass matrix has to be diagonalized. Diagonalizing this mass matrix yields the fields,

$$A_\mu = \frac{(g' W_{3\mu} + g B_\mu)}{\sqrt{g^2 + g'^2}}$$

$$Z_\mu = \frac{(-g W_{3\mu} + g' B_\mu)}{\sqrt{g^2 + g'^2}}.$$

A_μ is massless and identified as the photon field. The other field has a mass,

$$M_Z^2 = \frac{g^2 + g'^2}{2}(v_u^2 + v_d^2). \tag{8.22b}$$

Defining the weak mixing angle by $\tan\theta_W \equiv g'/g$, we recover the SM relation $M_W = M_Z \cos\theta_W$.

Instead of working with H_u and H_d, we could have equally well worked with the linear combinations,

$$\phi = \sin\beta H_u + \cos\beta H_d^*,$$

$$\phi' = \cos\beta H_u - \sin\beta H_d^*.$$

The doublet ϕ acquires a VEV $v \equiv \sqrt{v_u^2 + v_d^2} \simeq 174 \, \text{GeV}$ for its neutral component and can be identified with the SM Higgs doublet. The field ϕ' does not acquire a VEV and is just an additional scalar field that has nothing to do with symmetry breaking.

8.3.2 *Matter fermions*

Matter fermions acquire masses via Yukawa interactions in the superpotential. Specifically, these masses arise from the terms

$$\mathcal{L} \ni -\frac{1}{2}\sum_{i,j} \bar{\psi}_i \left[\left(\frac{\partial^2 \hat{f}}{\partial \hat{\mathcal{S}}_i \partial \hat{\mathcal{S}}_j}\right)_{\hat{\mathcal{S}}=\mathcal{S}} \frac{1 - \gamma_5}{2} + \left(\frac{\partial^2 \hat{f}}{\partial \hat{\mathcal{S}}_i \partial \hat{\mathcal{S}}_j}\right)_{\hat{\mathcal{S}}=\mathcal{S}}^\dagger \frac{1 + \gamma_5}{2} \right] \psi_j$$

in our master formula. We will focus on the mass of the electron; the calculation of other SM fermion masses follows along identical lines.

We first note that, since the superpotential contains the term $\hat{f} \ni f_e \hat{e} \hat{h}_d^0 \hat{E}^c$, we find

$$\frac{\partial^2 \hat{f}}{\partial \hat{e} \partial \hat{E}^c}\bigg|_{\hat{S}=S} = f_e \hat{h}_d^0 \big|_{\hat{h}_d^0 = h_d^0} = f_e h_d^0,$$

so that

$$\mathcal{L} \ni -\frac{1}{2}\bar{\psi}_e \left[f_e h_d^0 \frac{1-\gamma_5}{2} + f_e h_d^{0*} \frac{1+\gamma_5}{2} \right] \psi_{E^c}$$

$$-\frac{1}{2}\bar{\psi}_{E^c} \left[f_e h_d^0 \frac{1-\gamma_5}{2} + f_e h_d^{0*} \frac{1+\gamma_5}{2} \right] \psi_e$$

$$= -\left[\bar{\psi}_{E^c} f_e h_d^0 \frac{1-\gamma_5}{2} \psi_e + \bar{\psi}_e f_e h_d^{0*} \frac{1+\gamma_5}{2} \psi_{E^c} \right],$$

where in the last step we have used the Majorana bilinear relations to combine terms. Using the definition (8.3) of the Dirac electron field, and replacing the field h_d^0 by its VEV, the reader can easily check that these terms reduce to a mass term for the Dirac electron. Specifically,

$$\mathcal{L} \ni -f_e v_d \bar{e} e = -m_e \bar{e} e, \tag{8.23}$$

with $m_e \equiv f_e v_d$. Thus, as in the SM, the electron acquires a mass via its coupling to the Higgs field. This justifies our calling the superpotential coupling f_e as the electron Yukawa coupling. Note that in the MSSM the electron mass comes from $\langle h_d^0 \rangle$. The same is true for the other charged leptons and down-type quarks that couple just to the doublet H_d via superpotential interactions. A similar calculation for the masses of $T_3 = +1/2$ fermions of the SM finds their masses proportional to v_u. The neutrino, of course, remains massless just as in the SM, since we have not introduced a Yukawa coupling for it.

Exercise *Verify that the fermion Yukawa couplings can be written as,*

$$f_i = \frac{g m_i}{\sqrt{2} M_W} 1/\sin\beta, \text{ if } T_{3f} = \frac{1}{2}, \tag{8.24a}$$

and

$$f_i = \frac{g m_i}{\sqrt{2} M_W} 1/\cos\beta, \text{ if } T_{3f} = -\frac{1}{2}. \tag{8.24b}$$

Notice that these expressions for the MSSM Yukawa couplings f_i in terms of the fermion masses are different from the corresponding expressions for the SM Yukawa couplings λ_i.

We remark that since the mass of the fermions arises from the superpotential, it must be supersymmetric, i.e. the scalar superpartners will get an identical contribution to the mass from the superpotential Yukawa couplings. This should not be surprising since we have already seen that we cannot have soft SUSY breaking masses for chiral fermions.

8.3.3 Higgs bosons

Before turning to the masses of the superpartners, let us examine the spectrum of physical particles from the electroweak symmetry breaking sector. Within the SM with just one complex doublet, we know that a single neutral spin zero particle – the Higgs boson – is left in the spectrum as a relic of the spontaneous breakdown of $SU(2)_L \times U(1)_Y \rightarrow U(1)_{em}$. This is because the charged component of the doublet and one of the neutral components are the three would-be Goldstone bosons that become the longitudinal components of the $W^{\pm}$ and Z^0 after the Higgs mechanism. Since the symmetry breaking pattern of the MSSM is the same as that of the SM, we expect the same set of would-be Goldstone bosons: however, since we now start with two sets of complex doublets, one charged and three neutral spin zero bosons remain in the physical spectrum of the MSSM.

In order to identify these states and compute their masses , we must examine the Higgs potential:

$$
\begin{aligned}
V^{\text{Higgs}} &= (m_{H_u}^2 + \mu^2)(|h_u^0|^2 + |h_u^+|^2) + (m_{H_d}^2 + \mu^2)(|h_d^0|^2 + |h_d^-|^2) \\
&\quad - B\mu(h_u^+ h_d^- + h_u^0 h_d^0 + \quad \text{h.c.}) \\
&\quad + \frac{g^2}{8}\left\{(|h_u^+|^2 - |h_u^0|^2 + |h_d^0|^2 - |h_d^-|^2)^2 + 4|h_u^+|^2|h_u^0|^2 + 4|h_d^0|^2|h_d^-|^2 \right. \\
&\quad \left. - 4(h_u^{+*}h_d^{-*}h_u^0 h_d^0 + h_u^{0*}h_d^{0*}h_u^+ h_d^-)\right\} \\
&\quad + \frac{g'^2}{8}\left[|h_u^+|^2 + |h_u^0|^2 - |h_d^0|^2 - |h_d^-|^2\right]^2 .
\end{aligned} \tag{8.25}
$$

The neutral fields may be broken up into real and imaginary components $h_u^0 = \frac{h_{uR}^0 + ih_{uI}^0}{\sqrt{2}}$ and $h_d^0 = \frac{h_{dR}^0 + ih_{dI}^0}{\sqrt{2}}$, so that the scalar potential can be regarded as a function $V(h_{uR}^0, h_{uI}^0, h_{dR}^0, h_{dI}^0, h_u^+, h_u^{+*}, h_d^-, h_d^{-*})$ of eight independent fields. Since we are interested in excitations of the vacuum, we expand the Higgs potential about its minimum as,

$$V^{\text{Higgs}} = V_{\min} + \sum_{h_i} \left. \frac{\partial V}{\partial h_i} \right|_{h_i = \langle h_i \rangle} (h_i - \langle h_i \rangle)$$

$$+ \frac{1}{2} \sum_{h_i, h_j} \left. \frac{\partial^2 V}{\partial h_i \partial h_j} \right|_{h_{i,j} = \langle h_{i,j} \rangle} (h_i - \langle h_i \rangle)(h_j - \langle h_j \rangle) + \cdots, \quad (8.26)$$

where the h_i are the eight arguments of V as listed above, and the only non-vanishing VEVs are $\langle h_{dR}^0 \rangle = \sqrt{2} v_d$ and $\langle h_{uR}^0 \rangle = \sqrt{2} v_u$. The coefficients of the linear terms should all vanish, since the derivatives are evaluated at the minimum of the potential; the quadratic terms will then be Higgs boson mass terms, and since in general there will be mixing, these will form mass matrices. The conservation of electric charge means that there can be no mixing between charged and neutral Higgs fields, so that there is one mass matrix for the charged sector and a different one in the neutral sector. Moreover, because of the (assumed) CP invariance of the Higgs sector, the real and imaginary components of the neutral Higgs bosons do not mix either, so that the 4×4 mass matrix in the neutral sector decomposes into two 2×2 blocks.

First, let us construct the mass matrices that contain the would-be Goldstone bosons. These reside in the charged sector and in the CP-odd sector (i.e. the imaginary components) of the neutral fields. The states orthogonal to the Goldstone boson will automatically be the physical states in these sectors.

We begin with the charged fields. The Lagrangian will have the form

$$\mathcal{L} \ni \left(h_u^{+*} \; h_d^- \right) \mathcal{M}_{h^\pm}^2 \begin{pmatrix} h_u^+ \\ h_d^{-*} \end{pmatrix}, \quad (8.27)$$

where

$$\mathcal{M}_{h^\pm}^2 = \begin{pmatrix} \left. \frac{\partial^2 V}{\partial h_u^+ \partial h_u^{+*}} \right|_{h_i \to v_i} & \left. \frac{\partial^2 V}{\partial h_u^{+*} \partial h_d^{-*}} \right|_{h_i \to v_i} \\ \left. \frac{\partial^2 V}{\partial h_u^+ \partial h_d^-} \right|_{h_i \to v_i} & \left. \frac{\partial^2 V}{\partial h_d^- \partial h_d^{-*}} \right|_{h_i \to v_i} \end{pmatrix}.$$

The derivatives are simple to compute if we remember that we want to evaluate these at the VEV of the Higgs fields; then we can drop terms that are proportional to h_u^+, h_d^-, h_{uI}^0 or h_{dI}^0 (after the derivatives are taken) as these fields vanish in the vacuum. For instance,

$$\left. \frac{\partial^2 V}{\partial h_u^+ \partial h_u^{+*}} \right|_{h_i \to v_i} = (m_{H_u}^2 + \mu^2) + \frac{g^2}{4} \left(v_u^2 + v_d^2 \right) + \frac{g'^2}{4} (v_u^2 - v_d^2)$$

$$= B\mu \cot\beta + \frac{g^2}{2} v_d^2,$$

where in the last step we have used the first of the minimization conditions (8.16a) to eliminate $m_{H_u}^2 + \mu^2$ in favor of $B\mu$. The mass squared matrix in the charged sector is found to be,

$$\mathcal{M}_{h^\pm}^2 = \begin{pmatrix} B\mu\cot\beta + \frac{g^2}{2}v_d^2 & -B\mu - \frac{g^2}{2}v_u v_d \\ -B\mu - \frac{g^2}{2}v_u v_d & B\mu\tan\beta + \frac{g^2}{2}v_u^2 \end{pmatrix}, \tag{8.28}$$

where we have used (8.16b) to eliminate $m_{H_d}^2 + \mu^2$ from the lower right entry of the matrix. Its eigenvalues are given by

$$m_{G^\pm} = 0 \quad \text{and} \quad m_{H^\pm}^2 = B\mu(\cot\beta + \tan\beta) + M_W^2. \tag{8.29}$$

The zero eigenvalue merely confirms that, but for the Higgs mechanism, $G^\pm$ would have been the Goldstone boson. In the unitarity gauge, these do not appear in the Lagrangian with massive W bosons. The other state, $H^\pm$, remains in the spectrum. The mixing matrix takes the form,

$$\begin{pmatrix} G^+ \\ H^+ \end{pmatrix} = \begin{pmatrix} \cos\beta & \sin\beta \\ -\sin\beta & \cos\beta \end{pmatrix} \begin{pmatrix} h_d^{-*} \\ h_u^+ \end{pmatrix}. \tag{8.30}$$

Let us now turn to the neutral sector, focussing for the moment on the mass terms for the imaginary components of the neutral fields. These may be written as,

$$\mathcal{L} \ni \frac{1}{2} \begin{pmatrix} h_{uI}^0 & h_{dI}^0 \end{pmatrix} \mathcal{M}_{h_{iI}^0}^2 \begin{pmatrix} h_{uI}^0 \\ h_{dI}^0 \end{pmatrix}, \tag{8.31}$$

with

$$\mathcal{M}_{h_{iI}^0}^2 = \begin{pmatrix} \dfrac{\partial^2 V}{\partial h_{uI}^{02}}\Big|_{h_i \to v_i} & \dfrac{\partial^2 V}{\partial h_{uI}^0 \partial h_{dI}^0}\Big|_{h_i \to v_i} \\ \dfrac{\partial^2 V}{\partial h_{uI}^0 \partial h_{dI}^0}\Big|_{h_i \to v_i} & \dfrac{\partial^2 V}{\partial h_{dI}^{02}}\Big|_{h_i \to v_i} \end{pmatrix}.$$

A computation similar to that for the charged sector gives,

$$\mathcal{M}_{h_{iI}^0}^2 = \begin{pmatrix} B\mu\cot\beta & B\mu \\ B\mu & B\mu\tan\beta \end{pmatrix}. \tag{8.32}$$

The eigenvalues are,

$$m_{G^0} = 0 \quad \text{and} \quad m_A^2 = B\mu(\cot\beta + \tan\beta). \tag{8.33}$$

From the eigenvalue corresponding to $m_{H^\pm}^2$, we see that

$$m_{H^\pm}^2 = m_A^2 + M_W^2, \tag{8.34}$$

so that, at least at tree level, $m_{H^\pm} \geq M_W$ and $m_{H^\pm} \geq m_A$.

Exercise *We have already argued that the real and imaginary components of the neutral fields cannot mix. Explicitly verify that this is indeed the case. Also, verify the mass matrix is indeed given by (8.32). To obtain the diagonal entries, you will once again have to use the minimization conditions.*

Again, after a gauge transformation to the unitarity gauge, G^0 disappears from the Lagrangian which now includes a mass for the Z^0 boson. The massive A particle remains as a *pseudoscalar* Higgs boson, as will be seen when we calculate its couplings to matter fermions.[3] The mixing matrix for G^0 and A is

$$\begin{pmatrix} G^0 \\ A \end{pmatrix} = \begin{pmatrix} \sin\beta & -\cos\beta \\ \cos\beta & \sin\beta \end{pmatrix} \begin{pmatrix} h^0_{uI} \\ h^0_{dI} \end{pmatrix}. \tag{8.35}$$

Finally, let us turn to the mass matrix for the remaining neutral scalars involving the h^0_{uR} and h^0_{dR}. The mass squared matrix of the real components of the neutral Higgs scalars occurs in the Lagrangian as,

$$\mathcal{L} \ni \frac{1}{2} \begin{pmatrix} h^0_{uR} & h^0_{dR} \end{pmatrix} \mathcal{M}^2_{h^0_{iR}} \begin{pmatrix} h^0_{uR} \\ h^0_{dR} \end{pmatrix}, \tag{8.36}$$

with

$$\mathcal{M}^2_{h^0_{iR}} = \begin{pmatrix} \left.\frac{\partial^2 V}{\partial h^{02}_{uR}}\right|_{h_i \to v_i} & \left.\frac{\partial^2 V}{\partial h^0_{uR} \partial h^0_{dR}}\right|_{h_i \to v_i} \\ \left.\frac{\partial^2 V}{\partial h^0_{uR} \partial h^0_{dR}}\right|_{h_i \to v_i} & \left.\frac{\partial^2 V}{\partial h^{02}_{dR}}\right|_{h_i \to v_i} \end{pmatrix}$$

$$= \begin{pmatrix} m^2_A \cos^2\beta + M^2_Z \sin^2\beta & -(m^2_A + M^2_Z)\sin\beta\cos\beta \\ -(m^2_A + M^2_Z)\sin\beta\cos\beta & m^2_A \sin^2\beta + M^2_Z \cos^2\beta \end{pmatrix}, \tag{8.37}$$

where to obtain the last step we have used manipulations very similar to those used to obtain the other mass matrices above. The eigenvalues of this mass matrix are

$$m^2_{h,H} = \frac{1}{2}\left[(m^2_A + M^2_Z) \mp \sqrt{(m^2_A + M^2_Z)^2 - 4m^2_A M^2_Z \cos^2 2\beta} \right], \tag{8.38}$$

where h and H denote the lighter and heavier of the neutral scalar mass eigenstates.

Exercise *The masses of h and H respect several important bounds. To see this, recall that the expectation value of the matrix (8.37) – for any vector $(\cos\theta, \sin\theta)^T$ – must lie between the eigenvalues m^2_h and m^2_H. Verify that setting $\theta = \beta$ yields,*

$$m_h \le m_A |\cos 2\beta| \le m_H, \tag{8.39a}$$

---

[3] Because parity is not conserved in weak interactions, the attentive reader may wonder whether A remains an eigenstate beyond tree level. However, CP is conserved and the CP-odd A is precluded from mixing with the CP-even scalar Higgs bosons that we consider shortly.

while setting $\theta = \pi/2 - \beta$ *yields,*

$$m_h \leq M_Z|\cos 2\beta| \leq m_H. \tag{8.39b}$$

Notice that this implies that $m_h = 0$ *if* $\tan \beta = 1$.

Note that these bounds hold only at tree level. Radiative corrections that we alluded to earlier allow h to be significantly heavier than M_Z. This is fortunate since otherwise the non-observation of h in experiments at LEP2 would have excluded the MSSM!

Finally, we may write the physical Higgs scalars in terms of $h_{u\mathrm{R}}^0$ and $h_{d\mathrm{R}}^0$ as

$$\begin{pmatrix} h \\ H \end{pmatrix} = \begin{pmatrix} \cos\alpha & \sin\alpha \\ -\sin\alpha & \cos\alpha \end{pmatrix} \begin{pmatrix} h_{u\mathrm{R}}^0 \\ h_{d\mathrm{R}}^0 \end{pmatrix}, \tag{8.40a}$$

with α the Higgs scalar mixing angle being given by

$$\tan\alpha = \frac{(m_A^2 - M_Z^2)\cos 2\beta + \sqrt{(m_A^2 + M_Z^2)^2 - 4m_A^2 M_Z^2 \cos^2 2\beta}}{(m_A^2 + M_Z^2)\sin 2\beta}. \tag{8.40b}$$

Let us now turn to the mass spectrum of the superpartners. We first discuss masses of gauge and Higgs fermions, and then turn to the partners of the matter fermions.

8.3.4 Gluinos

The gluino $\tilde{g}$, the gaugino partner of the gluon, is the only color octet fermion. Since $SU(3)_C$ is not broken, the gluino cannot mix with any other fermion, and must be a mass eigenstate. Its mass term then arises just from the soft supersymmetry breaking gaugino mass term,[4]

$$\mathcal{L} \ni -\frac{1}{2}M_3\bar{\tilde{g}}\tilde{g} \tag{8.41}$$

so that its mass at tree level is simply $m_{\tilde{g}} = |M_3|$. If the real parameter M_3 is negative, following the discussion in the Technical Aside of Chapter 7, we can always redefine the gluino field $\tilde{g} \to -i\gamma_5\tilde{g}$. The new gluino field then has positive mass and retains its Majorana character. For later convenience, we will write this redefinition as $\tilde{g} \to (-i\gamma_5)^{\theta_{\tilde{g}}}\tilde{g}$, where $\theta_{\tilde{g}} = 0$ (1) for $M_3 > 0$ ($M_3 < 0$).

Exercise *Show that the transformation* $\psi \to \psi' = -i\gamma_5\psi$ *changes the sign of the mass term in the Lagrangian for a free Majorana fermion, but not the kinetic energy term. Show also that if* ψ *is Majorana, then so is* ψ'.

---

[4] Recall that we have already discussed how the CP-violating mass M_3' can be removed by field redefinition.

8.3.5 Charginos and neutralinos

Spontaneous breakdown of $SU(2)_L \times U(1)_Y$ implies that states with the same electric charge, color, and spin will mix. This means that gauginos and higgsinos cannot be the physical particles with definite mass. Rather, the neutral fermion fields $\psi_{h_u^0}$, $\psi_{h_d^0}$, λ_3 and λ_0 mix to form neutral fermion mass eigenstates, the *neutralinos*, while the negatively charged fields $\psi_{h_u^+ R}$, $\psi_{h_d^- L}$, and the linear combination $\frac{\lambda_1 + i\lambda_2}{\sqrt{2}}$ (this is just the superpartner of the field W_μ^- defined earlier) mix to form the negative *charginos*.[5]

We first work out the form of the chargino and neutralino mass matrices, and then diagonalize them to identify the physical charginos and neutralinos. These mass matrices receive a supersymmetric contribution from the superpotential higgsino mass term μ, a SUSY breaking one from gaugino masses, and finally a contribution from electroweak symmetry breaking. This last contribution is also SUSY breaking unless $v_u = v_d$ because D-term contributions to the potential from the Higgs field do not vanish in the vacuum.

The supersymmetric contribution, which arises from the superpotential terms,

$$\hat{f} \ni \mu \left(\hat{h}_u^0 \hat{h}_d^0 + \hat{h}_u^+ \hat{h}_d^- \right) \tag{8.42}$$

gives rise to fermion bilinear terms,

$$\mathcal{L} \ni -\frac{1}{2} \sum_{i,j} \bar{\psi}_i \left(\frac{\partial^2 \hat{f}}{\partial \hat{\mathcal{S}}_i \partial \hat{\mathcal{S}}_j} \right)_{\hat{\mathcal{S}} = \mathcal{S}} P_L \psi_j + \quad \text{h.c.},$$

which take the form,

$$\mathcal{L}_{\text{mass}} \ni -\frac{\mu}{2} \left[\bar{\psi}_{h_u^0} \psi_{h_d^0} + \bar{\psi}_{h_d^0} \psi_{h_u^0} \right]$$
$$-\frac{\mu}{2} \left[\bar{\psi}_{h_u^+} \psi_{h_d^-} + \bar{\psi}_{h_d^-} \psi_{h_u^+} \right]. \tag{8.43}$$

Gaugino–higgsino bilinear terms coming from electroweak breaking arise from,

$$\mathcal{L} \ni -\sqrt{2} \sum_i g \mathcal{S}_i^\dagger t_A \bar{\lambda}_A P_L \psi_i + \quad \text{h.c.}, \tag{8.44}$$

---

[5] Recall that $\psi_{h_u^+}$ is a Majorana spinor whose left-chiral component is positively charged while the right-chiral component is negatively charged.

when S_i are the Higgs fields that develop VEVs. These contributions can be written as,

$$\mathcal{L} \ni - \sqrt{2}\left(h_u^{+\dagger},\ h_u^{0\dagger}\right)\frac{1}{2}\begin{bmatrix} g\bar{\lambda}_3 + g'\bar{\lambda}_0 & g\bar{\lambda}_1 - ig\bar{\lambda}_2 \\ g\bar{\lambda}_1 + ig\bar{\lambda}_2 & -g\bar{\lambda}_3 + g'\bar{\lambda}_0 \end{bmatrix} P_L \begin{pmatrix} \psi_{h_u^+} \\ \psi_{h_u^0} \end{pmatrix}$$
$$- \sqrt{2}\left(h_d^{-\dagger},\ h_d^{0\dagger}\right)\frac{1}{2}\begin{bmatrix} -g\bar{\lambda}_3 - g'\bar{\lambda}_0 & -g\bar{\lambda}_1 - ig\bar{\lambda}_2 \\ -g\bar{\lambda}_1 + ig\bar{\lambda}_2 & g\bar{\lambda}_3 - g'\bar{\lambda}_0 \end{bmatrix} P_L \begin{pmatrix} \psi_{h_d^-} \\ \psi_{h_d^0} \end{pmatrix}$$
$$+ \quad \text{h.c.} \tag{8.45}$$

Electroweak symmetry breaking contributions to gaugino–higgsino masses arise when the Higgs boson fields develop VEVs. The corresponding terms in (8.45) involving charged higgsinos are,

$$-\frac{gv_u}{\sqrt{2}}\bar{\psi}_{h_u^+} P_R(\lambda_1 - i\lambda_2) - \frac{gv_d}{\sqrt{2}}(-\bar{\lambda}_1 + i\bar{\lambda}_2)P_L\psi_{h_d^-} + \quad \text{h.c.}$$
$$= -\frac{gv_u}{\sqrt{2}}(\bar{\lambda}_1 - i\bar{\lambda}_2)P_R\psi_{h_u^+} + \frac{gv_d}{\sqrt{2}}(\bar{\lambda}_1 - i\bar{\lambda}_2)P_L\psi_{h_d^-} + \quad \text{h.c.},$$

where the first term in the first line comes from the Hermitian conjugate part of (8.45), and in the second step we have used the Majorana bilinear identities to swap the order of the spinors. This then leads us to define Dirac fields for the negatively charged gaugino,

$$\lambda = \frac{\lambda_1 + i\lambda_2}{\sqrt{2}} \tag{8.46a}$$

and a negatively charged higgsino,

$$\tilde{\chi} = P_L\psi_{h_d^-} - P_R\psi_{h_u^+} \tag{8.46b}$$

in terms of which the charged and neutral gaugino–higgsino mass terms in (8.45) can then be written as,

$$\mathcal{L}_{\text{mass}} = gv_u\bar{\lambda}\frac{1 + \gamma_5}{2}\tilde{\chi} + gv_d\bar{\lambda}\frac{1 - \gamma_5}{2}\tilde{\chi} + \quad \text{h.c.}$$
$$+ \frac{gv_u}{\sqrt{2}}\bar{\lambda}_3\psi_{h_u^0} - \frac{g'v_u}{\sqrt{2}}\bar{\lambda}_0\psi_{h_u^0} - \frac{gv_d}{\sqrt{2}}\bar{\lambda}_3\psi_{h_d^0} + \frac{g'v_d}{\sqrt{2}}\bar{\lambda}_0\psi_{h_d^0}. \tag{8.47}$$

Exercise *Verify that the charged higgsino mass term in the Lagrangian (8.43) simply becomes* $+\mu\bar{\tilde{\chi}}\tilde{\chi}$.

Finally, the Lagrangian contribution from the soft SUSY breaking gaugino masses is,

$$\mathcal{L}_{\text{mass}} = -\frac{1}{2}M_1\bar{\lambda}_0\lambda_0 - M_2\frac{1}{2}\bar{\lambda}_3\lambda_3 - M_2\bar{\lambda}\lambda. \tag{8.48}$$

The gaugino–higgsino mass terms (8.43), (8.47), and (8.48) can be written as

$$\mathcal{L}_{\text{neutralino}} = -\frac{1}{2} \left(\bar{\psi}_{h_u^0}, \, \bar{\psi}_{h_d^0}, \, \bar{\lambda}_3, \, \bar{\lambda}_0 \right) \mathcal{M}_{\text{neutral}} \begin{pmatrix} \psi_{h_u^0} \\ \psi_{h_d^0} \\ \lambda_3 \\ \lambda_0 \end{pmatrix},$$

with

$$\mathcal{M}_{\text{neutral}} = \begin{pmatrix} 0 & \mu & -\frac{g v_u}{\sqrt{2}} & \frac{g' v_u}{\sqrt{2}} \\ \mu & 0 & \frac{g v_d}{\sqrt{2}} & -\frac{g' v_d}{\sqrt{2}} \\ -\frac{g v_u}{\sqrt{2}} & \frac{g v_d}{\sqrt{2}} & M_2 & 0 \\ \frac{g' v_u}{\sqrt{2}} & -\frac{g' v_d}{\sqrt{2}} & 0 & M_1 \end{pmatrix} \tag{8.49a}$$

and

$$\mathcal{L}_{\text{chargino}} = - \left(\bar{\lambda}, \, \bar{\tilde{\chi}} \right) \left(\mathcal{M}_{\text{charge}} P_{\text{L}} + \mathcal{M}_{\text{charge}}^T P_{\text{R}} \right) \begin{pmatrix} \lambda \\ \tilde{\chi} \end{pmatrix},$$

with

$$\mathcal{M}_{\text{charge}} = \begin{pmatrix} M_2 & -g v_d \\ -g v_u & -\mu \end{pmatrix}. \tag{8.49b}$$

The physical charginos and neutralinos are eigenstates of these mass matrices. The neutralino mass matrix is real and Hermitian, and so can be diagonalized by an orthogonal transformation as usual. The chargino mass matrix is not symmetric, so that the chargino mass terms are "γ_5-dependent". The diagonalization of charginos is performed as described in the Technical Note of Chapter 7.

Diagonalization of neutralinos

The neutralino mass matrix $\mathcal{M}_{\text{neutral}}$ is guaranteed to have real eigenvalues since it is Hermitian. It can be diagonalized by a unitary (in fact, real orthogonal) matrix V_n such that,

$$V_n^\dagger \mathcal{M}_{\text{neutral}} V_n = \mathcal{M}_D$$

where $\mathcal{M}_D$ is the diagonal matrix of eigenvalues which, though real, are not necessarily positive. The matrix V_n is the matrix whose columns are the eigenvectors of $\mathcal{M}_{\text{neutral}}$. The neutral higgsino and gaugino fields are related to the mass eigenstate fields by,

$$
\begin{pmatrix} \psi_{h_u^0} \\ \psi_{h_d^0} \\ \lambda_3 \\ \lambda_0 \end{pmatrix} = \begin{pmatrix} v_1^{(1)} & v_1^{(2)} & v_1^{(3)} & v_1^{(4)} \\ v_2^{(1)} & v_2^{(2)} & v_2^{(3)} & v_2^{(4)} \\ v_3^{(1)} & v_3^{(2)} & v_3^{(3)} & v_3^{(4)} \\ v_4^{(1)} & v_4^{(2)} & v_4^{(3)} & v_4^{(4)} \end{pmatrix} \begin{pmatrix} \widetilde{Z}_1' \\ \widetilde{Z}_2' \\ \widetilde{Z}_3' \\ \widetilde{Z}_4' \end{pmatrix}.
\tag{8.50}
$$

It is customary to define mass eigenstate fields with positive eigenvalues. We thus define mass eigenstates such that

$$
\widetilde{Z}_i = (-i\gamma_5)^{\theta_i} \widetilde{Z}_i',
\tag{8.51}
$$

with θ_i equals 0 (1) if the eigenvalue corresponding to $\widetilde{Z}_i'$ is positive (negative). The neutralinos are labeled according to increasing mass, with $\widetilde{Z}_1$ being the lightest neutralino and $\widetilde{Z}_4$ the heaviest.

The neutralino mass matrix can be diagonalized analytically, but the resulting formulae are lengthy and not particularly illuminating. Usually, the eigenvalues and eigenvectors are calculated numerically.

Diagonalization of charginos

The chargino mass terms are γ_5-dependent and, as discussed in the Technical Note of Chapter 7, can be diagonalized by different unitary transformations of the left- and right-handed components of the fields. We can write

$$
P_L \begin{pmatrix} \lambda \\ \tilde{\chi} \end{pmatrix} = U P_L \begin{pmatrix} \widetilde{W}_2 \\ \widetilde{W}_1 \end{pmatrix} \; ; \; P_R \begin{pmatrix} \lambda \\ \tilde{\chi} \end{pmatrix} = V P_R \begin{pmatrix} \widetilde{W}_2 \\ \widetilde{W}_1 \end{pmatrix},
\tag{8.52}
$$

with U and V being 2×2 unitary matrices. Then,

$$
\mathcal{L} \ni - \left(\overline{\widetilde{W}}_2 \; \overline{\widetilde{W}}_1 \right) V^\dagger \mathcal{M}_{\text{charge}} U P_L \begin{pmatrix} \widetilde{W}_2 \\ \widetilde{W}_1 \end{pmatrix}
$$
$$
- \left(\overline{\widetilde{W}}_2, \; \overline{\widetilde{W}}_1 \right) U^\dagger \mathcal{M}_{\text{charge}}^T V P_R \begin{pmatrix} \widetilde{W}_2 \\ \widetilde{W}_1 \end{pmatrix}.
$$

We construct matrices U and V so that these mass terms are diagonal, i.e.

$$
V^\dagger \mathcal{M}_{\text{charge}} U = \begin{pmatrix} m_{\widetilde{W}_2} & 0 \\ 0 & m_{\widetilde{W}_1} \end{pmatrix} \equiv \mathcal{M}_D \quad \text{and}
$$
$$
U^\dagger \mathcal{M}_{\text{charge}}^T V = \begin{pmatrix} m_{\widetilde{W}_2} & 0 \\ 0 & m_{\widetilde{W}_1} \end{pmatrix} \equiv \mathcal{M}_D^\dagger,
\tag{8.53}
$$

with $m_{\widetilde{W}_1}$ and $m_{\widetilde{W}_2}$ as real (but not necessarily positive) numbers. U is simply the unitary matrix that diagonalizes the Hermitian matrix $\mathcal{M}_{\text{charge}}^T \mathcal{M}_{\text{charge}}$, while V is the corresponding matrix that diagonalizes $\mathcal{M}_{\text{charge}} \mathcal{M}_{\text{charge}}^T$. The eigenvalues of the matrix $\mathcal{M}_{\text{charge}}^T \mathcal{M}_{\text{charge}}$ (which are the same as those of the matrix $\mathcal{M}_{\text{charge}} \mathcal{M}_{\text{charge}}^T$) are of course real and positive. Since

$$M_D^\dagger M_D = U^\dagger \left(\mathcal{M}_{\text{charge}}^T \mathcal{M}_{\text{charge}} \right) U,$$

these eigenvalues are just $m_{\tilde{W}_{2,1}}^2$, and are given by

$$m_{\tilde{W}_{1,2}}^2 = \frac{1}{2} \left[(\mu^2 + M_2^2 + 2M_W^2) \mp \zeta \right], \tag{8.54}$$

with

$$\zeta^2 = (\mu^2 - M_2^2)^2 + 4M_W^2 \left[M_W^2 \cos^2 2\beta + \mu^2 + M_2^2 - 2\mu M_2 \sin 2\beta \right].$$

We define $\tilde{W}_1$ to be the lighter chargino mass eigenstate, and $\tilde{W}_2$ the heavier one.

It is easy to see that U, the matrix of eigenvectors of $\mathcal{M}_{\text{charge}}^T \mathcal{M}_{\text{charge}}$, is

$$U = \begin{pmatrix} \frac{1}{\sqrt{1+x_2^2}} & \frac{1}{\sqrt{1+x_1^2}} \\ \frac{x_2}{\sqrt{1+x_2^2}} & \frac{x_1}{\sqrt{1+x_1^2}} \end{pmatrix},$$

where

$$x_{2/1} = \frac{\mu^2 - M_2^2 + 2M_W^2 \cos 2\beta \pm \zeta}{2\sqrt{2}M_W(-M_2 \cos \beta + \mu \sin \beta)}. \tag{8.55}$$

Likewise, the matrix V, constructed from the eigenvectors of $\mathcal{M}_{\text{charge}} \mathcal{M}_{\text{charge}}^T$, is given by

$$V = \begin{pmatrix} \frac{1}{\sqrt{1+y_2^2}} & \frac{1}{\sqrt{1+y_1^2}} \\ \frac{y_2}{\sqrt{1+y_2^2}} & \frac{y_1}{\sqrt{1+y_1^2}} \end{pmatrix},$$

with

$$y_{2/1} = \frac{\mu^2 - M_2^2 - 2M_W^2 \cos 2\beta \pm \zeta}{2\sqrt{2}M_W(-M_2 \sin \beta + \mu \cos \beta)}. \tag{8.56}$$

It is straightforward to check that $x_1 x_2 = y_1 y_2 = -1$, as expected from the orthogonality of the eigenvectors. Using this to eliminate x_2 and y_2, the U and V matrices can be recast as,

$$U = \begin{pmatrix} \theta_{x_1} \cos \gamma_L & \sin \gamma_L \\ -\theta_{x_1} \sin \gamma_L & \cos \gamma_L \end{pmatrix} \tag{8.57a}$$

and

$$V = \begin{pmatrix} \theta_y \cos \gamma_R & \sin \gamma_R \\ -\theta_y \sin \gamma_R & \cos \gamma_R \end{pmatrix}, \tag{8.57b}$$

with $\theta_x = \text{sign}(x_1)$ and $\theta_y = \text{sign}(y_1)$. The mixing angles γ_L and γ_R lie in the range $0 \leq \gamma_L, \gamma_R \leq 180°$, and are given by,

$$\tan \gamma_L = 1/x_1 \quad \text{and} \quad \tan \gamma_R = 1/y_1. \tag{8.58}$$

From Eq. (8.53) we see that *unsquared* chargino masses are given by,

$$m_{\tilde{W}_1} = \sin \gamma_R \left(M_2 \sin \gamma_L - \sqrt{2} M_W \cos \beta \cos \gamma_L \right)$$
$$- \cos \gamma_R \left(\sqrt{2} M_W \sin \beta \sin \gamma_L + \mu \cos \gamma_L \right) \tag{8.59a}$$

and

$$m_{\tilde{W}_2} = \theta_x \theta_y \left[\cos \gamma_R \left(M_2 \cos \gamma_L + \sqrt{2} M_W \cos \beta \sin \gamma_L \right) \right.$$
$$\left. + \sin \gamma_R \left(\sqrt{2} M_W \sin \beta \cos \gamma_L - \mu \sin \gamma_L \right) \right]. \tag{8.59b}$$

If either of the $m_{\tilde{W}_i}$ is negative, we replace $\tilde{W}_i \to \gamma_5 \tilde{W}_i$, and work with fields with positive mass eigenvalues.

In general, the chargino and neutralino mixing patterns are complex, and depend on the parameters, μ, M_1, M_2, and $\tan \beta$. However, if $|\mu| \gg |M_{1,2}|$, M_W, then $\tilde{W}_2$ and $\tilde{Z}_{3,4}$ are approximately higgsinos with squared masses of about μ^2, while the lighter chargino and the two lighter neutralinos are gaugino-like. If $|M_{1,2}| \gg |\mu|$, M_W, the situation is reversed, and the heavier chargino, and the two heavy neutralinos are gaugino-like, while the lighter chargino and the lighter neutralinos are approximately higgsino-like. These properties will be useful in understanding sparticle decay patterns discussed in Chapter 13.

Exercise *From the "squared mass" matrices of charginos and neutralinos, show that,*

$$m_{\tilde{W}_1}^2 + m_{\tilde{W}_2}^2 - 2M_W^2 = \mu^2 + M_2^2,$$

and

$$m_{\tilde{Z}_1}^2 + m_{\tilde{Z}_2}^2 + m_{\tilde{Z}_3}^2 + m_{\tilde{Z}_4}^2 - 2M_Z^2 = 2\mu^2 + M_1^2 + M_2^2.$$

These are, of course, tree-level relations.

Exercise *If soft SUSY breaking gaugino masses are zero, show that the lightest neutralino is a massless photino, $\tilde{\gamma} \equiv \sin \theta_W \lambda_3 + \cos \theta_W \lambda_0$. In this case, show that $\tilde{W}_1$ and $\tilde{Z}_2$ are lighter than M_W and M_Z, respectively.*

Although we note this in the context of the MSSM, this result is much more general, in the sense that it does not depend on the details of the electroweak symmetry breaking sector.

Incidently, note also that if $M_1 = M_2 = M$, the photino, defined above, is an eigenstate of the neutralino mass matrix with mass M.

Exercise *We have just seen that the lightest neutralino is a massless photino if gaugino masses are zero. Show that it is a massless higgsino* $\cos\beta\,\bar{\psi}_{h_u^0} + \sin\beta\,\bar{\psi}_{h_d^0}$ *if, instead, μ vanishes.*

Show that a massless neutralino can also occur if

$$\mu + M_W^2 \sin 2\beta \left(\frac{1}{M_2} + \frac{\tan^2\theta_W}{M_1} \right) = 0.$$

Find the appropriate eigenvector in this case.

8.3.6 Squarks and sleptons

Now we turn to squark and slepton masses. Unlike matter fermions whose masses only arise from superpotential Yukawa interactions, squarks and sleptons (collectively referred to as sfermions) have four distinct sources for these mass terms. For definiteness, we will write these terms for top squarks, but it will be obvious how to write the corresponding terms for other squarks as well as sleptons.

Superpotential terms

We expect that sfermions must get a mass contribution equal to the corresponding fermion mass. The relevant part of the superpotential is,

$$\hat{f} \ni \mu \hat{h}_u^0 \hat{h}_d^0 + f_t \hat{t}\hat{h}_u^0 \hat{T}^c.$$

Since $\mathcal{L} \ni -\sum_i |\partial \hat{f}/\partial \hat{S}_i|^2_{\hat{S}=S}$, we see that the squares of $\partial \hat{f}/\partial \hat{t} = f_t \hat{h}_u^0 \hat{T}^c$ and of $\partial \hat{f}/\partial \hat{T}^c = f_t \hat{t}\hat{h}_u^0$, upon the replacement $h_u^0 \to v_u$, give the anticipated terms,

$$\mathcal{L} \ni -m_t^2 \tilde{t}_L^\dagger \tilde{t}_L - m_t^2 \tilde{t}_R^\dagger \tilde{t}_R. \tag{8.60a}$$

This is, however, not the only t-squark bilinear that can come from the superpotential because the cross terms from $|\partial \hat{f}/\partial \hat{h}_u^0|^2$, upon the replacement $h_d^0 \to v_d$, yield an intra-generational mixing contribution to the $\tilde{t}$ mass,

$$\mathcal{L} \ni - (\mu m_t \cot\beta) \left(\tilde{t}_L^\dagger \tilde{t}_R + \tilde{t}_R^\dagger \tilde{t}_L \right). \tag{8.60b}$$

Notice that both these contributions will vanish if the corresponding *quark* Yukawa coupling is zero.

Soft SUSY breaking scalar masses

These terms arise from

$$\mathcal{L} \ni -\tilde{Q}_i^\dagger \mathbf{m}_{Qij}^2 \tilde{Q}_j - \tilde{u}_{Ri}^\dagger \mathbf{m}_{Uij}^2 \tilde{u}_{Rj}$$
$$\ni -m_{\tilde{t}_L}^2 \tilde{t}_L^\dagger \tilde{t}_L - m_{\tilde{t}_R}^2 \tilde{t}_R^\dagger \tilde{t}_R. \tag{8.61}$$

Remember that there is just one soft SUSY breaking squark (slepton) mass for each generation of left-squarks (left-sleptons); i.e.

$$m_{\tilde{t}_L} = m_{\tilde{b}_L} = m_{Q3}, \; m_{\tilde{e}_L} = m_{\tilde{\nu}_e} = m_{L1}, \quad \text{etc.}$$

Clearly, these terms come from SUSY breaking and are present regardless of whether or not electroweak symmetry is spontaneously broken.

Soft SUSY breaking trilinear terms

Soft SUSY breaking interactions of squarks with neutral Higgs bosons,

$$\mathcal{L} \ni A_t f_t \tilde{t}_L h_u^0 \tilde{t}_R^\dagger + \quad \text{h.c.},$$

give rise to intra-generational squark mixing terms

$$\mathcal{L} \ni -(-A_t m_t)(\tilde{t}_L^\dagger \tilde{t}_R + \tilde{t}_R^\dagger \tilde{t}_L), \tag{8.62}$$

when the Higgs field is replaced by its VEV. That these terms appear proportional to m_t is an artifact of writing a_t as $A_t f_t$. Nevertheless, like the superpotential terms, these terms are absent if the electroweak symmetry is unbroken.

D-term contributions

We write these terms which come from

$$\mathcal{L} \ni -\frac{1}{2} \sum_A |\sum_i S_i^\dagger g_\alpha t_{\alpha A} S_i|^2$$

$$\ni -\frac{1}{2} g^2 |\tilde{Q}^\dagger T_{3Q} \tilde{Q} + H_u^\dagger \frac{\tau_3}{2} H_u + H_d^\dagger (-\frac{\tau_3}{2}) H_d|^2$$

$$-\left(\frac{g'}{2}\right)^2 |H_u^\dagger Y_{H_u} H_u + H_d^\dagger Y_{H_d} H_d + \tilde{Q}^\dagger Y_Q \tilde{Q} + \tilde{u}_{Ri}^\dagger Y_{U^c} \tilde{u}_{Ri} + \tilde{d}_{Ri}^\dagger Y_{D^c} \tilde{d}_{Ri}|^2,$$

for both top and bottom squarks. Squark mass contributions arise from cross terms between squark and Higgs boson fields. The $SU(2)$ D-term gives,

$$\mathcal{L} \ni -\frac{1}{2}\left[2(\frac{g}{2})^2(v_d^2 - v_u^2)(\tilde{t}_L^\dagger \tilde{t}_L - \tilde{b}_L^\dagger \tilde{b}_L)\right]$$

$$= -M_W^2 \cos 2\beta T_{3Q_i} \tilde{Q}_{Li}^\dagger \tilde{Q}_{Li}, \tag{8.63a}$$

while the hypercharge D-term gives,

$$\mathcal{L} \ni \sin^2 \theta_W \cos 2\beta\, M_Z^2 \left(\tilde{t}_L^\dagger \frac{Y_Q}{2} \tilde{t}_L + \tilde{b}_L^\dagger \frac{Y_Q}{2} \tilde{b}_L + \tilde{t}_R^\dagger \frac{Y_{U^c}}{2} \tilde{t}_R + \tilde{b}_R^\dagger \frac{Y_{D^c}}{2} \tilde{b}_R \right). \qquad (8.63b)$$

Note that the hypercharges that appear in the terms involving right-handed fields are those for the corresponding *left*-handed antiquark fields that appear in Table 8.1. Eliminating the hypercharge in favor of the electric charge, the D-term contribution to any MSSM sfermion squared mass can be written as,

$$m_{D\text{-term}}^2 = M_Z^2 \cos 2\beta \left(T_3 - Q \sin^2 \theta_W \right). \qquad (8.64)$$

We can now assemble the mass squared matrices for the sfermions. For top squarks, we have

$$\mathcal{L} \ni - \left(\tilde{t}_L^\dagger,\ \tilde{t}_R^\dagger \right) \mathcal{M}_{\tilde{t}}^2 \begin{pmatrix} \tilde{t}_L \\ \tilde{t}_R \end{pmatrix},$$

where the matrix $\mathcal{M}_{\tilde{t}}^2$ is given by

$$\begin{pmatrix} m_{\tilde{t}_L}^2 + m_t^2 + D(\tilde{t}_L) & m_t(-A_t + \mu \cot \beta) \\ m_t(-A_t + \mu \cot \beta) & m_{\tilde{t}_R}^2 + m_t^2 + D(\tilde{t}_R) \end{pmatrix}, \qquad (8.65a)$$

and

$$D(\tilde{t}_L) = M_Z^2 \cos 2\beta (\frac{1}{2} - \frac{2}{3} \sin^2 \theta_W),$$

$$D(\tilde{t}_R) = M_Z^2 \cos 2\beta (+\frac{2}{3} \sin^2 \theta_W),$$

are the hypercharge D-term contributions (8.64) to the squared masses of $\tilde{t}_L$ and $\tilde{t}_R$. The eigenvalues of this matrix are,

$$m_{\tilde{t}_{1,2}}^2 = \frac{1}{2} \left(m_{\tilde{t}_L}^2 + m_{\tilde{t}_R}^2 \right) + \frac{1}{4} M_Z^2 \cos 2\beta + m_t^2$$

$$\mp \left\{ \left[\frac{1}{2}(m_{\tilde{t}_L}^2 - m_{\tilde{t}_R}^2) + M_Z^2 \cos 2\beta (\frac{1}{4} - \frac{2}{3} x_W) \right]^2 + m_t^2 (\mu \cot \beta - A_t)^2 \right\}^{\frac{1}{2}},$$

$$(8.65b)$$

with $\tilde{t}_1$ the lighter top squark mass eigenstate, and $\tilde{t}_2$ the heavier one, and $x_W \equiv \sin^2 \theta_W$. The top squark mixing matrix is defined by

$$\begin{pmatrix} \tilde{t}_1 \\ \tilde{t}_2 \end{pmatrix} = \begin{pmatrix} \cos \theta_t & -\sin \theta_t \\ \sin \theta_t & \cos \theta_t \end{pmatrix} \begin{pmatrix} \tilde{t}_L \\ \tilde{t}_R \end{pmatrix}, \qquad (8.65c)$$

with the top squark mixing angle θ_t given by,

$$\tan \theta_t = \frac{m_{\tilde{t}_L}^2 + m_t^2 + M_Z^2 \cos 2\beta \left(\frac{1}{2} - \frac{2}{3}x_W\right) - m_{\tilde{t}_1}^2}{m_t \left(-A_t + \mu \cot \beta\right)}. \tag{8.65d}$$

For bottom squarks, we find the mass matrix $\mathcal{M}_{\tilde{b}}^2$ to be

$$\begin{pmatrix} m_{\tilde{b}_L}^2 + m_b^2 + D(\tilde{b}_L) & m_b(-A_b + \mu \tan \beta) \\ m_b(-A_b + \mu \tan \beta) & m_{\tilde{b}_R}^2 + m_b^2 + D(\tilde{b}_R) \end{pmatrix}, \tag{8.66a}$$

with $m_{\tilde{b}_L} = m_{\tilde{t}_L}$ by $SU(2)$ symmetry, and

$$D(\tilde{b}_L) = M_Z^2 \cos 2\beta \left(-\frac{1}{2} + \frac{1}{3}\sin^2 \theta_W\right),$$

$$D(\tilde{b}_R) = M_Z^2 \cos 2\beta \left(-\frac{1}{3}\sin^2 \theta_W\right).$$

The corresponding eigenvalues are,

$$m_{\tilde{b}_{1,2}}^2 = \frac{1}{2}\left(m_{\tilde{b}_L}^2 + m_{\tilde{b}_R}^2\right) - \frac{1}{4}M_Z^2 \cos 2\beta + m_b^2$$

$$\mp \left\{\left[\frac{1}{2}(m_{\tilde{b}_L}^2 - m_{\tilde{b}_R}^2) - M_Z^2 \cos 2\beta \left(\frac{1}{4} - \frac{1}{3}x_W\right)\right]^2 + m_b^2(\mu \tan \beta - A_b)^2\right\}^{\frac{1}{2}}, \tag{8.66b}$$

and the bottom squark mixing angle (defined the same way as in Eq. (8.65c)) is

$$\tan \theta_b = \frac{m_{\tilde{b}_L}^2 + m_b^2 + M_Z^2 \cos 2\beta \left(-\frac{1}{2} + \frac{1}{3}x_W\right) - m_{\tilde{b}_1}^2}{m_b \left(-A_b + \mu \tan \beta\right)}. \tag{8.66c}$$

For tau sleptons we have,

$$\begin{pmatrix} m_{\tilde{\tau}_L}^2 + m_\tau^2 + D(\tilde{\tau}_L) & m_\tau(-A_\tau + \mu \tan \beta) \\ m_\tau(-A_\tau + \mu \tan \beta) & m_{\tilde{\tau}_R}^2 + m_\tau^2 + D(\tilde{\tau}_R) \end{pmatrix}, \tag{8.67a}$$

with

$$D(\tilde{\tau}_L) = M_Z^2 \cos 2\beta \left(-\frac{1}{2} + \sin^2 \theta_W\right),$$

$$D(\tilde{\tau}_R) = M_Z^2 \cos 2\beta \left(-\sin^2 \theta_W\right),$$

and

$$m_{\tilde{\tau}_{1,2}}^2 = \frac{1}{2}\left(m_{\tilde{\tau}_L}^2 + m_{\tilde{\tau}_R}^2\right) - \frac{1}{4}M_Z^2 \cos 2\beta + m_\tau^2$$

$$\mp \left\{ \left[\frac{1}{2}(m_{\tilde{\tau}_L}^2 - m_{\tilde{\tau}_R}^2) - M_Z^2 \cos 2\beta(\frac{1}{4} - x_W) \right]^2 + m_\tau^2(\mu \tan \beta - A_\tau)^2 \right\}^{\frac{1}{2}},$$

(8.67b)

and

$$\tan \theta_\tau = \frac{m_{\tilde{\tau}_L}^2 + m_\tau^2 + M_Z^2 \cos 2\beta \left(-\frac{1}{2} + x_W\right) - m_{\tilde{\tau}_1}^2}{m_\tau \left(-A_\tau + \mu \tan \beta\right)}.$$

(8.67c)

Since we have ignored neutrino masses, the MSSM only contains the scalar partner for the left-handed neutrino, one for each flavor. Also, because lepton flavor has been assumed to be conserved, the three sneutrinos cannot mix with one another, and hence, must be mass eigenstates. For the third generation, we thus have

$$m_{\tilde{\nu}_\tau}^2 = m_{L3}^2 + \frac{1}{2} M_Z^2 \cos 2\beta,$$

(8.68)

where the first term is the soft SUSY breaking mass for the third generation scalar lepton doublet, and the second term comes from the D-term contribution to the sneutrino mass. Since there are only superpartners of left-handed neutrinos in the MSSM, we will henceforth drop the subscript L on the sneutrinos.

The masses of the first and second generation squarks and sleptons can be obtained in exactly the same fashion. However, since first and second generation quark and lepton masses are small compared to the soft SUSY breaking masses, intra-generation mixing effects can be neglected so that $\tilde{f}_L$ and $\tilde{f}_R$ are essentially mass eigenstates. To a very good approximation, the masses of the first generation of sfermions are given by

$$m_{\tilde{u}_L}^2 = m_{Q_1}^2 + m_u^2 + M_Z^2 \cos 2\beta(\frac{1}{2} - \frac{2}{3}\sin^2 \theta_W)$$

(8.69a)

$$m_{\tilde{d}_L}^2 = m_{Q_1}^2 + m_d^2 + M_Z^2 \cos 2\beta(-\frac{1}{2} + \frac{1}{3}\sin^2 \theta_W)$$

(8.69b)

$$m_{\tilde{u}_R}^2 = m_{U_1}^2 + m_u^2 + M_Z^2 \cos 2\beta(\frac{2}{3}\sin^2 \theta_W)$$

(8.69c)

$$m_{\tilde{d}_R}^2 = m_{D_1}^2 + m_d^2 + M_Z^2 \cos 2\beta(-\frac{1}{3}\sin^2 \theta_W)$$

(8.69d)

$$m_{\tilde{e}_L}^2 = m_{L_1}^2 + m_e^2 + M_Z^2 \cos 2\beta(-\frac{1}{2} + \sin^2 \theta_W)$$

(8.69e)

$$m_{\tilde{\nu}_e}^2 = m_{L_1}^2 + M_Z^2 \cos 2\beta(\frac{1}{2})$$

(8.69f)

$$m_{\tilde{e}_R}^2 = m_{E_1}^2 + m_e^2 + M_Z^2 \cos 2\beta(-\sin^2 \theta_W),$$

(8.69g)

where the first terms on the right-hand side of these expressions are the soft SUSY breaking masses for the first generation of sfermions. There are analogous expressions for second generation masses. Notice that we are abusing notation here in that sometimes we use $m_{\tilde{\tau}_L}$ to denote the entire entry in the squark mass matrix (as implied by these equations), while at other times we use it to denote just the corresponding soft SUSY breaking mass. We trust that the meaning will be clear from the context.

We remind the reader that in deriving these MSSM mass spectra, we have ignored the possibility of **c**-terms. If such terms are present, like *a*-terms, they will contribute to intra-generation sfermion mixing, and possibly also to flavor physics.

Finally, we stress that (8.68) follows only from $SU(2)_L$ gauge symmetry, so that its analogue for the first two generations of sleptons and squarks (whose Yukawa couplings are negligible) gives a *model independent* relation between the physical masses of the up and down components of the slepton/squark doublets. Most importantly, it tells us that the mass gap between $\tilde{\ell}_L$ and the corresponding sneutrino ($\ell = e, \mu$), and likewise for the left-squarks, can never be too large. This is clearly relevant for collider searches for SUSY.

Exercise *The alert reader may wonder why the sfermion masses do not equal the corresponding fermion mass even if we take the "SUSY limit" in the sfermion mass squared matrix, and set the soft-masses and A-parameters to zero, and take* $\tan \beta = 1$ *so that the Higgs field D-terms vanish in the vacuum. The point is that within the MSSM, electroweak symmetry is unbroken unless we introduce soft SUSY breaking masses for the Higgs fields. Then, fermion and sfermion masses become equal as both vanish!*

There is, however, an interesting extension of the MSSM that leads to a SUSY limit in which electroweak symmetry is spontaneously broken. We need to introduce a SM "singlet Higgs" superfield $\hat{N}$, *and choose the superpotential as,*

$$\hat{f} = \hat{f}_{MSSM} + \lambda \hat{H}_u \hat{H}_d \hat{N} - K \hat{N} \tag{8.70a}$$

where the parameter $K > 0$ *has dimensions of mass squared, and appropriate group contractions are implied. Show that the scalar potential is given by,*

$$V = |\lambda h_d^0 N + \mu h_d^0|^2 + |\lambda h_u^0 N + \mu h_u^0|^2 + |\lambda h_u^0 h_d^0 - K|^2 + \cdots , \tag{8.70b}$$

where the ellipsis refers to terms involving charged Higgs boson or squark and slepton fields. Again assuming that these do not develop any VEV, show that this potential can have a minimum with $v_u = v_d \neq 0$ *and* $\langle N \rangle \neq 0$ *with* $\lambda \langle N \rangle + \mu = 0$. *Notice that because N condenses, there is effectively an additional contribution to* μ *equal to* $\lambda \langle N \rangle$. *In other words, the total "effective* μ *term" vanishes!*

Work out the top squark mass squared matrix for this model. Show that the off-diagonal terms vanish, while the diagonal terms are just m_t^2. In the sfermion sector, we thus have what looks like a SUSY limit of the MSSM but with non-vanishing masses for the fermions and sfermions. This model with the extra gauge singlet superfield is referred to as the Next to Minimal Supersymmetric Standard Model, or the NMSSM.

8.4 Interactions in the MSSM

In order to work out the phenomenological implications of the MSSM, we must first evaluate the interactions of the various superpartners, i.e. the mass eigenstates, with SM particles. This is done in two steps. First, we write down the interactions of the primitive fields of the MSSM (the fields with definite $SU(3)_C \times SU(2)_L \times U(1)_Y$ quantum numbers) using our master formula, and then transform these to the interactions of the mass eigenstates by performing the "rotations" (and, in the case of the Higgs sector, also a shift) of these fields discussed in the last section.

As in any gauge theory, before proceeding further we must fix a gauge. Since our attention will be mainly on tree level processes, we will write these in the unitarity gauge, where only physical fields are present. For many loop calculations, it is more convenient to work in the renormalizable R_ξ gauges, in which the propagator has better high energy behavior. Then, additional couplings involving Dewitt–Fadeev–Popov ghosts and unphysical Goldstone bosons must be included. We do not work these couplings out in this book.

In the following, we first evaluate the interactions in supersymmetric QCD. Next, we work out the interactions between matter fermions, sfermions, electroweak gauge bosons and the charginos and neutralinos. We then list the couplings of the MSSM Higgs bosons to other particles and sparticles. Finally, we list some "hybrid" interactions of matter sfermions.

8.4.1 QCD interactions in the MSSM

We begin by showing that we can recover the SM QCD Lagrangian written in Chapter 1 using our master formula. Clearly, the gluon field kinetic energy term $\mathcal{L} = -\frac{1}{4}F_{\mu\nu A}F_A^{\mu\nu}$ in the master formula has the usual form, and leads to the three and four gluon interactions listed in Eq. (1.7).

The kinetic energies and gauge couplings of quarks are contained in the terms,

$$\mathcal{L} \ni \frac{i}{2}\sum_i \bar{\psi}_i \, \displaystyle{\not}D \psi_i,$$

of the master formula, where the ψ_i are the fermion components of the quark superfields $\hat{q}$ and $\hat{Q}^c$. Just as for the electron field in Eq. (8.4), manipulation of the kinetic energy terms for ψ_q and ψ_{Q^c} leads to canonically normalized kinetic energy terms for the Dirac quark field q defined by,

$$q = P_L \psi_q + P_R \psi_{Q^c}.$$

To obtain the coupling between quarks and gluons, we must examine the interaction terms. Using (3.8c) and (3.8d) it is easy to see that

$$\bar{\psi}_q t_A^* \mathcal{G}_A P_R \psi_q = -\bar{\psi}_q t_A \mathcal{G}_A P_L \psi_q,$$

. so that

$$
\begin{aligned}
\mathcal{L}_{gq\bar{q}} &\ni -g_s \bar{\psi}_q t_A \mathcal{G}_A P_L \psi_q - g_s \bar{\psi}_{Q^c} t_A \mathcal{G}_A P_R \psi_{Q^c} \\
&= -g_s \bar{q} \gamma_\mu \frac{\lambda_A}{2} G_{A\mu} q,
\end{aligned}
\tag{8.71}
$$

which is just the interaction in Eq. (1.7).

The gauge invariant kinetic energy term for any flavor of left- or right-type squark field is,

$$
\begin{aligned}
\mathcal{L} &\ni (D^\mu \tilde{q})^\dagger (D_\mu \tilde{q}) \\
&= (\partial^\mu \tilde{q}^\dagger - i g_s \tilde{q}^\dagger t_A G_A^\mu)(\partial_\mu \tilde{q} + i g_s t_A G_{A\mu} \tilde{q}).
\end{aligned}
$$

The cross terms lead to

$$\mathcal{L}_{g\tilde{q}\tilde{q}} = -i g_s \left(\tilde{q}^\dagger \frac{\lambda_A}{2} \partial_\mu \tilde{q} - \partial_\mu \tilde{q}^\dagger \frac{\lambda_A}{2} \tilde{q} \right) G_A^\mu, \tag{8.72}$$

while the remaining interaction term yields,

$$\mathcal{L}_{gg\tilde{q}\tilde{q}} = g_s^2 \tilde{q}^\dagger \frac{\lambda_A}{2} \frac{\lambda_B}{2} \tilde{q} G_{A\mu} G_B^\mu, \tag{8.73}$$

where matrix multiplication is implied.

Exercise *We have obtained the interactions of gluons with $\tilde{q}_L$ and $\tilde{q}_R$. Show that the interactions of the squark mass eigenstates $\tilde{q}_1$ and $\tilde{q}_2$ with gluons have the same forms as in (8.72) and (8.73). This is just the familiar GIM (Glashow–Iliopoulos–Maiani) mechanism in a different setting.*

The gluino–quark–squark interaction comes from the Lagrangian term

$$\mathcal{L} \ni -\sqrt{2} \sum_{i,A} \mathcal{S}_i^\dagger g t_A \bar{\lambda}_A \frac{1 - \gamma_5}{2} \psi_i + \quad \text{h.c.,}$$

with $(\mathcal{S}_i, \psi_i) = (\tilde{q}_L, \psi_q)$ or $(\tilde{q}_R^\dagger, \psi_{Q^c})$. For the contribution from the superfield $\hat{Q}^c$, we write the term involving the right projector from the Hermitian conjugate part, and then use the Majorana bilinear identities to get the interaction,

$$\mathcal{L} \ni -\sqrt{2}\tilde{q}_{La}^\dagger \left(\frac{g\lambda_A}{2}\right)_{ab} \bar{\lambda}_A \frac{1-\gamma_5}{2}\psi_{qb} - \sqrt{2}\tilde{q}_{Ra}^\dagger \left(-\frac{g\lambda_A}{2}\right)_{ab} \bar{\lambda}_A \frac{1+\gamma_5}{2}\psi_{Q^c b},$$

where $\tilde{q}_R^\dagger$ is the field that *annihilates* the scalar partner of the weak singlet antiquark, or *creates* the scalar partner of the right-handed quark. To obtain this form, we must remember that the superfields $\hat{q}$ and $\hat{Q}^c$ belong to the **3** and **3*** representations, respectively, and write the generator t_A accordingly. We can allow for the possibly negative value of M_3 by replacing the gaugino λ_A by $(+i\gamma_5)^{\theta_{\tilde{g}}}\tilde{g}_A$ (rather than just $\tilde{g}_A$). Making the additional replacements of $P_L\psi_q = P_L q$ and $P_R\psi_{Q^c} = P_R q$ to write the interaction in terms of the Dirac quark field q leads to

$$\mathcal{L}_{\tilde{g}q\tilde{q}} = -\sqrt{2}g_s(-i)^{\theta_{\tilde{g}}}\tilde{q}_L^\dagger \bar{\tilde{g}}_A \frac{\lambda_A}{2}P_L q + \sqrt{2}g_s(i)^{\theta_{\tilde{g}}}\tilde{q}_R^\dagger \bar{\tilde{g}}_A \frac{\lambda_A}{2}P_R q + \quad \text{h.c.} \quad (8.74)$$

We can take into account intra-generation squark mixing by writing $\tilde{q}_L$ and $\tilde{q}_R$ in terms of the squark mass eigenstates $\tilde{q}_1$ and $\tilde{q}_2$ defined as in (8.65c). The quark–squark–gluino interaction then depends on the squark mixing angle, and we have

$$\mathcal{L}_{\tilde{g}q\tilde{q}_i} = -\sqrt{2}g_s\tilde{q}_1^\dagger \bar{\tilde{g}}_A \frac{\lambda_A}{2}\left[(-i)^{\theta_{\tilde{g}}}\cos\theta_q P_L + (i)^{\theta_{\tilde{g}}}\sin\theta_q P_R\right]q$$
$$-\sqrt{2}g_s\tilde{q}_2^\dagger \bar{\tilde{g}}_A \frac{\lambda_A}{2}\left[(-i)^{\theta_{\tilde{g}}}\sin\theta_q P_L - (i)^{\theta_{\tilde{g}}}\cos\theta_q P_R\right]q + \quad \text{h.c.}$$
$$(8.75)$$

Although we have written this for generic squarks, in practice, mixing angle effects are usually only important for the third generation.

We have a gluon–gluino–gluino interaction arising from the minimal coupling of the color octet gluino,

$$\mathcal{L} \ni \frac{i}{2}\bar{\lambda}_A \slashed{D}\lambda_A \ni -\frac{1}{2}g_s\bar{\tilde{g}}_A (t_B^{\text{adj}}\mathbb{G}_B)_{AC}\tilde{g}_C,$$

which leads to

$$\mathcal{L}_{g\tilde{g}\tilde{g}} = i\frac{g_s}{2}f_{ABC}\bar{\tilde{g}}_A\gamma_\mu\tilde{g}_B G_C^\mu. \quad (8.76)$$

Notice that this interaction is not altered by the transformation, $\tilde{g}_A \to (-i\gamma_5)^{\theta_{\tilde{g}}}\tilde{g}_A$.

Finally, supersymmetry necessarily implies the existence of four squark interactions. These arise from the D-terms on the third line of our master formula, and take the form,

$$\mathcal{L}_{4\tilde{q}} = -\frac{g_s^2}{8} \sum_A \left(\sum_i \tilde{q}_{Li}^\dagger \lambda_A \tilde{q}_{Li} - \sum_i \tilde{q}_{Ri}^\dagger \lambda_A \tilde{q}_{Ri} \right)^2, \tag{8.77}$$

where i denotes the flavor of the squark. Notice that the cross terms in the sum over flavors and types imply vertices such as $\tilde{u}_R^\dagger \tilde{u}_R \tilde{b}_L^\dagger \tilde{b}_L$, where the squark pairs could have different flavors and/or types. Moreover, for the same reason as in the last exercise, we see that writing this in terms of mass eigenstates ($\tilde{q}_1$ and $\tilde{q}_2$) does not lead to "cross terms" (such as $\tilde{q}_1^\dagger \tilde{q}_2$) in this coupling.

8.4.2 Electroweak interactions in the MSSM

Standard Model interactions

The triple and quartic vector boson gauge self-couplings arise from the squared field strength term in the master formula Eq. (6.44) and so are exactly as given by (1.19a) and (1.19b). Next, we turn to the SM electroweak interactions of quarks and leptons from the master formula. We will first evaluate the couplings of the up and down quarks to the gauge bosons $W^\pm$, Z^0 and γ. The starting point in the master formula is the term,

$$\mathcal{L} \ni \frac{i}{2} \sum_i \bar{\psi}_i \not{D} \psi_i$$

where $D_\mu = \partial_\mu + ig(t \cdot V_\mu)P_L - ig(t^* \cdot V_\mu)P_R$ and $i = \hat{Q}$, $\hat{U}^c$, and $\hat{D}^c$. We will leave it to the reader to verify that the second and third terms of the covariant derivative yield identical contributions to the Lagrangian. The $SU(2)_L$ and $U(1)_Y$ gauge boson interactions take the form,

$$\mathcal{L} \ni -\frac{g}{2} \left(\bar{\psi}_{u_L} \ \bar{\psi}_{d_L} \right) \begin{pmatrix} W_3 & W_1 - i\, W_2 \\ W_1 + i\, W_2 & -W_3 \end{pmatrix} \begin{pmatrix} \psi_{u_L} \\ \psi_{d_L} \end{pmatrix}$$

$$- \frac{1}{3}\frac{g'}{2} \bar{\psi}_u \not{B} P_L \psi_u - \frac{1}{3}\frac{g'}{2} \bar{\psi}_d \not{B} P_L \psi_d - \frac{4}{3}\frac{g'}{2} \bar{\psi}_{U^c} \not{B} P_R \psi_{U^c}$$

$$+ \frac{2}{3}\frac{g'}{2} \bar{\psi}_{D^c} \not{B} P_R \psi_{D^c}.$$

To write these in terms of the Dirac quark fields u and d, we substitute $P_L\psi_u = P_L u$, $P_L\psi_d = P_L d$, $P_R\psi_{U^c} = P_R u$, and $P_R\psi_{D^c} = P_R d$ and, finally, we eliminate the fields W_i and B in favor of the gauge boson mass eigenstates. The resulting Lagrangian is,

$$\mathcal{L}_{W\bar{u}d} = -\frac{g}{\sqrt{2}} \bar{u} \gamma^\mu P_L d\, W_\mu^+ + \quad \text{h.c.} \tag{8.78}$$

for the charged gauge bosons, and

$$\mathcal{L} = -e(+\frac{2}{3})\bar{u}\gamma_\mu u A^\mu + e\bar{u}\gamma_\mu \left[(-\frac{5}{12}t + \frac{1}{4}c) + (-\frac{1}{4}c - \frac{1}{4}t)\gamma_5\right] u Z^{0\mu} \quad (8.79)$$

for the electromagnetic and Z-boson interactions with u-quarks. Aside from inter-generational mixing between the quarks, these results are in accord with the SM interactions that we obtained in Chapter 1. The gauge interactions of other quarks and leptons can be obtained in the same fashion. These interactions have all been listed in Eq. (1.16a) and Eq. (1.16b), with coupling constants defined in Table 1.2.

Gauge boson couplings to matter scalars

The interactions of gauge bosons with sfermions originate in the gauge invariant kinetic terms,

$$\mathcal{L} \ni (D_\mu S_i)^\dagger (D^\mu S_i),$$

for the scalars. Notice that in addition to the coupling of a vector boson to a sfermion pair, these terms also include a two-gauge boson–two-sfermion interaction.

Three-point couplings: $W^\pm$ bosons do not couple to the $SU(2)$ singlet sfermions $\tilde{f}_R$. The coupling of $W^\pm$ to doublet sfermions of the first generation takes the form,

$$\mathcal{L} \ni -\frac{ig}{\sqrt{2}} \left(\tilde{u}_L^\dagger \partial_\mu \tilde{d}_L - \tilde{d}_L \partial_\mu \tilde{u}_L^\dagger\right) W^{+\mu} - \frac{ig}{\sqrt{2}} \left(\tilde{v}_e^\dagger \partial_\mu \tilde{e}_L - \tilde{e}_L \partial_\mu \tilde{v}_e^\dagger\right) W^{+\mu} + \quad \text{h.c.}$$

(8.80)

Except for intrageneration sfermion mixing, other sfermion generations couple to W in exactly the same way. For third generation squarks and sleptons, mixing effects can be important. These couplings can be readily obtained from Eq. (8.80) via the replacement,

$$\tilde{f}_L = \cos\theta_f \tilde{f}_1 + \sin\theta_f \tilde{f}_2,$$

where $f = t, b$ or τ. In addition to these three-point couplings the kinetic energy term for sfermions also includes a two-gauge boson–two-sfermion interaction. We will list these couplings shortly.

The interaction of a photon with a sfermion pair is given by,

$$\mathcal{L} \ni -ieq_f \left(\tilde{f}_i^\dagger \partial_\mu \tilde{f}_i - \tilde{f}_i \partial_\mu \tilde{f}_i^\dagger\right) A^\mu, \quad (8.81)$$

where $\tilde{f}$ is any squark or slepton, q_f is the electric charge of the sfermion (which is, of course, the same as the charge of the corresponding fermion), and $i = $ L or R. Notice that the photon couples just to left- or to right-sfermion pairs, i.e. there is no $\tilde{f}_L \tilde{f}_R \gamma$ interaction. Intra-generational (or for that matter, inter-generational) mixing does not alter the form of (8.81).

Exercise *Show that the conservation of electric current for the coupling,*

$$\mathcal{L} = J_\mu A^\mu,$$

implies that the photon cannot couple two sfermions with different masses. This explains why there is no $\tilde{f}_1 \tilde{f}_2 \gamma$ interaction.

The interactions of sfermions with a Z^0 boson are given by,

$$\mathcal{L} \ni ie\left[(\alpha_f - \beta_f)\,\tilde{f}_L^\dagger \partial_\mu \tilde{f}_L + (\alpha_f + \beta_f)\,\tilde{f}_R^\dagger \partial_\mu \tilde{f}_R\right] Z^\mu + \quad \text{h.c.,} \qquad (8.82)$$

where again $\tilde{f}_i$ is any squark or slepton of type i and α_f and β_f, which also determine the couplings of Z^0 to matter *fermions*, are given in Table 1.2. Like the photon, Z^0 interactions do not couple left- and right-type sfermions to each other. This should not be surprising since gauge bosons do not couple left-handed and right-handed *fermions* to each other. Supersymmetry then implies that they cannot couple the respective superpartners to one another either.

Exercise *In the presence of intra-generational mixing show that the couplings of Z^0 to sfermions are modified to,*

$$\mathcal{L} \ni ie\left[(\alpha_f - \beta_f \cos 2\theta_f)\,\tilde{f}_1^\dagger \partial_\mu \tilde{f}_1 + (\alpha_f + \beta_f \cos 2\theta_f)\,\tilde{f}_2^\dagger \partial_\mu \tilde{f}_2 \right.$$
$$\left. - \beta_f \sin 2\theta_f \left(\tilde{f}_1^\dagger \partial_\mu \tilde{f}_2 + \tilde{f}_2^\dagger \partial_\mu \tilde{f}_1\right)\right] Z^\mu + \quad \text{h.c.} \qquad (8.83)$$

Notice that unlike the photon, Z^0 does couple sfermions of different masses together.

Four-point Couplings: We now work out the two-vector boson–two-sfermion couplings that are also contained in the gauge invariant kinetic energy terms. The covariant derivative for squark fields can be written as,

$$D_\mu \tilde{u}_L = \partial_\mu \tilde{u}_L + i\left(eq_u A_\mu - e(\alpha_u - \beta_u)Z_\mu + g_s \frac{\lambda_A}{2} G_{A\mu}\right)\tilde{u}_L + \frac{ig}{\sqrt{2}} W_\mu^+ \tilde{d}_L,$$

$$D_\mu \tilde{d}_L = \partial_\mu \tilde{d}_L + i\left(eq_d A_\mu - e(\alpha_d - \beta_d)Z_\mu + g_s \frac{\lambda_A}{2} G_{A\mu}\right)\tilde{d}_L + \frac{ig}{\sqrt{2}} W_\mu^+ \tilde{u}_L,$$

$$D_\mu \tilde{u}_R = \partial_\mu \tilde{u}_R + i\left(eq_u A_\mu - e(\alpha_u + \beta_u)Z_\mu + g_s \frac{\lambda_A}{2} G_{A\mu}\right)\tilde{u}_R,$$

$$D_\mu \tilde{d}_R = \partial_\mu \tilde{d}_R + i\left(eq_d A_\mu - e(\alpha_d + \beta_d)Z_\mu + g_s \frac{\lambda_A}{2} G_{A\mu}\right)\tilde{d}_R,$$

where q_f, α_f, and β_f are defined in Table 1.2. Here, $\tilde{u}$ and $\tilde{d}$ denote any up- or down-type squark. Except for obvious replacements and the absence of the gluon field, $\tilde{\ell}_L$, $\tilde{\ell}_R$, and sneutrino covariant derivatives are identical to those for $\tilde{d}_L$, $\tilde{d}_R$ and $\tilde{u}_L$, respectively.[6]

The quartic interactions that we mentioned are now easy to work out. The interactions with photons, Z^0, and gluons can be written as

$$\mathcal{L}_{VV\tilde{f}\tilde{f}} = \tilde{f}_{L/R}^\dagger \left(eq_f A_\mu - e(\alpha_f \mp \beta_f)Z_\mu + \xi_f g_s \frac{\lambda_A}{2} G_{A\mu} \right)$$

$$\times \left(eq_f A^\mu - e(\alpha_f \mp \beta_f)Z^\mu + \xi_f g_s \frac{\lambda_B}{2} G_B^\mu \right) \tilde{f}_{L/R}, \quad (8.84a)$$

where the minus sign in the terms involving Z^0 is for $\tilde{f}_L$ and the plus sign for $\tilde{f}_R$ and $\xi_f = 1$ for squarks and $\xi_f = 0$ for charged sleptons and sneutrinos. Notice that in addition to just electroweak interactions, squarks also have QCD–electroweak hybrid interactions. Quartic interactions involving $W^\pm$ bosons can be written as

$$\mathcal{L}_{WW\tilde{f}\tilde{f}} = \frac{1}{2} g^2 \tilde{f}_L^\dagger \tilde{f}_L W_\mu^\pm W^{\mp\mu} \quad (8.84b)$$

where $\tilde{f}_L = \tilde{u}_L$, $\tilde{d}_L$, $\tilde{\ell}_L$ or $\tilde{\nu}$. Finally, the interactions involving both neutral and charged gauge bosons are,

$$\mathcal{L}_{VW\tilde{u}\tilde{d}} = \frac{g}{\sqrt{2}} \tilde{u}_L^\dagger \left(e(q_u + q_d)A_\mu - e(\alpha_u + \alpha_d)Z_\mu + g_s \frac{\lambda_A}{2} G_{A\mu} \right) W^{+\mu} \tilde{d}_L$$

$$+ \frac{g}{\sqrt{2}} \tilde{\nu}_L^\dagger \left(eq_\ell A_\mu - e(\alpha_\ell + \alpha_\nu)Z_\mu \right) W^{+\mu} \tilde{\ell}_L + \quad \text{h.c.} \quad (8.84c)$$

Left-type squark pairs have a contact interaction with the W-boson gluon pair.

In writing Eq. (8.84a)–(8.84c) we have ignored intragenerational mixing of sfermions. This can be easily included by writing $\tilde{f}_L$ and $\tilde{f}_R$ in terms of the mass eigenstates. Clearly, the four-point interactions involving just gluons and photons will couple just $\tilde{f}_1 \tilde{f}_1$ and $\tilde{f}_2 \tilde{f}_2$ pairs, while the others will couple $\tilde{f}_1 \tilde{f}_2$ pairs as well.

Chargino and neutralino couplings to matter

Because these are dimension four interactions, these interactions are unaffected by the soft SUSY breaking terms. There are just two sources of these couplings. First, the gaugino components of charginos and neutralinos couple to fermions and sfermions via the term

$$\mathcal{L} \ni -\sqrt{2} \sum_{i,A} g S_i^\dagger t_A \bar{\lambda}_A \frac{1 - \gamma_5}{2} \psi_i + \quad \text{h.c.},$$

---

[6] These covariant derivatives give an alternative way to write down the coupling of any gauge boson to a sfermion pair.

in Eq. (6.44). These couplings are completely determined by gauge interactions and various sparticle mixing matrices. The higgsino components of the charginos and neutralinos also contribute to these couplings via superpotential Yukawa interactions contained in

$$\mathcal{L} \ni -\frac{1}{2}\bar{\psi}_i \left(\frac{\partial^2 \hat{f}}{\partial \hat{S}_i \partial \hat{S}_j}\right)_{\hat{S}=S} P_L \psi_j + \quad \text{h.c.}$$

For most purposes, these couplings are only important for the third generation.

We begin by evaluating the neutralino–quark–squark couplings arising from gaugino interactions. The relevant terms are contained in

$$\mathcal{L} \ni -\frac{1}{\sqrt{2}} \left\{ \left(\tilde{u}_L^\dagger \ \tilde{d}_L^\dagger\right) \begin{pmatrix} g\bar{\lambda}_3 + \frac{g'}{3}\bar{\lambda}_0 & g(\bar{\lambda}_1 - i\bar{\lambda}_2) \\ g(\bar{\lambda}_1 + i\bar{\lambda}_2) & -g\bar{\lambda}_3 + \frac{g'}{3}\bar{\lambda}_0 \end{pmatrix} P_L \begin{pmatrix} \psi_u \\ \psi_d \end{pmatrix} \right.$$
$$\left. + \tilde{u}_R^\dagger g' \left(-\frac{4}{3}\right) \bar{\lambda}_0 P_R \psi_{U^c} + \tilde{d}_R^\dagger g' \left(+\frac{2}{3}\right) \bar{\lambda}_0 P_R \psi_{D^c} \right\} + \quad \text{h.c.,} \quad (8.85)$$

where, for convenience, we have written the Hermitian conjugate of the terms involving the $SU(2)$ singlet antiquarks. The interactions of charged sleptons and sneutrinos can be obtained by replacing $\tilde{u}_L \to \tilde{v}$, $\tilde{d}_L \to \tilde{\ell}$, $\tilde{d}_R \to \tilde{\ell}_R$, dropping the term involving $\tilde{u}_R$, and replacing 2/3, the weak hypercharge of the $SU(2)$ singlet $\bar{d}$, by 2, the hypercharge of the antilepton. We also need to replace the quark hypercharges that multiply $\bar{\lambda}_0$ by corresponding lepton or neutrino hypercharges, and also appropriately replace the quark Majorana spinors by those of the lepton/neutrino. We proceed, however, to extract the quark–squark–neutral gaugino interactions, and eliminate the Majorana fields in favor of the Dirac quark fields using $P_L\psi_u = P_Lu$, $P_L\psi_d = P_Ld$, $P_R\psi_{U^c} = P_Ru$ and $P_R\psi_{D^c} = P_Rd$. Finally, using Eq. (8.50) and (8.51), we substitute $\lambda_3 = \sum_i v_3^{(i)}(i\gamma_5)^{\theta_i}\widetilde{Z}_i$, and $\lambda_0 = \sum_i v_4^{(i)}(i\gamma_5)^{\theta_i}\widetilde{Z}_i$ to write,

$$\mathcal{L}_{\bar{f}f\widetilde{Z}_i} = \sum_{f=u,d,\ell,\nu} \left[iA_{\widetilde{Z}_i}^f \, \tilde{f}_L^\dagger \widetilde{Z}_i P_L f + iB_{\widetilde{Z}_i}^f \, \tilde{f}_R^\dagger \widetilde{Z}_i P_R f + \quad \text{h.c.}\right], \quad (8.86)$$

where

$$A_{\widetilde{Z}_i}^u = \frac{(-i)^{\theta_i-1}}{\sqrt{2}} \left[g v_3^{(i)} + \frac{g'}{3} v_4^{(i)}\right], \quad (8.87a)$$

$$A_{\widetilde{Z}_i}^d = \frac{(-i)^{\theta_i-1}}{\sqrt{2}} \left[-g v_3^{(i)} + \frac{g'}{3} v_4^{(i)}\right], \quad (8.87b)$$

$$B_{\widetilde{Z}_i}^u = \frac{4}{3\sqrt{2}} g'(i)^{\theta_i-1} v_4^{(i)} \quad \text{and} \quad (8.87c)$$

$$B_{\widetilde{Z}_i}^d = -\frac{2}{3\sqrt{2}} g'(i)^{\theta_i-1} v_4^{(i)}. \quad (8.87d)$$

The couplings of leptons and sleptons to neutralinos have the same form as in (8.86) above, but with couplings given by

$$A^{\ell}_{\widetilde{Z}_i} = -\frac{(-i)^{\theta_i-1}}{\sqrt{2}}\left[gv_3^{(i)} + g'v_4^{(i)}\right],$$
(8.88a)

$$A^{\nu}_{\widetilde{Z}_i} = \frac{(-i)^{\theta_i-1}}{\sqrt{2}}\left[gv_3^{(i)} - g'v_4^{(i)}\right],$$
(8.88b)

$$B^{\ell}_{\widetilde{Z}_i} = -(i)^{\theta_i-1}\sqrt{2}g'v_4^{(i)} \quad\text{and}$$
(8.88c)

$$B^{\nu}_{\widetilde{Z}_i} = 0.$$
(8.88d)

Next, we turn to the contribution to fermion–sfermion–neutralino interactions that arise from the superpotential terms,

$$\mathcal{L} \ni -\frac{1}{2}\bar{\psi}_i\left(\frac{\partial^2\hat{f}}{\partial\hat{\mathcal{S}}_i\partial\hat{\mathcal{S}}_j}\right)_{\mathcal{S}=\mathcal{S}}P_L\psi_j + \quad\text{h.c.},$$

with

$$\hat{f} \ni f_u\hat{u}\hat{h}_u^0\hat{U}^c + f_d\hat{d}\hat{h}_d^0\hat{D}^c + f_e\hat{e}\hat{h}_d^0\hat{E}^c + \cdots,$$

where the ellipsis denotes Yukawa couplings for the second and third generations. For up- (down-)type (s)fermions, we have contributions when one of ψ_i, ψ_j is $\psi_{h_u^0}$ ($\psi_{h_d^0}$), with the other one being ψ_f or ψ_{F^c}. It is straightforward to check that these contributions can be written as,

$$\mathcal{L} \ni -f_f v_a^{(i)}(-i)^{\theta_i}\,\tilde{f}_R^\dagger\overline{\widetilde{Z}}_i P_L f - f_f v_a^{(i)}(i)^{\theta_i}\,\tilde{f}_L^\dagger\overline{\widetilde{Z}}_i P_R f,$$

with $a = 1$ for up-type (s)fermions, and $a = 2$ for down-type ones. Combining this with the contributions (8.86) from the gaugino components of neutralinos, we have,

$$\mathcal{L}_{\widetilde{Z}_i f\tilde{f}} \ni \tilde{f}_L^\dagger\overline{\widetilde{Z}}_i\left(iA^f_{\widetilde{Z}_i}P_L - (i)^{\theta_i}f_f v_a^{(i)}P_R\right)f$$
$$+ \tilde{f}_R^\dagger\overline{\widetilde{Z}}_i\left(iB^f_{\widetilde{Z}_i}P_R - (-i)^{\theta_i}f_f v_a^{(i)}P_L\right)f + \quad\text{h.c.}$$
(8.89)

Finally, eliminating $\tilde{f}_L$ and $\tilde{f}_R$ in favor of the sfermion mass eigenstates $\tilde{f}_1$ and $\tilde{f}_2$, we arrive at

$$\mathcal{L}_{\widetilde{Z}_i f\tilde{f}} = \tilde{f}_j^\dagger\overline{\widetilde{Z}}_i\left[\alpha^{\tilde{f}_j}_{\widetilde{Z}_i}P_L + \beta^{\tilde{f}_j}_{\widetilde{Z}_i}P_R\right]f + \quad\text{h.c.},$$
(8.90)

with

$$\alpha^{\tilde{f}1}_{\tilde{Z}_i} = iA^f_{\tilde{Z}_i}\cos\theta_f + (-i)^{\theta_i}f_f v^{(i)}_a \sin\theta_f, \qquad (8.91a)$$

$$\beta^{\tilde{f}1}_{\tilde{Z}_i} = -iB^f_{\tilde{Z}_i}\sin\theta_f - (i)^{\theta_i}f_f v^{(i)}_a \cos\theta_f, \qquad (8.91b)$$

$$\alpha^{\tilde{f}2}_{\tilde{Z}_i} = iA^f_{\tilde{Z}_i}\sin\theta_f - (-i)^{\theta_i}f_f v^{(i)}_a \cos\theta_f, \qquad (8.91c)$$

$$\beta^{\tilde{f}2}_{\tilde{Z}_i} = iB^f_{\tilde{Z}_i}\cos\theta_f - (i)^{\theta_i}f_f v^{(i)}_a \sin\theta_f. \qquad (8.91d)$$

Again, $a = 1$ if f is an up-type quark, and $a = 2$ if it is a down-type quark or a charged lepton. Since we do not have a right-handed neutrino superfield, the neutrino–sneutrino–neutralino coupling is given by (8.86).

The interactions of charginos with either squarks and quarks or sleptons and leptons can be calculated in a similar fashion. For chargino–quark–squark interactions, using (8.85) we find that

$$\mathcal{L} \ni -g\tilde{u}^{\dagger}_{\mathrm{L}}\bar{\lambda}P_{\mathrm{L}}d \ - g\tilde{d}^{\dagger}_{\mathrm{L}}\overline{\lambda^c}P_{\mathrm{L}}u + \quad \text{h.c.}$$

Here, λ^c is the charge conjugate of the charged Dirac gaugino λ. Eliminating λ and λ^c in favor of the chargino mass eigenstates, we find

$$\mathcal{L} \ni iA^d_{\tilde{W}_i}\tilde{u}^{\dagger}_{\mathrm{L}}\overline{\tilde{W}_i}P_{\mathrm{L}}d + iA^u_{\tilde{W}_i}\tilde{d}^{\dagger}_{\mathrm{L}}\overline{\tilde{W}^c_i}P_{\mathrm{L}}u + \quad \text{h.c.,} \qquad (8.92)$$

where

$$A^d_{\tilde{W}_1} = i(-1)^{\theta_{\tilde{w}_1}}g\sin\gamma_{\mathrm{R}}, \qquad (8.93a)$$

$$A^d_{\tilde{W}_2} = i(-1)^{\theta_{\tilde{w}_2}}\theta_y g\cos\gamma_{\mathrm{R}}, \qquad (8.93b)$$

$$A^u_{\tilde{W}_1} = ig\sin\gamma_{\mathrm{L}}, \qquad (8.93c)$$

$$A^u_{\tilde{W}_2} = i\theta_x g\cos\gamma_{\mathrm{L}}. \qquad (8.93d)$$

These couplings, which originate in the gauge interactions, are generation independent; i.e. u and d ($\tilde{u}_{\mathrm{L}}$ and $\tilde{d}_{\mathrm{L}}$) respectively refer to any up- and down-type quark (squark). Moreover, the coupling of charginos to leptons and sleptons is identical, with the identification $u \to \nu$ and $d \to \ell$.

There are also superpotential contributions to these chargino interactions that can be worked out in the same way as for neutralinos. We will leave it to the reader to work out that including these leads to the couplings,

$$\mathcal{L}_{\tilde{u}d\tilde{W}_i} = \tilde{u}^{\dagger}_1\overline{\tilde{W}_i}\left[(iA^d_{\tilde{W}_i}\cos\theta_u - B_{\tilde{W}_i}\sin\theta_u)P_{\mathrm{L}} + B'_{\tilde{W}_i}\cos\theta_u P_{\mathrm{R}}\right]d$$

$$+ \tilde{u}^{\dagger}_2\overline{\tilde{W}_i}\left[(iA^d_{\tilde{W}_i}\sin\theta_u + B_{\tilde{W}_i}\cos\theta_u)P_{\mathrm{L}} + B'_{\tilde{W}_i}\sin\theta_u P_{\mathrm{R}}\right]d + \quad \text{h.c.,}$$

$$\qquad (8.94)$$

where

$$B_{\widetilde{W}_1} = -(-1)^{\theta_{\widetilde{W}_1}} f_u \cos \gamma_R, \tag{8.95a}$$

$$B_{\widetilde{W}_2} = (-1)^{\theta_{\widetilde{W}_2}} \theta_y f_u \sin \gamma_R, \tag{8.95b}$$

$$B'_{\widetilde{W}_1} = -f_d \cos \gamma_L, \tag{8.95c}$$

$$B'_{\widetilde{W}_2} = f_d \theta_x \sin \gamma_L. \tag{8.95d}$$

For chargino–sbottom–top interactions, we have

$$\mathcal{L}_{\tilde{d}u\widetilde{W}_i} = \tilde{d}_1^\dagger \overline{\widetilde{W}_i^c} \Big[(iA_{\widetilde{W}_i}^u \cos \theta_d - B'_{\widetilde{W}_i} \sin \theta_d) P_L + B_{\widetilde{W}_i} \cos \theta_d P_R \Big] u$$
$$+ \tilde{d}_2^\dagger \overline{\widetilde{W}_i^c} \Big[(iA_{\widetilde{W}_i}^u \sin \theta_d + B'_{\widetilde{W}_i} \cos \theta_d) P_L + B_{\widetilde{W}_i} \sin \theta_d P_R \Big] u + \quad \text{h.c.}$$

$$\tag{8.96}$$

Finally, the chargino–slepton–neutrino and chargino–sneutrino–lepton interactions can be obtained by replacing $u \to \nu$ and $d \to \ell$ everywhere *including* in the definitions of the couplings in (8.95a)–(8.95d). We then have,

$$\mathcal{L}_{\tilde{\tau}\nu_\tau \widetilde{W}_i} = \tilde{\tau}_1^\dagger \overline{\widetilde{W}_i^c} \Big[(iA_{\widetilde{W}_i}^\nu \cos \theta_\tau - B''_{\widetilde{W}_i} \sin \theta_\tau) P_L \nu_\tau \Big]$$
$$+ \tilde{\tau}_2^\dagger \overline{\widetilde{W}_i^c} \Big[(iA_{\widetilde{W}_i}^\nu \sin \theta_\tau + B''_{\widetilde{W}_i} \cos \theta_\tau) P_L \Big] \nu_\tau$$
$$+ \tilde{\nu}_\tau^\dagger \overline{\widetilde{W}}_i \Big[iA_{\widetilde{W}_i}^\tau P_L + B''_{\widetilde{W}_i} P_R \Big] \tau + \quad \text{h.c.,} \tag{8.97}$$

with

$$A_{\widetilde{W}_i}^\nu = A_{\widetilde{W}_i}^u, \tag{8.98a}$$

$$A_{\widetilde{W}_i}^\tau = A_{\widetilde{W}_i}^d, \tag{8.98b}$$

$$B''_{\widetilde{W}_1} = -f_\tau \cos \gamma_L, \tag{8.98c}$$

$$B''_{\widetilde{W}_2} = f_\tau \theta_x \sin \gamma_L. \tag{8.98d}$$

Gauge boson interactions with charginos and neutralinos

These interactions arise from two sources, both of which are supersymmetric. First, there is the contribution from gaugino kinetic energy terms,

$$\mathcal{L} \ni \frac{i}{2} \bar{\lambda} \, \slashed{D} \lambda,$$

in the master formula, with the covariant derivative involving gauge group generators in the adjoint representation: $(\slashed{D}\lambda)_A = \slashed{\partial}\lambda_A + ig(t_B^{\text{adj}} \slashed{W}_B)_{AC}\lambda_C$, with

$[t_B^{adj}]_{AC} = -i\epsilon_{ACB}$. The $SU(2)_L$ gauginos thus have a coupling of the form

$$\mathcal{L} \ni \frac{-ig}{2}\left(-\bar{\lambda}_1 \not{W}_3\lambda_2 + \bar{\lambda}_1 \not{W}_2\lambda_3 + \bar{\lambda}_2 \not{W}_3\lambda_1 - \bar{\lambda}_2 \not{W}_1\lambda_3\right.$$
$$\left. -\bar{\lambda}_3 \not{W}_2\lambda_1 + \bar{\lambda}_3 \not{W}_1\lambda_2\right),$$
$$= g\left[\bar{\lambda} \not{W}_3\lambda - (\bar{\lambda} \not{W}^-\lambda_3 + \text{h.c.})\right],$$

while there is no coupling to the hypercharge gaugino. To obtain the last step, we have used $\bar{\lambda}_3\gamma_\mu\lambda^c = -\bar{\lambda}\gamma_\mu\lambda_3$, as the reader can readily verify.

There are also higgsino contributions

$$\mathcal{L} \ni \frac{i}{2}\left[\left(\bar{\psi}_{h_u^+}\ \bar{\psi}_{h_u^0}\right)\frac{i}{2}\begin{bmatrix} g\not{W}_3 + g'\not{B} & g\not{W}_1 - ig\not{W}_2 \\ g\not{W}_1 + ig\not{W}_2 & -g\not{W}_3 + g'\not{B} \end{bmatrix}P_L\begin{pmatrix}\psi_{h_u^+} \\ \psi_{h_u^0}\end{pmatrix}\right.$$
$$\left. +\left(\bar{\psi}_{h_d^-}\ \bar{\psi}_{h_d^0}\right)\frac{i}{2}\begin{bmatrix} -g\not{W}_3 - g'\not{B} & -g\not{W}_1 - ig\not{W}_2 \\ -g\not{W}_1 + ig\not{W}_2 & g\not{W}_3 - g'\not{B} \end{bmatrix}P_L\begin{pmatrix}\psi_{h_d^-} \\ \psi_{h_d^0}\end{pmatrix}\right] + \text{h.c.}$$

Exercise *Verify that we can write the gaugino and higgsino contributions as:*

$$\mathcal{L} \ni g\left\{\bar{\lambda} \not{W}_3\lambda - (\bar{\lambda} \not{W}^-\lambda_3 + \text{h.c.})\right\}$$
$$+ \frac{1}{2}\bar{\tilde{\chi}}\left(g\not{W}_3 + g'\not{B}\right)\tilde{\chi}$$
$$+ \frac{1}{4}\sqrt{g^2 + g'^2}\left(\bar{\psi}_{h_u^0}\gamma_\mu\gamma_5\psi_{h_u^0} - \bar{\psi}_{h_d^0}\gamma_\mu\gamma_5\psi_{h_d^0}\right)Z^\mu$$
$$- \frac{g}{\sqrt{2}}(\bar{\tilde{\chi}} \not{W}^- P_R\psi_{h_u^0} - \bar{\tilde{\chi}} \not{W}^- P_L\psi_{h_d^0} + \text{h.c.}).$$

Here, the first line clearly comes from the couplings of the gauginos to gauge bosons, while the rest comes from the gauge interactions of higgsinos.

We can now write these in terms of the chargino and neutralino mass eigenstates to obtain the following couplings to the photon and Z^0 boson:

$$\mathcal{L} = e\left(\overline{\widetilde{W}}_1\gamma_\mu\widetilde{W}_1 + \overline{\widetilde{W}}_2\gamma_\mu\widetilde{W}_2\right)A^\mu$$
$$- e\cot\theta_W\overline{\widetilde{W}}_1\gamma_\mu(x_c - y_c\gamma_5)\widetilde{W}_1 Z^\mu - e\cot\theta_W\overline{\widetilde{W}}_2\gamma_\mu(x_s - y_s\gamma_5)\widetilde{W}_2 Z^\mu$$
$$+ (-1)^{(\theta_{\widetilde{W}_1} + \theta_{\widetilde{W}_2})}\frac{e}{2}(\cot\theta_W + \tan\theta_W)$$
$$\times \left[\overline{\widetilde{W}}_1\gamma_\mu(x\gamma_5 - y)(\gamma_5)^{(\theta_{\widetilde{W}_1} + \theta_{\widetilde{W}_2})}\widetilde{W}_2 Z^\mu + \text{h.c.}\right],$$

$$(8.99)$$

where

$$x_c = 1 - \frac{1}{4}\sec^2\theta_W(\cos^2\gamma_L + \cos^2\gamma_R), \tag{8.100a}$$

$$y_c = \frac{1}{4}\sec^2\theta_W(\cos^2\gamma_R - \cos^2\gamma_L), \tag{8.100b}$$

$$x_s = 1 - \frac{1}{4}\sec^2\theta_W(\sin^2\gamma_L + \sin^2\gamma_R), \tag{8.100c}$$

$$y_s = \frac{1}{4}\sec^2\theta_W(\sin^2\gamma_R - \sin^2\gamma_L), \tag{8.100d}$$

$$x = \frac{1}{2}(\theta_x\sin\gamma_L\cos\gamma_L - \theta_y\sin\gamma_R\cos\gamma_R), \quad\text{and} \tag{8.100e}$$

$$y = \frac{1}{2}(\theta_x\sin\gamma_L\cos\gamma_L + \theta_y\sin\gamma_R\cos\gamma_R). \tag{8.100f}$$

Notice that the photon does not couple to the $\widetilde{W}_1^+\widetilde{W}_2^-$ pair, as may be expected from the conservation of electromagnetic current.

The couplings of Z^0 with the neutralinos arise only via their higgsino components, and are given by,

$$\mathcal{L} = \frac{1}{4}\sqrt{g^2 + g'^2}\sum_{i,j}(-\mathrm{i})^{\theta_i}(\mathrm{i})^{\theta_j}(v_1^{(i)}v_1^{(j)} - v_2^{(i)}v_2^{(j)})\,\overline{\widetilde{Z}}_i\gamma_\mu(\gamma_5)^{\theta_i+\theta_j+1}\widetilde{Z}_j Z^\mu$$

$$\equiv \sum_{ij}W_{ij}\overline{\widetilde{Z}}_i\gamma_\mu(\gamma_5)^{\theta_i+\theta_j+1}\widetilde{Z}_j Z^\mu. \tag{8.101}$$

In models where $|\mu| \gg (\ll)|M_{1,2}|$, the neutralinos $\widetilde{Z}_1$ and $\widetilde{Z}_2$ ($\widetilde{Z}_3$ and $\widetilde{Z}_4$) are mainly gaugino-like so that their couplings to Z^0 are strongly suppressed by mixing angles. The couplings of neutralino pairs to gauge bosons are, therefore, very sensitive to model parameters. This is not the case for charginos. Their couplings to the photon are fixed by their electric charge. Moreover, chargino pairs couple to Z^0 via both their gaugino as well as their higgsino components, so that their couplings to vector bosons are much more robust.

Exercise *If* $\tan\beta = 1$ *show that the higgsino* $\frac{1}{\sqrt{2}}(\psi_{h_u^0} + \psi_{h_d^0})$ *has mass* $|\mu|$ *but that* Z^0 *does not couple to a pair of these higgsinos.*

Finally for charged vector bosons, substituting in terms of the mass eigenstates, we obtain,

$$\mathcal{L} = -g(-\mathrm{i})^{\theta_j}\sum_{i,j}\overline{\widetilde{W}}_i\left(X_i^j + Y_i^j\gamma_5\right)\gamma_\mu\widetilde{Z}_j W^\mu + \quad\text{h.c.}, \tag{8.102}$$

with

$$X_1^j = \frac{1}{2}\left[(-1)^{\theta_{\tilde{w}_1}+\theta_j}\left(\frac{\cos\gamma_R}{\sqrt{2}}v_1^{(j)} + \sin\gamma_R v_3^{(j)}\right)\right.$$
$$\left. -\frac{\cos\gamma_L}{\sqrt{2}}v_2^{(j)} + \sin\gamma_L v_3^{(j)}\right], \tag{8.103a}$$

$$X_2^j = \frac{1}{2}\left[(-1)^{\theta_{\tilde{w}_2}+\theta_j}\theta_y\left(\frac{-\sin\gamma_R}{\sqrt{2}}v_1^{(j)} + \cos\gamma_R v_3^{(j)}\right)\right.$$
$$\left. +\theta_x\left(\frac{\sin\gamma_L}{\sqrt{2}}v_2^{(j)} + \cos\gamma_L v_3^{(j)}\right)\right]. \tag{8.103b}$$

The $Y_{1,2}^j$ can be obtained from the $X_{1,2}^j$ by changing the sign of just the first term inside the square brackets. We see that W bosons couple to the chargino–neutralino system via both gaugino and higgsino components. In this sense, $W\tilde{W}_i\tilde{Z}_j$ couplings should, like the couplings of Z^0 to charginos, also be quite robust. Only if $|M_1| \ll |M_2|$ and $|\mu|$ (in which case the neutralino is dominantly a hypercharge gaugino) is this coupling dynamically suppressed.

8.4.3 Interactions of MSSM Higgs bosons

Higgs boson couplings to SM fermions

The interactions of Higgs bosons with SM fermions arise directly from the terms,

$$\mathcal{L} \ni -\frac{1}{2}\sum_{i,j}\bar{\psi}_i \left.\frac{\partial^2\hat{f}}{\partial\hat{\mathcal{S}}_i\partial\hat{\mathcal{S}}_j}\right|_{\hat{\mathcal{S}}=\mathcal{S}} P_L\psi_j + \quad \text{h.c.},$$

in our master formula. We have already examined a portion of these terms when we discussed masses for the SM fermions. Our present discussion proceeds along the same lines. The superpotential contains

$$\hat{f} \ni f_u(\hat{u}\hat{h}_u^0 - \hat{d}\hat{h}_u^+)\hat{U}^c + f_d(\hat{u}\hat{h}_d^- + \hat{d}\hat{h}_d^0)\hat{D}^c + f_e(\hat{\nu}_\tau\hat{h}_d^- + \hat{e}\hat{h}_d^0)\hat{E}^c + \cdots$$

We can easily work out the coupling of Dirac fermions to the scalar components in $\hat{h}_u$ to obtain,

$$\mathcal{L} \ni -f_u\bar{u}P_L u h_u^0 - f_u\bar{u}P_R u h_u^{0\dagger}$$

We can now eliminate h_u^0 in favor of the Higgs mass eigenstates using (8.35) and (8.40a). Recalling that $f_u = gm_u/\sqrt{2}M_W \sin\beta$ we find the required Lagrangian density,

$$\mathcal{L} \ni -\frac{gm_u}{2M_W\sin\beta}[\cos\alpha\,\bar{u}u h - \sin\alpha\,\bar{u}u H - i\cos\beta\,\bar{u}\gamma_5 u A]. \tag{8.104}$$

A similar calculation for the down-type quark and charged lepton Yukawa interactions yields,

$$\mathcal{L} \ni -\frac{g m_d}{2 M_W \cos\beta} \left[\sin\alpha\, \bar{d}dh + \cos\alpha\, \bar{d}dH - i\sin\beta\, \bar{d}\gamma_5 d\, A \right]$$
$$- \frac{g m_e}{2 M_W \cos\beta} \left[\sin\alpha\, \bar{e}eh + \cos\alpha\, \bar{e}eH - i\sin\beta\, \bar{e}\gamma_5 e A \right]. \tag{8.105}$$

The interactions with charged Higgs bosons can be similarly obtained by eliminating h_u^+ and h_d^- using (8.30):

$$\mathcal{L} \ni \frac{g}{2\sqrt{2} M_W} H^+ \left[(m_u \cot\beta + m_d \tan\beta)\bar{u}d + (m_d \tan\beta - m_u \cot\beta)\bar{u}\gamma_5 d \right.$$
$$\left. + m_e \tan\beta\, \bar{\nu}_e (1+\gamma_5)e \right] + \quad \text{h.c.} \tag{8.106}$$

Higgs boson couplings to vector bosons

As in any Yang–Mills theory, the coupling of vector bosons to Higgs boson pairs is fixed by the minimal coupling prescription; i.e. these arise from cross terms in the scalar field kinetic energy terms

$$\mathcal{L} \ni (D_\mu \mathcal{S}_i)^\dagger (D^\mu \mathcal{S}_i),$$

where $\mathcal{S}_i = H_u$ and H_d. Expanding these terms and substituting for the physical vector boson and Higgs fields yields the expected photon coupling to the charged Higgs boson pair,

$$\mathcal{L} \ni ie \left(H^+ \partial_\mu H^- - H^- \partial_\mu H^+ \right) A^\mu. \tag{8.107}$$

The Z^0 boson couples to both charged as well as neutral Higgs fields, with couplings given by,

$$\mathcal{L} \ni \frac{i}{2}(g' \sin\theta_W - g\cos\theta_W) \left(H^+ \partial_\mu H^- - H^- \partial_\mu H^+ \right) Z^{0\mu}, \tag{8.108}$$

and

$$\mathcal{L} \ni \frac{1}{2}(g'\sin\theta_W + g\cos\theta_W) \left[\cos(\alpha+\beta) \left(h\partial_\mu A - A\partial_\mu h \right) \right.$$
$$\left. - \sin(\alpha+\beta) \left(H\partial_\mu A - A\partial_\mu H \right) \right] Z^{0\mu}. \tag{8.109}$$

Notice that Z^0 only couples the pseudoscalar boson to a scalar boson. Couplings of Z^0 to hh, hH, and HH pairs are forbidden by the assumed CP invariance of the Higgs boson sector.

The couplings of W bosons to pairs of Higgs bosons are given by,

$$\mathcal{L} \ni i\frac{g}{2}\left[\cos(\alpha+\beta)\left(h\partial_\mu H^- - H^-\partial_\mu h\right) - \sin(\alpha+\beta)\left(H\partial_\mu H^- - H^-\partial_\mu H\right)\right.$$
$$\left. + i\left(A\partial_\mu H^- - H^-\partial_\mu A\right)\right]W^{+\mu} + \quad \text{h.c.} \tag{8.110}$$

The gauge kinetic term for the Higgs fields also contains two-vector boson–two-Higgs boson couplings. These are given by,

$$\mathcal{L} \ni H^+ H^- \left[e^2 A^\mu A_\mu + \frac{1}{4}(g'\sin\theta_W - g\cos\theta_W)^2 Z^{0\mu}Z^0_\mu\right.$$
$$\left. + e(g'\sin\theta_W - g\cos\theta_W)A^\mu Z^0_\mu + \frac{g^2}{2}W^{+\mu}W^-_\mu\right], \tag{8.111a}$$

$$\mathcal{L} \ni \left(\frac{g^2}{4}W^{+\mu}W^-_\mu + \frac{1}{8}(g\cos\theta_W + g'\sin\theta_W)^2 Z^{0\mu}Z^0_\mu\right)\left[h^2 + H^2 + A^2\right], \tag{8.111b}$$

and

$$\mathcal{L} \ni \frac{1}{2}eg\left(A^\mu + \tan\theta_W Z^{0\mu}\right)W^-_\mu H^+$$
$$\times \left[\cos(\alpha+\beta)h - \sin(\alpha+\beta)H - iA\right] + \quad \text{h.c.} \tag{8.111c}$$

Finally, a vector boson–vector boson–Higgs boson coupling can also arise from the four-point interactions in the case when one of the neutral Higgs fields is replaced by its vacuum expectation value. Instead of starting over, we can get these couplings from the four-point couplings that we have just obtained in (8.111b) and (8.111c), and simply set one of the neutral fields to their VEV using Eq. (8.40a):

$$\langle h \rangle = \sqrt{2}\left(\cos\alpha\, v_u + \sin\alpha\, v_d\right),$$
$$\langle H \rangle = \sqrt{2}\left(\cos\alpha\, v_d - \sin\alpha\, v_u\right),$$
$$\langle A \rangle = 0.$$

The resulting interaction is,

$$\mathcal{L} \ni gM_W\left(W^{+\mu}W^-_\mu + \frac{Z^{0\mu}Z^0_\mu}{2\cos^2\theta_W}\right)\left[\sin(\alpha+\beta)\, h + \cos(\alpha+\beta)\, H\right]. \tag{8.112}$$

Notice that there is no $Z^0 W^- H^+$ coupling and, by electromagnetic gauge invariance, also no $\gamma W^- H^+$ coupling.

Higgs boson self-couplings

We have already remarked at the end of Section 8.2 that in the MSSM, quartic interactions of Higgs fields arise only from D-terms, and so are completely determined by gauge couplings. Since we have already worked out the complete potential in (8.25), it is straightforward to write the quartic couplings in terms of mass eigenstates. We find,

$$
\begin{aligned}
\mathcal{L} \ni -\frac{1}{8} \Big\{ & 2g^2 H^+ H^- \left[\cos^2(\beta - \alpha)h^2 + \sin^2(\beta - \alpha)H^2 \right. \\
& \left. + \sin 2(\beta - \alpha)hH + \cos^2 2\beta A^2\right] \\
& + (g^2 + g'^2)\cos^2 2\beta (H^+ H^-)^2 + \frac{1}{4}(g^2 + g'^2) \\
& \times \left[\cos 2\alpha \, (h^2 - H^2) - 2\sin 2\alpha \, hH + \cos 2\beta \, A^2\right]^2 \\
& - (g^2 - g'^2)\cos 2\beta \, H^+ H^- \left[\cos 2\alpha(h^2 - H^2) - 2\sin 2\alpha hH + \cos 2\beta A^2\right] \Big\}.
\end{aligned}
$$

$$(8.113)$$

We see that the Higgs quartic scalar self-couplings are all fixed by gauge interactions. This is the origin of the tree-level bounds on m_h in (8.39a) and (8.39b), respectively. That these bounds are special to the MSSM is exemplified by the following exercise.

Exercise *Show that if the Higgs sector of the MSSM is extended by the inclusion of an extra $SU(3)_C \times SU(2)_L \times U(1)_Y$ singlet (as in the exercise at the end of Section 8.3), the quartic self interactions of Higgs bosons are no longer determined by just the gauge couplings. Convince yourself that the tree-level bounds on m_h are not valid in this case.*

The D-terms also result in trilinear couplings amongst the Higgs fields. As before, we can obtain these by setting one of the neutral Higgs fields to their VEV. The result is,

$$
\begin{aligned}
\mathcal{L} \ni -\frac{1}{8} \Big\{ & H^+ H^- [8g M_W (\sin(\alpha + \beta)h + \cos(\alpha + \beta)H) \\
& + \frac{4g M_Z \cos 2\beta}{\cos \theta_W} (\sin(\beta - \alpha) h - \cos(\beta - \alpha) H)] \\
& + \frac{2g M_Z}{\cos \theta_W} [\sin(\beta - \alpha) h - \cos(\beta - \alpha) H] \\
& \times \left[\cos 2\alpha \, h^2 - \cos 2\alpha \, H^2 - 2\sin 2\alpha \, hH + \cos 2\beta \, A^2\right] \Big\}.
\end{aligned}
$$

$$(8.114)$$

Notice that although these are all dimension 3 operators, there are no explicit soft SUSY breaking contributions to these interactions. This is because there is no gauge-invariant combination of three Higgs field doublets.

Higgs boson couplings to charginos and neutralinos

Supersymmetry dictates that Higgs bosons must interact with charginos and neutralinos. Since trilinear Higgs boson terms in the superpotential are forbidden by gauge invariance, these interactions can arise only from the couplings of Higgs bosons and higgsinos to $SU(2) \times U(1)$ gauginos. Letting $\mathcal{S}_i = H_u$ and H_d in the terms

$$\mathcal{L} \ni -\sqrt{2} \sum_{i,A} \mathcal{S}_i^\dagger g t_A \bar{\lambda}_A \frac{1-\gamma_5}{2} \psi_i + \quad \text{h.c.}$$

in the master formula, and eliminating the original fields in favor of the mass eigenstate fields leads to the required interactions. Since we have already done several similar calculations, we will simply present the final results.

The couplings of the light Higgs scalar to charginos and neutralinos are given by

$$\mathcal{L} = g\sqrt{2} S_1^h \overline{\widetilde{W}}_1 \widetilde{W}_1 h + g\sqrt{2} S_2^h \overline{\widetilde{W}}_2 \widetilde{W}_2 h + \left[\frac{g}{\sqrt{2}} \overline{\widetilde{W}}_1 (S^h + P^h \gamma_5) \widetilde{W}_2 h + \quad \text{h.c.} \right]$$
$$+ \sum_{i,j} X_{ij}^h \overline{\widetilde{Z}}_i (-i\gamma_5)^{\theta_i + \theta_j} \widetilde{Z}_j h, \tag{8.115}$$

where

$$S_1^h = \frac{1}{2}(-1)^{\theta_{\widetilde{W}_1}} [\sin\alpha \sin\gamma_R \cos\gamma_L + \cos\alpha \sin\gamma_L \cos\gamma_R], \tag{8.116a}$$

$$S_2^h = \frac{1}{2}(-1)^{\theta_{\widetilde{W}_2}+1} \theta_x \theta_y [\sin\alpha \cos\gamma_R \sin\gamma_L + \cos\alpha \cos\gamma_L \sin\gamma_R], \tag{8.116b}$$

$$S^h = \frac{1}{2} \big[-(-1)^{\theta_{\widetilde{W}_1}} \theta_x \sin\gamma_R \sin\gamma_L \sin\alpha + (-1)^{\theta_{\widetilde{W}_1}} \theta_x \cos\gamma_L \cos\gamma_R \cos\alpha$$
$$- (-1)^{\theta_{\widetilde{W}_2}} \theta_y \sin\gamma_L \sin\gamma_R \cos\alpha + (-1)^{\theta_{\widetilde{W}_2}} \theta_y \cos\gamma_L \cos\gamma_R \sin\alpha \big],$$
$$\tag{8.116c}$$

and P^h is the same as S^h except that the signs of the first two terms are reversed. Finally,

$$X_{ij}^h = -\frac{1}{2}(-1)^{\theta_i+\theta_j} \left(v_2^{(i)} \sin\alpha - v_1^{(i)} \cos\alpha \right) \left(g v_3^{(j)} - g' v_4^{(j)} \right). \tag{8.117}$$

The couplings of the heavy scalar H can be obtained from those of h by replacing $\cos \alpha \rightarrow -\sin \alpha$ and $\sin \alpha \rightarrow \cos \alpha$.

The corresponding couplings of the pseudoscalar A are given by

$$
\mathcal{L} \ni ig\sqrt{2}S_1^A \overline{\widetilde{W}}_1 \gamma_5 \widetilde{W}_1 A + ig\sqrt{2}S_2^A \overline{\widetilde{W}}_2 \gamma_5 \widetilde{W}_2 A
$$
$$
+ \left[\frac{-ig}{\sqrt{2}} \overline{\widetilde{W}}_1 (S^A + P^A \gamma_5) \widetilde{W}_2 A + \quad \text{h.c.} \right]
$$
$$
+ \sum_{i,j} X_{ij}^A \overline{\widetilde{Z}}_i (-i\gamma_5)^{\theta_i + \theta_j + 1} \widetilde{Z}_j A, \tag{8.118}
$$

where

$$
S_1^A = \frac{1}{2}(-1)^{\theta_{\widetilde{W}_1}} [\sin \gamma_R \cos \gamma_L \sin \beta + \sin \gamma_L \cos \gamma_R \cos \beta], \tag{8.119a}
$$
$$
S_2^A = -\frac{1}{2}(-1)^{\theta_{\widetilde{W}_2}} \theta_x \theta_y [\cos \gamma_R \sin \gamma_L \sin \beta + \cos \gamma_L \sin \gamma_R \cos \beta], \tag{8.119b}
$$
$$
S^A = \frac{1}{2} \left[-(-1)^{\theta_{\widetilde{W}_1}} \theta_x \sin \gamma_R \sin \gamma_L \sin \beta + (-1)^{\theta_{\widetilde{W}_1}} \theta_x \cos \gamma_L \cos \gamma_R \cos \beta \right.
$$
$$
\left. + (-1)^{\theta_{\widetilde{W}_2}} \theta_y \sin \gamma_L \sin \gamma_R \cos \beta - (-1)^{\theta_{\widetilde{W}_2}} \theta_y \cos \gamma_L \cos \gamma_R \sin \beta \right], \tag{8.119c}
$$

and P^A is obtained by reversing the sign of the first two terms of the expression for S^A. The coupling of A to neutralinos is,

$$
X_{ij}^A = \frac{1}{2}(-1)^{\theta_i + \theta_j} \left(v_2^{(i)} \sin \beta - v_1^{(i)} \cos \beta \right) \left(g v_3^{(j)} - g' v_4^{(j)} \right). \tag{8.120}
$$

Note that h and H couple to the scalar combination of $\widetilde{W}_i \widetilde{W}_i$ or $\widetilde{Z}_i \widetilde{Z}_i$ while A couples to the pseudoscalar combination. It is for this reason that we refer to h and H as scalars, and to A as a pseudoscalar.

Finally, the interactions of the charged Higgs bosons are given by,

$$
\mathcal{L} = \sum_k (i)^{\theta_k} \left[\cos \beta A_1^{(k)} \theta_y (-1)^{\theta_{\widetilde{W}_2}} \overline{\widetilde{Z}}_k P_R \widetilde{W}_2 + \cos \beta A_2^{(k)} (-1)^{\theta_{\widetilde{W}_1}} \overline{\widetilde{Z}}_k P_R \widetilde{W}_1 \right.
$$
$$
\left. - \sin \beta A_3^{(k)} \theta_x (-1)^{\theta_k} \overline{\widetilde{Z}}_k P_L \widetilde{W}_2 - \sin \beta A_4^{(k)} (-1)^{\theta_k} \overline{\widetilde{Z}}_k P_L \widetilde{W}_1 \right] H^+ + \quad \text{h.c.}
$$
$$
\tag{8.121}
$$

with

$$A_1^{(k)} = -\frac{1}{\sqrt{2}} \left(g v_3^{(k)} + g' v_4^{(k)} \right) \sin \gamma_R - g v_1^{(k)} \cos \gamma_R, \qquad (8.122a)$$

$$A_2^{(k)} = \frac{1}{\sqrt{2}} \left(g v_3^{(k)} + g' v_4^{(k)} \right) \cos \gamma_R - g v_1^{(k)} \sin \gamma_R, \qquad (8.122b)$$

$$A_3^{(k)} = -\frac{1}{\sqrt{2}} \left(g v_3^{(k)} + g' v_4^{(k)} \right) \sin \gamma_L + g v_2^{(k)} \cos \gamma_L, \qquad (8.122c)$$

$$A_4^{(k)} = \frac{1}{\sqrt{2}} \left(g v_3^{(k)} + g' v_4^{(k)} \right) \cos \gamma_L + g v_2^{(k)} \sin \gamma_L. \qquad (8.122d)$$

Higgs boson couplings to squarks and sleptons

In addition to the couplings that we have listed, there are several four scalar interactions in the MSSM. Since these are dimension four operators, there are no explicit soft-SUSY breaking contributions to these.

The D-term contributions from the term

$$\mathcal{L} \ni -\frac{1}{2} \sum_A \left| \sum_i \mathcal{S}_i^\dagger g_\alpha t_{\alpha A} \mathcal{S}_i \right|^2$$

in the master formula can be written as,

$$
\begin{aligned}
\mathcal{L} \ni -\frac{1}{2} \Bigg\{ &\frac{g^2}{4} \Big[(h_u^{+\dagger} h_u^0 + h_u^{0\dagger} h_u^+) - (h_d^{-\dagger} h_d^0 + h_d^{0\dagger} h_d^-) \\
&+ (\tilde{v}_e^\dagger \tilde{e}_L + \tilde{e}_L^\dagger \tilde{v}_e) + (\tilde{u}_L^\dagger \tilde{d}_L + \tilde{d}_L^\dagger \tilde{u}_L) + \cdots \Big]^2 \\
-&\frac{g^2}{4} \Big[(h_u^{+\dagger} h_u^0 - h_u^{0\dagger} h_u^+) + (h_d^{-\dagger} h_d^0 - h_d^{0\dagger} h_d^-) \\
&+ (\tilde{v}_e^\dagger \tilde{e}_L - \tilde{e}_L^\dagger \tilde{v}_e) + (\tilde{u}_L^\dagger \tilde{d}_L - \tilde{d}_L^\dagger \tilde{u}_L) + \cdots \Big]^2 \\
+&\frac{g^2}{4} \Big[(h_u^{+\dagger} h_u^+ - h_u^{0\dagger} h_u^0) - (h_d^{-\dagger} h_d^- - h_d^{0\dagger} h_d^0) \\
&+ (\tilde{v}_e^\dagger \tilde{v}_e - \tilde{e}_L^\dagger \tilde{e}_L) + (\tilde{u}_L^\dagger \tilde{u}_L - \tilde{d}_L^\dagger \tilde{d}_L) + \cdots \Big]^2 \\
+& g'^2 \Big[\frac{H_u^\dagger H_u - H_d^\dagger H_d - \tilde{L}_e^\dagger \tilde{L}_e + \frac{1}{3} \tilde{Q}_1^\dagger \tilde{Q}_1 + \cdots}{2} \\
&+ \tilde{e}_R^\dagger \tilde{e}_R - \frac{2}{3} \tilde{u}_R^\dagger \tilde{u}_R + \frac{1}{3} \tilde{d}_R^\dagger \tilde{d}_R + \cdots \Big]^2 \Bigg\} \\
-&\frac{g_s^2}{8} \sum_A \left(\sum_i \tilde{q}_{Li}^\dagger \lambda_A \tilde{q}_{Li} - \sum_i \tilde{q}_{Ri}^\dagger \lambda_A \tilde{q}_{Ri} \right)^2. \qquad (8.123)
\end{aligned}
$$

The ellipses denote sfermion terms from the second and third generations. In the first term of the last square parenthesis, an $SU(2)$ matrix product is implied; i.e. $h_u^\dagger h_u \equiv h_u^{+\dagger} h_u^+ + h_u^{0\dagger} h_u^0$, etc. and $\tilde{L}_e$ and $\tilde{Q}_1$ denote the first generation slepton and squark doublets, respectively. The last term is just the squark D-terms from supersymmetric QCD discussed previously.

We have already seen some of these terms before. For instance, terms involving the squares of bilinears in just the Higgs fields lead to the quartic self-interactions in (8.113). We see that there are several other quartic self-interactions that originate in these D-terms:

1. The cross terms between the Higgs and scalar matter bilinears lead to four-point vertices involving a pair of Higgs bosons and a pair of scalars (squarks or sleptons). These Higgs boson couplings are fixed by gauge interactions and, hence, *are generation-independent*. In the case where both the Higgs bosons are neutral, a quick examination shows that there is no $hA\tilde{q}\tilde{q}$ or $HA\tilde{q}\tilde{q}$ coupling or, for that matter, the corresponding slepton couplings.

2. The squares of the sfermion bilinears lead to several new quartic interactions amongst squarks and sleptons. These include four squark interactions, four slepton interactions, and also two squark two slepton contact interactions. All these couplings are again fixed by gauge interactions. Notice that the sfermions participating in these interactions may be of the same or different type (L or R) and of the same or different flavor. Note also that although some of the four squark couplings, for instance, the four $\tilde{u}_R$ couplings from the hypercharge D-term, superficially resemble that from the QCD interaction, the color structure of these interactions is quite different.

Trilinear superpotential terms also yield four scalar interactions determined by the Yukawa couplings. Clearly there are many such terms – even for just one generation there are $7 + 4 = 11$ terms corresponding to taking the derivative of the superpotential with respect to any of the seven chiral matter fields ($\hat{u}$, $\hat{d}$, $\hat{e}$, $\hat{v}$, $\hat{U}^c$, $\hat{D}^c$, and $\hat{E}^c$) or the four Higgs fields. We will leave it to the interested reader to enumerate all the terms which are straightforward to list, but only illustrate the form of the result with just one term arising from the derivative with respect to $\hat{h}_d^-$. This yields the interactions,

$$\mathcal{L} = -\left| f_d \tilde{u}_L \tilde{d}_R^\dagger + f_e \tilde{v} \tilde{e}_R^\dagger + \cdots \right|^2, \tag{8.124}$$

where the ellipsis denotes contributions from the second and third generations. The following features of the four-point interactions from D-terms and F-terms might be worth noting.

- The superpotential F-terms do not contribute to Higgs potential.
- Both D- and F-terms yield four scalar interactions as well as two sfermion two Higgs boson couplings. However, unlike the generation-independent D-terms, the superpotential couplings are important only for the third generation. In particular, four scalar couplings from the superpotential that involve sfermions of different generations are small.
- The bilinears that enter the D-terms always involve matter sfermions of the same type (L or R). In contrast, the corresponding F-term bilinears always couple L and R sfermions together. The form of the couplings in (8.123) and in (8.124) is, therefore, quite different.

It is now straightforward to write the interactions in (8.123) and (8.124) in terms of the mass eigenstate fields. However, since we will not have any occasion to use these couplings in the remainder of this book, we have chosen not to list the lengthy and cumbersome formulae that result upon doing so.

The quartic interactions of Higgs and sfermion fields also lead to $H\tilde{f}\tilde{f}$ couplings if one of the Higgs fields acquires a VEV. In addition, soft SUSY breaking scalar trilinear couplings (A-terms) are an additional source of these interactions. The process of obtaining the couplings of the physical Higgs fields to the left- and right-squark fields is lengthy but straightforward. We present here the results for a single generation of squarks. Of course, $\tilde{q}_L$ and $\tilde{q}_R$ need to be replaced by the corresponding mass eigenstates to obtain the coupling to physical particles.

The couplings of squarks to charged Higgs bosons are given by,

$$
\begin{aligned}
\mathcal{L}_{H^+\tilde{q}\tilde{q}} \ni g &\left[-\frac{M_W}{\sqrt{2}} \sin 2\beta + \frac{m_d^2 \tan\beta + m_u^2 \cot\beta}{\sqrt{2}M_W} \right] \left(\tilde{u}_L^\dagger \tilde{d}_L H^+ + \tilde{d}_L^\dagger \tilde{u}_L H^- \right) \\
&+ \left[\frac{g m_u m_d (\cot\beta + \tan\beta)}{\sqrt{2}M_W} \right] \left(\tilde{u}_R^\dagger \tilde{d}_R H^+ + \tilde{d}_R^\dagger \tilde{u}_R H^- \right) \\
&+ \left[\frac{-g m_d}{\sqrt{2}M_W} (A_d \tan\beta + \mu) \right] \left(\tilde{u}_L^\dagger \tilde{d}_R H^+ + \tilde{d}_R^\dagger \tilde{u}_L H^- \right) \\
&+ \left[\frac{-g m_u}{\sqrt{2}M_W} (A_u \cot\beta + \mu) \right] \left(\tilde{u}_R^\dagger \tilde{d}_L H^+ + \tilde{d}_L^\dagger \tilde{u}_R H^- \right).
\end{aligned}
\tag{8.125a}
$$

Here, and in the following, we have eliminated the Yukawa couplings in favor of the corresponding *quark* mass.

The couplings to the lighter scalar h are,

$$
\begin{aligned}
\mathcal{L}_{h\tilde{q}\tilde{q}} \ni g &\left[M_W(T_{3\hat{u}_L} - \tfrac{1}{2}Y_{\hat{u}_L} \tan^2\theta_W)\sin(\beta - \alpha) - \frac{m_u^2 \cos\alpha}{M_W \sin\beta} \right] \tilde{u}_L^\dagger \tilde{u}_L h \\
&+ g \left[M_W(T_{3\hat{d}_L} - \tfrac{1}{2}Y_{\hat{d}_L} \tan^2\theta_W)\sin(\beta - \alpha) - \frac{m_d^2 \sin\alpha}{M_W \cos\beta} \right] \tilde{d}_L^\dagger \tilde{d}_L h
\end{aligned}
$$

$$+ g \left[M_W (T_{3\hat{U}^c} - \frac{1}{2} Y_{\hat{U}^c} \tan^2 \theta_W) \sin(\beta - \alpha) - \frac{m_u^2 \cos \alpha}{M_W \sin \beta} \right] \tilde{u}_R^\dagger \tilde{u}_R h$$

$$+ g \left[M_W (T_{3\hat{D}^c} - \frac{1}{2} Y_{\hat{D}^c} \tan^2 \theta_W) \sin(\beta - \alpha) - \frac{m_d^2 \sin \alpha}{M_W \cos \beta} \right] \tilde{d}_R^\dagger \tilde{d}_R h$$

$$+ \frac{g m_d}{2 M_W \cos \beta} (-\mu \cos \alpha + A_d \sin \alpha) \left(\tilde{d}_L^\dagger \tilde{d}_R + \tilde{d}_R^\dagger \tilde{d}_L \right) h$$

$$+ \frac{g m_u}{2 M_W \sin \beta} (-\mu \sin \alpha + A_u \cos \alpha) \left(\tilde{u}_L^\dagger \tilde{u}_R + \tilde{u}_R^\dagger \tilde{u}_L \right) h, \qquad (8.125\text{b})$$

while the corresponding couplings to H are given by,

$$\mathcal{L}_{H\tilde{q}\tilde{q}} \ni g \left[-M_W (T_{3\hat{u}_L} - \frac{1}{2} Y_{\hat{u}_L} \tan^2 \theta_W) \cos(\beta - \alpha) + \frac{m_u^2 \sin \alpha}{M_W \sin \beta} \right] \tilde{u}_L^\dagger \tilde{u}_L H$$

$$+ g \left[-M_W (T_{3\hat{d}_L} - \frac{1}{2} Y_{\hat{d}_L} \tan^2 \theta_W) \cos(\beta - \alpha) - \frac{m_d^2 \cos \alpha}{M_W \cos \beta} \right] \tilde{d}_L^\dagger \tilde{d}_L H$$

$$+ g \left[-M_W (T_{3\hat{U}^c} - \frac{1}{2} Y_{\hat{U}^c} \tan^2 \theta_W) \cos(\beta - \alpha) + \frac{m_u^2 \sin \alpha}{M_W \sin \beta} \right] \tilde{u}_R^\dagger \tilde{u}_R H$$

$$+ g \left[-M_W (T_{3\hat{D}^c} - \frac{1}{2} Y_{\hat{D}^c} \tan^2 \theta_W) \cos(\beta - \alpha) - \frac{m_d^2 \cos \alpha}{M_W \cos \beta} \right] \tilde{d}_R^\dagger \tilde{d}_R H$$

$$+ \frac{g m_d}{2 M_W \cos \beta} (\mu \sin \alpha + A_d \cos \alpha) \left(\tilde{d}_L^\dagger \tilde{d}_R + \tilde{d}_R^\dagger \tilde{d}_L \right) H$$

$$+ \frac{g m_u}{2 M_W \sin \beta} (-\mu \cos \alpha - A_u \sin \alpha) \left(\tilde{u}_L^\dagger \tilde{u}_R + \tilde{u}_R^\dagger \tilde{u}_L \right) H. \qquad (8.125\text{c})$$

Note that the isospin and hypercharge values that appear in (8.125b) and (8.125c) refer to the corresponding quantities for the MSSM fields in Table 8.1.

Finally, the couplings to the pseudoscalar neutral Higgs field are given by,

$$\mathcal{L}_{A\tilde{q}\tilde{q}} \ni i \frac{g m_d}{2 M_W} (\mu + A_d \tan \beta) \left(\tilde{d}_R^\dagger \tilde{d}_L - \tilde{d}_L^\dagger \tilde{d}_R \right) A$$

$$+ i \frac{g m_u}{2 M_W} (\mu + A_u \cot \beta) \left(\tilde{u}_R^\dagger \tilde{u}_L - \tilde{u}_L^\dagger \tilde{u}_R \right) A. \qquad (8.125\text{d})$$

As already noted, especially for the third generation squarks and sleptons, mixing effects must be included by substituting for the appropriate mass eigenstates.

The corresponding couplings to sleptons can be obtained by substituting $m_d \rightarrow m_e, m_u \rightarrow 0, A_d \rightarrow A_e, A_u \rightarrow 0, \tilde{u}_L \rightarrow \tilde{\nu}_L, \tilde{d}_L \rightarrow \tilde{\ell}_L, \tilde{u}_R \rightarrow 0$, and $\tilde{d}_R \rightarrow \tilde{\ell}_R$, and by making appropriate weak isospin and hypercharge assignments.

8.5 Radiative corrections

Up to now, we have focussed our attention on the tree-level masses and couplings of MSSM particles. Since MSSM couplings are all assumed to be in the perturbative regime, this should be a good approximation to the true masses and couplings. There are, however, some situations where radiative corrections are very important. The best known of these is in the Higgs boson sector where the tree-level bound (8.39b), if applicable, would already exclude the model! Clearly, such a correction cannot be neglected. In this section we briefly discuss the radiative corrections that cannot be neglected in phenomenological analyses of SUSY. This discussion is not meant to be complete, but is included as a caution, and to provide the reader a flavor of the issues involved. For a comprehensive discussion, we refer the reader to the original literature.

8.5.1 Higgs boson masses

We have already mentioned that radiative corrections to Higgs boson masses can be large, and are especially important for the lightest Higgs scalar h. The biggest corrections arise from the top (quark and squark) Yukawa coupling to Higgs field H_u. For large values of $\tan\beta$ b-Yukawa, and to a lesser degree τ-Yukawa, contributions are also significant. Smaller corrections also arise from gauge interactions of the Higgs bosons.

   These radiative corrections can be included diagrammatically, by calculating the relevant Higgs boson self-energy graphs, and by identifying the location of the pole in the propagator. An alternative procedure involves analyzing the one-loop corrected effective potential. The form of the one-loop correction to the scalar potential can be written as

$$\Delta V = \sum_i \frac{(-1)^{2s_i}}{64\pi^2} Tr\left((\mathcal{M}_i\mathcal{M}_i^\dagger)^2\left[\log\frac{\mathcal{M}_i\mathcal{M}_i^\dagger}{Q^2} - \frac{3}{2}\right]\right), \qquad (8.126)$$

where the sum over i runs over all fields that couple to Higgs fields, $\mathcal{M}_i^2$ is the *Higgs field dependent* mass squared matrix (defined as the second derivative of the tree-level Lagrangian) of each of these fields, and the trace is over the internal as well as any spin indices. The function ΔV depends on the Higgs fields through $\mathcal{M}$, and must be added to the tree-level potential. It is this corrected effective potential that must be used to obtain the vacuum state as well as the masses and mixings of the physical particles in the Higgs sector. Here, we illustrate how to obtain the dominant corrections arising from top Yukawa couplings. To keep things simple, we also ignore intra-generational mixing.

Exercise *Show that the neutral Higgs field dependent mass matrix for stops in the*
$(\tilde{t}_L, \tilde{t}_R)$ *basis is given by,*

$$\begin{pmatrix} m_{\tilde{t}_L}^2 + f_t^2 |h_u^0|^2 & 0 \\ 0 & m_{\tilde{t}_R}^2 + f_t^2 |h_u^0|^2 \end{pmatrix},$$

*while the corresponding top quark mass is given by $f_t h_u^{0*}$ where, for simplicity, we
have ignored any $\tilde{t}_L$–$\tilde{t}_R$ mixing. Use these to show that the one-loop correction to
the effective Higgs potential due to top Yukawa couplings is given by,*

$$\Delta V \simeq \frac{3}{32\pi^2} \left[(m_{\tilde{t}_L}^2 + f_t^2 |h_u^0|^2)^2 \log(m_{\tilde{t}_L}^2 + f_t^2 |h_u^0|^2) \right.$$
$$+ (m_{\tilde{t}_R}^2 + f_t^2 |h_u^0|^2)^2 \log(m_{\tilde{t}_R}^2 + f_t^2 |h_u^0|^2)$$
$$\left. - 2 f_t^4 |h_u^0|^4 \log(f_t^2 |h_u^0|^2) - \frac{3}{2} \right]. \tag{8.127}$$

*To obtain this, we have to remember that in Eq. (8.126) the contribution from scalar
loops is written for real scalar fields. Since $\tilde{t}_L$ and $\tilde{t}_R$ are complex, their contribution
needs to be doubled. The factor 3 is a color factor.*

*Finally, we remark that to obtain the effective potential for the charged as well
as neutral Higgs fields, we must allow both top and bottom quarks and squarks in
the loops. Technically, this means that we have to construct a 4×4 field-dependent
mass matrix for the squarks, and 2×2 mass matrix for the fermions. Even for our
simplified calculation, these matrices are no longer diagonal. The trace can be
evaluated by evaluating the (field-dependent) eigenvalues of these squared mass
matrices, inserting these in place of $\mathcal{M}_i$ in (8.126) and summing. Carry out these
steps, and show that you obtain an $SU(2) \times U(1)$ invariant effective potential.*

We can use this effective potential to construct corrected Higgs boson mass
matrices in the same way as before. We will now have additional contributions
from the top quark Yukawa coupling f_t, and involving the top quark and top squark
masses. The result for the scalar Higgs bosons is simple to write down in this
approximation:

$$m_{h,H}^2 = \frac{1}{2} \left[(m_A^2 + M_Z^2 + \delta) \mp \xi^{1/2} \right] \tag{8.128}$$

where

$$\xi = \left[(m_A^2 - M_Z^2) \cos 2\beta + \delta \right]^2 + \sin^2 2\beta (m_A^2 + M_Z^2)^2,$$

and

$$\delta = \frac{3g^2 m_t^4}{16\pi^2 M_W^2 \sin^2 \beta} \log\left[(1 + \frac{m_{\tilde{t}_L}^2}{m_t^2})(1 + \frac{m_{\tilde{t}_R}^2}{m_t^2})\right]. \qquad (8.129)$$

In addition, the relation

$$m_{H^\pm}^2 = m_A^2 + M_W^2$$

is unaltered as long as bottom quark Yukawa couplings are neglected. Finally, the Higgs scalar mixing angle α is modified to

$$\tan \alpha = \frac{(m_A^2 - M_Z^2)\cos 2\beta + \delta + \xi^{1/2}}{\sin 2\beta(m_A^2 + M_Z^2)}. \qquad (8.130)$$

In these formulae, we have eliminated f_t using (8.24a). Notice that the presence of δ in the expression for m_h allows h to exceed M_Z as seems to be required by the LEP data discussed earlier.

Although we have illustrated these corrections keeping only top quark Yukawa couplings and neglecting intra-generation stop mixing, many phenomenological analyses include mixing effects as well as the corrections due to b and τ Yukawa couplings (these are important if $\tan \beta$ is large), and also gauge couplings. Much effort has gone into making as precise predictions as possible for Higgs boson masses, especially m_h. At the present time, state-of-the-art calculations including dominant two-loop effects indicate that the value of m_h can be as high as about 130 GeV, well beyond the reach of the LEP2 e^+e^- collider at CERN, which ran at a maximum energy of ~ 208 GeV, and even larger if $m_t > 175$ GeV.

Gluino mass

It has been noted by Martin and Vaughn that the tree-level gluino mass suffers large corrections – up to 25% – due to loop corrections.[7] In this case, one must compute the gluino self-energy diagrams, and look for the pole position in the gluino propagator. Including loop graphs with gluon exchange and quark–squark loops, they find

$$m_{\tilde{g}} = M_3(Q)\left(1 + \frac{\alpha_s}{4\pi}[15 + 6\log(Q/M_3) + \sum A_{\tilde{q}}]\right) \qquad (8.131)$$

in the $\overline{DR}$ regularization scheme.[8] Here,

$$A_{\tilde{q}} = \int_0^1 dx\, x \log[x m_{\tilde{q}}^2/M_3^2 + (1-x)m_q^2/M_3^2 - x(1-x)], \qquad (8.132)$$

---

[7] S. Martin and M. Vaughn, *Phys. Lett.* **B318**, 331 (1993).
[8] The calculation is performed in the $\overline{DR}$ renormalization scheme: see Chapter 9.

where the sum is over the 12 different quark–squark multiplets, and squark mixing has been neglected.

8.5.2 Squark mass

The dominant corrections to squark masses come from strong interactions, and so are the same for $\tilde{q}_L$ and $\tilde{q}_R$, and also independent of flavor. The radiatively corrected squark mass is given by,

$$
\delta m_{\tilde{q}}^2 = m_{\tilde{q}}^2 - m_{\tilde{q}}^2(Q)
$$

$$
= \frac{2\alpha_s(Q)}{3\pi} m_{\tilde{q}}(Q)^2 \left\{ 1 + 3x + (x-1)^2 \ln|x-1| - x^2 \ln x + 2x \ln \frac{Q^2}{m_{\tilde{q}}^2} \right\}.
$$

(8.133)

If intra-generation squark mixing is not negligible, the form of the corrections is more complicated, and we refer the reader to Pierce *et al.* for the complete result.[9]

8.5.3 Chargino and neutralino masses

By and large the corrections to these masses are not very large, but there are regions of parameter space where they can be several percent. Nevertheless, there are important circumstances where inclusion of these corrections could be important. We will see later that the phenomenology is to a great extent determined by what the lightest supersymmetric particle (LSP) is. This is largely because (as long as R-parity is conserved) all sparticle decays terminate in the LSP. In many models, the LSP is the lightest neutralino or the lighter stau, depending on the values of model parameters. In the case where these sparticles are approximately degenerate at tree level, the radiative corrections might prove to be crucial in identifying the LSP.[10]

Another scenario where radiative corrections are crucial occurs when $|M_2|$ is much smaller than other soft SUSY breaking parameters so that the $SU(2)$ gauginos are the lightest of the sparticles. It is then the radiative corrections that break the degeneracy between the charged and neutral partners, making the latter slightly lighter. A realization of such a scenario occurs in the so-called anomaly-mediated SUSY breaking model discussed in Chapter 11.

---

[9] See, D. Pierce *et al.*, *Nucl. Phys.* **B491**, 3 (1997).

[10] We have oversimplified the discussion here. In gauge-mediated SUSY breaking models that we will discuss in Chapter 11 the LSP is a gravitino: since couplings of sparticles to the gravitino are extremely weak, all other sparticles cascade decay to the next lightest super particle (NLSP) which then decays to the gravitino. It is very important to correctly identify the NLSP to obtain the correct phenomenology.

These radiative corrections have been analyzed in the literature. Complete formulae can be found in Pierce *et al.*, where details are provided.

8.5.4 Yukawa couplings and SM fermion masses

At tree level, the Yukawa couplings that enter the superpotential are simply related to the corresponding SM fermion masses via (8.24a) and (8.24b). This is because $\hat{h}_u^0$ ($\hat{h}_d^0$) couple only to up-(down-)type fermions. At one-loop level, the field h_u^0 can also couple to down-type quarks via its couplings to up- and down-type squarks.

Exercise *Draw a Feynman diagram involving a gluino and down-type squark, or a chargino and an up-type squark, in a loop to show that h_u^0 can couple to the down quarks at the one-loop level.*

Thus, beyond tree level, down-type quarks can obtain contributions to their masses proportional to v_u. Although these contributions are loop-suppressed, they can be comparable to the tree-level contribution (proportional to v_d) if $\tan\beta \gg 1$. Clearly, then the relation between Yukawa couplings and the corresponding quark mass is considerably modified. We refer the interested reader to the paper by Pierce *et al.* for details regarding these corrections.

8.6 Should the goldstino be part of the MSSM?

The MSSM is the simplest viable supersymmetric extension of the SM. Within this framework, our ignorance of the underlying mechanism of SUSY breaking is reflected in the 178 parameters discussed in Section 8.1.2. We should regard the MSSM as an effective theory that will someday be obtained from a more fundamental theory, once we understand the principles behind the physics of SUSY breaking. Presumably, this will result in a dramatic reduction in the set of parameters that one will regard as fundamental, in the sense that most soft SUSY breaking parameters will be derived from more basic considerations.

Indeed despite the many suggestions for how SUSY breaking effects are felt by the superpartners of SM particles, no compelling theory has as yet emerged. There are two common themes to all models of SUSY breaking.

• First, it appears that the SUSY breaking occurs in a sector of the theory that differs from that containing the SM particles and their superpartners. We are forced into considering such theories because models where SUSY breaking occurs in the SM sector run into phenomenological troubles with the sum rules such as (7.35) that led to light scalars as discussed in Chapter 7. This then raises an additional question: even if we can dynamically break SUSY, how do we convey this information to the observable sector of SM particles and their superpartners? The answer to this question will be taken up in Chapter 11 where we discuss various models.

• Second, supersymmetry is broken spontaneously rather than explicitly. Clearly, this is the more appealing route, and also affords a rationale for why SUSY breaking is soft in the MSSM: since SUSY breaking operators are proportional to a VEV, dimensional analysis tells us that dimension four SUSY breaking interactions are forbidden at least in a renormalizable theory.

The attentive reader will, however, be disturbed by the fact that spontaneous breaking of SUSY should be accompanied by a massless Goldstone fermion in the low energy spectrum. This should then be the LSP. Yet, our discussion of the MSSM has made no mention of this. Indeed the MSSM (as we have formulated it with explicit SUSY breaking terms) does not contain a goldstino.

The naive reason that we can get away with doing so is that in most models we promote SUSY to a local supersymmetry. This, as we will discuss in Chapter 11, results in a theory that necessarily incorporates gravity, and requires the introduction of the (spin 2) graviton and its superpartner, a spin 3/2 fermion, the gravitino. Then, when SUSY is spontaneously broken, the would-be Goldstone fermion combines with the (originally massless) gravitino to form a massive gravitino and disappears from the physical spectrum, while the graviton (which is protected from acquiring a mass by the unbroken reparametrization invariance) remains massless. This phenomenon is analogous to what happens in spontaneously broken local gauge theories: the would-be Goldstone bosons combine to form the longitudinal components of a massive gauge field, and no massless spin zero excitations remain in the spectrum.

In principle, if the gravitino is light enough we ought to include it as part of the low energy theory. It is, generally speaking, not usual to do so for the same reason that we do not include the graviton: like the graviton, the gravitino typically couples too weakly to matter for particle physics.[11] Thus the MSSM is a parametrization of the effective low energy theory, but with some prejudices thrown in.

---

[11] We will discuss an exception to this in Chapter 11 when we discuss gauge-mediated SUSY breaking. If the SUSY breaking scale is low enough, we will see that the couplings of the longitudinal components of the gravitino (i.e. the goldstino) play an important role for collider signals.

9

Implications of the MSSM

In this chapter, we discuss various implications of the MSSM relevant to low energy experiments in particle physics and to cosmology. We will postpone examination of signals from direct production of sparticles at high energy colliders to later chapters.

In any theory (like the MSSM) with many scalar fields, there are potentially new sources of flavor-changing neutral currents (FCNC). Experiment tells us that such flavor-violating effects are strongly suppressed. Experimental constraints on these restrict the form of soft SUSY breaking masses and couplings in the MSSM. As we will discuss in more detail, viable models may be classified by the pattern (universality, alignment or decoupling) of scalar mass matrices. We also discuss constraints from potentially large CP-violating processes such as the electric dipole moment of the electron and neutron.

We then proceed to study the effects of renormalization in the MSSM, which differ from corresponding effects in the SM because of the presence of weak scale superpartners. The prediction of gauge coupling unification in the MSSM – but not in the SM – is the best known, and perhaps the most spectacular, of these differences. It is possible to view the MSSM as a theory *defined* at the scale $M_{\text{SUSY}} \sim M_{\text{weak}}$, but with > 100 additional parameters that have well-defined values at that scale. However, since the MSSM is stable against radiative corrections, it may be valid up to much larger energy scales, perhaps as high as those associated with grand unification or string phenomena. New physics at these scales may provide an organizing principle that determines the multitude of weak scale SUSY parameters in terms of a few more fundamental parameters. The renormalization group equations (RGEs) then provide a link between the parameters of the theory at these ultra-high energy scales, and the weak scale, where superpartners are expected to be observed. In particular, we show that the breakdown of electroweak symmetry may be a *derived* consequence of the breakdown of supersymmetry, resulting from the large top quark Yukawa coupling. This picture fits in neatly with the recent discovery of the top quark with $m_t \simeq 175$ GeV. To avoid fine-tuning of SUSY

190

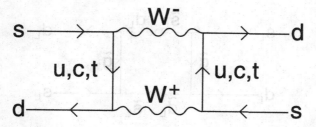

Figure 9.1 A SM box diagram contributing to the K_L–K_S mass difference.

parameters associated with electroweak symmetry breaking, a "naturalness" constraint suggests that SUSY particles that couple directly to the Higgs sector ought to have masses below ~1 TeV, and hence ought to be accessible to collider search experiments in the near future.

Having set up the framework, we proceed to quantify how various observations restrict the values of the soft SUSY breaking masses in the MSSM. These include measurements of the rare decays $b \to s\gamma$ and $B_s \to \mu^+\mu^-$, the anomalous magnetic moment $a_\mu = (g - 2)_\mu/2$ of the muon, and determination of the amount of relic neutralino "dark matter" left over from the Big Bang. If R-parity is indeed conserved (as we assume in the MSSM), then the lightest supersymmetric particle (LSP) should be absolutely stable, and LSPs produced in the early Universe should pervade all space, and could form the bulk of the dark matter that is required to exist by astrophysical measurements. If these cosmological relics are gravitationally clumped in our galactic halo, they may be detectable by both direct and indirect dark matter search experiments.

9.1 Low energy constraints on the MSSM

9.1.1 The SUSY flavor problem

Flavor-changing neutral current processes are forbidden at tree level in the SM due to the GIM mechanism. However, non-zero FCNC rates do occur in the SM at the loop level. A famous example occurs in the neutral K-meson system, where the $K_L - K_S$ mass difference can be calculated from box diagrams such as the one listed in Fig. 9.1. The contribution from the charm quark dominates, and the SM contribution to Δm_K is approximately given by

$$\Delta m_K \simeq \frac{G_F}{\sqrt{2}} \frac{\alpha}{6\pi} \frac{f_K^2 m_K}{\sin^2 \theta_W} \cos^2 \theta_C \sin^2 \theta_C \frac{m_c^2}{M_W^2}, \tag{9.1}$$

where f_K is the kaon decay constant, θ_C is the Cabibbo mixing angle, and m_c is the charm quark mass. The kaon mass difference was, in fact, used to predict the charm quark mass shortly before the discovery of charm.

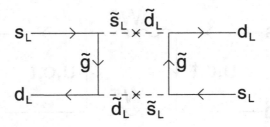

Figure 9.2 An MSSM box diagram contributing to the K_L–K_S mass difference.

In the MSSM, additional contributions from box diagrams involving squarks and gluinos are also present, such as the one shown in Fig. 9.2. (Other diagrams involving chargino and neutralino loops also contribute.) The cross in Fig. 9.2 represents an off-diagonal entry in the squark mass squared matrix

$$\mathcal{L}_{\text{soft}} \ni \tilde{d}_L^\dagger (\mathbf{m}_Q^2)_{12} \tilde{s}_L,$$

which naively is expected to be comparable to the corresponding diagonal entry: 100–1000 GeV. In this case, SUSY contributions to Δm_K violate limits from experiment, and the model is excluded. This is an example of what is referred to as the *SUSY flavor problem*. It occurs because the transformation that diagonalizes the quark mass matrices does not simultaneously diagonalize the corresponding squark mass squared matrices. It is up to theorists to devise models that restrict soft SUSY breaking mass matrices in such a way that bounds from FCNC processes are not violated.

Diagonalization of scalar mass matrices is always possible, but the large off-diagonal mass matrix elements will then lead to non-degenerate squarks that *all* couple via the gluino to *both* s and d quarks. If U denotes the unitary matrix that diagonalizes the quark mass matrix, and $\tilde{U}$ the unitary matrix that diagonalizes the squark mass matrix, the $\tilde{g} - \tilde{q} - q$ interaction (in the mass basis for quarks and squarks) will be proportional to $U\tilde{U}^\dagger$. A calculation of the complete box diagram shows that the contribution to Δm_K is proportional to

$$\sum_{\alpha,\beta=\tilde{d}_L,\tilde{s}_L,\tilde{b}_L} (U\tilde{U}^\dagger)_{i\alpha} (U\tilde{U}^\dagger)^*_{j\alpha} (U\tilde{U}^\dagger)_{i\beta} (U\tilde{U}^\dagger)^*_{j\beta} f(m_\alpha^2, m_\beta^2), \qquad (9.2)$$

where i and j label the external quarks, α and β label the internal squarks, and f is some function of the left-squark mass eigenvalues. A necessary condition for flavor-changing processes is that there are large off-diagonal entries in the $U\tilde{U}^\dagger$ matrix.

Exercise *Convince yourself that if the left-squark mass matrix has degenerate eigenvalues so that $f(m_\alpha^2, m_\beta^2)$ is independent of the squark indices, the gluino*

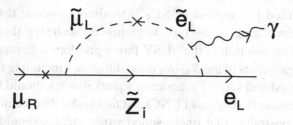

Figure 9.3 Feynman diagram contributing to $\mu \to e\gamma$ decay via a SUSY loop.

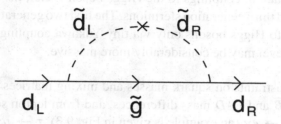

Figure 9.4 An MSSM contribution to the down quark self-energy.

contribution to Δm_K vanishes. The same argument clearly holds for the neutralino contribution. Work out the corresponding argument for chargino contributions.

Flavor violation is not confined to the kaons. A large off-diagonal entry $(\mathbf{m}_Q^2)_{23}$ or $(\mathbf{m}_D^2)_{23}$ (or worse, $(\mathbf{a_d})_{23}$, discussed below) would result in large flavor-violating gluino vertices, and an unacceptable rate for $b \to s\gamma$ decays via diagrams involving squark and gluino loops, analogous to those in Fig. 9.3 with (s)leptons replaced by s(quarks) and the neutralino by the gluino, or in Fig. 9.4 with a photon attached to the squark line.

Eq. (9.2) then suggests three distinct mechanisms to avoid large FCNCs in the MSSM, and thus to solve the SUSY flavor problem. The first two suppress the off-diagonal entries in (9.2) while the third suppresses loop effects by making sparticles very heavy.

1. Arrange for *degeneracy* amongst the masses of squarks with the same quantum numbers.
2. Arrange the SUSY breaking mechanism so that squark and quark mass matrices are diagonalized by the same unitary transformation, so that the matrix $U \tilde{U}^\dagger \simeq \mathbf{1}$. Squarks can be non-degenerate. In this case, the quark and squark mass matrices are said to be *aligned*.[1] Such an alignment can be arranged in models which include so-called *horizontal* symmetries, linking the various generations.

[1] Y. Nir and N. Seiberg, *Phys. Lett.* **B309**, 337 (1993).

3. The third method to suppress FCNCs is to simply assume that the masses of the squarks circulating in the box diagrams are so heavy that the diagram is suppressed. This solution to the SUSY flavor problem is known as *decoupling*. Detailed computations of the K_L–K_S mass difference including QCD corrections indicate that first and second generation squark masses should be larger than $\sim$ 40 TeV to adequately suppress FCNCs. At first sight, this seemingly contradicts *naturalness* constraints that imply superpartner masses should lie at or below the TeV scale. We should note though that these apply most directly to sparticles that have substantial couplings to the Higgs boson sector, i.e. to charginos and neutralinos and third generation sfermions. The first two generations of sfermions which couple to Higgs bosons only via tiny Yukawa couplings or indirectly at the two-loop level may be considerably more massive.

Additional constraints on squark masses and mixing matrices come from measurements of B-$\overline{B}$ and D-$\overline{D}$ mass differences, and from lepton sector FCNC processes such as $\mu \to e\gamma$ (an example is given in Fig. 9.3), $\tau \to \mu\gamma$, and $\tau \to e\gamma$.[2] The constraints that can be extracted vary in their severity, but all can be fulfilled by implementing one or a combination of the solutions of degeneracy, alignment or decoupling.

Constraints from FCNCs also restrict the form of the soft SUSY breaking trilinear terms $(\mathbf{a}_u)_{ij}$, $(\mathbf{a}_d)_{ij}$, and $(\mathbf{a}_e)_{ij}$. For instance, when a Higgs field develops a VEV, then off-diagonal mass terms such as

$$(\mathbf{a}_d)_{12} \tilde{Q}_1^a H_{da} \tilde{d}_{R2}^\dagger \to (\mathbf{a}_d)_{12} v_d \tilde{d}_L \tilde{s}_R^\dagger \tag{9.3a}$$

or flavor-diagonal masses such as

$$(\mathbf{a}_d)_{11} \tilde{Q}_1^a H_{da} \tilde{d}_{R1}^\dagger \to (\mathbf{a}_d)_{11} v_d \tilde{d}_L \tilde{d}_R^\dagger \tag{9.3b}$$

will be induced. The first of these will again be restricted by processes such as K–$\overline{K}$ mixing, so that we must require the off-diagonal entries of the $\mathbf{a}$ matrix to be small. The second of these terms can make (flavor-conserving) contributions to *fermion masses*, such as the down quark via gluino–squark loops (see Fig. 9.4). Requiring these contributions to fermion masses to be smaller (in order of magnitude) than the fermion masses themselves leads to tight constraints on the magnitudes of terms such as $(\mathbf{a}_u)_{ii}$ and $(\mathbf{a}_d)_{ii}$ and $(\mathbf{a}_e)_{ii}$, for generations $i = 1$ and 2.

[2] For a general analysis of FCNC and *CP*-violating effects, see F. Gabbiani, E. Gabrielli, A. Masiero and L. Silvestrini, *Nucl. Phys.* **B477**, 321 (1996).

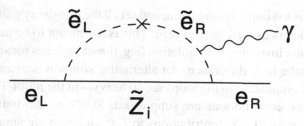

Figure 9.5 A supersymmetric contribution to the electric dipole moment of the electron.

9.1.2 The SUSY CP violation problem

Since, as discussed in Chapter 8, soft SUSY breaking parameters are in general complex, one should expect TeV scale imaginary components to these, which would correspond to the presence of large CP-violating phases. Many constraints on these imaginary components can also be extracted from low energy data. For instance, SUSY contributions to the parameter $\epsilon_K = \frac{1}{2}\frac{\text{Im}\langle K|\mathcal{H}_{eff}|\bar{K}\rangle}{\text{Re}\langle K|\mathcal{H}_{eff}|\bar{K}\rangle}$ can be used to set bounds on the *imaginary* part of squark mass squared matrix elements. The flavor-violating contributions have been parametrized by Gabbiani *et al.* as complex "mass insertions" $(\Delta m^2)_{ij}^{ab}$ (i, j label the flavor and $a, b = $ L, R the squark type), and the constraints are expressed as bounds on dimensionless quantities $\delta_{ij}^{ab} = (\Delta m^2)_{ij}^{ab}/\tilde{m}^2$. Measurements of Δm_K and ϵ_K put constraints on the δ_{12}^{ab}, but further assumptions are needed to extract these. For instance, assuming that the real parts of the δs dominate their imaginary parts, and further that the real part is at its upper bound, one can obtain an upper limit on the imaginary part. As an example: for $m_{\tilde{q}} \sim m_{\tilde{g}} = \tilde{m}$, the general analysis from Gabbiani *et al.* shows that for the down squark sector, $\sqrt{\text{Im}(\Delta m^2)_{12}^{LL}} \leq 0.01\tilde{m}$.

Limits on the imaginary parts of the soft SUSY breaking **a** terms can be obtained from experimental upper limits on the electric dipole moments (EDMs) of both the electron and the neutron. These contributions come from diagrams such as those depicted in Fig. 9.5. For instance, the quantity $\sqrt{|\text{Im}(\mathbf{a}_e)_{11}v_d|}$ is restricted to be less than $6 \times 10^{-4}m_{\tilde{e}}$. Likewise, the bound on the neutron electric dipole moment restricts $\sqrt{|\text{Im}(\mathbf{a}_d)_{11}v_d|}$ to be less than $\sim 0.002m_{\tilde{d}}$. In the case of R-parity-violating scenarios, phases of R-parity-violating couplings are also restricted by similar considerations.

Finally, measurements of the CP-violating decays $K_L \rightarrow \pi\pi$ are related to the CP-violating parameter ϵ'_K. These measurements further restrict $\sqrt{|\text{Im}(\mathbf{a}_d)_{12}v_d|}$ to be smaller than $\sim 0.004m_{\tilde{d}}$.

Determining the physical principle behind why so many of the CP-violating phases are so small is known as the *SUSY CP problem*. Motivated by the stringent limits on the magnitude of CP-violating phases, an ad hoc but frequent assumption in the literature is to simply ignore them, and set all the imaginary parts of soft SUSY breaking parameters as well as μ to zero. This is not meant to be taken literally. In many analyses not involving CP violation (e.g. direct searches for sparticles), these *small* phases make little difference. An alternative solution is to again assume the SUSY particles circulating in the loops are so heavy – in the multi-TeV range – that the CP-violating contributions are suppressed. SUSY model builders, however, have to explain why SUSY contributions to CP violation are small, and perhaps to make predictions for the patterns of CP violation for the third generation where data are not yet in.

A common but stronger assumption which solves both the SUSY flavor and CP problems is to assume *universality* and *reality* of soft SUSY breaking masses:

$$\mathbf{m}_Q^2 = m_Q^2 \mathbf{1}; \quad \mathbf{m}_U^2 = m_U^2 \mathbf{1}; \quad \mathbf{m}_D^2 = m_D^2 \mathbf{1}; \quad \mathbf{m}_L^2 = m_L^2 \mathbf{1}; \quad \mathbf{m}_E^2 = m_E^2 \mathbf{1}. \quad (9.4)$$

In addition, the trilinear **a** terms are assumed proportional to their corresponding Yukawa matrices:

$$\mathbf{a}_u = A_u \mathbf{f}_u; \quad \mathbf{a}_d = A_d \mathbf{f}_d; \quad \mathbf{a}_e = A_e \mathbf{f}_e. \quad (9.5)$$

In this case, almost all FCNC contributions will be well below experimental limits (a super-GIM mechanism operates), and all CP-violating phases other than in the usual CKM mixing matrix will be vanishing. The universality assumption, however, goes beyond just fulfilling experimental constraints. In particular, it should be kept in mind that restrictions on many of the soft SUSY breaking mass parameters are very loose or even non-existent, and it remains for experimental measurements to determine or limit these parameters.

9.1.3 Large CP-violating parameters in the MSSM?

As we have discussed, measurements of the EDM of the electron and the neutron and, most recently, of atoms such as mercury have placed stringent constraints on CP-violating phenomena in the MSSM. However, these constraints do *not* guarantee that many of the CP-violating MSSM parameters are small. It could be that flavor-blind CP-violating parameters (such as the gaugino masses M_i') and CP-violating phases associated with the first two generations are small, but those associated with the third generation are large. Or, it could be that sparticle masses are in the multi-TeV range so that SUSY loop contributions to the various EDMs are suppressed even if the CP-violating parameters are relatively large. Finally, it could be that individual phases/parameters are large, but that there exist cancellations amongst the amplitudes which contribute to the various EDMs. Any theory of SUSY breaking would then have to explain the origin of these cancellations if they are not to be attributed to mere accident. We caution the reader that there is no clear consensus in the literature as to just how well these cancellation mechanisms work. In part, this is because the conversion of the stringent experimental bound on the EDM of mercury to limits on the EDM of quarks and electrons (or its QCD analogue, the chromo electric dipole moment of quarks) that are predicted by any high scale theory is not entirely straightforward. A careful assessment of the associated subtleties is beyond the scope of this text, and we refer the interested reader to the literature for a discussion of these issues.[3]

If indeed the CP-violating parameters are large, these can lead to observable effects in the sparticle and Higgs boson sectors, even in non-CP-violating phenomena. One simple example comes from the chargino mass matrix, exemplified in the exercise at the end of this section. In Chapter 8, we noted that while one of the CP-violating phases associated with gaugino masses (we chose the case of M_3') could be rotated away, the remaining two could not. Thus, M_2' and also a complex phase in the μ parameter (where $\mu = |\mu|e^{i\phi_\mu}$) would enter the chargino mass matrix and alter the corresponding physical chargino mass eigenvalues. The CP-violating parameters would also modify the chargino mixing matrices, and hence could modify chargino production cross sections and branching fractions. Likewise, neutralino mass eigenvalues, production rates, and decay rates can depend on M_1', M_2', and ϕ_μ. Squark and slepton observables can depend on these parameters, as well as on possibly large CP-violating phases in the **a**-parameters: within the mSUGRA-like framework with a universal A-parameter introduced in Section 9.2.2, this effect is most pronounced for the third generation sfermions. If SUSY sources of CP violation are large, Higgs boson phenomenology may also be significantly altered: in particular, as already alluded to in Section 8.3.3, neutral Higgs bosons would no longer be mass eigenstates, and their phenomenology would be correspondingly altered. These new sources of CP violation can also obviously lead to novel effects in the physics of K and B mesons, since not all CP-violating effects would then be described by the Kobayashi–Maskawa phase.[4] These SUSY sources of CP violation may also have a significant impact on cosmology at early times, most notably on baryogenesis.

---

[3] The EDM of mercury was evaluated within the MSSM framework by T. Falk, K. Olive, M. Pospelov and R. Roiban, *Nucl. Phys.* **B560**, 3 (1999). For an overview of the potential uncertainties in this evaluation, see e.g. the reviews by Ibrahim and P. Nath, hep-ph/0210251 (2002) and D. Chung *et al.*, hep-ph/0312378 (2003), and references cited therein. An overview of how SUSY CP violation affects sparticle phenomenology is also contained in these reviews. See also, J. Erler and M. Ramsey-Musolf, hep-ph/0404291 for a general discussion of the evaluation of EDMs within extensions of the SM.

[4] For instance, determination of the angles of the so-called unitarity triangles in different processes will not yield consistent values for these. Especially topical at the time of this writing is the discrepancy in the decays $B \rightarrow \phi K_s$ vis-à-vis the decays $B \rightarrow \psi K_s$, reported by the BELLE experiment. This discrepancy has, however, not been seen by the BaBar experiment. For textbook discussions of these questions see G. Branco, L. Lavoura and J. Silva, *CP Violation*, Oxford (1999); I. Bigi and A. Sanda, *CP Violation*, Cambridge University Press (2000).

We will for simplicity of discussion ignore the possibility of large CP-violating parameters in the remainder of this book. We expect that the interested reader will be able to carry out the necessary modifications to include their effects in the discussion.

Exercise *To illustrate the potential impact of CP-violating SUSY parameters on sparticle masses and couplings, generalize the derivation of the chargino masses and mixing patterns in Section 8.3.5 to the case where the CP-violating gaugino mass term M'_2 does not vanish, and the supersymmetric parameter μ is complex.*

Working in the convention that the Higgs scalars have real VEVs (this may lead to CP-violating phases in the interactions of Higgs bosons with other scalars) show that the mass terms for the charginos can now be written as,

$$\mathcal{L}_{\text{chargino}} = -\left(\bar{\lambda},\ \bar{\tilde{\chi}}\right)\left(\mathcal{M}_{\text{charge}}P_{\text{L}} + \mathcal{M}^{\dagger}_{\text{charge}}P_{\text{R}}\right)\begin{pmatrix}\lambda \\ \tilde{\chi}\end{pmatrix},$$

where now

$$\mathcal{M}_{\text{charge}} = \begin{pmatrix} M_2 - \mathrm{i}M'_2 & -gv_d \\ -gv_u & -\mu \end{pmatrix}.$$

Many authors who use the two-component notation introduce instead a complex mass $\mathcal{M}_2 \equiv M_2 - \mathrm{i}M'_2$ and write $\mathcal{M}_2 \equiv |\mathcal{M}_2|e^{\mathrm{i}\phi_2}$, and work with the real numbers $|\mathcal{M}_2|$ and ϕ_2 instead of our M_2 and M'_2, as alluded to at the very end of Chapter 7.

Show that the squared chargino masses, the eigenvalues of the matrix $\mathcal{M}_{\text{charge}}\mathcal{M}^{\dagger}_{\text{charge}}$, are now given by,

$$m^2_{\widetilde{W}_{1,2}} = \frac{1}{2}\left[(|\mu|^2 + |\mathcal{M}_2|^2 + 2M^2_W) \mp \zeta\right],$$

with

$$\zeta^2 = (|\mu|^2 - |\mathcal{M}_2|^2)^2 + 4M^2_W\big[M^2_W \cos^2 2\beta + |\mu|^2 + |\mathcal{M}_2|^2$$
$$-2|\mu||\mathcal{M}_2|\sin 2\beta \cos(\phi_\mu + \phi_2)\big].$$

Notice that if $|\mathcal{M}_2| = \sqrt{M^2_2 + M'^2_2}$ and/or $|\mu|$ are much larger than M_W, once again the two charginos are dominantly gauginos and higgsinos.

The matrices U and V that enter the chargino couplings via the diagonalization of charginos (see Sec. 8.3.5) will now depend on these additional CP-violating parameters, and potentially cause CP violation in processes involving charginos.

The effect of CP-violating gaugino masses and complex μ term on the neutralino sector can be similarly worked out.

9.2 Renormalization group equations

Since the MSSM is free of quadratic divergences, mass parameters of order the weak scale remain stable under radiative corrections. In particular, if the MSSM is embedded in a larger framework, such as a GUT or string model, parameters of order the weak scale will remain that same order even after radiative corrections involving the ultra-high energy scales associated with these models. This stabilization of mass hierarchies allows the possibility of reliably extending the predictions of the MSSM up to very high energy scales. Logarithmic divergences, however, remain, and perturbative calculations involving energies $Q \sim M_{\mathrm{GUT}}$ will typically contain powers of $\frac{\alpha_i}{4\pi} \log(M_{\mathrm{GUT}}/M_Z)$, where α_i is a gauge coupling. Fortunately, these large logarithms which would invalidate the perturbative expansion in α_i can be summed by using renormalization group methods. The coupling constants and mass parameters of the theory are replaced by running couplings and masses, with values depending on the energy scale. The scale dependence of the parameters of the theory is given by the renormalization group equations.

9.2.1 Gauge couplings and unification

In quantum field theory, perturbative calculations beyond tree level are usually performed using *renormalized perturbation theory* (RPT), as opposed to bare perturbation theory. The bare fields, mass terms and coupling constants that occur in the original Lagrangian are (perturbatively) divergent quantities. In RPT, these are replaced by finite, renormalized fields, masses and coupling constants, and divergent quantities are formally shuffled into *counterterms*. The form of the counterterms is determined by specifying renormalization conditions at some arbitrarily chosen energy scale Q, referred to as the renormalization scale. While Green functions of the bare theory are independent of the renormalization scale, Green functions calculated in RPT are dependent on the renormalization scale. The dependence of Green functions of RPT on shifts in the renormalization scale Q is governed by the Callan–Symanzik equation. As the renormalization scale shifts, so too do the fields, coupling constants and masses of the theory. The evolution of a coupling constant g with renormalization scale, in particular, is governed by the Callan–Symanzik beta function $\beta(g)$, defined as

$$\beta(g) = Q \frac{\partial g}{\partial Q}.$$

The procedure for evaluating β is to calculate the logarithmically divergent parts of diagrams which contribute to the coupling constant renormalization, and then take the logarithmic derivative with respect to the renormalization scale Q. For non-supersymmetric non-Abelian gauge theories, the calculation of the one-loop

β-function is performed in many texts.[5] The result, generalized to include scalars and left- and right-handed fermions in different representations, is

$$\beta(g) = -\frac{g^3}{16\pi^2}\left[\frac{11}{3}C(G) - \frac{2}{3}n_F S(R_F) - \frac{1}{3}n_H S(R_H)\right], \tag{9.6}$$

where $C(G)$ is the quadratic Casimir for the adjoint representation of the associated Lie algebra, $S(R_F)$ is the Dynkin index for representation R_F of the fermion fields, $S(R_H)$ is the Dynkin index for representation R_H of the scalar fields, n_F is the number of fermion species, and n_H is the number of complex scalars. For an $SU(N)$ gauge theory, for fermions or scalars in the fundamental N-dimensional representation, $S(R) = 1/2$, while $C(G) = N$. For small values of n_F, the β-function is negative, resulting in the well-known phenomenon of asymptotic freedom.

For $SU(3)_C$, with fermions u_L, d_L, u_R and d_R, $n_F = 4n_g$, where n_g is the number of generations. For $SU(2)_L$, with three colors of left doublet quarks and a single left doublet of leptons, n_F is again $4n_g$. Finally, for $U(1)_Y$, $S(R) = 1$, and we simply sum over the squared hypercharges of a complete generation: $\sum(Y/2)^2 = 10/3$. The final result for the SM, at one-loop, is

$$\beta_i = \frac{g_i^3}{16\pi^2}b_i, \tag{9.7}$$

where the b_i ($i = 1, 2, 3$) are given by

$$\begin{pmatrix} b_1 \\ b_2 \\ b_3 \end{pmatrix} = \begin{pmatrix} 0 \\ -\frac{22}{3} \\ -11 \end{pmatrix} + n_g \begin{pmatrix} \frac{4}{3} \\ \frac{4}{3} \\ \frac{4}{3} \end{pmatrix} + n_H \begin{pmatrix} \frac{1}{10} \\ \frac{1}{6} \\ 0 \end{pmatrix}, \tag{9.8}$$

n_g is the number of generations, and n_H is the number of (complex) Higgs doublets ($n_H = 1$ in the SM). The expression for b_1 holds for the evolution of the rescaled charge $g_1 = \sqrt{5/3}g'$ appropriate for a GUT model.

In the MSSM, there will be additional loop contributions to the various counterterms involving gauginos, matter scalars, Higgs scalars and higgsinos. To calculate the various loop diagrams, a suitable regularization scheme must be chosen. For SM calculations, dimensional regularization (DREG) is most frequently chosen, since it preserves gauge symmetry and hence the validity of Ward identities in loop calculations. In models with supersymmetry, DREG is usually not the regulator of choice, since it violates supersymmetry. The reason is that, by modifying the dimensionality of spacetime, one introduces a mismatch between the number of degrees of freedom in vector fields versus their supersymmetric counterpart gauginos.

---

[5] See, for instance, M. Peskin and D. V. Schroeder, *Introduction to Quantum Field Theory*, Chapter 16, Addison-Wesley (1995).

A modification of DREG known as dimensional reduction (DRED) also modifies the dimensionality of spacetime, but maintains four-vectors as four-component objects.[6] DRED thus preserves supersymmetry, at least for one-loop calculations. Calculations using DRED versus DREG differ only in the finite parts of one-loop diagrams, but differ even in the divergent parts of two-loop diagrams. Thus, the RGEs calculated via DREG or via DRED will be equivalent to one-loop order.

In the MSSM, the β-functions are modified by superpartner contributions from gauginos, higgsinos and matter scalars. These can be readily computed from (9.6). Using $S(R) = N$ for the adjoint representation in $SU(N)$, it is straightforward to show that the one-loop β-functions for the MSSM take the form,

$$\beta(g) = -\frac{g^3}{16\pi^2} \left[3C(G) - S(R) \right], \qquad (9.9)$$

where the Dynkin index $S(R)$ is summed over all the matter and Higgs fields, and their superpartners, in the model. This then yields,

$$\begin{pmatrix} b_1 \\ b_2 \\ b_3 \end{pmatrix} = \begin{pmatrix} 0 \\ -6 \\ -9 \end{pmatrix} + n_{\rm g} \begin{pmatrix} 2 \\ 2 \\ 2 \end{pmatrix} + n_{\rm H} \begin{pmatrix} \frac{3}{10} \\ \frac{1}{2} \\ 0 \end{pmatrix}. \qquad (9.10)$$

Exercise *Using Eq. (9.6) to obtain the contributions of the superpartner gauginos, higgsinos, and sfermions, verify that the gauge β-functions for the MSSM are indeed as given by (9.9) and (9.10).*

The RGEs for the gauge couplings can now be simply integrated. The constant of integration can be fixed using the experimentally measured value of the gauge coupling at some reference scale Q_0. We then find,

$$\frac{1}{g_i(Q)^2} - \frac{1}{g_i(Q_0)^2} = -\frac{b_i}{8\pi^2} \ln \left(\frac{Q}{Q_0} \right). \qquad (9.11)$$

In Fig. 9.6, we show how the gauge couplings, given by Eq. (9.11), evolve with the scale choice Q. It is customary to plot the inverse of $\alpha_i = g_i^2/4\pi$, beginning with the values of α_1, α_2, and α_3 which are known to high precision at the value $Q = M_Z$. The evolution in the SM is shown in Fig. 9.6a. The three gauge coupling constants evolve in a generally convergent direction towards higher energy scales, becoming roughly the same at $Q \sim 10^{13}$–10^{17} GeV. In Fig. 9.6b, the case for the MSSM is shown. Remarkably, the three gauge couplings unify with impressive precision at $Q \simeq 2 \times 10^{16}$ GeV! In this case, we have evolved the gauge couplings according to

---

[6] W. Siegel, *Phys. Lett.* **B84**, 193 (1979).

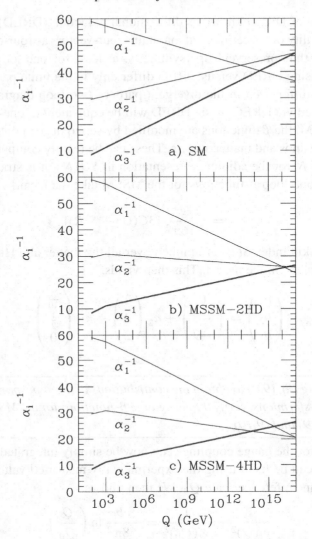

Figure 9.6 Evolution of the $SU(3)_C \times SU(2)_L \times U(1)_Y$ gauge coupling constants from the weak scale to the GUT scale for the case of (*a*) the SM, (*b*) the MSSM with two Higgs doublets, and (*c*) the MSSM with four Higgs doublets.

SM RGEs between $Q = 10^2$ and $Q = 10^3$ GeV, and switched to MSSM evolution equations for higher Q values. The gauge coupling unification in the MSSM is startling, and strongly suggests that the MSSM may be the remnant of some sort of supersymmetric grand unified theory, with superpartners around the TeV scale. In Fig. 9.6*c*, we show the same gauge coupling evolution, but this time we include four Higgs doublets in the supersymmetric model. For this case, gauge coupling unification is once again off the mark.

The successful prediction of gauge coupling unification is viewed by many as indirect evidence for weak scale supersymmetry. It motivates us to consider the possibility that there may indeed be no new physics all the way up to a very high scale, and that the weakly coupled MSSM is valid up to $Q \sim M_{\text{GUT}}$. This is an assumption. Above $Q = M_{\text{GUT}}$, there may be new physics: gauge coupling unification suggests grand unification with a desert as the simplest possibility, but this need not be the case.

It is possible that an examination of the parameters of the MSSM renormalized at a very high scale (obtained from their measured values) might provide clues as to what this new physics might be, in the same way that the unification of gauge couplings points to grand unification. But such a bottom-up approach is clearly not possible today since we do not know the weak scale values of the SUSY parameters or even the superpotential Yukawa couplings. Instead, what is usually done is to make *simple* ansätze about the values of these soft SUSY breaking parameters at the high scale, and then evolve these down to the weak scale relevant for phenomenology using renormalization group equations. These ansätze serve as boundary conditions for the evolution.[7] It should be stressed that the evolution does not involve any new physics beyond the MSSM.

Typically, the simple boundary conditions can be expressed in terms of just a handful of parameters, from which all the parameters of the MSSM may be computed. In this sense, the MSSM augmented by the boundary conditions is a very predictive framework. As discussed in Chapter 7, we hope that an understanding of the physics of supersymmetry breaking, and its mediation to the observable sector, will yield the correct boundary conditions. In Chapter 11, we will discuss various models for SUSY breaking, but for the present we will regard the specification of the boundary values of the soft supersymmetry breaking parameters as an additional ansatz.

We have already studied the renormalization group equations for the gauge couplings. The one-loop RGEs for third generation Yukawa couplings of the MSSM are given by

$$\frac{\mathrm{d}f_t}{\mathrm{d}t} = \frac{f_t}{16\pi^2} \left(-\sum_{i=1-3} c_i g_i^2 + 6f_t^2 + f_b^2 \right), \tag{9.12a}$$

$$\frac{\mathrm{d}f_b}{\mathrm{d}t} = \frac{f_b}{16\pi^2} \left(-\sum_{i=1-3} c_i' g_i^2 + f_t^2 + 6f_b^2 + f_\tau^2 \right), \tag{9.12b}$$

$$\frac{\mathrm{d}f_\tau}{\mathrm{d}t} = \frac{f_\tau}{16\pi^2} \left(-\sum_{i=1-3} c_i'' g_i^2 + 3f_b^2 + 4f_\tau^2 \right), \tag{9.12c}$$

---

[7] Of course, care must be taken to ensure that these boundary conditions lead to weak scale parameters consistent with all experimental constraints.

where $c_i = (13/15, 3, 16/3)$, $c_i' = (7/15, 3, 16/3)$, $c_i'' = (9/5, 3, 0)$ and $t = \log(Q)$. Effects of Yukawa couplings of the first two generations should be negligible, and are usually neglected in calculations, unless one is attempting to match the entire mass spectrum of SM fermions. To find the weak scale boundary conditions on the Yukawa couplings, one starts with running fermion masses (evaluated at the scale of the fermion mass) that are usually extracted in the $\overline{MS}$ (modified minimal subtraction) scheme and then converts these to corresponding masses at a scale M_Z (or m_t) in the $\overline{DR}$ scheme ($\overline{DR}$ = modified minimal subtraction using dimensional reduction). Once the running fermion masses are known at the weak scale, they can be converted to running Yukawa couplings. If $\tan\beta$ is large, it is important to include supersymmetric loop contributions to fermion masses for a reliable extraction of weak scale Yukawa couplings, especially for f_b. The Yukawa couplings can then be evolved to any other scale where the MSSM is valid using the RGEs given above. Notice that unlike the RGEs for gauge couplings that form a closed system, a knowledge of gauge couplings (but not the sparticle spectrum) is necessary to determine the Yukawa coupling evolution.

9.2.2 Evolution of soft SUSY breaking parameters

Like the gauge and Yukawa couplings, the various soft SUSY breaking parameters as well as the superpotential Higgs mass μ, evolve with energy scale. The RGE for the gaugino mass can be obtained from the generalization of the expression for the gluino mass in Eq. (8.131). Taking the derivative with respect to $t = \log(Q)$ gives,

$$\frac{dM_i}{dt} = \frac{g_i^2}{16\pi^2} M_i \left(-6C(G) + 2S(R)\right). \tag{9.13}$$

Exercise *Noting that the β-function for the gaugino mass is proportional to the β-function for the corresponding gauge coupling, show that*

$$\frac{M_i(Q)}{g_i^2(Q)} = K_i, \tag{9.14}$$

where the constant K_i is independent of the scale Q.

In models where the K_i defined in the last exercise are the same for each gauge group factor, we would have the relation

$$\frac{\alpha_1}{M_1} = \frac{\alpha_2}{M_2} = \frac{\alpha_3}{M_3}. \tag{9.15}$$

Such a relation is natural in many simple SUSY GUT theories. In this case, the three couplings as well as the three gaugino masses must unify at $Q = M_{\text{GUT}}$, and the K_i's in Eq. (9.14) are independent of i. The relation (9.15) is therefore often referred to as the GUT relation for gaugino masses. We should remind the reader that the coupling α_1 is related to the conventionally normalized weak hypercharge gauge coupling α' by

$$\alpha' = \frac{3}{5}\alpha_1.$$

The one-loop RGEs for the soft SUSY breaking parameters and for μ can most easily be worked out using the analysis of Falck, and are listed below.[8] Two-loop RGEs have also been worked out; for these, we refer the reader to the original literature.[9] In writing the RGEs, we neglect any inter-generation mixing. Also, following our earlier discussion, we write the trilinear soft SUSY breaking coupling a_i as $a_i = f_i A_i$. Finally, we write the RGEs only for third generation sfermion masses and A-parameters. The corresponding RGEs for the first two generations may be obtained by self-evident replacement of the Yukawa couplings and "X" parameters (defined below). With these assumptions, the RGEs are given by,

$$\frac{dM_i}{dt} = \frac{2}{16\pi^2} b_i g_i^2 M_i, \tag{9.16a}$$

$$\frac{dA_t}{dt} = \frac{2}{16\pi^2}\left(-\sum_i c_i g_i^2 M_i + 6f_t^2 A_t + f_b^2 A_b\right), \tag{9.16b}$$

$$\frac{dA_b}{dt} = \frac{2}{16\pi^2}\left(-\sum_i c_i' g_i^2 M_i + 6f_b^2 A_b + f_t^2 A_t + f_\tau^2 A_\tau\right), \tag{9.16c}$$

$$\frac{dA_\tau}{dt} = \frac{2}{16\pi^2}\left(-\sum_i c_i'' g_i^2 M_i + 3f_b^2 A_b + 4f_\tau^2 A_\tau\right), \tag{9.16d}$$

$$\frac{dB}{dt} = \frac{2}{16\pi^2}\left(-\frac{3}{5}g_1^2 M_1 - 3g_2^2 M_2 + 3f_b^2 A_b + 3f_t^2 A_t + f_\tau^2 A_\tau\right), \tag{9.16e}$$

$$\frac{d\mu}{dt} = \frac{\mu}{16\pi^2}\left(-\frac{3}{5}g_1^2 - 3g_2^2 + 3f_t^2 + 3f_b^2 + f_\tau^2\right), \tag{9.16f}$$

$$\frac{dm_{Q_3}^2}{dt} = \frac{2}{16\pi^2}\left(-\frac{1}{15}g_1^2 M_1^2 - 3g_2^2 M_2^2 - \frac{16}{3}g_3^2 M_3^2 + \frac{1}{10}g_1^2 S + f_t^2 X_t + f_b^2 X_b\right), \tag{9.16g}$$

$$\frac{dm_{\tilde{t}_R}^2}{dt} = \frac{2}{16\pi^2}\left(-\frac{16}{15}g_1^2 M_1^2 - \frac{16}{3}g_3^2 M_3^2 - \frac{2}{5}g_1^2 S + 2f_t^2 X_t\right), \tag{9.16h}$$

$$\frac{dm_{\tilde{b}_R}^2}{dt} = \frac{2}{16\pi^2}\left(-\frac{4}{15}g_1^2 M_1^2 - \frac{16}{3}g_3^2 M_3^2 + \frac{1}{5}g_1^2 S + 2f_b^2 X_b\right), \tag{9.16i}$$

[8] N. K. Falck, *Z. Phys.* **C30**, 247 (1986).
[9] S. Martin and M. Vaughn, *Phys. Rev.* **D50**, 2282 (1994); Y. Yamada, *Phys. Rev.* **D50**, 3537 (1994); I. Jack and D. R. T. Jones, *Phys. Lett.* **B333**, 372 (1994).

$$\frac{dm_{L_3}^2}{dt} = \frac{2}{16\pi^2} \left(-\frac{3}{5} g_1^2 M_1^2 - 3 g_2^2 M_2^2 - \frac{3}{10} g_1^2 S + f_\tau^2 X_\tau \right), \tag{9.16j}$$

$$\frac{dm_{\tilde{\tau}_R}^2}{dt} = \frac{2}{16\pi^2} \left(-\frac{12}{5} g_1^2 M_1^2 + \frac{3}{5} g_1^2 S + 2 f_\tau^2 X_\tau \right), \tag{9.16k}$$

$$\frac{dm_{H_d}^2}{dt} = \frac{2}{16\pi^2} \left(-\frac{3}{5} g_1^2 M_1^2 - 3 g_2^2 M_2^2 - \frac{3}{10} g_1^2 S + 3 f_b^2 X_b + f_\tau^2 X_\tau \right), \tag{9.16l}$$

$$\frac{dm_{H_u}^2}{dt} = \frac{2}{16\pi^2} \left(-\frac{3}{5} g_1^2 M_1^2 - 3 g_2^2 M_2^2 + \frac{3}{10} g_1^2 S + 3 f_t^2 X_t \right), \tag{9.16m}$$

where m_{Q_3} and m_{L_3} denote the mass term for the third generation $SU(2)$ squark and slepton doublet respectively, and

$$X_t = m_{Q_3}^2 + m_{\tilde{t}_R}^2 + m_{H_u}^2 + A_t^2, \tag{9.17a}$$

$$X_b = m_{Q_3}^2 + m_{\tilde{b}_R}^2 + m_{H_d}^2 + A_b^2, \tag{9.17b}$$

$$X_\tau = m_{L_3}^2 + m_{\tilde{\tau}_R}^2 + m_{H_d}^2 + A_\tau^2, \quad \text{and} \tag{9.17c}$$

$$S = m_{H_u}^2 - m_{H_d}^2 + Tr \left[\mathbf{m}_Q^2 - \mathbf{m}_L^2 - 2\mathbf{m}_U^2 + \mathbf{m}_D^2 + \mathbf{m}_E^2 \right]. \tag{9.17d}$$

Here, the trace denotes a sum over generations. In many models (including the model with "universal" mass parameters to be introduced shortly), $S = 0$.

Exercise *Obtain the one-loop RGE for S and show that if S vanishes at one scale, it vanishes at all scales. We therefore do not have to worry about the S-term in the class of models where the boundary conditions ensure that S vanishes.*

Notice that the RGE for μ is completely decoupled from the soft SUSY breaking parameters, as is appropriate for a parameter occurring in the superpotential.

The complete set of 26 RGEs can be solved easily numerically as follows. Given initial values of the gauge couplings, Yukawa couplings, soft breaking terms, and μ parameter at some scale Q_0, we can plug into the right-hand side of each of the RGEs to calculate the slope, and then make a linear extrapolation along a small step size ΔQ to find new values of each of these parameters. By iterating this approach, the trajectories of each parameter can be found. In practice, more sophisticated numerical methods such as Runge–Kutta integration are used.

As an example, inspired by the apparent gauge coupling unification at the grand unified scale together with the constraints from FCNCs and electric dipole moments, we can adopt the *universality* hypothesis at the scale $Q = M_{\text{GUT}} \simeq 2 \times$

10^{16} GeV:

$$g_1 = g_2 = g_3 \equiv g_{\text{GUT}}, \tag{9.18a}$$

$$M_1 = M_2 = M_3 \equiv m_{1/2}, \tag{9.18b}$$

$$m_{Q_i}^2 = m_{U_i}^2 = m_{D_i}^2 = m_{L_i}^2 = m_{E_i}^2 = m_{H_u}^2 = m_{H_d}^2 \equiv m_0^2, \tag{9.18c}$$

$$A_t = A_b = A_\tau \equiv A_0, \tag{9.18d}$$

where all off-diagonal soft SUSY breaking scalar masses and A parameters are set to zero. Inter-generation mixing then occurs only via superpotential Yukawa couplings. Many analyses *not involving flavor physics* can be simplified by ignoring Yukawa interactions for the first two generations. Then, each generation number is separately conserved, and off-diagonal sfermion masses or A-parameters will not be generated by renormalization group evolution. Notice also that we use two notations interchangeably: m_{E_1} and $m_{\tilde{e}_R}$ both denote the soft SUSY breaking mass for the right-handed selectron, while $m_{L_1} = m_{\tilde{e}_L} = m_{\tilde{\nu}_{eL}}$ denotes the common soft SUSY breaking mass parameter of the selectron, and the electron sneutrino, and similarly for squarks. The assumption that the MSSM is valid between the weak scale and GUT scale, and that the "boundary conditions" (9.18a)–(9.18d) hold is often referred to as the *mSUGRA*, or the minimal supergravity model. We will see in Chapter 11 that these boundary conditions are obtained in the simplest supergravity GUT models, assuming that below $Q = M_{\text{GUT}}$, the field content is that of the MSSM.[10]

An example of the evolution of soft SUSY breaking parameters is shown in Fig. 9.7. We take $m_0 = 100$ GeV, $m_{1/2} = 200$ GeV, $A_0 = 0$, and $\tan \beta = 4$. The short dashed lines depict the running of the three gaugino masses from their common GUT scale value. The value of M_3 increases, since it has a negative β-function, while M_1 and M_2 both decrease. In the mSUGRA model, we thus expect at the weak scale the ratio of gaugino masses $M_1 : M_2 : M_3 \sim 1 : 2 : \sim 7$, in accord with the values of the weak scale gauge couplings. Thus, the gluino should be far heavier than the lighter chargino or the two lighter neutralinos.

The evolution of first generation squark and slepton mass parameters is shown by the solid lines. The evolution is solely due to their gauge interactions (Yukawa couplings are neglected) which always *increase* these masses as we run from the high scale down to the weak scale. Squark masses evolve the most because of strong interaction loop contributions to their RGEs. The small intra-generational mass splitting is due to differences in their electroweak interaction. Sleptons, because they

---

[10] Many phenomenological analyses of weak scale supersymmetry have been performed within this framework. Its popularity can be judged by the number of acronyms associated with it: MSSM, CMSSM, MGUM, MSGM, SSC, ... We will see in Chapter 11 that supergravity does not necessarily lead to high scale universality as was originally thought. So, rather than associating SUGRA with supergravity, one may instead consider that mSUGRA stands for *m*inimal *s*upersymmetric model with *u*niversality, *g*auge coupling unification, and *r*adiative electroweak symmetry breaking.

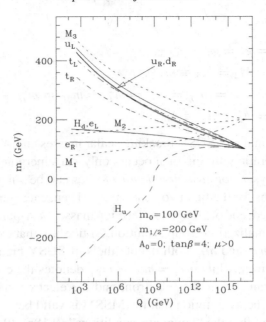

Figure 9.7 Evolution of gaugino masses, Higgs boson mass parameters, and first generation scalar mass parameters, versus energy scale in the mSUGRA model. For the scalars, we actually plot sign $(m^2) \cdot \sqrt{|m^2|}$, so that the negative values on the dashed H_u curve correspond to negative values of $m^2_{H_u}$. We use two-loop RGEs for this figure.

have just electroweak interactions, evolve much less than squarks, with $m_{\tilde{e}_R}$ evolving less than $m_{\tilde{e}_L}$ because the $SU(2)$ gauge coupling is larger than the hypercharge gauge coupling. We can see from the RGEs that the sfermion mass evolution depends on the gaugino masses which, in turn, are all proportional to $m_{1/2}$. Thus, if $m_0 \gg m_{1/2}$ most of the mass at the weak scale comes from m_0, and the sfermions will be approximately degenerate and much heavier than the gluino. If, on the other hand, $m_{1/2} \gg m_0$ as in our illustration, radiative corrections to the sfermion masses are large, and sleptons will be much lighter than the squarks, with the right selectron being the lightest of the first generation matter scalars. These important features of the soft SUSY breaking masses of the first two sfermion generations are captured by the following simple approximations to the soft terms:

$$m^2_{\tilde{q}} \simeq m^2_0 + (5-6)m^2_{1/2}, \qquad (9.19a)$$

$$m^2_{\tilde{e}_L} \simeq m^2_0 + 0.5m^2_{1/2}, \qquad (9.19b)$$

$$m^2_{\tilde{e}_R} \simeq m^2_0 + 0.15m^2_{1/2}, \qquad (9.19c)$$

where D-term contributions given by (8.64) have been neglected for simplicity. Notice that these relations (together with the relation between M_3 and $m_{1/2}$) imply that the first two generations of squarks can never be much lighter than the gluino. The effect of Yukawa coupling contributions to the third generation scalar RGEs can be seen by the dot-dashed lines for the top squark soft SUSY breaking mass parameters. Yukawa interactions have an opposite effect compared to gauge interactions: they *reduce* scalar masses as we run from a high scale down to the weak scale. Indeed we see that the stop mass parameters are considerably smaller than the corresponding first generation squark masses. Remember that by $SU(2)$ invariance, $m_{\tilde{b}_L} = m_{\tilde{t}_L}$. For the low value of $\tan \beta$ used in Fig. 9.7, the bottom Yukawa coupling is small, and we expect that $m_{\tilde{b}_R} \simeq m_{\tilde{d}_R} > m_{\tilde{b}_L}$. Even though top scalars have an additional contribution m_t^2 (see (8.65a)), generally speaking these will be lighter than their first and second generation counterparts. In fact, care must be exercised to ensure that these masses do not become negative, since then charge and color breaking minima may occur in the scalar potential.

Finally, we note that because H_d and $\tilde{e}_L$ have the same gauge quantum numbers, and if bottom quark Yukawa interactions are negligible, the evolution of $m_{H_d}^2$ is virtually identical to that of $m_{\tilde{e}_L}^2$. The evolution of the Higgs mass parameter $m_{H_u}^2$ is very different and particularly noteworthy: it begins at the common GUT scale value but, because it has large top quark Yukawa interactions, it evolves to *negative* values. At first sight, something appears terribly wrong! However, as we will now see, this turns out to be just what is needed for an appropriate breakdown of electroweak symmetry.

Exercise *Notice from Fig. 9.7 that the Yukawa coupling contributions to the evolution of $m_{\tilde{t}_R}^2$, which reduce it relative to $m_{\tilde{u}_R}^2$, are larger than the corresponding contributions that reduce $m_{\tilde{t}_L}^2$. This is because the correction to $m_{\tilde{t}_R}^2$ can come from either a t_L or a b_L (or the corresponding squarks) and, respectively, the neutral or the charged component of the higgsino (scalar) component of $\hat{h}_u$, while the correction to $m_{\tilde{t}_L}^2$ can come only from the singlet top quark (or squark) in the loop. Since all the vertices are determined by just the superpotential top quark Yukawa coupling, we expect that the Yukawa coupling correction to $m_{\tilde{t}_R}^2$ is twice that of $m_{\tilde{t}_L}^2$. Identify this factor of two in the RGEs. Now consider the diagrams that give rise to Yukawa coupling corrections to $m_{H_u}^2$. Relative to the correction to the stop masses, how big do you expect this correction to be? Identify the corresponding term in the RGE, and check whether your answer is consistent with this.*

9.2.3 Radiative breaking of electroweak symmetry

In the SM, electroweak symmetry is spontaneously broken if a scalar field that transforms non-trivially under $SU(2)_L \times U(1)_Y$ acquires a VEV. The situation is no different in the MSSM. If the scalar field potential, evaluated at the weak scale, has a minimum for non-zero field values of h_u^0 or h_d^0 with zero values for other fields, we would have the desired symmetry breaking pattern. In the MSSM with arbitrary values for each of the soft SUSY breaking parameters this can be trivially arranged by choosing $m_{H_u}^2$ or $m_{H_d}^2$ to be negative. Of course, the conditions (8.17a) requiring that the origin be a maximum of the potential, and (8.17b) requiring the potential to be bounded from below need to be satisfied. The remarkable thing that we have just seen is that *even with universal mass parameters at the high scale, renormalization group evolution can cause $m_{H_u}^2$ to turn negative at the weak scale,* leaving squark and slepton squared masses to be positive. We stress that although the scalar potential with parameters renormalized at a scale $Q \gg M_{\mathrm{weak}}$ has only positive squared masses, this does not imply that its true minimum is at the origin in field space. Radiative corrections can be substantial because of the large value of $\log(Q/M_{\mathrm{weak}})$, and can qualitatively change this picture. Evolving the parameters of this potential to the weak scale sums these large logs, and a more reliable picture of the true potential is obtained.[11] We will, therefore, refer to this mechanism, wherein $m_{H_u}^2$ turns negative due to its renormalization group evolution, as radiative electroweak symmetry breaking (REWSB).

In Chapter 8 we minimized the tree-level scalar potential of the MSSM, and found two conditions necessary for spontaneous breaking of electroweak symmetry:

$$B = \frac{(m_{H_u}^2 + m_{H_d}^2 + 2\mu^2)\sin 2\beta}{2\mu} \quad \text{and} \quad (9.20a)$$

$$\mu^2 = \frac{m_{H_d}^2 - m_{H_u}^2 \tan^2\beta}{(\tan^2\beta - 1)} - \frac{M_Z^2}{2}. \quad (9.20b)$$

The first of these can be used to determine B (or equivalently B_0) in terms of $\tan\beta$, μ, and the Higgs mass parameters. Since B never enters the RGEs for the other parameters, its value is not needed for computing their evolution. The second of these minimization conditions determines the value of μ^2 in terms of the Higgs mass parameters and $\tan\beta$, possibly at the expense of some fine-tuning.

REWSB, which was discovered in the early 1980s,[12] occurs over a wide range of model parameters if the top quark Yukawa coupling is large enough, corresponding to $m_t \sim 100 - 200$ GeV. The subsequent discovery of the top quark with mass $m_t \simeq 175$ GeV lends credence to this mechanism. As mentioned above there are, however, regions of parameter space where charge- and color-breaking minima

---

[11] In practice, one usually also includes higher loop calculations and computes the minima using the effective potential as discussed in the previous chapter.

[12] L. E. Ibáñez and G. G. Ross, *Phys. Lett.* **B110**, 215 (1982); K. Inoue *et al.*, *Prog. Theor. Phys.* **68**, 927 (1982) and **71**, 413 (1984); L. Ibáñez, *Phys. Lett.* **B118**, 73 (1982); J. Ellis, J. Hagelin, D. Nanopoulos and M. Tamvakis, *Phys. Lett.* **B125**, 275 (1983); L. Alvarez-Gaumé, J. Polchinski and M. Wise, *Nucl. Phys.* **B221**, 495 (1983).

may occur. Thus although it is fair to say that REWSB links electroweak symmetry breaking with the breakdown of supersymmetry at some higher scale, it is premature to conclude that the *pattern* of electroweak symmetry breaking is explained.

Since REWSB is driven by the top quark Yukawa coupling, we have $m_{H_u}^2 < m_{H_d}^2$ (assuming that we start with universal masses), which implies $\tan\beta > 1$. Furthermore, in order for REWSB to be driven by the top quark Yukawa coupling, $\tan\beta$ has to be bounded above. This follows because $f_t > f_b$ implies that $m_t/m_b = f_t v_u/f_b v_d > \tan\beta$, where the Yukawa couplings, and hence the quark mass parameters, are to be evaluated at the weak scale. The bound $\tan\beta \lesssim 60$ thus obtained should be regarded as qualitative because it would be modified by radiative corrections.

The evolution of the soft SUSY breaking masses from M_{GUT} to the weak scale now allows us to determine the weak scale values of the soft SUSY breaking masses that are needed to determine all the sparticle masses as well as their couplings in terms of just a handful of parameters. Note that the minimization condition for REWSB specifies the value of μ^2, but not the sign of μ. It is convenient to eliminate the high scale parameter B_0 in favor of $\tan\beta$. The mSUGRA model is completely specified by the parameter set:

$$m_0, \; m_{1/2}, \; A_0, \; \tan\beta, \; \mathrm{sign}(\mu). \tag{9.21}$$

A selection of sparticle masses for the same mSUGRA model parameters used in Fig. 9.7 is shown in Table 9.1.

9.2.4 Naturalness constraint on superparticle masses

It is often stated loosely that sparticles ought to have masses typically below $\sim 1\,\mathrm{TeV}$ so that the hierarchy between the SUSY breaking and weak scales can be maintained without resorting to too much fine tuning. But how much fine tuning is too much fine tuning?

To gain a better handle on how heavy sparticles can be, many groups have tried to quantify a measure of fine tuning, in order to decide which values of SUSY model parameters are natural. First, one has to decide on such a measure, and then one must decide how much fine tuning is too much. Clearly, there is a good deal of subjectivity built into constraints from naturalness. We should also keep in mind that it is possible that what appears to be fine tuning from the vantage point of the low energy theory could be the result of particular relationships in the (unknown) high energy theory. Thus, while we would regard the fine tuning required by the SM as indicative of new physics, we would not necessarily be alarmed by what appears to be fine tuning at, for instance, a part per mille level.

Table 9.1 *Weak scale sparticle masses and parameters (GeV)*
for the mSUGRA model with $m_0 = 100$ GeV, $m_{1/2} = 200$ GeV,
$A_0 = 0$, $\tan\beta = 4$, and $\mu > 0$. These results were obtained
from the computer program ISAJET version 7.69.

parameter	value (GeV)
$m_{\tilde{g}}$	500.5
$m_{\tilde{u}_L}$	463.4
$m_{\tilde{d}_R}$	451.7
$m_{\tilde{t}_1}$	324.7
$m_{\tilde{b}_1}$	426.9
$m_{\tilde{e}_L}$	176.0
$m_{\tilde{e}_R}$	131.0
$m_{\tilde{\tau}_1}$	129.6
$m_{\tilde{W}_1}$	135.3
$m_{\tilde{Z}_2}$	136.5
$m_{\tilde{Z}_1}$	72.8
m_h	104.4
m_A	343.1
μ	292.5

A particularly simple measure of fine tuning can be extracted from the second of the electroweak symmetry breaking relations (9.20b) listed in the previous subsection. Naively, if $|\mu| \gg M_Z$, the term involving the Higgs mass parameters must also be large so that these two terms may combine and largely cancel to give M_Z. It is possible to argue that models with $|\mu| \gg M_Z$ would be fine tuned, and the weak scale value of $|\mu|$ itself can be used as a measure of fine tuning. This naive example may be too simplistic, and many authors would also argue that many parameter choices leading to $|\mu| \sim M_Z$ are also fine tuned.

As an example of a more sophisticated measure of fine tuning, we can discuss naturalness constraints in the mSUGRA model. The fundamental parameters associated with SUSY breaking are,

$$a_i = \{m_0, \ m_{1/2}, \ A_0, \ B_0, \ \text{and} \ \mu_0\}. \tag{9.22}$$

Quantities associated with the weak scale can be calculated in terms of these fundamental parameters. We have included the GUT scale superpotential μ parameter in this list since we will regard the value of M_Z (which is given by the second of the electroweak symmetry breaking conditions (9.20b) mentioned above) as an output. Of course, only certain sets of GUT scale input parameters $\{a_i\}$ will give the correct value of M_Z. For variations of the input parameters $a_i \rightarrow a_i + \Delta a_i$, we can derive a corresponding value of $M_Z^2 + \Delta M_Z^2$. The fine-tuning requirement is that,

$$\frac{\Delta M_Z^2}{M_Z^2} < c_i \frac{\Delta a_i}{a_i},$$

where c_i is the *fine-tuning* parameter.[13] It is up to the reader to decide what constitutes an acceptable choice of c_i. Typical values quoted in the literature range from $c_i = 10$–100. Thus, if a tiny change in input parameters a_i leads to a big change in the derived value of M_Z^2 (or some other weak scale observable), then the model is said to be fine-tuned. In terms of derivatives, the fine-tuning requirement is written as

$$\left| \frac{\partial \log M_Z^2}{\partial \log a_i} \right| < c_i. \tag{9.23}$$

As an example, we show in Fig. 9.8 the m_0 vs. $m_{1/2}$ plane of the mSUGRA model, taking $A_0 = 0$, $\tan \beta = 10$, $\mu > 0$, and $m_t = 175$ GeV. The gray region in the upper left is excluded if we require that the lightest SUSY particle be electrically neutral (to fulfill cosmological constraints). The gray regions on the far right are excluded by a lack of appropriate REWSB (using the one-loop corrected scalar potential). The dark gray region for low $m_{1/2}$ is excluded in that the lightest chargino mass falls below limits from LEP2 searches: $m_{\widetilde{W}_1} < 100$ GeV. For reference, we plot also contours of $m_{\tilde{g}} = 1000$ and 2000 GeV, and $m_{\tilde{u}_L} = 1000$ and 2000 GeV. To illustrate how subjective fine-tuning considerations can be, we show examples of fine-tuning bounds obtained using various criteria for fine-tuning limits from the literature. First, we show a contour of the weak scale value of $\mu = 500$ GeV. Taking the value of μ as a fine-tuning parameter generally restricts the parameter plane to values of $m_{1/2}$ below about 400 GeV, unless m_0 is very large, in which case very large values of $m_{1/2}$ can yield "natural" models. The trajectory of constant μ is known as the hyperbolic branch (HB), and all parameter space points with low $|\mu|$ may be regarded as natural.[14] A second contour (labeled AC) was obtained by Anderson and Castaño.[15] These authors include the top Yukawa coupling f_t in the list of fundamental parameters, and use a weighted average of fine-tuning parameters to obtain their contour. Their result clearly prefers low values of both m_0 and $m_{1/2}$ to obtain natural models. Finally, a fine-tuning contour calculated by Feng, Matchev and Moroi neglects the top Yukawa coupling on the basis that it is associated with the flavor sector, and not the SUSY breaking sector.[16] Their contour (labeled FMM) excludes large $m_{1/2}$ values, but does admit solutions with

[13] R. Barbieri and G. Guidice, *Nucl. Phys.* **B306**, 63 (1988).
[14] See K. Chan, U. Chattopadhyay and P. Nath, *Phys. Rev.* **D58**, 096004 (1998).
[15] G. Anderson and D. Castaño, *Phys. Rev.* **D53**, 2403 (1996).
[16] J. L. Feng, K. T. Matchev and T. Moroi, *Phys. Rev.* **D61**, 075005 (2000).

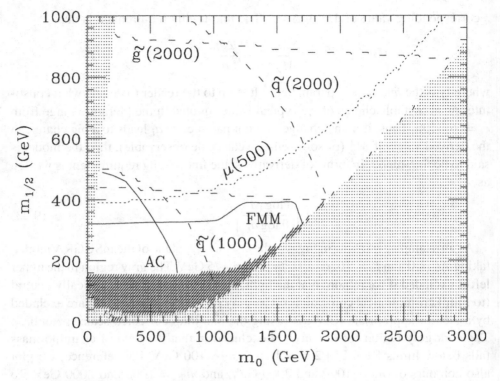

Figure 9.8 A plot of mSUGRA parameter space in the m_0 vs. $m_{1/2}$ plane, for $A_0 = 0$, $\tan \beta = 10$, and $\mu > 0$. We show contours of gluino and squark mass for 1000 and 2000 GeV. We also show sample fine-tuning contours i) $\mu = 500$ GeV, plus contours extracted from ii) Anderson and Castaño (AC), and iii) Feng, Matchev and Moroi (FMM). The proposed acceptable regions are below the fine-tuning contours.

large m_0. The large m_0 solutions have been referred to as *focus point* supersymmetry (FP), since (for $m_{1/2} \ll m_0$) the value of $m_{H_u}^2$ evolves to a fixed weak scale value independent of its GUT scale value, i.e. it is focussed in its RG trajectory. The focus point solutions offer the possibility of solving, at least partially, the SUSY flavor and *CP* problems, since in this case all the squark and slepton masses can be beyond 1 TeV while "maintaining naturalness".

9.3 Constraints from $b \to s\gamma$ decay

We have already mentioned that the agreement between the observed rate for the decay $b \to s\gamma$ and SM expectation yields significant constraints on off-diagonal squark mass squared matrix elements and **a** parameters. This should not be surprising because, as we saw in the exercise at the end of Section 6.6, the radiative decay of the b quark is a probe of the supersymmetry breaking sector.[17] Even in the mSUGRA model with universal GUT scalar masses and a universal A_0-parameter, significant constraints are obtained. Within this framework flavor violation, which is the other essential ingredient for this decay to occur, occurs only via Yukawa couplings. The flavor-violating matrix elements can then be calculated in terms of known quark masses and Kobayashi–Maskawa (KM) matrix elements (see exercise below). The point is that we can always go to a quark basis (at the weak scale) where the fields u_R, d_R, and d_L are the same as in the mass basis, while the field u_L is related to the corresponding field in the mass basis by the KM matrix. The Yukawa coupling matrices are known in this basis, and can be evolved to the high scale where the mSUGRA boundary conditions are specified. At this scale, if squarks with the same gauge quantum numbers have a common mass (the mSUGRA framework, where all squarks have a universal mass is a special case), we do not need to know the basis of squarks since all bases are equivalent. The mSUGRA boundary conditions, together with the *known* Yukawa coupling matrices thus specify the framework completely, and flavor-violating effects can be unambiguously computed.

Exercise *Unlike in the Standard Model, flavor physics is not always specified by just the Kobayashi–Maskawa matrix. A general two doublet Higgs model serves to illustrate this point. Let two $SU(2)$ doublets (H^+, H^0) and (K^+, K^0) couple the quark doublet to d_R via "Yukawa coupling matrices" $\mathbf{f_H}$ and $\mathbf{f_K}$, respectively while the conjugate doublets $(H^{0*}, -H^-)$ and $(K^{0*}, -K^-)$ respectively couple the quark doublet to u_R with Yukawa coupling matrices $\widetilde{\mathbf{f}}_H$ and $\widetilde{\mathbf{f}}_K$. Show that (1) the couplings of H^0 and K^0 as well as those of H^+ and K^+ to the quarks depend on all four matrices $V_L(u)$, $V_L(d)$, $V_R(u)$, and $V_R(d)$ that connect the weak current and mass bases for u_L, d_L, u_R, and d_R type quarks (the Kobayashi–Maskawa matrix, which is determined by the couplings of quarks to $W^\pm$ bosons, is just $V_L(u)^\dagger V_L(d)$), and (2) the interactions of H^0 and K^0 do not conserve flavor. Verify that for the "MSSM-like case" where $\mathbf{f_K}$ and $\widetilde{\mathbf{f}}_H$ vanish, the charged boson couplings are completely determined by the KM matrix and quark masses and, further, that the flavor-violating couplings of H^0 and K^0 vanish.*

For the case of the MSSM, some couplings involving squarks may also depend on matrices that diagonalize the squarks. It is, therefore, noteworthy that in models

---

[17] In practice, it is the inclusive decay $B \to X_s\gamma$ that is bounded by the experiment. The transition magnetic dipole moment operator that we argued to be absent in the SUSY limit is the operator of lowest dimensionality that could have a contribution to this decay. In principle, higher dimensional operators involving additional gluons may contribute to the inclusive decay of B *mesons* through non-renormalizable terms in the Kähler potential, but these contributions would be extremely suppressed, and quite likely smaller than the theoretical uncertainty in the calculation.

where squarks with the same quantum numbers have a common mass at some high
scale, flavor physics effects are fixed by just the KM matrix and quark masses.

In the SM the decay $b \to s\gamma$ proceeds at lowest order via a $t W^-$ loop. In supersymmetric models, there are additional contributions from the $t H^-$ loop, as well as sparticle loops containing $\tilde{u}_i \widetilde{W}_j$, $\tilde{d}_i \tilde{g}$, and $\tilde{d}_i \widetilde{Z}_j$. Since SM as well as SUSY contributions both occur at the one-loop level, it is reasonable to expect that if sparticles are at the weak scale, SUSY contributions to the decay amplitude will be comparable to the SM contribution, so that the experimental determination of the branching ratio will provide strong constraints on the parameters of supersymmetric models.

The $b \to s\gamma$ decay rate is usually calculated by evaluating lowest order matrix elements of effective theory operators at a scale $Q \sim m_b$. The complete calculation is complicated by the fact that QCD corrections are large. These are included via renormalization group resummation of leading logs (LL) which arise due to a disparity between the scale at which new physics enters the $b \to s\gamma$ loop corrections (usually taken to be $Q \sim M_W$), and the scale at which the $b \to s\gamma$ decay rate is evaluated ($Q \sim m_b$). The resummation is most easily performed within the framework of effective field theories. Above the scale $Q = M_W$ (all scales $Q \sim M_W$ are equivalent in LL perturbation theory), calculations are performed within the full theory. Below $Q = M_W$, particles heavier than M_W are integrated out, leading to an effective Hamiltonian,

$$H_{\text{eff}} = -\frac{4G_F}{\sqrt{2}} V_{tb} V_{ts}^* \sum_{i=1}^{8} C_i(Q) O_i(Q). \tag{9.24}$$

Matching the two theories at $Q = M_W$ yields the values of the so-called Wilson coefficients $C_i(Q = M_W)$. The O_i in Eq. (9.24) are a complete set of operators that mix via QCD; their form can be found in the literature.[18] The logs are summed by solving the renormalization group equations (RGEs) for the Wilson coefficients

$$Q \frac{\mathrm{d}}{\mathrm{d}Q} C_i(Q) = \gamma_{ji} C_j(Q), \tag{9.25}$$

where γ is the 8×8 anomalous dimension matrix. The matrix elements of the operators O_i are finally calculated at a scale $Q \sim m_b$ and multiplied by the appropriately evolved Wilson coefficients to obtain the decay amplitude. The LL QCD corrections just described yield enhancements in the $b \to s\gamma$ decay rate of factors of 2–5. Variation of the scale choice between $m_b/2 < Q < 2m_b$ yields approximately a 25% uncertainty in the theoretical calculation. This is reduced to about 9% by working at next-to-leading order. We also note that the SUSY

---

[18] See e.g., G. Buchalla, A. Buras and M. Lautenbacher, *Rev. Mod. Phys.* **68**, 1125 (1996).

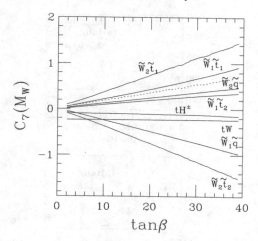

Figure 9.9 Contributions to the Wilson coefficient $C_7(M_W)$ versus $\tan\beta$ from various loop contributions. We take $m_0 = 100$ GeV, $m_{1/2} = 200$ GeV, $A_0 = 0$, and $\mu > 0$. The dotted line shows the total contribution from sparticle loops. Reprinted with permission from H. Baer, M. Brhlik, D. Castaño and X. Tata, *Phys. Rev.* **D58**, 015007 (1998), copyright (1998) by the American Physical Society.

calculation has larger uncertainty, especially if $\tan\beta$ is large. Our discussion here is to give the reader a flavor of the ingredients that go into such a calculation. These calculations are sophisticated, and the reader who is actually interested in performing the calculation is well advised to consult the original literature.

The most important of the above operators is the magnetic operator $O_7 \sim \bar{s}_L\sigma^{\mu\nu}b_R F_{\mu\nu}$. In Fig. 9.9, we show the magnitude of the Wilson coefficient $C_7(M_W)$ versus $\tan\beta$ from various contributions involving tW, tH^-, and $\widetilde{W}_i\tilde{q}_j$ loops for one choice of mSUGRA parameters. The total contribution from gluino and neutralino loops is negligible. We see from the figure that large cancellations are possible between the various contributions.

In Fig. 9.10, we show the $b \rightarrow s\gamma$ branching fraction contours in the m_0 vs. $m_{1/2}$ plane, for $\tan\beta = 10$, $A_0 = 0$, and a) $\mu < 0$ and b) $\mu > 0$. Data from the CLEO, BELLE, and ALEPH experiments, roughly speaking, restrict $2 \times 10^{-4} \lesssim B(b \rightarrow s\gamma) \lesssim 5 \times 10^{-4}$, if we conservatively factor in theoretical uncertainties. Clearly, the mSUGRA model with $\mu < 0$ is only consistent with data for large values of $m_{1/2} > 300$ GeV. For $\mu > 0$, virtually the entire plane is allowed. Qualitatively similar results are obtained for even larger values of $\tan\beta$.

9.4 $B_s \rightarrow \mu^+\mu^-$ decay

Within the framework of the Minimal Supersymmetric Standard Model (MSSM), FCNC conservation is ensured (at tree level) by requiring that the matter multiplets

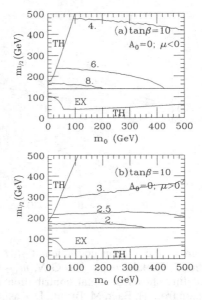

Figure 9.10 Contours of constant branching fraction $B(b \to s\gamma)$ in the m_0 vs. $m_{1/2}$ plane for $\tan \beta = 10$, $A_0 = 0$. The number labeling each contour must be multiplied by 10^{-4} to obtain the branching fraction. The region labeled EX is excluded by the constraint $m_{\widetilde{W}_1} > 100$ GeV while the region marked TH is not allowed for theoretical reasons.

with weak isospin $T_3 = 1/2$ couple only to the Higgs superfield $\hat{H}_u$, while those with $T_3 = -1/2$ couple just to the Higgs superfield $\hat{H}_d$.

At the one-loop level, however, a coupling of $\hat{H}_u$ to down-type fermions is induced. This induced coupling leads to a new contribution, proportional to v_u, to the down-type fermion mass matrix. Although this contribution is suppressed by a loop factor relative to the tree-level contribution, this suppression is (partially) compensated if $\tan \beta$ is sufficiently large. As a result, down-type Yukawa interactions and down-type quark mass matrices are no longer diagonalized by the same transformation, and flavor-violating couplings of neutral Higgs scalars h, H, and A emerge. In the limit of large m_A, the Higgs sector becomes equivalent to the Standard Model (SM) Higgs sector with a light Higgs boson $h \simeq H_{\rm SM}$, and the effects of flavor violation decouple from the low energy theory . The interesting feature is that the flavor-violating couplings of h, H, and A *do not decouple for large superparticle mass parameters*: being dimensionless, these couplings depend only on ratios of these mass parameters, and so remain finite even for very large values of SUSY mass parameters.[19] This flavor-violating neutral Higgs boson coupling results in a potentially observable branching fraction for the decay $B_s \to \mu^+ \mu^-$

---

[19] These not only include sparticle masses, but also the superpotential parameter μ and also the soft SUSY breaking A-parameters.

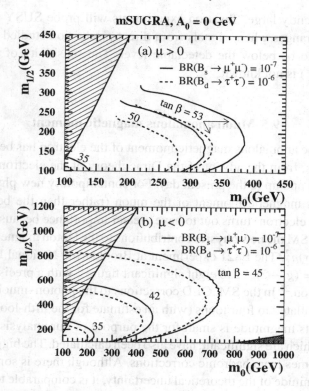

Figure 9.11 Contours of $B(B_s \to \mu^+\mu^-) = 10^{-7}$ (solid) and $B(B_d \to \tau^+\tau^-) = 10^{-6}$ (dashed) in the m_0 vs. $m_{1/2}$ plane of the mSUGRA model for several values of $\tan\beta$ and (a) $\mu > 0$, and (b) $\mu < 0$. In frame (a), the contours end where $\tilde{Z}_1$ is no longer the LSP. The region where this occurs for $\tan\beta = 35$ is shaded. Reprinted with permission from J. K. Mizukoshi, X. Tata, and Y. Wang, *Phys. Rev* **D66**, 115003 (2002), copyright (2002) by the American Physical Society.

mediated by the neutral Higgs bosons, h, H, and A, and possibly also the decay $B_d \to \tau^+\tau^-$.[20] The former might be probed at the Tevatron (the CDF experiment has already limited it to be $< 5.8 \times 10^{-7}$), while the latter might be detectable at B-factories. Within the MSSM, this branching fraction – which depends sensitively on $\tan\beta$ and m_A and less sensitively on other sparticle masses – can be more than 1000 times its SM value.

In Fig. 9.11 we illustrate the branching fraction for these Higgs-mediated leptonic decays of B_s and B_d mesons within the mSUGRA framework for a) $\mu > 0$, and b) $\mu < 0$. The solid lines show contours of $B(B_s \to \mu^+\mu^-) = 10^{-7}$, a level that Tevatron experiments should probe with an integrated luminosity ~ 2 fb^{-1}, for the values of $\tan\beta$ shown on the contours. The dashed lines are contours of $B(B_d \to \tau^+\tau^-) = 10^{-6}$. The contours in frame a) end where $\tilde{Z}_1$ is no longer the LSP. If

---

[20] See, e.g., K. S. Babu and C. Kolda, *Phys. Rev. Lett.* **84**, 228 (2000).

$\tan \beta$ is sufficiently large, Tevatron experiments will probe SUSY via B_s decays for parameter ranges where signals from direct production studied in Chapter 15 are predicted to be below the detectable level. The sensitivity of B-factories to $B(B_d \rightarrow \tau^+\tau^-)$ is not known.

9.5 Muon anomalous magnetic moment

Historically, the anomalous magnetic moment of the electron has been a harbinger of new physics, from the advent of the Dirac theory of the electron to the formulation of QED, and up to the present day. For contemporary new physics searches, the anomalous magnetic moment of the muon (rather than the better measured moment of the electron) turns out to have greater importance because for many extensions of the SM the new physics contributions to the lepton magnetic moment are proportional to m_ℓ^2. The E821 experiment at Brookhaven National Laboratory has measured $a_\mu = (g - 2)_\mu/2$ to eight significant figures, with a precision better than a part per million.[21] In the SM, QED corrections to the photon–muon–muon vertex have been calculated to four loops (with an estimate for the fifth-loop contribution, showing that its magnitude is small for the purpose of our analysis). Electroweak corrections, which are significant, have also been calculated. The biggest theoretical uncertainty comes from hadronic corrections. Although there is some controversy about the magnitude of the theoretical uncertainty, it is comparable to or better than the experimental uncertainty. If weak scale SUSY exists, then there will also be SUSY contributions to a_μ via the $\widetilde{W}_i - \tilde{\nu}_\mu$ and $\widetilde{Z}_i - \tilde{\mu}_j$ loops shown in Fig. 9.12. The SUSY contribution gives

$$\Delta a_\mu^{\text{SUSY}} \propto \frac{m_\mu^2 \mu M_i \tan \beta}{M_{\text{SUSY}}^4}, \tag{9.26}$$

where M_i $(i = 1, 2)$ is a gaugino mass and M_{SUSY} is a characteristic sparticle mass circulating in the loop. The complete one-loop result is given, for instance, by Moroi.[22] We see that $\Delta a_\mu^{\text{SUSY}}$ grows with $\tan \beta$ and, for models with a positive gaugino mass, has the same sign as the superpotential Higgs mass term μ. Depending on SUSY parameters, its magnitude may be comparable to that of the weak contribution, so that the sensitivity of the E821 experiment is at a level where it can probe these SUSY contributions.

[21] See e.g. G. W. Bennett *et al.* (Muon $g - 2$ Collaboration), hep-ex/0401008.
[22] T. Moroi, *Phys. Rev.* **D53**, 6565 (1996). That the SUSY contributions to a_μ cancel the corresponding SM contributions in the SUSY limit of the MSSM – this limit is discussed in the exercise at the end of Section 8.3.6 – in accord with the general result demonstrated in the exercise at the end of Section 6.6 has been explicitly demonstrated at the one-loop level by T. Ibrahim and P. Nath, *Phys. Rev.* **D61**, 095008 (2000).

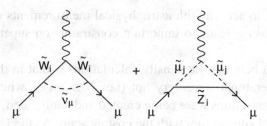

Figure 9.12 Supersymmetric contributions to $g - 2$ of the muon.

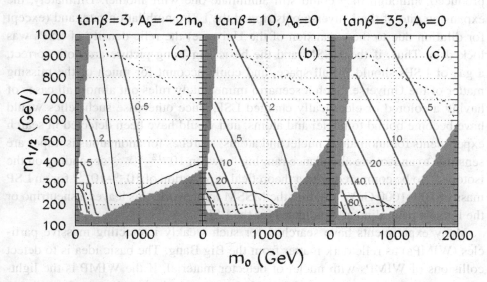

Figure 9.13 Contours of $a_\mu \times 10^{10}$ in the mSUGRA model for $\mu > 0$. The Fermilab Tevatron (dashes) and CERN LHC (dot-dashed) reach contours are also shown.

Contours of $\Delta a_\mu^{\text{SUSY}} \times 10^{10}$ are shown in Fig. 9.13 for three $\tan \beta$ values in the mSUGRA model, and $\mu > 0$. Regions of parameter space where $\Delta a_\mu^{\text{SUSY}} \gtrsim 60 \times 10^{-10}$ and $\Delta a_\mu^{\text{SUSY}} < 0$ are currently disfavored. As the experimental error reduces, and theoretical calculations improve, the results will more definitively point to preferred and excluded regions of SUSY model parameter space.

9.6 Cosmological implications

Since R-parity is assumed to be conserved in the MSSM, the lightest SUSY particle is absolutely stable. This has profound implications for cosmology and, in particular, may imply that relic LSPs left over from the Big Bang could account for the bulk of the matter in the Universe. Moreover, the requirement that the relic density of LSPs should be in accord with astrophysical measurements of the dark matter density of the Universe leads to important constraints on supersymmetric model parameters.

The central idea behind relic density calculations is that in the very early Universe, when temperatures were very hot (i.e. $T \gg m_{LSP}$, where the Boltzmann constant $k = 1$), neutralinos were being created and annihilated, but that they were in a state of thermal equilibrium with the cosmic soup. As the Universe expanded, and cooled, its temperature dropped below the level where LSPs could be pair-produced, although they could still annihilate one with another. Ultimately, the expansion rate of the Universe outstripped the LSP annihilation rate, and (except for dilution due to the expansion of the Universe) the relic density of LSPs was locked in. Thus, if the MSSM and the basic Big Bang picture are both correct, a gas of LSPs should fill all space, and could account for much of the missing matter of the Universe. Such a scenario immediately rules out almost all cases of having a colored or electrically charged LSP, since otherwise such relics would have become bound to nuclei and atoms, and would have been detected in search experiments for anomalous nuclei and atoms: searches for anomalous isotopes are sensitive to an isotope abundance ranging between 10^{-12}–10^{-29} depending on the isotope,[23] to be compared with a theoretical expectation of 10^{-6}–10^{-10} for an LSP mass of 100–1000 GeV.[24] Within the MSSM framework, this leaves a sneutrino or the lightest neutralino as candidates for the LSP.

Many experiments have searched for such weakly interacting massive particles (WIMPs) as relic dark matter from the Big Bang. The basic idea is to detect collisions of WIMPs with nuclei of detector material. If the WIMP is the lightest neutralino with a mass ~ 100 GeV, then a typical neutralino-nucleus elastic scattering will involve energies of a few keV. To detect such tiny energy depositions, detector materials are frequently cooled to ultra-low temperatures, so that phonons, ionization or superconducting phase transitions can be detected. If instead the WIMP is a sneutrino heavier than about 25 GeV, then it should have been seen already by such direct dark matter detection experiments. Sneutrinos lighter than ~ 25 GeV are excluded by measurements of the properties of the Z boson at LEP (and also by the non-observation of energetic solar neutrinos in the Kamiokande detector). Thus, cosmological considerations point to the lightest neutralino, $\widetilde{Z}_1$, as being the LSP for the MSSM. It is satisfying that in most model calculations (see Chapter 11) involving the MSSM, the lightest neutralino is in fact the LSP.

---

[23] See e.g. T. Hemmick *et al.*, *Phys. Rev.* **D41**, 2074 (1990) and references therein.
[24] S. Wolfram, *Phys. Lett.* **B82**, 65 (1979); C. B. Dover, T. Gaisser and G. Steigman, *Phys. Rev. Lett.* **42**, 1117 (1979).

Another possibility is that the LSP is the gravitino.[25] In this case, gravitinos could account for the cold dark matter (CDM) in the Universe, but direct or indirect gravitino detection would likely be impossible.

9.6.1 Relic density of neutralinos

The total matter/energy density $\Omega = \rho/\rho_c$ of the Universe is usually written as a fraction in terms of the critical closure density $\rho_c = 3H_0^2/8\pi G_N \simeq 1.88 \times 10^{-29}h^2$ g cm^{-3}. Here, $H_0 \simeq 71$ km s^{-1} Mpc^{-1} is the value of the Hubble parameter today, and G_N is Newton's gravitational constant. H_0 is frequently parametrized as $H_0 \equiv 100h$ km s^{-1} Mpc^{-1}, where h is a dimensionless scaling constant.

The past decade has witnessed increasingly precise measurements of the anisotropies of the cosmic microwave background (CMB) radiation left over from the Big Bang. Recent results come from the Wilkinson Microwave Anisotropy Probe (WMAP) satellite measurements. Astonishingly, an analysis of their results pinpoints the age of the Universe to be 13.7 ± 0.2 Gyrs.[26] In addition, the geometry of the Universe is flat, consistent with simple inflationary models. The dark energy content of the Universe is found to be about 73%, while the matter content is about 27%. A best fit of WMAP and other data sets to cosmological parameters in the ΛCDM cosmological model yields a determination of baryonic matter density $\Omega_b h^2 = 0.0224 \pm 0.0009$, which is in excellent agreement with estimates from Big Bang nucleosynthesis, a total matter density of $\Omega_m h^2 = 0.135^{+0.008}_{-0.009}$, and a very low density of hot dark matter (relic neutrinos). From these values the cold dark matter density of $\Omega_{CDM} h^2 = 0.1126^{+0.0161}_{-0.0181}$ (at 2σ) can be inferred.

The discrepancy between baryonic and total matter density may come from CDM particles (that is, non-relativistic matter that does not radiate light), while the remaining energy density may come from a non-zero cosmological constant, as was first suggested by measurements of type Ia supernovae at the highest red shifts, and then strikingly confirmed by the CMB data. We will see that the lightest neutralino of supersymmetry can be an excellent candidate for CDM in the Universe.

The relic density of neutralinos predicted by the MSSM can be found by solving the Boltzmann equation as formulated for a Friedmann–Robertson–Walker (FRW) Universe:

$$\frac{dn}{dt} = -3Hn - \langle \sigma v_{rel} \rangle (n^2 - n_0^2). \tag{9.27}$$

Here, n is the number density of neutralinos, t is time, n_0 is the thermal equilibrium number density, and $\langle \sigma v_{rel} \rangle$ is the thermally averaged neutralino annihilation cross

---

[25] For a discussion of this possibility, see J. Feng, S. Su and F. Takayama, hep-ph/0404231, and references therein.
[26] See e.g. D. N. Spergel *et al.* (WMAP Collaboration), *Astrophys. J. Suppl.* **148**, 175 (2003).

section times relative velocity. (Remember that except for s-wave scattering, σv_{rel} depends on v_{rel}, so that the thermal average depends on the temperature.) The first term on the right represents a diminution of number density as the Universe expands, while the second term represents the change due to annihilation of neutralinos into SM particles.

Using conservation of entropy and the kinematics of a FRW Universe, it is convenient to reparametrize the Boltzmann equation in terms of temperature rather than time. In the radiation dominated era, the entropy density $\propto T^3$, so that the size of the Universe $R \propto 1/T$, and $t \simeq 1/(2H) = \sqrt{\frac{45}{16\pi^3 g_* G_N} \frac{1}{T^2}}$, where $g_* \sim 80$ counts the total number of relativistic degrees of freedom.[27] Defining $f = n/T^3$ and rescaling the temperature in terms of particle mass, $x = T/m$, the Boltzmann equation can be recast in the form

$$\frac{\mathrm{d}f}{\mathrm{d}x} = m\sqrt{\frac{45}{4\pi^3 g_* G_N}} \langle \sigma v_{\text{rel}} \rangle (f^2 - f_0^2). \tag{9.28}$$

The Boltzmann equation can be solved in several steps.

1. At very early times the last term in (9.27) dominates, and n is close to its equilibrium value, so that $f \simeq f_0$. For non-relativistic particles (including their rest mass), $E \simeq m + p^2/2m$, and the equilibrium number density is given by

$$
\begin{aligned}
f_0(x) &= \frac{n_0}{T^3} = \frac{1}{T^3} \frac{g}{(2\pi)^3} \int \mathrm{d}^3 p \, e^{-E/T} \\
&= \frac{1}{T^3} \frac{4\pi g}{(2\pi)^3} e^{-m/T} \int_0^\infty p^2 \mathrm{d}p \, e^{-p^2/2mT} \\
&= \frac{g}{2}\sqrt{\frac{1}{2\pi^3}} \left(\frac{m}{T}\right)^{\frac{3}{2}} e^{-\frac{m}{T}} \\
&= \frac{g}{2}\sqrt{\frac{1}{2\pi^3}} x^{-\frac{3}{2}} e^{-\frac{1}{x}}, \tag{9.29}
\end{aligned}
$$

where $g = 2$ is the number of spin degrees of freedom for a neutralino.

2. As the Universe cools to temperatures below $T = m$, the number density of neutralinos falls exponentially. However, if this would continue, neutralinos would no longer be able to annihilate efficiently, and the first term on the right-hand side of the Boltzmann equation would begin to dominate. In this regime, we would have

$$\frac{1}{n}\frac{\mathrm{d}n}{\mathrm{d}t} = -3\frac{1}{R}\frac{\mathrm{d}R}{\mathrm{d}t},$$

---

[27] E. W. Kolb and M. S. Turner, *The Early Universe*, Addison-Wesley (1990).

so that $n \propto 1/R^3$: i.e. the number density of neutralinos would reduce only due to the expansion of the Universe, and no longer drop exponentially. In other words, the number of neutralinos would be much larger than expected from thermal equilibrium. This is referred to as freeze out. The temperature at which this occurs may be estimated by using the equilibrium f on the left-hand side of (9.28) and setting $f^2 - f_0^2 \simeq f_0^2$ on its right-hand side. This then gives the freeze out temperature,

$$1/x_F = \log \left[\frac{m}{2\pi^3} \sqrt{\frac{45}{2g_* G_N}} \langle \sigma v_{rel} \rangle \sqrt{x_F} \right]. \tag{9.30}$$

This equation can be solved iteratively, and typically yields $T_F \simeq m/20$.

3. With the relic density locked in at a value much larger than its value in thermal equilibrium, $f^2 \gg f_0^2$, we can integrate the Boltzmann equation to obtain the relic density *today* as,

$$n(T_0) = \frac{1}{m} \left(\frac{T_0}{T_\gamma} \right)^3 (T_\gamma)^3 \sqrt{\frac{4\pi^3 g_* G_N}{45}} \left[\int_0^{x_F} \langle \sigma v_{rel} \rangle dx \right]^{-1}, \tag{9.31}$$

where $T_\gamma = 2.75$ K is today's cosmic microwave background temperature. We see that the relic number density $\propto T^3$ showing that it is indeed dropping only due to the expansion of the Universe as discussed above. The reason for writing the relic density in this form is that we do not know the neutralino temperature T_0: but for the fact that photons are reheated as species decouple, these two temperatures would be the same. Since the reheating process is isentropic, and $s = gT^3$, $(T_\gamma/T_0)^3$ can be obtained from the ratio of the number of degrees of freedom at freeze out to the effective number of degrees of freedom today and is approximately equal to 19.4.

The neutralino relic density can be recast in the form

$$\Omega_{\tilde{Z}_1} h^2 = \frac{\rho_{\tilde{Z}_1}(T_0)}{8.1 \times 10^{-47} \text{ GeV}^4}, \tag{9.32}$$

with $\rho_{\tilde{Z}_1} = mn(T_0)$ given by,

$$\rho_{\tilde{Z}_1}(T_0) \simeq \frac{1.66}{M_{Pl}} \left(\frac{T_0}{T_\gamma} \right)^3 T_\gamma^3 \sqrt{g_*} \frac{1}{\int_0^{x_F} \langle \sigma v_{rel} \rangle dx}. \tag{9.33}$$

Central to the calculation is the evaluation of the thermally averaged neutralino annihilation cross section times velocity. This has been simplified to a one-dimensional integral by Gondolo and Gelmini:[28]

$$\langle \sigma v_{\text{rel}} \rangle = \frac{\int \sigma v_{\text{rel}} e^{-E_1/T} e^{-E_2/T} d^3 p_1 d^3 p_2}{\int e^{-E_1/T} e^{-E_2/T} d^3 p_1 d^3 p_2}$$

$$= \frac{1}{4x K_2^2(\frac{1}{x})} \int_2^\infty da \sigma(a) a^2 (a^2 - 4) K_1(\frac{a}{4}), \qquad (9.34)$$

where $a = \sqrt{s}/m_{\tilde{Z}_1}$ and K_i are modified Bessel functions of order i. Evaluation of the relic density thus requires the knowledge of all neutralino annihilation cross sections $\tilde{Z}_1 \tilde{Z}_1 \rightarrow f_1 f_2$, where f_1 and f_2 are SM particles. How to compute cross sections, starting with the interactions derived in the last chapter, will be discussed in Chapter 12.

If there are other sparticles with mass close to the LSP mass, these will also be present in the thermal bath right up to the time that the LSP decouples. In this case, it is necessary to take into account SUSY processes involving annihilation of pairs of these sparticles as well as co-annihilation of these sparticles and the LSP to accurately obtain the neutralino relic density. Although the number density of the heavier sparticles is suppressed by the Boltzmann factor $\exp(-m/T)$, this may be compensated for by the fact that the cross sections for co-annihilation or pair annihilation may be much larger than the LSP annihilation cross section. For instance, if the $\tilde{\tau}_1$ is close in mass to a gaugino-like $\tilde{Z}_1$, its annihilation rate may be much larger than the annihilation rate for $\tilde{Z}_1$ pairs. Alternatively, in models with small values of μ, $m_{\tilde{Z}_1} \sim m_{\tilde{Z}_2} \sim m_{\tilde{W}_1}$, and $\sigma_{\tilde{Z}_1 \tilde{Z}_2}$ or $\sigma_{\tilde{W}_1 \tilde{W}_1}$ (which are not P-wave suppressed at threshold) may be much larger than the annihilation cross section for two $\tilde{Z}_1$s.

The WMAP determination $\Omega_{\text{CDM}} h^2 = 0.1126^{+0.0161}_{-0.0181}$ implies an upper limit $\Omega_{\text{WIMP}} h^2 < 0.129$ (2σ) on the relic density of any stable WIMP. Only if we further assume that the cold dark matter consists solely of a single component can we infer that the relic density of any particular WIMP (in our case $\Omega_{\tilde{Z}_1} h^2$) will saturate the WMAP value. In Fig. 9.14, we show regions of relic density $\Omega_{\tilde{Z}_1} h^2$ in the m_0 vs. $m_{1/2}$ plane for $a)$ $\tan \beta = 10$ and $\mu > 0$ and (b) $\tan \beta = 45$ for $\mu < 0$, where $A_0 = 0$ and $m_t = 175$ GeV. The very dark gray regions on the right and far left are excluded by either not having a neutralino LSP, or not having the correct EWSB pattern. The white regions for both $\tan \beta$ values have $\Omega_{\tilde{Z}_1} h^2 > 1$, so that the Universe would be younger than 10 billion years. The appropriately labeled light gray region is where $\Omega_{\tilde{Z}_1} h^2 \leq 0.1$, and can be regarded as the theoretically favored region. Four regions of parameter space emerge with $\Omega_{\tilde{Z}_1} h^2 < 0.129$, as determined by the WMAP analysis. In frame (a), we see:

[28] P. Gondolo and G. Gelmini, *Nucl. Phys.* **B360**, 145 (1991). Formulae including co-annihilation effects can be found in J. Edsjo and P. Gondolo, *Phys. Rev.* **D56**, 1879 (1997).

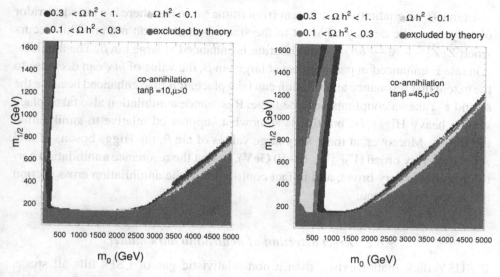

Figure 9.14 Predictions for neutralino relic density $\Omega_{\widetilde{Z}_1} h^2$ in the m_0–$m_{1/2}$ plane of the mSUGRA model for $\mu > 0$ and two values of $\tan\beta$. We thank A. Belyaev for supplying this figure.

- the bulk annihilation region,
- the stau co-annihilation region, and
- the HB/FP region.

In addition, in frame (*b*), we see

- the *A*-annihilation funnel.

The bulk annihilation region occurs at low m_0 and low $m_{1/2}$ where neutralino annihilation mainly occurs via $\widetilde{Z}_1\widetilde{Z}_1 \to \ell\bar{\ell}$, via *t*-channel slepton exchange. As m_0 increases, the slepton masses also increase, suppressing the neutralino annihilation rate and increasing the relic density. The stau co-annihilation region is the narrow corridor of favored relic density adjacent to the region where $\tilde{\tau}_1$ becomes the LSP; this is where $\widetilde{Z}_1\tilde{\tau}_1$ and $\tilde{\tau}_1\bar{\tilde{\tau}}_1$ co-annihilation can take place. The HB/FP region occurs at large m_0 along the lack of REWSB excluded region. In this area, since $|\mu|$ is becoming small, the $\widetilde{Z}_1$ becomes increasingly higgsino-like, and annihilation into WW, ZZ, and Zh states becomes large. Directly adjacent to the REWSB excluded region, where $\mu \to 0$, co-annihilation of $\widetilde{Z}_1$ with $\widetilde{W}_1$ and $\widetilde{Z}_2$ is also important. There is also a narrow strip of low relic density at $m_{1/2} \sim 160\,\mathrm{GeV}$ just beyond the reach of LEP2 where neutralino annihilation via the narrow light Higgs *h* resonance occurs: $\widetilde{Z}_1\widetilde{Z}_1 \to h \to b\bar{b}, \ \tau\bar{\tau}$.

Frame (b) is qualitatively different from frame (a) in that there is a broad corridor of very low relic density adjacent to the stau co-annihilation region. This occurs when $\widetilde{Z}_1 \widetilde{Z}_1 \to A \to b\bar{b}$, $\tau\bar{\tau}$ annihilation is enhanced at large $\tan \beta$. The annihilation rate is enhanced in part because at large $\tan \beta$, the value of m_A can decrease to the extent that resonance annihilation can take place. It is also enhanced because the b- and τ-Yukawa couplings become large. Resonance annihilation also takes place via the heavy Higgs H, but this is somewhat suppressed relative to annihilation through A. Moreover, at these very large values of $\tan \beta$, the Higgs bosons H and A become very broad ($\Gamma_{H,A} \sim 10$–50 GeV), so that the resonance annihilation corridor becomes very broad, and in fact contributes to the annihilation cross section across the entire plane.

9.6.2 Direct detection of neutralino dark matter

If SUSY dark matter exists, then a non-relativistic gas of LSPs fills all space. Moreover, the LSPs are gravitationally clumped to form a galactic dark matter halo. A number of direct WIMP detection experiments have been built or are under construction to detect this halo. The general idea behind these experiments is that as the earth moves through this halo, relic WIMPs, be they neutralinos or something else, will scatter off the nuclei in some material, depositing typically tens of keV of energy. The energy that is deposited could be detected via: (i) changes in resistance due to a slight temperature increase (bolometry), (ii) a magnetic flux change due to a superconducting granule phase transition, (iii) ionization, or (iv) phonons.[29] Sneutrinos have a large scattering cross section, and it is the lack of a signal in such experiments that disfavors the sneutrino as the LSP. Neutralino cross sections are much smaller, and require higher sensitivity for their detection. Gravitinos are essentially undetectable. The technical challenge is to build detectors that could pick out the relatively rare, low energy neutralino scattering events from backgrounds mainly due to cosmic rays and radioactivity in surrounding matter. Future detectors are aiming to reach a sensitivity of 0.01–0.001 events kg^{-1} day^{-1}. It is possible that the first evidence for SUSY may come from direct neutralino detection rather than from accelerator experiments, though identifying the SUSY origin of the signal may require other analyses.

The first step involved in a neutralino–nucleus scattering calculation is to calculate the effective four-particle neutralino–quark and neutralino–gluon interactions. The neutralino–quark axial vector interaction leads, in the non-relativistic limit, to a neutralino–nucleon spin-spin interaction, which involves the measured quark spin content of the nucleon. To obtain the neutralino–nucleus scattering cross section, a

---

[29] For a review, see G. Jungman, M. Kamionkowski and K. Griest, *Phys. Rep.* **267**, 195 (1996).

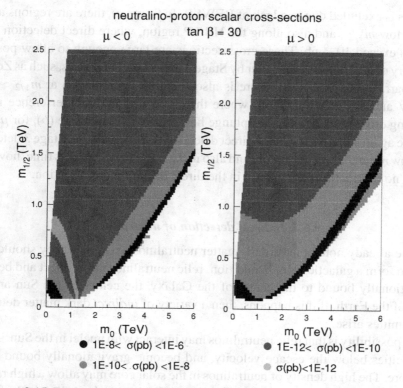

Figure 9.15 Regions of scalar neutralino–proton cross sections in the mSUGRA model, in units of pb. The blank regions are excluded by theoretical and experimental considerations. We thank J. O'Farrill for supplying this figure.

convolution with nuclear spin form factors must be performed. Neutralino–quark and neutralino–gluon interactions (via loop diagrams) can also resolve into scalar and tensor components. These interactions can then be converted into an effective scalar neutralino–nucleon interaction involving quark and gluon parton distribution functions. The neutralino–nucleus scattering cross section can be obtained by convoluting with suitable scalar nuclear form factors. The final neutralino detection rate is obtained by multiplying by the *local* neutralino relic density (estimates are obtained from galaxy formation modeling), and appropriate functions involving the velocity distribution of relic neutralinos and the Earth's velocity around the Sun and around the galactic center. When the Earth's velocity around the Sun is aligned with the Sun's galactic velocity, the scattering rate should increase, leading to a seasonal modulation of these direct detection rates.

In Fig. 9.15, we show regions of scalar neutralino–nucleus cross section in the mSUGRA model for $\tan \beta = 30$, $A_0 = 0$, and (a) $\mu < 0$ or (b) $\mu > 0$. The left-hand side of the plots is excluded because $\tilde{\tau}_1$ becomes the LSP, and the lower right side of the plots is excluded due to a lack of REWSB. In frame a), there are regions at low m_0 and low $m_{1/2}$, and also along the HB/FP region, where direct detection cross sections exceed 10^{-9} pb. These cross sections are large enough to allow possible discovery of neutralino dark matter by Stage 3 dark matter detectors, such as Zeplin-4, Cryoarray, and, XENON. There is also a region in frame (a) at $m_{1/2} \sim 0.3$–0.8 TeV and $m_0 \sim 0.2$–1.2 TeV, where there is a destructive interference in the scattering cross section, and rates plunge below 10^{-12} pb. In frame (b), for $\mu > 0$, there are again sufficient rates for direct detection of dark matter at Stage 3 detectors in the low m_0 and low $m_{1/2}$ region, and also in the HB/FP region. This time, however, there is no destructive interference in the direct detection cross section.

9.6.3 Indirect detection of neutralinos

We have already noted that if dark matter neutralinos exist, then they should condense to form a galactic halo. In addition, relic neutralinos may collect and become gravitationally bound to the center of the Galaxy, the center of the Sun and the center of the Earth. If this happens, then a variety of *indirect* dark matter detection opportunities arise.

One possibility is that relic neutralinos may interact with nuclei in the Sun, scatter to velocities below the escape velocity, and become gravitationally bound in the solar core. The high density of neutralinos in the solar core may allow a high rate for neutralino annihilation into SM particles. (Neutralinos may also collect in the core of the Earth and experience enhanced annihilation, but rates are typically smaller than from the Sun.) Most SM annihilation products will be immediately absorbed by the solar material. However, high energy neutrinos arising from neutralino annihilation may escape the Sun, and be detected by neutrino telescopes such as Antares (a water Cherenkov device in the Mediterranean) or IceCube (an array of phototubes deployed in the ice at the South Pole). Muon neutrinos would convert to muons in the water or ice, and Cherenkov radiation from the muons could be detected. The rate for neutralino annihilation in the solar core is given by

$$\Gamma = \frac{1}{2} C \tanh^2(\sqrt{CA}t_\odot), \tag{9.35}$$

where C is the solar capture rate, A is the total annihilation rate times relative velocity per unit volume, and $t_\odot$ is the present age of the Sun. For the Sun, the age of the Solar System exceeds the equilibration time, so $\Gamma \sim C/2$. Thus, highest rates for neutrinos from solar core annihilations occur in parameter space regions where the neutralino–nucleus scattering cross section is largest. From Fig. 9.15, this would mean the bulk annihilation region or the HB/FP region. It is intriguing that these regions also have low relic densities in accord with the WMAP analysis.

Another possibility for indirect neutralino dark matter detection occurs if neutralinos annihilate in the galactic core or halo to SM particles. High energy photons can be produced as part of the annihilation products, and can be detected by gamma ray observatories. In this case, the highest rates for gamma ray detection occur in regions of parameter space where the neutralino annihilation cross section times velocity is largest: i.e. in the bulk annihilation region, the HB/FP region or in the A-annihilation funnel. Since the neutralino density is expected to be high around the galactic core, a directional signal may be found emanating from this source. Neutralinos may also annihilate via loop diagrams as $\widetilde{Z}_1 \widetilde{Z}_1 \rightarrow \gamma\gamma$. In this case, the rate would be quite low, but the signature spectacular, since the gamma ray energy would be essentially equal to $m_{\widetilde{Z}_1}$.

Finally, neutralinos present in the galactic halo may also annihilate, leading to positrons or antiprotons, which may be detected by cosmic ray detectors. In this case, the e^+s or $\bar{p}$s would likely be non-directional, since their path of flight would be bent by galactic magnetic fields. Again, the highest rates are to be expected where the neutralino annihilation cross section is highest: in the bulk region, the HB/FP region or the A-annihilation funnel. The rates for detection of the indirect signals depend on assumptions regarding the density profile of neutralinos in the galactic core and halo. Clearly, if clumping of dark matter occurs, then rates may be higher than expected. Alternatively, if the galactic halo neutralino density profile has been overestimated, then signal rates may be lower than expected.

In principle, the results of direct and indirect dark matter searches may pinpoint the mechanism responsible for depletion of neutralinos in the early Universe so that the current value of the relic density is consistent with WMAP results. The bulk annihilation region at low m_0 and low $m_{1/2}$ may give rise to large signal rates in *all* direct and indirect search experiments. However, this region is largely disallowed due to large contributions to $(g - 2)_\mu$, $BF(b \rightarrow s\gamma)$ and a value of m_h lower than bounds from LEP2. The stau co-annihilation region is likely to give *no* signals for neutralino direct or indirect detection, while *all* signals for direct and indirect neutralino detection may be possible in the HB/FP region. If neutralino annihilation through the broad A and H resonances is the main sink for neutralinos in the early Universe, then direct neutralino detection and also detection of neutrinos from neutralino annihilation in the core of the Sun are unlikely. However, detection of γs, e^+s, and $\bar{p}$s from neutralino annihilation in the galactic core and halo may occur at detectable rates.

9.7 Neutrino masses

Data from solar and atmospheric neutrino experiments provide unambiguous evidence for neutrino oscillations, and strongly suggest an interpretation in terms of neutrino masses and mixings. These data are consistent with a hierarchical structure of neutrino masses, with $m_{\nu_e} \ll m_{\nu_\mu} \ll m_{\nu_\tau}$ and $m_{\nu_\tau} \sim 0.05$ eV and near-maximal neutrino mixing, though other mass patterns are certainly possible. Cosmological data also tell us that neutrinos are all lighter than a few eV.

We have seen, however, that like the SM, the MSSM (which after all is essentially a direct supersymmetrization of the SM) does not allow for neutrino masses. As in the SM, one can allow for lepton number conserving Dirac neutrino masses by introducing neutrino Yukawa couplings into the superpotential. This necessarily entails the introduction of new right-handed neutrino (RHN) superfields. Since neutrinos are electrically neutral, it is also possible to introduce lepton number violating Majorana mass terms for these. Within the SM, the well-known see-saw mechanism provides an elegant way of obtaining the small values of neutrino masses indicated by the data;[30] in the supersymmetric context, this again requires the introduction of RHN superfields. Within the supersymmetric framework, Majorana masses for neutrinos are also obtained if the superpotential includes lepton number and *R*-parity violating interactions, *without* the need for any new RHN superfields. This is discussed in detail in Chapter 16. Here, we will confine our attention to the simplest extension of the MSSM that accommodates the incorporation of neutrino masses via the see-saw mechanism. This, of course, requires us to extend the superfield content of the MSSM by the RHN superfields, one for each generation.

9.7.1 The MSSM plus right-handed neutrinos

In order to implement the see-saw mechanism for neutrino masses, we are led to introduce three additional gauge singlet left-chiral scalar superfields $\hat{N}_i^c$ ($i = 1$–3 denotes the generation),

$$\hat{N}_i^c = \tilde{v}_{Ri}^\dagger(\hat{x}) + i\sqrt{2}\bar{\theta}\,\psi_{N_i^c L}(\hat{x}) + i\bar{\theta}\theta_L \mathcal{F}_{N_i^c}(\hat{x}),$$

whose Majorana fermion component destroys left-handed $SU(2)$ singlet anti-neutrinos, or create the corresponding right-handed neutrinos (v_{Ri}). These singlet superfields are coupled to other MSSM superfields via the superpotential

$$\hat{f} = \hat{f}_{MSSM} + (\mathbf{f}_v)_{ij}\epsilon_{ab}\hat{L}_i^a \hat{H}_u^b \hat{N}_j^c + \frac{1}{2}M_{Ni}\hat{N}_i^c \hat{N}_i^c, \tag{9.36}$$

---

[30] The see-saw mechanism has an interesting history. To our knowledge, the see-saw formula for the neutrino mass first appears in H. Fritzsch and P. Minkowski, *Phys. Lett.* **B62**, 72 (1976) and P. Minkowski, *Phys. Lett.* **B67**, 421 (1977). The mechanism was independently invented and cast into its modern form by T. Yanagida, KEK Report No. 79-18 (1979); M. Gell-Mann, P. Ramond and R. Slansky, in *Supergravity*, D. Freedman *et al.*, Editors, North-Holland, Amsterdam (1980); S. Glashow, in *Quarks and Leptons, Cargèse 1979*, M. Lévy *et al.*, Editors, Plenum (1980); R. Mohapatra and G. Senjanovic, *Phys. Rev. Lett.* **44**, 912 (1980). For a recent review of the original idea and its variants, see e.g. R. Mohapatra, hep-ph/9910365.

where summation over generation indices i and j as well as $SU(2)$ indices a and b is implied. Notice that the superpotential includes lepton number violating Majorana mass parameters M_{Ni} for these right-handed neutrinos. In (9.36) we have, without loss of generality, chosen a basis for the RHN superfields so that the superpotential mass terms are diagonal. Since the mass terms for these gauge singlet superfields are not forbidden by symmetry considerations (other than ad hoc global symmetries such as lepton number conservation), these are naturally expected to be large – $\sim M_{\text{Planck}}$ in the present framework, or comparable to the $SO(10)$ breaking scale if the model is embedded into an $SO(10)$ GUT, as discussed in Chapter 11. Indeed, values of M_{Ni} well beyond the weak scale and ranging up to M_{GUT} are favored by SUSY GUT models which seek to explain neutrino oscillation data. When electroweak symmetry is broken, Dirac neutrino mass entries $(m_D)_{ij}$ are also induced. The resulting 6×6 neutrino mass matrix must be diagonalized to obtain the masses of the physical neutrinos. Assuming that $(m_D)_{ij} \ll M_i$ for all i and j, we know that there must be three nearly sterile, heavy Majorana neutrinos with masses very close to M_i. Since these essentially saturate the trace, there must be three light active Majorana neutrinos with masses that depend on the details of the Dirac mass matrix, but whose values vary inversely as the M_i. In the limit where we ignore the mixing of active neutrino flavors (not a good approximation to the data), the formulae become simple and we have $m_{\nu_i} \simeq m_{Di}^2/M_{Ni}$ as the mass of the active neutrino of generation i. Though the neutrino masses would be different in the case of mixed neutrinos, we would expect that this simple formula reproduces their order of magnitude.

The soft SUSY breaking terms must now also be augmented to include

$$\mathcal{L} \ni \mathcal{L}_{\text{MSSM}} - \tilde{\nu}_{Ri}^\dagger \mathbf{m}_{\tilde{\nu}_R ij}^2 \tilde{\nu}_{Rj} + \left[(\mathbf{a}_\nu)_{ij} \epsilon_{ab} \tilde{L}_i^a \tilde{H}_u^b \tilde{\nu}_{Rj}^\dagger + \frac{1}{2} b_{\nu ij} \tilde{\nu}_{Ri} \tilde{\nu}_{Rj} + \text{h.c.} \right],$$

$$(9.37)$$

where once again a summation over repeated indices is implied. The parameters $(\mathbf{m}_{\tilde{\nu}_R})_{ij}$, $(\mathbf{a}_\nu)_{ij}$, and $b_{\nu ij}$ are assumed to be of order the weak scale. Assuming that $M_i \gg M_W$, the right-handed sneutrinos have masses $\sim M_{Ni}$ and, like the ν_{Ri}'s, decouple from the low energy theory. When considering the renormalization group evolution of couplings and SSB parameters, one must remember that for energy scales above $Q = M_{Ni}$, the effective theory is the MSSM augmented by the corresponding right-handed neutrinos and sneutrinos, while at scales below the smallest of the M_{Ni}, these are all integrated out, leaving the MSSM as the effective field theory. Indeed, the RGEs of the MSSM must be augmented to include potentially significant effects of the new neutrino Yukawa couplings.[31]

---

[31] The RGEs for the MSSM augmented by a RHN are listed, for instance, in H. Baer *et al.*, *JHEP* **04**, 016 (2000).

The inclusion of the new neutrino-sector superpotential and soft SUSY breaking parameters can in general lead to lepton-flavor-violating processes (LFV).[32] Even in the case where one assumes mSUGRA-like conditions at $Q = M_{GUT}$ on the new parameters, neutrino Yukawa coupling contributions to renormalization group evolution between M_{GUT} and M_{Ni} can induce off-diagonal slepton mass matrix entries that lead to LFV processes like $\mu \to e\gamma$, $\mu \to e$ conversion, $\tau \to \mu\gamma$ or $\mu^- \to e^+ e^- e^-$ at potentially observable rates. The stringent experimental limits on these rare decays strongly constrain the neutrino sector parameters. In the future, discovery of LFV processes may help pin down the parameters associated with the right-handed neutrinos. Of course, LFV could also show up in the direct decays of sparticles, for instance $\widetilde{Z}_j \to \ell_1^+ \ell_2^- \widetilde{Z}_i$ or $\tilde{\ell}_1 \to \ell_2 \widetilde{Z}_i$, if these are produced at future colliders. We should also mention that renormalization effects from neutrino Yukawa couplings would cause small inter-generation splitting between the sneutrinos, in much the same way that the tau Yukawa interaction splits $m_{\tilde{e}_L}$ from $m_{\tilde{\tau}_L}$. These splittings may provide a direct test of the see-saw mechanism if sneutrino masses are precisely measured in the future.

Finally, we mention that the existence of long-lived, heavy right-handed Majorana neutrinos and sneutrinos offers a novel solution of the baryogenesis problem via leptogenesis, provided that neutrino Yukawa interactions also violate CP conservation. This is possible in the same way that SM quark interactions do not conserve CP. Assuming that CP is violated, there will be a difference in the rates for the decay of ν_R into leptons and antileptons at the one-loop level, so that a leptonic matter–antimatter asymmetry can be induced at temperature $T \lesssim M_{Ni}$, below which the right-handed neutrinos and sneutrinos fall out of thermal equilibrium. This lepton asymmetry is then converted to a baryon asymmetry at lower temperatures via sphaleron interactions, as discussed in Chapter 16.[33]

[32] For an overview, see Y. Kuno and Y. Okada, *Rev. Mod. Phys.* **73**, 151 (2001).
[33] See M. Fukugita and T. Yanagida, *Phys. Lett.* **B174**, 45 (1986).

10

Local supersymmetry

We know that the superpartners of SM particles must acquire SUSY breaking masses, since otherwise they would have been produced in experiments via their gauge interactions. This requires an understanding of the mechanism of supersymmetry breaking. A variety of models for supersymmetry breaking have been postulated in the literature. The general consensus seems to be that the SM superpartners cannot acquire tree-level masses via spontaneous breaking of *global supersymmetry* at the TeV scale: we have seen in Chapter 7 that this leads to phenomenological problems with tree-level sum rules which imply that some sfermions must be lighter than fermions. Within the framework of the MSSM our ignorance of the SUSY breaking mechanism is parametrized by 178 soft SUSY breaking parameters.

The MSSM is, therefore, regarded as a low energy effective theory to be derived from a theory that incorporates supersymmetry breaking. In the next chapter, we will discuss various models for the generation of soft SUSY breaking parameters that have been suggested in the literature. These models circumvent the problems with the sum rules in one of two different ways. Either the models are based on local supersymmetry, or the soft SUSY breaking parameters are generated only at the loop level. As preparation for a discussion of the first of these classes of models, in this chapter we present a short discussion of locally supersymmetric theories where the parameters of SUSY transformations depend on the spacetime co-ordinates. Since supersymmetry is a spacetime symmetry, local supersymmetry necessarily involves gravitation. Local supersymmetry is, therefore, also referred to as supergravity. Supergravity is a large and complex subject in its own right, and its elaboration is beyond the scope of this book. Our purpose here is only to provide the reader with the basic ideas so as to facilitate the development of particle physics models based on it. We begin by reviewing general relativity, the classical theory of gravitation whose supersymmetric extension naturally leads to supergravity.

10.1 Review of General Relativity

Before proceeding to discuss supergravity, it will be useful to review the classical theory of gravitation, as embodied in Einstein's General Relativity (GR). In GR, physics is formulated on a curved four-dimensional spacetime manifold, and gravitation is a manifestation of this curvature.

The principle of *special* relativity states that the laws of physics are the same for all *inertial* observers. This is bothersome obviously because we can evidently discern the laws of physics, even though we live on Earth in an accelerating frame. Einstein generalized the principle of special relativity to include *all* observers, including those in accelerating frames.

Einstein was deeply impressed by demonstrations such as the Eötvös experiment that gravitational and inertial mass were equal to very high precision. He reasoned that in a freely falling elevator, one would not be able to discern any effects of gravitation via any experiment confined to a sufficiently small region of measurement. This led to the formulation of the *principle of equivalence*, which is one of the cornerstones of GR. It states that in an arbitrary gravitational field one can always transform co-ordinates to a freely falling (locally Lorentz) frame, where effects of gravitation are locally eliminated. In this freely falling frame, the laws of physics take their special relativistic form. Einstein described the equivalence principle as "the happiest thought of my life".

The effects of gravitation can be incorporated by starting with (local) equations that we know to hold in the absence of gravitation, and generalizing these to be form invariant under general co-ordinate transformations. This is so because the equivalence principle tells us that we can always transform to a co-ordinate system (the freely falling frame) in which the effects of gravity are locally absent. To make the equations form-invariant, we will see that we are led to introduce new "fields" (the affine connection introduced below) that incorporate the effects of gravitation. The situation is quite analogous to that in local gauge theories where, to maintain the invariance of the field equations under local gauge transformations, one is forced to introduce the vector fields and the related field strength tensors.

10.1.1 General co-ordinate transformations

General relativity requires the laws of physics to be the same for *any* observer, be they in a co-ordinate system which is rotating, accelerating, or whatever. Whether we use a co-ordinate system x^μ or $x'^\mu = x'^\mu(x)$, we should arrive at the same physical equations, except that the quantities would appear in a different co-ordinate system. This means that the equations describing the laws of physics take the *tensor* form.

Denoting a general co-ordinate transformation (GCT) by

$$x^\mu \to x'^\mu = x'^\mu(x) \quad \text{(GCT)}, \tag{10.1}$$

the differential line element dx^μ transforms under GCTs as

$$dx^\mu \to dx'^\mu = \frac{\partial x'^\mu}{\partial x^\nu} dx^\nu. \tag{10.2}$$

By the chain rule for differentiation, we note that $\partial x^\mu / \partial x'^\nu$ is the inverse of the transformation matrix $\partial x'^\nu / \partial x^\rho$ that appears in the GCT of the line element in (10.2). The differential volume element

$$d^4 x' = dx'^0 dx'^1 dx'^2 dx'^3 = J dx^0 dx^1 dx^2 dx^3, \tag{10.3}$$

where the Jacobian is the determinant of the transformation matrix $J = |\partial x'^\mu / \partial x^\nu|$.

An object is a *contravariant* vector under GCTs if its components transform as,

$$V^\mu \to V'^\mu = \frac{\partial x'^\mu}{\partial x^\nu} V^\nu. \tag{10.4a}$$

The differential line element dx^μ is thus a contravariant vector. Contravariant tensors of rank n are objects with n indices whose components transform as,

$$A^{\mu_1 \mu_2 \dots \mu_n} \to A'^{\mu_1 \mu_2 \dots \mu_n} = \frac{\partial x'^{\mu_1}}{\partial x^{\rho_1}} \frac{\partial x'^{\mu_2}}{\partial x^{\rho_2}} \cdots \frac{\partial x'^{\mu_n}}{\partial x^{\rho_n}} A^{\rho_1 \rho_2 \dots \rho_n}, \tag{10.4b}$$

while scalars transform as

$$\phi \to \phi' = \phi. \tag{10.4c}$$

A scalar may thus be thought of as a tensor of rank zero, and a vector as a tensor of rank one.

The derivative of a scalar function $\phi(x)$, which under a GCT becomes $\phi'(x')$, transforms as

$$\frac{\partial \phi}{\partial x^\mu} \to \frac{\partial \phi'}{\partial x'^\mu} = \frac{\partial \phi}{\partial x'^\mu} = \frac{\partial \phi}{\partial x^\nu} \frac{\partial x^\nu}{\partial x'^\mu}. \tag{10.5a}$$

The transforming matrix is the inverse of the transformation matrix for contravariant vectors. Objects which transform like $\partial \phi / \partial x^\mu$, i.e. as

$$V_\mu \to V'_\mu = \frac{\partial x^\nu}{\partial x'^\mu} V_\nu \tag{10.5b}$$

are known as covariant vectors. Covariant tensors of rank n are defined to be objects with n indices whose components transform as,

$$A_{\mu_1 \mu_2 \dots \mu_n} \to A'_{\mu_1 \mu_2 \dots \mu_n} = \frac{\partial x^{\rho_1}}{\partial x'^{\mu_1}} \frac{\partial x^{\rho_2}}{\partial x'^{\mu_2}} \cdots \frac{\partial x^{\rho_n}}{\partial x'^{\mu_n}} A_{\rho_1 \rho_2 \dots \rho_n}. \tag{10.5c}$$

Notice that the indices corresponding to contravariant components are written as superscripts, while those corresponding to covariant components are written as subscripts. In this sense, it is convenient to write $\partial \phi / \partial x^\mu$ as $\partial_\mu \phi$.

Mixed tensors with n contravariant and m covariant indices are analogously defined.

Exercise *If $A^{\mu_1 \mu_2 \ldots \mu_n}$ and $B^{\nu_1 \nu_2 \ldots \nu_m}$ are contravariant components of tensors with rank n and m, respectively, show that the entity S with $n + m$ indices defined by $S^{\mu_1 \mu_2 \ldots \mu_n \nu_1 \nu_2 \ldots \nu_m} = A^{\mu_1 \mu_2 \ldots \mu_n} B^{\nu_1 \nu_2 \ldots \nu_m}$ transforms as a contravariant tensor of rank $n + m$. An analogous result also holds for covariant as well as mixed tensors.*

Exercise *If $A^{\mu_1 \mu_2 \ldots \mu_n}_{\nu_1 \nu_2 \ldots \nu_m}$ is a mixed tensor with n contravariant and m covariant indices, show that $A^{\mu_1 \mu_2 \ldots \mu_n}_{\mu_1 \nu_2 \ldots \nu_m}$ (where the index μ_1 is summed over) is a mixed tensor with $n - 1$ contravariant and $m - 1$ covariant indices.*

Exercise *Verify that if a tensor is zero in one frame, it is zero in all frames. Convince yourself that this implies that tensor equations retain their form under GCTs. This is why we required that the equations of GR should take the tensorial form.*

Exercise *Let $A^{\mu\nu \ldots \sigma} B_{\alpha\beta \ldots \sigma} = T^{\mu\nu \ldots}_{\alpha\beta \ldots}$ where A and T transform as tensors of the appropriate rank. Show that B transforms as a tensor. We will use this result to show that the "metric tensor" indeed transforms as a tensor.*

10.1.2 Covariant differentiation, connection fields, and the Riemann curvature tensor

We have just seen that the derivative of a scalar function gives us a vector function. It is, therefore, reasonable to ask whether the derivative of a tensor function results in a tensor with rank higher by one. To check this, we consider how the derivative of a (first rank) tensor transforms under a GCT: $\partial V^\mu / \partial x^\nu$. Under a GCT, this transforms as

$$\frac{\partial V^\mu}{\partial x^\nu} \to \frac{\partial V'^\mu}{\partial x'^\nu} = \frac{\partial}{\partial x'^\nu} \left(\frac{\partial x'^\mu}{\partial x^\rho} V^\rho \right)$$

$$= \frac{\partial x^\sigma}{\partial x'^\nu} \frac{\partial}{\partial x^\sigma} \left(\frac{\partial x'^\mu}{\partial x^\rho} V^\rho \right)$$

$$= \frac{\partial x^\sigma}{\partial x'^\nu} \frac{\partial x'^\mu}{\partial x^\rho} \frac{\partial V^\rho}{\partial x^\sigma} + \frac{\partial x^\sigma}{\partial x'^\nu} \frac{\partial^2 x'^\mu}{\partial x^\sigma \partial x^\rho} V^\rho. \tag{10.6}$$

The presence of the second term in the last line shows that $\partial V^\mu / \partial x^\nu$ does *not* transform as a tensor.[1] The situation is reminiscent of that encountered in local gauge theory. If the field transformed according to some representation of the gauge group, the ordinary derivative of the field did not transform properly. In the same spirit, we introduce a covariant derivative,

$$D_\nu V^\mu \equiv \partial_\nu V^\mu + \Gamma^\mu_{\rho\nu} V^\rho, \tag{10.7}$$

where $\Gamma^\mu_{\rho\nu}(x)$ is a *connection* field,[2] analogous to the vector potential in the covariant derivative of gauge theories. We require that $D_\nu V^\mu$ transforms as a tensor under GCT:

$$D_\nu V^\mu \rightarrow D'_\nu V'^\mu = \frac{\partial x^\sigma}{\partial x'^\nu} \frac{\partial x'^\mu}{\partial x^\rho} D_\sigma V^\rho.$$

This then implies that the connection must transform as

$$\Gamma'^\mu_{\rho\nu} = \frac{\partial x^\sigma}{\partial x'^\nu} \frac{\partial x'^\mu}{\partial x^\tau} \frac{\partial x^\lambda}{\partial x'^\rho} \Gamma^\tau_{\lambda\sigma} - \frac{\partial x^\sigma}{\partial x'^\nu} \frac{\partial x^\tau}{\partial x'^\rho} \frac{\partial^2 x'^\mu}{\partial x^\sigma \partial x^\tau}. \tag{10.8}$$

Evidently the connection field does not transform as a tensor; its transformation property is that of an *affine connection*. If we construct the symmetric and antisymmetric parts of the affine connection under interchange of the lower indices,

$$\Gamma^\mu_{\rho\nu} = \frac{1}{2} \left(\Gamma^\mu_{\rho\nu} + \Gamma^\mu_{\nu\rho} \right) + \frac{1}{2} \left(\Gamma^\mu_{\rho\nu} - \Gamma^\mu_{\nu\rho} \right) \equiv S^\mu_{\rho\nu} + A^\mu_{\rho\nu}, \tag{10.9}$$

it is easy to see that the antisymmetric piece $A^\mu_{\rho\nu}$ transforms as a tensor. This tensor is known as the torsion tensor. The torsion tensor, usually taken to be zero in GR, does not vanish in supergravity theories when gravitinos (see below) are present.

Since the gradient of a scalar field transforms as a vector, the covariant derivative of a scalar is the same as its ordinary derivative: $\partial_\mu \phi = D_\mu \phi$. If we require that the covariant derivative satisfy the usual product rule, then

$$D_\nu \left(V^\mu W_\mu \right) = (D_\nu V^\mu) W_\mu + V^\mu \left(D_\nu W_\mu \right) = \partial_\nu \left(V^\mu W_\mu \right),$$

for any contravariant vector V^μ and any covariant vector W_μ. This is only possible if $D_\nu W_\mu = \partial_\nu W_\mu - \Gamma^\rho_{\mu\nu} W_\rho$, i.e. the connection field enters with a minus sign for derivatives of covariant vectors. Covariant derivatives of higher rank tensors can be made by simply introducing a connection field term for each index: e.g. $D_\mu A^\rho_\nu = \partial_\mu A^\rho_\nu + \Gamma^\rho_{\sigma\mu} A^\sigma_\nu - \Gamma^\sigma_{\nu\mu} A^\rho_\sigma$.

Unlike ordinary derivatives, covariant derivatives (except when they act on scalar functions) do not commute. We had already noted this when we considered gauge theories, where we had seen that the commutator of covariant derivatives yields the

---

[1] Note that if the transformation $x \rightarrow x'$ is linear (as is the case for special relativity), this offending second term would be absent.

[2] Manifolds on which a continuous connection field can be defined are known as *affine manifolds*.

field strength tensor $F_{\mu\nu A}$ (see Eq. (6.46)). We can perform a similar exercise in GR:

$$\left[D_\mu, D_\nu\right] V^\rho = R^\rho_{\ \tau\mu\nu} V^\tau + 2A^\tau_{\mu\nu} D_\tau V^\rho, \tag{10.10}$$

where

$$R^\rho_{\ \tau\mu\nu} = \partial_\mu \Gamma^\rho_{\tau\nu} - \partial_\nu \Gamma^\rho_{\tau\mu} + \Gamma^\rho_{\sigma\mu} \Gamma^\sigma_{\tau\nu} - \Gamma^\rho_{\sigma\nu} \Gamma^\sigma_{\tau\mu} \tag{10.11}$$

defines the Riemann curvature tensor, and $A^\tau_{\mu\nu}$ is the torsion tensor.

Exercise *This exercise illustrates the use of the equivalence principle described at the beginning of this section.*

In the absence of gravitation (and any other forces) the equation of motion for a spinless particle is

$$\frac{\mathrm{d}^2 x^\mu}{\mathrm{d}\tau^2} = 0. \tag{10.12a}$$

Even in the presence of gravitation, this equation still holds true in the freely falling frame, according to the principle of equivalence. A GCT into any other (non-freely falling) frame with co-ordinates $x'^\mu(x)$ implies that

$$\frac{\mathrm{d}^2 x^\mu}{\mathrm{d}\tau^2} \rightarrow \frac{\partial x'^\mu}{\partial x^\nu} \frac{\mathrm{d}^2 x^\nu}{\mathrm{d}\tau^2} + \frac{\partial^2 x'^\mu}{\partial x^\lambda \partial x^\nu} \frac{\partial x^\nu}{\partial \tau} \frac{\partial x^\lambda}{\partial \tau}.$$

Notice that the second term (whose presence tells us that $\mathrm{d}^2 x^\mu / \mathrm{d}\tau^2$ is not a vector under GCTs) is the same as the corresponding term in the transformation (10.8). Hence deduce that the equation

$$\frac{\mathrm{d}^2 x'^\mu}{\mathrm{d}\tau^2} + \Gamma'^\mu_{\rho\nu} \frac{\mathrm{d}x'^\rho}{\mathrm{d}\tau} \frac{\mathrm{d}x'^\nu}{\mathrm{d}\tau} = 0 \tag{10.12b}$$

is covariant under GCTs. Since Γ' vanishes in the frame in which there is no gravity, this equation then reduces to (10.12a). Hence, the equivalence principle tells us that (10.12b) describes the motion of a particle in an external gravitational field.

Note that torsion makes no contribution to the motion of the particle.

10.1.3 The metric tensor

In the previous section, we have made no mention of the metric tensor in our discussion of the covariant derivative, the connection or even the curvature tensor. Even the equation of motion for a particle in a gravitational field can be stated in terms of just the connection fields. Indeed, there are non-metric theories of gravity, e.g. theories with torsion, but these violate the equivalence principle as we will show.

From now on, we will focus our attention on standard GR where it is assumed that spacetime is a Riemannian manifold.

Riemannian manifolds, which are manifolds on which a metric (introduced below) can be defined, form a natural setting for formulating GR. On any sufficiently small patch of such a manifold, it is possible to find a Cartesian co-ordinate system for which the separation between two points is given by a Pythagorean-type law. On such a manifold, the differential line element is given by

$$ds^2 = g_{\mu\nu}(x)dx^\mu dx^\nu, \tag{10.13a}$$

and accordingly the length squared of any four vector is given by

$$V^2 = g_{\mu\nu}(x)V^\mu V^\nu. \tag{10.13b}$$

Since the left-hand side is a scalar and the line elements on the right-hand side are vectors, by one of the previous exercises the quantity $g_{\mu\nu}$ transforms as a covariant second rank tensor known as the metric tensor. The metric tensors $g_{\mu\nu}(x)$ and $g^{\mu\nu}(x)$ can be used to raise and lower indices in GR.

We will assume a four-dimensional spacetime with one time-like direction. The principle of equivalence then tells us that we can always transform to a freely falling co-ordinate frame where the metric tensor is locally flat (Minkowski), i.e.

$$g_{\mu\nu}(x) \to \eta_{\mu\nu} = \begin{pmatrix} 1 & 0 & 0 & 0 \\ 0 & -1 & 0 & 0 \\ 0 & 0 & -1 & 0 \\ 0 & 0 & 0 & -1 \end{pmatrix}. \tag{10.14}$$

In this frame, the derivative of the metric vanishes. This can be covariantly written as:

$$D_\mu g_{\nu\lambda} = \partial_\mu g_{\nu\lambda} - \Gamma^\rho_{\mu\nu}g_{\lambda\rho} - \Gamma^\rho_{\mu\lambda}g_{\nu\rho} = 0. \tag{10.15}$$

Using the transformation property of the metric tensor together with (10.8), it is straightforward to check that

$$\Gamma^\tau_{\mu\lambda} - \frac{1}{2}g^{\nu\tau}\left(\partial_\mu g_{\nu\lambda} + \partial_\lambda g_{\mu\nu} - \partial_\nu g_{\lambda\mu}\right) \tag{10.16}$$

transforms as a tensor. The part of this tensor *symmetric* under $\mu \leftrightarrow \lambda$ vanishes in the frame where the metric is locally Minkowskian, and hence must vanish in all frames. We thus obtain,

$$\Gamma^\tau_{\mu\lambda} = \frac{1}{2}g^{\nu\tau}\left(\partial_\mu g_{\nu\lambda} + \partial_\lambda g_{\mu\nu} - \partial_\nu g_{\lambda\mu}\right), \tag{10.17}$$

for the components of the connection that are *symmetric* under $\mu \leftrightarrow \lambda$. The corresponding antisymmetric components of the connection are *not* determined by the metric, but depend on the torsion tensor introduced above.

10.1.4 Einstein Lagrangian and field equations

To obtain the field equations of GR from an action principle, we can try to find an appropriate Lagrangian density, and vary the corresponding action $S = \int \mathcal{L} d^4 x$. For $\mathcal{L}$, we can construct a scalar by performing successive contractions on the Riemann tensor:

$$R_{\nu\tau} = R^{\rho}_{\nu\rho\tau} \quad \text{(Ricci tensor),} \quad \text{and} \tag{10.18a}$$

$$R = g^{\nu\tau} R_{\nu\tau} \quad \text{(Ricci scalar).} \tag{10.18b}$$

The Ricci scalar R is a candidate Lagrangian density, but we also know that the measure $d^4 x$ is not invariant under GCTs. However, $\sqrt{-g}\, d^4 x$ is invariant, where $g = det(g_{\mu\nu})$. Thus, $\mathcal{L} = \sqrt{-g}\, R$ is a candidate Lagrangian density for GR, and is known as the Einstein Lagrangian. Since the Lagrangian density must have mass dimension four, it must be multiplied by a constant with dimensions of M^2. Hence, we write the Lagrangian density for the gravitational field as,

$$\mathcal{L}_G = -\frac{1}{2\kappa^2} \sqrt{-g}\, R \tag{10.19}$$

where κ^{-2} has dimensions of mass squared.

Exercise *Using the transformation properties of $g_{\mu\nu}$ and $d^4 x$, show that $\sqrt{-g}\, d^4 x$ is invariant under GCTs.*

Exercise *Show that the Ricci tensor obtained by contracting the Riemann curvature tensor is symmetric.*

Variation of the Einstein action with respect to the fields $g_{\mu\nu}$ is a lengthy calculation, but can be made simpler using the Palatini formalism wherein the connection fields $\Gamma^{\tau}_{\mu\nu}$ and their derivatives are regarded as independent fields along with $g_{\mu\nu}(x)$. Either approach leads to Einstein's field equations in a vacuum:

$$R_{\mu\nu} - \frac{1}{2} g_{\mu\nu} R = 0. \tag{10.20}$$

This equation is generally covariant, and contains at most the second derivative of the metric. We could have included higher powers of R into the action but these would have led to higher derivatives in the equations of motion.

We may also add the effects of matter and/or energy to the Einstein Lagrangian. For instance, including a real scalar field ϕ with Lagrangian $\mathcal{L}_M = \sqrt{-g}(g_{\mu\nu}\partial^{\mu}\phi\partial^{\nu}\phi - m^2\phi^2)$ into the action will bring a source term involving the symmetric energy momentum tensor $T_{\mu\nu}$ into the equations of motion. Although we have illustrated this for coupling to scalar fields, the same is true for coupling to *all* matter fields. The constant κ that we introduced for dimensional relations determines the gravitational coupling of matter. It must be chosen to obtain agreement with Newtonian gravity in the non-relativistic, weak field limit. It turns out that $\kappa^2 = 8\pi G_N/c^4$, with G_N being Newton's constant. In natural units with $\hbar = c = 1$, the Planck mass $M_{Pl} = G_N^{-1/2}$. The reduced Planck mass is defined by $M_P = M_{Pl}/\sqrt{8\pi}$ so that $\kappa = 1/M_P$, with $M_P \simeq 2.4 \times 10^{18}$ GeV. It is common to use units in which M_P is also set to unity.

Finally, we can also include the term $\mathcal{L}_\Lambda = \frac{\sqrt{-g}}{\kappa^2}\Lambda$ into the Lagrangian density without bringing higher derivatives of the metric into the field equations. Here, Λ is known as the cosmological constant. Indeed there is evidence for a small but non-zero cosmological constant ($\Lambda \sim (3 \text{ meV})^4$ in natural units) in Einstein's equations, indicative of a dark energy that pervades the Universe. Including matter as well as the cosmological constant, Einstein's field equations become,

$$R_{\mu\nu} - \frac{1}{2}g_{\mu\nu}R - g_{\mu\nu}\Lambda = 8\pi G_N T_{\mu\nu}. \tag{10.21}$$

Notice that both sides of this equation are symmetric under interchange of tensor indices.

10.1.5 Spinor fields in General Relativity

The preceding formulation of GR can admit fields transforming as scalars, vectors, and tensors. In supersymmetry, we must necessarily include spinor fields as well, but there exists no generalization of spinorial Lorentz transformation rules to general co-ordinate transformations: mathematically speaking, the group $GL(4)$ has no finite dimensional spinor representations. What is done instead is to define, for every point on the curved spacetime, a tangent space with a flat Minkowski metric in which the spinors may transform. Thus, the action we construct should be invariant under GCTs $x^\mu \to x'^\mu$ on the curved manifold, *and* invariant under local Lorentz transformations (LLTs) on the flat tangent space:

$$\xi^a \to \xi'^a = \Lambda^a_b(x)\xi^b. \tag{10.22}$$

For each spacetime point, $\xi^a(x)$ define a (locally inertial) co-ordinate system in the flat tangent space. It is customary to take Greek indices $\mu = 0\text{-}3$ for objects transforming under GCTs, and Latin indices $a = 0\text{-}3$ for objects transforming under LLTs. The transformation from local Lorentz co-ordinates to general co-ordinates is given by

$$\frac{\partial\xi^a}{\partial x^\mu} \equiv e^a_\mu, \tag{10.23}$$

where e_μ^a is known as the *vierbein*.

The vierbein transforms under GCTs as

$$e_\mu^a(x) \to e_\mu'^a(x') = \frac{\partial x^\nu}{\partial x'^\mu} e_\nu^a(x), \qquad (10.24)$$

and under LLTs as

$$e_\mu^a \to e_\mu'^a = \Lambda_b^a e_\mu^b(x). \qquad (10.25)$$

The vierbein allows us to connect one co-ordinate system with the other. Thus, an object v^μ which transforms as a vector under GCTs can be related to an object V^a which transforms as a vector under LLTs via

$$V^a = e_\mu^a v^\mu. \qquad (10.26)$$

In particular, the metric tensor in each space is related as

$$g_{\mu\nu}(x) = e_\mu^a e_\nu^b \eta_{ab}, \qquad (10.27)$$

where η_{ab} is the usual Minkowski metric. From the above relation, knowledge of the vierbein completely determines the form of the metric tensor, and it is sometimes convenient to think of the vierbein as a "square root" of the metric tensor. The Minkowski metric tensor η_{ab} (η^{ab}) can be used to lower (raise) Latin indices, just as $g_{\mu\nu}$ ($g^{\mu\nu}$) can be used to lower (raise) Greek indices. Thus, we also have

$$g^{\mu\nu} = e_a^\mu e_b^\nu \eta^{ab}. \qquad (10.28)$$

Taking the determinant of Eq. (10.27), we are able to replace the Jacobian factor $\sqrt{-g}$ by $\mathbf{e} \equiv det(e_\mu^a)$.

Spinors transform under LLTs as

$$\psi_m(x) \to \psi_m'(x') = \Lambda_{\frac{1}{2}mn} \psi_n(x) \qquad (10.29)$$

where $\Lambda_{\frac{1}{2}mn} = \left[e^{-i\epsilon_{rs}(x)\sigma_{rs}} \right]_{mn}$, and the spinor index $m = 1-4$, and $\sigma_{rs} = \frac{1}{2}[\gamma_r, \gamma_s].$[3] The Dirac matrices satisfy $\{\gamma_r, \gamma_s\} = 2\eta_{rs}$ in local Lorentz space. They are related to the curved space gamma matrices via $\gamma^\mu = e_r^\mu \gamma^r$, and where

$$\{\gamma^\mu(x), \gamma^\nu(x)\} = 2g^{\mu\nu}(x). \qquad (10.30)$$

The transformation parameter ϵ_{rs} is antisymmetric on rs and includes six parameters: three rotations and three boosts.

In order to define a covariant derivative for spinor fields $D_\mu \psi$ such that

$$D_\mu \psi \to D_\mu' \psi' = \Lambda_{\frac{1}{2}} \left(D_\mu \psi \right), \qquad (10.31)$$

---

[3] We are economizing notation here by not writing the transformation matrix as $\mathcal{D}(\Lambda_{1/2})_{mn}$, as is the practice by many authors.

we introduce *spin connection* fields ω_μ^{rs} such that

$$D_\mu \psi = \partial_\mu \psi - \frac{i}{4} \omega_\mu^{rs} \sigma_{rs} \psi, \qquad (10.32)$$

and, as usual, require these to transform so that (10.31) is satisfied. The covariant derivative of the vierbein will involve both connection and spin connection fields:

$$D_\mu e_\nu^a = \partial_\mu e_\nu^a - \Gamma_{\mu\nu}^\lambda e_\lambda^a + \omega_{\mu b}^a e_\nu^b. \qquad (10.33)$$

A field strength tensor can be computed from the spinor field covariant derivative, just as from a vector field covariant derivative.

Exercise *Evaluate the commutator of spinor covariant derivatives and show that it can be written as*

$$[D_\mu, D_\nu]\psi = -\frac{i}{4} \sigma_{uv} R_{\mu\nu}^{uv} \psi \qquad (10.34a)$$

where

$$R_{\mu\nu}^{uv} = \partial_\mu \omega_\nu^{uv} - \partial_\nu \omega_\mu^{uv} + \omega_{\mu r}^u \omega_\nu^{rv} - \omega_{\mu r}^v \omega_\nu^{ur}. \qquad (10.34b)$$

This quantity is related to the Riemann curvature tensor via

$$R_{\mu\nu}^{uv} = e_\rho^u e_\sigma^v R_{\mu\nu}^{\rho\sigma}. \qquad (10.34c)$$

Hint: Recall the generators $M_{ab} = \sigma_{ab}/2$ of the Lorentz group obey the algebra (4.6).

We can again apply the principle of equivalence as we did to obtain $D_\mu g_{\nu\lambda} = 0$, but this time for the vierbein: $D_\mu e_\nu^a = 0$. This gives $4 \times 6 = 24$ constraints, the number of independent components of the spin connection, which can be eliminated as an independent field. Indeed, the spin connection fields ω_μ^{ab} can be constructed from knowledge of the vierbein via,

$$\omega_\mu^{ab} = \frac{1}{2} e^{av} (\partial_\mu e_\nu^b - \partial_\nu e_\mu^b) + \frac{1}{4} e^{a\rho} e^{b\sigma} (\partial_\sigma e_\rho^c - \partial_\rho e_\sigma^c) e_{c\mu} - (a \leftrightarrow b). \qquad (10.35)$$

10.2 Local supersymmetry implies (super)gravity

Our next goal is to examine what happens when we allow the parameters α that characterize SUSY transformations to be spacetime dependent; i.e. when we allow SUSY to be a local symmetry. Such local SUSY transformations are known as *supergravity* transformations since, as we will see presently, a consistent implementation of local SUSY transformations necessarily brings a massless spin 2 field into the theory. Moreover, this spin 2 field couples to the energy-momentum tensor for matter, just as in general relativity, and its quanta are identified with gravitons. The spin $\frac{3}{2}$ Rarita–Schwinger field is needed since SUSY requires that the gravitons must have fermionic partners with spin differing by $1/2$. Its quanta are referred to as gravitinos.

An aside on the spin $\frac{3}{2}$ Rarita–Schwinger field *We briefly discuss the basics of massive spin $\frac{3}{2}$ fields, since after supersymmetry breaking the gravitinos acquire a mass. A free massive gravitino may be described by a "vector-spinor" field $\psi_\lambda(x)$, each of whose Majorana spinor components (the spinor index is suppressed) satisfies the Dirac equation,*

$$(i\partial\!\!\!/ - m)\psi_\lambda = 0, \tag{10.36a}$$

and is subject to the subsidiary condition,

$$\gamma^\lambda \psi_\lambda = 0. \tag{10.36b}$$

Contracting (10.36a) with γ^λ, it is easy to see that,

$$\partial^\lambda \psi_\lambda = 0. \tag{10.36c}$$

To understand why ψ_λ describes a spin $\frac{3}{2}$ particle, let us examine the plane wave solutions $\psi_\lambda(x) = u_\lambda(k)e^{-ikx}$ of (10.36a) in the rest frame of the particle. Eq. (10.36a) then implies

$$\gamma^0 u_\lambda = u_\lambda. \tag{10.37a}$$

It is most convenient to do the analysis using the standard representation for the gamma matrices. Exactly as for the case of a massive spin $\frac{1}{2}$ particle in its rest frame, we find that the lower two components of all four u_λ must vanish. The subsidiary condition (10.36b) implies that

$$u_0 = \vec{\gamma} \cdot \vec{u}, \tag{10.37b}$$

where $\vec{u}$ has as its components the three four-spinors $u_1, u_2,$ and $u_3,$ all of whose lower components vanish, and whose three upper components are the three two-spinors $\chi_1, \chi_2,$ and χ_3. Using the explicit form of the $\vec{\gamma}$ matrices, we see from (10.37b) that,

$$u_0 = 0$$
$$\vec{\sigma} \cdot \vec{\chi} = 0. \tag{10.37c}$$

The two constraints (10.37c) imply that just four of the six components of $\vec{\chi}$ are truly independent. Since the spinors ψ_λ are completely fixed by $\vec{\chi}$, we see that these *are specified by four independent components, just the right number to describe a massive spin $\frac{3}{2}$ particle in its rest frame.*

Exercise *Show that the Lagrangian density*

$$\mathcal{L} = -\frac{1}{2}\epsilon^{\mu\nu\rho\sigma}\,\bar{\psi}_\mu\gamma^5\gamma_\nu\partial_\rho\psi_\sigma - \frac{1}{4}m\bar{\psi}_\mu[\gamma^\mu,\gamma^\nu]\psi_\nu \tag{10.38}$$

yields the Dirac equation (10.36a) as well as the constraint conditions (10.36b) and (10.36c), assuming $m \neq 0$. You may find the identity

$$\gamma^5\gamma^\nu = \frac{i}{3!}\epsilon^{\nu\rho\sigma\tau}\gamma_\rho\gamma_\sigma\gamma_\tau$$

useful.

Notice that the Lagrangian for the massless case is invariant under the transformation $\psi_\mu \to \psi_\mu + \partial_\mu\alpha$. For this case, the constraints do not follow from the Lagrangian, but have to be imposed as gauge fixing conditions.

To obtain a locally supersymmetric theory, we will adopt the Noether procedure, which was used to derive the simplest supergravity Lagrangians. The Noether procedure is a systematic technique for obtaining a theory invariant under a local symmetry transformation, starting from a theory that is invariant under the corresponding global transformation.

QED serves as an illustrative example. We may start with the simple Dirac Lagrangian for an electron $\mathcal{L} = i\bar{\psi}\partial\!\!\!/\psi$ which is invariant under a global phase transformation $\psi \to e^{i\alpha}\psi$, where α is a constant. If we make the transformation local, so that $\alpha \to \alpha(x)$, then this Lagrangian is no longer invariant, changing by an amount $\delta\mathcal{L} = -\bar{\psi}\gamma^\mu\psi\partial_\mu\alpha$. Invariance can be restored by adding a gauge field term to $\mathcal{L}$ given by $\mathcal{L}' = -e\bar{\psi}\gamma^\mu A_\mu\psi$, where the gauge field transforms as $A_\mu \to A_\mu - \frac{1}{e}\partial_\mu\alpha$, and e in this case is the magnitude of the electric charge. The final QED Lagrangian is obtained by adding the gauge field kinetic term $-\frac{1}{4}F_{\mu\nu}F^{\mu\nu}$ and an electron mass term $-m\bar{\psi}\psi$, which are separately locally gauge invariant.

To illustrate why local supersymmetry necessarily implies gravity, we apply the Noether procedure to the Wess–Zumino model introduced in Chapter 3. To simplify our analysis, we will examine only the free, massless case with the fields "on shell", meaning that these satisfy their equations of motion. Then we do not have to worry about the auxiliary fields which can be set to zero. Furthermore, from (3.7d) and (3.7e), we see that the SUSY transforms of the auxiliary fields also vanish as long as the fermion field satisfies its equation of motion, $\partial\!\!\!/\psi = 0$.

The Lagrangian for this very simplified model takes the form,

$$\mathcal{L} = \mathcal{L}_{\text{kin}} = \frac{1}{2}(\partial_\mu A)^2 + \frac{1}{2}(\partial_\mu B)^2 + \frac{i}{2}\bar{\psi}\partial\!\!\!/\psi, \tag{10.39}$$

and is invariant under

$$\delta A = i\bar{\alpha}\gamma_5\psi, \tag{10.40a}$$

$$\delta B = -\bar{\alpha}\psi, \tag{10.40b}$$

$$\delta\psi = -i\partial(-B + i\gamma_5 A)\alpha. \tag{10.40c}$$

If we now let $\alpha \to \alpha(x)$, and define the local transformation so that the derivative in (10.40c) acts only on the fields, a straightforward calculation shows that the Lagrangian no longer transforms as a total derivative, but instead as,

$$\delta\mathcal{L}_{\mathrm{kin}} = \partial^\mu \left(\frac{1}{2}\bar{\alpha}\gamma_\mu\partial(-B + i\gamma_5 A)\psi \right) + (\partial^\mu\bar{\alpha}) \left(\partial\gamma_\mu(-B + i\gamma_5 A) \right) \psi. \tag{10.41}$$

The additional term can be cancelled by adding to the Lagrangian a term given by

$$\mathcal{L}_1 = -\kappa\bar{\psi}_\mu\partial_\nu(-B + i\gamma_5 A)\gamma^\nu\gamma^\mu\psi, \tag{10.42}$$

where ψ_μ is a spin $\frac{3}{2}$ field. It has mass dimensionality $[\psi_\mu] = \frac{3}{2}$, so that a dimensional constant κ with $[\kappa] = -1$ must be included to give a dimension four Lagrangian term. The field ψ_μ is effectively a gauge field for the local supersymmetry transformation, just as A_μ was the gauge field for a local phase transformation in the QED example. If ψ_μ transforms under local SUSY as $\bar{\psi}_\mu \to \bar{\psi}_\mu + \frac{1}{\kappa}\partial_\mu\bar{\alpha}$, then the transformation term involving $\partial^\mu\bar{\alpha}$ will cancel!

This of course does not mean that the action corresponding to the Lagrangian density $\mathcal{L}_{\mathrm{kin}} + \mathcal{L}_1$ is supersymmetric because we must now apply the local SUSY transformation laws to the additional Lagrangian term $\mathcal{L}_1$ as well. Clearly, the terms resulting from this transformation are $\mathcal{O}(\kappa)$. Indeed a somewhat lengthy calculation shows that,

$$\delta(\mathcal{L}_{\mathrm{kin}} + \mathcal{L}_1) = -2i\kappa\bar{\psi}_\mu\gamma_\nu T^{\mu\nu}\alpha + \cdots, \tag{10.43a}$$

where the ellipsis denotes terms involving derivatives of α or total derivatives, and

$$T^{\mu\nu} = (\partial^\mu A)(\partial^\nu A) + (\partial^\mu B)(\partial^\nu B) - \frac{1}{2}\eta^{\mu\nu}\left[(\partial_\rho A)^2 + (\partial_\rho B)^2\right] + \frac{i}{2}\bar{\psi}\gamma^\mu\partial^\nu\psi, \tag{10.43b}$$

is the canonical energy–momentum tensor for the WZ model.

Exercise *Verify the transformation (10.43a).*

To obtain the $T^{\mu\nu}$ term on the right-hand side of (10.43a) which is written up to derivatives of the parameter α, we need to consider only "global" SUSY transformations when performing the variation. We need to Fierz transform the fermion quartic term so that the transformation parameter α is contracted with the gravitino spinor: only the vector and axial-vector combinations survive. Moreover, since we write this variation only up to a total derivative, and for "on-shell fields", *many terms vanish due to (10.36a)–(10.36c). Finally, we note that the $\eta^{\mu\nu}$ terms for the scalar field contributions to $T^{\mu\nu}$ vanish when the fields are on-shell.*

This term can now be cancelled by adding another term,

$$\mathcal{L}_2 = -g_{\mu\nu} T^{\mu\nu}, \tag{10.44}$$

to the Lagrangian density. We see that the Noether procedure forces us to introduce a massless spin 2 field $g_{\mu\nu}$ that couples to the energy momentum tensor as in General Relativity. The quanta of this field are the gravitons. We require this spin 2 field $g_{\mu\nu}$ to transform as

$$\delta g_{\mu\nu} = -i\kappa\bar{\alpha}(\gamma_\nu\psi_\mu + \gamma_\mu\psi_\nu). \tag{10.45}$$

We see that local supersymmetry implies gravity. The dimensionful coupling constant κ that we have been forced to introduce can be related to Newton's gravitational constant.

The procedure we have outlined was for the simple case of the massless, non-interacting on-shell WZ model. The locally supersymmetric couplings of the (on-shell) scalar supermultiplet of the Wess–Zumino model can be found in Ferrara *et al.*, and includes many more terms.[4] One must, of course, also include kinetic terms for both the graviton and gravitino fields, and derivatives must be made covariant with respect to general co-ordinate and local Lorentz transformations. A complete derivation is beyond the scope of this text, and we will simply present the answer. The relevant Lagrangian is given by a sum of a pure (supersymmetrized) gravity piece together with a second piece that describes the supersymmetrized gravitational couplings of matter:

$$\mathcal{L} = \mathcal{L}_{\rm G} + \mathcal{L}_{\rm M}. \tag{10.46}$$

Here, $\mathcal{L}_{\rm G}$ is given by a sum of the Einstein Lagrangian and the kinetic term (10.38) for the massless Rarita–Schwinger field:

$$\mathcal{L}_{\rm G} = -\frac{\mathbf{e}}{2\kappa^2}R - \frac{1}{2}\epsilon^{\lambda\rho\mu\nu}\bar{\psi}_\lambda\gamma_5\gamma_\mu D_\nu\psi_\rho, \tag{10.47}$$

where $\mathbf{e}$, the determinant of the vierbein, is the Jacobian factor $\sqrt{-g}$ that appears in the Einstein Lagrangian. A comparison with Eq. (10.19) shows that the constant κ introduced in our discussion of local supersymmetry transformations indeed coincides with the same constant that appears in our discussion of general relativity.

---

[4] S. Ferrara, D. Freedman, P. van Nieuwenhuizen, P. Breitenlohner, F. Gliozzi and J. Scherk, *Phys. Rev.* **D15**, 1013 (1977).

The (super)gravitational interactions of the matter supermultiplet take the form,

$$
\begin{aligned}
\mathcal{L}_{\mathrm{M}} = {} & \mathbf{e} g_{\mu\nu} \left(\frac{1}{2} \partial^{\mu} A \partial^{\nu} A + \frac{1}{2} \partial^{\mu} B \partial^{\nu} B \right) + \mathbf{e} \, \frac{\mathrm{i}}{2} \bar{\psi} \, \slashed{D} \psi \\
& - \frac{\kappa}{2} \mathbf{e} \, \bar{\psi}_{\mu} \partial_{\nu} (-B + \mathrm{i}\gamma_5 A) \gamma^{\nu} \gamma^{\mu} \psi - \frac{\kappa^2}{16} \mathbf{e} \, (\bar{\psi}\gamma_5\gamma_{\mu}\psi)(\bar{\psi}\gamma_5\gamma^{\mu}\psi) \\
& - \mathrm{i}\frac{\kappa^2}{8} (B \overset{\leftrightarrow}{\partial}_{\sigma} A) \left[\epsilon^{\mu\nu\rho\sigma} \bar{\psi}_{\mu}\gamma_{\nu}\psi_{\rho} - \mathrm{i}\mathbf{e} \, \bar{\psi}\gamma_5\gamma^{\sigma}\psi \right] \\
& + \frac{\kappa^2}{16} \bar{\psi}\gamma_5\gamma_{\sigma}\psi \left[\mathrm{i}\epsilon^{\mu\nu\rho\sigma} \bar{\psi}_{\mu}\gamma_{\nu}\psi_{\rho} + \mathbf{e} \, \bar{\psi}_{\mu}\gamma_5\gamma^{\sigma}\psi_{\mu} \right],
\end{aligned} \tag{10.48}
$$

which includes relativistically covariant kinetic energy terms for scalar and spinor fields together with interaction terms involving the gravitational coupling constant κ. At low energies, these terms are suppressed by inverse powers of M_{P}. The covariant derivatives that appear in (10.47) and (10.48) are given by,

$$
\slashed{D}\psi = \gamma^{\mu}(\partial_{\mu} - \frac{\mathrm{i}}{4}\omega_{\mu}^{rs}\sigma_{rs})\psi \quad \text{and} \tag{10.49a}
$$

$$
D_{\nu}\psi_{\rho} = \partial_{\nu}\psi_{\rho} - \frac{\mathrm{i}}{4}\omega_{\nu}^{rs}\sigma_{rs}\psi_{\rho} - \Gamma_{\rho\nu}^{\sigma}\psi_{\sigma}. \tag{10.49b}
$$

Of course, the last term of $D_{\nu}\psi_{\rho}$ makes no contribution to the kinetic energy of the gravitino once the connection is written as a function of the metric (using the equations of motion) so that it is symmetric in its lower indices.

The local SUSY transformation laws are given by

$$
\delta A = \mathrm{i}\bar{\alpha}\gamma_5\psi, \tag{10.50a}
$$

$$
\delta B = -\bar{\alpha}\psi, \tag{10.50b}
$$

$$
\delta\psi = -\mathrm{i}\slashed{\partial}(-B + \mathrm{i}\gamma_5 A)\alpha + \mathrm{i}\frac{\kappa}{2}(\bar{\psi}_{\mu}\psi)\gamma^{\mu}\alpha + \mathrm{i}\frac{\kappa}{2}(\bar{\psi}_{\mu}\gamma_5\psi)\gamma^{\mu}\gamma_5\alpha
$$
$$
+ \frac{1}{4}\kappa^2 \left[\bar{\alpha}(-B + \mathrm{i}\gamma_5 A)\gamma_5\psi\right]\gamma_5\psi, \tag{10.50c}
$$

$$
\delta e_{\mu}^{a} = -\mathrm{i}\kappa\bar{\alpha}\gamma^{a}\psi_{\mu}, \quad \text{and} \tag{10.50d}
$$

$$
\delta\psi_{\mu} = \frac{2}{\kappa}D_{\mu}\alpha + \frac{\mathrm{i}}{2}\kappa(B \overset{\leftrightarrow}{\partial}_{\mu} A)\gamma_5\alpha - \frac{\kappa^2}{4}[\bar{\alpha}(-B + \mathrm{i}\gamma_5 A)\gamma_5\psi]\gamma_5\psi_{\mu}. \tag{10.50e}
$$

Notice that the transformation law for the vierbein reproduces the SUSY transformation (10.45) for the metric that we had obtained above. We do not write the transformation law for the connection fields as these are complicated, and are not needed for our discussion. The gravitino and vierbein fields can be combined into what is called the metric superfield. This is the gravitational analogue of the gauge superfield and, hence, is a *real* superfield. Since the gravitino carries a vector index, the metric superfield is a real vector superfield.

The reader may have noticed that the supergravity Lagrangian contains non-renormalizable terms. This was also true of the Lagrangian for Einsteinian gravity. Such non-renormalizable terms enter because the gravitational coupling constant κ has dimensions of inverse mass. The situation is analogous to Fermi's theory of β-decay which though non-renormalizable was practically useful, and which has since been understood as the low energy limit of a more fundamental theory (the Standard Model). In the same vein, we will regard the non-renormalizable supergravity Lagrangian as the low energy limit of an as yet unformulated locally supersymmetric fundamental theory (perhaps, superstring theory) to be discovered in the future.

10.3 The supergravity Lagrangian

We have seen that the construction of locally supersymmetric field theories forces us to consider non-renormalizable interactions. If we give up the restriction of renormalizability the globally supersymmetric Lagrangian for gauge theories in (6.47) can be generalized to,

$$\mathcal{L} = -\frac{1}{4} \int d^4\theta\, K\left(\hat{S}^\dagger e^{-2g t_A \hat{\Phi}_A}, \hat{S}\right) - \frac{1}{2} \left[\int d^4 x\, d^2\theta_L\, \hat{f}(\hat{S}) + \quad \text{h.c.} \right]$$
$$- \frac{1}{4} \int d^2\theta_L\, f_{AB}(\hat{S}) \overline{\hat{W}^c_A} \hat{W}_B. \tag{10.51}$$

In particular, the Kähler potential and the superpotential functions are no longer restricted to be quadratic and cubic polynomials, though the latter is still required to be an analytic function of the fields. Moreover, we have introduced the *gauge kinetic function* $f_{AB}(\hat{S})$ which, like the superpotential $\hat{f}(\hat{S})$, is an analytic function of the chiral superfields $\hat{S}_i$ so that, like the superpotential term, the last term is also an F-term of a chiral superfield (and hence supersymmetric). Renormalizability (and gauge invariance) restricted $f_{AB} = \delta_{AB}$ in (6.47), but now the more general form is possible. To preserve gauge invariance, f_{AB} must transform as the symmetric product of two adjoints of the gauge group. As before, choosing the Kähler potential $K(\hat{S}^\dagger, \hat{S})$ and the superpotential $\hat{f}(\hat{S})$ to be invariant under *global* gauge transformations guarantees local gauge invariance of (10.51). Except for these restrictions from gauge invariance, the Kähler potential, the superpotential and the gauge kinetic function are arbitrary functions of all chiral superfields.

Although it is possible in principle to obtain the complete Lagrangian for locally supersymmetric gauge theories by applying the Noether procedure to the globally supersymmetric Lagrangian (10.51), in practice, more efficient techniques involving tensor calculus of local supersymmetry have been developed to obtain the complete result including all auxiliary fields. A discussion of these techniques is beyond the scope of this text. The final result, analogous to our master formula, but for local supersymmetry, was first obtained in 1982 by Cremmer *et al.*[5] We simply present it here, in terms of component fields, after all auxiliary fields have been eliminated. It is customary to factor out the Jacobian term e, and to write the result in units with $M_P = 1$. The reduced Planck mass can be re-inserted term-by-term by requiring the dimensionality of each term be equal to four.

Although the Lagrangian for a general non-renormalizable supersymmetric theory depends on three independent functions, K, $\hat{f}$, and f_{AB}, the remarkable feature of the supergravity Lagrangian is that it depends on the gauge kinetic function and just one combination,

$$G(\hat{S}^\dagger, \hat{S}) = K(\hat{S}^\dagger, \hat{S}) + \log |\hat{f}(\hat{S})|^2, \tag{10.52}$$

of the Kähler potential and superpotential. We will refer to G as the Kähler function, not to be confused with the Kähler potential K.[6] In what follows, derivatives of the Kähler function with respect to chiral superfields are denoted by,

$$G^i = \left.\frac{\partial G}{\partial \hat{S}_i}\right|_{\hat{S}=S} \quad \text{and} \quad G_j = \left.\frac{\partial G}{\partial \hat{S}^{j\dagger}}\right|_{\hat{S}=S}. \tag{10.53a}$$

Also,

$$G^i_j = \left.\frac{\partial^2 G}{\partial \hat{S}_i \partial \hat{S}^{j\dagger}}\right|_{\hat{S}=S} \tag{10.53b}$$

defines the Kähler metric.[7] Higher derivatives of G are analogously defined. Finally, we define the inverse of the metric by,

$$(G^{-1})^i_j G^j_k = \delta^i_k. \tag{10.53c}$$

Exercise *If the Lagrangian depends only on the combination G rather than separately on K and $\hat{f}$, the choice of the superpotential is not unique. Show that (classically) the transformations,*

$$K(\hat{S}^\dagger, \hat{S}) \to K - [h(\hat{S})]^\dagger - h(\hat{S})$$

$$\hat{f}(\hat{S}) \to \exp\left(h(\hat{S})\right)$$

---

[5] E. Cremmer, S. Ferrara, L. Girardello and A. van Proeyen, *Nucl. Phys.* **B212**, 413 (1983).
[6] Some authors refer to G as the Kähler potential. Moreover, what we call K is sometimes denoted by d and, to make matters worse, a different K is defined by $d = -3\log(-K/3)$.
[7] The use of j, the index labeling the adjoint of the jth field, as a superscript is merely conventional and should not cause confusion. It allows for contraction of upper and lower indices according to "usual rules" of tensor calculus. For notational clarity, we write the gauge generator matrix with only lower indices.

leave G (and hence the Lagrangian) invariant. This means that we can move all the analytic terms in the Kähler potential to the superpotential if we wish or, alternatively, that we may choose the superpotential to be a positive constant.

We are now in a position to write down the locally supersymmetric Lagrangian for a Yang–Mills gauge theory coupled to gravity. We break up this Lagrangian into purely bosonic terms $\mathcal{L}_B$, and terms with fermions $\mathcal{L}_F$. We further divide each of these terms into two parts: one part ($\mathcal{L}_B^C$) independent of the gauge kinetic function, and the other ($\mathcal{L}_B^G$) containing all the dependence on f_{AB}. The latter piece is, of course, absent in a theory without gauge fields. The purely bosonic Lagrangian can be written as,

$$\mathcal{L}_B = \mathcal{L}_B^C + \mathcal{L}_B^G \tag{10.54}$$

with (in units where the coupling $\kappa = 1$)

$$\mathbf{e}^{-1}\mathcal{L}_B^C = -\frac{R}{2} + G_j^i D_\mu \mathcal{S}_i D^\mu \mathcal{S}^{j*} - \mathbf{e}^G \left(G_i (G^{-1})_j^i G^j - 3 \right) \tag{10.55a}$$

and

$$\mathbf{e}^{-1}\mathcal{L}_B^G = -\frac{1}{4}(\mathrm{Re}\, f_{AB})F_{A\mu\nu}F_B^{\mu\nu} - \frac{1}{4}(\mathrm{Im}\, f_{AB})F_{A\mu\nu}\tilde{F}_B^{\mu\nu}$$

$$- \frac{g^2}{2}(\mathrm{Re}\, f_{AB}^{-1})G^i t_{Aij}\mathcal{S}_j G^k t_{Bk\ell}\mathcal{S}_\ell, \tag{10.55b}$$

where $\tilde{F}_B^{\mu\nu} = \frac{1}{2}\epsilon^{\mu\nu\rho\sigma} F_{B\rho\sigma}$. Here, $\mathcal{L}_B^G$ has been written as though the gauge group is simple: if the gauge group has several factors, a sum over each of these factors is implied. The first term in $\mathcal{L}_B^C$ is the Einstein Lagrangian (10.19). The second term contains the kinetic energy terms for the scalar components of the chiral superfields (hence the superscript C on this Lagrangian) while the last term in $\mathcal{L}_B^C$ is the part of the scalar potential that originates in the superpotential. Notice that, unlike the scalar potential for globally supersymmetric theories, this term may be negative. The kinetic terms for the gauge fields are contained in $\mathcal{L}_B^G$.

The part of the Lagrangian involving fermions is more complicated, and for convenience of writing we further split it into terms which give the kinetic energy terms $\mathcal{L}_{F,\mathrm{kin}}$ and other terms that only contain interactions, i.e.

$$\mathcal{L}_F = \mathcal{L}_{F,\mathrm{kin}} + \mathcal{L}_{F,\mathrm{Int}}. \tag{10.56a}$$

As before, each of these is further split into pieces, depending on whether or not there is dependence on the gauge kinetic function. We then have,

$$\mathcal{L}_{F,\mathrm{kin}} = \mathcal{L}_{F,\mathrm{kin}}^C + \mathcal{L}_{F,\mathrm{kin}}^G, \tag{10.56b}$$

and

$$\mathcal{L}_{F,\text{Int}} = \mathcal{L}_{F,\text{Int}}^C + \mathcal{L}_{F,\text{Int}}^G. \tag{10.56c}$$

The terms that appear in $\mathcal{L}_{F,\text{kin}}$ are given by,

$$
\mathbf{e}^{-1}\mathcal{L}_{F,\text{kin}}^C = -\frac{\mathbf{e}^{-1}}{2}\epsilon^{\mu\nu\rho\sigma}\bar{\psi}_{\mu}\gamma_5\gamma_{\nu}D_{\rho}\psi_{\sigma} + \left(\frac{i}{2}G_j^i\bar{\psi}_{iR}\gamma^{\mu}D_{\mu}\psi_R^j + \quad\text{h.c.}\right)
$$

$$
+ \left(\frac{\mathbf{e}^{-1}}{8}\epsilon^{\mu\nu\rho\sigma}\bar{\psi}_{\mu}\gamma_{\nu}\psi_{\rho}G^i D_{\sigma}\mathcal{S}_i + \quad\text{h.c.}\right)
$$

$$
+ \left(\frac{i}{2}\bar{\psi}_{iR}\,\slashed{D}\mathcal{S}_j\psi_R^k(-G_k^{ij} + \frac{1}{2}G_k^iG^j)\right.
$$

$$
\left. + \frac{i}{\sqrt{2}}G_i^j\bar{\psi}_{\mu R}\,\slashed{D}\mathcal{S}^{i\dagger}\gamma^{\mu}\psi_{jL} + \quad\text{h.c.}\right) \tag{10.57a}
$$

and

$$
\mathbf{e}^{-1}\mathcal{L}_{F,\text{kin}}^G = \left[\frac{1}{2}\text{Re}(f_{AB})\left(\frac{i}{2}\bar{\lambda}_A\,\slashed{D}\lambda_B + \frac{1}{4}\bar{\lambda}_A\gamma^{\mu}\sigma^{\nu\rho}\psi_{\mu}F_{B\nu\rho}\right)\right.
$$

$$
- \frac{i}{2}G^iD^{\mu}\mathcal{S}_i\bar{\lambda}_{AL}\gamma_{\mu}\lambda_{BL}\Big) + \frac{1}{8}\text{Im}\,(f_{AB})\mathbf{e}^{-1}D_{\mu}(\mathbf{e}\bar{\lambda}_A\gamma_5\gamma^{\mu}\lambda_B)
$$

$$
\left. - \frac{1}{4\sqrt{2}}\frac{\partial f_{AB}}{\partial\mathcal{S}_i}\bar{\psi}_{iR}\sigma^{\mu\nu}F_{A\mu\nu}\lambda_{BL}\right] + \quad\text{h.c.} \tag{10.57b}
$$

The first two terms in (10.57a) contain the kinetic energies of the gravitino and the chiral fermions, while the first term of (10.57b) contains the kinetic energy of the gauginos. Finally, the pieces of $\mathcal{L}_{F,\text{Int}}$ are given by

$$
\mathbf{e}^{-1}\mathcal{L}_{F,\text{Int}}^C = \left[\frac{i}{2}\mathbf{e}^{G/2}\bar{\psi}_{\mu L}\sigma^{\mu\nu}\psi_{\nu R} + \frac{1}{2}gG^it_{Aij}\mathcal{S}_j\bar{\psi}_{\mu R}\gamma^{\mu}\lambda_{AR}\right.
$$

$$
- g\sqrt{2}G_i^jt_{Ajk}\mathcal{S}_k\bar{\lambda}_{AL}\psi_R^i
$$

$$
- \frac{1}{2}\mathbf{e}^{G/2}(-G^{ij} - G^iG^j + G_k^{ij}(G^{-1})_{\ell}^kG^{\ell})\bar{\psi}_{iR}\psi_{jL}
$$

$$
- \frac{1}{\sqrt{2}}\mathbf{e}^{G/2}G^i\bar{\psi}_{\mu L}\gamma^{\mu}\psi_{iL}
$$

$$
+ \frac{i}{16}G_i^j\bar{\psi}_{iL}\gamma_d\psi_{jL}\,(\epsilon^{abcd}\bar{\psi}_a\gamma_b\psi_c - i\bar{\psi}^a\gamma^5\gamma^d\psi_a)
$$

$$
\left. + \left(\frac{1}{8}G_{kl}^{ij} - \frac{1}{8}G_m^{ij}(G^{-1})_n^mG_{kl}^n - \frac{1}{16}G_k^iG_l^j\right)\bar{\psi}_{iR}\psi_{jL}\bar{\psi}_L^k\psi_R^l\right]
$$

$$
+ \quad\text{h.c.} \tag{10.58a}
$$

and

$$
\begin{aligned}
e^{-1}\mathcal{L}^G_{F,\text{Int}} = \Bigg[& \frac{1}{4}e^{G/2}\frac{\partial f^*_{AB}}{\partial S^{j*}}(G^{-1})^j_k G^k \bar{\lambda}_{AL}\lambda_{BR} \\
& + \frac{g}{2\sqrt{2}}(\text{Re } f_{AB})^{-1}\frac{\partial f_{BC}}{\partial S_k} G^i t_{Aij} S_j \bar{\psi}_{kR}\lambda_{CL} \\
& - \frac{1}{32}(G^{-1})^k_l \frac{\partial f_{AB}}{\partial S_l}\frac{\partial f^*_{CD}}{\partial S^{k*}} \bar{\lambda}_{AR}\lambda_{BL}\bar{\lambda}_{CL}\lambda_{DR} \\
& + \frac{3}{32}\big[\text{Re}\,(f_{AB})\bar{\lambda}_{AR}\gamma_\mu \lambda_{BR}\big]^2 + \frac{i}{16}\text{Re}\,(f_{AB})\bar{\lambda}_A \gamma^\mu \sigma^{\rho\sigma}\psi_\mu \bar{\psi}_\rho \gamma_\sigma \lambda_B \\
& + \frac{i}{4\sqrt{2}}\frac{\partial f_{AB}}{\partial S_i}\left(\bar{\psi}_{iR}\sigma^{\mu\nu}\lambda_{AL}\bar{\psi}_{\nu R}\gamma_\mu \lambda_{BR} + \frac{i}{2}\bar{\psi}_{\mu L}\gamma^\mu \psi_{iL}\bar{\lambda}_{AR}\lambda_{BL}\right) \\
& + \frac{1}{16}\bar{\psi}_{iR}\gamma^\mu \psi^j_R \bar{\lambda}_{DL}\gamma_\mu \lambda_{CL}\left[G^i_j \text{Re}\,(f_{CD}) + \frac{1}{2}\text{Re}\left(f^{-1}_{AB}\frac{\partial f_{AC}}{\partial S_i}\frac{\partial f^*_{BD}}{\partial S^{j*}}\right)\right] \\
& - \frac{1}{16}\bar{\psi}_{iR}\psi_{jL}\bar{\lambda}_{CR}\lambda_{DL} \\
& \times \left(-2G^{ij}_k(G^{-1})^k_l\frac{\partial f_{CD}}{\partial S_l} + 2\frac{\partial^2 f_{CD}}{\partial S_i \partial S_j} - \frac{1}{2}\text{Re}\,f^{-1}_{AB}\frac{\partial f_{AC}}{\partial S_i}\frac{\partial f_{BD}}{\partial S_j}\right) \\
& - \frac{1}{128}\bar{\psi}_{iR}\sigma_{\mu\nu}\psi_{jL}\bar{\lambda}_{CR}\sigma^{\mu\nu}\lambda_{DL}\text{Re}\left(f^{-1}_{AB}\frac{\partial f_{AC}}{\partial S_i}\frac{\partial f_{BD}}{\partial S_j}\right)\Bigg] + \text{ h.c.}
\end{aligned}
$$

$$\text{(10.58b)}$$

The transformation laws of local supersymmetry are given by,

$$\delta S_i = -i\sqrt{2}\bar{\alpha}\psi_{iL}, \tag{10.59a}$$

$$
\begin{aligned}
\delta\psi_{iL} = &\sqrt{2}\,\slashed{D}S_i\alpha_R + i\sqrt{2}e^{G/2}(G^{-1})^j_i G_j\alpha_L \\
& - \frac{i}{2\sqrt{2}}\alpha_L\bar{\lambda}_A\lambda_{BR}(G^{-1})^j_i \frac{\partial f^*_{AB}}{\partial S^{*j}} + \cdots,
\end{aligned}
\tag{10.59b}
$$

$$\delta e^a_\mu = -i\bar{\alpha}\gamma^a\psi_\mu, \tag{10.59c}$$

$$\delta\psi_\mu = 2D_\mu\alpha + ie^{G/2}\gamma_\mu\alpha + \cdots, \tag{10.59d}$$

$$\delta V^\mu_A = -i\bar{\alpha}\gamma^\mu\lambda_A, \tag{10.59e}$$

$$\delta\lambda_{AR} = \frac{i}{2}\sigma^{\mu\nu}F_{A\mu\nu}\alpha_R - ig\text{Re}\,(f^{-1}_{AB})G^i(t_B)_{ij}S_j\alpha_R + \cdots \tag{10.59f}$$

The ellipses represent additional terms that we will not need for our subsequent discussion. These terms contain products of fermion fields, or, as in (10.59d), contain derivatives of scalar fields.

Except for the restrictions from gauge invariance and analyticity already mentioned, there is no known principle for the choice of the Kähler potential, the superpotential, and the gauge kinetic function in a general non-renormalizable theory.

Supergravity couplings, however, depend only on the gauge kinetic function and the Kähler function G. Choosing the Kähler potential and the gauge kinetic function to be what they are in renormalizable theories,

$$K = \sum_i \hat{\mathcal{S}}^{i\dagger}\hat{\mathcal{S}}_i \qquad (10.60a)$$

and

$$f_{AB}(\hat{\mathcal{S}}) = \delta_{AB}, \qquad (10.60b)$$

leads to canonical kinetic energy terms for "matter" (scalar and fermion) fields and for gauginos, respectively. The theory that is obtained from the general supergravity Lagrangian (10.54)–(10.58b) for this choice of the Kähler potential and the gauge kinetic function is sometimes referred to as "minimal supergravity".[8]

Exercise *Verify that for any gauge-invariant superpotential $\hat{f}(\hat{\mathcal{S}})$,*

$$\frac{\partial \hat{f}}{\partial \hat{\mathcal{S}}_i} t_{Aij}\hat{\mathcal{S}}_j = 0.$$

Exercise (Recovering the Lagrangian for global SUSY) *The locally supersymmetric Lagrangian must reduce to the globally supersymmetric Lagrangian in our master formula (6.44) if we take the limit $M_P \to \infty$.*

(a) *Identify the kinetic energy terms for all the fields.*

(b) *Convince yourself that the coupling of matter and gauge fields with gravitons and gravitinos, as well as all contributions from non-minimal terms in K and f_{AB}, all result in non-renormalizable interactions suppressed by powers of the reduced Planck mass. We can thus confine ourselves to the minimal supergravity choice,*

$$K = \sum_i \frac{\hat{\mathcal{S}}^{i\dagger}\hat{\mathcal{S}}_i}{M_P^2},$$

and

$$f_{AB} = \delta_{AB}$$

for these functions in the remainder of this exercise. Notice that we have inserted appropriate powers of the reduced Planck mass required to make K dimensionless.

---

[8] This should be distinguished from the minimal supergravity model discussed in the next chapter, where the same Kähler potential and a related gauge kinetic function are used. The reader should also note that the field-independent choice of the gauge kinetic function leaves gauginos massless at the tree level.

(c) *Verify that the last term of (10.55a) reduces to* $-\sum_i \left|\frac{\partial \hat{f}}{\partial S_i}\right|^2$, *the part of the scalar potential that originates in the superpotential on the third line of the master formula. Using the result of the previous exercise, show that the last term of (10.55b) reduces to the remainder of the scalar potential in our master formula. Remember that we have written the supergravity Lagrangian in units where* $\kappa = 1/M_P = 1$. *You will have to reinsert this factor on the various terms using dimensional analysis.*

(d) *Finally, convince yourself that the terms on the second and third lines of (10.58a) reduce to the couplings of gauginos to the sfermion–fermion pair and to chiral fermion superpotential Yukawa couplings, respectively.*

10.4 Local supersymmetry breaking

In Chapter 7, we showed that in order for global SUSY to be broken, the variation of a spinorial operator had to be non-zero. The same holds for models with local supersymmetry, i.e. we may have either $\langle 0|\delta\psi_i|0\rangle \neq 0$, or $\langle 0|\delta\lambda_A|0\rangle \neq 0$.

When we considered the spontaneous breaking of global supersymmetry without also breaking the Poincaré symmetry, we were led to just two possibilities: F-type SUSY breaking with $\langle 0|\mathcal{F}_i|0\rangle \neq 0$, or D-type breaking with $\langle 0|\mathcal{D}_A|0\rangle \neq 0$. For both cases, some auxiliary fields acquired a vacuum expectation value. For the case of local supersymmetry, the same is true although we cannot see this because we have written these supersymmetry transformations, (10.59a)–(10.59f), with the auxiliary fields already eliminated. If we assume that fermion fields cannot acquire vacuum expectation values, the condition for local SUSY breaking from (10.59b) is

$$\langle 0|e^{G/2}(G^{-1})_i^j G_j|0\rangle \neq 0 \tag{10.61a}$$

or from (10.59f),

$$\langle 0|\mathrm{Re}\,(f_{AB})^{-1}G^i t_{Bij} S_j|0\rangle \neq 0. \tag{10.61b}$$

The terms denoted by ellipses in the supergravity transformations (10.59a)–(10.59f) cannot acquire VEVs, and so are not relevant for this discussion. These conditions are the generalization of the conditions for global supersymmetry breaking that we found in Chapter 7.

These conditions take a simpler form for the minimal supergravity case introduced earlier. Then $(G^{-1})_i^j = \delta_i^j$, and $G_i = S_i + \frac{1}{\hat{f}^*}\partial \hat{f}^*/\partial S^{i*}$, and the F-type breaking condition reduces to,

$$\frac{\partial \hat{f}}{\partial S_i} + \frac{S^{i*}\hat{f}}{M_P^2} \neq 0. \tag{10.62}$$

Clearly, this condition reduces to Eq. (7.3a) in the limit $M_P \to \infty$. It is easy to see that (10.61b) is, similarly, a generalization of the D-term SUSY breaking condition Eq. (7.3b).

Our discussion of local supersymmetry breaking up to now has omitted one important possibility that was actually mentioned in Section 7.5. Supersymmetry may also be broken if the last term in Eq. (10.59b) acquires a VEV. This is not possible if gauge couplings remain perturbative. There may, however, be gauge interactions (not contained in the MSSM) that become strong at a high scale, and cause the associated gauginos to condense.[9]

It is also instructive to calculate the form of the scalar potential for minimal supergravity. In this case, from the $\mathcal{L}_B$ terms in the supergravity Lagrangian, we obtain

$$V = e^{\frac{\mathcal{S}^{i\dagger}\mathcal{S}_i}{M_P^2}} \left(-\frac{3}{M_P^2}|\hat{f}|^2 + \left| \frac{\partial \hat{f}}{\partial \mathcal{S}_i} + \frac{\mathcal{S}^{i\dagger}\hat{f}}{M_P^2} \right|^2 \right) + \frac{g^2}{2}\mathcal{S}^{i\dagger}t_{Aij}\mathcal{S}_j\mathcal{S}^{k\dagger}t_{Ak\ell}\mathcal{S}_\ell. \quad (10.63)$$

The negative term above offers the possibility of a small or even zero cosmological constant in supergravity theories (even if supersymmetry is broken), whereas in global SUSY the scalar potential was always positive semi-definite. There is no known reason though why the negative and positive terms should (almost) cancel, and a small cosmological constant is only possible by severe fine-tuning.

10.4.1 Super-Higgs mechanism

Recall that Goldstone bosons are the relics of spontaneous breaking of global symmetries: corresponding to every symmetry generator that does not annihilate the ground state, there is a massless boson (with derivative couplings) in the physical spectrum. If instead the spontaneously broken symmetry is local, the Goldstone boson is "eaten by the gauge fields", in that it becomes the longitudinal component of a gauge field which then acquires a mass. This is the well-known Higgs mechanism.

The situation for supersymmetry is quite similar. We have already seen that when global SUSY is spontaneously broken we obtain a massless Goldstone fermion, the goldstino, in the spectrum. In supergravity theories, where we have invariance under local SUSY transformations, the gravitino plays the same role that gauge fields play in local gauge theories. If SUSY is spontaneously broken, it is then natural to examine whether the goldstino degrees of freedom become the longitudinal degrees of freedom of the gravitino, the gauge field of supergravity, thereby endowing it

---

[9] In this context, we should mention that condensation of chiral fermions associated with new gauge interactions is also a possibility. Indeed, if there are chiral fermions in the adjoint representation of the gauge group, hybrid $\bar{\psi}\lambda$ condensates may also be possible. In these cases, the terms denoted by the ellipses in the supergravity transformations may be relevant.

with a mass. Although we do not analyze the details of the "supersymmetric Higgs mechanism" here, we see from the first term of Eq. (10.58a) that the gravitino becomes massive if the Kähler function G acquires a VEV:

$$\frac{i}{2} e^{\frac{G}{2}} \bar{\psi}_\mu \sigma^{\mu\nu} \psi_\nu \rightarrow \frac{i}{2} e^{\frac{G_0}{2}} \bar{\psi}_\mu \sigma^{\mu\nu} \psi_\nu, \tag{10.64}$$

where G_0 is the VEV of G. Thus the gravitino mass can be identified as

$$m_{3/2}^2 = e^{G_0} M_P^2. \tag{10.65}$$

The goldstino associated with either D- or F-type SUSY breaking is absorbed by the gravitino, and does not appear in the physical spectrum, while the gravitino becomes massive. Indeed, with an appropriate (field-dependent) choice of the local SUSY transformation parameter (unitarity gauge choice), the goldstino field can be completely eliminated from the Lagrangian.

We should also mention that the supertrace formula (7.35) that we obtained in Chapter 7 is also modified if the supersymmetry is local. For the case of minimal supergravity with N chiral supermultiplets, from Cremmer *et al.* we have

$$STr\mathcal{M}^2 = 2 \sum_A \mathcal{D}_A Tr(gt_A) + (N-1)(2m_{3/2}^2 - \frac{\mathcal{D}_A \mathcal{D}_A}{M_P^2}). \tag{10.66}$$

The first term is the same as the case for global SUSY but the last term is new. This term will play an important role in the next chapter where we consider realistic supergravity models of particle physics.

Exercise *For the flat Kähler metric show that the gravitino mass, assuming that the cosmological constant vanishes, can be written as*

$$m_{3/2}^2 = \frac{\langle F_i F^{i*} \rangle}{3M_P^2}, \tag{10.67a}$$

where

$$F_i = e^{\frac{G}{2}} (G^{-1})_i^j G_j \tag{10.67b}$$

is the auxiliary field whose VEV (10.61a) breaks supersymmetry.

An illustrative example: the Polonyi superpotential

A particularly simple illustration of the ideas that we have just discussed is obtained for the minimal supergravity model with a single chiral scalar superfield coupled via the Polonyi superpotential $\hat{f}$ given by,

$$\hat{f} = m^2 (\hat{S} + \beta), \tag{10.68}$$

where m^2 and β are real constants.

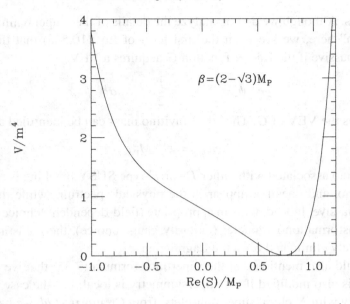

Figure 10.1 The scalar potential of the Polonyi model (in units of m^4) versus $\text{Re}\,(\mathcal{S})/M_P$ for the choice $\beta = (2 - \sqrt{3})M_P$ with $\text{Im}\,\mathcal{S}$ set to zero. Supersymmetry is necessarily broken, and the cosmological constant is zero for this choice of parameters.

It is straightforward to obtain the scalar potential which is given by,

$$V = m^4 e^{\mathcal{S}^*\mathcal{S}} \left(\left| 1 + \mathcal{S}^*(\mathcal{S} + \beta) \right|^2 - 3 \,|\mathcal{S} + \beta|^2 \right), \qquad (10.69a)$$

while the condition (10.62) for SUSY to remain unbroken becomes,

$$1 + \mathcal{S}^*(\mathcal{S} + \beta) = 0. \qquad (10.69b)$$

It is easy to see that SUSY is broken if $\beta^2 < 4$ (in Planck units).

The scalar potential has several extrema. In the following, we confine ourselves to those minima with $V = 0$, which implies

$$\left| 1 + \mathcal{S}^*(\mathcal{S} + \beta) \right|^2 = 3 \,|\mathcal{S} + \beta|^2 .$$

We can see that $\mathcal{S} = (\sqrt{3} - 1)M_P$ is one such minimum if $\beta = (2 - \sqrt{3})M_P$.[10] The shape of the scalar potential is shown in Fig. 10.1 for $\text{Im}\,\mathcal{S} = 0$. In this minimum, the gravitino mass is given by,

$$m_{3/2} = e^{G_0/2} M_P = e^{(2-\sqrt{3})} \frac{m^2}{M_P^2} M_P. \qquad (10.70)$$

Thus, if the parameter $m \sim 10^{10}$ GeV, then $m_{3/2} \sim 100$ GeV.

---

[10] This is an incredible fine-tuning. For any other value of β the cosmological constant would be very large.

Realistic supersymmetric models

It should be clear that, without further assumptions, the MSSM is not a tractable framework for SUSY phenomenology: there are just too many free parameters. This is not to say that we cannot do *any* phenomenology with the MSSM. First, assuming only R-parity conservation, we know that all sparticles must decay into other sparticles, until the decay chain terminates in the stable LSP. We have already seen that cosmological considerations require that the LSP cannot have electromagnetic or strong interactions. Since it couples to quarks and electrons only via *weak interactions*, it behaves like a neutrino in that it escapes the experimental apparatus undetected. As a result, the production of SUSY particles in high energy collisions is generically signaled by events with apparent "missing energy and momentum" (carried off by the undetected LSPs). With some mild assumptions of sparticle mass ordering, other relatively robust inferences may also be possible. For instance, if we assume that $\tilde{\mu}_R$ is the only charged sparticle that is accessible at an e^+e^- collider, and that the lightest neutralino is the LSP, we can conclude that

- smuons will be pair produced in e^+e^- collisions with cross sections that are fixed in terms of $m_{\tilde{\mu}_R}$ by *known* couplings to the photon and the Z-boson;
- both smuons will dominantly decay via $\tilde{\mu}_R \rightarrow \mu \tilde{Z}_1$.

Smuon pair production will thus be signaled by a calculable rate for missing energy events with acolinear muon pairs. We will see later that this is the way that the current bound on $m_{\tilde{\mu}_R}$ is obtained from experiments at LEP2.

If instead $\widetilde{W}_1$ is the lightest charged sparticle, there are additional complications from the fact that it is a model-dependent mixture of charged gauginos and higgsinos. The decays of other heavier sparticles are sensitively dependent on the sparticle (and Higgs boson) mass ordering, as well as on the sparticle mixing matrices discussed in Chapter 8. The size of the parameter space makes a general analysis of heavy sparticle decay patterns quite intractable. The analysis of

SUSY loop-induced contributions to low energy processes is also complicated for the same reason. Indeed, as we saw in Chapter 9, phenomenologically well-motivated, but theoretically ad hoc, universality assumptions are invoked for these analyses.

Clearly, the problem is that we do not have any theoretical principle for determining the soft SUSY breaking parameters of the MSSM. We speculate that the MSSM is the low energy approximation to an underlying fundamental theory in which SUSY is spontaneously broken by some as yet unknown dynamics. We hope that experimental data on sparticle properties will guide us to this dynamics once these are discovered. In the absence of such guidance, we adopt a "top-down" approach based on theoretical assumptions about how superpartners of SM particles acquire masses, resulting in different models of supersymmetry.

The first attempts to construct supersymmetric theories of particle physics were based on global supersymmetry, with the supersymmetry broken at the weak scale. As we saw in Chapter 7 these attempts typically run into problems with the tree-level mass sum rule (7.35). These problems can be avoided if

1. supersymmetry is promoted to a local symmetry, in which case the sum rule is modified to (10.66); the term proportional to $m_{3/2}$ on the right-hand side means that the scalar masses can all be shifted up, thereby evading the phenomenologically unacceptable existence of scalars lighter than the fermions.
2. Alternatively, the tree-level sum rule of global supersymmetry can be evaded if superpartners of SM particles get their masses only at the loop level.

Models that exploit both these alternatives have been constructed. A common feature of all realistic supersymmetric models of particle physics is the necessity of assuming a "hidden sector" whose dynamics somehow breaks supersymmetry. This sector is dubbed hidden because it couples only indirectly (and very weakly) to the "observable sector" of SM particles and their superpartners. The details of how supersymmetry is broken in this sector are, as we will see, unimportant for the physics of the observable sector. What is important is "the agent" that couples the hidden and observable sectors, and communicates the effects of SUSY breaking to the superpartners of SM particles, which then acquire soft SUSY breaking masses and couplings. The idea is that supersymmetry is broken at a scale $F = M_{SUSY}^2 \gg M_W^2$ in the hidden sector where the goldstino resides. This sector is assumed to interact with the observable sector only via the exchange of superheavy particles X. The couplings of the goldstino to the observable particles are suppressed by (a power of) M_{SUSY}/M_X, and the effective mass gap in the observable sector is $M_{eff} \sim M_{SUSY}^2/M_X$ (or, more generally, M_{SUSY}^{n+1}/M_X^n, $n = 1, 2, \ldots$). It is this gap that is required to be comparable to the weak scale, even though the fundamental SUSY breaking scale may be much larger.[1]

The reason that this approach makes any sense at all is that the radiative corrections in the observable sector correspond to the scale M_{eff}. The effective potential of the low energy theory (the observable sector) does not contain any terms of $\mathcal{O}(M_{\text{SUSY}}^4, M_{\text{SUSY}}^3 M_{\text{eff}}, M_{\text{SUSY}}^2 M_{\text{eff}}^2, M_{\text{SUSY}} M_{\text{eff}}^3)$ which would render the whole approach inconsistent. This was first analyzed by Polchinski and Susskind in a toy model, and later by others in more realistic scenarios.[2]

Supersymmetric models are characterized by the agent that communicates supersymmetry breaking effects in the hidden SUSY breaking sector to the observable world. Since gravity couples universally to energy, gravitational interactions are one obvious choice for mediating SUSY breaking effects. Indeed in a truly supersymmetric world, gravity mediation *must be present*. Whether its effects are swamped by other things is the relevant issue. Of course, gravity-mediated SUSY breaking requires that supersymmetry is local (as it must be if *all* interactions are supersymmmetric), so that these models are based on supergravity. Supersymmetry breaking effects may also be communicated by the usual SM gauge interactions. In these so-called gauge-mediated supersymmetry breaking (GMSB) scenarios, new "messenger fields" that couple directly to the hidden sector, but which also have SM gauge couplings, act as mediators of SUSY breaking effects: MSSM gauginos and sfermions get supersymmetry breaking masses and couplings only at the loop level, thereby evading the tree-level mass sum rule. We will see that both supergravity models as well as GMSB models have the general structure of geometric hierarchy models discussed above: since gravity is the mediator of SUSY breaking in supergravity models, the scale M_X is identified with the Planck scale, and $M_{\text{SUSY}} \sim 10^{10}$ GeV. In GMSB scenarios, M_{SUSY} is identified with the mass of messenger fields; if these are relatively light, the underlying SUSY breaking scale can be much smaller than in gravity-mediated scenarios. More recently other mediation mechanisms, which could most naturally occur if the world had additional (compactified) spatial dimensions, have also been considered.

In this chapter, we will introduce the physical ideas behind these various models, focussing on the differences in their phenomenological implications. We emphasize that each of these models is based on untested assumptions about physics at scales well above the weak scale. It may be that these assumptions will prove to be incorrect. The important thing, however, is that the models make characteristic predictions which will be subject to test in experiments at high energy colliders. A

---

[1] Indeed there were attempts to make realistic globally supersymmetric models based on this idea, with $M_X \gg M_W$ and $n = 2$. These models were dubbed geometric hierarchy models for obvious reasons.

[2] J. Polchinski and L. Susskind, *Phys. Rev.* **D26**, 3661 (1982); L. Hall, J. Lykken and S. Weinberg, *Phys. Rev.* **D27**, 2359 (1983), and references therein.

common feature of all these scenarios is that sfermions with the same $SU(3)_C \times SU(2)_L \times U(1)_Y$ quantum numbers will turn out to have the same mass parameters and the same A-parameters, renormalized at some high scale, so that unwanted flavor-violating effects discussed in Chapter 9 are absent. The reader should keep an eye open for how this comes about in each of these cases.

11.1 Gravity-mediated supersymmetry breaking

We begin our discussion of models by considering the case where gravitational interactions mediate the effects of supersymmetry breaking in the hidden supersymmetry breaking sector to the superpartners of SM particles. We will concentrate on the *form* of the Lagrangian for the low energy theory that is obtained as a result of coupling the hidden and SM sectors via supergravity, but will not write down the most general formulae for the coefficients of the various terms in the resulting low energy theory. These formulae are cumbersome and, in the absence of a compelling high energy theory, not particularly useful for the abstraction of the low energy phenomenology.

11.1.1 Hidden sector origin of soft supersymmetry breaking terms

We begin our construction of supergravity models by grouping the left-chiral supermultiplets of the model $\{\hat{S}_i\}$ into observable sector fields $\hat{C}_i$ (these include all the MSSM fields) and the "hidden sector" fields $\hat{h}_m$. The $\hat{h}_m$ fields are assumed to be gauge singlets under the observable sector gauge symmetry group, which may be taken to be G_{SM} or a grand unifying group. The observable sector gauge superfields are correspondingly chosen. The superpotential is chosen to be a sum of two independent parts, with no "superpotential couplings" between them:[3]

$$\hat{f}(\hat{S}_i) = \hat{f}_o(\hat{C}_n) + \hat{f}_h(\hat{h}_m). \tag{11.1}$$

New gauge interactions (under which the SM particles are singlets), and the associated degrees of freedom may also be present in the hidden sector.

The locally supersymmetric Lagrangian corresponding to these sets of fields can be worked out using the general results of the previous chapter. With our assumptions, (super)gravity is the only coupling between the hidden sector and the observable sector. We assume that the dynamics of the hidden sector somehow breaks supersymmetry. This could be by any of the mechanisms discussed in Chapter 7. The goldstino degrees of freedom are absorbed by the gravitino which obtains

---

[3] We may imagine that a symmetry forbids superpotential couplings between the hidden and observable sectors if the superpotential is restricted to be polynomial in the fields. Since supergravity is not renormalizable, the division into the two sectors appears to require an alternative explanation.

a mass $m_{3/2}$. The low energy effective field theory valid below the Planck scale is obtained by taking the limit as $M_P \to \infty$, keeping $m_{3/2}$ fixed. This will turn out to be a renormalizable supersymmetric Yang–Mills theory based on the low energy gauge group together with a slew of soft SUSY breaking (SSB) masses and couplings, with magnitudes $\sim m_{3/2}$. It should be kept in mind that in this framework, higher dimensional, non-renormalizable operators (consistent with low energy symmetries) suppressed by appropriate powers of M_P will also be present. These operators are referred to as "Planck slop" in the literature.

To illustrate this procedure, we will adopt a simple model wherein the observable sector consists of the fields of the MSSM, and the gauge symmetry is $SU(3)_C \times SU(2)_L \times U(1)_Y$. We take the hidden sector to consist of a single left-chiral superfield $\hat{h}$ whose dynamics breaks local supersymmetry. The hidden sector superpotential $\hat{f}_h$ might be the Polonyi superpotential, although we will be somewhat more general than that.

We first work out the scalar potential for the case of the flat Kähler metric, with

$$K(\hat{S}, \hat{S}^\dagger) = \hat{h}^\dagger \hat{h} + \sum_n \hat{C}^{\dagger n} \hat{C}_n.$$

We will return to the more general case later. This potential is given by,

$$
\begin{aligned}
V &= e^{(S^{i*}S_i)/M_P^2} \left[\left| \frac{\partial \hat{f}}{\partial S_i} + \frac{S^{i*} \hat{f}}{M_P^2} \right|^2 - \frac{3}{M_P^2} |\hat{f}|^2 \right] \\
&= e^{(h^*h + C^{n*}C_n)/M_P^2} \left[\left| \frac{\partial \hat{f}_h}{\partial h} + \frac{h^* \hat{f}}{M_P^2} \right|^2 + \left| \frac{\partial \hat{f}_o}{\partial C_n} + \frac{C^{n*} \hat{f}}{M_P^2} \right|^2 - \frac{3}{M_P^2} |\hat{f}|^2 \right],
\end{aligned}
$$

$$(11.2)$$

where a sum over fields is implied. We assume that the F-component of the hidden sector field $\hat{h}$ develops a VEV $\sim m^2$ which breaks local SUSY and, further, that its scalar component h develops a VEV of order M_P as well. The VEVs of the scalar components of the observable sector fields are assumed to be negligible compared to M_P. Accordingly, we parametrize these VEVs by,

$$\langle h \rangle = a M_P, \tag{11.3a}$$

and

$$\langle \hat{f}_h \rangle = b m^2 M_P \quad \text{and} \quad \left\langle \frac{\partial \hat{f}_h}{\partial h} \right\rangle = m^2, \tag{11.3b}$$

with a and b being dimensionless coefficients of order 1. For the specific choice of the Polonyi model, $a = \sqrt{3} - 1$ and $b = 1$. The gravitino mass is given by,

$$m_{3/2} = \frac{bm^2}{M_P} e^{a^2/2}. \tag{11.3c}$$

The ratio m/M_P is arbitrary at this point.

The next step in the calculation of the effective scalar potential of the "light" observable sector fields valid at an energy scale $Q \ll M_P$ is to evaluate the scalar potential with the "heavy" hidden sector field (whose quanta cannot be excited at this low energy scale) set to its VEV. Finally, we take the flat space limit, $M_P \to \infty$ while keeping $m_{3/2}$ fixed. The effective scalar potential reduces to,

$$V_{\text{eff}} = m^4 e^{a^2} \left[(1 + ab)^2 - 3b^2 \right] + \left| \frac{\partial \tilde{f}_0}{\partial C_n} \right|^2 + V_{\text{ssb}}, \tag{11.4a}$$

where $\tilde{f}_0$ is the rescaled scalar superpotential $\tilde{f}_0 = e^{a^2/2} \hat{f}_0$,

$$V_{\text{ssb}} = \sum_n \left[1 + \left(a + \frac{1}{b} \right)^2 - 3 \right] m_{3/2}^2 |C_n|^2$$

$$+ m_{3/2} \sum_n \left[\frac{\partial \tilde{f}_0}{\partial C_n} C_n + A \tilde{f}_0 + \quad \text{h.c.} \right], \tag{11.4b}$$

and $A = a \left(\frac{1}{b} + a \right) - 3$.

The first term of V_{eff} in (11.4a) is the cosmological constant, which may be fine-tuned to zero by adopting $(1 + ab)^2 = 3b^2$. The second term of V_{eff} is the "usual" superpotential contribution to the scalar potential, as in theories with global SUSY, where $\tilde{f}_0$ is now identified as the superpotential of the effective theory. The term V_{ssb} evidently contains various soft SUSY breaking terms. The first of these are mass terms for the scalar components of the visible sector superfields: if the cosmological constant is fine-tuned to zero, they are *all* given by $m_{\text{scalar}} = m_{3/2}$ in this simple case, and the desired universality is obtained. We can choose m^2 so that $m_{3/2} \sim M_{\text{weak}}$, the size required for low energy supersymmetry to stabilize the electroweak scale. The smallness of the ratio m/M_P still needs to be explained. The remaining terms in V_{ssb} correspond to bilinear and trilinear soft SUSY breaking terms, and are also of order $m_{3/2}$. These correspond to the b and $\mathbf{a}$ terms in (8.10) of Chapter 8. Notice that the $\mathbf{c}$ terms discussed in Chapter 8 are absent; this is because of our simple choice of the Kähler potential.

Before turning to the case of the general Kähler metric, we note that SUSY breaking in the hidden sector may also lead to soft SUSY breaking gaugino masses in the observable sector. This can clearly be seen from the first term in $\mathcal{L}_{F,\text{Int}}^G$ in (10.58b):

$$\mathcal{L}_F^G \ni \frac{1}{4} e^{G/2} \frac{\partial f_{AB}^*}{\partial \mathcal{S}^{j*}} (G^{-1})_k^j G^k \bar{\lambda}_A \lambda_B. \tag{11.5}$$

In order to obtain non-zero gaugino masses, the gauge kinetic function must be a non-trivial function of hidden sector fields, and SUSY must be broken; i.e. $\langle G^k \rangle \neq 0$. Since $\langle G^k \rangle \sim m^2$, and $\partial f^*_{AB}/\partial \mathcal{S}^{j*} \sim 1/M_P$, we expect that the resulting gaugino mass $m_{1/2} \sim m^2/M_P \sim m_{3/2}$, and is also of order the weak scale. Clearly, without any assumptions about unification of gauge interactions, we will obtain *independent* masses for $SU(3)_C$, $SU(2)_L$, and $U(1)_Y$ gauginos.

We should keep in mind that the soft SUSY breaking effective Lagrangian for the low energy theory that we have just obtained will still obtain radiative corrections from the interactions in this *low energy* theory. Although this low energy theory may be weakly coupled, these radiative corrections would be expected to depend on $k = \frac{g^2}{8\pi^2} \ln(M_P/m_{3/2})$, where g is a typical coupling in the low energy theory (it could be one of the gauge couplings.) Since k is not small, it is important to sum these logarithms. This is done by using "running parameters" obtained by solving the renormalization group equations discussed in Chapter 9. In other words, the parameters in the "low energy theory" should be regarded as renormalized at some high scale $\sim M_P$.

The careful reader will have traced the reason for universal masses in (11.4b) to our assumption of a flat Kähler metric. More general forms of the Kähler potential lead to non-universal SSB masses and couplings as can be seen from the exercise at the end of this section. This was first pointed out by Soni and Weldon.[4] Rather than list the rather complicated expressions for these parameters, we will just note several important features:

1. Although universality is not a generic feature of supergravity models, the scale of SSB masses and couplings is generally still set by $m_{3/2}$.
2. The trilinear **a** terms, in general, are not proportional to the corresponding superpotential Yukawa couplings.
3. Trilinear **c** terms are possible, but are suppressed by higher powers of M_P.

Regarding the first point, we should mention that even if we do arrange for a model with universal scalar masses at tree level (for instance, by the minimal choice for the Kähler potential), loop corrections will in general spoil the degeneracy. Indeed, assuming universal scalar masses is tantamount to assuming a $U(n)$ symmetry amongst the n observable sector superfields. However, this symmetry is explicitly broken by superpotential Yukawa couplings in the observable sector. Moreover, there is no theoretical argument for such a symmetry. Thus, while it might be possible to accommodate universality by making technical assumptions of the underlying physics, it seems fair to say that universality is not a generic feature of supergravity

[4] S. Soni and H. A. Weldon, *Phys. Lett.* **126B**, 215 (1983). Convenient general expressions for the soft SUSY breaking parameters are given by A. Brignole *et al.* in *Perspectives on Supersymmetry*, edited by G. Kane, World Scientific (1998).

models. This has provided motivation for the construction of alternative scenarios where the scalar mass degeneracies needed for solving the SUSY flavor and CP problems occur for different reasons.

Exercise (Non-universal scalar masses) *Show that the low energy effective potential leads to non-universal scalar mass terms if the Kähler potential has the form,*

$$K(\hat{S}, \hat{S}^{\dagger}) = \hat{h}^{\dagger}\hat{h} + \sum_{n} \tilde{K}_{n}\left(\frac{\hat{h}\hat{h}^{\dagger}}{M_{P}^{2}}\right)\hat{C}^{\dagger n}\hat{C}_{n},$$

where $\tilde{K}_{n}$ are dimensionless functions of the hidden sector field. Remember that for this Kähler potential the kinetic energy terms of the scalar components of $\hat{C}_{n}$ do not have the canonical form, so that these fields have to be rescaled to obtain canonical kinetic energies. Notice that this redefinition implies that we would have obtained universal mass terms if K_{n} had just been some constants rather than field-dependent functions.

Check also the form of supersymmetry breaking trilinear interactions in the low energy theory. Are the A terms universal?

Exercise *We found that **c**-type trilinears were absent in the minimal model. Convince yourself that this type of SUSY trilinears can arise if one allows trilinear terms in the Kähler potential. Remember that these terms will always be suppressed by powers of M_{P}.*

Although the details of the hidden sector are unimportant for low energy phenomenology, it is gratifying to see such a hidden sector is present in many theoretical frameworks. For instance, in heterotic string models, there is a natural, built-in hidden sector comprised of the dilaton field S, arising from the gravitational sector of the theory, and the moduli fields T_{i}, which parametrize the size and shape of the compactification. If the auxiliary fields of these multiplets provide the seeds for SUSY breaking, then the resulting effective theory below the Planck scale may be just a four-dimensional supersymmetric gauge theory with weak scale soft SUSY breaking terms.

11.1.2 Why is the μ parameter small?

We have just seen that supergravity models provide a rationale for why the scale of SUSY breaking parameters of the MSSM is much smaller than M_{P}. On dimensional grounds we would expect though that the supersymmetry conserving μ parameter to be of order M_P rather than $m_{3/2}$, which would destroy the mechanism for electroweak symmetry breaking in SUSY models. This is known as the μ problem.

Supergravity models provide an elegant mechanism for generating the μ term with the right magnitude, provided that we assume that μ is forbidden by a symmetry that is violated only by interactions with the hidden sector.[5] Although it would then be absent in the tree-level superpotential of the observable sector, an effective μ term would develop via the gravitational interactions with the hidden sector. To see this, we note that there is nothing to forbid a (non-renormalizable) term,

$$K(\hat{h}, \hat{H}_u, \hat{H}_d) \ni \frac{\lambda \hat{h}^\dagger \hat{H}_u \hat{H}_d}{M_P}, \tag{11.6a}$$

in the Kähler potential, where $\hat{h}$ is a hidden sector left-chiral superfield and $\hat{H}_u$ and $\hat{H}_d$ the Higgs superfields of the MSSM. Since the F-component of $\hat{h}$ develops a VEV $\sim m^2$, the action of the low energy effective theory includes a term,

$$\int d^4x \mathcal{L} \ni -\frac{1}{4} \int d^4x d^4\theta \, K(\hat{h}, \hat{H}_u, \hat{H}_d) + \quad \text{h.c.}$$

$$\sim \frac{m^2 \lambda}{M_P} \int d^4x d^2\theta \, \hat{H}_u \hat{H}_d + \quad \text{h.c.} \tag{11.6b}$$

The reader will recognize this as the superpotential μ term of the MSSM, with a magnitude $|\mu| \sim m_{3/2}$ as phenomenologically required.

11.1.3 Supergravity Grand Unification (SUGRA GUTs)

Minimal supergravity (mSUGRA) model

We have already encountered the minimal supergravity (mSUGRA) model in Chapter 9 where we adopted the universality hypothesis for gaugino masses, scalar masses and the various A-parameters at a high scale $Q \sim M_{GUT}$. This scenario can be obtained within the framework with gravity mediated SUSY breaking.[6] The choice of a flat Kähler metric leads to a common mass for all scalars of $m_0^2 = m_{3/2}^2 + V_0/M_P^2$, where V_0 is the minimum of the scalar potential. It is this technical assumption of the "minimal" choice of the Kähler potential that the "minimal" in minimal supergravity refers to. In this case though universality is likely to hold closer to $Q \sim M_P$. Common gaugino masses at $Q = M_{GUT}$ may arise because of grand unification of gauge interactions. But these may also be obtained by assuming that the gauge kinetic function has the same field dependence on the

[5] G. F. Giudice and A. Masiero, *Phys. Lett.* **206B**, 480 (1988).

[6] A. Chamseddine, R. Arnowitt and P. Nath, *Phys. Rev. Lett.* **49**, 970 (1982); R. Barbieri, S. Ferrara and C. Savoy, *Phys. Lett.* **B 119**, 343 (1982); N. Ohta, *Prog. Theor. Phys.* **70**, 542 (1983); L. Hall, J. Lykken and S. Weinberg, *Phys. Rev.* **D27**, 2359 (1983).

hidden sector fields, for each factor of gauge symmetry: e.g. $f_{AB}^a = c^a \delta_{AB} f(h_m)$, where a labels the different factors of the gauge group, and c^a are real numbers. The constants c^a disappear from the mass term upon canonical normalization of the gaugino kinetic energy terms.

The fundamental parameters of the model are the set (9.22). We have already seen that radiative EWSB (discussed in Chapter 9) allows us to trade the bilinear SSB parameter B (equivalently B_0) in favor of $\tan\beta$, and also to fix the value of μ^2 to reproduce the experimental value of M_Z. Then, the parameter space of the model is given by

$$m_0, \ m_{1/2}, \ A_0, \ \tan\beta, \text{sign}\,(\mu). \tag{11.7}$$

It is common to assume that the universality of SSB parameters holds at M_{GUT} rather than M_{P}.

SU(5) supergravity GUT model

We will assume that the reader is familiar with the motivations for grand unification. Grand unified theories are especially attractive when combined with supergravity. The simplest model, based on $SU(5)$ gauge symmetry,

- allows for unification of the gauge symmetries of the SM into a single Lie group,
- provides a group theoretic explanation for the ad hoc hypercharge assignments of the SM or MSSM fields.

It is usually assumed that supersymmetric $SU(5)$ grand unification is valid at mass scales $Q > M_{\text{GUT}} \simeq 2 \times 10^{16}$ GeV, extending at most to the reduced Planck scale M_{P} where gravitational effects become sizable. Below $Q = M_{\text{GUT}}$, the $SU(5)$ model (with a minimal matter content) breaks down to the MSSM with the usual $SU(3)_{\text{C}} \times SU(2)_{\text{L}} \times U(1)_{\text{Y}}$ gauge symmetry.

In the $SU(5)$ model, the $\hat{D}^c$ and $\hat{L}$ superfields are members of a $\bar{\mathbf{5}}$ superfield $\hat{\phi}$, while the $\hat{Q}$, $\hat{U}^c$, and $\hat{E}^c$ superfields occur in the $\mathbf{10}$ representation $\hat{\psi}$. There is a replication of generations. The Higgs sector of the minimal $SU(5)$ model is comprised of three super-multiplets: $\hat{\Sigma}(\mathbf{24})$ which is responsible for breaking $SU(5)$, together with $\hat{\mathcal{H}}_1(\bar{\mathbf{5}})$ and $\hat{\mathcal{H}}_2(\mathbf{5})$ which contain the MSSM Higgs doublet superfields $\hat{H}_d$ and $\hat{H}_u$ respectively.[7] The superpotential is given by,

$$\hat{f} = \mu_\Sigma \text{Tr}\hat{\Sigma}^2 + \frac{1}{6}\lambda_\Sigma \text{Tr}\hat{\Sigma}^3 + \mu_H \hat{\mathcal{H}}_1 \hat{\mathcal{H}}_2 + \lambda \hat{\mathcal{H}}_1 \hat{\Sigma} \hat{\mathcal{H}}_2$$

$$+ \frac{1}{4} f_t \epsilon_{ijklm} \hat{\psi}^{ij} \hat{\psi}^{kl} \hat{\mathcal{H}}_2^m + \sqrt{2} f_b \hat{\psi}^{ij} \hat{\phi}_i \hat{\mathcal{H}}_{1j} + \cdots, \tag{11.8a}$$

---

[7] This is the primary reason why we assigned $\hat{H}_d$ to transform as the $\mathbf{2}^*$ representation of $SU(2)_{\text{L}}$.

where we neglect the Yukawa couplings of the first two generations, and retain just f_t and f_b, the top and bottom quark Yukawa couplings. The couplings λ and λ_Σ are GUT Higgs sector self-couplings, and μ_Σ and μ_H are superpotential Higgs mass terms. Note that in this model $f_b = f_\tau$ when the $SU(5)$ symmetry is unbroken.

Proton decay is the smoking gun signature of grand unification. In non-supersymmetric GUTs, this occurs via dimension 6 operators involving X and Y gauge bosons. In supersymmetric GUTs, proton decay can also be mediated by color-triplet higgsinos which, being fermions, lead to dimension 5 baryon-number-violating operators which are potentially much more dangerous.[8] Furthermore, higgsino-mediated proton decay depends on Yukawa couplings. As a result, in SUSY GUTs, the proton preferentially decays to kaons rather than to pions as in standard GUTs.

Soft supersymmetry breaking terms are induced by hidden sector local SUSY breaking, and are parametrized by:

$$\mathcal{L}_{\text{soft}} = -m_{\mathcal{H}_1}^2|\mathcal{H}_1|^2 - m_{\mathcal{H}_2}^2|\mathcal{H}_2|^2 - m_\Sigma^2\text{Tr}\{\Sigma^\dagger\Sigma\} - m_5^2|\phi|^2 - m_{10}^2\text{Tr}\{\psi^\dagger\psi\}$$
$$-\frac{1}{2}M_5\bar{\lambda}_\alpha\lambda_\alpha$$
$$+\left[B_\Sigma\mu_\Sigma\text{Tr}\Sigma^2 + \frac{1}{6}A_{\lambda_\Sigma}\lambda_\Sigma\text{Tr}\Sigma^3 + B_H\mu_H\mathcal{H}_1\mathcal{H}_2 + A_\lambda\lambda\mathcal{H}_1\Sigma\mathcal{H}_2 \right.$$
$$\left. + \frac{1}{4}A_t f_t\epsilon_{ijklm}\psi^{ij}\psi^{kl}\mathcal{H}_2^m + \sqrt{2}A_b f_b\psi^{ij}\phi_i\mathcal{H}_{1j} + \text{h.c.} \right]. \quad (11.8b)$$

The various SSB parameters and the gauge and Yukawa couplings evolve with energy scale according to 15 renormalization group equations (the first two generations are degenerate). Assuming universality at M_P, one imposes

$$m_{10} = m_5 = m_{\mathcal{H}_1} = m_{\mathcal{H}_2} = m_\Sigma \equiv m_0$$
$$A_t = A_b = A_\lambda = A_{\lambda_\Sigma} \equiv A_0, \quad (11.9a)$$

and evolves all the soft masses from M_P to M_{GUT}. Although there are no large logarithms or couplings, there is substantial evolution due to large group theory factors arising from the fact that the representations contain many particles. The MSSM soft breaking masses at M_{GUT} are specified via

$$m_Q^2 = m_U^2 = m_E^2 \equiv m_{10}^2, \quad m_D^2 = m_L^2 \equiv m_5^2,$$
$$m_{H_d}^2 = m_{\mathcal{H}_1}^2, \quad m_{H_u}^2 = m_{\mathcal{H}_2}^2. \quad (11.10)$$

---

[8] Indeed, if $\lambda \lesssim 0.7$, triplet higgsinos are too light and one runs into trouble with constraints from proton decay. Some authors have argued that these constraints, in fact, rule out minimal SUSY $SU(5)$. It is clear, however, that the proton decay rate depends on the unknown details of GUT scale physics, and can be altered by complicating the GUT sector. For this reason, we will not consider such constraints here.

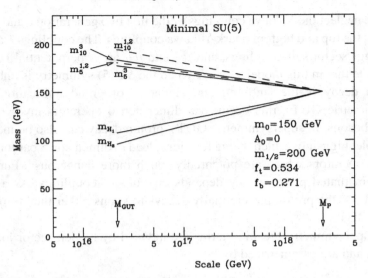

Figure 11.1 Evolution of SSB masses for a typical case study in SUSY $SU(5)$ GUT model with $\tan \beta = 35$, which allows for b-τ Yukawa unification. We choose $\lambda = 0.7$ and $\lambda_\Sigma = 0.1$. Although not shown, A_t and A_b evolve to -88 GeV and -78 GeV at $Q = M_{\mathrm{GUT}}$. Reprinted from H. Baer, M. Diáz, P. Quintana and X. Tata, *JHEP* **04**, 016 (2000).

The evolution of SSB masses in SUSY $SU(5)$ is shown in Fig. 11.1. A striking feature is the sizable GUT scale splitting between the Higgs and matter scalar mass parameters, arising from the large λ Yukawa coupling contribution to the running of Higgs boson mass parameters. The masses of the **10** and **5*** multiplets evolve differently, as do those of multiplets of the different generations. In particular, third generation multiplet masses are somewhat suppressed compared to their counterparts of the first two generations owing to Yukawa coupling contributions to the RGE running. Thus, even in an $SU(5)$ SUGRA GUT model, we would expect non-universality. However, scalar masses for multiplets in the first two generations are still highly degenerate, so that FCNCs are suppressed. We have also checked that even if we start with $A_0 = 0$ at $Q = M_P$, sizable values of A_t and A_b are generated at $Q = M_{\mathrm{GUT}}$. This could have a significant effect on the phenomenology of third generation sparticles.

Exercise *Draw a Feynman diagram involving triplet higgsino exchange for the baryon number violating process,*

$$\tilde{d}\tilde{u} \rightarrow \bar{s}\bar{\nu}_\mu.$$

Notice that the squarks in the initial state can be obtained from u and d quarks by exchanging a chargino. This "dressed" diagram mediates the process $du \rightarrow \bar{s}\bar{\nu}_\mu$

which in turn can cause $p \rightarrow K^+ \bar{\nu}_\mu$. *Convince yourself that the amplitude for this process is suppressed only by a single power of* M_{GUT} *and, hence, is much larger than that of the baryon number violating amplitude mediated by exchange of GUT gauge bosons.*

Non-universal gaugino masses

Since there is no reason to expect the gauge kinetic function to be field independent, gaugino masses are a generic prediction of supergravity models (or for that matter any non-renormalizable theory with broken supersymmetry). Moreover, (11.5) shows that the gaugino mass scale in the effective low energy theory is expected to be $\sim m_{3/2}$. Gauge invariance dictates that the gauge kinetic function must transform as the symmetric product of two adjoints under the gauge symmetry. If any of the auxiliary fields that break supersymmetry transform non-trivially under the grand unifying gauge group (but of course as an MSSM gauge singlet), non-universal MSSM gaugino masses are obtained. These may then be parametrized by,

$$\mathcal{L} \supset \frac{\langle F_h \rangle_{AB}}{M_P} \bar{\lambda}_A \lambda_B + \cdots \qquad (11.11)$$

where the λ_A are the gaugino fields, and F_h is the auxiliary field component of $\hat{h}$ that acquires a SUSY, and possibly GUT symmetry, breaking VEV. It is only in the special case where the fields F_h which break supersymmetry are GUT singlets that universal gaugino masses are obtained.

In the context of $SU(5)$ grand unification, F_h belongs to an $SU(5)$ representation which appears in the symmetric product of two adjoints:

$$(\mathbf{24} \times \mathbf{24})_{\text{symmetric}} = \mathbf{1} \oplus \mathbf{24} \oplus \mathbf{75} \oplus \mathbf{200}, \qquad (11.12)$$

where only $\mathbf{1}$ yields universal masses. If instead F_h transforms as any other irreducible representation that appears in (11.12), the MSSM gaugino mass parameters at $Q = M_{\text{GUT}}$, though different, are related by group theory. The weak scale gaugino masses are then obtained by renormalization group evolution, starting from these non-universal values, as discussed in Chapter 9. The relative GUT scale $SU(3)$, $SU(2)$, and $U(1)$ gaugino masses M_3, M_2, and M_1 are listed in Table 11.1 along with the approximate masses after RGE evolution to $Q \sim M_Z$. These scenarios represent the predictive subset of the more general (and less predictive) case of an arbitrary superposition of these representations. The model parameters may be chosen to be,

$$m_0, \ M_3^0, \ A_0, \ \tan\beta, \ \text{and sign}\,(\mu), \qquad (11.13)$$

where M_i^0 is the $SU(i)$ gaugino mass at scale $Q = M_{\text{GUT}}$. M_2^0 and M_1^0 can be obtained in terms of M_3^0 via Table 11.1. Notice that the nature of the neutralino LSP

Table 11.1 *Relative gaugino mass parameters at*
$Q = M_{GUT}$ *and their relative values evolved to* $Q = M_Z$ *in*
the four possible F_h *irreducible representations in* $SU(5)$
SUSY GUTS.

F_h	M_{GUT}			M_Z		
	M_3	M_2	M_1	M_3	M_2	M_1
1	1	1	1	~6	~2	~1
24	2	−3	−1	~12	~ − 6	~ − 1
75	1	3	−5	~6	~6	~ − 5
200	1	2	10	~6	~4	~10

as well as the mass gap between the LSP and $\tilde{Z}_2$ depend on the gauge transformation properties of F_h: as a result, SUSY phenomenology changes significantly in the different scenarios.[9]

$SO(10)$ *supergravity GUT models*

We saw in Section 9.7 that it is necessary to introduce right-handed neutrino superfields in order to give neutrinos a mass without spoiling the conservation of R-parity. Within the MSSM, or within the $SU(5)$ GUT framework just discussed, these gauge singlet superfields had to be introduced ad hoc. The body of evidence in support of neutrino mass, however, makes the grand unified group $SO(10)$ especially appealing because the minimal $SO(10)$ model contains three generations of matter superfields, with each generation together with a singlet neutrino superfield $\hat{N}^c$ included in the **16**-dimensional spinorial representation $\hat{\psi}_{16}$. Thus, $SO(10)$ allows not only for gauge group unification, but also for unification of matter in each generation into a single irreducible representation. Morover, singlet neutrinos essential for the implementation of the see-saw mechanism occur automatically in this framework. The Higgs bosons $\hat{H}_u$ and $\hat{H}_d$ lie within a **10**-dimensional fundamental representation $\hat{\phi}_{10}$. The superpotential for the model includes the term

$$\hat{f} \ni f\hat{\psi}_{16}\hat{\psi}_{16}\hat{\phi}_{10} + \cdots , \tag{11.14}$$

responsible for quark and lepton masses, with f being the single Yukawa coupling per generation in the GUT scale theory. The ellipsis represents terms including for instance higher dimensional Higgs representations and interactions responsible for the breaking of $SO(10)$.

[9] G. Anderson *et al.*, *Phys. Rev.* **D61**, 095005 (2000).

The unification of Yukawa couplings of each generation means that the third generation neutrino Yukawa coupling f_ν is large: $f_\nu = f_t$. The third generation neutrino Yukawa coupling f_ν splits $\tilde{\tau}_L$ and $\tilde{\nu}_\tau$ masses from their first and second generation cousins, in the same way that tau Yukawa couplings split the staus from other slepton masses. In some models, this splitting is potentially measurable in linear collider experiments.

The soft SUSY breaking terms will include a common mass m_{16} for all matter scalars and a mass m_{10} for the Higgs scalars, along with a universal gaugino mass $m_{1/2}$, and common trilinear and bilinear SSB masses A_0 and B. Motivated by apparent gauge coupling unification in the MSSM, it is common to assume that $SO(10)$ breaks directly to the gauge group $SU(3)_C \times SU(2)_L \times U(1)_Y$ at $Q = M_{GUT} = 2 \times 10^{16}$ GeV, though $SO(10)$ could well have broken to $SU(5)$ at a yet higher scale.

A novel feature arises because the rank of $SO(10)$ (the rank is the largest number of mutually commuting generators) is one higher than that of the MSSM gauge group. This effectively means that $SO(10)$ includes a (broken) $U(1)_X$ factor that is not a low energy symmetry. Naively, one would suppose that if the $U(1)_X$ breaking scale M_X is sufficiently large, $U(1)_X$ would be negligible for TeV scale physics. To see that this is not the case, let us consider a simple toy model, where $U(1)_X$ is broken by VEVs of a pair of MSSM gauge singlet fields Φ and $\bar{\Phi}$ with $U(1)_X$ charges $+1$ and -1, respectively. If we denote the scalar components of the MSSM fields by S_i and their $U(1)_X$ charges by x_i, we can write the scalar potential that determines $\langle\Phi\rangle$ and $\langle\bar{\Phi}\rangle$ as,

$$V = V_{\text{symm}}(\Phi, \bar{\Phi}) + m^2|\Phi|^2 + \overline{m}^2|\bar{\Phi}|^2 + \frac{g_X^2}{2}\left[|\Phi|^2 - |\bar{\Phi}|^2 + x_i|S_i|^2\right]^2. \quad (11.15)$$

The term V_{symm} comes from the superpotential for the heavy fields and is chosen to be symmetric under $\Phi \leftrightarrow \bar{\Phi}$. The next two terms are SSB masses for the heavy fields Φ and $\bar{\Phi}$: they are of order of the weak scale, but otherwise unrelated. The last term, which is the usual $U(1)_X$ D-term contribution to the potential, forces the minimum to be along the nearly D-flat direction $\langle\Phi\rangle \approx \langle\bar{\Phi}\rangle$. However, if $m \neq \overline{m}$, the minimum of the potential deviates from the D-flat direction by,

$$\langle\Phi\rangle^2 - \langle\bar{\Phi}\rangle^2 \simeq \frac{\overline{m}^2 - m^2}{2g_X^2}. \quad (11.16)$$

The last term in (11.15) then shows that the MSSM scalars S_i receive an additional contribution to the mass proportional to their $U(1)_X$ charge,

$$\Delta m_i^2 = \frac{x_i}{2} \times (\overline{m}^2 - m^2), \quad (11.17)$$

which by (11.16) is of order the weak scale. Thus $U(1)_X$ leaves its imprint on the MSSM sfermion mass spectrum even if M_X is very large.

Returning to the $SO(10)$ model, the scalar field squared mass parameters at $Q = M_{\text{GUT}}$ are then given by

$$m_Q^2 = m_E^2 = m_U^2 = m_{16}^2 + M_D^2$$
$$m_D^2 = m_L^2 = m_{16}^2 - 3M_D^2$$
$$m_{H_{u,d}}^2 = m_{10}^2 \mp 2M_D^2$$
$$m_N^2 = m_{16}^2 + 5M_D^2, \tag{11.18}$$

where M_D^2 parametrizes the magnitude of the $U(1)_X$ D-terms just discussed, and can, owing to our ignorance of the gauge symmetry breaking mechanism, be taken as a free parameter of order of the weak scale, with either positive or negative values. Thus, the model is characterized by the following free parameters

$$m_{16}, \ m_{10}, \ M_D^2, \ m_{1/2}, \ A_0, \ \text{sign}\,(\mu). \tag{11.19}$$

Since

$$\frac{m_t}{m_b} \sim \frac{f_t v_u}{f_b v_d},$$

solutions with unification of Yukawa couplings are possible only for large values of $\tan \beta$. This argument is only qualitative because radiative corrections to m_b are very important. In practice, the value of $\tan \beta$ is restricted by the requirement of Yukawa coupling unification, and so is tightly constrained to a narrow range around $\tan \beta \sim 50$–55.

Inverted hierarchy models

A phenomenologically interesting class of models referred to as *inverted mass hierarchy* (IMH) models, have the matter sfermion mass order inverted from the order of the corresponding fermions. Specifically, scalars of the first and second generation are expected to have masses at the multi-TeV scale so that a decoupling solution to the SUSY flavor and CP problems may be invoked. Because these sparticles have very tiny couplings to the Higgs sector, they do not lead to unnaturally large fine-tuning. On the other hand, third generation sfermions (which have large couplings to the Higgs sector) are expected to be in the sub-TeV mass range to accommodate constraints from naturalness.

One class of IMH models has the inverted mass hierarchy generated *radiatively*. In this case, all the scalars begin with multi-TeV masses at the GUT scale, while gaugino masses are in the sub-TeV range. In models with Yukawa unification and

$SO(10)$-like GUT scale boundary conditions of

$$4A_0^2 = 2m_{10}^2 = m_{16}^2, \tag{11.20}$$

the large third generation Yukawa coupling acts to drive third generation and Higgs scalars to sub-TeV values, while leaving multi-TeV first and second generation scalars. A positive D-term contribution with $M_D \sim m_{16}/3$ is needed for radiative symmetry breaking.[10] When fully implemented, including constraints from radiative EWSB, it turns out that first and second generation scalars as heavy as 2–3 TeV can be allowed. This is sufficient to suppress many CP-violating processes, but is not enough to fully suppress FCNCs. Such a model might be viable if it is coupled with a partial degeneracy solution for first and second generation scalars. Yukawa coupling unification is also possible for first and second generation scalar masses ~8–10 TeV, but third generation sfermions then have masses around 3–5 TeV.

A second class of IMH models arises if one assumes an IMH already in place at the GUT scale. This may be possible in non-minimal gravity-mediated models. In evaluating sparticle mass spectra from these GUT scale IMH models, it is crucial to use two-loop RGEs. The form of the two-loop RGEs for SSB masses is given by

$$\frac{dm_i^2}{dt} = \frac{1}{16\pi^2}\beta_{m_i^2}^{(1)} + \frac{1}{(16\pi^2)^2}\beta_{m_i^2}^{(2)}, \tag{11.21}$$

where $t = \ln Q$, $i = Q_j$, U_j, D_j, L_j, and E_j, and $j = 1\text{–}3$ is a generation index. The one-loop β-function for the evolution of (the initially sub-TeV) third generation scalar masses depends only on these scalar masses and the (also sub-TeV) gaugino masses. Two-loop terms are formally suppressed relative to one-loop terms by the square of a coupling constant as well as an additional loop factor of $16\pi^2$. However, these two-loop terms include contributions from *all* scalars. Specifically, the two-loop β functions include,

$$\beta_{m_i^2}^{(2)} \ni a_i g_3^2 \sigma_3 + b_i g_2^2 \sigma_2 + c_i g_1^2 \sigma_1, \tag{11.22}$$

where

$$\sigma_1 = \frac{1}{5}g_1^2 \left\{ 3(m_{H_u}^2 + m_{H_d}^2) + Tr[\mathbf{m}_Q^2 + 3\mathbf{m}_L^2 + 8\mathbf{m}_U^2 + 2\mathbf{m}_D^2 + 6\mathbf{m}_E^2] \right\},$$

$$\sigma_2 = g_2^2 \left\{ m_{H_u}^2 + m_{H_d}^2 + Tr[3\mathbf{m}_Q^2 + \mathbf{m}_L^2] \right\}, \quad \text{and}$$

$$\sigma_3 = g_3^2 Tr[2\mathbf{m}_Q^2 + \mathbf{m}_U^2 + \mathbf{m}_D^2],$$

and the $\mathbf{m}_i^2$ are squared mass matrices in generation space. The numerical coefficients a_i, b_i, and c_i are related to the quantum numbers of the scalar fields, but are

---

[10] Such a D-term reduces $m_{H_u}^2$ relative to $m_{H_d}^2$, which facilitates EWSB. Indeed, slightly better Yukawa coupling unification is obtained if the D-term splitting is applied to just the Higgs scalars rather than to all the sparticles, but a qualitatively similar hierarchy is obtained.

all positive quantities. Thus, incorporation of multi-TeV masses for the first and second generation scalars leads to an overall positive, *possibly dominant*, contribution to the slope of SSB mass trajectories versus energy scale. Although formally a two-loop effect, the smallness of the couplings is compensated by the much larger values of masses of the first two generations of scalars. In running from M_{GUT} to M_{weak}, this results in an overall *reduction* of scalar masses, and is most important for the sub-TeV third generation scalar masses which may be driven tachyonic. That this not occur then constrains the size of the hierarchy. For values of SSB masses which fall short of these constraints, a sort of see-saw effect amongst scalar masses occurs: the higher the value of first and second generation scalar masses, the larger will be the two-loop suppression of third generation and Higgs scalar masses. In this class of models, first and second generation scalars with masses of order 10–15 TeV may co-exist with sub-TeV third generation scalars, thus giving a very large suppression to both FCNC and CP-violating processes.

11.2 Anomaly-mediated SUSY breaking

In supergravity models, MSSM soft SUSY breaking parameters are thought to arise from tree-level gravitational interactions of observable sector superfields with gauge singlet hidden sector fields that can acquire a Planck scale VEV. It was subsequently recognized that there is an additional one-loop contribution to SSB parameters that is always present when SUSY is broken.[11] Usually this latter contribution, which originates in the super-Weyl anomaly (and is, therefore, called the anomaly-mediated supersymmetry breaking (AMSB) contribution), only makes a loop suppressed correction to the leading tree-level SSB parameters, so that the pattern of sparticle masses is qualitatively unchanged from what we have described in the last section. However, in models without SM gauge singlet superfields that can acquire a Planck scale VEV, the usual supergravity contribution to gaugino masses is suppressed by an additional factor $M_{\text{SUSY}}/M_{\text{P}}$ relative to $m_{\frac{3}{2}} = M_{\text{SUSY}}^2/M_{\text{P}}$, and the anomaly-mediated contribution can dominate. Extra dimensional theories potentially offer an alternative way to suppress supergravity couplings between the observable sector and the hidden sector (goldstino) field which lead to tree-level MSSM SSB parameters $\sim m_{\frac{3}{2}}$: these supergravity contributions may be exponentially suppressed if the SUSY breaking and visible sectors reside on different branes that are "sufficiently separated" in a higher dimensional space.[12] In this case, the suppression is the result of geometry and not a symmetry, though then one has to wonder about the dynamics that results in such a geometry. Moreover, it has been

---

[11] L. Randall and R. Sundrum, *Nucl. Phys.* **B557**, 79 (1999); G. Giudice *et al.*, *JHEP* **12**, 027 (1998).
[12] The term brane means a lower dimensional spatial slice of the entire space.

argued that while it is possible to find models where AMSB terms may dominate, their construction appears to require more than just spatial separation between the observable sector and SUSY breaking branes.[13]

A derivation of the AMSB contribution to SSB parameters would require techniques beyond those that we have developed. We will, therefore, simply list the relevant results and proceed to discuss their implications. Before doing so, we note that these contributions are determined just by the super-conformal anomaly. Since anomalies depend only on the low energy theory, the AMSB contributions to SSB parameters are insensitive to (unknown) physics at the high scale. These contributions, which can be written in terms of the β-functions and anomalous dimensions of the theory with *unbroken* supersymmetry, can be explicitly checked to be invariant under renormalization group evolution, consistent with their insensitivity to ultra-violet physics.[14]

The AMSB contribution to the gaugino mass is given by,

$$M_i = \frac{\beta_{g_i}}{g_i} m_{\frac{3}{2}},$$ (11.23)

where β_{g_i} is the corresponding beta function, defined by $\beta_{g_i} \equiv \mathrm{d}g_i/\mathrm{d}\ln\mu$. The gaugino masses are not universal, but given by the ratios of the respective β-functions.

The anomaly-mediated contribution to the scalar mass parameter is given by,

$$m_{\tilde{f}}^2 = -\frac{1}{4}\left\{\frac{\mathrm{d}\gamma}{\mathrm{d}g}\beta_g + \frac{\mathrm{d}\gamma}{\mathrm{d}f}\beta_f\right\} m_{\frac{3}{2}}^2,$$ (11.24)

where β_f is the β-function for the corresponding superpotential Yukawa coupling, and $\gamma = \partial \ln Z/\partial \ln \mu$, with Z the wave function renormalization constant. Finally, the anomaly-mediated contribution to the trilinear SUSY breaking scalar coupling is given by,

$$A_f = \frac{\beta_f}{f} m_{\frac{3}{2}}.$$ (11.25)

The following features of the AMSB contributions to the SSB parameters are worth noting.

1. AMSB contributions to gaugino and sfermion masses as well as A-parameters are all of the same scale, $m_{3/2}/16\pi^2$. Requiring this to be the weak scale puts the gravitino mass in a cosmologically safe range.[15]

---

[13] See A. Anisimov, M. Dine, M. Graesser and S. Thomas, *Phys. Rev.* **D65**, 105011 (2002).

[14] Indeed the AMSB expressions for scalar masses and A-parameters were first obtained via this route. See I. Jack, D.R.T. Jones and A. Pickering, *Phys. Lett.* **B426**, 73 (1998); L. Avdeev, D. Kazakov and I. Kondrashuk, *Nucl. Phys.* **B510**, 289 (1998).

[15] See e.g. S. Weinberg, *The Quantum Theory of Fields, Vol. III*, p 198, Cambridge University Press (2000).

2. Since Yukawa interactions are negligible for the first two generations, the anomaly-mediated contributions to the masses of the corresponding matter scalars with the same gauge quantum numbers are essentially equal. This solves the SUSY flavor problem if the AMSB contribution is dominant. Indeed the ultra-violet insensitivity of the AMSB scenario guarantees that no flavor violation results from high scale physics as long as AMSB contributions dominate.

3. The anomaly contribution turns out to be negative for sleptons, necessitating additional sources for the squared masses of scalars. Since the masses are insensitive to high scale physics, we cannot ameliorate this within this framework by adding new fields at the high scale. There are several proposals in the literature, but phenomenologically it suffices to add a universal contribution m_0^2 (which, of course, preserves the desired degeneracy between the first two generations of scalars) to Eq. (11.24), and regard m_0 as an additional parameter. It is assumed that the ad hoc introduction of m_0^2 in Eq. (11.24) does not affect the other parameters. This is referred to as the minimal AMSB model which we now examine.

11.2.1 The minimal AMSB (mAMSB) model

As we have just mentioned, the mAMSB model is *defined* by assuming that gaugino masses and A-parameters are given by (11.23) and (11.25), respectively while the expression for SSB scalar masses is amended by the addition of a (sufficiently large) universal mass parameter m_0^2 to make slepton masses positive. It is assumed that the AMSB mass relations hold at $Q = M_{\text{GUT}}$, and weak scale parameters are obtained from these via RGE evolution.

At one-loop level, with the field content of the MSSM at low energy, gaugino masses are given by,

$$M_1 = \frac{33}{5} \frac{g_1^2}{16\pi^2} m_{3/2}, \tag{11.26a}$$

$$M_2 = \frac{g_2^2}{16\pi^2} m_{3/2}, \quad \text{and} \tag{11.26b}$$

$$M_3 = -3 \frac{g_3^2}{16\pi^2} m_{3/2}. \tag{11.26c}$$

Notice the differing sign on the gluino mass term. This has implications for the sign of the SUSY contribution to the anomalous magnetic moment of the muon. Third generation scalar masses are given by

$$m_{U_3}^2 = \left(-\frac{88}{25}g_1^4 + 8g_3^4 + 2f_t\hat{\beta}_{f_t}\right)\frac{m_{3/2}^2}{(16\pi^2)^2} + m_0^2, \qquad (11.27a)$$

$$m_{D_3}^2 = \left(-\frac{22}{25}g_1^4 + 8g_3^4 + 2f_b\hat{\beta}_{f_b}\right)\frac{m_{3/2}^2}{(16\pi^2)^2} + m_0^2, \qquad (11.27b)$$

$$m_{Q_3}^2 = \left(-\frac{11}{50}g_1^4 - \frac{3}{2}g_2^4 + 8g_3^4 + f_t\hat{\beta}_{f_t} + f_b\hat{\beta}_{f_b}\right)\frac{m_{3/2}^2}{(16\pi^2)^2} + m_0^2, (11.27c)$$

$$m_{L_3}^2 = \left(-\frac{99}{50}g_1^4 - \frac{3}{2}g_2^4 + f_\tau\hat{\beta}_{f_\tau}\right)\frac{m_{3/2}^2}{(16\pi^2)^2} + m_0^2, \qquad (11.27d)$$

$$m_{E_3}^2 = \left(-\frac{198}{25}g_1^4 + 2f_\tau\hat{\beta}_{f_\tau}\right)\frac{m_{3/2}^2}{(16\pi^2)^2} + m_0^2, \qquad (11.27e)$$

$$m_{H_u}^2 = \left(-\frac{99}{50}g_1^4 - \frac{3}{2}g_2^4 + 3f_t\hat{\beta}_{f_t}\right)\frac{m_{3/2}^2}{(16\pi^2)^2} + m_0^2, \qquad (11.27f)$$

$$m_{H_d}^2 = \left(-\frac{99}{50}g_1^4 - \frac{3}{2}g_2^4 + 3f_b\hat{\beta}_{f_b} + f_\tau\hat{\beta}_{f_\tau}\right)\frac{m_{3/2}^2}{(16\pi^2)^2} + m_0^2. \qquad (11.27g)$$

The A-parameters are given by,

$$A_t = \frac{\hat{\beta}_{f_t}}{f_t}\frac{m_{3/2}}{16\pi^2}, \qquad (11.28a)$$

$$A_b = \frac{\hat{\beta}_{f_b}}{f_b}\frac{m_{3/2}}{16\pi^2}, \quad \text{and} \qquad (11.28b)$$

$$A_\tau = \frac{\hat{\beta}_{f_\tau}}{f_\tau}\frac{m_{3/2}}{16\pi^2}. \qquad (11.28c)$$

The quantities $\hat{\beta}_{f_i}$ that enter the expressions for scalar masses and A-parameters are given by,

$$\hat{\beta}_{f_t} = 16\pi^2\beta_t = f_t\left(-\frac{13}{15}g_1^2 - 3g_2^2 - \frac{16}{3}g_3^2 + 6f_t^2 + f_b^2\right), \quad (11.29a)$$

$$\hat{\beta}_{f_b} = 16\pi^2\beta_b$$

$$= f_b\left(-\frac{7}{15}g_1^2 - 3g_2^2 - \frac{16}{3}g_3^2 + f_t^2 + 6f_b^2 + f_\tau^2\right), \quad (11.29b)$$

$$\hat{\beta}_{f_\tau} = 16\pi^2\beta_\tau = f_\tau\left(-\frac{9}{5}g_1^2 - 3g_2^2 + 3f_b^2 + 4f_\tau^2\right). \qquad (11.29c)$$

The first two generations of squark and slepton masses are given by the corresponding formulae above with the Yukawa couplings set to zero. Eq. (11.26a)–(11.28c) serve as RGE boundary conditions at $Q = M_{GUT}$. We evolve the MSSM parameters to the weak scale and, as usual obtain B and μ^2 in accord with the constraint from radiative electroweak symmetry breaking. The model is, therefore, characterized by the parameter set,

$$m_0, \ m_{3/2}, \ \tan\beta, \ \text{and sign}(\mu). \tag{11.30}$$

The most notable feature of this framework is the hierarchy of gaugino masses. The gluino is (as in models with unified gaugino mass parameters) much heavier than the electroweak gauginos, but the novel feature is that $M_1/M_2 \sim 3$, so that the wino is lighter than the bino. Ignoring gaugino–higgsino mixing, the charged and neutral components of the $SU(2)$ gauginos would be degenerate: it is important to include radiative corrections to decide which of these is the LSP. Happily, these make the neutralino lighter than the chargino (else the model would be in trouble with cosmology). The near degeneracy of the chargino and the wino LSP have implications for particle phenomenology as well as cosmology. In particular, for the evaluation of the relic neutralino density, charginos and neutralinos coexist at the neutralino decoupling temperature, and co-annihilation effects are very important.

In Table 11.2, we show sparticle masses in the minimal AMSB model for two values of m_0, with other parameters being the same. Note that the parameter $m_{3/2}$ should be selected typically above 30–35 TeV to evade constraints from LEP experiments. From the spectra in the table, we see that for the smaller value of m_0, sleptons can be very light, though for very large values of m_0 they will be degenerate with squarks. We observe several characteristics of the AMSB spectrum. Most notable is that the $\widetilde{W}_1$ and $\widetilde{Z}_1$ are nearly degenerate in mass, so that in addition to the usual leptonic decay modes $\widetilde{W}_1 \to \widetilde{Z}_1 \ell\nu$, the only other kinematically allowed (and in these cases dominant) decay of the chargino is $\widetilde{W}_1^{\pm} \to \widetilde{Z}_1 \pi^{\pm}$. The chargino has a very small width, corresponding to a lifetime $\sim 1.5 \times 10^{-9}$ s, so that it would be expected to travel a significant fraction of a meter before decaying. We also see that the $\tilde{\ell}_L$ and $\tilde{\ell}_R$ are nearly mass degenerate. This degeneracy, which seems fortuitous, is much tighter than expected in the mSUGRA framework.

In the minimal AMSB framework, $m_{\widetilde{W}_1} - m_{\widetilde{Z}_1}$ is typically bigger than 160 MeV, so that $\widetilde{W}_1 \to \widetilde{Z}_1 \pi$ is always allowed and the chargino decays within the detector. The chargino would then manifest itself only as missing energy, unless the decay length is a few tens of centimeters, so that the chargino track can be established in the detector. The track would then seem to disappear since the presence of the soft pion would be very difficult to detect. Some parameter regions with $m_{\widetilde{W}_1} - m_{\widetilde{Z}_1} < m_{\pi^{\pm}}$ may be possible; in this case, the chargino would mainly decay via $\widetilde{W}_1 \to \widetilde{Z}_1 e\nu$ and its decay length (depending on the mass difference) may then be larger than several meters. It would then show up via a search for long-lived charged exotics.

Table 11.2 *Model parameters and weak scale
sparticle masses in GeV for two minimal
anomaly-mediated SUSY breaking case studies.*

parameter	AMSB(200)	AMSB(500)
m_0	200	500
$m_{3/2}$	35,000	35,000
$\tan\beta$	5	5
μ	> 0	> 0
$m_{\tilde{g}}$	804	818
$m_{\tilde{u}_L}$	775	894
$m_{\tilde{t}_1}$	542	611
$m_{\tilde{b}_1}$	683	774
$m_{\tilde{\ell}_L}$	149	481
$m_{\tilde{\ell}_R}$	136	477
$m_{\tilde{\tau}_1}$	118	471
$m_{\tilde{\tau}_2}$	160	484
$m_{\widetilde{W}_1}$	109	110
$m_{\widetilde{Z}_2}$	313	316
$m_{\widetilde{W}_1} - m_{\widetilde{Z}_1}$	0.171	0.172
m_h	114	113
m_A	658	813
μ	634	643
θ_τ	0.96	0.98
θ_b	0.08	0.05

Exercise *Verify by explicit computation that the one-loop expressions for the gaugino masses and A-parameters are scale invariant. For the hypercharge gaugino mass, for example, this means that*

$$M_1(Q) = \frac{33}{5} \frac{g_1(Q)^2}{16\pi^2} m_{3/2}$$

is true at **all scales**, *not just as a boundary condition. Thus you need to verify that,*

$$Q \frac{dM_1}{dQ} = \frac{33}{5} \frac{m_{3/2}}{16\pi^2} Q \frac{dg_1^2}{dQ},$$

etc. are consistent with the RGEs of the MSSM listed in Section 9.2.2 together with the RGEs for gauge and Yukawa couplings.

Verify also that the expressions (11.27a)–(11.27g) for scalar masses are similarly scale invariant only if $m_0^2 = 0$.

11.2.2 D-term improved AMSB model

While the addition of a common term m_0^2 to all scalar squared masses solves the tachyonic slepton mass problem, it destroys the scale invariance of the soft parameters with respect to renormalization group evolution, which renders the predictions of AMSB models insensitive to high scale physics. Indeed a variety of ways have been suggested to solve the tachyon mass problem, many of which do not maintain the scale invariance of the AMSB soft SUSY breaking parameters. Instead of studying all these various alternatives, we will focus on a modification of the AMSB relation that preserves this scale invariance.

The key observation is that additional contributions to soft SUSY breaking scalar masses that arise from Fayet–Iliopoulos D-terms, introduced in Section 6.5.1, automatically preserve this scale invariance property, as long as the charges of the corresponding $U(1)$ symmetries have no mixed anomalies with the MSSM gauge group.[16] In other words, as long as the extra contributions to scalar mass squared parameters take the form,

$$\delta m_i^2 = m_0^2 \sum_a k_a Y_{ai},$$

where Y_a are the generators of (mixed anomaly-free) $U(1)$ symmetries, and k_a are constants (one k_a for each such $U(1)$ factor), the scale invariance of AMSB scalar masses is maintained.[17] Moreover, since this invariance holds for arbitrarily small values of the corresponding "gauge coupling", it is not necessary for these $U(1)$s to survive as gauge symmetries of the low energy theory for this mechanism to work: i.e. global $U(1)$ symmetries of the low energy superpotential are sufficient. Notice that these D-term contributions to scalar mass parameters are the same for all sparticles with the same gauge quantum numbers so that flavor-changing neutral current constraints are satisfied.

The MSSM symmetries already include the hypercharge $U(1)$. Unfortunately, the corresponding D-term contributions cannot solve the slepton mass problem since the superfields $\hat{L}$ and $\hat{E}^c$ have opposite signs of hypercharge: the hypercharge D-term can make only one of $m^2(\tilde{\ell}_L)$ or $m^2(\tilde{\ell}_R)$ positive, but not both. We need at least one other D-term. Assuming that lepton flavor is not separately conserved by the superpotential (i.e neutrinos mix), there are only two independent anomaly-free $U(1)$ symmetries in the MSSM. These are the usual hypercharge symmetry, and $U(1)_{B-L}$ (or combinations thereof). Their D-term contributions to sparticle masses (at the weak scale) can thus be parametrized in terms of two parameters, D_Y and

---

[16] See I. Jack and D. R. T. Jones, *Phys. Lett.* **B482**, 167 (2000).

[17] Another possibility that also preserves the scale invariance has been proposed by I. Jack and D. R. T. Jones, *Phys. Lett.* **B491**, 151 (2000).

D_{B-L}, as[18]

$$\delta m_U^2 = -\frac{4}{3}D_Y - \frac{1}{3}D_{B-L}, \tag{11.31a}$$

$$\delta m_D^2 = \frac{2}{3}D_Y - \frac{1}{3}D_{B-L}, \tag{11.31b}$$

$$\delta m_Q^2 = \frac{1}{3}D_Y + \frac{1}{3}D_{B-L}, \tag{11.31c}$$

$$\delta m_E^2 = 2D_Y + D_{B-L}, \tag{11.31d}$$

$$\delta m_L^2 = -D_Y - D_{B-L}, \tag{11.31e}$$

$$\delta m_{H_u}^2 = D_Y, \tag{11.31f}$$

$$\delta m_{H_d}^2 = -D_Y. \tag{11.31g}$$

The values of D_Y and D_{B-L} must be of order of the weak scale squared. The value of D_{B-L} may possibly be the only imprint of the additional $U(1)$ symmetry. A necessary (but not sufficient) condition for a viable spectrum is,

$$0 < D_Y < -D_{B-L} < 2D_Y.$$

To summarize, the negative slepton mass problem can be solved maintaining the attractive ultra-violet insensitivity characteristic of the AMSB framework if there is an additional source of SUSY breaking that results in non-vanishing D-terms of a $U(1)$ symmetry with charges that are free of any mixed anomalies with the MSSM gauge group factors.[19]

11.3 Gauge-mediated SUSY breaking

As the name indicates, in gauge-mediated SUSY breaking (GMSB), SM gauge interactions communicate the effects of SUSY breaking to the superpartners of SM particles.[20] In addition to the fields of the SUSY breaking and the observable sectors that we have already discussed, there is a third set of fields that has both SM gauge interactions, as well as couplings to the hidden sector: these couplings may originate in the superpotential, or in new gauge interactions with the hidden sector (under which SM particles are neutral). Through these couplings, SUSY breaking effects are first felt by the fields in the new sector, and then

---

[18] The reader can easily check that this parametrization is equivalent to that in the original paper of I. Jack and D. R. T. Jones, with their parameters ζ_1 and ζ_2 given by $\zeta_1 = 2D_Y + \frac{8}{11}D_{B-L}$ and $\zeta_2 = \frac{1}{11}D_{B-L}$.

[19] For an explicit model that realizes this scenario, see N. Arkani-Hamed, D. E. Kaplan, H. Murayama and Y. Nomura, *JHEP* **0102**, 041 (2001).

[20] Interest in this picture was rekindled by M. Dine and A. Nelson, *Phys. Rev.* **D48**, 1277 (1993) and M. Dine, A. Nelson, Y. Nir and Y. Shirman, *Phys. Rev.* **D53**, 2658 (1996); see also references therein.

communicated to the observable sector by SM gauge interactions. The third sector that links the SUSY breaking and observable fields is referred to as the "messenger sector".

At tree level, SUSY is unbroken in the MSSM sector. MSSM sparticles feel SUSY breaking effects only via their couplings to messenger particles in loops, and so evade the fatal tree-level mass sum rule. These loop effects, which involve the usual SM gauge couplings, again lead to SSB masses of the geometric hierarchy form,

$$m_i \propto \frac{g_i^2}{16\pi^2} \frac{\langle F_S \rangle}{M}$$

where $\langle F_S \rangle$ is the *induced* SUSY breaking VEV of some (elementary or composite) gauge singlet superfield in the messenger sector, M is the messenger sector mass scale, g_i is the SM gauge coupling constant for the corresponding sparticle, and $16\pi^2$ is a loop factor.[21] We thus conclude that colored superpartners are heavier than their uncolored counterparts and, likewise, uncolored particles that have just hypercharge gauge interactions are lighter than their cousins which also couple to $SU(2)_L$. Such a spectrum is the hallmark of the GMSB scenario.

The induced SUSY breaking scale in the messenger sector should be distinguished from the corresponding scale $\langle F \rangle$ in the SUSY breaking sector. If the sectors are perturbatively coupled, we would expect $\langle F_S \rangle < \langle F \rangle$, while if they are strongly coupled, $\langle F_S \rangle \sim \langle F \rangle$. The gravitino mass, however, is determined by the fundamental SUSY breaking scale $\langle F \rangle$, and by (10.67a) is,

$$m_{3/2} = \frac{\langle F \rangle}{\sqrt{3} M_P}.$$

We note that the SSB masses of MSSM superpartners are suppressed by just the messenger mass scale M, and not M_P as in gravity-mediated scenarios. If $M \ll M_P$, the underlying scale of SUSY breaking can be much lower in GMSB models as compared to gravity-mediated scenarios.[22] In this case of low energy SUSY breaking the gravitino mass, which is suppressed relative to other sparticle masses by a factor $\sim M/M_P$, may be very small. Indeed, the gravitino may be the LSP. Since the lightest MSSM particle can now decay to the gravitino, the phenomenological implications of such a scenario may differ dramatically from corresponding expectations in mSUGRA.

---

[21] If the sparticle has coupling to more than one factor of $SU(3)_C \times SU(2)_L \times U(1)_Y$, there will be one such contribution for each coupling.

[22] If SUSY is local, there will be a gravity mediated contribution also, but this is negligible compared to the corresponding gauge-mediated contribution.

11.3.1 The minimal GMSB model

The messenger sector is assumed to consist of n_5 vector-like multiplets of messenger lepton and messenger quark superfields that carry the $SU(3)_C \times SU(2)_L \times U(1)_Y$ quantum numbers,

$$\hat{\ell} \sim (\mathbf{1}, \mathbf{2}, 1) \qquad \hat{\ell}' \sim (\mathbf{1}, \mathbf{2}^*, -1)$$
$$\hat{q} \sim (\mathbf{3}, \mathbf{1}, -\frac{2}{3}) \qquad \hat{q}' \sim (\mathbf{3}^*, \mathbf{1}, \frac{2}{3}), \tag{11.32a}$$

coupled via the superpotential,

$$\hat{f}_M = \lambda_\ell \hat{S}\hat{\ell}'\hat{\ell} + \lambda_q \hat{S}\hat{q}'\hat{q}. \tag{11.32b}$$

Here $\hat{S}$ is a gauge singlet field that also couples to the SUSY breaking sector. We assume that this coupling induces a VEV for both its scalar and its auxiliary component. Notice that the messenger sector forms complete vector multiplets of $SU(5)$. This ensures that the apparent unification of gauge couplings is not altered by their inclusion.

It is straightforward to see that the messenger quarks (and likewise, messenger leptons) combine to form a Dirac quark ($SU(2)$ doublet lepton) with a mass $m_{q_M} = \lambda_q \langle S \rangle$ ($m_{\ell_M} = \lambda_\ell \langle S \rangle$) where $\langle S \rangle$ is the VEV of the scalar component of the singlet field $\hat{S}$. In addition to this supersymmetric mass contribution, the scalar partners of the messenger quarks (leptons) acquire a SUSY breaking mass from the VEV $\langle F_S \rangle$ of the auxiliary component of $\hat{S}$ that mix scalar components of $\hat{q}$ and $\hat{q}'$ ($\hat{\ell}$ and $\hat{\ell}'$). Diagonalizing the messenger squark and slepton mass matrices, we find that these acquire masses,

$$m_{\tilde{\ell}_M}^2 = |\lambda_\ell \langle S \rangle|^2 \pm |\lambda_\ell \langle F_S \rangle|, \tag{11.33a}$$
$$m_{\tilde{q}_M}^2 = |\lambda_q \langle S \rangle|^2 \pm |\lambda_q \langle F_S \rangle|. \tag{11.33b}$$

Notice that $\langle F_S \rangle / \lambda_i \langle S \rangle^2$ cannot be arbitrarily large – otherwise the messengers will be too light or even tachyonic. We will denote the messenger mass scale by $M \equiv \lambda \langle S \rangle$, where $\lambda \simeq \lambda_\ell \simeq \lambda_q$. If $\langle F_S \rangle \to 0$, we recover a supersymmetric spectrum in the messenger sector.

Exercise *Using the master formula, compute the mass spectrum of the messenger quarks. Note that the supersymmetry breaking contribution to messenger squark masses comes from the F-term of the superpotential,*

$$\lambda_q \hat{S}\hat{q}\hat{q}'\big|_F \ni (\lambda_q \tilde{q}\tilde{q}' + \text{h.c.})\langle F_s \rangle.$$

Combine this with the supersymmetric contribution to messenger squark masses to obtain their masses.

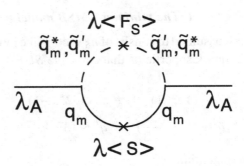

Figure 11.2 Diagram leading to gluino and hypercharge gaugino masses in GMSB models. A similar diagram with messenger leptons and messenger sleptons in the loop will also contribute to the $SU(2)$ gaugino mass. Messenger leptons and sleptons also contribute to the hypercharge gaugino mass. The dashed line denotes messenger sfermions while the solid line denotes the messenger fermion. A contribution arises only from the messenger fermion mass term indicated by the cross on the fermion line. The cross on the sfermion line indicates the SSB mixing term between the sfermions.

It is now possible to compute the SSB mass parameters induced in the visible sector via gauge interactions with messenger sector fields. Gauginos obtain masses from one-loop diagrams including messenger fields as indicated in Fig. 11.2. In the approximation $\langle F_S \rangle \ll \lambda \langle S \rangle^2$ (i.e. the SUSY breaking scale is smaller than the messenger mass scale), the gaugino for gauge group i gets a mass

$$M_i = \frac{\alpha_i}{4\pi} n_5 \Lambda \tag{11.34}$$

where

$$\Lambda = \frac{\langle F_S \rangle}{\langle S \rangle}. \tag{11.35}$$

The factor n_5 arises because each messenger generation makes the same contribution to the gaugino mass.

MSSM scalars do not couple directly to the messenger sector, so that their squared masses are induced only via two-loop diagrams such as the ones depicted in Fig. 11.3. Now, the squared mass scales with the number of messenger multiplets, and in the same approximation as in (11.34) we obtain,

$$m_i^2 = 2n_5 \Lambda^2 [C_1^i (\frac{\alpha_1}{4\pi})^2 + C_2^i (\frac{\alpha_2}{4\pi})^2 + C_3^i (\frac{\alpha_3}{4\pi})^2]. \tag{11.36a}$$

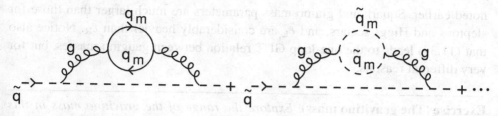

Figure 11.3 Examples of two-loop Feynman diagrams leading to scalar masses in GMSB models.

Here, C_i are quadratic Casimirs given by,

$$C_1^i = \frac{3}{5} Y_i^2,$$

$$C_2^i = \begin{cases} 3/4 \text{ for doublets} \\ 0 \text{ for singlets} \end{cases} \qquad (11.36b)$$

$$C_3^i = \begin{cases} 4/3 \text{ for triplets} \\ 0 \text{ for singlets.} \end{cases}$$

Note that the unknown messenger sector Yukawa couplings drop out from (11.34) and (11.36a). These formulae are rather general, in that if the messengers can be grouped into a **10** or **10*** of $SU(5)$, then $n_5 \to n_{10}$ where $n_{10} = 3$ for each set of **10** and **10*** messenger fields. The value of n_5 (and hence n_{10}) cannot be too large since the gauge couplings will then diverge in their running from the weak to the GUT scale, and perturbative unification will be spoiled. Typically, $n_5 \leq 4$ is a valid choice, though if the messenger scale is large higher values of n_5 are allowed. As the parameter n_5 increases, the gaugino masses increase at a greater rate than the scalars, since $M_i \propto n_5$, while $m_i \propto \sqrt{n_5}$.

We note that these formulae for gaugino and scalar masses are simply modified by multiplication by threshold functions if our approximation $x \equiv \langle F_S \rangle / \lambda \langle S \rangle^2 \ll 1$ in which we have written them ceases to be valid.[23] Our formulae for the gaugino (scalar) masses are good approximations for x as large as 0.9 (0.5).

We see from (11.36a) that scalars with the same gauge quantum numbers will receive identical masses. This gives a natural explanation for the scalar mass degeneracy needed to solve the SUSY flavor problem, and provides strong motivation for this class of models.[24] We see also the characteristic pattern of sparticle masses

---

[23] S. Martin, *Phys. Rev.* **D55**, 3177 (1997).

[24] We remark, however, that since messenger field $\tilde{\ell}$ and $\hat{H}_u$ carry the same gauge quantum numbers, in any Yukawa coupling involving h_u, we can replace the Higgs field by a messenger slepton. This would lead to flavor violation unless such couplings are forbidden by a global symmetry, or the messenger scale is sufficiently large. Thus it is really the squark loop contributions to FCNC effects that are naturally suppressed in these scenarios.

noted earlier. Squark and gluino mass parameters are much larger than those for sleptons and Higgs scalars, and $\tilde{\ell}_L$ are considerably heavier than $\tilde{\ell}_R$. Notice also that (11.34) leads to the one-loop GUT relation between gaugino masses, but for very different reasons.

Exercise (The gravitino mass) *Explore the range of the gravitino mass in this framework. To do so, write*

$$m_{3/2} = \frac{\langle F \rangle}{\lambda \langle F_S \rangle} \times \frac{\Lambda M}{\sqrt{3} M_P} \equiv C_{\mathrm{grav}} \frac{\Lambda M}{\sqrt{3} M_P}, \tag{11.37}$$

where λ is the messenger sector Yukawa coupling, taken to be common for messenger quarks and leptons. Since we want sparticles at the weak scale ~ 100 GeV, we must have Λ to be few tens of TeV. For given values of Λ and M, the gravitino is lightest when $\langle F_S \rangle$ is close to the fundamental SUSY breaking scale and when the messenger scale is not very different from Λ. Show that for reasonable values of parameters the gravitino mass may be in the eV range. How heavy can it be?

Like scalar masses, the **a**-terms only arise via two-loop diagrams. Remember, however, that it is the *squared* scalar masses that arise at two loops, so that scalar mass parameters have the same order of magnitude as gaugino masses. In comparison to this, **a**-terms which are suppressed by an extra loop factor, are small. As an approximation, they are frequently taken to be

$$\mathbf{a}_u = \mathbf{a}_d = \mathbf{a}_e = 0. \tag{11.38}$$

It should be remembered that the formulae (11.34), (11.36a), and (11.38) for the MSSM parameters in GMSB models hold at the scale $Q \sim M$ where the heavy messenger fields are integrated out. As for SUGRA models, these parameters must be evolved to the weak scale for the extraction of phenomenology using the MSSM.

The bilinear b term is also generated at two loops and so is tiny. In principle, this means that the requirement of radiative EWSB should fix $\tan \beta$ since the weak scale $B\mu$ is fixed by the condition $b_0 = 0$. This is not what is usually done in practice. The reason is that it is difficult to generate μ in these scenarios. The rationale then is any modification to the model that allows for μ affects the Higgs sector and so will presumably also affect the b term. In practice, therefore, μ and b are treated as free weak scale parameters: as usual μ^2 is fixed to reproduce M_Z^2, and b_0 is traded in for $\tan \beta$. There is one difference in the radiative symmetry breaking mechanism from gravity-mediated models that seems worth mentioning. In mSUGRA, $m_{H_u}^2$ turns negative because of the large logarithm that arises due to the disparity between the GUT and weak scales. In the GMSB scenario $m_{H_u}^2$ turns negative even if the messenger scale is close to the weak scale because the colored squarks are much heavier than Higgs scalars, i.e. large t-squark masses drive $m_{H_u}^2$ to negative values.

We have already noted that if the messenger scale $M \ll M_P$, the gravitino may be very light. But if gravitinos couple with gravitational strength, why do we care? The point is that since gravitinos get masses via the super-Higgs mechanism the couplings of their longitudinal components (essentially the goldstinos) are enhanced by a factor $E/m_{3/2}$ in exactly the same way that longitudinal W couplings are enhanced by a factor of E/M_W. In other words, "the effective dimensionless coupling" of longitudinal gravitinos to a particle–sparticle pair is $\sim E/M_P \times E/m_{3/2}$, where the first factor is the usual coupling of gravity to energy and the second factor the enhancement just discussed. It is easy to check that for $E \sim 100$ GeV and $m_{3/2} \sim 1$ eV, this coupling is $\sim 10^{-6}$. Dimensional analysis gives the lifetime of a 100 GeV particle decaying via this coupling as $\sim 10^{-12}$ seconds! *Thus interactions of very light longitudinal gravitinos may be relevant for particle physics, and even for collider phenomenology.* We will return to this in later chapters.

If gravitinos are light, sparticles can decay via $\tilde{p} \to p\tilde{G}$ with a decay rate that depends on the gravitino mass. It is more convenient to use C_{grav} introduced in (11.37) to parametrize this decay rate. Notice that, by construction, $C_{\text{grav}} \geq 1$. The parameter space of GMSB models can thus be specified by,

$$\Lambda, \ M, \ n_5, \ \tan\beta, \ \text{sign}(\mu), \ C_{\text{grav}}. \tag{11.39}$$

For a given number n_5 of messenger multiplets, the mass scale of MSSM superpartners is set by Λ. The second entry, M ($M > \Lambda$) is the mass scale associated with the messenger fields, and specifies the scale at which the mass formulae (11.34) and (11.36a) as well as $\mathbf{a} = 0$, hold. The SSB parameters relevant for phenomenology are then obtained by evolving these from M to the weak scale where radiative EWSB determines the magnitude but not the sign of μ. MSSM sparticle masses are, therefore, only logarithmically sensitive to M, and, of course, independent of C_{grav}. Increasing C_{grav} only increases the lifetime of sparticles which decay mainly to the gravitino, but does not affect MSSM sparticle masses.

An example of the renormalization group evolution that fixes the sparticle spectrum is shown in Fig. 11.4. While the gaugino masses are related as in mSUGRA, sfermion masses are very different. In particular, we have $m_{\tilde{q}} \gg m_{\tilde{e}_L} \sim m_{\tilde{e}_R}$.

For GMSB models, the parameter Λ should be ~ 10–150 TeV in order for sparticles to obtain masses of order of the weak scale. The messenger scale $M \geq \Lambda$. If the SUSY breaking scale is small so that the gravitino is the LSP, GMSB phenomenology may differ dramatically from phenomenology of models with a weak scale gravitino and a neutralino LSP. If the gravitino is the LSP and other

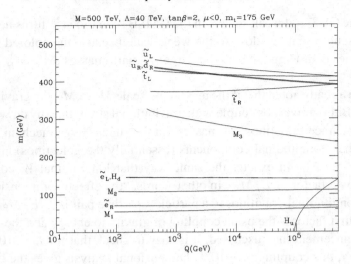

Figure 11.4 Renormalization group trajectories for the soft SUSY breaking masses versus renormalization scale Q from the messenger scale ($M = 500$ TeV) to the weak scale. In this example, we take $\Lambda = 40$ TeV, $n_5 = 1$, $\tan\beta = 2$, $\mu < 0$, and $m_t = 175$ GeV. We will see later that this scenario is excluded by lower bounds on both the selectron as well as h masses. The point of this figure is only to illustrate the RG evolution, for which the LEP exclusion is not relevant. Reprinted with permission from H. Baer, M. Brhlik, C.-H. Chen and X. Tata, *Phys. Rev* **D55**, 4463 (1997), copyright (1997) by the American Physical Society.

sparticles can decay to it in a lifetime short compared to the age of the Universe, then the cosmological considerations that require the lightest MSSM sparticle to be only weakly interacting no longer apply, and this next-to-lightest SUSY particle (NLSP) may be charged. Typically, the NLSP is the lightest neutralino or the lighter stau (which would be very close in mass to $\tilde{e}_R$ and $\tilde{\mu}_R$ for small to moderate values of $\tan\beta$).

In the case of a neutralino NLSP, collider phenomenology is most different when the gravitino is very light, so that the NLSP decays *inside the experimental apparatus*.[25] The main decay modes for a neutralino NLSP are $\widetilde{Z}_1 \rightarrow \gamma\tilde{G}$, $Z\tilde{G}$ or $h\tilde{G}$. For a stau NLSP, the decay mode would be $\tilde{\tau}_1 \rightarrow \tau\tilde{G}$. Heavier sparticles cascade decay to the NLSP which subsequently decays into the gravitino LSP, and SUSY event topologies are very sensitive to the nature of the NLSP.

Since sparticle masses are only weakly dependent on the messenger scale M, the $\Lambda - \tan\beta$ plane provides a convenient panorama for displaying the various phenomenological possibilities. These are illustrated in Fig. 11.5 where we show

---

[25] Of course, heavier sparticles can then also decay into gravitinos but, as we will see in Chapter 13, the branching fractions for these decays are negligible.

this plane for values of n_5 ranging from 1–4. The gray region is excluded because electroweak symmetry is not correctly broken, while the various shaded regions are excluded by constraints from LEP experiments. In the region labeled 1, the neutralino is the NLSP and decays into the gravitino. In region 2, $m_{\tilde{\tau}_1} < m_{\widetilde{Z}_1}$, with all other sleptons heavier than $\widetilde{Z}_1$, so that cascade decays terminate in $\tilde{\tau}_1$ (which decays via $\tilde{\tau}_1 \to \tau\tilde{G}$), except very close to the boundary between regions 1 and 2 where $\widetilde{Z}_1 \to \tilde{\tau}_1\tau$ is forbidden. In regions 3 and 4, in addition to $\widetilde{Z}_1 \to \tilde{\tau}_1\tau$, the decays $\widetilde{Z}_1 \to \tilde{\ell}_R\ell$ ($\ell = e, \mu$) are also allowed. In region 3, however, $m_{\tilde{\ell}_1} < m_{\tilde{\tau}_1} + m_\tau$, while just the opposite is the case in region 4. We will discuss the implications of this in Chapter 13.

The lifetimes for NLSP decay depend on C_{grav} and range from essentially instantaneous to very long. NLSPs produced in collider detectors may have a long lifetime, and decay with a displaced vertex, or possibly even decay outside the detector. In the latter case, a neutralino NLSP would escape undetected as in gravity-mediated models. A stau (or charged slepton) NLSP would behave as a stable charged particle in the apparatus, and leave an ionizing track which may be detectable. NLSP decays will be considered in more detail in Chapter 13.

11.3.2 Non-minimal GMSB models

While the minimal GMSB framework leads to strong correlations between various sparticle masses, it is possible to conceive of extensions where the correlations are relaxed. Examples of things that have been considered include:

- Additional interactions needed to generate μ and b parameters may split the SSB mass parameters of the Higgs and lepton doublets at $Q = M$, even though these have the same gauge quantum numbers.
- Allowing incomplete messenger representations can effectively result in different numbers of messengers n_{5_i} for each factor of the low energy gauge group.
- If the hypercharge D-term has a non-vanishing VEV in the messenger sector, there would be additional contributions to the scalar masses that may be parametrized by $\delta m_{\tilde{f}}^2 = Y_{\tilde{f}} K_Y$, where K_Y is the D-term VEV with the gauge coupling absorbed into it.

We mention these variations to make the reader aware that although the minimal GMSB framework is well motivated and constrained, the implications that we have drawn from it are based on a number of unstated assumptions about physics at the messenger scale and beyond.

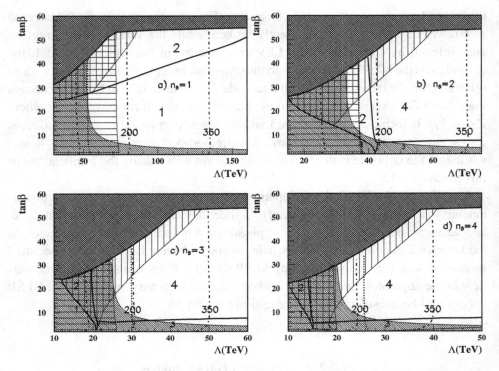

Figure 11.5 The four regions of the $\Lambda - \tan\beta$ parameter plane of the mGMSB model discussed in the text. The heavy solid lines denote the boundaries between these regions. The gray region is excluded because electroweak symmetry is not correctly broken. The shaded regions are excluded by various constraints from LEP experiments: $m_{\tilde{\tau}} > 76$ GeV (vertical shading), $m_{\tilde{Z}_1} > 95$ GeV (horizontal shading), and $m_h > 110$ GeV (diagonal shading). The dot-dashed contours are where the chargino mass is 100, 200 or 350 GeV, while the dotted line is the contour of $m_{\tilde{e}_R} = 100$ GeV. We thank Dr. Yili Wang for supplying this figure which appears in her doctoral dissertation.

11.4 Gaugino-mediated SUSY breaking

Gaugino-mediated SUSY breaking is a model based on extra dimensions that provides a novel solution to the SUSY flavor problem.[26] Within this framework, chiral supermultiplets of the observable sector reside on one brane whereas the SUSY breaking sector is confined to a different, spatially separated brane. Gravity and gauge superfields, which propagate in the bulk, directly couple to fields on both the branes. As a result of their direct coupling to the SUSY breaking sector, gauginos acquire a mass. Direct couplings between the observable and SUSY breaking

---

[26] D. E. Kaplan, G. D. Kribs and M. Schmaltz, *Phys. Rev.* **D62**, 035010 (2000); Z. Chacko *et al.*, *JHEP* **01**, 003 (2000); M. Schmaltz and W. Skiba, *Phys. Rev.* **D62**, 095004 (2000) and **D62**, 095005 (2000).

sectors are exponentially suppressed, and MSSM scalars dominantly acquire SUSY breaking masses via their interactions with gauginos (or gravity) which directly feel the effects of SUSY breaking. As a result, scalar SSB mass parameters are suppressed relative to gaugino masses, and may be neglected in the first approximation. The same is true for the A- and B-parameters.

In a specific realization, to preserve the success of the unification of gauge couplings, it is assumed that there is grand unification (either $SU(5)$ or $SO(10)$) and, further, that the compactification scale M_c, below which there are no Kaluza–Klein excitations, is larger than M_{GUT}. Furthermore, since the construction ensures flavor-blind interactions for just light bulk fields, we require that the scale $M_c \lesssim M_{Planck}/10$ in order to sufficiently suppress other flavor-violating scalar couplings from heavy bulk fields that would be generically present. Based on the discussion in the last paragraph, the boundary conditions for the soft SUSY breaking parameters of the MSSM are taken to be $m_0 = A_0 = B_0 = 0$ at the scale M_c. The condition $B_0 = 0$ fixes $\tan\beta$. In both $SU(5)$ and $SO(10)$ models, this value of $\tan\beta$ is found to be too small to be compatible with the unification of bottom and tau Yukawa couplings in the MSSM, which requires $\tan\beta \geq 30$. For this reason, and because the value of $B_0\mu_0$ may also depend on how the μ problem is solved, we will ignore the $B_0 = 0$ constraint and, as usual, choose $\tan\beta$ instead of B_0 as a free parameter.[27] The MSSM parameters can then be obtained from the parameter set,

$$m_{1/2}, \quad M_c, \quad \tan\beta, \quad \text{and} \operatorname{sign}(\mu) \tag{11.40}$$

where it is the grand unification assumption that leads to a universal gaugino mass above $Q = M_{GUT}$, and $|\mu|$ is fixed assuming radiative EWSB. The gravitino can be made heavier than gauginos and, as in the mSUGRA framework, is irrelevant for collider phenomenology. The LSP may be the stau or the lightest neutralino, though cosmological considerations exclude the former (unless R-parity is not conserved).

For illustration we choose the GUT group to be $SU(5)$. This model is then a special case of our earlier discussion of $SU(5)$, except that the SSB parameters now "unify" at the scale M_c (rather than M_P) where they take on values specific to the model. The unification of the τ and b Yukawa couplings constrain $\tan\beta \sim 30$–50. In Fig. 11.6, we show the evolution of the various SSB parameters of the MSSM, starting with the inoMSB boundary conditions. Here, the unified gaugino mass is taken to be 300 GeV at $Q = M_{GUT}$. The compactification scale is taken to be $M_c = 10^{18}$ GeV. We see that although these start from zero, RG evolution results in GUT scale scalar masses and A-parameters that are not negligible compared to $m_{1/2}$: although there is no large logarithm, large group theory coefficients are the cause

---

[27] It is also possible that Higgs fields reside in the bulk, in which case they would directly feel SUSY breaking effects, resulting in a non-vanishing value for B_0 as well as other SSB parameters in the Higgs sector. Such scenarios are, of course, less predictive than the minimal one that we consider here.

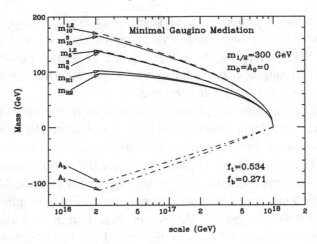

Figure 11.6 Renormalization group evolution of soft SUSY breaking $SU(5)$ masses versus scale in the minimal gaugino mediation model. We take $\tan \beta = 35$ and $\mu < 0$ to achieve $b - \tau$ Yukawa coupling unification. Reprinted from H. Baer, M. Diáz, P. Quintana and X. Tata, *JHEP* **04**, 016 (2000).

of this sizable renormalization group evolution. While the inter-generation splitting is small, the splittings between the **5** and the **10** dimensional matter multiplets, as well as between these and the Higgs multiplets, is substantial.

In Table 11.3 we show a sample spectrum for this model. We choose $m_{1/2} = 300$ GeV, $\tan \beta = 35$, and other parameters as in Fig. 11.6. The spectrum is not unlike that in the mSUGRA framework with small m_0, so that sleptons are relatively light and squarks are lighter than the gluino.

11.5 An afterword

The reader will have noticed that we have not constructed a complete supersymmetric model in the sense of the SM. Instead, we have assumed that SUSY is broken in some sector, and discussed several mechanisms for how this is communicated to MSSM superpartners. As mentioned at the start of this chapter, MSSM phenomenology depends more upon this messenger mechanism and not so much upon the dynamics of SUSY breaking.

This is not to say that the question of SUSY breaking is not important. Indeed, a complete model must address the μ problem, and at the same time generate $b \equiv B\mu$ and other SSB parameters so that (8.19b), with radiative corrections included, yields the correct value of M_Z, and a sparticle spectrum consistent with experimental constraints. The value of $\tan \beta$ as given by (8.19a) would then be a prediction. We stress that these EWSB conditions (8.19a) and (8.19b) only depend on our assumption of the MSSM field content of the low energy theory, and therefore

Table 11.3 *Input and output parameters for
the Minimal Gaugino Mediation model
case study described in the text. Mass
parameters are in GeV units.*

parameter	scale	value
m_0	M_c	0
A_0	M_c	0
$m_{1/2}$	M_{GUT}	300
g_5	M_{GUT}	0.717
f_t	M_{GUT}	0.534
$f_b = f_\tau$	M_{GUT}	0.271
λ	M_{GUT}	1
λ'	M_{GUT}	0.1
$\tan\beta$	M_{weak}	35
μ	M_{weak}	< 0
$m_{\tilde{g}}$	M_{weak}	737.2
$m_{\tilde{u}_L}$	M_{weak}	668.5
$m_{\tilde{d}_R}$	M_{weak}	633.1
$m_{\tilde{t}_1}$	M_{weak}	482.8
$m_{\tilde{b}_1}$	M_{weak}	541.5
$m_{\tilde{\ell}_L}$	M_{weak}	258.6
$m_{\tilde{\ell}_R}$	M_{weak}	210.0
$m_{\tilde{\tau}_1}$	M_{weak}	143.3
$m_{\widetilde{W}_1}$	M_{weak}	240.2
$m_{\widetilde{Z}_2}$	M_{weak}	240.0
$m_{\widetilde{Z}_1}$	M_{weak}	124.8
m_h	M_{weak}	115.6
m_A	M_{weak}	311.2
μ	M_{weak}	-411.5

should be valid as long as the underlying fundamental theory reduces to the MSSM at low energy. This is not to say that every high energy theory will necessarily lead to an acceptable model. For instance, while there is an elegant mechanism for generating μ in gravity-mediated SUSY breaking scenarios (where the SUSY breaking scale is large), it is not straightforward (see Section 11.3.2) to generate acceptable values for both μ and b in GMSB scenarios with a low SUSY breaking scale. We circumvent the complications associated with the underlying mechanism of SUSY breaking and the associated μ problem because whatever the underlying physics is, it must be consistent with the EWSB conditions (8.19a) and (8.19b) as long as the low energy theory is the MSSM. Fortunately, TeV scale phenomenology depends more on how SUSY breaking is felt by weak scale superpartners and not so much on the underlying dynamics of SUSY breaking.

12

Sparticle production at colliders

The interaction Lagrangian for the physical particles of the MSSM presented in Chapter 8 can be used to compute the S-matrix elements for any physical process, and production cross sections and decay rates can then be obtained. In this chapter, we focus on the evaluation of tree-level superparticle production cross sections in high energy collisions, and present sparticle production rates at currently operating colliders, as well as at colliding beam facilities under construction, or those being considered for construction in the future. We first examine production reactions at hadron colliders such as the Fermilab Tevatron $p\bar{p}$ collider, which is currently operating at a center of mass (CM) energy $\sqrt{s} \simeq 2$ TeV. Negative results of SUSY searches at the Tevatron have been interpreted by the CDF and DØ collaborations as a lower limit $m_{\tilde{g}} \geq 195$ GeV ($m_{\tilde{g}} \geq 260$–300 GeV if squarks are degenerate and have a mass equal to $m_{\tilde{g}}$) on the gluino mass. We also show example cross sections for the CERN Large Hadron Collider (LHC), a pp collider, which is scheduled to operate at a CM energy around 14 TeV. The CERN LHC will have sufficient energy to either establish or rule out many models of weak scale supersymmetry. The evaluation of sparticle production rates by hadronic collisions is complicated by the fact that hadrons are not elementary, but composed of quarks and gluons.

In the second section of this chapter, we discuss sparticle production reactions at e^+e^- colliders. Since electrons, unlike protons, are elementary particles, the production processes are much simpler. Searches for supersymmetric matter at the CERN LEP2 e^+e^- collider, which concluded operation in November 2000, have provided significant lower limits on several sparticle masses. The clean environment of e^+e^- scattering events, together with the well-defined energy of the initial state, make these machines ideal for precision measurements of sparticle properties. Designs for linear e^+e^- colliders operating at $\sqrt{s} \simeq 0.5$–1.5 TeV and beyond usually include the possibility of longitudinal electron beam polarization and possibly even positron beam polarization. Beam polarization can be a valuable tool, both for eliminating SM backgrounds, as well as for separating signals from different

SUSY reactions. We will, therefore, consider sparticle production from polarized initial beams: results for unpolarized (or partially polarized) beams can be obtained by suitable averaging over polarization.

Leading order formulae for cross sections for sparticle pair production are collected in Appendix A.

12.1 Sparticle production at hadron colliders

Since superpartners are assumed to be heavy, sparticle pair production is a high Q^2 process, and at hadron colliders occurs predominantly via collisions between the constituents of hadrons: the quarks, antiquarks, and gluons. Production cross sections are calculated within the framework of the parton model.[1] Suppose parton a is a constituent of hadron A, and parton b is a constituent of hadron B. Parton a carries fractional longitudinal momentum x_a of hadron A, and parton b carries fractional longitudinal momentum x_b of hadron B. We let $f_{a/A}(x_a, Q^2)$ denote the probability density of finding parton a with fractional momentum x_a in hadron A, where Q^2 is the squared four-momentum transfer of the underlying elementary process. Its magnitude is the typical energy scale of this reaction. $f_{a/A}(x_a, Q^2)$ is the parton distribution function (PDF). For a hadronic reaction,

$$A + B \to c + d + X,$$

where c and d are superpartners and X represents assorted hadronic debris, we have an associated subprocess reaction

$$a + b \to c + d,$$

whose cross section can be computed using the Lagrangian for the MSSM. To obtain the final cross section, we must convolute the appropriate subprocess production cross section $d\hat{\sigma}$ with the parton distribution functions:

$$d\sigma(AB \to cdX) = \sum_{a,b} \int_0^1 dx_a \int_0^1 dx_b\, f_{a/A}(x_a, Q^2)\, f_{b/B}(x_b, Q^2)\, d\hat{\sigma}(ab \to cd),$$

(12.1)

where the sum extends over all initial partons a, b whose collisions produce the final state $c + d$.

Notice that the longitudinal momentum $\mathbf{p}_a + \mathbf{p}_b$ of the initial state is not known. It is for this reason that complete kinematic reconstruction is usually not possible at hadron colliders. The initial partons, however, have negligible transverse

---

[1] See, e.g., V. Barger and R. J. N. Phillips, *Collider Physics*, Addison-Wesley (1987).

momentum. Constraints from transverse momentum balance, therefore, play a central role in hadron collider physics.

Once the interactions of sparticles are known, the computation of the hard scattering cross section for any sparticle production process is straightforward. One way is to develop the Feynman rules for the MSSM and use these to obtain the production amplitudes and then the cross section.[2] The presence of Majorana neutralinos is an additional complication that leads to somewhat unusual Feynman rules. Instead of following this route, we will describe a procedure for evaluating the invariant matrix element starting from the interaction Lagrangian. In effect, this procedure involves doing exactly what one would do to derive the Feynman rules, and so is not new. We find it convenient to use because all particles are treated uniformly, the relative signs between various amplitudes are automatically obtained, and no new rules have to be committed to memory.

The invariant amplitude $\mathcal{M}$ that enters the computation of the cross section for the process $i \rightarrow f$, where i and f denote the initial and final states, respectively, arises from the non-trivial part of the S-matrix element

$$\langle f|S|i \rangle = \langle f|T \left(\exp[-i \int d^4 x \mathcal{H}_{int}] \right) |i\rangle, \qquad (12.2a)$$

where

$$S = 1 + i(2\pi)^4 \delta^4(P_f - P_i)\mathcal{M}. \qquad (12.2b)$$

We assume that the reader is familiar with the evaluation of $\mathcal{M}$ using covariant perturbation theory and, in particular, with how various numerical factors coming from different ways of Wick contracting to obtain the same Feynman diagram usually cancel (or sometimes give the so-called combinatorial factor). Once the matrix element $\mathcal{M}$ has been computed, the cross section for the hard scattering process can be readily obtained using,

$$d\hat{\sigma} = \frac{1}{2\hat{s}} \frac{1}{(2\pi)^2} \int \frac{d^3 p_c}{2E_c} \frac{d^3 p_d}{2E_d} \delta^4(p_a + p_b - p_c - p_d) \cdot F_{color} F_{spin} \sum |\mathcal{M}|^2, \qquad (12.2c)$$

where F_{color} and F_{spin} are factors arising from averaging over the colors and spins in the initial state (assuming it to be unpolarized) and the sum extends over the colors and spins of the initial and final states.

---

[2] See, e.g., M. E. Peskin and D. V. Schroeder, *Introduction to Quantum Field Theory*, Chapter 4, Perseus Press (1995).

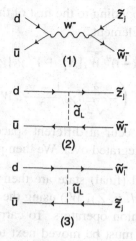

Figure 12.1 Feynman diagrams for chargino–neutralino pair production from quark–antiquark annihilation at hadron colliders.

12.1.1 Chargino–neutralino production

Cross section for $d\bar{u} \to \widetilde{W}_i^- \widetilde{Z}_j$: a worked example

As an illustration of the method we will work out the cross section for chargino–neutralino production which dominantly occurs by annihilation of quarks and antiquarks at hadron colliders: $d\bar{u} \to \widetilde{W}_i^- \widetilde{Z}_j$. Subdominant contributions from other flavors can be analogously included. In the next chapter, we will see that the subsequent decays of the chargino and the neutralino can lead to a final state with three hard (high p_T), isolated leptons (e's or μ's) plus large missing transverse momentum carried off by the LSPs. This may be one of the best discovery modes for gravity-mediated SUSY breaking models at the Fermilab Tevatron.

The subprocess $d\bar{u} \to \widetilde{W}_i^- \widetilde{Z}_j$ takes place at second order in the perturbation expansion via the three Feynman diagrams listed in Fig. 12.1. The relevant vertices can be obtained from the interaction terms,

$$\mathcal{L}_{W\bar{u}d} = -\frac{g}{\sqrt{2}} \bar{u}\gamma_\mu \frac{1-\gamma_5}{2} d\, W^{+\mu} + \quad \text{h.c.}$$

$$\mathcal{L}_{W\widetilde{W}_i\widetilde{Z}_j} = -g(-\mathrm{i})^{\theta_j} \overline{\widetilde{W}}_i [X_i^j + Y_i^j \gamma_5]\gamma_\mu \widetilde{Z}_j W^{-\mu} + \quad \text{h.c.}$$

$$\mathcal{L}_{q\bar{q}\widetilde{W}_i} = \mathrm{i}A_{\widetilde{W}_i}^d \tilde{u}_L^\dagger \overline{\widetilde{W}}_i \frac{1-\gamma_5}{2} d + \mathrm{i}A_{\widetilde{W}_i}^u \tilde{d}_L^\dagger \overline{\widetilde{W}}_i^c \frac{1-\gamma_5}{2} u + \quad \text{h.c.}$$

and

$$\mathcal{L}_{q\bar{q}\widetilde{Z}_j} = \mathrm{i}A_{\widetilde{Z}}^q \tilde{q}_L^\dagger \overline{\widetilde{Z}}_j \frac{1-\gamma_5}{2} q + \quad \text{h.c.},$$

listed in Chapter 8.

The amplitude $\mathcal{M}_1$ corresponding to the first of the diagrams in Fig. 12.1 obviously depends on the matrix element,

$$\langle \widetilde{W}_i \widetilde{Z}_j | T \left[\left(-g(-\mathrm{i})^{\theta_j} \overline{\widetilde{W}}_i [X_i^j + Y_i^j \gamma_5] \gamma_\mu \widetilde{Z}_j W^{-\mu}(x) \right) \right.$$

$$\left. \times \left(-\frac{g}{\sqrt{2}} \bar{u} \gamma_\nu \frac{1 - \gamma_5}{2} d W^{+\nu}(y) \right) \right] |d\bar{u}\rangle,$$

where the two interactions occur at different spacetime points x and y whose coordinates are ultimately integrated over. We then proceed as follows:

1. The particles in the initial (final) state are then "reduced" in any chosen order (which we take to be d, $\bar{u}$, $\widetilde{Z}_j$, $\widetilde{W}_i$) using the annihilation (creation) pieces of the corresponding fermion operators. To carry out this reduction, the corresponding field operator must be moved next to the state in question. Since fermion fields anticommute with other fermion fields, this process can lead to minus signs. In the present case, the reduction of the quarks in the prescribed order does not lead to any sign, but to reduce the neutralino in the final state one has to anticommute the $\widetilde{Z}_j(x)$ past $\overline{\widetilde{W}}_i(x)$, giving an additional minus sign for this amplitude. The reduction of the Dirac quarks [antiquarks] in the initial state and the chargino in the final state, as usual, leaves wave function factors $u(p_d)\exp(-\mathrm{i}p_d \cdot y)$ $[\bar{v}(p_{\bar{u}})\exp(-\mathrm{i}p_{\bar{u}} \cdot y)]$, and $\bar{u}(p_{\widetilde{W}_j})\exp(\mathrm{i}p_{\widetilde{W}_j} \cdot x)$. Notice that because the neutralino is Majorana, it can be reduced by the operator $\widetilde{Z}_j(x)$ (as opposed to its Dirac conjugate) even though it is in the final state. In other words, the neutralino is treated as an anti-particle, and the associated wave function factor is, $v(p_{\widetilde{Z}_j})exp(\mathrm{i}p_{\widetilde{Z}_j} \cdot x)$. This is also the reason for the reversed direction of the arrow (which denotes the flow of fermion number) on the neutralino line in diagram (1) of Fig. 12.1.

2. Once the external particles are all reduced, aside from c-number wave function and coupling constant factors, we are left with

$$\langle 0 | T(W^{-\mu}(x) W^{+\nu}(y)) | 0 \rangle$$

which is of course the propagator for the W-boson between the spacetime points x and y. For the final step, it is convenient to write this propagator (in the unitary gauge) in terms of its momentum space expansion with the four-momentum variable q_W as

$$\langle 0 | T \left(W^{-\mu}(x) W^{+\nu}(y) \right) | 0 \rangle = \mathrm{i} \int \frac{\mathrm{d}^4 q_W}{(2\pi)^4} \frac{-g^{\mu\nu} + \frac{q_W^\mu q_W^\nu}{M_W^2}}{q_W^2 - M_W^2 + \mathrm{i}M_W\Gamma_W} e^{-\mathrm{i}q_W \cdot (x-y)}.$$

3. Finally, integration over x and y leads to four-momentum conservation at each vertex (so that the propagator momentum $q_W = p_d + p_{\bar{u}}$), leaving us with an

overall four-momentum conserving δ function as in the last term of (12.2b). Neglecting quark masses, the $q_W^\mu q_W^\nu$ term in the propagator cannot contribute because

$$\not{p}_d u(p_d) = \bar{v}(p_{\bar{u}})\not{p}_{\bar{u}} = 0,$$

by the Dirac equation. All factors of (2π) cancel and there is no combinatorial factor. We are left with[3]

$$\mathcal{M}_1 = \frac{g^2}{\sqrt{2}}(-\mathrm{i})^{\theta_j} D_W(\hat{s})\bar{u}(\widetilde{W}_i)[X_i^j + Y_i^j \gamma_5]\gamma^\mu v(\widetilde{Z}_j)\, \bar{v}(\bar{u})\gamma_\mu \frac{1-\gamma_5}{2}u(d). \quad (12.3a)$$

Here particle labels denote the corresponding four-momenta, $\hat{s} = (d+\bar{u})^2$, and $D_W(\hat{s}) = (\hat{s} - M_W^2 + \mathrm{i}M_W\Gamma_W)^{-1}$.

The amplitude for the $\tilde{d}_L$ exchange diagram (2) in Fig. 12.1 depends on the matrix element,

$$\langle \widetilde{W}_i \widetilde{Z}_j | T \left[\left(\mathrm{i}A_{\widetilde{Z}_j}^d \tilde{d}_L^\dagger \overline{\widetilde{Z}}_j \frac{1-\gamma_5}{2}d \right) \left(-\mathrm{i}A_{\widetilde{W}_i}^{u*} \tilde{d}_L \bar{u} \frac{1+\gamma_5}{2} \widetilde{W}_i^c \right) \right] |d\bar{u}\rangle.$$

The reduction of the d and $\bar{u}$ quarks in the initial state gives the usual Dirac wave functions for these. This time, the neutralino is reduced by the operator $\overline{\widetilde{Z}}_j$ (so that it is treated as a particle rather than as an antiparticle as in the evaluation of $\mathcal{M}_1$). Finally, the chargino is reduced by the creation part of the $\widetilde{W}_i^c$ operator (which destroys a positive chargino or creates a negative chargino), and by the expansion analogous to (3.33) we obtain the wave function $v(\widetilde{W}_i)\exp(\mathrm{i}\widetilde{W}_i \cdot x)$ for the chargino. Notice that the directions of the arrows on the chargino and neutralino lines in diagram (2) of Fig. 12.1 are in accord with this assignment. The scalar field operators contract together to form the $\tilde{d}_L$ propagator, and the corresponding amplitude can be written as,

$$\mathcal{M}_2 = -A_{\widetilde{Z}_j}^d A_{\widetilde{W}_i}^{u*} \bar{u}(\widetilde{Z}_j)\frac{1-\gamma_5}{2}u(d)\frac{1}{(\widetilde{W}_i - \bar{u})^2 - m_{\tilde{d}_L}^2}\bar{v}(\bar{u})\frac{1+\gamma_5}{2}v(\widetilde{W}_i), \quad (12.3b)$$

where, once again, there is an additional minus sign from anticommuting fermion field operators. We will leave it to the reader to work out that the amplitude for the

---

[3] In writing (12.3a) we have left out a factor $(\mathrm{i})^3$ where two powers of i come from the fact that we are doing second order perturbation theory, and the third power of i comes from the propagator. Since all three diagrams come from second order perturbation theory, and each of these contains one propagator, this amounts to leaving out an irrelevant phase in the overall amplitude from the way it is conventionally written. Moreover, from (12.2b) we see that what we have evaluated is really $\mathrm{i}\mathcal{M}_1$ rather than $\mathcal{M}_1$; again, this only changes the overall phase. We will omit these irrelevant phase factors in the rest of this book. We warn the reader that one should be careful in doing so. In calculations where the amplitude comes from contributions with different numbers of propagators, or from different orders of expansion (although with the same powers of couplings, of course) of the time evolution operator, these phases must be retained.

$\tilde{u}_L$ exchange diagram (3) that depends on the matrix element

$$\langle \widetilde{W}_i \widetilde{Z}_j | T \left[\left(-iA_{\widetilde{Z}_j}^{u*} \bar{u} \frac{1+\gamma_5}{2} \widetilde{Z}_j \tilde{u}_L \right) \left(iA_{\widetilde{W}_i}^d \tilde{u}_L^\dagger \overline{\widetilde{W}_i} \frac{1-\gamma_5}{2} d \right) \right] | d\bar{u} \rangle,$$

takes the form.

$$\mathcal{M}_3 = A_{\widetilde{W}_i}^d A_{\widetilde{Z}_j}^{u*} \bar{v}(\bar{u}) \frac{1+\gamma_5}{2} v(\widetilde{Z}_j) \frac{1}{(\widetilde{Z}_j - \bar{u})^2 - m_{\tilde{u}_L}^2} \bar{u}(\widetilde{W}_i) \frac{1-\gamma_5}{2} u(d). \quad (12.3c)$$

Note that constructing the amplitudes in this fashion allows us to keep track of the relative signs between them.

The amplitudes $\mathcal{M}_1$, $\mathcal{M}_2$, and $\mathcal{M}_3$ can now be squared and summed over initial and final spin states using standard trace techniques. We find,

$$\sum_{\text{spins}} |\mathcal{M}_1|^2 = 8g^4 |D_W(\hat{s})|^2 \left\{ [X_i^{j2} + Y_i^{j2}](\widetilde{Z}_j \cdot d\, \widetilde{W}_i \cdot \bar{u} + \widetilde{Z}_j \cdot \bar{u}\, \widetilde{W}_i \cdot d) \right.$$

$$+ 2(X_i^j Y_i^j)(\widetilde{Z}_j \cdot d\, \widetilde{W}_i \cdot \bar{u} - \widetilde{Z}_j \cdot \bar{u}\, \widetilde{W}_i \cdot d)$$

$$\left. + [X_i^{j2} - Y_i^{j2}] m_{\widetilde{W}_i} m_{\widetilde{Z}_j} d \cdot \bar{u} \right\}, \quad (12.4a)$$

$$\sum_{\text{spins}} |\mathcal{M}_2|^2 = \frac{4|A_{\widetilde{W}_i}^u|^2 |A_{\widetilde{Z}_j}^d|^2}{[(\widetilde{W}_i - \bar{u})^2 - m_{\tilde{d}_L}^2]^2} d \cdot \widetilde{Z}_j\, \widetilde{W}_i \cdot \bar{u} \quad (12.4b)$$

and

$$\sum_{\text{spins}} |\mathcal{M}_3|^2 = \frac{4|A_{\widetilde{W}_i}^d|^2 |A_{\widetilde{Z}_j}^u|^2}{[(\widetilde{Z}_j - \bar{u})^2 - m_{\tilde{u}_L}^2]^2} \bar{u} \cdot \widetilde{Z}_j\, \widetilde{W}_i \cdot d. \quad (12.4c)$$

Next, we turn to the interference terms between these amplitudes. Here we will often find a "mismatch" of spinors. For instance, in computing $\sum_{\text{spins}}(\mathcal{M}_1 \mathcal{M}_2^\dagger)$, we find

$$\sum_{\text{spins}} \mathcal{M}_1 \mathcal{M}_2^\dagger = -(-i)^{\theta_j} \frac{g^2}{\sqrt{2}} D_W(\hat{s}) \frac{1}{(\widetilde{W}_i - \bar{u})^2 - m_{\tilde{d}_L}^2} A_{\widetilde{Z}_j}^{d*} A_{\widetilde{W}_i}^u$$

$$\times \bar{u}(\widetilde{W}_i)(X_i^j + Y_i^j \gamma_5)\gamma^\mu v(\widetilde{Z}_j) \cdot \bar{v}(\bar{u})\gamma_\mu \frac{1-\gamma_5}{2} u(d)$$

$$\times \bar{u}(d) \frac{1+\gamma_5}{2} u(\widetilde{Z}_j) \cdot \bar{v}(\widetilde{W}_i) \frac{1-\gamma_5}{2} v(\bar{u}),$$

so that the chargino and neutralino spinors are not in the proper format for us to evaluate the spin sums using as usual

$$\sum_{\text{spins}} u(p)\bar{u}(p) = \not{p} + m,$$

etc. In order to do the spin sums using the spinor completeness relations, we may use the relations $u = C\bar{v}^T$ and $v = C\bar{u}^T$ to write,

$$
\begin{aligned}
\bar{u}(\widetilde{W}_i)(X_i^j + Y_i^j \gamma_5)\gamma^\mu v(\widetilde{Z}_j) &= v^T(\widetilde{W}_i)C(X_i^j + Y_i^j \gamma_5)\gamma^\mu C\bar{u}^T(\widetilde{Z}_j) \\
&= v^T(\widetilde{W}_i)(X_i^j + Y_i^j \gamma_5)^T \gamma^{\mu T}\bar{u}^T(\widetilde{Z}_j) \\
&= \bar{u}(\widetilde{Z}_j)\gamma^\mu(X_i^j + Y_i^j \gamma_5)v(\widetilde{W}_i).
\end{aligned}
$$

Now we may apply the spinor completeness relations and follow the usual trace techniques to obtain,

$$
\begin{aligned}
\sum_{\text{spins}}(\mathcal{M}_1\mathcal{M}_2^* + \text{c.c.}) ={} & \frac{-\sqrt{2}g^2\mathrm{Re}[A_{\widetilde{Z}_j}^{d*}A_{\widetilde{W}_i}^u(-\mathrm{i})^{\theta_j}](\hat{s} - M_W^2)|D_W(\hat{s})|^2}{(\widetilde{W}_i - \bar{u})^2 - m_{\tilde{d}_L}^2} \\
& \times \left\{8(X_i^j + Y_i^j)\widetilde{Z}_j \cdot d\, \bar{u} \cdot \widetilde{W}_i + 4(X_i^j - Y_i^j)m_{\widetilde{W}_i}m_{\widetilde{Z}_j}d \cdot \bar{u}\right\}.
\end{aligned}
$$

(12.4d)

Similarly, we find that

$$
\begin{aligned}
\sum_{\text{spins}}(\mathcal{M}_1\mathcal{M}_3^* + \text{c.c.}) ={} & \frac{\sqrt{2}g^2\mathrm{Re}[A_{\widetilde{W}_i}^{d*}A_{\widetilde{Z}_j}^u(-\mathrm{i})^{\theta_j}](\hat{s} - M_W^2)|D_W(\hat{s})|^2}{(\widetilde{Z}_j - \bar{u})^2 - m_{\tilde{u}_L}^2} \\
& \times \left\{8(X_i^j - Y_i^j)\widetilde{Z}_j \cdot \bar{u}d \cdot \widetilde{W}_i + 4(X_i^j + Y_i^j)m_{\widetilde{W}_i}m_{\widetilde{Z}_j}d \cdot \bar{u}\right\},
\end{aligned}
$$

(12.4e)

and

$$
\sum_{\text{spins}}(\mathcal{M}_2\mathcal{M}_3^* + \text{c.c.}) = -\frac{4\mathrm{Re}[A_{\widetilde{Z}_j}^d A_{\widetilde{W}_i}^{u*} A_{\widetilde{W}_i}^{d*} A_{\widetilde{Z}_j}^u]m_{\widetilde{W}_i}m_{\widetilde{Z}_j}d \cdot \bar{u}}{[(\widetilde{W}_i - \bar{u})^2 - m_{\tilde{d}_L}^2][(\widetilde{Z}_j - \bar{u})^2 - m_{\tilde{u}_L}^2]}.
$$

(12.4f)

The hard subprocess cross section is obtained using,

$$
\mathrm{d}\hat{\sigma} = \frac{1}{2\hat{s}}\frac{1}{(2\pi)^2}\int \frac{\mathrm{d}^3 p_{\widetilde{W}_i}}{2E_{\widetilde{W}_i}}\frac{\mathrm{d}^3 p_{\widetilde{Z}_j}}{2E_{\widetilde{Z}_j}}\delta^4(\bar{u} + d - \widetilde{W}_i - \widetilde{Z}_j)\cdot\frac{1}{3}\frac{1}{4}\sum_{\text{spins}}|\mathcal{M}|^2,
$$

(12.5a)

or

$$
\frac{\mathrm{d}\hat{\sigma}}{\mathrm{d}\cos\theta} = \frac{p_{\widetilde{W}_i}}{16\pi\hat{s}^{3/2}}\frac{1}{12}\sum_{\text{spins}}|\mathcal{M}|^2,
$$

(12.5b)

where

$$
p_{\widetilde{W}_i} = p_{\widetilde{Z}_j} = \lambda^{1/2}(\hat{s}, m_{\widetilde{W}_i}^2, m_{\widetilde{Z}_j}^2)/2\sqrt{\hat{s}},
$$

(12.6a)

with

$$\lambda(x, y, z) = x^2 + y^2 + z^2 - 2xy - 2xz - 2yz. \tag{12.6b}$$

The factor $1/3$ $(1/4)$ in (12.5a) comes from averaging over color (spin) in the initial state. A sum over colors in the initial and (when applicable) final states is implied. Since the squared matrix element given by the sum of (12.4a)–(12.4f) is Lorentz invariant, we can evaluate it in any frame. It is convenient to evaluate it in the CM frame of the colliding partons. There is no loss of generality if we choose their directions to be along the $\pm z$-axis, and take the chargino and neutralino to lie in the xz plane. Their four vectors can thus be written as:

$$d = \frac{\sqrt{\hat{s}}}{2}(1, \ 0, \ 0, \ 1), \tag{12.7a}$$

$$\bar{u} = \frac{\sqrt{\hat{s}}}{2}(1, \ 0, \ 0, \ -1), \tag{12.7b}$$

$$\widetilde{W}_i = (E_{\widetilde{W}_i}, \ p_{\widetilde{W}_i} \sin\theta, \ 0, \ p_{\widetilde{W}_i} \cos\theta), \tag{12.7c}$$

$$\widetilde{Z}_j = (E_{\widetilde{Z}_j}, \ -p_{\widetilde{Z}_j} \sin\theta, \ 0, \ -p_{\widetilde{Z}_j} \cos\theta). \tag{12.7d}$$

We can now evaluate all the scalar products that appear in the squared matrix element in terms of the scattering angle θ in the parton CM frame, and obtain our result for the differential scattering cross section for the hard process $d\bar{u} \to \widetilde{W}_i \widetilde{Z}_j$ in terms of $z = \cos\theta$ as,

$$\frac{d\hat{\sigma}}{dz}(d\bar{u} \to \widetilde{W}_i \widetilde{Z}_j) = \frac{p_{\widetilde{W}_i}}{16\pi \hat{s}^{3/2}} \frac{1}{12} (M_1 + M_2 + M_3 + M_{12} + M_{13} + M_{23}), \tag{12.8}$$

where

$$M_1 = g^4 |D_W(\hat{s})|^2 \left\{ (X_i^{j2} + Y_i^{j2}) \left[\hat{s}^2 - (m_{\widetilde{W}_i}^2 - m_{\widetilde{Z}_j}^2)^2 + 4\hat{s} p_{\widetilde{W}_i}^2 z^2 \right] \right.$$
$$\left. + 8X_i^j Y_i^j \hat{s}^{3/2} pz + 4(X_i^{j2} - Y_i^{j2})\hat{s} m_{\widetilde{W}_i} m_{\widetilde{Z}_j} \right\} \tag{12.9a}$$

$$M_2 = \frac{1}{4}|A_{\widetilde{Z}_j}^d|^2 |A_{\widetilde{W}_i}^u|^2 \ G(m_{\widetilde{Z}_j}, m_{\widetilde{W}_i}, m_{\bar{d}_L}, -z) \tag{12.9b}$$

$$M_3 = \frac{1}{4}|A_{\widetilde{W}_i}^d|^2 |A_{\widetilde{Z}_j}^u|^2 \ G(m_{\widetilde{W}_i}, m_{\widetilde{Z}_j}, m_{\bar{u}_L}, z) \tag{12.9c}$$

$$M_{12} = \frac{\frac{g^2}{\sqrt{2}}\text{Re}[(-i)^{\theta_j} A_{\widetilde{W}_i}^u A_{\widetilde{Z}_j}^{d*}](\hat{s} - M_W^2)|D_W(\hat{s})|^2}{[\frac{1}{2}(\hat{s} - m_{\widetilde{Z}_j}^2 - m_{\widetilde{W}_i}^2) + \sqrt{\hat{s}} pz + m_{\bar{d}_L}^2]} \left\{ (X_i^j + Y_i^j) \right.$$
$$\left. [\hat{s}^2 - (m_{\widetilde{W}_i}^2 - m_{\widetilde{Z}_j}^2)^2 + 4\hat{s}^{3/2} pz + 4\hat{s} p^2 z^2] + 4(X_i^j - Y_i^j)\hat{s} m_{\widetilde{W}_i} m_{\widetilde{Z}_j} \right\} \tag{12.9d}$$

$$
M_{13} = \frac{-\frac{g^2}{\sqrt{2}}\mathrm{Re}[(-\mathrm{i})^{\theta_j} A^u_{\tilde{Z}_j} A^{d*}_{\tilde{W}_i}](\hat{s} - M^2_W)|D_W(\hat{s})|^2}{[\frac{1}{2}(\hat{s} - m^2_{\tilde{Z}_j} - m^2_{\tilde{W}_i}) - \sqrt{\hat{s}}pz + m^2_{\tilde{u}_L}]} \left\{ (X^j_i - Y^j_i) \right.
$$

$$
\left. [\hat{s}^2 - (m^2_{\tilde{W}_i} - m^2_{\tilde{Z}_j})^2 - 4\hat{s}^{3/2}pz + 4\hat{s}p^2z^2] + 4(X^j_i + Y^j_i)\hat{s}m_{\tilde{W}_i}m_{\tilde{Z}_j} \right\} \quad (12.9e)
$$

$$
M_{23} =
$$

$$
\frac{-2\mathrm{Re}[A^d_{\tilde{Z}_j} A^{u*}_{\tilde{W}_i} A^{d*}_{\tilde{W}_i} A^u_{\tilde{Z}_j}]\hat{s}m_{\tilde{W}_i}m_{\tilde{Z}_j}}{[\frac{1}{2}(\hat{s} - m^2_{\tilde{Z}_j} - m^2_{\tilde{W}_i}) - \sqrt{\hat{s}}pz + m^2_{\tilde{u}_L}][\frac{1}{2}(\hat{s} - m^2_{\tilde{Z}_j} - m^2_{\tilde{W}_i}) + \sqrt{\hat{s}}pz + m^2_{\tilde{d}_L}]},
$$

$$(12.9f)$$

where

$$
G(m_1, m_2, M, z) = \frac{\hat{s}^2 - (m^2_1 - m^2_2)^2 - 4\hat{s}^{3/2}pz + 4\hat{s}p^2z^2}{[\frac{1}{2}(\hat{s} - m^2_1 - m^2_2) - \sqrt{\hat{s}}pz + M^2]^2}. \quad (12.10)
$$

It is not difficult to integrate the subprocess cross section over scattering angles to obtain the total cross section. For the purposes of event generation at hadron colliders, discussed in Chapter 14, this is not especially useful and, although expressions for these total cross sections are available, we do not reproduce these here. In any event, to obtain the total cross section at a hadron collider, we must convolute this subprocess cross section with appropriate PDFs. This is done numerically. Throughout this book, we use CTEQ5L PDFs.[4] Here, we take the renormalization and factorization scales equal, and equal to $Q^2 = \hat{s}$. As an example, various chargino–neutralino production cross sections are shown in Fig. 12.2 versus $m_{\tilde{g}}$ for the CERN LHC pp collider. We have assumed that all flavors of $\tilde{q}_L$ that enter via the t-channel propagators have a common mass.[5] In this figure, we have taken the superpotential parameter $\mu = m_{\tilde{g}} = m_{\tilde{q}}$ and $\tan\beta = 5$. We also assume the gaugino mass unification condition that relates weak scale gaugino mass parameters according to

$$
\frac{M_1}{\alpha_1} = \frac{M_2}{\alpha_2} = \frac{M_3}{\alpha_3},
$$

where $\alpha_i = g^2_i/4\pi$ for $i = 1, 2, 3$ and $g_1 = \sqrt{5/3}g'$, $g_2 = g$, and $g_3 = g_s$. The region to the left of the vertical line is excluded by experiments at LEP2, since they require $m_{\tilde{W}_1} \gtrsim 100$ GeV.

By far the dominant cross section in this class of models occurs for $\tilde{W}_1 \tilde{Z}_2$ production. Gaugino mass unification implies roughly $M_1 : M_2 : M_3 \simeq 1 : 2 : 7$. Since

---

[4] H. L. Lai *et al.* (CTEQ Collaboration), *Eur. Phys. J.* **C12**, 375 (2000).
[5] For the purpose of illustrating the various cross sections, in this chapter, we will take all 12 flavors of squarks to be degenerate.

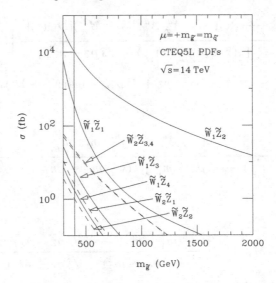

Figure 12.2 Cross sections for chargino plus neutralino production at the CERN LHC *pp* collider for tan $\beta = 5$, and assuming gaugino mass unification at M_{GUT}. The vertical line corresponds to $m_{\widetilde{W}_1} = 100$ GeV.

$\mu \gg M_2$, M_1, the $\widetilde{Z}_1$ will be mainly bino-like (i.e. $\widetilde{Z}_1 \simeq \lambda_0$), while $\widetilde{Z}_2$ and $\widetilde{W}_1$ will be wino-like. Electroweak gauge symmetry implies that the W boson cannot couple to the bino, so that $\widetilde{Z}_1$ couples to W only via its small wino and higgsino components. The wino-like $\widetilde{Z}_2$ and $\widetilde{W}_1$, on the other hand, have large $SU(2)_L$ gaugino components, and so have large couplings to the W as well as to the quark–squark system. The states $\widetilde{Z}_3$, $\widetilde{Z}_4$, and $\widetilde{W}_2$ are mainly higgsino-like and so have smaller isodoublet (rather than the larger isotriplet) coupling to W; this, as well as kinematics, suppresses their production compared to their gaugino-like cousins. This explains why $\widetilde{W}_1\widetilde{Z}_2$ production has the largest cross section in Fig. 12.2. Even for values of $m_{\tilde{g}}$ as high as 2000 GeV, over 1000 $\widetilde{W}_1\widetilde{Z}_2$ events are expected at the CERN LHC, assuming 100 fb^{-1} of integrated luminosity. At the Fermilab Tevatron collider, $\widetilde{W}_1\widetilde{Z}_2$ production could be the dominant SUSY production reaction because production of colored particles is kinematically suppressed in many models. If the branching ratios for the decays $\widetilde{W}_1 \to \ell\bar{\nu}_\ell\widetilde{Z}_1$ and $\widetilde{Z}_2 \to \ell\ell\widetilde{Z}_1$ are large enough then, as already noted, isolated trilepton plus missing energy events may provide a distinctive signature for the discovery of SUSY at the Fermilab Tevatron.

12.1.2 Chargino pair production

At leading order, chargino pair production occurs by $d\bar{d}$ annihilation via the diagrams shown in Fig. 12.3; there are corresponding contributions from annihilation

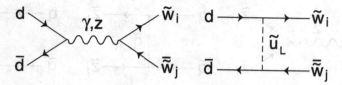

Figure 12.3 Feynman diagrams for leading order chargino pair production via $d\bar{d}$ annihilation at hadron colliders. There are analogous diagrams from the annihilation of other quark flavors.

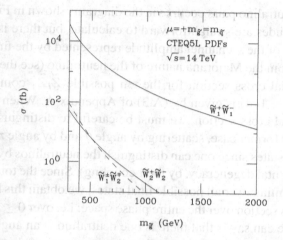

Figure 12.4 Cross sections for chargino pair production at the CERN LHC pp collider for $\tan\beta = 5$, and assuming gaugino mass unification at M_{GUT}.

of other quark flavors. The possible final states consist of $\widetilde{W}_1\overline{\widetilde{W}}_1$, $\widetilde{W}_2\overline{\widetilde{W}}_2$, and $\widetilde{W}_1\overline{\widetilde{W}}_2 + \overline{\widetilde{W}}_1\widetilde{W}_2$. The first two of these occur via γ or Z exchange in the s-channel and t-channel squark exchange, while $\widetilde{W}_1\overline{\widetilde{W}}_2 + \overline{\widetilde{W}}_1\widetilde{W}_2$ production occurs only via Z exchange in the s-channel and t-channel squark exchange. This is because conservation of the electromagnetic current forbids the coupling of the photon to particles of unequal mass. The relevant couplings are listed in Chapter 8, and can be used to construct the production amplitudes as in the previous section. These can be squared using the same techniques described in the last sub-section. The resulting subprocess cross sections are listed in (A.1)–(A.2) of Appendix A. As before, we convolute with CTEQ5L PDFs, and illustrate the total production cross sections for chargino pair production at the LHC in Fig. 12.4. We see that $\widetilde{W}_1\overline{\widetilde{W}}_1$ production is the largest of this set, and is comparable in magnitude to the cross section for $\widetilde{W}_1\widetilde{Z}_2$ pair production shown in Fig. 12.2.

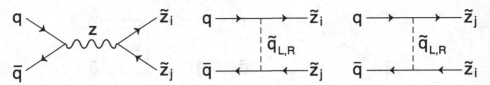

Figure 12.5 Feynman diagrams for leading order neutralino pair production processes at hadron colliders.

12.1.3 Neutralino pair production

Production of neutralino pairs occurs via the diagrams shown in Fig. 12.5. The four t-channel amplitudes are straightforward to calculate, but there is a small subtlety in the evaluation of the s-channel amplitude represented by the first diagram in the figure coming from the Majorana nature of the neutralino (see the exercise below).

The differential cross section for the ten possible $\widetilde{Z}_i\widetilde{Z}_j$ combinations (corresponding to $i, j = 1$–4) is given by (A.3) of Appendix A. When integrating these to obtain the total cross section, we must be careful to distinguish between $i \neq j$ and $i = j$. In the former case, scattering by angle θ and by angle $\pi - \theta$ correspond to *different* final states since one can distinguish the neutralinos by their mass (or, if there is an accidental degeneracy, by their coupling). Since the total cross section is obtained by summing over all possible final states, we obtain this by integrating the differential cross section over the entire phase space: i.e. over $0 \leq \theta \leq \pi$. However, for $i = j$, all one can say is that there is one neutralino at an angle θ (with respect to the quark beam) and a second neutralino at an angle $\pi - \theta$, but there is no way to tell, even in principle, which of the two neutralinos is at θ. In other words, *the state with scattering angle θ is the same state as the one with scattering angle $\pi - \theta$*, and so, to obtain the total cross section we should integrate over just half the phase space (since otherwise we would double count the final states). We can write the total neutralino cross section as,

$$\sigma_{\text{tot}}(q\bar{q} \rightarrow \widetilde{Z}_i\widetilde{Z}_j) = \Delta_{ij} \int_{-1}^{1} \frac{d\sigma}{dz}(q\bar{q} \rightarrow \widetilde{Z}_i\widetilde{Z}_j)dz \qquad (12.11a)$$

with

$$\Delta_{ij} = 1 - \frac{1}{2}\delta_{ij}. \qquad (12.11b)$$

Neutralino pair production rates (particularly for the gaugino-like neutralinos) are more sensitive to model parameters than corresponding rates for $\widetilde{W}_i\widetilde{Z}_j$ and $\widetilde{W}_i^-\widetilde{W}_j^+$ production. This is because they couple to Z only via their small higgsino components (so that the s-channel amplitude is suppressed) while the t-channel amplitude is obviously sensitive to the squark mass. This is in sharp contrast to $\widetilde{W}_1\widetilde{Z}_2$ production for which we saw that (as long as $|\mu| \gg M_2 \simeq 2M_1$) the

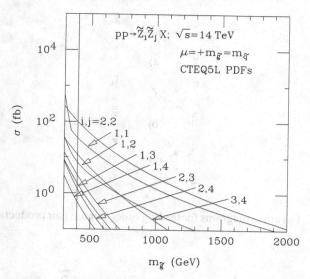

Figure 12.6 Cross sections for neutralino pair production at the CERN LHC *pp* collider for $\tan \beta = 5$, and assuming gaugino mass unification at M_{GUT}.

W amplitude is always large. The s-channel contributions are also always sizable for the case of $\widetilde{W}_1^- \widetilde{W}_1^+$ production: the chargino obviously always couples to the photon, and the Z has large weak-isovector couplings to the gaugino-like chargino.

Sample cross sections for the CERN LHC are shown in Fig. 12.6. For the parameters in this figure, the gaugino-like neutralino states, $\widetilde{Z}_2$ and $\widetilde{Z}_1$, are most strongly produced. In models with a $\widetilde{Z}_1$ LSP and R-parity conservation, the $\widetilde{Z}_1$ is absolutely stable, and will escape detection at collider detectors. Thus, the $\widetilde{Z}_1 \widetilde{Z}_1$ reaction would be invisible, aside from any initial state QCD radiation into instrumented regions of the detector. Many of these reactions occur at low rates and do not lead to distinctive signatures at hadron colliders.

Exercise *The amplitude for the first diagram in Fig. 12.5 depends on the matrix element*

$$\langle \widetilde{Z}_j \widetilde{Z}_i | e \bar{q} \gamma_\mu (\alpha_q + \beta_q \gamma_5) q \, Z^\mu \sum_{ab} W_{ab} \overline{\widetilde{Z}}_a \gamma_\nu (\gamma_5)^{\theta_a + \theta_b + 1} \widetilde{Z}_b Z^\nu | q \bar{q} \rangle.$$

The matrix element is non-zero only when either $a = i$ with $b = j$ or $a = j$ with $b = i$. Both these contributions must be included to correctly obtain the amplitude. Evaluate these contributions and, using $W_{ij} = (-1)^{\theta_j - \theta_i} W_{ji}$ together with the charge conjugation properties $u = C \bar{v}^T$ and $v = C \bar{u}^T$ of the solutions to the Dirac equation, show that the two contributions are equal.

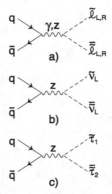

Figure 12.7 Feynman diagrams for leading order slepton pair production at hadron colliders.

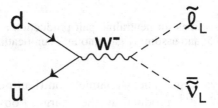

Figure 12.8 Feynman diagram for slepton–sneutrino associated production at hadron colliders.

12.1.4 Slepton and sneutrino pair production

At leading order, charged sleptons and sneutrinos may be produced in pairs via the diagrams in Fig. 12.7, or charged sleptons can be produced in association with their sneutrino partner via the W exchange diagram shown in Fig. 12.8. The former correspond to the supersymmetric analogue of the Drell–Yan process, whereas the latter is the analogue of the classic process via which the W boson was discovered at the CERN $p\bar{p}$ collider. Only like-type (L or R) slepton pairs can be produced for the first two generations of charged sleptons though intrageneration mixing also allows $\tilde{\tau}_1\bar{\tilde{\tau}}_2 + \bar{\tilde{\tau}}_1\tilde{\tau}_2$ production via Z exchange. Since W couples only to left-handed leptons and their superpartners, $\tilde{\ell}_R\tilde{\nu}_L$ production is forbidden. Both $\tilde{\tau}_1$ and $\tilde{\tau}_2$ can be produced in association with $\tilde{\nu}_\tau$; the state with the large admixture of $\tilde{\tau}_L$ ($\tilde{\tau}_2$ in many models) has the bigger coupling to W.

The computation of the various amplitudes differs from what we have already seen only because of the derivative coupling of sleptons to gauge bosons. To illustrate how these are handled, we write the amplitude for the associated slepton–sneutrino production process in Fig. 12.8. We need to evaluate the matrix element,

$$\langle \tilde{\ell} \tilde{v}_{\mathrm{L}} \Big| T \left[\left(-\frac{g}{\sqrt{2}} \bar{u} \gamma^\mu \frac{1 - \gamma_5}{2} d\, W_\mu^- \right) \left(-\frac{ig}{\sqrt{2}} (\tilde{\ell}_{\mathrm{L}}^\dagger \partial^\nu \tilde{v}_{\mathrm{L}} - \tilde{v}_{\mathrm{L}} \partial^\nu \tilde{\ell}_{\mathrm{L}}^\dagger) W_\nu^+ \right) \right] \Big| d\bar{u} \rangle.$$

In reducing the sleptons in the final state, we get the derivative of the sneutrino (charged slepton) wave function $\exp(i\tilde{v} \cdot x)$ ($\exp(i\tilde{\ell}_{\mathrm{L}} \cdot x)$) which gives an i times the momentum factors in the amplitude. The contraction of the W fields gives us the W propagator and, as before, integration over the spacetime points where the interactions occur give us momentum conservation at each vertex. We are then left with the matrix element

$$\mathcal{M} = -\frac{1}{2} g^2 \bar{v}(\bar{u}) \gamma^\mu \frac{1 - \gamma_5}{2} u(d) D_W(\hat{s}) (\tilde{e}_{\mathrm{L}} - \tilde{v}_{\mathrm{L}})_\mu, \tag{12.12}$$

which is now straightforward to square to obtain the differential cross section listed in (A.14).[6] The cross sections for the charged slepton (including stau) and sneutrino pair production processes can be similarly obtained and are given in (A.15a)–(A.15b). Note that the cross sections for the production of the first two generations of charged sleptons and sneutrinos are completely determined by their masses, and so are model-independent. For staus, model dependence enters via the stau mixing angle.

In Fig. 12.9, we show slepton pair production cross sections as a function of slepton mass for the Fermilab Tevatron and for the CERN LHC.[7] These results include next-to-leading order corrections (mentioned below) in the limit of very heavy squark masses. The negative results of slepton searches at LEP2 require $m_{\tilde{e}}$ ($m_{\tilde{\mu}}$) to be greater than about 100 (85) GeV. In the region $m_{\tilde{\ell}} \simeq 100$–200 GeV, the cross sections for the Fermilab Tevatron are always below 100 fb, and simulation studies indicate that sleptons beyond the reach of LEP2 would be very difficult to detect.[8] Detection of slepton pairs via their direct production seems possible at the CERN LHC if slepton masses are below ~ 300–400 GeV.

---

[6] Instead of writing this as a differential cross section $d\sigma/dz$ as before, we have written it as a differential cross section over the Mandelstam variable $\hat{t} = (d - \tilde{\ell})^2$ using

$$\frac{d\sigma}{d\hat{t}} = \frac{1}{16\pi \hat{s}^2} \frac{1}{12} |\mathcal{M}|^2,$$

where the factor $1/12$ comes from color and spin averaging over the initial state.

[7] Sometimes in the subsequent discussion of sparticle pair production, we will for convenience use $\tilde{\ell}$ to collectively denote both sleptons and antisleptons, or $\tilde{q}$ to denote both squarks and antisquarks. It should be clear from the context when this occurs.

[8] H. Baer *et al.*, *Phys. Rev.* **D49**, 3283 (1994).

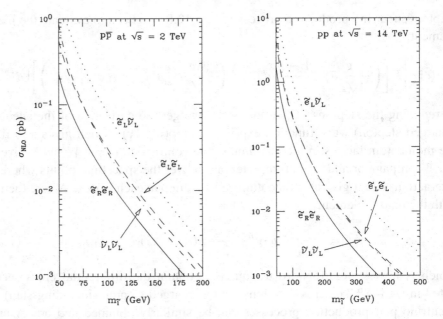

Figure 12.9 Cross sections for production of slepton pairs at the Tevatron and the CERN LHC.

12.1.5 Production of gluinos and squarks

Gluino and squark production at hadron colliders occurs dominantly via strong interactions. Thus, their production rate may be expected to be considerably larger than that for sparticles with just electroweak interactions whose production we have been considering up to now. This is tempered by the fact that in many models colored sparticles are expected to be the heaviest of all the superparticles, so that their production may be kinematically suppressed.

Gluino production at hadron colliders mainly occurs via the diagrams listed in Fig. 12.10. Since the gluon luminosity in hadron collisions falls off rapidly with $\hat{s}$, gluino production from the gg initial state is usually dominant for lower values of $m_{\tilde{g}}$, while $q\bar{q}$ annihilation dominates if $m_{\tilde{g}}$ is large. The differential cross sections for gluino pair production by gg scattering and by $q\bar{q}$ scattering is given by (A.5a) and (A.5b), respectively.[9] Gluino pair production leads to a large rate for multi-jet events with apparent E_T^{miss} carried off by the daughter LSPs from the decay of the gluinos. Other distinctive gluino signatures will be discussed in subsequent chapters.

---

[9] The derivative coupling at the three gluon vertex can be handled as explained in the previous subsection. In the present case, the derivative may also act on the gluon propagator but this can be dealt with exactly as before.

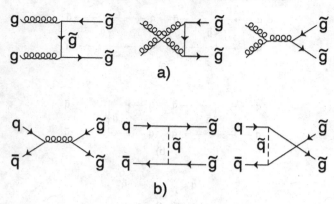

Figure 12.10 Feynman diagrams for leading order gluino pair production processes at hadron colliders.

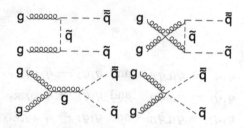

Figure 12.11 Feynman diagrams for squark pair production via gluon scattering at hadron colliders.

Pair production of squarks via gg scattering takes place via the diagrams listed in Fig. 12.11. These scattering reactions lead to particle–antiparticle pairs of the same flavor and type, e.g. $\tilde{u}_L \bar{\tilde{u}}_L$, $\tilde{u}_R \bar{\tilde{u}}_R$, etc. This is also true for t-squark pair production: only $\tilde{t}_i \bar{\tilde{t}}_i$ ($i = 1, 2$) pairs can be produced because gluons do not couple to $\tilde{t}_1 \tilde{t}_2$ pairs. In addition, as shown in Fig. 12.12, squark pairs can also be produced via quark–quark or quark–antiquark scattering. These contributions are important only for those flavors with significant luminosity in the colliding hadron beams. Not only do different Feynman diagrams contribute to the production of different flavors and types of squarks, as we will see in the next chapter these different squarks have their distinct decay patterns. Thus the cross section magnitudes, angular distributions, and the final decay products all depend on which pair of squarks is being produced. For simulations of superparticle production at colliders it is, therefore, important to separate out the production of different types of squark pairs. The component

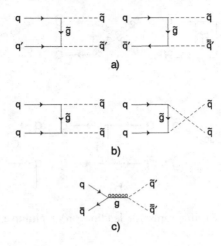

Figure 12.12 Feynman diagrams for squark pair production via quark scattering at hadron colliders.

reactions are

$$q_1\bar{q}_2 \to \tilde{q}_{1L}\bar{\tilde{q}}_{2R} \quad \text{and} \quad q_1\bar{q}_2 \to \tilde{q}_{1R}\bar{\tilde{q}}_{2L}, \tag{12.13a}$$

$$q_1\bar{q}_2 \to \tilde{q}_{1L}\bar{\tilde{q}}_{2L} \quad \text{and} \quad q_1\bar{q}_2 \to \tilde{q}_{1R}\bar{\tilde{q}}_{2R}, \tag{12.13b}$$

$$q_1 q_2 \to \tilde{q}_{1L}\tilde{q}_{2R} \quad \text{and} \quad q_1 q_2 \to \tilde{q}_{1R}\tilde{q}_{2L}, \tag{12.13c}$$

$$q_1 q_2 \to \tilde{q}_{1L}\tilde{q}_{2L} \quad \text{and} \quad q_1 q_2 \to \tilde{q}_{1R}\tilde{q}_{2R}, \tag{12.13d}$$

$$q\bar{q} \to \tilde{q}_L\bar{\tilde{q}}_R \quad \text{and} \quad q\bar{q} \to \tilde{q}_R\bar{\tilde{q}}_L, \tag{12.13e}$$

$$q\bar{q} \to \tilde{q}_L\bar{\tilde{q}}_L \quad \text{and} \quad q\bar{q} \to \tilde{q}_R\bar{\tilde{q}}_R, \tag{12.13f}$$

$$q\bar{q} \to \tilde{q}'_L\bar{\tilde{q}}'_L \quad \text{and} \quad q\bar{q} \to \tilde{q}'_R\bar{\tilde{q}}'_R, \tag{12.13g}$$

$$qq \to \tilde{q}_L\tilde{q}_L \quad \text{and} \quad qq \to \tilde{q}_R\tilde{q}_R, \tag{12.13h}$$

$$q\bar{q} \to \tilde{q}_L\bar{\tilde{q}}_R \quad \text{and} \quad q\bar{q} \to \tilde{q}_R\bar{\tilde{q}}_L. \tag{12.13i}$$

The differential cross sections for these various squark pair production reactions are listed in (A.7a)–(A.7j) of Appendix A. While most of the necessary amplitudes can be straightforwardly calculated using techniques that we have already described, in the evaluation of the amplitude for the processes $q_1 q_2 \to \tilde{q}_{1L}\tilde{q}_{2R}$ which occur via gluino exchanges in the t-channel, and also u-channel if $q_1 = q_2$, (see Fig. 12.12) we encounter a new complication. The relevant amplitude depends on the matrix element

$$\langle \tilde{q}_{1Lc}\tilde{q}_{2Rd}| - 2g_s^2 T \left[\left(\tilde{q}_L^\dagger \frac{\bar{\tilde{g}}_A \lambda_A}{2} \frac{1-\gamma_5}{2} q \right)(x) \left(\tilde{q}_R^\dagger \frac{\bar{\tilde{g}}_B \lambda_B}{2} \frac{1+\gamma_5}{2} q \right)(y) \right] |q_{1a}q_{2b}\rangle,$$

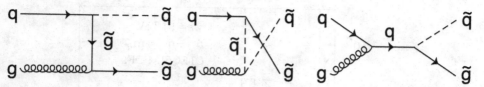

Figure 12.13 Feynman diagrams for leading order gluino–squark associated production at hadron colliders.

where *a–d* are color indices. Reducing the particles in the external states leaves us with,

$$-2g_s^2 \langle 0|T\left[\left(e^{i\tilde{q}_{1L}\cdot x}(\frac{\bar{\tilde{g}}_A(x)\lambda_A}{2})_{ca}\frac{1-\gamma_5}{2}u(q_1)e^{-iq_1\cdot x}\right)\right.$$
$$\left.\times \left(e^{i\tilde{q}_{2R}\cdot y}(\frac{\bar{\tilde{g}}_B(y)\lambda_B}{2})_{db}\frac{1+\gamma_5}{2}u(q_2)e^{-iq_2\cdot y}\right)\right]|0\rangle. \qquad (12.14a)$$

We now see that because we have two Dirac-conjugated gluino fields, the vacuum expectation value of their time-ordered product is *not* the Feynman propagator for the gluino. To bring it to this form, we recall that the Majorana nature of the gluino means that the spinor $\tilde{g} \equiv \lambda_A \tilde{g}_A/2$ is a Majorana spinor so that,

$$\bar{\tilde{g}}(x)\frac{1-\gamma_5}{2}u(q_1) = \tilde{g}(x)^T C\frac{1-\gamma_5}{2}C\bar{v}(q_1)^T = -\bar{v}(q_1)\frac{1-\gamma_5}{2}\tilde{g}(x).$$

If we substitute this into (12.14a), we see that the matrix element contains the gluino propagator as expected, but that we obtain a v-spinor for the wave function of the quark q_1.[10] As usual, we can now write the gluino propagator as a Fourier integral over the four-momentum $p_{\tilde{g}}$; also, integration over the co-ordinates x and y gives four-momentum conservation at each vertex, and the matrix element for $q_1 q_2 \to \tilde{q}_{1L}\tilde{q}_{2R}$ reduces to,

$$\mathcal{M} = 2g_s^2\bar{v}(q_1)\frac{1-\gamma_5}{2}(\frac{\lambda_A}{2})_{ca}\frac{1}{(\not{p}_{\tilde{g}} - m_{\tilde{g}})}(\frac{\lambda_A}{2})_{db}\frac{1+\gamma_5}{2}u(q_2), \qquad (12.14b)$$

which can be now squared using usual trace techniques.

Finally, gluinos and squarks may also be produced in association with each other via gluon–quark scattering, as shown in Fig. 12.13. The corresponding cross section is given by (A.6).

---

[10] This is equivalent to saying that we write the Lagrangian at point x in terms of the anti-quark field, i.e. a field $\psi_{\bar{q}}$ that destroys an antiquark in the initial state or creates a quark in the final state, and the "anti-gluino" field. The reader may also recall that we encountered a similar manipulation in Chapter 3, when we examined the quadratic divergences in the corrections to the two-point function of the field A. See also Eq. (3.37a) and (3.37b).

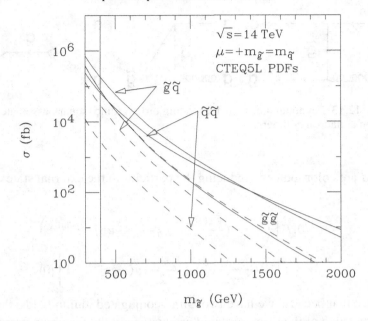

Figure 12.14 Cross sections for squark and gluino production at the CERN LHC pp collider for $m_{\tilde{q}} = m_{\tilde{g}}$ (solid) and for $m_{\tilde{q}} = 2m_{\tilde{g}}$ (dashes).

It is worth emphasizing that because there are no third generation partons in the initial state, squark and gluino production rates are fixed by SUSY QCD in terms of just the squark and gluino masses, and do not depend upon the details of any model. In Fig. 12.14, we show sample cross sections for gluino and squark pair production at the CERN LHC, assuming six flavors of mass degenerate left- and right-squarks. In this example, we take $m_{\tilde{q}} = m_{\tilde{g}}$ (solid lines) and $m_{\tilde{q}} = 2m_{\tilde{g}}$ (dashed lines). The renormalization and factorization scale is chosen to be half the average mass of the sparticles produced, which yields results in accord with next-to-leading order predictions. For the case of $m_{\tilde{q}} = m_{\tilde{g}}$, $\tilde{g}\tilde{q}$ associated production dominates over most of the range of $m_{\tilde{g}}$, until $\tilde{q}\tilde{q}$ pair production dominates at the highest values of $m_{\tilde{g}}$. This behavior is in part a reflection of the PDFs, where production via gluons is dominant for small x values, but production via valence quark scattering dominates for large x values and large sparticle masses. In the case of $m_{\tilde{q}} = 2m_{\tilde{g}}$, $\tilde{g}\tilde{g}$ production is dominant, since these are the lightest mass pairs of sparticles. We see that even for gluinos as heavy as 1 TeV, $\mathcal{O}(10^3$–$10^4)$ gluino and squark events are expected at the LHC for an integrated luminosity of just 10 fb^{-1}. It is in this sense that the LHC will be a sparticle factory.

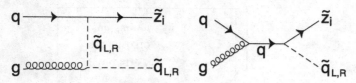

Figure 12.15 Feynman diagrams for leading order squark–neutralino associated production at hadron colliders.

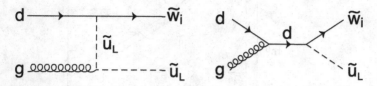

Figure 12.16 Feynman diagrams for leading order squark–chargino associated production at hadron colliders.

Exercise *The reader will have noticed that gluinos can be produced only from gg and $q\bar{q}$ initial states, but not from the qq initial state. Argue that this must be the case by color symmetry. Use the reduction of $SU(3)$ tensor products,*

$$\mathbf{3} \otimes \mathbf{3} = \mathbf{3}^* \oplus \mathbf{6},$$
$$\mathbf{8} \otimes \mathbf{8} = \mathbf{1} \oplus \mathbf{8} \oplus \mathbf{8} \oplus \mathbf{10} \oplus \mathbf{10}^* \oplus \mathbf{27},$$

to make the argument.

12.1.6 Gluino or squark production in association with charginos or neutralinos

Gluinos and squarks may also be produced in association with charginos and neutralinos in a semi-strong reaction. Diagrams leading to squark production in association with neutralinos (charginos) are shown in Fig. 12.15 (Fig. 12.16). These reactions occur by quark–gluon scattering via u- and s-channel graphs, with cross sections given by (A.8)–(A.10). Sample reaction rates for the CERN LHC are shown in Fig. 12.17 versus $m_{\tilde{g}}$ for $\mu = m_{\tilde{g}} = m_{\tilde{q}}$, $\tan\beta = 5$, and assuming gaugino mass unification and degenerate squarks.

Feynman diagrams for gluino production in association with neutralinos (charginos) are shown in Fig. 12.18 (Fig. 12.19). In this case, production occurs via quark–antiquark scattering via t- and u-channel squark exchange. The relevant cross sections are given by (A.11)–(A.12). Example cross sections for the LHC are shown in Fig. 12.20 for the same parameters as in Fig. 12.17.

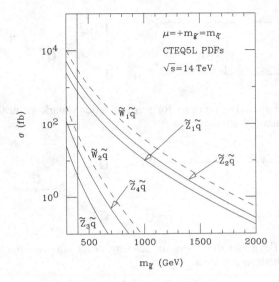

Figure 12.17 Cross sections for chargino or neutralino production in association with squarks at the CERN LHC pp collider for $\tan\beta = 5$, and assuming gaugino mass unification at M_{GUT} and degenerate squarks.

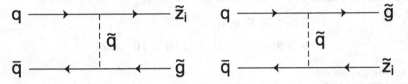

Figure 12.18 Feynman diagrams leading to gluino plus neutralino production at hadron colliders.

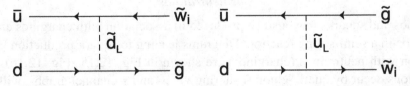

Figure 12.19 Feynman diagrams leading to gluino plus chargino production at hadron colliders.

Generally, the rates for all semi-strong associated production reactions are smaller than rates for direct pair production of gluinos and squarks at the LHC, or to chargino and neutralino pair production at the Fermilab Tevatron. The signatures are not especially distinctive from those arising from cascade decays of gluino and squark pair production, so that these processes appear to be less important for the search for supersymmetry.

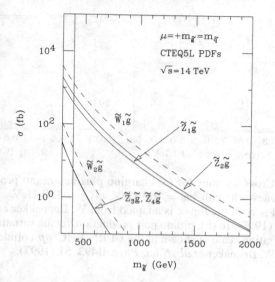

Figure 12.20 Cross sections for chargino or neutralino production in association with gluinos at the CERN LHC *pp* collider for tan $\beta = 5$, and assuming gaugino mass unification at M_{GUT}.

12.1.7 Higher order corrections

Next-to-leading order (NLO) QCD corrections to scattering cross sections are necessary to improve the accuracy of numerical predictions, and such calculations have been carried out for all the sparticle production mechanisms discussed above. The accuracy of leading order (LO) predictions can be ascertained by varying the renormalization and factorization scales inherent in the cross section calculations. For simplicity, we set these two scales equal to each other, and denote them by Q. In Fig. 12.21a, we show the variation of LO and NLO calculations of $\widetilde{W}_1\widetilde{Z}_2$ production at the CERN LHC with respect to variation in the scale choice, expressed as a ratio with the average mass of the produced sparticles. The uncertainty of the LO result is $\sim 30\%$, while the scale variation of the NLO result is minimal. Typically, for this reaction, the NLO result represents an enhancement of 20%–50%.

In Fig. 12.21b, the cross section variation versus scale choice is shown for gluino pair production. In this case, the LO cross section varies by a factor of ~ 3, while the NLO result varies only by about 30%. As noted above, for strongly produced SUSY particles a scale choice of 0.3–0.5 times the average sparticle mass will yield LO cross section predictions in accord with NLO results.[11]

---

[11] NLO sparticle pair production cross sections can be generated by the computer program PROSPINO: see W. Beenakker, R. Höpker and M. Spira, hep-ph/9611232.

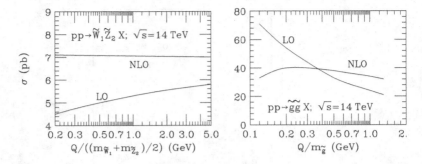

Figure 12.21 Cross sections for (a) chargino plus neutralino production for the mSUGRA framework with $m_0 = 100$ GeV, $m_{1/2} = 150$ GeV, $A_0 = 300$ GeV, $\tan \beta = 4$, and $\mu > 0$; the figure is adapted from W. Beenakker *et al.*, *Phys. Rev. Lett.* **83**, 3780 (1999). In (b), gluino pair production versus variation in renormalization/factorization scale is shown at the CERN LHC *pp* collider; this figure is adapted from W. Beenaker *et al.*, *Nucl. Phys.* **B492**, 51 (1997).

12.1.8 Sparticle production at the Tevatron and LHC

In Fig. 12.22, we show total cross sections for production of supersymmetric particles at the Fermilab Tevatron, for $p\bar{p}$ collisions at $\sqrt{s} = 2$ TeV, as a function of the physical gluino mass, assuming the squarks are all degenerate. In frame (a), for $m_{\tilde{q}} = m_{\tilde{g}}$, we see that chargino and neutralino production is the dominant production mechanism over the entire range of $m_{\tilde{g}}$ values shown. Strong production of gluinos and squarks never dominates, mainly because in this case the gluino and squark masses are so heavy compared to the charginos and neutralinos.[12] In frame (b), we show the corresponding cross sections for $m_{\tilde{q}} = 2m_{\tilde{g}}$. In this case, strong production cross sections are even more suppressed due to large squark masses, and production of charginos and neutralinos is dominant. We see that $\tilde{W}_1^+ \tilde{W}_1^-$ and $\tilde{W}_1^\pm \tilde{Z}_2$ production processes dominate sparticle production at the Tevatron.

Figure 12.23 illustrates sparticle production rates at the CERN LHC. In frame (*a*) for $\mu = m_{\tilde{g}} = m_{\tilde{q}}$, gluino and squark production dominates unless gluinos and squarks are heavier than 1.7 TeV, in which case chargino and neutralino production has the largest rate. For the heavy squark case in frame (*b*), gluino and squark production is dominant for $m_{\tilde{g}} \lesssim 800$ GeV. Associated production is never dominant.

12.2 Sparticle production at e^+e^- colliders

Since superpartners were not discovered at the CERN LEP2 e^+e^- collider, operating at $\sqrt{s} \simeq 200$ GeV, it seems likely that if weak scale supersymmetry exists

---

[12] This is not the case for $\mu = -m_{\tilde{g}}$ for which charginos and neutralinos tend to be heavier. Then strong production is dominant if $m_{\tilde{g}} \lesssim 300$ GeV (200 GeV) for $m_{\tilde{q}} = m_{\tilde{g}}$ ($m_{\tilde{q}} = 2m_{\tilde{g}}$).

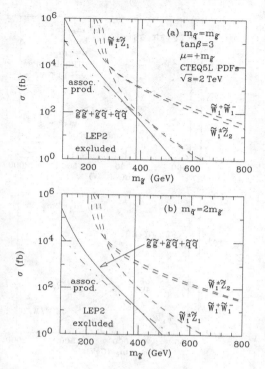

Figure 12.22 Cross sections for the production of gluinos, squarks, charginos, and neutralinos at the Fermilab Tevatron $p\bar{p}$ collider. We assume gaugino mass unification at $Q = M_{GUT}$, and also that all squarks have the same mass. To the left of the vertical line, the chargino is lighter than 100 GeV.

in nature, its discovery will take place at a hadron machine. Nevertheless, there is considerable interest in the construction of a linear e^+e^- collider to operate in the energy regime of $\sqrt{s} \sim 0.5$–1.5 TeV. Despite the lower energy, the advantages of such a machine (over hadron colliders) for the elucidation of weak scale supersymmetry are numerous:

- Unlike at hadron colliders where the energy available for the production of new particles is limited to that of the colliding partons, essentially all of the available center of mass energy may go into creating new states at an e^+e^- collider. This is because, unlike hadrons, electrons and positrons are elementary particles.
- For the same reason, the e^+e^- initial state has a well-defined energy and momentum, and allows detailed kinematic reconstruction of scattering events, facilitating precision measurements. Again for this same reason, e^+e^- scattering events are very clean because the hard scattering event is free of contamination from spectator jets and initial state QCD radiation that are necessarily present in hadron

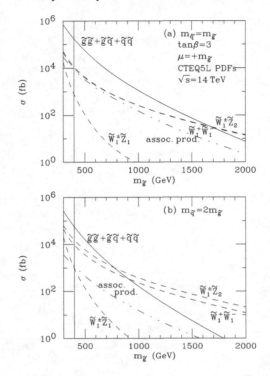

Figure 12.23 Cross sections for gluinos, squarks, charginos and neutralinos at the CERN LHC *pp* collider. As in Fig. 12.22, we assume gaugino mass unification and the degeneracy of squarks. The chargino is lighter than 100 GeV to the left of the vertical line.

scattering. The clean environment together with the simplicity of the initial state allows final states to be reconstructed with greater precision.

- Aside from kinematic suppression, all particles with non-trivial $SU(2)_L \times U(1)_Y$ quantum numbers are produced at comparable rates so that signal to background is never very small.
- The availability of a longitudinally polarized electron, and possibly also a positron, beam is a novel feature of electron–positron colliders. Since SUSY signals and SM backgrounds are both sensitive to beam polarization, polarized beams can be a very valuable tool, both for reducing SM backgrounds and for separating SUSY reactions from one another.
- The beam energy is tunable. Together with beam polarization capability, this will allow experimentalists to isolate particular SUSY processes, further facilitating determination of sparticle properties.

The biggest physics advantages of hadron colliders are (a) the higher beam energy, which makes them an ideal facility for a broad band search for new physics, and (b) the sizable cross section for SUSY processes which results in observable signal rates for luminosity and energy which is supposed to be well within the realm of current technology. In contrast, both signal and background cross sections tend to be small at high energy e^+e^- colliders, so that very high beam intensities are essential for physics. Thus, while sparticles may be discovered at a hadron collider, and many of their properties determined there, a TeV scale e^+e^- collider operating with polarizable beams will allow a systematic program of precision studies of all the superparticles with significant production cross sections.

12.2.1 Production of sleptons, sneutrinos, and squarks

Pair production of smuons, staus, and their corresponding sneutrinos takes place via the same Feynman diagrams as in Fig. 12.7, with $q\bar{q}$ replaced by e^-e^+. Squark pairs are also produced via the same Feynman diagrams as for charged slepton production, with the sleptons replaced by squarks. The relevant matrix elements can be evaluated as before. The one new element is that, for reasons explained at the start of this chapter, we present the cross sections for polarized electron/positron beams. This simply entails inserting corresponding chiral projectors $P_{L/R} = \frac{1 \mp \gamma_5}{2}$ to select out the desired polarization in front of the initial state electron/positron spinor wave functions when evaluating the various amplitudes.[13] The cross sections for squark pair production, as well as for charged sleptons and sneutrinos of the first two generations, are given by (A.21a)–(A.21c). The cross section for unpolarized beams, or for partially polarized beams, can be obtained from these using (A.28). We note that for the first two generations of squarks as well as for smuon and sneutrino production, the cross sections are determined by just the sfermion mass (together with known SM parameters), and so are model-independent.

In Fig. 12.24, we show the cross section for smuon and sneutrino ($\tilde{\nu}_\mu$ or $\tilde{\nu}_\tau$) pair production from unpolarized beams as a function of the sparticle mass, for an e^+e^- collider operating at $\sqrt{s} = 1$ TeV. The stau cross section depends on the stau mixing angle but typically has a similar magnitude. Linear colliders are currently being designed, and the projected luminosity for such a machine might be 10–50 fb^{-1} per year, or larger. The Technical Design Report of the TESLA collider being considered for construction quotes a luminosity of 3.4×10^{34} cm^{-2}s^{-1} at $\sqrt{s} = 500$ GeV, corresponding to a projected design luminosity in excess of 300 fb^{-1}/yr, assuming the machine runs about a third of the time. Depending on the luminosity that is ultimately attained, several hundred to several thousand smuon pair events might be expected annually for smuon masses heavy enough to be within 80% of the kinematic limit.

---

[13] At the energies of interest it is safe to neglect the electron mass so that there is no difference between chirality and helicity.

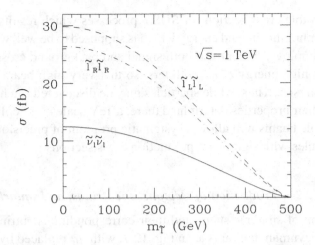

Figure 12.24 Cross sections for production of smuons and associated sneutrinos at a $\sqrt{s} = 1$ TeV e^+e^- collider with unpolarized beams.

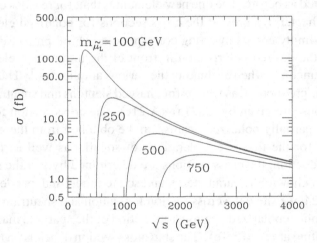

Figure 12.25 Cross sections for left-smuon pair production versus $\sqrt{s}$ for various smuon masses at an e^+e^- collider with unpolarized beams.

The variation of this cross section versus collider $\sqrt{s}$ is shown in Fig. 12.25, for various smuon masses. Slightly above threshold, the cross section attains a maximum, falling off as the energy escalates. The rapid rise of the cross section close to the kinematic end-point is characteristic of the β^3 p-wave threshold behavior evident in (A.21a).

The cross sections for various squark pair production processes as a function of squark mass are shown in Fig. 12.26 for a 1 TeV e^+e^- collider with unpolarized beams. For third generation squarks, as for staus, the cross sections will be modified

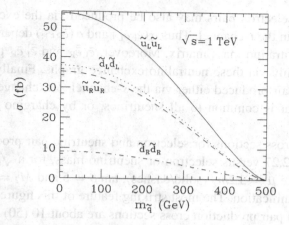

Figure 12.26 Cross sections for various types of squark pairs at a 1 TeV e^+e^- collider with unpolarized beams, versus $m_{\tilde{q}}$.

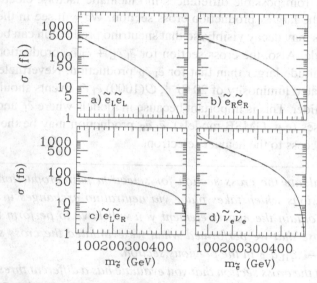

Figure 12.27 Cross sections for producing left- and right-selectron and electron sneutrinos at a $\sqrt{s} = 1$ TeV e^+e^- collider with unpolarized beams, versus the corresponding selectron or sneutrino mass for the parameters listed in the text. The solid (dashed) lines correspond to positive (negative) values of μ.

by mixing angle factors. Production of left-squarks is the largest of these cross sections. We note that these cross sections are much smaller than the corresponding cross sections at the LHC.

The mechanisms for the production of first generation sleptons and sneutrinos are more complicated. In addition to the first of the Feynman graphs of Fig. 12.7, left- and right-selectron pairs may also be produced via the exchange of any of the neutralinos in the t-channel. Thus $\sigma(\tilde{e}_L\tilde{e}_L)$ and $\sigma(\tilde{e}_R\tilde{e}_R)$ depend on parameters entering the neutralino mass matrix. Moreover, $\tilde{e}_L\bar{\tilde{e}}_R$ and $\bar{\tilde{e}}_L\tilde{e}_R$ pairs can also be produced but only via these neutralino exchange graphs. Finally, electron sneutrinos may be pair produced either via the s-channel Z exchange diagram shown in Fig. 12.7 that is common to all sneutrinos, or by chargino exchange in the t-channel.

Production cross sections for selectron and sneutrino pair production are illustrated in Fig. 12.27 versus selectron or sneutrino mass, for a $\sqrt{s} = 1$ TeV e^+e^- collider. We take $\mu = \pm 2m_{\tilde{l}}$ (solid/dashes), $\tan\beta = 5$ and $M_2 = m_{\tilde{l}}$, and assume gaugino mass unification. The most striking feature of this figure is that the selectron (sneutrino) pair production cross sections are about 10 (50) times larger than the corresponding cross sections for second generation sleptons. This is because of the t-channel contributions to their production. Notice also that for the first generation, aside from possible differences in kinematic factors, electron sneutrinos usually have the largest production cross section. We will see in the next chapter that sneutrinos may decay visibly, so that sneutrino production can be an important discovery mode. Also, the cross section for $\tilde{e}_L\tilde{e}_L + \tilde{e}_R\tilde{e}_R$ production is almost an order of magnitude larger than that for $\tilde{e}_L\tilde{e}_R$ production. Nevertheless, even for a modest integrated luminosity of 20 fb^{-1}, $\mathcal{O}(1000)$ $\tilde{e}_L\tilde{e}_R$ events should be expected at a linear collider. This is important because in models where $\tilde{e}_L$ and $\tilde{e}_R$ have very different masses (e.g. GMSB models), $\tilde{e}_L\tilde{e}_R$ production may be the only reaction which gives access to the heavier selectron.

Exercise *Evaluate the cross section for selectron pair production by electron–electron collisions which takes place via neutralino exchanges in the t- and u-channels. To obtain the matrix element you will have to perform manipulations similar to those that we performed when we evaluated the cross section for the process $q_1q_2 \rightarrow \tilde{q}_{1L}\tilde{q}_{2R}$ in the previous section.*

Notice that the cross section that you evaluate has a different threshold behavior from that for selectron production in e^+e^- collisions. This, together with the fact that lepton number conservation implies that we have no SM backgrounds from W^-W^- production, suggests that the selectron mass can be more precisely measured via this process than at e^+e^- colliders.

12.2.2 Production of charginos and neutralinos

Production of $\widetilde{W}_1\overline{\widetilde{W}}_1$ and $\widetilde{W}_2\overline{\widetilde{W}}_2$ pairs proceeds via the Feynman diagrams of Fig. 12.3, by replacing $d\bar{d}$ with e^-e^+, and $\tilde{u}_L$ by $\tilde{\nu}_{eL}$. The s-channel Z exchange

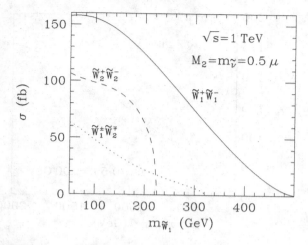

Figure 12.28 Cross sections for chargino pair production at a 1 TeV e^+e^- collider with unpolarized beams, versus $m_{\widetilde{W}_1}$, for $\tan \beta = 5$.

and t-channel sneutrino exchange graphs also lead to the production of $\widetilde{W}_2 \overline{\widetilde{W}}_1$ and $\widetilde{W}_1 \overline{\widetilde{W}}_2$ pairs. The differential cross sections for these various chargino production processes are given by (A.27a)–(A.27d). In many models, $|\mu| \gg M_2$, so that the lighter (heavier) chargino is gaugino-like (higgsino-like). Typically $\sigma(\widetilde{W}_1^+ \widetilde{W}_1^-)$ is large because of the enhanced isotriplet coupling of the charginos to Z^0. However, this cross section can be sensitive to the sneutrino mass because of the interference between the s- and t-channel amplitudes which *reduces* the cross section if $\sqrt{s} > M_Z$.

In Fig. 12.28 we illustrate the cross sections versus the lighter chargino mass for various chargino production processes at a $\sqrt{s} = 1$ TeV e^+e^- collider. We take $M_2 = m_{\widetilde{\nu}_e} = 0.5\mu$, $\tan \beta = 5$. For $M_2 = 0.5\mu \gg M_W$, $m_{\widetilde{W}_2} \sim 2m_{\widetilde{W}_1}$ and production of heavier chargino pairs is kinematically (as well as dynamically) suppressed relative to that of lighter chargino pairs, and cuts off at the kinematic limit which is close to $2m_{\widetilde{W}_1} \sim m_{\widetilde{W}_2} = 500$ GeV. The mixed process $\widetilde{W}_2 \overline{\widetilde{W}}_1 + \widetilde{W}_1 \overline{\widetilde{W}}_2$ always occurs at a lower rate. Note, however, that for an integrated luminosity of 100 fb^{-1} there should be several hundred $\widetilde{W}_1^\pm \widetilde{W}_2^\mp$ events beyond the kinematic limit for $\widetilde{W}_2^+ \widetilde{W}_2^-$ production.

In many supersymmetric models, $\widetilde{W}_1$ is the lightest of visibly decaying SUSY particles. If charginos are kinematically accessible, they should be produced at observable rates in e^+e^- collisions because of their unambiguous couplings to the photon and to the Z. As shown in Fig 12.29, this rate may be significantly smaller than its typical expectation if the sneutrino happens to be relatively light, but should nonetheless be observable.

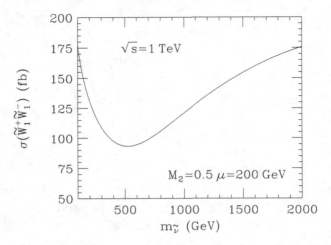

Figure 12.29 Cross sections for chargino pair production processes at a 1 TeV e^+e^- collider with unpolarized beams, versus $m_{\tilde{\nu}_e}$, for $M_2 = 0.5\mu = 200$ and $\tan\beta = 5$.

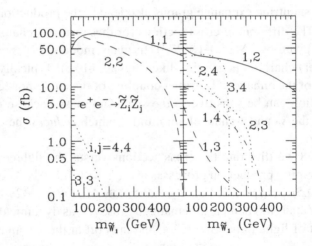

Figure 12.30 Cross sections for neutralino pair production at a 1 TeV e^+e^- collider with unpolarized beams, versus $m_{\tilde{W}_1}$, for $\tan\beta = 5$ and $M_2 = m_{\tilde{e}} = 0.5\mu$.

Neutralino pair production occurs at e^+e^- colliders via the diagrams of Fig. 12.5 with $q\bar{q}$ replaced by e^-e^+ and $\tilde{q}_{L,R}$ replaced by $\tilde{e}_{L,R}$. The corresponding differential cross section is given by (A.26). Sample neutralino production cross sections are shown in Fig. 12.30, for the same parameters as in Fig. 12.28 but with $m_{\tilde{\nu}_e}$ replaced by $m_{\tilde{e}_L} = m_{\tilde{e}_R}$. For the parameters selected, $\widetilde{Z}_1 \simeq \tilde{B}$ and $\widetilde{Z}_2 \simeq \tilde{W}$, so that by $SU(2)_L \times U(1)_Y$ gauge symmetry, these neutralinos have *small* couplings to Z bosons,[14] but couple to the $e\tilde{e}_{L,R}$ system via the corresponding *gauge coupling* (since the wino component does not couple to $\tilde{e}_R$, the $e\tilde{e}_R\tilde{Z}_2$ coupling is small). Since selectrons have been assumed to be relatively light, in this illustration, t-channel amplitudes are large: $\tilde{Z}_1\tilde{Z}_1$ pair production is large because of the large hypercharge of $\tilde{e}_R$. $\tilde{Z}_1\tilde{Z}_2$ production mainly occurs via $\tilde{e}_L$ exchange because $\tilde{e}_L$ couples to both the bino and the wino; however, this rate is smaller than that for $\tilde{Z}_1\tilde{Z}_1$ because of the smaller hypercharge of $\tilde{e}_L$ (in addition to kinematic suppression). For the same reason, $\tilde{Z}_2\tilde{Z}_2$ production mainly occurs via $\tilde{e}_L$ exchange. The neutralinos $\tilde{Z}_3$ and $\tilde{Z}_4$ are mainly higgsino-like, with the magnitude of each higgsino component being close to $\frac{1}{\sqrt{2}}$. Cross sections for $\tilde{Z}_3\tilde{Z}_3$ and $\tilde{Z}_4\tilde{Z}_4$ pair production are, however, very small because the $Z\tilde{Z}_{3(4)}\tilde{Z}_{3(4)}$ coupling in (8.101) is clearly strongly supressed, and the corresponding amplitudes for t-channel exchanges are supressed for dynamical reasons. The rate for $\tilde{Z}_3\tilde{Z}_4$ production, which mainly occurs via unsuppressed couplings to the Z boson is large, and dominates the kinematically favored (but dynamically suppressed) production of "light–heavy" neutralino pairs.

In R-parity conserving models with $\tilde{Z}_1$ as the LSP, the $\tilde{Z}_1\tilde{Z}_1$ final state will be invisible, except for initial state photon radiation. However, as in this example, in mSUGRA and mGMSB models with $m_{\tilde{Z}_2} \simeq m_{\tilde{W}_1} \simeq 2m_{\tilde{Z}_1}$, $\tilde{Z}_1\tilde{Z}_2$ production may be observable even if chargino pairs are not kinematically accessible. We should stress though that unlike chargino cross sections that are relatively robust, neutralino production cross sections are very sensitive to model parameters. In particular, if selectrons are very heavy and $|\mu| \gg |M_{1,2}|$ (as is possible in many models), $\tilde{Z}_1\tilde{Z}_1$, $\tilde{Z}_1\tilde{Z}_2$, and $\tilde{Z}_2\tilde{Z}_2$ production mainly occurs via Z exchange through the suppressed higgsino components of the neutralinos: in this case, these production cross sections can be very small even if neutralino production is kinematically unsuppressed.

12.2.3 *Effect of beam polarization*

We have already mentioned that the availability of longitudinally polarized beams at a linear e^+e^- collider serves as a powerful additional tool for signal analysis at these facilities. The degree of longitudinal beam polarization can be parametrized as

$$P_L(e^-) = f_L - f_R, \quad \text{where} \tag{12.15a}$$

$$f_L = \frac{n_L}{n_L + n_R} = \frac{1 + P_L}{2}, \quad \text{and} \tag{12.15b}$$

$$f_R = \frac{n_R}{n_L + n_R} = \frac{1 - P_L}{2}. \tag{12.15c}$$

---

[14] After all, this is the SUSY analogue of the three neutral vector boson coupling which is forbidden by gauge invariance.

Here, $n_{L,R}$ is the number of left-(right-)polarized electrons in the beam, and $f_{L,R}$ is the corresponding fraction. Thus, a 90% right-polarized beam would correspond to $P_L(e^-) = -0.8$, and a completely unpolarized beam corresponds to $P_L(e^-) = 0$.

In Appendix A we have collected the various SM and SUSY cross sections for polarized electron and positron beams. In practice, however, beams are always partially polarized, and the relevant cross sections can be obtained using

$$\sigma = f_L(e^-)f_L(e^+)\sigma_{LL} + f_L(e^-)f_R(e^+)\sigma_{LR}$$
$$+ f_R(e^-)f_L(e^+)\sigma_{RL} + f_R(e^-)f_R(e^+)\sigma_{RR}, \qquad (12.16)$$

where f_L and f_R are defined above, and σ_{ij} ($i, j = L, R$) is the cross section from $e_i^- e_j^+$ annihilation.

In Fig. 12.31, we show the production cross sections for various SM particle pair production processes at an e^+e^- collider operating at $\sqrt{s} = 500$ GeV, versus the electron beam polarization parameter $P_L(e^-)$, taking the positrons to be unpolarized. The most striking feature is the strong dependence of the W boson pair production cross section on $P_L(e^-)$. This is important because W^+W^- production, which is the SM process with the largest cross section (for unpolarized beams), can lead to events with "missing energy" and "missing momentum" carried off by neutrinos from leptonic decays of W, and so is an important background to the SUSY signal. Fortunately, this rate can be reduced to tiny values by using an increasingly right-handed electron beam (see the exercise below). The other SM processes have a less severe dependence on beam polarization, but generally have the largest rates for left-polarized beams.

The polarization dependence of SUSY particle production cross sections is illustrated in Fig. 12.32, for the mSUGRA model with parameters shown in the figure. We see that the production of first generation sleptons, $\widetilde{W}_1$ pairs and some neutralino pairs is strongly sensitive to $P_L(e^-)$. By adjusting the polarization of the electron beam, we see that it is possible to select out event samples that are rich in $\tilde{e}_L$ or $\tilde{e}_R$ (in addition to other sparticles). In addition to the fact that polarization can be used to reduce SM background, this intra-generational separation can also be important for detailed studies of these sparticles. Indeed we will see that electron beam polarization is a very useful tool when engaging in precision studies of the properties of SUSY particles.

While the degree of beam polarization that will be attained at future linear colliders is still uncertain, it is thought that 80%, or higher, polarization for the electron beam will certainly be possible. The situation for positron beams is less clear, but positron beam polarization of about 60% seems to be the target.

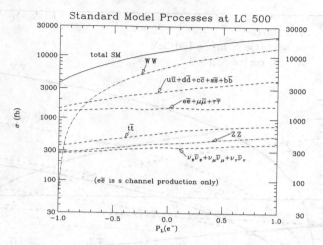

Figure 12.31 Cross sections for various SM pair production processes versus the electron beam polarization parameter $P_L(e^-)$, for e^+e^- collisions at $\sqrt{s} = 500$ GeV. We take the positrons to be unpolarized. Reprinted with permission from H. Baer, R. Munroe, and X. Tata, *Phys. Rev.* **D54**, 6735 (1996), copyright (1996) by the American Physical Society.

Exercise *We saw in Fig. 12.31 that the WW cross section showed a very strong dependence on the electron beam polarization. In view of the importance of eliminating this background, it is worthwhile to understand the smallness of the cross section for $P_L(e^-) = -1$.*

(a) *Draw the Feynman diagrams by which this process occurs. Since W's couple only to left-handed electrons, it is straightforward to see that the amplitude for the neutrino exchange diagram vanishes if $P_L(e^-) = -1$. Remember that electron masses are negligible at the energy that we are considering. For a purely right-handed electron beam, this leaves us with just the Z and photon exchange amplitudes.*

(b) *To analyze these s-channel amplitudes, it is convenient to work in terms of the original hypercharge and $SU(2)_L$ gauge bosons rather than in terms of the photon and the Z. Since right-handed electrons have no coupling to the $SU(2)_L$ gauge boson, the internal vector boson line in the s-channel Feynman diagram must start off as a hypercharge gauge boson at the electron positron vertex. Gauge invariance precludes any coupling between this boson and the W^+W^- pair. Thus, this amplitude would vanish but for mixing between the hypercharge and $SU(2)$ gauge bosons. This mixing originates in the gauge-covariant kinetic*

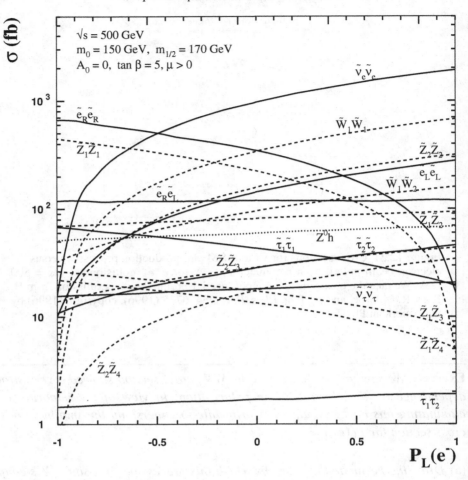

Figure 12.32 Cross sections for production of various sparticle pairs in the mSUGRA model versus the electron beam polarization parameter, for e^+e^- collisions at $\sqrt{s} = 500$ GeV. The positron beam is taken to be unpolarized. Reprinted with permission from H. Baer, C. Balázs, J. K. Mizukoshi and X. Tata, *Phys. Rev* **D63**, 055011 (2001), copyright (2001) by the American Physical Society.

energy,

$$|D_\mu \phi|^2 \ni \sim gg' \langle \phi \rangle^2 W_{3\mu} B^\mu,$$

of the field ϕ defined below (8.22b). Use this to show that the s-channel amplitude must be proportional to M_Z^2. For dimensional reasons it must, therefore, depend on M_Z^2/s and so becomes very small at high energy for right-handed electron beams.

Exercise *Explain the polarization dependence of the slepton and sneutrino pair production cross sections in Fig. 12.32. In particular, explain clearly why stau and tau sneutrino pair production is much less sensitive to $P_L(e^-)$ compared to first generation sleptons.*

12.2.4 Bremsstrahlung and beamstrahlung

Up to now, we have focussed on particle production cross sections at e^+e^- colliders where the full beam energy goes into the hard scattering. However, to properly describe signals and backgrounds to sparticle production at e^+e^- colliders operating in the TeV range, one must allow for forward initial state radiation of high energy photons or bremsstrahlung. An additional complication comes from energy loss due to beam–beam interactions, the so-called beamstrahlung effect. The photons from both these effects are lost down the beam pipe resulting in an unmeasurable loss in the energy of the beam. This reduces the CM energy of the colliding beams, and results in an (unknown) longitudinal momentum for the hard scattering initial state. It is essential to incorporate bremsstrahlung and beamstrahlung losses for precision studies that are possible at linear colliders.

The bremsstrahlung effect can be included by convoluting e^+e^- cross sections with an effective electron structure function. A simple parametrization is the Kuraev–Fadin distribution, given by[15]

$$D_e^{\text{brem}}(x, Q^2) = \frac{1}{2}\beta(1-x)^{\frac{\beta}{2}-1}(1+\frac{3}{8}\beta) - \frac{1}{4}\beta(1+x), \qquad (12.17a)$$

where

$$\beta \equiv \frac{2\alpha}{\pi}(\ln\frac{Q^2}{m_e^2} - 1), \qquad (12.17b)$$

x is the electron fractional momentum, and Q is the scale of hard scattering. The bremsstrahlung distribution is shown by the dashed curve in Fig. 12.33.

In addition, for the very dense, compact electron and positron beams that are essential to obtain the high luminosity needed for high energy linear colliders, one must account for beamstrahlung. In effect, the electron or positron beams are so compact that energy loss can occur due to beam interactions before the hard scattering. This energy loss can be calculated semi-classically, and gives rise to a beamstrahlung distribution function, $D_e^{\text{beam}}(x)$. A parametrization of the beamstrahlung distribution function is too complicated to present here, but it does depend on machine characteristics and beam profile. Stipulating a beamstrahlung

---

[15] E. Kuraev and V. Fadin, *Sov. J. Nucl. Phys.* **41**, 466 (1985).

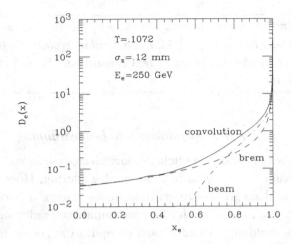

Figure 12.33 Distribution of electrons in the electron, due to bremsstrahlung, beamstrahlung and their convolution. Reprinted from H. Baer, T. Krupovnickas and X. Tata, *JHEP* **06**, 061 (2004).

parameter Υ along with σ_z related to the bunch length is sufficient to determine the beamstrahlung distribution as characterized in calculations by P. Chen.[16] The beamstrahlung distribution for a collider with $\Upsilon = 0.1072$ and beam size $\sigma_z = 0.12$ mm is also shown by the dot-dashed curve in Fig. 12.33, for beam energy $E_e = 250$ GeV. To account for both bremsstrahlung and beamstrahlung, a convolution,

$$D_e(x) = \int_x^1 \mathrm{d}z D_e^{\mathrm{brem}} \left(\frac{x}{z}, Q^2\right) D_e^{\mathrm{beam}}(z)/z, \qquad (12.18)$$

of the two distribution functions must be performed. The resulting beam energy distribution with both beamstrahlung and bremsstrahlung effects included is shown by the solid curve in Fig. 12.33. As can be seen, the highest probability is that electrons or positrons with $x \sim 1$ will interact. But there is a significant probability that energy loss can result from beamstrahlung and bremsstrahlung, so that the energy in the hard scattering process is considerably smaller. This is especially important when examining the reconstruction of SUSY processes with high precision, because the energy loss due to beamstrahlung/bremsstrahlung photons distorts final state distributions as well as the missing energy spectrum that is one of the key elements of sparticle production reactions.

As an example, we show the distribution in dimuon invariant mass in Fig. 12.34 for $e^+e^- \rightarrow \mu^+\mu^-$, at $\sqrt{s} = 500$ GeV, using the beamstrahlung parameters of

[16] See P. Chen, *Phys. Rev.* **D46**, 1186 (1992).

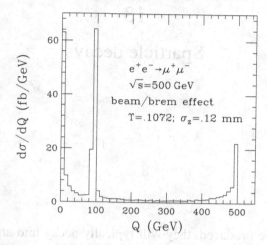

Figure 12.34 Differential cross section for muon pair production at a 500 GeV e^+e^- collider, as a function of dimuon mass. The Z and γ peaks are clearly evident, due to bremsstrahlung and beamstrahlung. The two-photon background discussed in the text is not included in this figure.

Fig. 12.33. A large fraction of events is produced with invariant mass $m(\mu^+\mu^-) \sim$ 500 GeV, as might be expected. However, the γ and Z poles in the production process lead to even larger dimuon rates at $m(\mu^+\mu^-) \sim 0$ and M_Z.[17] In addition, there are non-vanishing cross section contributions in the intermediate invariant mass regions.

We should also mention that for very low values of $m(\mu^+\mu^-)$ the cross section will actually be dominated by the higher order process $e^+e^- \rightarrow e^+e^-\mu^+\mu^-$, where the muons are mainly produced by collision of almost on-shell photons radiated by the electron and positron beams. These "two-photon processes" are a very important background if the observable final state is a pair of charged particles with low momentum and low invariant mass and the final state electrons and positrons escape undetected down the beam pipe. Within the context of supersymmetry, this occurs when the pair-produced charged sparticle is approximately degenerate with the LSP.

---

[17] This means that these machines are "self-scanning" for resonances that couple to e^+e^- pairs.

13

Sparticle decays

Once sparticles are produced, they will typically decay into another sparticle together with SM particles via many different channels. The daughter sparticles subsequently decay to yet lighter sparticles until the decay cascade terminates in the stable LSP. In this discussion we have implicitly assumed that R-parity is conserved: otherwise, sparticles may also decay into just SM particles, and the final state would be comprised of only SM particles. However, whether or not R-parity is conserved, sparticle production at colliders typically leads to a variety of final state topologies via which to search for SUSY. Signal rates into any particular topology are determined by sparticle production cross sections studied in the last chapter, and by the branching fractions for various decays of sparticles.

In this chapter, we examine sparticle decays in the context of the R-parity conserving MSSM. As just mentioned, R-parity conservation implies that any sparticle decay chain will end in a stable LSP which may be a neutralino, a sneutrino, or, in models with local supersymmetry, also a gravitino. We have already seen in Chapter 9 that a sneutrino LSP is disfavored. A weak scale gravitino is essentially decoupled as far as collider physics considerations go. Hence, for most of this chapter, we will assume the gravitino is unimportant for sparticle decay calculations. However, as we saw in Section 11.3.1, an important exception to this occurs if the scale of SUSY breaking is low so that gravitinos are very light. To cover this possibility, we address sparticle decays to gravitinos in the last section of this chapter.

Before proceeding with the detailed examination of the decay rates and various branching fractions for individual sparticle decays, we pause to estimate the expected lifetimes for unstable sparticles. The lifetimes of sparticles are relevant when considering collider signatures for SUSY.

- Sparticles with lifetimes much longer than the time they take to traverse the detector will appear to be stable for the purposes of collider physics. If these are color and electrically neutral, they escape the detector unseen and manifest themselves as apparent missing energy and momentum in SUSY events. If such sparticles are electrically charged, they would cause ionization (the extent of which would depend on their velocity) and leave tracks in the detector, and would reveal themselves in experiments searching for heavy charged exotics. If these are electrically neutral but have strong interactions their experimental signatures may be quite complicated.[1] A particularly striking possibility is that such a particle may intermittently change into its charged partner by charged pion exchanges with nucleons in the experimental apparatus, and then back to neutral!

- Neutral sparticles with lifetimes somewhat shorter than their traversal time in the experimental apparatus would result in displaced vertices. Such a sparticle would be produced at the primary vertex, but would travel a macroscopic distance before decaying at a secondary vertex, which may, depending on the lifetime, be quite distant from the primary interaction point. Experimentalists searching for new physics should keep this possibility in mind, and not discard such an exotic signal as due to background from secondary (cosmic ray) interactions or other noise. If the sparticle lifetime is comparable to B meson lifetimes, SUSY events would contain displaced vertices (with tracks not pointing back to the primary interaction point) that would be identified in specialized microvertex detectors that are an integral part of most contemporary general purpose detectors.

- Finally, sparticles with lifetimes too short to yield displaced vertices that can be resolved by the microvertex detectors would appear to decay promptly at the primary vertex. A familiar SM example of such a situation is the production and decay of the W or Z bosons. In this case, we can get a handle on sparticle properties only by studying their decay products.

The partial decay rate for a particle decaying via $A \to a_1 + a_2 + \cdots + a_n$ is given in the rest frame of A by,

$$\Gamma_n = (2\pi)^{4-3n} \frac{1}{2M_A} \int \frac{d^3 p_{a_1}}{2E_{a_1}} \cdots \frac{d^3 p_{a_n}}{2E_{a_n}} |\mathcal{M}(A \to a_1 a_2 \cdots a_n)|^2$$
$$\times \delta^4(P_A - P_{a_1} - P_{a_2} \cdots - P_{a_n}), \tag{13.1a}$$

where, for any sparticle A, the spin and color summed and averaged squared matrix element $|\mathcal{M}|^2$ for the decay is evaluated using the matrix element obtained using the sparticle interactions listed in Chapter 8. The *total decay rate* is then obtained by summing the partial decay rates for all possible decay modes of A. The lifetime of A is the inverse of this total decay rate,

$$\tau_A = \frac{1}{\sum_n \Gamma_n}. \tag{13.1b}$$

---

[1] The elementary sparticle may well be charged and colored, but may bind with SM quarks to produce an unconfined strongly interacting, electrically neutral "meson" that traverses the apparatus.

The mass dimension of the matrix element $\mathcal{M}$ that appears in (13.1a) can readily be checked to be $[\mathcal{M}] = 3 - n$. The matrix element for two-body decays has dimensions of mass, that for three-body decays is dimensionless, etc.

Before proceeding to evaluate the partial widths for the various decays of individual sparticles, let us estimate their order of magnitude. For two-body decays of unpolarized particles, Lorentz invariance implies that the squared matrix element, summed over final state spins, must be independent of final state momenta: i.e. it is constant.[2] This constant must generically be $\sim k^2 \times m_A^2$ where k is the coupling constant in the interaction responsible for the decay $A \to a_1 a_2$.[3] Using

$$\int \delta^4(P_A - P_{a_1} - P_{a_2})\frac{d^3 p_{a_1}}{2E_{a_1}}\frac{d^3 p_{a_2}}{2E_{a_2}} = \frac{\pi \lambda^{1/2}(m_A^2, m_{a_1}^2, m_{a_2}^2)}{2m_A^2}, \tag{13.2a}$$

it is easy to check that the partial width for the decay,

$$\Gamma(A \to a_1 a_2) \sim \frac{f}{4m_A}\frac{k^2}{4\pi}\lambda^{1/2}(m_A^2, m_{a_1}^2, m_{a_2}^2)$$

$$\simeq f\frac{k^2}{4\pi}\frac{m_A}{4}, \tag{13.2b}$$

where f includes spin and color factors, and in the last step we have ignored any phase space suppression for the decay. The point of this calculation is to show that if the coupling k is comparable to the electromagnetic coupling or larger, the typical width of a 100 GeV particle undergoing two-body decays is $\gtrsim 200$ MeV for a single channel, corresponding to a lifetime $\lesssim 10^{-23}$ seconds: frequently, the total decay rate is considerably larger because of color factors and also because there are several channels. Clearly, such lifetimes are orders of magnitude too short to be detectable by even the best vertex detectors. As shown in the exercise below, the same conclusion obtains if sparticles dominantly decay via three-body decays mediated by gauge interactions.

In the subsequent sections, we will see that essentially all MSSM sparticles can decay (at tree level) via two- or three-body decays mediated by SM gauge interactions. We conclude that, except in very special cases where there is severe phase space suppression, sparticles decay promptly in the experimental apparatus. Important exceptions may occur in GMSB models where the NLSP decays into a (longitudinal) gravitino via suppressed couplings as discussed in Section 11.3.1, or for the case of R-parity violating models where the lightest dominantly R-odd

---

[2] The matrix element can be a function of scalar products of various momenta which, by momentum conservation, can be written in terms of particle masses.

[3] We assume that the interaction does not have any special features that forbids the appearance of the parent's mass in the matrix element. An example where this is forbidden is the matrix element for charged pion decay which, because of chiral symmetry, has to be proportional to the final state fermion mass rather than m_π.

particle decays via very small *R*-parity violating couplings. These special situations will be treated separately.

In the rest of this chapter we will focus on the decay patterns of various sparticles since these determine the event topologies via which to search for supersymmetry at high energy colliders. We illustrate the calculation of partial decay widths by evaluating the width for three-body decays of the gluino. In Appendix B, we list formulae for widths of all tree-level two-body sparticle decay modes along with formulae for the important three-body decay widths.

Exercise *Estimate the order of magnitude of the partial width for a three-body decay of A and show that if this decay is mediated by gauge couplings, we should not expect a discernible secondary vertex in the experimental apparatus. Proceed by the following steps.*

(a) *Although the matrix element for three-body decays is not a constant but depends on the final state momenta, we may estimate its order of magnitude. If the decay is mediated by a virtual bosonic sparticle, the amplitude will contain a propagator of this heavy bosonic particle. Convince yourself that the order of magnitude of the matrix element (which we saw must be dimensionless) is given by,*

$$|\mathcal{M}|^2 \sim k_1^2 k_2^2 \left(\frac{m_A^2}{m_H^2} \right)^2,$$

where k_1 and k_2 are the dimensionless couplings at each of the two vertices involving the virtual heavy particle of mass m_H. Here, m_H^2 in the denominator comes from the propagator, and the m_A^2 is inserted to make the matrix element dimensionless.

(b) *Neglecting any masses for the final state particles, show that the partial width for the three-body decay is then given by,*

$$\Gamma(A \to a_1 a_2 a_3) \sim \frac{f}{32\pi} \left(\frac{k_1 k_2}{4\pi} \right)^2 \frac{m_A^5}{m_H^4},$$

where f again contains spin and color factors.

(c) *Assuming that the mass of the virtual sparticle is no more than an order of magnitude larger than that of the decaying parent, estimate the partial width for this decay, taking the couplings k_1 and k_2 to be comparable to gauge couplings.*

(d) *Frequently, each sparticle has several three-body decay modes, so that the total decay rate is enhanced by color and multiplicity factors. Convince yourself that the lifetime of a 100 GeV sparticle decaying via SM gauge interactions is typically smaller than $\sim 10^{-16}$ seconds.*

Note that if the virtual sparticle is a fermion, the matrix element may have just one power of m_H in the denominator, in which case the expected lifetime would be even smaller.

13.1 Decay of the gluino

If the gluino is heavy enough, it can decay via the strong interaction to quark plus squark. Neglecting intergenerational mixing, the possible two-body decays are:

$$\tilde{g} \rightarrow u\bar{\tilde{u}}_L, \ \bar{u}\tilde{u}_L, \ u\bar{\tilde{u}}_R, \ \bar{u}\tilde{u}_R, \tag{13.3a}$$

$$\rightarrow d\bar{\tilde{d}}_L, \ \bar{d}\tilde{d}_L, \ d\bar{\tilde{d}}_R, \ \bar{d}\tilde{d}_R, \tag{13.3b}$$

$$\rightarrow s\bar{\tilde{s}}_L, \ \bar{s}\tilde{s}_L, \ s\bar{\tilde{s}}_R, \ \bar{s}\tilde{s}_R, \tag{13.3c}$$

$$\rightarrow c\bar{\tilde{c}}_L, \ \bar{c}\tilde{c}_L, \ c\bar{\tilde{c}}_R, \ \bar{c}\tilde{c}_R, \tag{13.3d}$$

$$\rightarrow b\bar{\tilde{b}}_1, \ \bar{b}\tilde{b}_1, \ b\bar{\tilde{b}}_2, \ \bar{b}\tilde{b}_2, \quad \text{and} \tag{13.3e}$$

$$\rightarrow t\bar{\tilde{t}}_1, \ \bar{t}\tilde{t}_1, \ t\bar{\tilde{t}}_2, \ \bar{t}\tilde{t}_2. \tag{13.3f}$$

Each flavor combination must be separately calculated, since the different squark types will have different decay modes, and each decay chain can give rise to distinct final states and ensuing signatures. Unless they are kinematically suppressed these two-body decays generally dominate other decays. Their partial widths are given by (B.1a) and (B.1b) of Appendix B.

Since the gluino has only strong interactions, if these two-body decays to squarks are kinematically forbidden, then the gluino would dominantly decay to charginos and neutralinos via three-body decays mediated by virtual squarks. Again neglecting inter-generational mixing, the possible decays are,

$$\tilde{g} \rightarrow u\bar{u}\tilde{Z}_i, \ d\bar{d}\tilde{Z}_i, \ s\bar{s}\tilde{Z}_i, \ c\bar{c}\tilde{Z}_i, \ b\bar{b}\tilde{Z}_i, \ t\bar{t}\tilde{Z}_i, \tag{13.4a}$$

$$\rightarrow u\bar{d}\widetilde{W}_j^-, \ \bar{u}d\widetilde{W}_j^+, \ c\bar{s}\widetilde{W}_j^-, \ \bar{c}s\widetilde{W}_j^+, \ t\bar{b}\widetilde{W}_j^-, \ \bar{t}b\widetilde{W}_j^+, \tag{13.4b}$$

where $i = 1$–4 and $j = 1, \ 2$. Note that in all models with a neutralino LSP, the decays $\tilde{g} \rightarrow q\bar{q}\tilde{Z}_1$ are kinematically allowed ($q = u, d, s, c$). As an example calculation, we will illustrate gluino three-body decay to a pair of light quarks plus a chargino.

13.1.1 $\tilde{g} \rightarrow u\bar{d}\widetilde{W}_j$: a worked example

At leading order, the $\tilde{g} \rightarrow u\bar{d}\widetilde{W}_j$ decay occurs via the Feynman diagrams shown in Fig. 13.1. The decay amplitude for diagram (1) is constructed from

$$\langle u_a\bar{d}_b\widetilde{W}_j|T\left[\left(-\sqrt{2}g_s(\mathrm{i})^{\theta_{\tilde{g}}}\bar{\tilde{u}}_L\bar{u}P_R\frac{\lambda_B}{2}\tilde{g}_B(x)\right)\cdot\left(\mathrm{i}A_{\widetilde{W}_j}^d\tilde{u}_L^\dagger\overline{\widetilde{W}}_iP_Ld(y)\right)\right]|\tilde{g}_A\rangle,$$

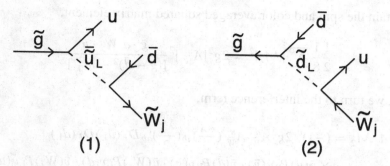

Figure 13.1 Feynman diagrams contributing to the decay $\tilde{g} \to u\bar{d}\widetilde{W}_j$.

where a, b, and A denote the color indices of the final state quarks and the decaying gluino. The matrix element can be evaluated as described in the last chapter. The external particles can be reduced using the creation/annihilation operators; again the exponential wave function factors lead to momentum conservation at each vertex. Finally, the $\tilde{u}_L$ and $\tilde{u}_L^\dagger$ fields contract together to yield a $\tilde{u}_L$ propagator factor $D_F(\tilde{u}_L) = \frac{1}{(\tilde{g}-u)^2 - m_{\tilde{u}_L}^2}$. Following the steps detailed in Chapter 12, we omit irrelevant factors of i, and find that the matrix element is given by,

$$\mathcal{M}_1 = -i(i)^{\theta_{\tilde{g}}} \sqrt{2} g_s A_{\widetilde{W}_j}^d \frac{\lambda_{Aab}}{2} \bar{u}(u) P_R u(\tilde{g}) \cdot D_F(\tilde{u}_L) \cdot \bar{u}(\widetilde{W}_j) P_L v(\bar{d}). \tag{13.5a}$$

The sum and average over color indices yields a factor

$$\frac{1}{8} \sum_A \frac{\lambda_{Aab}}{2} \frac{\lambda_{Aab}^*}{2} = \frac{1}{8} \text{Tr} \frac{\lambda_A}{2} \frac{\lambda_A}{2} = \frac{1}{8} \frac{1}{2} \delta_{AA} = \frac{1}{2},$$

where in the second step we have used the Hermiticity of the $SU(3)$ generators. Using usual trace techniques, the sum and average over colors and spins then yields the squared matrix element,

$$\frac{1}{2} \frac{1}{8} \sum |\mathcal{M}_1|^2 = 2 g_s^2 |A_{\widetilde{W}_j}^d|^2 \frac{\tilde{g} \cdot u \; \widetilde{W}_j \cdot \bar{d}}{[(\tilde{g} - u)^2 - m_{\tilde{u}_L}^2]^2}. \tag{13.5b}$$

A similar calculation for diagram (2) yields the matrix element,

$$\mathcal{M}_2 = -i(-i)^{\theta_{\tilde{g}}} \sqrt{2} g_s A_{\widetilde{W}_j}^{u*} \frac{\lambda_{Aab}}{2} \bar{v}(\tilde{g}) P_L v(\bar{d}) \cdot D_F(\tilde{d}_L) \cdot \bar{u}(u) P_R v(\widetilde{W}_j), \tag{13.6a}$$

where the chargino is treated as an antiparticle since its interaction with the down squark is written in terms of the field $\widetilde{W}_j^c$. This explains the direction of the arrow on the chargino line in diagram (2) of Fig. 13.1; we will leave it to the reader to check the reason for the reversal of the corresponding arrow on the gluino line. We then obtain the spin and color averaged squared matrix element,

$$\frac{1}{2}\frac{1}{8}\sum|\mathcal{M}_2|^2 = 2g_s^2|A_{\widetilde{W}_j}^u|^2\frac{\tilde{g}\cdot\bar{d}\ \widetilde{W}_j\cdot u}{[(\tilde{g}-\bar{d})^2 - m_{\bar{d}_L}^2]^2}. \tag{13.6b}$$

Finally, we turn to the interference term,

$$\mathcal{M}_1\mathcal{M}_2^\dagger = (-1)^{\theta_{\tilde{g}}}2g_s^2 A_{\widetilde{W}_j}^d A_{\widetilde{W}_j}^u (\frac{\lambda_A}{2})_{ab}(\frac{\lambda_A^*}{2})_{ab}D_F(\tilde{u}_L)D_F(\tilde{d}_L)$$
$$\times \bar{u}(u)P_Ru(\tilde{g})\cdot\bar{v}(\bar{d})P_Rv(\tilde{g})\cdot\bar{u}(\widetilde{W}_j)P_Lv(\bar{d})\cdot\bar{v}(\widetilde{W}_j)P_Lu(u).$$

Just as in the evaluation of the interference term following (12.4c), we find a mismatch between spinors involving the $\tilde{g}$ and also the $\widetilde{W}_j$. As before, this can be rectified using the relations $u = C\bar{v}^T$ and $v = C\bar{u}^T$, which yield:

$$\bar{v}(\bar{d})P_Rv(\tilde{g}) = u^T(\bar{d})C P_RC\bar{u}^T(\tilde{g}) = -\bar{u}(\tilde{g})P_Ru(\bar{d}), \quad\text{and}$$
$$\bar{u}(\widetilde{W}_j)P_Lv(\bar{d}) = v^T(\widetilde{W}_j)C P_LC\bar{u}^T(\bar{d}) = -\bar{u}(\bar{d})P_Lv(\widetilde{W}_j).$$

Then the spin and color summed and averaged interference term becomes,

$$\frac{1}{8}\frac{1}{2}\sum(\mathcal{M}_1\mathcal{M}_2^\dagger + \text{c.c.}) = -\frac{2g_s^2(-1)^{\theta_{\tilde{g}}}m_{\tilde{g}}m_{\widetilde{W}_j}\text{Re}(A_{\widetilde{W}_j}^d A_{\widetilde{W}_j}^u)u\cdot\bar{d}}{[(\tilde{g}-u)^2 - m_{\tilde{u}_L}^2][(\tilde{g}-\bar{d})^2 - m_{\bar{d}_L}^2]}. \tag{13.6c}$$

The width for the decay $\tilde{g}\rightarrow u\bar{d}\widetilde{W}_j$ can now be obtained using (13.1a) and integrating over the entire phase space. To integrate $\sum|\mathcal{M}_1|^2$, we first re-write the dot product $\widetilde{W}_j\cdot\bar{d} = (Q^2 - m_{\widetilde{W}_j}^2 - m_d^2)/2$, with $Q = \widetilde{W}_j+\bar{d} = \tilde{g} - u$, so that the integrand is independent of $\widetilde{W}_j$ and $\bar{d}$. The integration over the momenta of $\widetilde{W}_j$ and $\bar{d}$ can be easily performed using the invariant scalar integral (13.2a) leaving just the integral over the three momentum of the u quark to be performed. It is most convenient to write the integrand in the rest frame of the gluino. The measure $d^3u/2E_u = 2\pi|\vec{p}_u|dE_u$ so that the contribution to the partial width from $|\mathcal{M}_1|^2$ is

$$\Gamma_{11} = \frac{\alpha_s|A_{\widetilde{W}_j}^d|^2}{16\pi^2}\psi(m_{\tilde{g}}, m_{\tilde{u}_L}, m_{\widetilde{W}_j}), \tag{13.7a}$$

where

$$\psi(m_{\tilde{g}}, m_{\tilde{q}}, m) = \int dE\frac{E^2(m_{\tilde{g}}^2 - 2m_{\tilde{g}}E - m^2)^2}{(m_{\tilde{g}}^2 - 2m_{\tilde{g}}E - m_{\tilde{q}}^2)^2(m_{\tilde{g}}^2 - 2m_{\tilde{g}}E)}, \tag{13.7b}$$

and where the limits of integration (neglecting the u quark mass) range from $E_{\min} = 0$ to $E_{\max} = (m_{\tilde{g}}^2 - m^2)/2m_{\tilde{g}}$. Similarly, integrating $\sum|\mathcal{M}_2|^2$ over the phase space gives,

$$\Gamma_{22} = \frac{\alpha_s |A^u_{\widetilde{W}_j}|^2}{16\pi^2} \psi(m_{\tilde{g}}, m_{\tilde{d}_L}, m_{\widetilde{W}_j}).\tag{13.7c}$$

Finally, we must integrate over the interference term. Since this term involves $\tilde{g} \cdot u$ and $\tilde{g} \cdot \bar{d}$ dot products in the propagator denominators, we cannot use covariant scalar, vector or tensor integrals in its evaluation. Instead, we will evaluate the three-body phase space integral directly. Toward this end, we write

$$\frac{d^3 \widetilde{W}_j}{2E_{\widetilde{W}_j}} = d^4 \widetilde{W}_j \theta(\widetilde{W}^0_j)\delta(\widetilde{W}^2_j - m^2_{\widetilde{W}_j}),$$

and use the energy–momentum conserving δ-function to integrate over the chargino four-momentum. Since $\widetilde{W}_j = \tilde{g} - u - \bar{d}$, the step function $\theta(\widetilde{W}^0_j)$ is just one (because of limits on the particle energies obtained below). The remaining integrand can then be written in the rest frame of the gluino with the u quark direction chosen as the z-axis. The δ-function that specifies the chargino to be on its mass shell can be then written as,

$$\delta\left[(\tilde{g} - u - \bar{d})^2 - m^2_{\widetilde{W}_j}\right] = \frac{1}{2E_u E_{\bar{d}}}\delta\left[1 - \cos\theta + \frac{m^2_{\tilde{g}} - m^2_{\widetilde{W}_j} - 2m_{\tilde{g}}(E_u + E_{\bar{d}})}{2E_u E_{\bar{d}}}\right],$$

where θ is the angle between the up and down quark momenta. Neglecting quark masses, it is now straightforward to see that

$$\int \frac{u \cdot \bar{d}}{[(\tilde{g} - u)^2 - m^2_{\tilde{u}_L}][(\tilde{g} - \bar{d})^2 - m^2_{\tilde{d}_L}]}\delta^4(\tilde{g} - \widetilde{W}_j - u - \bar{d})\frac{d^3u}{2E_u}\frac{d^3\bar{d}}{2E_{\bar{d}}}\frac{d^3\widetilde{W}_j}{2E_{\widetilde{W}_j}}$$

$$= \pi^2 \int \frac{u \cdot \bar{d}\, dE_u dE_{\bar{d}}}{[(\tilde{g} - u)^2 - m^2_{\tilde{u}_L}][(\tilde{g} - \bar{d})^2 - m^2_{\tilde{d}_L}]}$$

$$= -\frac{\pi^2}{2}\int \frac{dE_u}{(m^2_{\tilde{g}} - 2m_{\tilde{g}}E_u - m^2_{\tilde{u}_L})}\int dE_{\bar{d}}\left(1 + \frac{m^2_{\tilde{d}_L} - m^2_{\widetilde{W}_j} - 2m_{\tilde{g}}E_u}{m^2_{\tilde{g}} - 2m_{\tilde{g}}E_{\bar{d}} - m^2_{\tilde{d}_L}}\right).$$

The integration over $dE_{\bar{d}}$ is simple, once the limits of integration are determined (see the exercise below). We then find that the contribution to the width from the interference term takes the form,

$$\Gamma_{12} = \frac{-\alpha_s(-1)^{\theta_{\tilde{g}}}\mathrm{Re}(A^u_{\widetilde{W}_j} A^d_{\widetilde{W}_j})}{8\pi^2}\phi(m_{\tilde{g}}, m_{\tilde{u}_L}, m_{\tilde{d}_L}, m_{\widetilde{W}_j}),\tag{13.8a}$$

where

$$\phi(m_{\tilde{g}}, m_{\tilde{u}_L}, m_{\tilde{d}_L}, m)$$

$$= \frac{m}{2} \int \frac{dE_u}{m_{\tilde{g}}^2 - m_{\tilde{u}_L}^2 - 2m_{\tilde{g}}E_u} \left[\frac{-E_u(m_{\tilde{g}}^2 - m^2 - 2m_{\tilde{g}}E_u)}{m_{\tilde{g}}(m_{\tilde{g}} - 2E_u)} \right.$$

$$\left. - \frac{2m_{\tilde{g}}E_u - m_{\tilde{d}_L}^2 + m^2}{2m_{\tilde{g}}} \log \frac{m_{\tilde{d}_L}^2(m_{\tilde{g}} - 2E_u) - m_{\tilde{g}}m^2}{(m_{\tilde{g}} - 2E_u)(m_{\tilde{d}_L}^2 - 2m_{\tilde{g}}E_u - m^2)} \right], \quad (13.8b)$$

with the range of integration from 0 to $(m_{\tilde{g}}^2 - m^2)/2m_{\tilde{g}}$. The partial decay width is then given by

$$\Gamma(\tilde{g} \to u\bar{d}\widetilde{W}_j) = \Gamma_{11} + \Gamma_{22} + \Gamma_{12}. \quad (13.9)$$

By CP invariance, $\Gamma(\tilde{g} \to u\bar{d}\widetilde{W}_j^-) = \Gamma(\tilde{g} \to d\bar{u}\widetilde{W}_j^+)$. This will lead to an important signature for gluinos. Moreover, these partial widths are generation-independent as long as quark Yukawa interactions can be neglected. For decays to third generation quarks the calculation is considerably more complicated mainly because the higgsino components of the charginos also couple via Yukawa interactions. Moreover, intra-generational squark mixing and final state quark masses also need to be taken into account. The formula for this partial width is given in Section B.1.4 of Appendix B.

Exercise *The requirement that $|\cos\theta| \leq 1$ determines the limits on the energy of the down quark. Using the value of $\cos\theta$ given by the chargino mass shell δ-function, show that*

$$E_{\bar{d}}(\min) = (m_{\tilde{g}}^2 - m_{\widetilde{W}_j}^2 - 2m_{\tilde{g}}E_u)/2m_{\tilde{g}},$$

$$E_{\bar{d}}(\max) = (m_{\tilde{g}}^2 - m_{\widetilde{W}_j}^2 - 2m_{\tilde{g}}E_u)/2(m_{\tilde{g}} - 2E_u).$$

The limits on the up quark energy are even easier to determine. In the gluino rest frame, if u is produced at rest, then $E_u(\min) = 0$, while if $\bar{d}$ is produced at rest, then $E_u(\max) = (m_{\tilde{g}}^2 - m_{\widetilde{W}_j}^2)/2m_{\tilde{g}}$.

Work out how these limits are modified if quarks have non-zero masses. This is relevant for the decay $\tilde{g} \to t\bar{b}\widetilde{W}_j$.

13.1.2 Other gluino decays

We have already mentioned that the decays $\tilde{g} \to c\bar{s}\widetilde{W}_j^-$ and $\tilde{g} \to t\bar{b}\widetilde{W}_j^-$ (along with the corresponding CP conjugate decays) may also occur. If the latter decay is kinematically unsuppressed, its partial width may be considerably larger than that

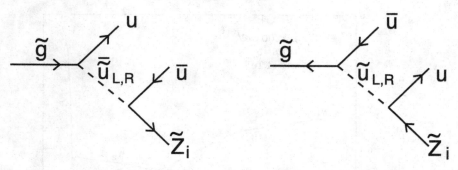

Figure 13.2 Feynman diagrams contributing to the decay $\tilde{g} \to u\bar{u}\tilde{Z}_i$. Decays to other flavors of squarks occur via similar diagrams.

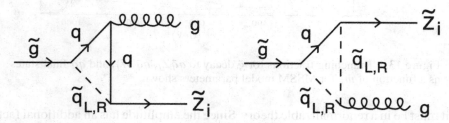

Figure 13.3 Feynman diagrams contributing to the decay $\tilde{g} \to g\tilde{Z}_i$. Since the gluino and the neutralino are Majorana particles, these same diagrams but with reversed arrows also contribute to the amplitude. This corresponds to distinct contractions in the evaluation of the decay matrix element.

for three-body decays to light squarks. This occurs in part because the top (and for large $\tan\beta$, bottom) quark Yukawa couplings are large, and also because in all models where squark mass parameters are (roughly) equal at some high scale, the physical masses of light bottom and top squarks are significantly smaller than first and second generation squark masses.

Gluinos can also decay via three-body mode to neutralinos. The diagrams contributing to $\tilde{g} \to u\bar{u}\tilde{Z}_i$ are shown in Fig. 13.2. The calculation of the decay width is very similar to the one illustrated for $\tilde{g} \to u\bar{d}\tilde{W}_j$. For decays to massless quarks, the chiral structure of the interaction ensures that there is no interference term between diagrams involving left- and right-squark exchange. The corresponding partial width is given by Eq. (B.4). The decays $\tilde{g} \to b\bar{b}\tilde{Z}_i$ and $\tilde{g} \to t\bar{t}\tilde{Z}_i$ may also occur. Once again, the evaluation of these partial widths is complicated because Yukawa couplings, squark mixing, and quark masses have all to be included. The relevant formulae can be found in Section B.1.3 of Appendix B.

It is also possible for the gluino to decay via loop diagrams as $\tilde{g} \to g\tilde{Z}_i$, as shown in Fig. 13.3. Each diagram is separately divergent but the summed amplitude is finite

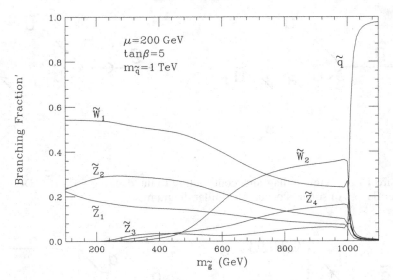

Figure 13.4 Branching fractions for $\tilde{g}$ decay to $q\bar{q}\widetilde{Z}_i$, $q\bar{q}'\widetilde{W}_j$, and $q\bar{q}$ final states as a function of $m_{\tilde{g}}$ for MSSM model parameters shown.

as it must be in a renormalizable theory. Since the amplitude has an additional factor of the strong coupling relative to the amplitudes for tree-level three-body decays as well as a loop suppression factor, the partial width for this decay is usually smaller than that for three-body tree-level decays. However, in some regions of MSSM parameter space, this decay mode can be significant, since it can be enhanced by third generation Yukawa couplings, and suffers less kinematic suppression. We do not list the formula for this decay here but refer the reader to the literature.[4]

In Fig. 13.4, we show gluino branching ratios to charginos and neutralinos as a function of $m_{\tilde{g}}$, for degenerate soft SUSY breaking squark masses of $m_{\tilde{q}} = 1$ TeV, with $\mu = 200$ GeV, and $\tan\beta = 5$, in the MSSM with gaugino mass unification. Values of $m_{\tilde{g}} \lesssim 550$ GeV are excluded by the LEP constraint on the chargino mass. However, we should understand that this figure is for illustrative purposes only. Two-body gluino decays are kinematically forbidden over most of the range of $m_{\tilde{g}}$ in the figure. For low values of $m_{\tilde{g}}$, $\widetilde{Z}_1$, $\widetilde{Z}_2$, and $\widetilde{W}_1$ are all extremely light, and the gluino decays mainly via three-body modes into $q\bar{q}'\widetilde{W}_1$, $q\bar{q}\widetilde{Z}_1$, and $q\bar{q}\widetilde{Z}_2$. Moreover, we see that the branching fraction to the kinematically favored $q\bar{q}\widetilde{Z}_1$ mode is smaller than that for gluino decays to the heavier neutralino $\widetilde{Z}_2$ or to the chargino. The reason is that for low values of $m_{\tilde{g}}$, $2M_1 \simeq M_2 \simeq m_{\tilde{g}}/3 \ll \mu$, so that the lightest neutralino is dominantly a bino while $\widetilde{Z}_2$ and $\widetilde{W}_1$ are dominantly winos. Since the $SU(2)_L$ gauge coupling is larger than the hypercharge gauge coupling, decays to the bino-like LSP are dynamically suppressed. The partial width for the decay to a chargino is almost

[4] See e.g., H. Baer, X. Tata and J. Woodside, *Phys. Rev.* **D42**, 1568 (1990).

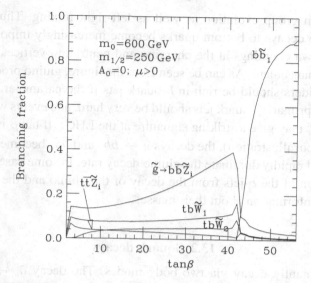

Figure 13.5 Branching fractions of the $\tilde{g}$ to final states involving third generation quarks versus $\tan\beta$ in the mSUGRA model. Also shown is the total branching fraction for decays to quarks of the first two generations.

twice that to $\tilde{Z}_2$; this is reasonable because there are two charged wino states and just one neutral wino state. Decays to heavy neutralinos and the heavier chargino (which are mainly higgsino-like) are kinematically and dynamically suppressed. As $m_{\tilde{g}}$ increases, decays to states including heavier charginos and neutralinos become possible. Ultimately, these dominate the branching fractions. This is because for fixed μ, M_1 and M_2 increase with $m_{\tilde{g}}$ so that for very heavy gluinos, it is the heavier chargino $\tilde{W}_2$ and the heavier neutralinos $\tilde{Z}_3$ and $\tilde{Z}_4$ that are mainly gaugino-like and so have large couplings to the quark–squark system: decays to the more higgsino-like $\tilde{W}_1$, $\tilde{Z}_1$, and $\tilde{Z}_2$, though kinematically favored are suppressed by mixing angles. That heavy gluinos decay to heavy charginos and neutralinos which subsequently decay to lighter charginos and neutralinos is quite a general feature of SUSY models. Of course, as we can see, if $m_{\tilde{g}} > (m_q + m_{\tilde{q}})$, then the two-body decays to quark plus squark become kinematically accessible and rapidly dominate the branching fraction. Since these occur via only strong interactions which are flavor independent, aside from mass effects, every flavor and type of squark will be democratically produced.

In Fig. 13.5, we show the $\tilde{g}$ branching fractions to states including third generation quarks, as a function of $\tan\beta$, in the mSUGRA model for $m_0 = 600$ GeV, $m_{1/2} = 250$ GeV, $A_0 = 0$, and $\mu > 0$. For small values of $\tan\beta$ just the top quark Yukawa coupling is important, but decays to t quarks are somewhat suppressed by phase space. As $\tan\beta$ increases, the magnitude of the bottom (and also tau) Yukawa coupling increases; as a result, $m_{\tilde{b}_1}$ is decreased both because of renormalization group evolution as well as left-right mixing. Thus as $\tan\beta$ increases, gluino decays to bottom quarks become increasingly important both due to direct Yukawa couplings at the chargino and neutralino vertices, as well as to propagator enhancement. As can be seen from the figure, gluino production events at hadron colliders should be rich in b-quark jets if the parameter $\tan\beta$ is large. Moreover, the primary b-quark jets should be very hard, and events with hard b-jets and large E_T^{miss} may give a striking signature at the LHC.[5] If $\tan\beta$ is large enough ($\tan\beta \gtrsim 42$ in our illustration), the decays $\tilde{g} \to b\bar{\tilde{b}}_1$ and $\bar{b}\tilde{b}_1$ become kinematically accessible, and rapidly dominate the gluino decay rate. In some cases, the momentum distribution of the b-jets from the decay of the gluino and the $\tilde{b}_1$ squark can even provide information about their masses.

13.2 Squark decays

Squarks dominantly decay via two-body modes. The decay $\tilde{q}_i \to q\tilde{Z}_1$ is kinematically accessible by assumption as long as the mass of the daughter quark is negligible. For the first generation, Yukawa couplings can be neglected and possible decay modes include,

$$\tilde{u}_L \to u\tilde{Z}_i,\ d\tilde{W}_j^+,\ u\tilde{g}, \tag{13.10a}$$

$$\tilde{d}_L \to d\tilde{Z}_i,\ u\tilde{W}_j^-,\ d\tilde{g}, \tag{13.10b}$$

$$\tilde{u}_R \to u\tilde{Z}_i,\ u\tilde{g}, \tag{13.10c}$$

$$\tilde{d}_R \to d\tilde{Z}_i,\ d\tilde{g}. \tag{13.10d}$$

Notice that right-squarks have no coupling to charginos, and so can only decay to $\tilde{g}$ or $\tilde{Z}_i$. The decay modes for $\tilde{c}_L$, $\tilde{c}_R$, $\tilde{s}_L$ and $\tilde{s}_R$ are similar. Unless they are kinematically suppressed, decays to gluinos dominate. Partial widths for two-body decays of squarks to gluinos, charginos, and neutralinos may be found in Appendix B.2.

For third generation squarks, squark mixing effects as well as non-negligible Yukawa couplings lead to more complicated decay patterns. Bottom squarks may decay via the following modes, if these are kinematically accessible:

$$\tilde{b}_{1,2} \to b\tilde{g},\ b\tilde{Z}_i,\ t\tilde{W}_j,\ W\tilde{t}_{1,2},\ H^-\tilde{t}_{1,2} \quad \text{and} \tag{13.11a}$$

$$\tilde{b}_2 \to Z\tilde{b}_1,\ h\tilde{b}_1,\ H\tilde{b}_1,\ A\tilde{b}_1. \tag{13.11b}$$

Unlike squarks of the first two generations, both light and heavy bottom squarks can potentially decay to charginos and W bosons, since they are mixtures of left- and

---

[5] For yet larger values of $m_{1/2}$ gluino decays to t-quarks are kinematically unsuppressed, and these serve as an additional source of b-jets.

right-squarks. Likewise, top squarks can decay via

$$\tilde{t}_{1,2} \to t\tilde{g}, \; t\tilde{Z}_i, \; b\tilde{W}_j, \; W\tilde{b}_{1,2}, \; H^+\tilde{b}_{1,2} \quad \text{and} \tag{13.12a}$$

$$\tilde{t}_2 \to Z\tilde{t}_1, \; h\tilde{t}_1, \; H\tilde{t}_1, \; A\tilde{t}_1. \tag{13.12b}$$

If top squarks are relatively light, they dominantly decay via $\tilde{t}_1 \to b\tilde{W}_1$, and possibly also via $\tilde{t}_1 \to t\tilde{Z}_1$. If both these modes are kinematically forbidden, then the $\tilde{t}_1$ can decay via usually suppressed modes

$$\tilde{t}_1 \to c\tilde{Z}_1, \; b\nu\tilde{\ell}_L, \; b\ell\tilde{\nu}_L, \; bW\tilde{Z}_1, \; \text{or} \quad bf\bar{f}'\tilde{Z}_1, \tag{13.13}$$

where f and $\bar{f}'$ are light SM fermions that couple to the W boson. The first of these decay modes can take place via the off-diagonal terms in the SUSY Lagrangian that give rise to flavor-violating interactions. Even if tree-level flavor-violating interactions are absent in the Lagrangian renormalized at high energy scales, radiative corrections can induce these at the weak scale, giving rise to the flavor-violating decay mode. We assume here that the decay $\tilde{t}_1 \to c\tilde{g}$ (which would be similarly induced) is kinematically forbidden. In models with universal squark masses at the high scale, it has been shown that the decay $\tilde{t}_1 \to c\tilde{Z}_1$ frequently dominates the four-body decay for rather light top squarks.[6] There are, however, regions of parameter space where the three- and even four-body decay modes can compete with, or even dominate, the $\tilde{t}_1 \to c\tilde{Z}_1$ decay.

In Fig. 13.6, we show the branching fractions for $\tilde{u}_L$ in the MSSM, for fixed values of $\mu = 200 \, \text{GeV}, m_{\tilde{g}} = 1000 \, \text{GeV}$, and $\tan\beta = 5$, versus $m_{\tilde{u}_L}$. Gaugino mass unification is also assumed. At a very low value of $m_{\tilde{u}_L}$, only the decay $\tilde{u}_L \to u\tilde{Z}_1$ is open, and hence dominates the branching fraction. As $m_{\tilde{u}_L}$ increases, new decay modes become accessible. In particular, when $\tilde{u}_L \to d\tilde{W}_1$ becomes accessible, it soon becomes dominant.[7] As $m_{\tilde{u}_L}$ increases even further, decays to the heavier charginos and neutralinos become kinematically accessible. Ultimately, decays to the $SU(2)_L$ gaugino-like $\tilde{W}_2$ and $\tilde{Z}_4$ dominate while decays to the higgsino-like $\tilde{Z}_3$ are dynamically suppressed. The heavy charginos and neutralinos will subsequently decay as described below so that heavy squarks, like heavy gluinos, will decay via a multi-step cascade that terminates in the LSP. Finally, at very high values of $m_{\tilde{u}_L}$, the decay to $u\tilde{g}$ becomes possible, and soon dominates the electroweak decays to charginos and neutralinos. Branching fractions for $\tilde{d}_L$ decays shown in Fig. 13.7 are qualitatively similar, except that the $\tilde{d}_L \to d\tilde{Z}_1$ decay is not as rapidly suppressed when other channels open up. Indeed the extremely rapid suppression of $\tilde{u}_L \to u\tilde{Z}_1$

---

[6] K. Hikasa and M. Kobayashi, *Phys. Rev.* **D36**, 724 (1987).

[7] To understand the branching fractions we note that, in this case, the neutralinos are fairly mixed with $\tilde{Z}_1$ dominantly bino-like, $\tilde{Z}_2$ equally mixed in all four components, $\tilde{Z}_3$ being essentially a higgsino, and $\tilde{Z}_4$ dominantly wino-like. The two charginos are substantial mixtures of gauginos and higgsinos, with $\tilde{W}_2$ being the more gaugino-like because $M_2 > \mu$.

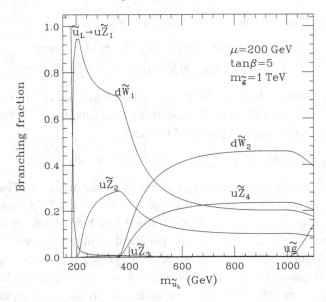

Figure 13.6 Branching fractions of the $\tilde{u}_L$ versus $m_{\tilde{u}_L}$ in the MSSM, assuming gaugino mass unification.

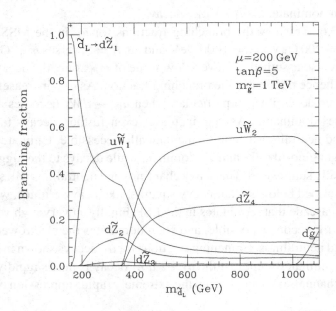

Figure 13.7 Branching fractions of the $\tilde{d}_L$ versus $m_{\tilde{d}_L}$ in the MSSM, assuming gaugino mass unification.

decays in Fig. 13.6 may be attributed to a somewhat accidental cancellation in the corresponding coupling.

In Fig. 13.8 and Fig. 13.9, we show branching fractions of the $\tilde{u}_R$ and $\tilde{d}_R$ squarks in the MSSM model versus the corresponding squark mass, for the same parameters as in Fig. 13.6. Since right-handed squarks are $SU(2)_L$ singlets, these can only decay to neutralinos, and (neglecting Yukawa couplings) only via their hypercharge gaugino components. The partial widths are, therefore, in the ratio of the corresponding $|v_4^{(i)}|^2$ for both types of squarks. Finally, for very high masses, the decay mode $\tilde{q}_R \to q\tilde{g}$ opens up, and soon dominates the branching fractions.

Exercise *Notice that the form of three boson couplings in Chapter 8 implies that the decays $\tilde{t}_2 \to \tilde{t}_1 Z$, $\tilde{t}_2 \to \tilde{b}_1 W$, and also $\tilde{t}_i \to \tilde{b}_j H^+$ (and the corresponding sbottom decays) may occur via gauge interactions. This would suggest that these decays may be relevant also for the first two generations of squarks. Verify that if Yukawa couplings can be ignored, the relevant coupling to Z vanishes, and further, that the decays of $\tilde{d}_L$ to W and $H^\pm$ bosons are kinematically forbidden assuming that $m_{\tilde{u}_L} + m_{\tilde{d}_L} > M_W$. Convince yourself that two-body decays to h, H, and A can only occur via Yukawa couplings.*

13.3 Slepton decays

First generation sleptons may decay via the following two-body modes, if kinematically allowed:

$$\tilde{e}_L \to e\tilde{Z}_i, \ v_e\tilde{W}_j^-, \tag{13.14a}$$

$$\tilde{v}_e \to v_e\tilde{Z}_i, \ e\tilde{W}_j^+, \tag{13.14b}$$

$$\tilde{e}_R \to e\tilde{Z}_i. \tag{13.14c}$$

Decays to W, Z and Higgs bosons are not possible for the same reasons as for first generation squarks. Smuons and muon sneutrinos have identical decay patterns and branching fractions as their first generation cousins. The partial widths for these decays are given by (B.53a)–(B.54b) of Appendix B.

We illustrate the branching fractions of the left-selectron, the right-selectron, and the sneutrino in Fig. 13.10, Fig. 13.11, and Fig. 13.12, respectively, as a function of the corresponding sparticle mass for the same MSSM parameters as Fig. 13.6. Except for the fact that these sleptons and sneutrinos never have two-body decays to gluinos, the decay patterns are qualitatively very similar to those of the corresponding squarks that we examined in the last section. In particular, while very light $SU(2)_L$ doublet sleptons $\tilde{e}_L$ and $\tilde{v}_e$ can only decay to the LSP, the branching

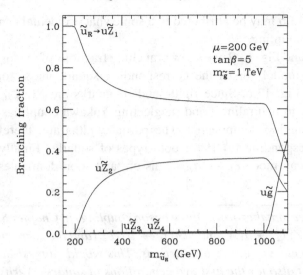

Figure 13.8 Branching fractions of the $\tilde{u}_\mathrm{R}$ versus $m_{\tilde{u}_\mathrm{R}}$ in the MSSM, assuming gaugino mass unification.

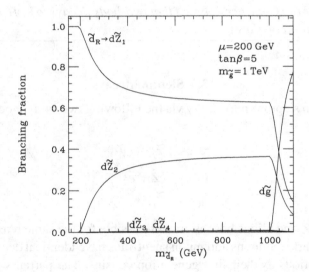

Figure 13.9 Branching fractions of the $\tilde{d}_\mathrm{R}$ versus $m_{\tilde{d}_\mathrm{R}}$ in the MSSM, assuming gaugino mass unification.

fractions for their decays to heavier charginos and neutralinos become dominant if these decays are not kinematically suppressed.[8] Thus a sneutrino heavier than the chargino is expected to have a significant branching fraction for visible decays.

---

[8] The strong suppression of the $\tilde{\nu}_e \rightarrow \tilde{Z}_2 \nu$ decay is accidental.

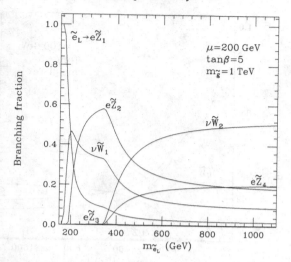

Figure 13.10 Branching fractions of the $\tilde{e}_L$ versus $m_{\tilde{e}_L}$ in the MSSM, assuming gaugino mass unification.

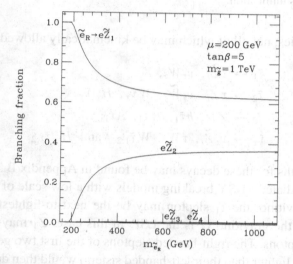

Figure 13.11 Branching fractions of the $\tilde{e}_R$ versus $m_{\tilde{e}_R}$ in the MSSM, assuming gaugino mass unification.

The right-selectron, like its squark cousin $\tilde{d}_R$, can only decay to neutralinos via the hypercharge gauge coupling: since $\widetilde{Z}_1$ has the largest bino component, this decay always dominates. As a result, $\tilde{\ell}_R$ pair production leads to events with opposite sign/same flavor dilepton pairs plus large missing energy.

Just as with third generation squarks, the decay possibilities of third generation sleptons are more complicated due to Yukawa coupling and mixing effects. The

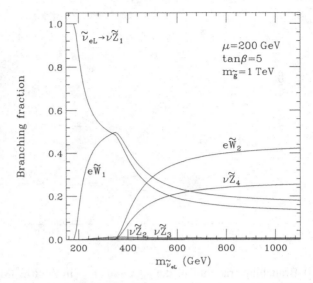

Figure 13.12 Branching fractions of the $\tilde{\nu}_{eL}$ versus $m_{\tilde{\nu}_{eL}}$ in the MSSM, assuming gaugino mass unification.

decay possibilities, not all of which may be kinematically allowed, include

$$\tilde{\tau}_1 \rightarrow \tau\tilde{Z}_i, \; \nu_\tau\tilde{W}_j, \tag{13.15a}$$

$$\tilde{\tau}_2 \rightarrow \tau\tilde{Z}_i, \; \nu_\tau\tilde{W}_j, \; W\tilde{\nu}_\tau, \; H^-\tilde{\nu}_\tau, \tag{13.15b}$$

$$\tilde{\tau}_2 \rightarrow Z\tilde{\tau}_1, \; h\tilde{\tau}_1, \; H\tilde{\tau}_1, \; A\tilde{\tau}_1, \tag{13.15c}$$

$$\tilde{\nu}_\tau \rightarrow \nu_\tau\tilde{Z}_i, \; \tau\tilde{W}_j, \; W\tilde{\tau}_{1,2} \quad \text{and} \quad H^+\tilde{\tau}_{1,2}. \tag{13.15d}$$

The partial widths for these decays may be found in Appendix B.3.

In gauge-mediated SUSY breaking models with a low scale of SUSY breaking and a light gravitino, the $\tilde{\tau}_1$ slepton may be the next-to-lightest SUSY particle (NLSP), while the gravitino $\tilde{G}$ is the LSP. In this case, $\tilde{Z}_1$ may be heavier than some of the sleptons. The right-handed sleptons of the first two generations (these would be much lighter than their left-handed sisters) would then dominantly decay via

$$\tilde{\ell}_R \rightarrow \tilde{\tau}_1^-\tau^+\ell \text{ and } \tilde{\ell}_R \rightarrow \tilde{\tau}_1^+\tau^-\ell \tag{13.16a}$$

mediated by neutralino exchange (recall that these couple to charginos only via tiny Yukawa couplings) which usually dominates the two-body decay $\tilde{\ell}_R \rightarrow \ell\tilde{G}$ (even for $C_{grav} = 1$) as long as these are not strongly suppressed by kinematics. For the case of the "co-NLSP" scenario where the sleptons of all three generations are almost degenerate, and $m_{\tilde{\ell}_1} - m_{\tilde{\tau}_1} < m_\tau$, the decay $\tilde{\ell}_1 \rightarrow \ell\tilde{G}$ dominates if $C_{grav} = 1$;

for larger values of C_{grav}, the decay

$$\tilde{\mu}_1 \to \tilde{\tau}_1 \bar{\nu}_\tau \nu_\mu \qquad (13.16b)$$

mediated by muon Yukawa couplings to a virtual chargino can potentially compete with the decay to the gravitino. In this case, the lifetimes of the NLSP may be large, and there may be displaced vertices or detectable charged sparticle tracks in the experimental apparatus.[9]

13.4 Chargino decays

Charginos decay only via electroweak interactions. They would dominantly decay via the following two-body modes if these are kinematically unsuppressed:

$$\widetilde{W}_j \to W\widetilde{Z}_i, \ H^-\widetilde{Z}_i, \qquad (13.17a)$$

$$\to \tilde{u}_L \bar{d}, \ \bar{\tilde{d}}_L u, \ \tilde{c}_L \bar{s}, \ \bar{\tilde{s}}_L c, \ \tilde{t}_{1,2} \bar{b}, \ \bar{\tilde{b}}_{1,2} t, \qquad (13.17b)$$

$$\to \tilde{\nu}_e \bar{e}, \ \bar{\tilde{e}}_L \nu_e, \ \tilde{\nu}_\mu \bar{\mu}, \ \bar{\tilde{\mu}}_L \nu_\mu, \ \tilde{\nu}_\tau \bar{\tau}, \ \bar{\tilde{\tau}}_{1,2} \nu_\tau, \ \text{and} \qquad (13.17c)$$

$$\widetilde{W}_2 \to Z\widetilde{W}_1, \ h\widetilde{W}_1, \ H\widetilde{W}_1, \ \text{and} \ A\widetilde{W}_1. \qquad (13.17d)$$

Partial widths for these decays are listed in Appendix B.5.1.

If all these modes are suppressed or forbidden (as may be the case for charginos in the mass range accessible to Tevatron searches), then three-body modes mediated by virtual bosons will dominate. Charginos may decay to a lighter neutralino via

$$\widetilde{W}_j \to \widetilde{Z}_i + f\bar{f}', \qquad (13.18a)$$

where f and $\bar{f}'$ are light SM fermions that couple to the W boson. For the lighter chargino, usually only the three-body decays to the $\widetilde{Z}_1$ are relevant. The heavy chargino may also decay via

$$\widetilde{W}_2 \to \widetilde{W}_1 f\bar{f} \qquad (13.18b)$$

as well.

Feynman diagrams for leading order contributions to $\widetilde{W}_1 \to e\bar{\nu}_e \widetilde{Z}_1$ decay are shown in Fig. 13.13. Three-body decays to other leptons or to quarks occur via analogous diagrams. For decays to the first two generations of fermions, Yukawa couplings, and hence also intragenerational sfermion mixings, are small; thus $\tilde{e}_R$ and H^+ exchange diagrams make negligible contributions. However, for $\widetilde{W}_1 \to \tau\bar{\nu}_\tau \widetilde{Z}_1$ decay, these contributions can be important if $\tan\beta$ is large. The partial width for the decay $\widetilde{W}_1 \to \tau\bar{\nu}_\tau \widetilde{Z}_1$ is given in Appendix B.5.2. The corresponding widths for

---

[9] For a discussion of three-body decays of sleptons, see S. Ambrosanio, G. Kribs and S. Martin, *Nucl. Phys.* **B516**, 55 (1998) and H. Baer, P. Mercadante, X. Tata and Y. Wang, *Phys. Rev.* **D60**, 055001 (1999).

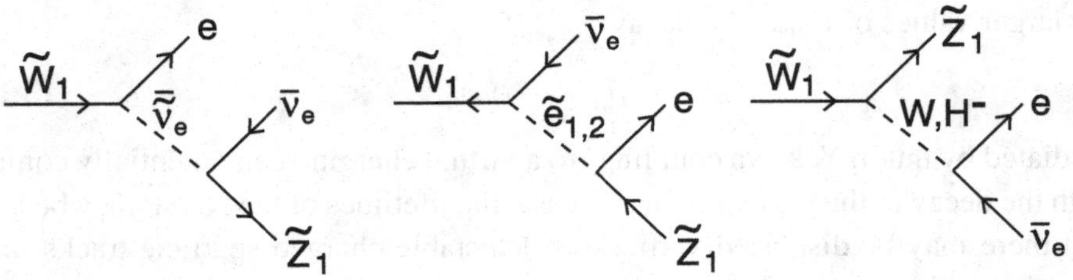

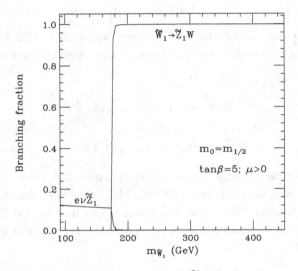

Figure 13.13 Feynman diagrams contributing to the decay $\widetilde{W}_1 \to e\bar{\nu}_e\widetilde{Z}_1$.

Figure 13.14 Branching fractions for decays of $\widetilde{W}_1$ versus $m_{\widetilde{W}_1}$ in the mSUGRA model. Below the threshold for $\widetilde{W}_1 \to W\widetilde{Z}_1$ decay, decays to other leptons families have essentially the same branching ratio as that for $\widetilde{W}_1 e\nu$. The rest of the time the chargino decays hadronically with these decays distributed essentially equally between the first two generations.

other decays can be obtained from this by setting the Yukawa coupling and the tau lepton mass to zero, and including appropriate color factors as spelled out there.

We illustrate the $\widetilde{W}_1$ decay branching ratio in Fig. 13.14 versus $m_{\widetilde{W}_1}$ for the mSUGRA model with parameters $m_0 = m_{1/2}$, $\tan\beta = 5$, $A_0 = 0$, and $\mu > 0$. In this case, squarks are much heavier than M_W and, except for the lowest values of the chargino mass, so are sleptons and sneutrinos. For $m_{\widetilde{W}_1} < M_W + m_{\widetilde{Z}_1}$ the amplitude for the decay is dominated by the virtual W boson exchange, resulting in a branching ratio $B(\widetilde{W}_1 \to \widetilde{Z}_1 f \bar{f}') \simeq B(W \to f \bar{f}')$, which is close to 11% for the decay $\widetilde{W}_1 \to \widetilde{Z}_1 e\nu$; the small increase in this branching for very low $m_{\widetilde{W}_1}$ values is due to contributions from slepton and sneutrino exchanges. Here, it is worth recalling the relative robustness of the $W\widetilde{W}_1\widetilde{Z}_1$ coupling that we mentioned below

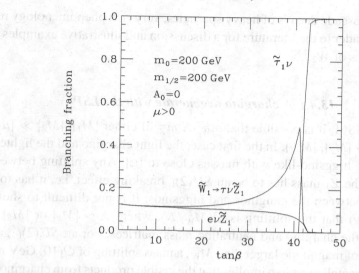

Figure 13.15 Branching fractions of the $\widetilde{W}_1$ versus $\tan\beta$ in the mSUGRA model with parameters as shown in the figure.

(8.103b): since this coupling is usually unsuppressed, the W exchange contribution tends to dominate chargino three-body decays if sfermions are heavy and $\tan\beta$ is not very large, so that chargino branching fractions to $\widetilde{Z}_1 f \bar{f}'$ are frequently close to those for $W \to f\bar{f}'$ decays. As $m_{\widetilde{W}_1}$ increases, the two-body mode $\widetilde{W}_1 \to W\widetilde{Z}_1$ opens up, and quickly dominates the branching fraction. The final state particles of the $\widetilde{W}_1$ decay (and the branching ratios) are the same as for low $m_{\widetilde{W}_1}$ values, but now the W boson is real instead of virtual.

The $\tan\beta$ dependence of the branching fractions of $\widetilde{W}_1$ is illustrated in Fig. 13.15 for the same mSUGRA model parameters as in the previous figure, but with $m_0 = m_{1/2} = 200$ GeV. For low values of $\tan\beta$, the chargino dominantly decays via $\widetilde{W}_1 \to f\bar{f}'\widetilde{Z}_1$ with branching fractions equal to those for $W \to f\bar{f}'$ as for the case of Fig. 13.14. As $\tan\beta$ increases, the τ Yukawa coupling grows, and the $\tilde{\tau}_1$ mass decreases due to Yukawa coupling contributions to RGE running, and due to non-negligible mixing effects. For $\tan\beta \sim 15$, the two branching fractions begin to separate and decays to τs become increasingly important; for large values of $\tan\beta$, contributions from the higgsino component of the chargino may also be relevant. The decay amplitude from the virtual $\tilde{\tau}_1$ Feynman diagram becomes comparable to and even larger than the virtual W contribution. For very large values of $\tan\beta$, the $\tilde{\tau}_1$ becomes so light that $\widetilde{W}_1 \to \tilde{\tau}_1\nu_\tau$ becomes accessible, and quickly dominates the branching fraction even though $\tilde{\tau}_1$ is dominantly $\tilde{\tau}_R$.

Heavy charginos usually decay via two-body modes. Their decay patterns are highly model and parameter-space dependent. The decay products of $\widetilde{W}_2$ frequently include W, Z, and Higgs bosons, and sometimes also sleptons. Indeed if $\widetilde{W}_2$s are produced via cascade decays of heavy sparticles, very rich phenomenology results. We refer the reader to the literature for a discussion and illustrative examples of the branching ratios of $\widetilde{W}_2$.[10]

13.4.1 A chargino degenerate with the LSP

Within the MSSM, it is possible that $m_{\widetilde{W}_1} \simeq m_{\widetilde{Z}_1}$ if either $|M_1|, |M_2| \gg |\mu|, M_W$ or $|M_1|, |\mu| \gg |M_2|, |M_W|$. In the first case, the light chargino and the lightest two neutralinos are higgsino-like with masses close to $|\mu|$. Any splitting between the chargino and the $\widetilde{Z}_1$ mass has to be an $SU(2)_L$ breaking effect, i.e. it has to come from mixing between the gauginos and higgsinos. It is not difficult to show (see exercise below) that the splitting is $\mathcal{O}(M_W^2/\Lambda)$, where $\Lambda \sim |M_1|$ or $|M_2|$ is the large scale in the chargino and neutralino mass matrices. For an $SU(2)_L$ gaugino mass an order of magnitude larger than M_W, a mass splitting of $\mathcal{O}(10)$ GeV may be expected. This small mass gap implies that the visible products from chargino decay will be rather soft compared to expectations in mSUGRA or mGMSB models, but the decay patterns of the charginos are qualitatively similar to those we have just discussed.[11]

In the second case where $|M_1|, |\mu| \gg |M_2|, M_W$, the $SU(2)_L$ gaugino would be lighter than the higgsinos or the hypercharge gauginos and in the absence of any gaugino–higgsino mixing we would expect that $\widetilde{Z}_1$ and $\widetilde{W}_1^{\pm}$ form a weak isotriplet with a mass $|M_2|$. The degeneracy again should not be surprising because any mass splitting between the charged and neutral winos has to be an $SU(2)_L$ breaking effect and, at tree level, gaugino–higgsino mixing is the only source of $SU(2)_L$ breaking. It is tedious but straightforward to show that in this case the tree-level mass splitting between the chargino and neutralino is $\mathcal{O}(M_W^4/\Lambda^3)$ where $\Lambda \sim |M_1|$ or $|\mu|$ is the large scale in the mass matrices. For an order of magnitude hierarchy between M_W and Λ, this corresponds to a sub-GeV mass gap. Then, the contribution to the mass splitting from radiative corrections can potentially be comparable to or even much larger than the tree-level splitting. These corrections have been evaluated,[12] and it has been shown that radiative corrections make the dominant contribution to the mass gap within the minimal anomaly-mediated SUSY breaking (AMSB) model which provides an example of just such a chargino–neutralino spectrum. Detailed calculation shows that the chargino–neutralino mass gap is typically 160–250 MeV. Fortunately, $m_{\widetilde{W}_1} > m_{\widetilde{Z}_1}$ so that the LSP is still neutral.

---

[10] See, e.g., H. Baer, A. Bartl, D. Karatas, W. Majerotto and X. Tata, *Int. J. Mod. Phys.* **A4**, 4111 (1989).

[11] Although $|\mu|$ is generically large within the mSUGRA framework, the recent determination of the relic dark matter density by the WMAP collaboration prefers selected regions of mSUGRA parameter space: in one of these regions, dubbed the hyperbolic branch region, $|\mu|$ may be much smaller than the gaugino masses.

[12] See e.g. D. Pierce and A. Papadopoulos, *Nucl. Phys.* **B430**, 278 (1994).

For such a small splitting, chargino decays are qualitatively altered from our discussion above. If $m_{\widetilde{W}_1} - m_{\widetilde{Z}_1} < m_\pi$, hadronic decays of the chargino are kinematically forbidden, and the chargino would dominantly decay via $\widetilde{W}_1^- \to e\nu\widetilde{Z}_1$, the mode with the largest phase space. The chargino could be rather long lived and could traverse a considerable distance before decaying, so there would be a charged particle track with a kink in the detector. If the decay $\widetilde{W}_1^- \to \pi^-\widetilde{Z}_1$ is allowed, the chargino decay length would be only a few centimeters, and the chargino track would then be more difficult to identify. For yet larger mass gaps, multi-pion decays would become possible and the lifetime would be even shorter.[13]

Exercise *For the case where the magnitude of the gaugino masses is much larger than $|\mu|$ or M_W, show that the eigenvalues of the neutralino mass matrix shift by:*

$$\mu \to \mu + \frac{1}{2}M_W^2(1 - \sin 2\beta)\left[\frac{1}{M_2} + \frac{\tan^2\theta_W}{M_1}\right], \tag{13.19a}$$

$$-\mu \to -\mu + \frac{1}{2}M_W^2(1 + \sin 2\beta)\left[\frac{1}{M_2} + \frac{\tan^2\theta_W}{M_1}\right], \tag{13.19b}$$

while the chargino mass (for $\mu > 0$) is given by,

$$m_{\widetilde{W}_1} = \mu + \frac{M_W^2}{M_2}\sin 2\beta. \tag{13.19c}$$

Hint: To find the shift of the neutralino eigenvalues, write the neutralino mass matrix in the basis where the higgsino sub-matrix is diagonal, and then treat the off-diagonal entries of the neutralino mass matrix in the new basis using standard second order perturbation theory. The chargino mass may be obtained using (8.54).

13.5 Neutralino decays

Like charginos, neutralinos dominantly decay via the following two-body modes if these are kinematically accessible:

$$\widetilde{Z}_i \to W\widetilde{W}_j, \ H^+\widetilde{W}_j, \ Z\widetilde{Z}_{i'}, \ h\widetilde{Z}_{i'}, \ H\widetilde{Z}_{i'}, \ A\widetilde{Z}_{i'} \tag{13.20a}$$

$$\to \widetilde{q}_{L,R}\bar{q}, \ \bar{\widetilde{q}}_{L,R}q, \ \widetilde{\ell}_{L,R}\ell, \ \bar{\widetilde{\ell}}_{L,R}\ell, \ \widetilde{\nu}_\ell\bar{\nu}_\ell, \ \bar{\widetilde{\nu}}_\ell\nu_\ell. \tag{13.20b}$$

Here, $i, i' = 1$–4 with $i > i'$, and q and ℓ denote all possible quark and lepton flavors. The partial widths for these decays are listed in Appendix B.4.1.

---

[13] Formulae for $\widetilde{W}_1$ decay for a tiny $m_{\widetilde{W}_1} - m_{\widetilde{Z}_1}$ mass difference can be found in C. H. Chen, M. Drees and J. F. Gunion, *Phys. Rev.* **D55**, 330 (1997), (erratum-ibid. **60**, 039901,1999).

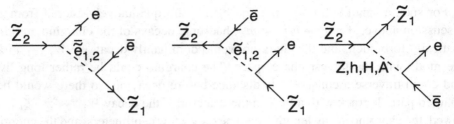

Figure 13.16 Feynman diagrams contributing to the decay $\widetilde{Z}_2 \to e\bar{e}\widetilde{Z}_1$.

If all these two-body modes are suppressed or kinematically forbidden, then the neutralino usually decays via

$$\widetilde{Z}_i \to \widetilde{Z}_{i'} + f\bar{f} \qquad (13.21a)$$

where f is a SM quark or lepton. The leading Feynman diagrams contributing to $\widetilde{Z}_2 \to e\bar{e}\widetilde{Z}_1$ decay at leading order are shown in Fig. 13.16, where $\tilde{e}_1$ and $\tilde{e}_2$ are selectron mass eigenstates (that essentially coincide with $\tilde{e}_R$ and $\tilde{e}_L$). Decays to other fermion flavors in (13.21a) as well as of other neutralinos occur via analogous diagrams. For decays to the first two generations, the three diagrams involving the Higgs bosons make a negligible contribution. The partial width for this decay is given in B.4.2. In addition, the three-body mode

$$\widetilde{Z}_i \to \widetilde{W}_j + f\bar{f}', \qquad (13.21b)$$

which occurs via diagrams analogous to those in Fig. 13.13 may also be relevant. Its partial width is given by Eq. (B.106) of Appendix B.

Neutralinos can also decay via

$$\widetilde{Z}_i \to \gamma\widetilde{Z}_{i'} \qquad (13.22)$$

at the one-loop level via diagrams involving charged sfermions/fermions and charginos/W or charged Higgs bosons in the loop. The branching fraction for this decay is usually small. However, it can be important if the widths of three-body modes are somehow suppressed. This suppression may occur either if one of the neutralinos is photino-like and the other higgsino-like since the photino (higgsino) does not couple to the Z boson (sfermion), or if both neutralinos are very close in mass because the strong three-body phase space suppression favors two-body decays. We do not list the partial width for this decay but will refer the interested reader to the original literature for this computation.[14]

[14] H. E. Haber and D. Wyler, *Nucl. Phys.* **B323**, 267 (1989); see also H. Baer and T. Krupovnickas, *JHEP* **0209**, 038 (2002).

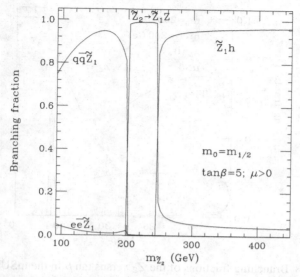

Figure 13.17 Branching fractions of the $\tilde{Z}_2$ versus $m_{\tilde{Z}_2}$ in the mSUGRA model. The branching ratios are almost generation independent for this low value of $\tan \beta$. The hadronic decays are summed over all quark flavors. Invisible decays make up the remainder of the branching fraction at low values of $m_{\tilde{Z}_1}$.

Since $\tilde{Z}_2$ is likely to be the most accessible visibly decaying neutralino, we show the branching fractions for its various decays in Fig. 13.17, assuming the mSUGRA model framework, and for the same model parameters as in Fig. 13.14. For low values of $m_{\tilde{Z}_2}$, the two-body decay modes are all inaccessible, and $\tilde{Z}_2$ mainly decays via three-body modes. If we compare these branching fractions to those for chargino decay in Fig. 13.14, we are immediately struck by the fact that while the branching fractions for chargino three-body decays were close to those for the W boson, the branching fractions for the neutralino decay $\tilde{Z}_2 \to \tilde{Z}_1 f \bar{f}$ differ considerably from those of $Z \to f \bar{f}$: i.e. even for sfermions considerably heavier than M_Z, the Z exchange graph does not dominate. This is because the couplings of Z to neutralinos are very sensitive to model parameters and, as we have discussed below (8.101), can be considerably suppressed. When this occurs, slepton exchange amplitudes remain important even for slepton masses of several hundred GeV. Over considerable regions of the MSSM parameter space, the leptonic three-body decays of $\tilde{Z}_2$ can be either enhanced or suppressed due to interference between scalar and Z boson exchange graphs, and neutralino branching fractions are quite different from those of the Z boson.[15] Neutralino decay patterns (and resulting signatures) are, therefore, much more sensitive to model parameters than those for chargino decays.

[15] For more details, see H. Baer and X. Tata, *Phys. Rev.* **D47**, 2739 (1993).

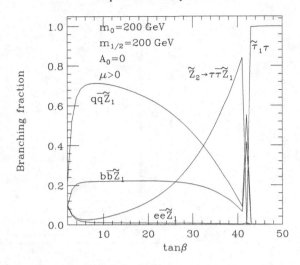

Figure 13.18 Branching fractions of the $\widetilde{Z}_2$ versus $\tan \beta$ in the mSUGRA model. Here $q = u, d, s, c$.

In Fig. 13.17, as $m_{\widetilde{Z}_2}$ increases, ultimately the two-body mode $\widetilde{Z}_2 \to Z\widetilde{Z}_1$ becomes accessible, and dominant. At even higher values of $m_{\widetilde{Z}_2}$, the decay mode $\widetilde{Z}_2 \to h\widetilde{Z}_1$ becomes accessible, and in this case quickly dominates. In SUSY particle cascade decays, we may expect an assortment of Higgs and vector bosons to be present.

In Fig. 13.18, we again show $\widetilde{Z}_2$ decay branching fractions in the mSUGRA model, but this time versus $\tan \beta$ and for the same parameters as in Fig. 13.15. At very low $\tan \beta$, $\widetilde{Z}_2$ decays via three-body modes with a large branching fraction into charged leptons. Decays into first, second, and third generation charged leptons occur at nearly the same rate. As $\tan \beta$ increases, the leptonic branching fraction drops and decays to quarks become increasingly dominant. The branching fraction into tau pairs begins diverging from that to electron (and muon) pairs around $\tan \beta \sim 5$. The decays to bottom quarks become more important relative to other hadronic decays but, in this example, decay to $\tau \bar{\tau} \widetilde{Z}_1$ becomes dominant for $\tan \beta \sim 30$, due to the enhanced tau lepton Yukawa coupling, and the gradual suppression of $m_{\widetilde{\tau}_1}$. Finally, around $\tan \beta \gtrsim 42$, two-body decays to $\widetilde{\tau}_1 \tau$ and $\bar{\widetilde{\tau}}_1 \tau$ turn on, and quickly dominate the branching fraction.

13.6 Decays of the Higgs bosons

Both the neutral and charged physical spin zero particles associated with the electroweak symmetry breaking sector dominantly decay via two-body modes into SM particles or, if they are heavy enough, also into lighter SUSY particles. The partial widths for the dominant tree level decays of Higgs bosons are listed in Appendix C.

13.6.1 Light scalar h

At tree level, the light scalar Higgs boson h can decay via the two-body modes,

$$h \to u\bar{u},\ d\bar{d},\ s\bar{s},\ c\bar{c},\ b\bar{b},\ e\bar{e},\ \mu\bar{\mu},\ \tau\bar{\tau}, \tag{13.23a}$$

$$h \to \widetilde{Z}_i \widetilde{Z}_{i'},\ \widetilde{W}_j^+ \widetilde{W}_{j'}^-,\ \tilde{f}\bar{\tilde{f}}, \tag{13.23b}$$

$$h \to AA \tag{13.23c}$$

where $i, i' = 1\text{–}4$ and $j, j' = 1, 2$. Since m_h is expected to be smaller than about 135 GeV within the MSSM framework with perturbative gauge couplings up to the GUT scale, its decays to $t\bar{t}$, W^+W^-, and ZZ are kinematically forbidden. Its decays to SUSY particles, possibly other than $\widetilde{Z}_1\widetilde{Z}_1$, are also expected to be suppressed. Over much of the parameter space, $h \to b\bar{b}$ decays dominate. For small to moderate values of $\tan\beta$, the bottom Yukawa coupling is small, and the h is narrow. In this case, especially the first of the three-body modes

$$h \to Wf\bar{f}'/Zf\bar{f} \tag{13.24}$$

may also be significant, particularly at the upper end of the m_h range. Since the h couples to mass, it dominantly decays to $b\bar{b}$ with a branching fraction of about 85%, and to $\tau\bar{\tau}$ pairs. The ratio of their branching ratios is fixed at tree level, but may be significantly affected by SUSY radiative corrections to the relation between the fermion mass and the corresponding Yukawa coupling. If neutralinos are light enough, h may also decay invisibly to $\widetilde{Z}_1\widetilde{Z}_1$. This decay, which occurs via gauge interactions, can potentially have a large branching fraction, although this is unlikely within constrained frameworks such as mSUGRA because of experimental limits on $m_{\widetilde{Z}_1}$.

Finally, h can also decay via

$$h \to gg,\ \gamma\gamma,\ Z\gamma, \tag{13.25}$$

through loops of gauge/Higgs sector fields and SM fermions, as well as their SUSY counterparts. Although the branching fractions for these decays are always suppressed by coupling and loop factors, the $h \to \gamma\gamma$ decay is an important search mode for LHC experiments which have excellent electromagnetic resolution. The $h \to \gamma\gamma$ branching fraction, which is $\mathcal{O}(10^{-3})$ for a SM-like h in the 100–120 GeV range, is enhanced for some ranges of SUSY parameters.[16]

---

[16] The h, H, and A can all decay via loop diagrams to $\gamma\gamma$ as well as to gg pairs. Formulae for these partial widths may be found in J. F. Gunion, H. E. Haber, G. Kane and S. Dawson, *The Higgs Hunter's Guide*, Addison-Wesley (1990); M. Bisset, U. of Hawaii thesis, UH-511-813-94 (1994).

13.6.2 Heavy scalar H

The heavy scalar Higgs boson H decays via the two-body modes

$$H \rightarrow u\bar{u},\ d\bar{d},\ s\bar{s},\ c\bar{c},\ b\bar{b},\ t\bar{t},\ e\bar{e},\ \mu\bar{\mu},\ \tau\bar{\tau}, \tag{13.26a}$$

$$\rightarrow WW,\ ZZ \tag{13.26b}$$

$$\rightarrow \widetilde{Z}_i\widetilde{Z}_{i'},\ \widetilde{W}_j^+\widetilde{W}_{j'}^-,\ \tilde{f}\bar{\tilde{f}}, \tag{13.26c}$$

$$\rightarrow hh,\ AA,\ H^+H^-,\ AZ, \tag{13.26d}$$

$$\rightarrow gg,\ \gamma\gamma,\ Z\gamma, \tag{13.26e}$$

as well as to (usually strongly suppressed) three-body modes, as does the h. If $m_A \gtrsim 200$ GeV, h is essentially a SM Higgs boson, and decays of H to vector bosons are suppressed by a factor $\cos^2(\alpha + \beta)$ (see the exercise below). Hence, the heavy scalar usually decays to $t\bar{t}$, $b\bar{b}$, hh or SUSY particles. As $\tan\beta$ increases, decays to $b\bar{b}$ and $\tau\bar{\tau}$ are enhanced relative to decays to $t\bar{t}$. SUSY decay modes of interest include the invisible $H \rightarrow \widetilde{Z}_1\widetilde{Z}_1$ channel, $H \rightarrow \widetilde{W}_1\widetilde{W}_1$, and $H \rightarrow \widetilde{Z}_2\widetilde{Z}_2$. This last decay results in gold-plated four isolated lepton events with missing energy if both neutralinos decay via $\widetilde{Z}_2 \rightarrow \ell\bar{\ell}\widetilde{Z}_1$.

Exercise *Starting from Eq. (8.40b) verify that* $\tan\alpha \rightarrow \cot\beta$ *as* $m_A \rightarrow \infty$, *so that* $\cos(\alpha + \beta) \rightarrow 0$ *in the same limit.*

13.6.3 Pseudoscalar A

The pseudoscalar Higgs boson A can decay via

$$A \rightarrow u\bar{u},\ d\bar{d},\ s\bar{s},\ c\bar{c},\ b\bar{b},\ t\bar{t},\ e\bar{e},\ \mu\bar{\mu},\ \tau\bar{\tau}, \tag{13.27a}$$

$$\rightarrow \widetilde{Z}_i\widetilde{Z}_{i'},\ \widetilde{W}_j^+\widetilde{W}_{j'}^-,\ \tilde{f}\bar{\tilde{f}}, \tag{13.27b}$$

$$\rightarrow hZ, \tag{13.27c}$$

$$\rightarrow gg,\ \gamma\gamma. \tag{13.27d}$$

Since A does not couple to vector boson pairs at tree level, its dominant decays are to $t\bar{t}$ or $b\bar{b}$ and $\tau\bar{\tau}$, unless its decays to hZ or SUSY particles are accessible: if this is the case, these latter decays usually dominate.

We remark that if CP is violated in the Higgs sector, A would mix with h and H, and its decay patterns would be qualitatively altered.

13.6.4 Charged scalar $H^\pm$

The charged Higgs H^+ dominantly decays via

$$H^+ \rightarrow u\bar{d},\ c\bar{s},\ t\bar{b},\ \nu_e\bar{e},\ \nu_\mu\bar{\mu},\ \nu_\tau\bar{\tau}, \tag{13.28a}$$
$$\rightarrow \widetilde{Z}_i\widetilde{W}_j^+,\ \tilde{f}\bar{\tilde{f}}', \tag{13.28b}$$
$$\rightarrow hW. \tag{13.28c}$$

Notice that, within the MSSM, the decay $H^+ \rightarrow W^+Z^0$ is absent at tree level. Thus, it dominantly decays to $t\bar{b}$, unless decays to hW or SUSY particles are open. If $H^+ \rightarrow t\bar{b}$ decay is also kinematically forbidden, H^+ preferentially decays via $H^+ \rightarrow \tau^+\nu_\tau$. In this case, the daughter tau dominantly has the opposite helicity from taus produced in W boson decays.

13.7 Top quark decays to SUSY particles

The top quark may be heavy enough for it to be able to decay to SUSY particles. However, branching fractions for its SUSY decays cannot be too large, as this would lead to inconsistencies between experimental measurements that agree well with SM predictions of top quark production and decay properties. In addition to its SM decay mode,

$$t \rightarrow bW^+, \tag{13.29a}$$

the decays

$$t \rightarrow bH^+, \tag{13.29b}$$
$$\rightarrow \tilde{t}_{1,2}\widetilde{Z}_i,\ \tilde{b}_{1,2}\widetilde{W}_j \tag{13.29c}$$

are also possible within the MSSM framework. The decay mode $t \rightarrow bH^+$ would then usually be followed by $H^+ \rightarrow \nu_\tau\bar{\tau}$, so an enhanced production of τ leptons would occur in top quark production events. If $t \rightarrow \tilde{t}_1\widetilde{Z}_1$, followed by $\tilde{t}_1 \rightarrow b\widetilde{W}_1 \rightarrow bf\bar{f}'\widetilde{Z}_1$, then the visible top quark decay products might be the same as in the SM, but with reduced energies, since some energy is taken by the mass of $\widetilde{Z}_1$. Such a decay chain may be almost excluded if we assume gaugino mass unification, but may be allowed if $|M_1| \ll |M_2|$. Alternatively, if $t \rightarrow \tilde{t}_1\widetilde{Z}_1$ is followed by $\tilde{t}_1 \rightarrow c\widetilde{Z}_1$, then a top quark decay would lead to a charm jet with an energy sensitively dependent upon $m_{\widetilde{Z}_1}$ and $m_{\tilde{t}_1}$.

13.8 Decays to the gravitino/goldstino

If the gravitino is the LSP, sparticles can decay to it. If these decays proceed only via the usual gravitational coupling (as do the decays to gravitinos with helicities $\pm\frac{3}{2}$), they would be completely irrelevant for the purposes of collider physics. In our discussion of the GMSB model we saw, however, that the amplitudes for decays to the longitudinal components of the gravitino with helicities $\pm\frac{1}{2}$ are enhanced by a factor $E/m_{3/2}$ which is very large if the gravitino is superlight. In this case, sparticle decays to the longitudinal components of the gravitino, which is essentially the goldstino, may be relevant. The NLSP, of course, can only decay into the gravitino. The considerations of this section most directly apply to the GMSB model with a low SUSY breaking scale.

13.8.1 Interactions

The couplings of the gravitino to the fermion–sfermion and to the gauge boson–gaugino system are given by the last term of (10.57a) and the second term of (10.57b), respectively. With $G^i_j = \delta^i_j + \cdots$ and $f_{AB} = \delta_{AB} + \cdots$ (the ellipsis denotes possible non-minimal terms in these), we find that these couplings can be written as,

$$\mathcal{L} \ni \frac{\mathrm{i}}{\sqrt{2}M_\mathrm{P}} \bar{\psi}_\mu \not{D} S^{i\dagger} \gamma^\mu \psi_{iL} + \frac{1}{8M_\mathrm{P}} \bar{\lambda}_A \gamma^\rho \sigma^{\mu\nu} \psi_\rho F_{A\mu\nu} + \quad \text{h.c.,} \qquad (13.30a)$$

where we have inserted the appropriate factors of M_P.

In principle, these couplings allow us to evaluate rates for sparticle decays to gravitinos. However, because of the unfamiliarity with manipulating the vector–spinor wave functions of spin $\frac{3}{2}$ particles, it is convenient to work only with the familiar spin $\frac{1}{2}$ goldstino that has been dynamically rearranged by the super-Higgs mechanism, and now forms the helicity $\pm\frac{1}{2}$ components of the gravitino. Then, just as W and Z interactions at high energies can be approximated by the interactions of their longitudinal components (the Goldstone bosons), so too can gravitino interactions be approximated by the interactions of the goldstino fields which they have absorbed by the super-Higgs mechanism.[17] But, we have already obtained the coupling of the goldstino to the chiral supermultiplet. Comparing the first term of (13.30a) with the goldstino coupling in (7.28), we see that the gravitino field can, in the high energy limit, be well approximated by

$$\psi_\mu \to \sqrt{\frac{2}{3}} \frac{1}{m_{3/2}} \partial_\mu \tilde{G}, \qquad (13.30b)$$

---

[17] The goldstino–gravitino equivalence, which was formally established by R. Casalbuoni *et al.*, *Phys. Lett.* **B215**, 313 (1988), ought to be an excellent approximation for decays of 100 GeV sparticles into eV, keV or even GeV scale gravitinos.

where we have used (10.67a) to eliminate the auxiliary field VEV in favor of the gravitino mass, and denoted the goldstino field (previously denoted by ψ_g) by $\tilde{G}$.[18] With this substitution, the interaction Lagrangian (13.30a) becomes

$$\mathcal{L} \ni \sqrt{\frac{2}{3}} \frac{1}{M_P m_{3/2}} \left[\frac{1}{8} \bar{\lambda}_A \gamma^\rho \sigma^{\mu\nu} (\partial_\rho \tilde{G}) F_{A\mu\nu} - \frac{i}{\sqrt{2}} \bar{\psi}_{iL} \gamma^\mu \not{D} S \partial_\mu \tilde{G} \right] + \quad \text{h.c.}$$

(13.30c)

The first term in (13.30c) clearly contains the coupling of the goldstino (or equivalently, helicity $\pm\frac{1}{2}$ gravitinos in the high energy limit) to gauginos and gauge bosons, while the second contains the corresponding couplings to the sfermion–fermion or the Higgs boson–higgsino pairs. Note, however, that when the Higgs fields are set equal to their VEV, even the second term contains (via the gauge covariant derivative) couplings of the goldstino to the vector boson–higgsino pair.[19]

These couplings can be used to obtain the interactions that are dominantly responsible for the decays $\tilde{Z}_i \rightarrow \gamma \tilde{G}$ (from the first term alone) or $\tilde{Z}_i \rightarrow Z\tilde{G}$, as well as the interactions that lead to the decay $\tilde{W}_i \rightarrow W\tilde{G}$. The second term yields interactions that lead to the decay of a neutralino (chargino) into a neutral (charged) Higgs boson and a gravitino, as well as to sfermion decays, $\tilde{f}_{1,2} \rightarrow f\tilde{G}$. Usually the branching fraction for these gravitino decay modes is significant only for the decay of the NLSP, with the gravitino being the LSP, as is the case in GMSB models with a low SUSY breaking scale.

To evaluate the couplings responsible for $\tilde{Z}_i \rightarrow \gamma \tilde{G}$, we write out the first term in (13.30c) for the neutral $U(1)_Y$ and neutral $SU(2)_L$ gauge and gaugino fields:

$$\mathcal{L} \ni \sqrt{\frac{2}{3}} \frac{1}{8 M_P m_{3/2}} \left[\bar{\lambda}_0 \gamma^\rho \sigma^{\mu\nu} \partial_\rho \tilde{G} (\partial_\mu B_\nu - \partial_\nu B_\mu) \right.$$
$$\left. + \bar{\lambda}_3 \gamma^\rho \sigma^{\mu\nu} \partial_\rho \tilde{G} (\partial_\mu W_{3\nu} - \partial_\nu W_{3\mu}) \right] + \quad \text{h.c.},$$

and substitute $B_\mu = \sin\theta_W Z_\mu + \cos\theta_W A_\mu$, $W_{3\mu} = \sin\theta_W A_\mu - \cos\theta_W Z_\mu$, $\lambda_0 = \sum_i v_4^{(i)} (i\gamma_5)^{\theta_i} \tilde{Z}_1$, and $\lambda_3 = \sum_i v_3^{(i)} (i\gamma_5)^{\theta_i} \tilde{Z}_i$ to obtain

$$\mathcal{L}_{\tilde{Z}_i \gamma \tilde{G}} = \sqrt{\frac{2}{3}} \frac{1}{4 M_P m_{3/2}} (v_4^{(i)} \cos\theta_W + v_3^{(i)} \sin\theta_W) \bar{\tilde{Z}}_i (i\gamma_5)^{\theta_i} \gamma^\rho \sigma^{\mu\nu} \partial_\rho \tilde{G} (\overset{\leftrightarrow}{\partial}_\mu A_\nu).$$

(13.31a)

In arriving at this we have used the fact that the Majorana properties of the goldstino and neutralinos imply that the Hermitian conjugate term is identical to the original term, accounting for a factor 2. For the $\tilde{Z}_i Z\tilde{G}$ interaction, both terms in Eq. (13.30c)

---

[18] This was first pointed out by P. Fayet, *Phys. Lett.* **B70**, 461 (1977).

[19] We will leave it to the reader to check that this contribution vanishes for the photon as it must since the VEVs leave the electromagnetic gauge invariance unbroken.

contribute, and the coupling is given by

$$
\mathcal{L}_{\tilde{Z}_i Z \tilde{G}} = \sqrt{\frac{2}{3}} \frac{1}{4 M_P m_{3/2}} \left[(v_4^{(i)} \sin \theta_W - v_3^{(i)} \cos \theta_W) \overline{\tilde{Z}}_i (i\gamma_5)^{\theta_i} \gamma^\rho \sigma^{\mu\nu} \partial_\rho \tilde{G} \overset{\leftrightarrow}{\partial}_\mu Z_\nu \right.
$$
$$
\left. + 2 M_Z (i)^{\theta_i} (\sin \beta v_1^{(i)} - \cos \beta v_2^{(i)}) \overline{\tilde{Z}}_i \Gamma \gamma^\mu \gamma^\nu \partial_\mu \tilde{G} Z_\nu \right], \tag{13.31b}
$$

where $\Gamma = 1 \ (\gamma_5)$ for $\theta_i = 0 \ (1)$.

The couplings of neutralinos to the goldstino and neutral Higgs bosons can be worked out from the second term in (13.30c) by substituting the higgsinos and the Higgs fields with definite hypercharges in terms of the corresponding mass eigenstate fields. We then find the neutralino–Higgs boson–goldstino interactions:

$$
\mathcal{L}_{\tilde{Z}_i \phi \tilde{G}} = \kappa_\phi \overline{\tilde{Z}}_i \frac{1 + \gamma_5}{2} \gamma^\mu \gamma^\nu \partial_\mu \tilde{G} \partial_\nu \phi + \quad \text{h.c.}, \tag{13.32a}
$$

where $\phi = h, \ H,$ and A, and

$$
\kappa_h = -\frac{(i)^{\theta_i + 1}}{\sqrt{6} M_P m_{3/2}} [v_1^{(i)} \cos \alpha + v_2^{(i)} \sin \alpha], \tag{13.32b}
$$

$$
\kappa_H = -\frac{(i)^{\theta_i + 1}}{\sqrt{6} M_P m_{3/2}} [-v_1^{(i)} \sin \alpha + v_2^{(i)} \cos \alpha], \quad \text{and} \tag{13.32c}
$$

$$
\kappa_A = -\frac{(i)^{\theta_i + 2}}{\sqrt{6} M_P m_{3/2}} [v_1^{(i)} \cos \beta + v_2^{(i)} \sin \beta]. \tag{13.32d}
$$

Exercise *Using the Majorana properties of the neutralino and goldstino fields, verify that these couplings can be rewritten as,*

$$
\mathcal{L}_{\tilde{Z}_i \phi \tilde{G}} = \overline{\tilde{Z}}_i \left[\frac{\kappa_\phi + \kappa_\phi^*}{2} + \frac{\kappa_\phi - \kappa_\phi^*}{2} \gamma_5 \right] \partial_\mu \tilde{G} \partial_\nu \phi. \tag{13.33}
$$

Notice that because κ_ϕ is either real or imaginary, the interaction is either scalar or pseudoscalar. This form of the coupling is, therefore, more convenient for evaluating the partial widths for the decays $\tilde{Z}_i \to \phi \tilde{G}$.

Finally, the last term in (13.30c) also gives the couplings of the goldstino to fermion–sfermion pairs. These can be written as

$$
\mathcal{L}_{f\tilde{f}\tilde{G}} = -\frac{i}{\sqrt{3}} \frac{1}{M_P m_{3/2}} \left[\bar{\psi}_f \frac{1 + \gamma_5}{2} \gamma^\mu \gamma^\nu \partial_\nu \tilde{f}_L + \bar{\psi}_{F^c} \frac{1 + \gamma_5}{2} \gamma^\mu \gamma^\nu \partial_\nu \tilde{f}_R^\dagger \right] \partial_\mu \tilde{G}
$$
$$
+ \quad \text{h.c.},
$$

where $\psi_f \ (\psi_{F^c})$ are, as usual, Majorana spinors whose left-handed components annihilate the left-handed $SU(2)_L$ doublet fermion, (left-handed $SU(2)_L$ singlet antifermion) and the SM Dirac fermion is given by

$$f = \frac{1 - \gamma_5}{2}\psi_f + \frac{1 + \gamma_5}{2}\psi_{F^c}.$$

Writing this Lagrangian with the Hermitian conjugate of the second term, and once again using the Majorana nature of the spinors we find that,

$$\mathcal{L}_{f\tilde{f}\tilde{G}} = -\frac{i}{\sqrt{3}}\frac{1}{M_P m_{3/2}}\left[\bar{f}\frac{1 + \gamma_5}{2}\gamma^\mu\gamma^\nu\partial_\nu\tilde{f}_L - \bar{f}\frac{1 - \gamma_5}{2}\gamma^\mu\gamma^\nu\partial_\nu\tilde{f}_R\right]\partial_\mu\tilde{G}$$
$$+ \quad \text{h.c.} \tag{13.34a}$$

Using this, we can readily obtain the goldstino interactions with the sfermion mass eigenstates,

$$\mathcal{L}_{f\tilde{f}_i\tilde{G}} = -\frac{i}{\sqrt{3}M_P m_{3/2}}\left[\bar{f}(\cos\theta_f P_R + \sin\theta_f P_L)\gamma^\mu\gamma^\nu\partial_\mu\tilde{G}\partial_\nu\tilde{f}_1\right.$$
$$\left. + \bar{f}(\sin\theta_f P_R - \cos\theta_f P_L)\gamma^\mu\gamma^\nu\partial_\mu\tilde{G}\partial_\nu\tilde{f}_2\right] + \quad \text{h.c.} \tag{13.34b}$$

The goldstino–tau–stau coupling leads to the dominant decay of the lighter stau in mGMSB models with the gravitino as the LSP and $\tilde{\tau}_1$ as the NLSP.

13.8.2 NLSP decay to a gravitino within the mGMSB model

Within the mGMSB framework, as we saw in Fig. 11.5 for the number of messenger generations $n_5 = 1$ and $\tan\beta$ not too large, the lightest neutralino tends to be the NLSP. Since gaugino masses scale with n_5 while scalar masses scale with $\sqrt{n_5}$, the lighter stau becomes the NLSP for larger values of n_5. If $\tan\beta$ is small to moderate, the tau Yukawa coupling is small and $\tilde{e}_R$ and $\tilde{\mu}_R$ are roughly degenerate with $\tilde{\tau}_1$, and we have the so-called co-NLSP scenario (region 3 of this figure).

The NLSP dominantly decays into a gravitino and a SM particle. It is straightforward to work out the partial widths for these two-body decays using the interactions presented in the last section. For a neutralino NLSP lighter than h or the Z boson, $\tilde{Z}_1 \to \gamma\tilde{G}$ is the only allowed two-body decay.

Exercise *Starting with the interaction in (13.31a), show that the width for the decay $\tilde{Z}_i \to \gamma\tilde{G}$ is given by,*

$$\Gamma(\tilde{Z}_i \to \gamma\tilde{G}) = \frac{(v_4^{(i)}\cos\theta_W + v_3^{(i)}\sin\theta_W)^2 m_{\tilde{Z}_i}^5}{48\pi m_{3/2}^2 M_P^2}. \tag{13.35}$$

Here, we have neglected the gravitino mass (except in the goldstino coupling, of course).

You may find it helpful to use the identity,

$$\gamma^\rho \sigma^{\mu\nu} = 2i(g^{\rho\mu}\gamma^\nu - g^{\rho\nu}\gamma^\mu) + \sigma^{\mu\nu}\gamma^\rho$$

and use $\tilde{G}_\mu \gamma^\mu u(\tilde{G}) = 0$ for the massless on-shell goldstino.

If $m_{\tilde{Z}_1}$ is large enough, its decays to Z as well as Higgs bosons may also be accessible. The partial widths for these two-body decays of the neutralino are listed in (B.67)–(B.69a) of Appendix B. Since this NLSP is mainly bino-like within the mGMSB model, it has large couplings to the hypercharge gauge boson, and as a result the decay $\tilde{Z}_1 \to \gamma\tilde{G}$ dominates the decay $\tilde{Z}_1 \to Z\tilde{G}$ for both dynamical as well as kinematic reasons. Decays to Higgs bosons are strongly suppressed. In non-minimal scenarios, the decays $\tilde{Z}_1 \to h\tilde{G}$ or $\tilde{Z}_1 \to Z\tilde{G}$ may be dominant.[20]

The $\tilde{Z}_1 \to \gamma\tilde{G}$ decay rate depends on $m_{3/2}$, which is independent of other sparticle masses. Recall that in the mGMSB framework, the gravitino mass, and hence the NLSP decay rate, is controlled by the parameter C_{grav}. If $m_{3/2}$ is large enough, then the $\tilde{Z}_1$ can be very long-lived. The mean decay length for a $\tilde{Z}_1$ with fractional velocity $\beta_{\tilde{Z}_1}$ is given by

$$d(\text{cm}) = \beta_{\tilde{Z}_1}\gamma_{\tilde{Z}_1}c\tau_{\tilde{Z}_1}$$

$$= \frac{10^{-2}}{(v_4^{(i)}\cos\theta_W + v_3^{(i)}\sin\theta_W)^2}(E^2/m_{\tilde{Z}_1}^2 - 1)^{1/2}\left(\frac{100\,\text{GeV}}{m_{\tilde{Z}_1}}\right)^5\left(\frac{\sqrt{\langle F\rangle}}{100\,\text{TeV}}\right)^4.$$

$$(13.36)$$

Remember that $\langle F\rangle$ is the true SUSY breaking scale (not the corresponding scale $\langle F_S\rangle$ in the messenger sector). For $m_{\tilde{Z}_1} \sim 100\,\text{GeV}$, the decay length mainly varies with the SUSY breaking scale $\langle F\rangle$ and can range from microns to kilometers and beyond, depending on $\langle F\rangle$. In a collider detector, the NLSP may have a decay vertex displaced from the interaction region, or may even decay outside of the detector. Thus, one of the signatures considered for GMSB models is the presence of hard isolated photons plus missing energy in collider events, where the photon induced EM shower may not point back to the interaction vertex. Indeed, a determination of the lifetime of the NLSP from its decay length distribution would yield the fundamental underlying SUSY scale. For this purpose, the higher order $\tilde{Z}_1 \to e^+e^-\tilde{G}$ decay may be more suitable for experimental reasons.

Finally, if the stau is the NLSP in the GMSB model, it would decay via,

$$\tilde{\tau}_1 \to \tau\tilde{G} \qquad (13.37)$$

---

[20] See, e.g., K. Matchev and S. Thomas, *Phys. Rev.* **D62**, 077702 (2000).

with a rate given by (B.60). If other flavors of sleptons are also only marginally lighter than the stau NLSP (region 3 of Fig. 11.5 where $m_{\tilde{\ell}_1} - m_{\tilde{\tau}_1} < m_\tau$), the decays (13.16a) are kinematically forbidden, and $\tilde{\ell}_1 \to \ell \tilde{G}$ or via (13.16b), depending on the value of C_{grav}. The rates for stau decays to gravitinos are comparable to the corresponding decay rate of a neutralino NLSP of the same mass. Hence, the charged NLSP might again be sufficiently long-lived, and (depending on its β) a highly ionizing track, terminating in a kink or a jet, may provide a characteristic signature.

14

Supersymmetric event generation

It is possible that the first indication of physics beyond the SM will come from indirect searches. These include direct or indirect detection of dark matter, $(g - 2)_\mu$, branching ratios (or event shapes) for various rare decays such as $B \to X_s\gamma$, $B \to X_s\ell^+\ell^-$, $B_s \to \ell^+\ell^-$ ($\ell = \mu, \tau$) or $\mu \to e\gamma$, or measurements of the electric dipole moment of the electron or the neutron. However, any such signal will likely be explainable by several new physics hypotheses, and not just supersymmetry. Thus, it is usually accepted that an unambiguous discovery of weak scale supersymmetry will have to occur at colliding beam experiments, where supersymmetric matter can be directly created, and the resultant scattering events can be scrutinized.

As we saw in Chapter 12 and Chapter 13, supersymmetric models can be used to predict various sparticle production rates and their subsequent decay patterns into final states containing quarks, leptons, photons, gluons (and LSPs in R-parity conserving models). However, quarks and gluons are never directly detected in any collider detector. Instead, detectors measure tracks of quasi-stable charged particles and their momenta as they bend in a magnetic field. They also measure energy deposited in calorimeter cells by hadrons, charged leptons, and photons. There is thus a gap between the predictions of supersymmetric models in terms of final states involving quarks, gluons, leptons and photons, and what is actually detected in the experimental apparatus. This gap is bridged by supersymmetric event generator computer programs. Once a collider type and supersymmetric model are specified, the event generator program can produce a complete simulation of the sorts of scattering events that are to be expected. The final state of any scattering event is composed entirely of electrons, muons, photons, and the long-lived hadrons (pions, kaons, nucleons, etc.) and their associated four-vectors that may be measured in a collider experiment.

The underlying idea of SUSY event generator programs is that for a specified collider type (e^+e^-, pp, $p\bar{p}$, ...) and center of mass energy, the event generator will, for any set of MSSM parameters, generate various sparticle pair production events in the ratio of their production cross sections, and with distributions as given by their differential cross sections discussed in Chapter 12. Moreover, the produced sparticles will undergo a (possibly multi-step cascade) decay into a partonic final state, according to branching ratios as fixed by the model.[1] Finally, this partonic final state is converted to one that is comprised of particles that are detected in an experimental apparatus. By generating a large number of "SUSY events" using these computer codes, the user can statistically simulate the various final states that are expected to be produced within the framework of any particular model. Although we have been focussing upon supersymmetry, we should mention that these programs also allow the user to simulate SM processes. This is essential for assessing SM backgrounds to new physics.

Several event generator programs that incorporate SUSY are currently available, including ISAJET, PYTHIA, HERWIG, and SUSYGEN. These include the $2 \rightarrow 2$ leading order SUSY production processes discussed in Chapter 12. In addition, specific $2 \rightarrow n$ ($n \leq 6$) SUSY reactions may be generated by such programs as CompHEP, Madgraph-II, and GRACE. The output of these latter programs must then be interfaced with one of the event generator programs to yield complete scattering event simulations. Ideally, event generator programs should be flexible enough to enable simulation of SUSY events from a variety of models such as mSUGRA, GMSB, etc. In other words, the user should be able to use the input parameters of these specific models (instead of the MSSM parameters) and generate the corresponding scattering events at any collider. In this way, different hypotheses about how MSSM superpartners obtain their masses may be directly tested by experiments at colliding beam facilities. In this connection, we also note that publicly available programs such as ISAJET, SPheno, SuSpect, and SOFTSUSY can be used to evaluate weak scale MSSM parameters and sparticle masses for several of the models that we have discussed in Chapter 11. Other than ISAJET, these programs do not generate sparticle production events, although the program SPheno will generate a table of sparticle decay branching fractions.

The simulation of hadron collider scattering events may be broken up into several steps, as illustrated in Fig. 14.1. The steps include:

- the perturbative calculation of the hard scattering subprocess in the parton model, and convolution with parton distribution functions (PDFs), as encapsulated by (12.1);
- inclusion of sparticle cascade decays;

---

[1] The user usually has the option to generate only a subset of SUSY production reactions or decays. This is useful if one wants to focus on a signal in a particular channel.

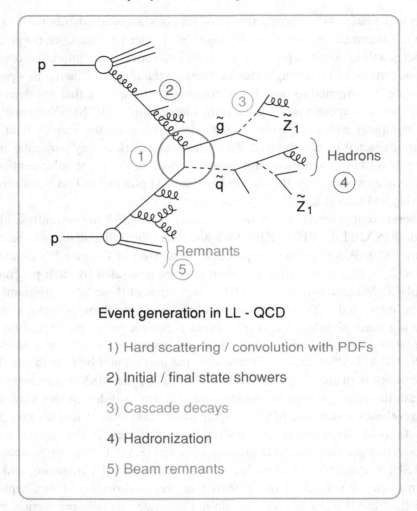

Event generation in LL - QCD

1) Hard scattering / convolution with PDFs

2) Initial / final state showers

3) Cascade decays

4) Hadronization

5) Beam remnants

Figure 14.1 Steps in any event generation procedure.

- implementation of perturbative parton showers for initial and final state colored particles, and for other colored particles which may be produced as decay products of heavier objects;
- implementation of a hadronization model which describes the formation of mesons and baryons from quarks and gluons. Also, unstable particles must be decayed to the (quasi-)stable daughters that are ultimately detected in the apparatus, with rates and distributions in accord with their measured or predicted values.
- Finally, the debris from the colored remnants of the initial beams must be modeled to obtain a valid description of physics in the forward regions of the collider detector.

Some of these steps are absent for simulations of electron–positron collisions which, as we saw in Chapter 12, are intrinsically simpler. However, for e^+e^- collider simulations, we have to allow for polarized initial beams.

In this chapter, we first briefly describe the physics involved in each of these steps. We then outline how this has been incorporated into some of the available event generator programs. Special attention is paid to the program ISAJET, since we have been involved with its development for describing supersymmetric processes.

14.1 Event generation

14.1.1 Hard scattering

The hard scattering and convolution with parton distributions forms the central calculation of event generator programs. The calculations are usually performed at lowest order in perturbation theory, so that the hard scattering is either a $2 \rightarrow 2$ or $2 \rightarrow 1$ scattering process.

For supersymmetric particle production at a high energy hadron collider such as the LHC, a large number of hard scattering subprocesses are likely to be kinematically accessible. Each subprocess reaction must be convoluted with parton distribution functions so that a total cross section for each reaction may be determined. The Q^2-dependent PDFs commonly used are constructed to be solutions of the Dokshitzer, Gribov, Lipatov, Altarelli, Parisi (DGLAP) QCD evolution equations, which account for multiple *collinear* emissions of quarks and gluons from the initial state in the leading log approximation. As Q^2 increases, more gluons are radiated, so that the distributions soften for large values of x, and correspondingly increase at small x values. Use of a running QCD coupling constant makes the entire calculation valid at leading log level.

Once the total cross sections are evaluated for all the allowed subprocesses, then reactions may be selected probabilistically (with an assigned weight) using a random number generator. This will yield sparticle events in the ratio predicted by the particular model being simulated.

For sparticle production at e^+e^- colliders, it may also be necessary to convolute with PDFs to incorporate bremsstrahlung and beamstrahlung effects as described in Chapter 12. In addition, if beam polarization is used, then each subprocess cross section will depend on beam polarization parameters as well.

14.1.2 Parton showers

For reactions occurring at both hadron and lepton colliders, to obtain a realistic portrait of supersymmetric (or Standard Model) events, it is necessary to account for multiple *non-collinear* QCD radiation effects. The evaluation of the cross section using matrix elements for multi-parton final states is prohibitively difficult. Instead, these multiple emissions are approximately included in an event simulation via a parton shower (PS) algorithm.[2] They give rise to effects such as jet broadening, radiation in the forward regions and energy flow into detector regions that are not described by calculations with only a limited number of final state partons.

In leading log approximation (LLA), the cross section for *single* gluon emission from a quark line is given by

$$d\sigma = \sigma_0 \frac{\alpha_s}{2\pi} \frac{dt}{t} P_{qq}(z) dz, \qquad (14.1)$$

where σ_0 is the overall hard scattering cross section, t is the intermediate state virtual quark mass, and $P_{qq}(z) = \frac{4}{3} \left(\frac{1+z^2}{1-z} \right)$ coincides with the Altarelli–Parisi splitting function for $q' \to qg$ for the fractional momentum of the final quark $z \equiv |\vec{p}_q|/|\vec{p}_{q'}| < 1$. Interference between various *multiple* gluon emission Feynman graphs, where the gluons are ordered differently, is a subleading effect which can be ignored. Thus, Eq. (14.1) can be applied successively, and gives a factorized probability for each gluon emission. The idea behind the PS algorithm is then to use these approximate emission probabilities (which are exact in the collinear limit), along with exact (non-collinear) kinematics to construct a program which describes multiple non-collinear parton emissions. Notice, however, that the cross section (14.1) is singular as $t \to 0$ and as $z \to 1$, i.e. in the regime of collinear and also soft gluon emission. These singularities can be regulated by introducing physically appropriate cut-offs. A cut-off on the value of $|t|$ of order $|t_c| \sim 1$ GeV corresponds to the scale below which QCD perturbation theory is no longer valid. A cut-off on z is also necessary, and physically corresponds to the limit beyond which the gluon is too soft to be resolved.

The PS algorithms available vary in their degree of sophistication. The simplest algorithm was created by Fox and Wolfram in 1979. Their method was improved to account for interference effects in the angle-ordered algorithm of Marchesini and Webber. In addition, parton emission from heavy particles results in a dead-cone effect, where emissions in the direction of the heavy particle are suppressed. Furthermore, it is possible to include spin correlations in the PS algorithm.

PS algorithms are also applied to the initial state partons. In this case, a backwards shower algorithm is most efficient, which develops the emissions from the hard scattering backwards in time towards the initial state. The backward shower algorithm developed by Sjöstrand makes use of the PDFs evaluated at different energy scales to calculate the initial state parton emission probabilities.

---

[2] For more detailed discussions beyond the scope of this text, see e.g. *Collider Physics*, V. Barger and R. J. N. Phillips, Addison-Wesley (1987), Chapter 9.

14.1.3 Cascade decays

We have already seen that not only are there many reactions available via which SUSY particles may be produced at colliders but, once produced, there exist many ways in which superparticles may decay. For the next-to-lightest SUSY particle (NLSP), there may be only one or at most a few ways to decay to the LSP. Thus, for a collider such as LEP or even the Fermilab Tevatron, where only the lightest sparticles will have significant production rates, we might expect that their associated decay patterns will be relatively simple. However, the number of possible final states increases rapidly if squarks and gluinos that can decay into the heavier charginos and neutralinos are accessible, and the book-keeping becomes correspondingly more complicated. Indeed, at the CERN LHC, where the massive strongly interacting sparticles such as gluinos and squarks are expected to be produced at large rates, sparticle cascade decay patterns can be very complex.[3] As an example, the many possible decay paths of a gluino in the mSUGRA model are shown in Fig. 14.2. Branching fractions to a variety of final states resulting from the cascade decay are also listed in the figure.

Monte Carlo event generators immensely facilitate the analysis of signals from such complex cascade decays, especially in the case where no single decay chain dominates. An event generator can select different cascade decay branches by generating a random number which picks out a particular decay choice, with a weight proportional to the corresponding branching fraction, at each step of the cascade decay. Quarks and gluons produced as the end products of cascade decays will shower off still more quarks and gluons, with probabilities determined by the PS algorithm.

The procedure that we have just described is exact for cascade decays of spinless particles into two other spinless particles at each step in the cascade. This is because the squared matrix element is just a constant, and there are no spin correlations possible. This is not true in general and in some cases it can be very important to include the decay matrix element and/or spin correlations in the calculation of cascade decays of sparticles. For instance, it has been suggested that the end point of the dilepton mass distribution from $\widetilde{W}_1 \widetilde{Z}_2 \rightarrow q\bar{q}'\widetilde{Z}_1 + \ell\bar{\ell}\widetilde{Z}_1$ production at hadron colliders yields a good measure of $m_{\widetilde{Z}_2} - m_{\widetilde{Z}_1}$. Frequently, interference between Z and slepton mediated amplitudes for $\widetilde{Z}_2$ decays suppresses this mass distribution near the kinematic end point, leading to greater uncertainties in its determination relative to the expectation with a constant matrix element. As an extreme example of the distortion due to effects from the decay matrix element, in Fig. 14.3 we show this distribution for $\widetilde{W}_1 \widetilde{Z}_2$ events at the Fermilab Tevatron collider for the choice of mSUGRA parameters $(m_0, m_{1/2}, A_0, \tan\beta, \text{sign}(\mu))$ shown on the figure. For this

---

[3] H. Baer, V. Barger, D. Karatas and X. Tata, *Phys. Rev.* **D36**, 96 (1987).

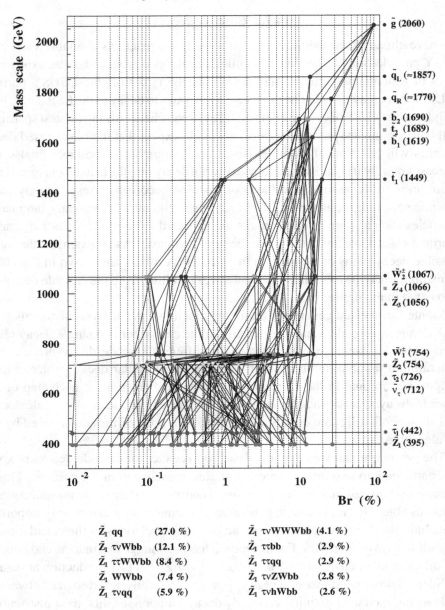

$\tilde{Z}_1$ qq   (27.0 %) $\tilde{Z}_1$ τνWWWbb (4.1 %)
$\tilde{Z}_1$ τνWbb   (12.1 %) $\tilde{Z}_1$ ττbb   (2.9 %)
$\tilde{Z}_1$ ττWWbb (8.4 %) $\tilde{Z}_1$ ττqq   (2.9 %)
$\tilde{Z}_1$ WWbb   (7.4 %) $\tilde{Z}_1$ τνZWbb   (2.8 %)
$\tilde{Z}_1$ τνqq   (5.9 %) $\tilde{Z}_1$ τνhWbb   (2.6 %)

Figure 14.2 An illustration of the branching fraction for various cascade decays of the gluino within the mSUGRA model with parameters $m_0 = 400$ GeV, $m_{1/2} = 900$ GeV, $\tan\beta = 35$, $A_0 = 0$, and $\mu > 0$. The masses of various sparticles are also shown. This figure is adapted from S. Abdullin and D. Denegri, hep-ph/9905510.

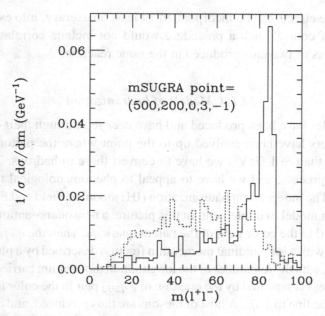

Figure 14.3 Distribution of opposite sign, same flavor dileptons from $\tilde{W}_1 \tilde{Z}_2$ production at the Fermilab Tevatron. The solid histogram shows the distribution including the exact matrix element, while the dashed histogram is the same distribution assuming that the decay matrix element is constant. Reprinted with permission from H. Baer, M. Drees, F. Paige, P. Quintana and X. Tata, *Phys. Rev.* **D61**, 095007 (2000), copyright (2000) by the American Physical Society.

parameter space point, $m_{\tilde{Z}_2} = 173$ GeV and $m_{\tilde{Z}_1} = 86$ GeV so one expects a dilepton mass distribution cut-off at $m_{\tilde{W}_1} - m_{\tilde{Z}_1} = 87$ GeV. The dilepton mass distribution including decay matrix element energy effects is denoted by the solid histogram, and is highly distorted by the pole of the virtual Z boson in the decay diagrams relative to the distribution using just phase space for the $\tilde{Z}_2$ decay (dashed histogram).

Spin correlation effects are especially important for precision measurements at e^+e^- linear colliders. While retaining spin correlations may be less crucial in many situations at a hadron collider, this is not always the case. For instance, relativistic τ^- leptons produced from W decay are always left-handed, while those produced from a charged Higgs decay are always right-handed. Likewise, the polarization of the taus from $\tilde{\tau}_1$ decays depends on the stau mixing angle. Since the undetectable energy carried off by ν_τ from tau decay depends sensitively on the parent tau helicity, it is necessary to include effects of tau polarization in any consideration involving the energy of "tau jets".[4] By evaluating the mean polarization of taus in any particular

---

[4] These effects are crucial for assessing the efficiency for identifying hadronically decaying taus at a hadron collider.

process, these effects can be incorporated, at least on average, into event generator programs. Of course, such a procedure would not include correlations between decay products of two taus produced in the same reaction.

14.1.4 Models of hadronization

Once sparticles have been produced and have decayed through their cascades, and parton showers have been evolved up to the point where the partons have virtuality smaller than ~ 1 GeV^2, we have to convert these to hadrons. This is a non-perturbative process, and we have to appeal to phenomenological models for its description. The independent hadronization (IH) model of Field and Feynman is the simplest such model to implement. In this picture, a new quark–antiquark pair $q_1 \bar{q}_1$ can be created in the color field of the parent quark q_0. Then the $q_0 \bar{q}_1$ pair can turn into a meson with a longitudinal momentum fraction described by a phenomenological function, with the remainder of the longitudinal momentum carried by the quark q_1. This process is repeated by the creation of a $q_2 \bar{q}_2$ pair in the color field of q_1, and so on down the line to $q_n \bar{q}_n$. A host of mesons are thus produced, and decayed to the quasi-stable π, K, ... mesons according to their experimental properties. The final residual quark q_n will have very little energy, and can be discarded without significantly affecting jet physics. Finally, a small transverse momentum can be added according to a pre-assigned Gaussian probability distribution to obtain a better description of the data. Quark fragmentation into baryons is also possible by creation of diquark pairs in its color field, and can be incorporated. The IH scheme will thus describe the bulk features of hadronization needed for event simulation programs.

The string model of hadronization developed at Lund is a more sophisticated model than IH, which treats hadron production as a universal process independent of the environment of the fragmenting quark. In the string model, a produced quark–antiquark pair is assumed to be connected by a color flux tube or string. As the quark–antiquark pair moves apart, more and more energy is stored in the string until it is energetically favorable for the string to break, creating a new quark–antiquark pair. Gluons are regarded as kinks in the string. The string model correctly accounts for color flow in the hadronization process, as opposed to the IH model. In $e^+ e^- \to q \bar{q} g$ (3-jet) events, the string model predicts fewer produced hadrons in the regions between jets than the IH model, in accord with observation.

A third scheme for hadronization is known as the cluster hadronization model. In this case, color flow is still accounted for, but quarks and antiquarks that are nearby in phase space will form a cluster, and will hadronize according to preassigned probabilities. This model avoids non-locality problems associated with the string hadronization model, where quarks and antiquarks separated by spacelike distances can affect the hadronization process.

14.1.5 Beam remnants

Finally, at a hadron collider the colored remnants of the nucleon that did not participate in the hard scattering must be accounted for. These beam remnant effects produce additional energy flow, especially in the far forward regions of the detector. A variety of approaches are available to describe these non-perturbative processes, including models involving Pomeron exchange and multiple scatterings. In addition, the beam remnants must be hadronized as well, and appear to require a different parametrization from "minimum bias" events where there are only beam jets but no hard scattering.

14.2 Event generator programs

Publicly available event generators for SUSY processes include

- **ISAJET:** (H. Baer, F. Paige, S. Protopopescu and X. Tata),
  http://www.phy.bnl.gov/~isajet/
- **PYTHIA:** (T. Sjöstrand, L. Lönnblad and S. Mrenna),
  http://www.thep.lu.se/~torbjorn/Pythia.html
- **HERWIG:** (G. Corcella, I. G. Knowles, G. Marchesini, S. Moretti, K. Odagiri, P. Richardson, M. Seymour and B. R. Webber),
  http://hepwww.rl.ac.uk/theory/seymour/herwig/
- **SUSYGEN:** (N. Ghodbane, S. Katsanevas, P. Morawitz and E. Perez),
  http://lyoinfo.in2p3.fr/susygen/susygen3.html

The event generator program ISAJET was originally developed in the late 1970s to describe scattering events at the ill-fated ISABELLE pp collider at Brookhaven National Laboratory. It was developed by F. Paige and S. Protopopescu to generate SM and beyond scattering events at hadron colliders and, to a lesser extent, e^+e^- colliders. ISAJET was the first event generator program developed to give a realistic portrayal of SUSY scattering events. ISAJET uses the IH model for hadronization, and the original Fox–Wolfram (Sjöstrand) PS shower algorithm for final state (initial state) parton showers. It includes an n-cut Pomeron model to describe beam-jet evolution.

The event generator PYTHIA was developed mainly by T. Sjöstrand in the early 1980s to implement the Lund string model for event generation. S. Mrenna contributed the inclusion of SUSY processes in PYTHIA.

The event generator HERWIG was developed in the mid-1980s to describe scattering events with angle-ordered parton showers, which accounted for interference effects neglected in the Fox–Wolfram shower approach. HERWIG implements a cluster hadronization model. Supersymmetric processes are now available in HERWIG.

The program SUSYGEN was developed by S. Katsanevas and others to generate $e^+e^- \to$ SUSY events for the LEP experiments. SUSYGEN interfaces with PYTHIA for hadronization and showering. SUSYGEN has since been upgraded to also generate events for hadron colliders.

A description of these codes, or a comparison of their relative virtues and shortcomings, is beyond the scope of this text. Moreover, any such discussion would rapidly become out of date as these programs are continually being upgraded. We refer the interested reader to the webpages cited above, both for how to use these codes, and also for a description of the physics underlying these event generators.

We have already noted that in addition to these event generator codes, there are several specialized codes (SPheno, SuSpect, SOFTSUSY) for the evaluation of sparticle spectra. In addition, there are publicly available codes for a careful evaluation (including loop effects) of the mass spectrum (FeynHiggs, FeynHiggsFast) and decay rates (HDECAY) of MSSM Higgs bosons. A careful evaluation of these is especially useful because, as we saw in Chapter 8, m_h is bounded above in a wide class of models, and experiments searching for a signal for h have already excluded significant portions of its allowed range.[5] Finally, we note that $2 \to n$ (with $n \leq 6$) hard scattering processes including SUSY particles may be generated by programs such as CompHEP (E. Boos *et al.*), Madgraph-II (T. Stelzer and W. F. Long), and GRACE (Minami–Tateya Collaboration, M. Jimbo *et al.*). These codes need to be interfaced with one of the event generators, and are especially useful if specific reactions need to be generated including, for instance, effects of spin correlations.

14.3 Simulating SUSY with ISAJET

In this section, we illustrate the use of SUSY event generators using ISAJET (with which we are most familiar) as an example. The interested reader can follow similar procedures for any of the other event generator codes.

14.3.1 Program set-up

ISAJET is a publicly available code, and can be obtained from the website http://www.phy.bnl.gov/~isajet/. ISAJET is written in Fortran 77, and the code is maintained by the Patchy code management system, which is included in the CERNLIB library of subroutines. The files available are

---

[5] We stress that one should interpret the excluded region with care, since it is sensitive to underlying assumptions. For instance, if MSSM parameters are complex, the excluded mass range may be significantly smaller.

`isajet.car` and a Unix `Makefile`. The `Makefile` program must be edited to suit the user's particular machine. Running the `Makefile` on `isajet.car` creates a number of programs, including `isasugra.x`, `isasusy.x`, `isajet.x`, and `isajet.tex`. The last is a LaTex file of the ISAJET manual.[6]

The program `isasusy.x` requires a weak scale MSSM parameter set as its input, and produces an output file with the corresponding physical sparticle masses along with (s)particle decay rates and branching fractions. The program `isasugra.x` accepts as an input parameters from the various SUSY breaking models described in Chapter 11, including SUGRA models with universal or non-universal SSB terms, GMSB models and AMSB models, and models including right-handed neutrinos ν_R. The program then uses the RGEs discussed in Chapter 9 to evolve these parameters, which are typically specified at some high scale, down to the weak scale relevant for phenomenology. These weak scale parameters are then used to evaluate the sparticle masses, decay rates and decay branching fractions which are written to a user-readable output file. The outputs of either `isasusy.x` or `isasugra.x` serve as input parameters for the program `isajet.x` which actually generates SUSY events corresponding to the particular model under study.

14.3.2 Models for SUSY in ISAJET

MSSM

The program `isasusy.x` calculates weak scale sparticle masses and their associated branching fractions in terms of a subset of weak scale MSSM input parameters. The relevant input parameters include:

$$m_t \tag{14.2a}$$

$$m_{\tilde{g}}, \quad \mu, m_A, \tan\beta \tag{14.2b}$$

$$m_{Q_{11}}, m_{D_{11}}, m_{U_{11}}, m_{L_{11}}, m_{E_{11}} \tag{14.2c}$$

$$m_{Q_{33}}, m_{D_{33}}, m_{U_{33}}, m_{L_{33}}, m_{E_{33}}, (A_u)_{33}, (A_d)_{33}, (A_\tau)_{33} \tag{14.2d}$$

plus optional inputs of

$$m_{Q_{22}}, m_{D_{22}}, m_{U_{22}}, m_{L_{22}}, m_{E_{22}} \tag{14.2e}$$

$$M_1, \quad M_2 \tag{14.2f}$$

$$m_{3/2}. \tag{14.2g}$$

ISAJET currently takes the soft-SUSY breaking sfermion mass squared matrices to be real and diagonal; also, only third generation diagonal trilinear A terms

---

[6] The ISAJET manual is also available on the hep-ph archive as H. Baer, F. Paige, S. Protopopescu and X. Tata, hep-ph/0001086.

are allowed. This corresponds to the simplified parameter space discussed in Section 8.1.2. If the optional second generation masses are not specified, then their values are set equal to the corresponding first generation masses. Also, the $U(1)_Y$ and $SU(2)_L$ gaugino masses are fixed by the gaugino mass unification relation

$$\frac{M_1}{\alpha_1} = \frac{M_2}{\alpha_2} = \frac{M_3}{\alpha_3} \tag{14.3}$$

unless the optional independent gaugino masses are specified. Finally, if the value of $m_{3/2}$ is not specified, it is assumed that the gravitino is heavy enough so that it effectively decouples from particle phenomenology.

The Higgs boson masses are computed using the RG improved one-loop effective potential evaluated at an optimized scale choice $Q = \sqrt{m_{\tilde{t}_L} m_{\tilde{t}_R}}$: using this high scale effectively accounts for some of the larger two-loop effects.

mSUGRA

For the mSUGRA model included in `isasugra.x`, the model inputs are

$$m_0, m_{1/2}, A_0, \tan\beta, \text{sign}(\mu), \text{ and } m_t. \tag{14.4}$$

Then ISAJET calculates the gauge and third generation Yukawa couplings at the weak scale in the $\overline{DR}$ scheme, and evolves the set of six gauge and Yukawa couplings via two-loop RGEs up to the GUT scale, which is defined as the Q value at which $g_1 = g_2$. At the scale M_{GUT}, all SSB scalar masses are set to m_0, all gaugino masses are set to $m_{1/2}$ and all third generation diagonal trilinear A terms are set to A_0. Then the set of 26 MSSM couplings and parameters are evolved via two-loop RGEs to the weak scale. More precisely, each SSB term is frozen out at the scale equal to its absolute value, except for Higgs sector parameters, which are frozen at $Q = \sqrt{m_{\tilde{t}_L} m_{\tilde{t}_R}}$. One-loop corrections are added to the scalar potential, which is then minimized to obtain the value of μ^2 and $B(Q)$, consistent with radiative breaking of electroweak symmetry with the correct value of M_Z. SUSY threshold corrections to m_t, m_b, and m_τ, which considerably modify the relation between SM fermion masses and the corresponding Yukawa couplings if $\tan\beta$ is large, are computed at this stage. The set of couplings and mass parameters are then evolved iteratively between M_{GUT} and M_{weak} until the solution to the RGEs converges to within a specified tolerance.

Non-universal SUGRA

To facilitate simulation of models with non-universal gaugino masses, or models with non-universal scalar masses (e.g. models with additional D-term contributions, or gaugino-mediated SUSY breaking models), ISAJET includes the "Non-universal

SUGRA" option which allows the user to set arbitrary values of the SSB parameters

$$M_1, \quad M_2, \quad M_3 \tag{14.5a}$$

$$A_t, \quad A_b, \quad A_\tau \tag{14.5b}$$

$$m_{H_d}, \quad m_{H_u} \tag{14.5c}$$

$$m_{Q_{11}}, \quad m_{D_{11}}, \quad m_{U_{11}}, \quad m_{L_{11}}, \quad m_{E_{11}} \tag{14.5d}$$

$$m_{Q_{33}}, \quad m_{D_{33}}, \quad m_{U_{33}}, \quad m_{L_{33}}, \quad m_{E_{33}} \tag{14.5e}$$

at $Q = M_{\text{GUT}}$, in place of the universal scalar mass and the universal A-parameter of the mSUGRA model. First and second generation SSB scalar masses are assumed equal. As in mSUGRA, these serve as boundary conditions for the RGEs which are again used to evaluate the weak scale values of MSSM SSB parameters from which sparticle masses, couplings, and decay branching fractions are obtained. The user also has the option to specify the scale Q at which the boundary conditions are to be implemented, allowing more accurate simulation of string-based scenarios.

GMSB models

ISAJET also allows for event generation in a variety of GMSB models. The set (11.39) of parameters of the mGMSB model

$$\Lambda, M, n_5, \tan\beta, \text{sign}(\mu), C_{\text{grav}} \tag{14.6a}$$

can directly be used as an input to ISAJET. As in mSUGRA, the gauge and Yukawa couplings at the weak scale are first evolved up in energy to $Q = M$, the messenger scale, where the calculated GMSB SSB mass parameters are used as boundary conditions for the RGEs. All parameters are then evolved back down to the weak scale, where the scalar potential is minimized and REWSB is imposed as usual.

ISAJET also allows event generation of many non-minimal GMSB models, by allowing several additional input parameters:

$$R, \delta m_{H_d}^2, \delta m_{H_u}^2, D_Y(M), n_5(1), n_5(2), n_5(3). \tag{14.6b}$$

In this set, R is a gaugino mass multiplier that decouples gaugino and scalar mass parameters at the messenger scale. This can occur if the scale for a $U(1)_R$ symmetry breaking differs from the SUSY breaking scale. For the minimal model, $R = 1$. The parameters $\delta m_{H_d}^2$ and $\delta m_{H_u}^2$ are additional contributions to the Higgs SSB masses at the messenger scale which may arise from additional interactions that generate the dimensional B and μ parameters. These additional contributions are zero in the mGMSB model. In (14.6b), $D_Y(M)$ is the VEV of the hypercharge D-term in the messenger sector, and can lead to additional contributions $\delta m^2(M) = g'Y D_Y(M)$ to scalar mass parameters at $Q = M$. Finally, allowing incomplete messenger representations can effectively yield differing numbers of messengers $(n_{5_1}, n_{5_2}, n_{5_3})$ for each factor of the gauge group.

AMSB models

ISAJET also has the minimal anomaly-mediated SUSY breaking model hardwired. It accepts the mAMSB parameter set

$$m_0, m_{3/2}, \tan\beta, \text{sign}(\mu), \tag{14.7}$$

as an input, and then evolves gauge and Yukawa couplings to M_{GUT}, where the mAMSB SSB masses are imposed as boundary conditions for the RGEs. The complete set of SSB masses and couplings are evolved to the weak scale where REWSB is imposed as usual.

Models with right-handed neutrinos

ISAJET allows the simulation of models with right-handed neutrino (RHN) superfields that are so topical today. In addition to other parameters, the user has to specify (see (9.37)),

$$M(\nu_3), M_N, A_\nu, m_{\tilde{\nu}_R}, \tag{14.8}$$

where $M(\nu_3)$ is the third generation neutrino mass, M_N is the heavy mass in the neutrino see-saw, A_ν is the new third generation neutrino trilinear A-parameter and $m_{\tilde{\nu}_R}$ the SSB mass of the RHNs. The parameter B_ν in (9.37) only affects the mass of the very heavy right-handed sneutrinos, and so is irrelevant for our purposes. As for other fermions, the masses and Yukawa couplings of the first two generations of neutrinos are neglected. Typically, we expect that $M_N \sim M_{\text{GUT}}$, while A_ν and $m_{\tilde{\nu}_R}$ are comparable to the weak scale. If one inputs a value of $M(\nu_3) = 0$, then ISAJET computes the third generation neutrino Yukawa coupling f_ν by imposing $f_\nu = f_t$ at M_{GUT}, as expected in $SO(10)$ SUSYGUT models.

14.3.3 Generating events with ISAJET

The programs `isasusy` and `isasugra` are useful for examining sparticle masses and branching fractions expected in different models. For generating collider events, `isajet` must be used. `isajet` takes its input from an input parameter file such as the file `sugra.par` as shown below. This file implements sparticle production events for the Fermilab Tevatron $p\bar{p}$ collider at $\sqrt{s} = 2$ TeV.

```
TEST SUGRA JOB
2000,5000,0,0/
SUPERSYM
```

```
BEAMS
'P','AP'/
SEED
9998871/
NTRIES
5000/
SUGRA
100,200,0,3,1/
TMASS
175,-1,-1/
JETTYPE1
'ALL'/
JETTYPE2
'ALL'/
PT
10,250,10,250/
END
STOP
```

The first line is a comment line, containing the job title. The second line is the collider CM energy, the number of events to be generated, the number of events to print out, and the number of events to skip between printing. The third line gives the class of reactions: in this case, supersymmetric ones. The fourth and fifth lines denote the beam types: here proton and antiproton. The fifth and sixth lines specify a random seed for event generation; by altering the seed, an independent set of events can be generated. NTRIES on the next two lines limits the number of tries (in this case, 5000) that the program makes to find a good event. The tenth line denotes the SUGRA model inputs (m_0, $m_{1/2}$, A_0, $\tan\beta$, sign (μ)) specified on the next line. Lines 12–13 show the top quark mass input, while lines 14-15 specify the types of sparticles to be produced in the $2 \to 2$ subprocess: in this case, all allowed reactions will occur. By limiting JETTYPE1 and JETTYPE2 to be specific sparticle(s), particular (sets of) SUSY reactions can be studied. Lines 18–19 show the p_T range of the final state particles in the $2 \to 2$ hard scattering process. Finally, the last two lines indicate the end of the file.

For generating scattering events at an e^+e^- linear collider, the input might look like this:

```
TEST SUGRA JOB FOR A LC WITH BEAM POLARIZATION
       AND BEAMSTRAHLUNG
500,5000,0,0/
```

```
E+E-
SEED
9998871/
NTRIES
5000/
SUGRA
2375,300,0,30,1/
EPOL
0.9,0/
EBEAM
400,500,.1072,.12/
TMASS
175,-1,-1/
JETTYPE1
'ALL'/
JETTYPE2
'ALL'/
END
STOP
```

In the above file, e^+e^- events are stipulated by the E+E- reaction card on line 3. Since the SUGRA model is stipulated, isajet will generate SUSY events. On lines 10–11, a left-polarized electron beam with $P_L(e^-) = 0.9$ is stipulated to scatter from an unpolarized positron beam. In lines 12–13, beamstrahlung is enabled, and the reaction subprocess energy is restricted to lie between 400–500 GeV. The beamstrahlung parameters $\Upsilon = 0.1072$ and $\sigma_z = 0.12$ mm (as defined in Section 12.2.4) must also be given.

There are several ISAJET output files. One will include various masses, and the sum total of all cross sections generated. Another will include the actual scattering events, which consists of dumping out various ISAJET common blocks, including PARTCL, which contains all final state particles, their identities, sources, and four-vectors. This output is in a form suitable for analysis, or for interface with detector simulation programs. Further details along with program updates can be found in the ISAJET manual, isajet.tex.

As an example of a supersymmetric scattering event, we show in Fig. 14.4 a $pp \rightarrow \tilde{g}\tilde{u}_L X$ event generated within the mSUGRA framework for the CMS detector at the CERN LHC collider with $\sqrt{s} = 14$ TeV. The response of various detector elements to the passage of particles through them was simulated by the program GEANT. The mSUGRA model parameters are also shown in the figure, along with several sparticle masses and the sparticle cascade decay chains. Six high E_T jets,

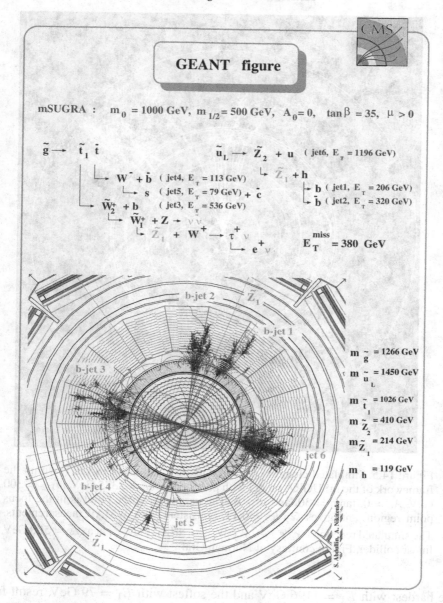

Figure 14.4 GEANT simulation of an ISAJET mSUGRA event for the CMS detector at the CERN LHC. Notice the large multiplicity of b-jets. Adapted from a figure, courtesy of Salavat Abdullin.

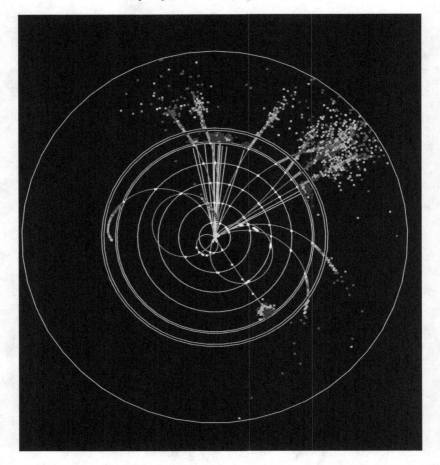

Figure 14.5 Simulated chargino pair production event using ISAJET, within the framework of the mSUGRA model with parameters $m_0 = 2375$ GeV, $m_{1/2} = 300$ GeV, $A_0 = 0$, $\tan \beta = 30$, and $\mu > 0$. The parameter space point is in the focus point region, and gives $\Omega_{\widetilde{Z}_1} h^2 = 0.11$, consistent with WMAP measurements. The simulated reaction is $e^+ e^- \rightarrow \widetilde{W}_1^+ \widetilde{W}_1^- \rightarrow u \bar{d} \widetilde{Z}_1 + e \bar{\nu}_e \widetilde{Z}_1$ at a $\sqrt{s} = 500$ GeV linear collider. Figure courtesy of Norman Graf.

the hardest with $E_T = 1196$ GeV and the softest with $E_T = 79$ GeV, result from these cascade decays. Moreover, four of the produced jets contain displaced vertices from B hadrons, though not all of these would be tagged in a real detector. Two of the four b-jets result from the decay of the Higgs boson h produced in the cascade decay of the $\tilde{u}_L$ squark. In such an event, it may be possible to reconstruct the h mass, but with large errors compared to the mass reconstruction from the $h \rightarrow \gamma \gamma$ signal. This event also contains 380 GeV of E_T^{miss} from the undetected neutralinos and neutrinos produced in the cascade decays. We also remark that there are two on-shell W bosons and an on-shell Z^0 boson in this event. Had the vector bosons decayed leptonically, they would have given rise to readily detectable hard electrons and muons that constitute the characteristic multilepton plus jet signature for SUSY.

A sample sparticle production event for a $\sqrt{s} = 500$ GeV linear e^+e^- collider is shown in Fig. 14.5. The event shown has $e^+e^- \rightarrow \widetilde{W}_1^+ \widetilde{W}_1^-$, where $\widetilde{W}_1^+ \rightarrow u \bar{d} \widetilde{Z}_1$ while $\widetilde{W}_1^- \rightarrow e \bar{\nu}_e \widetilde{Z}_1$. The electron track can be seen in the lower-right quadrant, where it deposits its energy in the EM calorimeter. The two quark jets are evident in the upper half plane. The event was generated in the mSUGRA model with parameters shown in the caption. The parameters are from the HB/FP region of mSUGRA parameter space (as discussed in Chapter 9), and give rise to a relic density $\Omega_{\widetilde{Z}_1} h^2 = 0.11$, in accord with WMAP measurements.

15

The search for supersymmetry at colliders

While the first clear hints of deviation from the SM may well come from any of a large variety of experiments, establishing precisely what the new physics is will be possible only by observations at energy scales close to, or beyond, the threshold for the new phenomena. Direct examination of the properties of any new states of matter associated with the new physics is probably the best way to study the new phenomena, if these degrees of freedom are kinematically accessible. If the new physics is supersymmetry, then the new states of matter will be the superpartners, and it is only by determining their quantum numbers and couplings that we can unambiguously establish that the new physics is actually supersymmetry. Of course, any new states of matter may be quite different from superpartners. For instance, if extra spatial dimensions exist which are accessible at the weak scale, the new degrees of freedom will be Kaluza–Klein excitations of SM particles. It is even possible that no new degrees of freedom are directly accessible, but that SM interactions acquire form factors that point to what the new physics might be. Our point here is that elucidation of new physics will only be possible at colliding beam facilities.

The purpose of this chapter is to examine what may be learned from a study of high energy collisions assuming that nature is supersymmetric at the weak scale. To start with, we review various searches for supersymmetry in previous collider and fixed target experiments. Up to now, no direct evidence for SUSY has been found. The negative searches have been interpreted as lower limits on sparticle masses, and as exclusion of regions of the parameter space of various specific models for MSSM sparticle masses. Next, we project the SUSY reach of the luminosity upgrade of the Fermilab Tevatron $p\bar{p}$ collider with $\sqrt{s} = 2$ TeV that has already begun operation, as well as of the CERN Large Hadron Collider (LHC), a 14 TeV $p\bar{p}$ collider scheduled to commence operation in 2007. We also discuss the capability of a high energy e^+e^- linear collider (LC) operating at $\sqrt{s} = 0.5$–1 TeV for SUSY studies; such a machine is being considered for construction in the not-too-distant future. In the latter part of the chapter we discuss how a determination of sparticle properties at the LHC and at a LC may be used to establish that the new physics is supersymmetry and, further, to zero in on the mechanism by which MSSM superpartners acquire SUSY breaking masses and couplings. Our discussion follows a "bottom-up" vision of a program for high energy physics over the next two decades that includes the following general steps:

- Establish the discovery of new physics.
- Figure out what the new physics is – here, we take it to be weak scale supersymmetry.
- From the experimentally determined values of sparticle masses and couplings, figure out the organizing principle(s) that lead to the observed supersymmetry breaking parameters.

Interpretation of negative results of searches for supersymmetric particles depends heavily on the assumptions made about the underlying supersymmetric model. Many sparticle search experiments try to be as model independent as possible, in which case limits can be placed on sparticle masses. But sometimes dependence on a model for a particular analysis is unavoidable. In these cases, null search results are often presented as limits on model parameter space in one or two-parameter space dimensions. Other model parameters can be scanned over, so that results apply for a wide range of model parameters. Alternatively, results can be presented for "typical" choices of other model parameters, such as $\tan \beta$. Finally, bounds can be presented as a function of the model parameter which gives the most conservative estimate of the reach into parameter space.

15.1 Early searches for supersymmetry

Experiments at the energy frontier have been searching for supersymmetry since the early 1980s when it was recognized that weak scale supersymmetry could protect the large hierarchy between the weak and GUT (or Planck) scales from large radiative corrections.

15.1.1 e^+e^- collisions

Searches for supersymmetry were performed in the early 1980s at the PEP e^+e^- collider at SLAC ($\sqrt{s} \simeq 29$ GeV) by the MAC and MARK 2 collaborations and at the PETRA e^+e^- collider at DESY ($\sqrt{s} \lesssim 47$ GeV) by the MARK J, CELLO, TASSO, and JADE collaborations. The center of mass energy was extended in the mid to late 1980s by the TOPAZ, VENUS, and AMY experiments at the Tristan e^+e^- collider operating at KEK at $\sqrt{s} \lesssim 60$ GeV.

Typically, the searches focussed on signals from the lightest charged sparticles since these could be produced with relatively large cross sections. It was assumed that any sparticles which were produced would promptly decay to the LSP (which was assumed to be photino-like). The relevant processes searched for were:

$$e^+e^- \rightarrow \tilde{\ell}^+\tilde{\ell}^- \rightarrow \ell^+\ell^-\widetilde{Z}_1\widetilde{Z}_1, \tag{15.1a}$$

$$e^+e^- \rightarrow \tilde{q}\bar{\tilde{q}} \rightarrow q\bar{q}\widetilde{Z}_1\widetilde{Z}_1, \tag{15.1b}$$

and

$$e^+e^- \rightarrow \widetilde{W}_1\overline{\widetilde{W}}_1 \rightarrow f_i\bar{f}'_i\widetilde{Z}_1 + f_j\bar{f}'_j\widetilde{Z}_1, \tag{15.2}$$

where f and f' are the upper and lower members of a weak $SU(2)$ doublet, and the subscripts i and j denote the fermion type. Since the neutralino LSPs would escape undetected in the experimental apparatus, the experimental signature for slepton (squark) production was taken to be a pair of acolinear leptons (jets) balanced by missing energy and missing transverse momentum carried off by the LSPs. Chargino pair production can lead to missing energy events with multiple jets, jets and a charged lepton, or a charged lepton pair, depending on how the charginos decay. Within the SM, missing energy can only arise if neutrinos are produced in the reaction in addition to the charged leptons and/or jets. The SM cross section for events with hard jets and/or leptons together with large missing energy is very small, and the non-observation of an excess of signal events above expected background levels is interpreted as a lower limit on $m_{\tilde{\ell}}$, $m_{\tilde{q}}$, and $m_{\widetilde{W}_1}$. In addition, there are non-physics backgrounds from experimental mismeasurements that can fake missing energy events. These backgrounds depend on the energy resolution of the experimental apparatus, and also on other details such as uninstrumented regions of this detector, etc. and so are detector-dependent. By selecting events to lie in a kinematic region with large E_T^{miss}, these backgrounds can be greatly suppressed. The residual background is usually evaluated using event simulation programs discussed in the previous chapter, interfaced with programs to simulate the response of the experimental apparatus. Because the background rate is small, the non-observation of a signal translates into a lower limit only a little below the beam energy. Mass limits on selectrons (whose cross section is enhanced by t-channel $\widetilde{Z}_1$ exchange) and charginos (whose cross section has an s-wave threshold compared to the p-wave threshold for sfermion production) were somewhat stronger than the limits placed on the other sfermions.

Assuming that the $\widetilde{Z}_1$ ($\tilde{\nu}_e$) is very light and escapes detection, the MAC, ASP, and AMY collaborations were able to obtain lower limits on $m_{\tilde{e}}$ ($m_{\widetilde{W}_1}$) beyond the kinematic limit for selectron (chargino) pair production by searching for single photon events coming from

$$e^+e^- \to \widetilde{Z}_1\widetilde{Z}_1\gamma \quad \text{or} \quad \tilde{\nu}_e\bar{\tilde{\nu}}_e\gamma, \tag{15.3}$$

where the first of these takes place via selectron exchange and the second via chargino exchange. The SM background from $e^+e^- \to \nu\bar{\nu}\gamma$ production is very small. Non-observation of these single photon events has also been re-interpreted as excluding portions of the selectron–goldstino mass plane. This search is also relevant in models with a very light gravitino.

15.1.2 Searches at the CERN S$p\bar{p}$S *collider*

Shortly after the inauguration in 1982 of the CERN S$p\bar{p}$S collider at $\sqrt{s} = 546$ GeV and the discovery of the W and Z bosons, a variety of anomalous collider events were reported by the UA1 and UA2 collaborations: these included events containing one or more jets plus missing transverse energy at UA1 and, at UA2, events containing a hard electron plus jets plus E_T^{miss}. The UA1 events were exactly the sort of events expected from pair production of gluinos or squarks with $m \sim 40$–50 GeV, depending on which is the lighter: the produced sparticles were assumed to decay directly to the LSP either via $\tilde{g} \to q\bar{q}\widetilde{Z}_1$, or via $\tilde{q} \to q\widetilde{Z}_1$. The excitement was short-lived as it was soon realized that SM processes such as $q\bar{q} \to Z + g$ or $q\bar{q}' \to W + g$ followed by $Z \to \nu\bar{\nu}$ or $W \to \tau\nu_\tau$ could give rise to these jet(s) $+E_T^{\mathrm{miss}}$ events at the observed rates.

Subsequently, the collider energy was raised to $\sqrt{s} = 630$ GeV and SM backgrounds to the E_T^{miss} data sample were carefully estimated. UA1 was able to place limits of $m_{\tilde{g}} > 53$ GeV and $m_{\tilde{q}} > 45$ GeV, assuming degenerate squark masses and a neutralino with mass less than 20 GeV. Their results could not exclude gluinos with mass less than 4 GeV, leaving open a window for a light gluino. The UA2 experiment, ultimately taking more data than UA1, was able to raise the lower bounds to $m_{\tilde{q}} > 74$ GeV and $m_{\tilde{g}} > 79$ GeV, but the light gluino window still remained open.

15.1.3 A light gluino window?

Gluinos with lifetimes long compared to the hadronization time will bind with a gluon or with $q\bar{q}$ pairs to form neutral or charged R-parity odd hadrons before decaying to the LSP. The lightest of these R-odd hadrons is expected to be neutral, and is denoted by R^0. R-hadrons in the mass range 1.5–7.5 GeV can be produced by strong interactions via collisions of protons on nuclear targets. During the late 1970s, several fixed target experiments obtained upper limits on the cross sections of a neutral hadron decaying into a final state containing a charged hadron, excluding various ranges of $m_{\tilde{g}}$ depending on its lifetime.

Gluinos were searched for but not found in neutrino beam dump experiments by looking for the re-interaction of the LSP produced via $\tilde{g} \to q\bar{q}\tilde{Z}_1$ by the WA-66, E-613 and CHARM collaborations. These searches, which exclude portions of the $m_{\tilde{g}} - \tau_{\tilde{g}}$ plane, become ineffective for large squark masses (long gluino lifetimes) because the neutralino interaction cross section falls as $1/m_{\tilde{q}}^4$. Light gluinos were also searched for by the WA-75 collaboration in π meson beam dumps onto emulsions, again with a negative result.

Light gluinos were also searched for in $\Upsilon \to \eta_{\tilde{g}}\gamma$ decays by the CUSB experiment, where $\eta_{\tilde{g}}$ is the pseudoscalar $\tilde{g}\tilde{g}$ bound state. The non-observation of monoenergetic photons excludes $1.5 \le m_{\tilde{g}} \le 3$ GeV, independent of the gluino lifetime. The ARGUS experiment searched for gluinos via $\chi_b \to \tilde{g}\tilde{g}g$ decays, with one of the R-hadrons decaying away from the interaction point, and excluded the mass range 1–4.5 GeV for an appropriate lifetime range.

Despite these efforts, a window for light gluinos still remained, where the R^0 hadron was expected to have a mass of 1–3 GeV and a lifetime of 10^{-10}–10^{-5} s. The R^0 was expected to decay mainly via $R^0 \to \rho\tilde{Z}_1 \to \pi^+\pi^-\tilde{Z}_1$ and at smaller rates into the C-violating mode $R^0 \to \pi^0\tilde{Z}_1$. In the late 1990s, the KTeV collaboration at Fermilab reported a null result from searches for the spontaneous appearance of $\pi^+\pi^-$ pairs, or a single π^0 consistent with the decay of a long-lived neutral particle produced by 800 GeV protons on a beryllium target. They excluded the interesting range $m_{R^0} \sim 1$–3 GeV for a lifetime between 3×10^{-10} and 10^{-3} s.

These experimental results leave little if any room for light gluinos with mass less than 5–10 GeV. When combined with limits from UA1 and UA2, it seems clear that $m_{\tilde{g}} \gtrsim 79$ GeV.

15.2 Search for SUSY at LEP and LEP2

In 1989, the CERN Large Electron Positron collider (LEP) began operating at and around the Z pole, $\sqrt{s} \simeq 91$ GeV. Data was collected at the Z pole by the four experiments ALEPH, DELPHI, L3 and OPAL through 1995. At that point, each experiment had accumulated over 4 million Z boson events, corresponding to an integrated luminosity of over 150 pb^{-1}. In 1995, the center of mass collider energy was raised to 136 GeV, and over subsequent years it was raised beyond WW and ZZ thresholds until a maximum energy of $\sqrt{s} \simeq 208$ GeV was reached in the year 2000, in an effort to flush out the Higgs boson.

15.2.1 SUSY searches at the Z pole

The four LEP experiments together accumulated a sample of about 17M Z^0 events allowing very precise determination of the Z^0 line-shape. In particular, $\Delta\Gamma_Z$ as well as $\Delta\Gamma_{\text{inv}}$, the non-SM contributions to the total and "invisible" widths of the

Z^0, are constrained to be smaller than a few MeV. The former leads to lower limits only slightly below $M_Z/2$ on masses of MSSM sparticles ($\tilde{f}$, $\widetilde{W}_i$) with significant couplings to Z^0. This limit is *independent* of the decay properties of the sparticles. The limit $\Delta\Gamma_{\text{inv}} < 2$ MeV strongly constrains the partial width for Z decays to $\widetilde{Z}_1$ pairs, but does not lead to a model-independent lower limit on $m_{\widetilde{Z}_1}$ because of the strong parameter dependence of the $Z\widetilde{Z}_1\widetilde{Z}_1$ coupling. The invisible width puts a bound very close to $M_Z/2$ on (quasi-)stable or invisibly decaying sneutrinos.

The large number of Z^0 boson events also gave lower limits essentially equal to $M_Z/2$ on $m_{\widetilde{W}_1}$, $m_{\tilde{q}_{\text{L,R}}}$, and $m_{\tilde{\ell}_{\text{L,R}}}$. These limits, which come from searches for final state configurations with low SM backgrounds, depend on how the sparticles decay, and so are somewhat model-dependent. These limits would be evaded if the parent sparticle had a mass close to the daughter LSP so that the visible decay products are very soft. Alternatively, limits on $\tilde{t}_1$ or $\tilde{b}_1$ masses could be evaded for values of the squark mixing angle such that the corresponding Z partial width is very small.

Searches for $Z^0 \to \widetilde{Z}_1\widetilde{Z}_2 \to \widetilde{Z}_1 + f\bar{f}\widetilde{Z}_1$ are of special interest because this reaction, which leads to distinctive events with acolinear leptons or jets and large missing transverse energy, may be kinematically accessible even if $Z \to \widetilde{W}_1^+\widetilde{W}_1^-$ is not. Unfortunately, the $Z\widetilde{Z}_i\widetilde{Z}_j$ coupling is very parameter-dependent, and vanishes if either neutralino is a gaugino. Even so the LEP experiments, which are able to exclude branching fractions for $Z \to \widetilde{Z}_1\widetilde{Z}_2$ larger than $(2-20) \times 10^{-6}$ (depending on the values of the neutralino masses), are able to exclude regions of parameter space that would otherwise not be accessible.

15.2.2 SUSY searches at LEP2

All four LEP experiments collected data for several center of mass energies ranging from $\sqrt{s} = M_Z$ up to $\sqrt{s} = 203$–208 GeV, where each experiment accumulated over 210 pb^{-1} of integrated luminosity. Non-observation of any signal in a large number of final states was interpreted as lower limits on many sparticle masses. The precise limits are somewhat model-dependent but, because of the clean experimental environment, are frequently close to the kinematic limit. We summarize these limits in Table 15.1.

Charged sleptons are searched for assuming that these decay via $\tilde{\ell} \to \ell\widetilde{Z}_1$, and that the neutralino LSP escapes detection. Since the cross section for selectron pair production is considerably larger than that for smuon or stau production, the limits on $m_{\tilde{e}_R}$ are somewhat stronger. The limit on $m_{\tilde{\tau}_1}$ also depends on θ_τ. Since third generation squarks are expected to be lighter than squarks of other generations, LEP2 experiments focussed on searches for $\tilde{t}_1$ and $\tilde{b}_1$ squarks. The limits obtained depend on the corresponding mixing angle as well as on their assumed decay patterns. Experiments at LEP2 have also searched for charginos produced via

Table 15.1 *Limits on various sparticle masses from the non-observation of any signal in experiments at LEP2. The limits on $m_{\tilde{t}_1}$ ($m_{\tilde{b}_1}$) are shown for two cases of squark mixing angle: no mixing, and mixing such that the coupling to Z^0 vanishes. The limit on the chargino mass for small mass gaps is obtained from a combination of results including searches for soft events with radiated photons from the initial state, for long-lived particles that manifest themselves by tracks with kinks or impact parameter off-sets, or for quasi-stable heavy charged particles.*

sparticle	mass bound (GeV)	comment
$\tilde{e}_R$	99	$\tilde{e}_R \rightarrow e\tilde{Z}_1$, $\Delta m > 10$ GeV
$\tilde{\mu}_R$	94	$\tilde{\mu}_R \rightarrow \mu\tilde{Z}_1$, $\Delta m > 10$ GeV
$\tilde{\tau}_1$	85	$\tilde{\tau}_R \rightarrow \tau\tilde{Z}_1$, $\Delta m > 10$ GeV
$\tilde{t}_1$	98 (94)	$\tilde{t}_1 \rightarrow c\tilde{Z}_1$, $\Delta m > 10$ GeV, $\theta_t = 0(56°)$
$\tilde{t}_1$	99 (95)	$\tilde{t}_1 \rightarrow b\ell\tilde{\nu}_L$, $\Delta m > 10$ GeV, $\theta_t = 0(56°)$
$\tilde{b}_1$	99 (95)	$\tilde{b}_1 \rightarrow b\tilde{Z}_1$, $\Delta m > 10$ GeV, $\theta_b = 0(68°)$
$\tilde{W}_1$	103.5	$m_{\tilde{\nu}_e} > 300$ GeV, $\Delta m > 10$ GeV, gaugino mass unification
$\tilde{W}_1$	91.9	$m_{\tilde{\nu}_e} \sim 500$ GeV

$e^+e^- \rightarrow \tilde{W}_1\overline{\tilde{W}}_1$, followed by $\tilde{W}_1 \rightarrow f\bar{f}'\tilde{Z}_1$, where f and f' are quarks or leptons. The signature channels include: (i) four-jet $+E^{\text{miss}}$ events, (ii) lepton + two-jets $+E^{\text{miss}}$ events, and (iii) lepton–antilepton $+E^{\text{miss}}$ events, where E^{miss} denotes the apparent missing energy in the event. Since the production cross section is not suppressed by the p-wave β^3 factor as for scalar pair production, the experimental limit is usually very close to the phase space boundary. Exceptions occur either when $m_{\tilde{W}_1} - m_{\tilde{Z}_1}$ is very small so that the energy of the visible decay products and the momentum carried off by the LSP are both small, or when the sneutrino is rather light and the contribution of the t-channel sneutrino exchange to the production amplitude (which interferes destructively with the s-channel contributions) is significant. Neutralino pair production was also searched for in the $e^+e^- \rightarrow \tilde{Z}_1\tilde{Z}_2$, $\tilde{Z}_2\tilde{Z}_2$ channels, where $\tilde{Z}_2 \rightarrow \tilde{Z}_1 + f\bar{f}$. The production cross sections and decay branching fractions are very parameter dependent, and no model-independent limit on neutralino masses can be extracted. Nonetheless, upper limits on the cross section for various event topologies restrict the parameter space of various models of MSSM sparticle masses.

Searches within mSUGRA

Many SUSY searches have been carried out within the mSUGRA framework, or the MSSM with additional assumptions about degeneracy of sfermions. The limits in Table 15.1 for $\Delta m > 10$ GeV are essentially those that would be obtained in mSUGRA. However, because this framework is very constrained, the limits on the chargino mass together with those on neutralino production cross sections imply a limit $m_{\widetilde{Z}_1} > 50$ GeV on the neutralino LSP for any set of mSUGRA parameters. This serves as an example of the interplay between collider experiments and searches for relic dark matter.

Searches within the mGMSB model

We have seen that SUSY signals may differ from those in the MSSM if the LSP is an ultra-light gravitino as may be the case within the mGMSB framework. In this case, searches would naturally focus on the next-to-lightest SUSY particle (NLSP) which, depending on n_5, is either $\widetilde{\tau}_1$ or the neutralino. The search strategy depends on the lifetime of the NLSP which, as we have seen, can vary over a wide range, depending on the gravitino mass. For the stau NLSP scenario, the negative result of the search for acoplanar tau pairs without any displaced vertices implies $m_{\widetilde{\tau}_1} > 87$ GeV. If the stau is very long-lived so that it decays outside the detector, searches for heavy stable charged particles imply $m_{\widetilde{\tau}_1} > 97$ GeV, while for intermediate lifetimes, searches for tracks with large impact parameters or tracks with kinks lead to a mass bound somewhere in between. For the co-NLSP case, corresponding searches imply $m_{\widetilde{\mu}_R} > 96$ GeV, independent of the smuon lifetime.

For the case of a neutralino LSP decaying outside the detector, sparticle masses are bounded as in Table 15.1. Stronger bounds can be obtained if the neutralino decays via $\widetilde{Z}_1 \rightarrow \gamma \widetilde{G}$ within the detector. Since the neutralino pair production cross section depends on the selectron mass, the limit obtained depends on n_5.[1] For $m_{\widetilde{e}_R} = 1.1 m_{\widetilde{Z}_1}(2m_{\widetilde{Z}_1})$ (this covers the range $n_5 = 1$–4) the negative results of a search for acolinear photon pairs at LEP2 implies that $m_{\widetilde{Z}_1} \gtrsim 92(96)$ GeV.

Searches within the AMSB model

In AMSB models, the chargino $\widetilde{W}_1$ and neutralino $\widetilde{Z}_1$ are expected to be nearly mass degenerate and, as discussed in Section 13.4.1, the visible decay products from chargino decay are very soft. In this case, the bound $m_{\widetilde{W}_1} > 91.9$ GeV in the last row of Table 15.1 applies since the sneutrino is typically quite heavy in this scenario.

15.2.3 SUSY Higgs searches at LEP2

The search for neutral Higgs scalars is especially interesting in the SUSY context because $m_h \lesssim 130$ GeV within the MSSM, and a Higgs boson in this mass range is

---

[1] Recall though that $\widetilde{Z}_1$ is typically the NLSP only for $n_5 = 1$.

what is expected from a global fit of LEP and other electroweak data to the SM.[2] A lighter Higgs boson h, within the kinematic reach of LEP2, could have been produced via

$$e^+e^- \rightarrow Zh \text{ or } Ah \tag{15.4}$$

processes, both of which occur via s-channel Z^0 exchange. Moreover, the two reactions are complementary in the sense that the ZZh and the ZAh coupling cannot both simultaneously vanish (at tree level). The first of these reactions is also the usual process for searching for the SM Higgs boson. While h and A are expected to dominantly decay into $b\bar{b}$ pairs, a variety of final states is possible, including the one with an "invisible h" if the decay $h \rightarrow \widetilde{Z}_1\widetilde{Z}_1$ is allowed.

Shortly before the termination of the LEP2 collider, an excess of events in the four-jet sample with displaced vertices and a "$b\bar{b}$ mass" ~ 114 GeV caused some excitement. However, a final dedicated run of LEP around $\sqrt{s} = 208$ GeV did not unearth any signal and a limit,

$$m_{H_{SM}} > 114.3 \text{ GeV}, \tag{15.5}$$

was obtained on the SM Higgs boson mass. Assuming CP is conserved in the Higgs sector, the same bound also applies to m_h for large values of m_A. However, the LEP collaborations also performed dedicated analyses to search for MSSM Higgs bosons in several channels, but found no signal. Since the masses and couplings of the Higgs bosons to SM particles are determined at tree level by $\tan\beta$ together with any *one* of the physical particle masses (taken to be m_A in Chapter 8), the results of these searches can be conveniently displayed in the $m_A - \tan\beta$ plane as shown in Fig. 15.1. Once radiative corrections are included, the Higgs sector depends also on other SUSY parameters: the excluded region shown is conservative in the sense that SUSY parameters are chosen to maximize m_h for a given value of $\tan\beta$.

15.3 Supersymmetry searches at the Tevatron

The Collider Detector at Fermilab (CDF) and DØ are the major general purpose experiments at the Fermilab Tevatron $p\bar{p}$ collider. During Run 1, when each of these experiments accumulated an integrated luminosity of ~ 100 pb^{-1} at $\sqrt{s} = 1.8$ TeV, the top quark was discovered and its mass determined to be $m_t = 174.3 \pm 5.1$ GeV. The experiments also searched for new physics, albeit with null results. Run 2 of the Tevatron began in 2001 at $\sqrt{s} \simeq 2$ TeV, featuring the Tevatron Main Injector along with upgraded detectors designed to handle the large increase in beam luminosity.

---

[2] If we assume that all couplings remain perturbative out to a very high energy scale, we obtain a model-independent bound $m_h \lesssim 160$ GeV as long as SUSY is broken at the weak scale, to be compared with the corresponding bound of about 200 GeV on the SM Higgs boson mass.

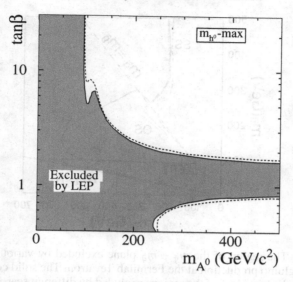

Figure 15.1 The shaded region shows the portion of the $m_A - \tan\beta$ plane excluded by the null results of the searches for MSSM Higgs bosons at LEP2. Here, $M_2 = -\mu = 200$ GeV, $m_{\tilde{g}} = 800$ GeV, all soft SUSY breaking sfermion masses set to 1 TeV and the top squark mixing adjusted to maximize m_h for a given value of $\tan\beta$. The LEP excluded region is sensitive to the value of the top quark mass which is taken to be 179.3 GeV. For this scenario, the LEP data exclude $0.9 \leq \tan\beta \leq 1.5$; this range is also sensitive to the choice of m_t. The dashed lines mark the boundaries of the region that would be expected to be excluded on the basis of Monte Carlo simulations, assuming no signal events. Throughout the analysis, it is assumed that there are no SUSY sources of CP violation. For details of the analysis, see LHWG-Note 2004-01. We thank P. Igo-Kemenes for supplying this figure.

Run 2 is expected to continue at least until 2007 when the CERN LHC pp collider is expected to commence operation. During this run, each experiment is currently expected to accumulate an integrated luminosity of 5–10 fb^{-1}, though higher values were initially anticipated.

15.3.1 Supersymmetry searches at run 1

Searches for gluinos and squarks

The CDF and DØ collaborations have continued the search for squarks and gluinos begun at CERN. The first searches focussed on the multijet $+E_T^{\text{miss}}$ signature from $\tilde{g}\tilde{g}$, $\tilde{g}\tilde{q}$, and $\tilde{q}\tilde{q}$ production, followed by the direct decays of squarks and gluinos to $\tilde{Z}_1$. It was subsequently realized that, as we saw in Chapter 13, heavier squarks and gluinos are more likely to decay via cascades than to decay directly to the LSP, so that the momentum of the LSPs, and hence the E_T^{miss}, is somewhat degraded. The cascade decay patterns are model-dependent, and the mSUGRA model began

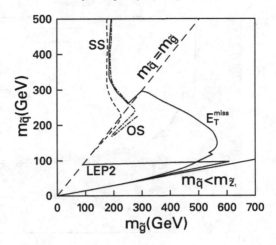

Figure 15.2 The region of the $m_{\tilde{g}} - m_{\tilde{q}}$ plane excluded by various searches for squark and gluino production at the Fermilab Tevatron. The solid contour labeled E_T^{miss} shows the boundary of the regions excluded by different searches in multijet $+E_T^{miss}$ channels, while the dashed (dashed-dotted) contours labeled SS (OS) mark the boundaries of the regions excluded by the CDF (DØ) search in the SS dilepton (OS dielectron) channel. We caution that these searches have been performed in somewhat different models, and refer the reader interested in details to the original papers. We note that the dot-dashed contour is our transcription of the original contour that was presented in the $m_0 - m_{1/2}$ plane of the mSUGRA model. Finally, the region marked LEP2 is excluded by searches for squark pair production at LEP2.

to be adopted for many phenomenological analyses. Within this framework, as we have already seen, squarks can never be much lighter than gluinos. The Tevatron collaborations, quite rightly, disregard this model-dependent restriction, and also perform a search for squark production, assuming that the gluino is heavy. For this purpose, they adopt the MSSM with gaugino mass unification, assuming a mass degeneracy for the three generations of squarks.

The analysis of the E_T^{miss} signal is complicated. To enhance the signal over backgrounds from SM events with neutrinos, or from mismeasurements of jets, carefully designed selection cuts are applied to the data.[3] Moreover, these cuts are optimized, depending on the mass of squarks and gluinos being searched for. The CDF and DØ collaborations have already performed several analyses, using different sets of cuts, but have found no evidence for any excess of events above SM expectations. The region of the $m_{\tilde{g}} - m_{\tilde{q}}$ plane excluded by these searches is summarized by the solid contour labeled E_T^{miss} in Fig. 15.2. This contour is a composite from several Tevatron searches with different selection cuts. In the upper

---

[3] These backgrounds mainly come from W and Z production, vector boson pair production (WW, WZ, and ZZ) and heavy flavor production ($c\bar{c}$, $b\bar{b}$, and $t\bar{t}$).

portion of the plane, the analysis is performed within the mSUGRA framework, but for the lower portion (where squarks much lighter than gluinos are not allowed in the mSUGRA model) the MSSM with gaugino mass unification, and ten flavors of degenerate squarks is used. Since different analyses are used for mSUGRA and the MSSM, the excluded region does not match up when $m_{\tilde{q}} = m_{\tilde{g}}$. The reason that the range of $m_{\tilde{q}}$ excluded by the E_T^{miss} search cuts off for large values of $m_{\tilde{g}}$ is that the LSP mass increases with $m_{\tilde{g}}$, and the transverse momentum carried off by the LSPs is correspondingly reduced. We see from the figure that within the mSUGRA framework, gluinos lighter than 195 GeV are excluded (95% CL) for any value of $m_{\tilde{q}}$ while, if $m_{\tilde{g}} \simeq m_{\tilde{q}}$, the mass limit extends to as much as 300 GeV, depending on the analysis.

Although cascade decays degrade the reach of CDF and DØ E_T^{miss} searches because they soften the E_T^{miss} spectrum, they also lead to novel signatures for gluino and squark production. If daughters $\widetilde{W}_1$ and $\widetilde{Z}_2$ decay leptonically, gluino and squark production leads to events with several jets together with n hard, isolated leptons and E_T^{miss}. Within the SM there is a substantial background from high p_T $W \to \ell\nu$ production if $n = 1$ but, for $n \geq 2$, SM backgrounds are rather small. One important background comes from high p_T $Z^0 \to \ell^+\ell^-$ events which contain opposite sign lepton pairs with the same flavor. These can be easily vetoed by requiring that the dilepton mass not reconstruct to M_Z within some error. Especially interesting are events with same sign dileptons from gluino pair production that we had mentioned in our discussion just below Eq. (13.9) (these may also come from $\tilde{g}\tilde{q}$ or $\tilde{q}\tilde{q}$ production), or events with $n \geq 3$ leptons because SM backgrounds to these event topologies are very small. The cross section for multilepton topologies is suppressed by branching fractions for leptonic decays of charginos and neutralinos and so requires data samples with significant integrated luminosities to obtain a handful of signal events. The dashed contour labeled SS in Fig. 15.2 shows the region excluded by a CDF search in the same sign dilepton channel, while the dot-dashed contour labeled OS shows the corresponding region from the dielectron analysis by the DØ collaboration. These analyses of course depend on the cascade decay patterns which are somewhat model-dependent. For instance, the OS contour, which was obtained within the mSUGRA model framework, terminates at the boundary of parameter space when $m_0 = 0$, while the wedge in it occurs because cascade decay patterns are altered when sleptons and/or sneutrinos become light enough to be produced as decay products of charginos and neutralinos. Our main point, however, is that these leptonic searches, even with an integrated luminosity of just 100 pb^{-1}, are already competitive with the E_T^{miss} search. With the much larger data sample anticipated in Run 2, it may be the case that the rate limited but cleaner same sign dilepton and trilepton event channels will lead to a better reach than the E_T^{miss} channel.

Search for charginos and neutralinos

Charginos and neutralinos are produced via electroweak interactions and so have cross sections comparable to those for pair production of W and Z^0 bosons. Signals from their hadronic decays are buried under QCD backgrounds, so that searches are forced to focus on events containing isolated leptons. Signals from $\widetilde{W}_1\widetilde{Z}_1$ production where $\widetilde{W}_1 \to \ell\nu\widetilde{Z}_1$ are buried under background from the resonantly produced $W \to \ell\nu$ decays. Indeed the most promising signal for chargino and neutralino production comes from hadronically quiet (except for jet activity from QCD radiation) isolated trilepton events expected from $(\widetilde{W}_1 \to \ell\nu_\ell\widetilde{Z}_1) + (\widetilde{Z}_2 \to \ell\bar{\ell}\widetilde{Z}_1)$ production. We have already seen in Fig. 12.22 that for models with gaugino mass unification, $\widetilde{W}_1\widetilde{Z}_2$ production may be the dominant production mechanism for SUSY particle production at the Tevatron, and also that if sleptons are sufficiently light, then the $\widetilde{Z}_2$ leptonic branching fraction may be significantly enhanced. Since leptons from $Z^0 \to \ell^+\ell^-$ can be readily identified, the most serious SM background comes from $W(\to \ell\nu_\ell) + Z(\to \tau\bar{\tau})$ followed by leptonic τ decays, and from $W^{(*)}\gamma^*$ and $W^{(*)}Z^{(*)}$ production, where the off-shell vector bosons "decay" leptonically.

Searches for isolated trilepton events from SUSY have been performed by both CDF and DØ for the Run 1 data sample. If the leptonic branching fractions for chargino and neutralino decays are similar to those of the W and Z^0 boson, the chargino mass bound obtained is well below the corresponding LEP2 limit, but exceeds it if leptonic chargino and neutralino decays are enhanced by the presence of light sleptons. These searches are, however, a proof of principle and will yield interesting results when the integrated luminosity levels associated with Run 2 are achieved.[4]

Search for top and bottom squarks

Since third generation squarks are expected to be lighter than other squarks, dedicated searches for these have been performed at the Tevatron by both the CDF and DØ collaborations. If $\tilde{t}_1\bar{\tilde{t}}_1$ production occurs at a large rate at the Tevatron, and $m_{\tilde{t}_1} < m_b + m_{\widetilde{W}_1}$, then $\tilde{t}_1$ is likely to decay dominantly via $\tilde{t}_1 \to c\widetilde{Z}_1$, resulting in a $c\bar{c} + E_T^{\text{miss}}$ final state. For $m_{\widetilde{Z}_1} \lesssim 50$ GeV, the CDF search excludes $m_{\tilde{t}_1}$ up to ~ 110 GeV, extending beyond the reach of LEP2.[5] Searches have also been performed for the case when $\tilde{t}_1 \to b\widetilde{W}_1$. In this case, a reach beyond the current LEP2 bound is obtained only if the leptonic branching fraction of $\widetilde{W}_1$ is large. Assuming that $\widetilde{W}_1 \to \ell\tilde{\nu}$, the combined result of the two collaborations implies that $m_{\tilde{t}_1} \gtrsim 125$–$140$ GeV for $m_{\tilde{\nu}} = 60$–85 GeV and $m_{\widetilde{W}_1}$ beyond the LEP2 bound.

---

[4] We should also emphasize that chargino and neutralino searches are independent of gluino searches via E_T^{miss} events, and in models without gaugino mass unification yield completely independent information.

[5] Since $\sigma(\tilde{t}_1\bar{\tilde{t}}_1)$ is completely determined by $m_{\tilde{t}_1}$, this excluded region is completely determined by $m_{\tilde{t}_1}$ and $m_{\widetilde{Z}_1}$, and is independent of other model parameters.

Searches have also been performed for $t\bar{t}$ production where $t \to \tilde{t}_1 \widetilde{Z}_1$ decay assuming $\tilde{t}_1 \to b\widetilde{W}_1 \to b\ell\tilde{v}$, but significant bounds on $m_{\tilde{t}_1}$ are obtained only if the branching fraction for the SUSY decay of t is in excess of $\sim 45\%$.

Both the DØ and CDF collaborations have also searched for $p\bar{p} \to \tilde{b}_1\bar{\tilde{b}}_1 X$ production assuming $\tilde{b}_1 \to b\widetilde{Z}_1$. The absence of a signal has been interpreted as an exclusion of a portion of the $m_{\tilde{b}_1}$ *vs.* $m_{\widetilde{Z}_1}$ plane. Values of $m_{\tilde{b}_1} \lesssim 130$ GeV are excluded if $m_{\tilde{b}_1} - m_{\widetilde{Z}_1} \gtrsim 50$–$60$ GeV, and $m_{\tilde{b}_1}$ as large as 145 GeV is excluded for low values of $m_{\widetilde{Z}_1}$.

Searches for SUSY in GMSB models

In GMSB models with a neutralino that decays via $\widetilde{Z}_1 \to \gamma\tilde{G}$ as the NLSP, we would expect sparticle production to lead to $\gamma\gamma + \text{jets} + \text{leptons} + E_T^{\text{miss}}$ events. The DØ collaboration found no excess above SM expectation in their inclusive $\gamma\gamma + E_T^{\text{miss}}$ sample and set a limit $m_{\widetilde{Z}_1} > 77$ GeV, corresponding to $m_{\widetilde{W}_1} > 150$ GeV at the 95% CL. From the null result of a search for gravitino pair production tagged by a high E_T jet from the initial state, the CDF collaboration concluded that the gravitino mass must be heavier than about 1.1×10^{-5} eV corresponding to the SUSY breaking scale $\sqrt{F} \geq 215$ GeV.[6] They also concluded that there was no signal in the $\ell + \gamma + E_T^{\text{miss}}$ as well as the b-jet $+ E_T^{\text{miss}}$ channels, although there was a small excess in the first of these channels.

15.3.2 Prospects for future SUSY searches

Run 2 of the Fermilab Tevatron began in 2001. Current expectation is that an integrated luminosity of 5–10 fb^{-1} will be expected before the LHC begins to operate, down from 15–25 fb^{-1} that had been originally anticipated. It is interesting to project the SUSY reach of Tevatron experiments for this vastly larger data sample.

The E_T^{miss} channel

The current limits on charginos from LEP2, that $m_{\widetilde{W}_1} > 103$ GeV given a reasonable mass gap between $\widetilde{W}_1$ and $\widetilde{Z}_1$, imply that $m_{\tilde{g}} \gtrsim 330$–$400$ GeV (depending on the sign of μ) in models with gaugino mass unification, if $\tan\beta \gtrsim 1.5$ as suggested by Fig. 15.1. As already mentioned, in addition to the E_T^{miss} signal, multilepton signals are also potentially important. The size of these signals is sensitive to which sparticles are dominantly produced, and on how they decay, and so are model-dependent. The E_T^{miss} signal is somewhat more robust.

In Fig. 15.3, we show regions of mSUGRA parameter space where the somewhat more robust $E_T^{\text{miss}} + \text{jets}$ signal ought to be visible above SM backgrounds at at least

---

[6] Recall from our discussion of goldstino interactions that this cross section is fixed by the gravitino mass.

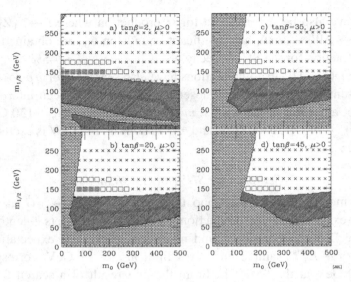

Figure 15.3 A plot of points accessible at 5σ level at Tevatron Run 2 for 2 fb^{-1} (gray squares) and 25 fb^{-1} (white squares) of data in searches for mSUGRA via $E_T^{miss}+$ multijet events. Points with a $\times$ symbol are inaccessible at Run 2 via the $E_T^{miss}+$ jets signal. Reprinted with permission from H. Baer, C. H. Chen, M. Drees, F. Paige and X. Tata, *Phys. Rev.* **D58**, 075008 (1998), copyright (1998) by the American Physical Society.

the 5σ level. The bricked areas are disallowed by a lack of radiative EWSB (low $m_{1/2}$ region) or a charged (slepton) LSP (low m_0 region). The hatched region is excluded by LEP2 searches. The gray squares denote model points where a 5σ signal is expected with 2 fb^{-1} of data, while points denoted by open squares are accessible only for an integrated luminosity of 25 fb^{-1}. We stress that, although it appears that the LEP limit on the chargino excludes much of the parameter plane accessible to experiment, this E_T^{miss} search is still important because it can probe squark and gluino masses without any assumption about gaugino mass unification.

Multilepton channels

Tevatron experiments have already shown that searches via multilepton $+ E_T^{miss}$ events are competitive with the traditional E_T^{miss} search. The signals can naturally be sorted according to the number of isolated leptons contained in each event: $E_T^{miss} +$ jets, $1\ell + E_T^{miss}+$ jets, opposite sign (OS) or same sign (SS) dileptons $+ E_T^{miss} +$ jets and $3\ell + E_T^{miss} +$ jets. We will focus our attention on the trilepton signal which, for large data samples, yields the largest reach in models with gaugino mass unification. Here, we focus on this signal within the mSUGRA framework, but it should be kept in mind that both OS and SS dilepton signals may also be observable.

Within the mSUGRA framework, LEP bounds imply that $\widetilde{W}_1 \widetilde{Z}_2$ and $\widetilde{W}_1 \widetilde{W}_1$ production have the largest sparticle production cross sections at the Tevatron if SUSY is accessible at all. It makes sense to focus on the trilepton signals arising from the former reaction since these have rather low SM backgrounds. The background size can be gauged from the fact that the inclusion of backgrounds from $W^{(*)}\gamma^*$ and $W^{(*)}Z^*$ sources of trileptons is important.

It might seem that the signal appears as trilepton events free from jet activity. Detailed studies, however, show that the largest reach is obtained in the inclusive trilepton channel after suitable cuts, since the production of heavy sparticles is frequently associated with jets from initial state QCD radiation. Moreover, gluino and squark production, which leads to jetty trilepton events, also makes a subdominant contribution to this signal.

For small to intermediate values of $\tan\beta$, the leptons from chargino and neutralino decays are relatively hard and readily detectable. We have seen in Chapter 13 that if $\tan\beta$ is large, decays to third generation leptons and neutrinos are enhanced at the expense of those to the experimentally detectable e and μ. For large $\tan\beta$, the leptons in the $\ell\ell\ell'$ signal ($\ell, \ell' = e, \mu$) arise as secondary daughters from τ decay, and so tend to be soft. It was shown, however, that using a special set of soft lepton cuts, the trilepton signal should be detectable above backgrounds over a wide range of $\tan\beta$.[7]

The region of the m_0 vs. $m_{1/2}$ plane where the trilepton signal is observable at the Tevatron is illustrated in Fig. 15.4 for a moderate and a high value of $\tan\beta$. The dark shaded region on the left is excluded because the stau is the LSP, while the right-hand side is excluded because electroweak symmetry is improperly broken, since $\mu^2 < 0$. Just to the left of this latter boundary, μ^2 is small, and $m_{\widetilde{W}_1} \sim m_{\widetilde{Z}_1} \sim |\mu|$. For small $m_{1/2}$, this is the so-called focus point (FP) region, while for larger $m_{1/2}$ values this has been referred to as the hyperbolic branch (HB).

The light-shaded region is excluded by constraints from LEP2. Below the band, $m_h < 114.1$ GeV. Below the solid (dashed) contours, Tevatron experiments should be able to see the trilepton signal at the 5σ (3σ) level with an integrated luminosity of 10 (25) fb^{-1}. For the $\tan\beta = 10$ case, we see a large signal at low values of m_0 for which charginos and neutralinos decay into real sleptons, so that their leptonic branching ratio is nearly 100%. In this case, the reach extends to $m_{1/2}$ as high as 240–260 GeV. As m_0 increases, these decays are no longer kinematically accessible, and the reach drops sharply. For $m_0 \sim 200$ GeV, $B(\widetilde{Z}_2 \to \ell\ell\widetilde{Z}_1)$ is very small because of the negative interference between the Z and slepton-mediated amplitudes for $\widetilde{Z}_2$ decay and there is no reach via this channel. As m_0 is increased further, the

---

[7] See, e.g., S. Abel *et al.*, Report of the SUGRA Working Group at the *Physics at Run II: SUSY and Higgs Workshop*, hep-ph/0003154, and references cited therein.

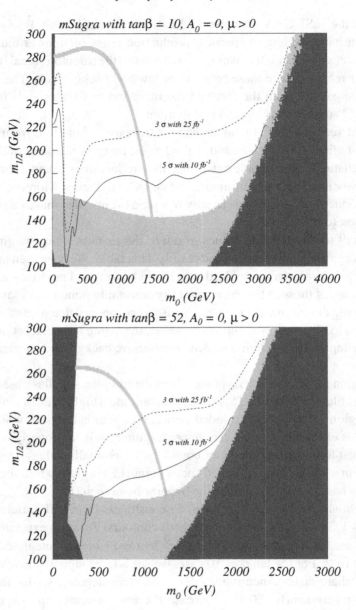

Figure 15.4 The region of the $m_0 - m_{1/2}$ plane where the inclusive trilepton signal with soft leptons is detectable at Tevatron Run 2. The dark-shaded region is excluded by theoretical constraints discussed in the text, while the light-shaded region is excluded by experimental constraints from LEP2. Below the thick light gray line, $m_h < 114.1$ GeV. Reprinted from H. Baer, T. Krupovnickas and X. Tata, *JHEP* **07**, 020 (2003).

slepton-mediated amplitudes become unimportant, and the leptonic branching ratio of $\widetilde{Z}_2$ becomes equal to that of the Z^0 boson, and the contours level off. Finally, for very large values of m_0, we enter the HB/FP region and the chargino becomes increasingly higgsino-like and light, and the contours extend to larger values of $m_{1/2}$. It is important to note that the signal becomes difficult to see because the mass gap between $\widetilde{W}_1$ or $\widetilde{Z}_2$ and the LSP becomes small, and the visible decay products become too soft to pass the experimental cuts. A signal might escape detection even if charginos and neutralinos are well within the kinematic reach of the Tevatron. For larger values of $\tan\beta$, the small m_0 region where the trilepton signal is observable shrinks because the chargino preferentially decays to staus, until this region completely disappears as illustrated for $\tan\beta = 52$ in the second frame of Fig. 15.4. Once again the contours rise in the HB/FP region where the chargino becomes relatively light.

The trilepton signal is important from another point of view. Since the like flavor, opposite sign lepton pair in an $\ell^+\ell^-\ell'^{\pm}$ event arises from $\widetilde{Z}_2 \to \ell^+\ell^-\widetilde{Z}_1$ decay, the $m(\ell^+\ell^-)$ distribution must kinematically be bounded by $m_{\widetilde{Z}_2} - m_{\widetilde{Z}_1}$. If the trilepton signal is sufficiently large, it will be possible to determine this dilepton edge which would then serve as a starting point for reconstructing SUSY particle masses.

Top and bottom squarks

Bottom squark pair production can be searched for at CDF and DØ via the $p\bar{p} \to \tilde{b}_1\bar{\tilde{b}}_1 X \to b\bar{b} + E_T^{\text{miss}}$ reaction. Values of $m_{\tilde{b}_1} \sim 210$ (245) GeV can be probed with 2 (25) fb^{-1} of data, assuming $\tilde{b}_1 \to b\widetilde{Z}_1$, and a large $m_{\tilde{b}_1} - m_{\widetilde{Z}_1}$ mass gap. If $\tilde{b} \to b\widetilde{Z}_2$ also occurs at a significant rate, then the reach will be reduced, but this degradation is typically smaller than 30–40 GeV.

The reaction $p\bar{p} \to \tilde{t}_1\bar{\tilde{t}}_1 X$ followed by $\tilde{t}_1 \to b\widetilde{W}_1$ can be searched for at Run 2 in the $b\bar{b}\ell\nu_\ell q\bar{q}'$ final state, which also occurs in direct $t\bar{t}$ production. This search was not possible at Run 1 due to low cross sections and large backgrounds. Mass values of $m_{\tilde{t}_1} \sim 160$–190 GeV can be probed with 2–20 fb^{-1} of data.

If instead $\tilde{t}_1 \to c\widetilde{Z}_1$ is the dominant decay mode, then the Run 1 search for $c\bar{c} + E_T^{\text{miss}}$ final states can be extended. It is estimated that $m_{\tilde{t}_1} \gtrsim 200$ GeV may be probed in 20 fb^{-1}. Alternatively, if $\tilde{t}_1 \to b\ell\tilde{\nu}_\ell$ dominates, then $m_{\tilde{t}_1}$ as large as 240 GeV can be explored if $m_{\tilde{\nu}}$ is as low as 45 GeV.

Search for SUSY Higgs bosons

One of the most intriguing predictions in the MSSM is that the scalar h is lighter than about 135 GeV. This is in the range favored by analyses of electroweak radiative corrections, and possibly within range of discovery at CDF and DØ. Indeed the lack of any excess of $p\bar{p} \to \phi b\bar{b} \to b\bar{b}b\bar{b}$ events ($\phi = h$, H or A) in their data sample has already allowed the CDF collaboration to exclude a portion of the $m_A - \tan\beta$

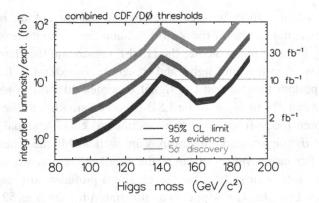

Figure 15.5 Projections for the integrated luminosity required per experiment after combining the data from the CDF and DØ experiments to detect/exclude a SM-like Higgs boson via the channels discussed in the text for a Higgs mass smaller than 130 GeV. For heavier Higgs bosons, not relevant to our discussion, other channels are used. This figure appears in M. Carena *et al.*, Report of the Tevatron Higgs Working Group, FERMILAB-CONF-00-279-T.

plane in Fig. 15.1 with $\tan\beta \geq 50\text{–}100$ for values of m_A not excluded at LEP2. The most important reaction for searching for h at the Tevatron is

$$p\bar{p} \to WhX, \quad W \to \ell\nu_\ell, \quad \text{and} \quad h \to b\bar{b}. \tag{15.6}$$

The signal is an isolated lepton together with two b-jets and E_T^{miss}, where the b-jets are tagged by displaced vertices owing to the long B meson lifetime. The major SM backgrounds come from $Wb\bar{b}$ and $t\bar{t}$ production. The signal is not large, but can be enhanced relative to background because the jet–jet mass is expected to cluster around m_h. Its statistical significance is sensitive to the efficiency for b-tagging and the jet–jet mass resolution that will be attained. This significance can be further enhanced by including signals in other event topologies from Zh, $t\bar{t}h$, and $b\bar{b}h$ production. The results of a detailed analysis (including neural net improvement) of the integrated luminosity required per experiment, after combining the signals in the $\ell b\bar{b} + E_T^{\text{miss}}$, $b\bar{b} + E_T^{\text{miss}}$, and $b\bar{b}\ell^+\ell^-$ channels, is shown in Fig. 15.5 for a SM Higgs boson.[8]

One striking implication of Fig. 15.5 is that given an integrated luminosity of 30–40 fb^{-1}, Tevatron experiments would have an excellent chance of discovering h, or apparently ruling out the MSSM. We should, however, be careful before jumping to such a strong conclusion. First, the upper bound on m_h depends on assumptions about how large third generation squark masses and A-parameters might be. Second,

[8] See M. Carena *et al.*, Report of the Tevatron Higgs Working Group, *Physics at Run II: Supersymmetry/Higgs Workshop*, hep-ph/0010338.

the reach shown in these plots depends on projections for jet–jet mass resolutions and b-tagging efficiencies during Run 2. In addition, recent projections indicate that Tevatron experiments will accumulate an integrated luminosity of 5–10 fb^{-1}, in which case the reach is considerably smaller. In this case, tantalizing 3σ effects may be observable if $m_h \lesssim 125$ GeV.

It is worth noting that other neutral Higgs bosons may be accessible to Tevatron searches if $\tan\beta$ is very large. Because the bottom Yukawa coupling increases with $\tan\beta$, the reactions

$$p\bar{p} \to Ab\bar{b}X, \; Hb\bar{b}X \tag{15.7}$$

may probe m_A as large as 160–200 GeV, with 25 fb^{-1} of data. Charged Higgs bosons are generally more difficult to detect.

GMSB models

If an ultra-light gravitino is the LSP, SUSY signals at colliders are sensitive to the identity of the NLSP, which is either the neutralino or the lighter stau (possibly with other sleptons essentially degenerate with the stau) within the mGMSB framework. The decay of the NLSP leads to isolated photons, leptons, or even Z^0 and Higgs bosons, as we saw in Section 13.8.2, in addition to jets, leptons, and E_T^{miss} expected within the MSSM. Moreover, the NLSP decay may be either prompt or delayed: the latter leads to a variety of novel handles for enhancing the SUSY signal. These include: displaced vertices, tracks with kinks, and tracks corresponding to charged quasi-stable heavy exotics, in addition to the visible daughters from NLSP decays.

In order to assess the Tevatron reach for GMSB models, it is expedient to analyze various "model lines" characterized by the decay properties of the NLSP. For each of these model lines, the reach is evaluated in terms of Λ (which can then be translated to the mass of any sparticle, for instance, the gluino), assuming that the NLSP decays promptly. This is a conservative assumption since delayed decays would serve to enhance the reach. All the model lines have $M = 3\Lambda$ and $C_{\text{grav}} = 1$, and are characterized by:

- **A.** A bino-like NLSP that mainly decays via $\widetilde{Z}_1 \to \gamma\widetilde{G}$, for model parameters $n_5 = 1$, $\tan\beta = 2.5$, and $\mu > 0$. SUSY events typically contain two isolated photons in addition to jets, leptons, and E_T^{miss}.
- **B.** A stau NLSP in models with $n_5 = 2$, $\tan\beta = 15$, and $\mu > 0$. In this case, sparticles cascade decay to $\tilde{\tau}_1$, which then decays via $\tilde{\tau}_1 \to \tau\widetilde{G}$.
- **C.** A stau–selectron–smuon co-NLSP for model parameters $n_5 = 3$, $\tan\beta = 3$, and $\mu > 0$. SUSY events are then expected to be rich in relatively easily detectable leptons from the decay of the NLSPs in this scenario.

- **D**. A higgsino-like NLSP model line where the NLSP mainly decays via $\widetilde{Z}_1 \rightarrow h\widetilde{G}$ as long as it is not kinematically suppressed. This does not occur in the mGMSB model where the NLSP tends to be bino-like. However, since the signals are so sensitively dependent on the decay of the NLSP, it is worthwhile to explore this non-canonical scenario and study just how the reach of Tevatron experiments is affected.[9] The model line examined has $n_5 = 2$, $\tan\beta = 3$ with $\mu = -\frac{3}{4}M_1$ to obtain a light higgsino. The Higgs boson yields SUSY events rich in b-jets.
- **E**. A higgsino-like NLSP which dominantly decays via $\widetilde{Z}_1 \rightarrow Z^0\widetilde{G}$ as long as the decay is not kinematically suppressed. It has the same parameters as model line **D**, except that $\mu = +\frac{3}{4}M_1$.

The reach of the Fermilab Tevatron for an integrated luminosity of 25 fb^{-1} is summarized in Table 15.2 where, in addition to the reach in Λ, we have shown the corresponding value of $m_{\tilde{g}}$ to compare with the reach in other models. We have also listed the event topology that yields the largest reach. We stress again that the reach shown is conservative in that if the NLSP has a long lifetime, the reach may be significantly larger. For instance, by searching for highly ionizing tracks from $\tilde{\tau}_1$ in model line **B**, or tracks with displaced kinks if $\tilde{\tau}_1$ decays within the detector but far from the production point, Λ values as high as ~ 85 TeV can be probed for 30 fb^{-1} of data.

15.4 Supersymmetry searches at supercolliders

The CERN LHC pp collider is scheduled to begin operation in 2007, at $\sqrt{s} \simeq 14$ TeV. Initial runs are expected to accumulate 10 fb^{-1} of integrated luminosity, while several hundred fb^{-1} of data are ultimately expected to be recorded. For gluino and squark masses smaller than ~ 1 TeV, we can see from Fig. 12.14 that several hundred thousand SUSY events would be expected in this data sample!

There is a developing consensus in the high energy physics community that the next big accelerator project should be an electron–positron linear collider operating at a center of mass energy $\sqrt{s} = 500$ GeV which would be upgradeable to $\sqrt{s} = 0.8$–1 TeV in the second stage. At the start of Section 12.2 we have already discussed the special advantages of these machines for studying new physics, and also the sense in which these could complement the data from the LHC.

In a discussion of supersymmetry at supercolliders, we need to address two conceptually distinct issues.

---

[9] It is worth noting that additional interactions needed to generate μ and $B\mu$ in this framework could alter the relation between μ and the gaugino masses making such a scenario more plausible.

Table 15.2 *A comparison of the SUSY reach of the Tevatron luminosity upgrade and the LHC for the various model lines of the GMSB framework that were introduced in the text, with the reach in the mSUGRA and AMSB models. For the GMSB model lines, we also show the dominant decay of the NLSP together with the channel that yields the largest reach. For the mSUGRA model, a significantly higher reach in $m_{\tilde{g}}$ is possible, both at the Tevatron as well as at the LHC, if $m_0 \ll m_{1/2}$. For the mAMSB model, the corresponding reach is also larger when m_0 is smaller than in the case that is shown. Studies of the Tevatron reach within the AMSB model are not available.*

Model line	NLSP	Tevatron (25 fb^{-1})	LHC (10 fb^{-1})
A	$\tilde{Z}_1 \sim \tilde{B}$ $\tilde{Z}_1 \to \gamma \tilde{G}$	$\Lambda \cong 115$ TeV, $m_{\tilde{g}/\tilde{q}} \sim 0.87$ TeV, $ll\gamma\gamma + E_T^{\mathrm{miss}}$	$\Lambda \cong 400$ TeV $m_{\tilde{g}/\tilde{q}} \sim 2.8$ TeV, $\gamma\gamma + E_T^{\mathrm{miss}}$
B	$\tilde{\tau}_1$	$\Lambda \cong 53$ TeV, $m_{\tilde{g}/\tilde{q}} \sim 0.82$ TeV, Clean channels $3l + 1\tau 2l + 1\tau 3l$ $+2\tau 1l + 3\tau 2l$	$\Lambda \cong 150$ TeV $m_{\tilde{g}/\tilde{q}} \sim 2.0$ TeV, $3l + E_T^{\mathrm{miss}}$
C	$\tilde{\tau}_1, \tilde{e}_R, \tilde{\mu}_R$	$\Lambda \cong 60$ TeV, $m_{\tilde{g}/\tilde{q}} \sim 1.3$ TeV, $\geq 4l + E_T^{\mathrm{miss}}$	$\Lambda \cong 155$ TeV $m_{\tilde{g}/\tilde{q}} \sim 3.0$ TeV, $4l + E_T^{\mathrm{miss}}$
D	$\tilde{Z}_1 \sim \tilde{h}$ $\tilde{Z}_1 \to h\tilde{G}$	$\Lambda \cong 105$ TeV, $m_{\tilde{g}/\tilde{q}} \sim 1.5$ TeV, $\geq 3b\text{-jets} + E_T^{\mathrm{miss}}$	$\Lambda \cong 140$ TeV $m_{\tilde{g}/\tilde{q}} \sim 2.0$ TeV, $\geq 2b\text{-jets} + E_T^{\mathrm{miss}}$
E	$\tilde{Z}_1 \sim \tilde{h}$ $\tilde{Z}_1 \to Z\tilde{G}$	$\Lambda \cong 120$ TeV, $m_{\tilde{g}/\tilde{q}} \sim 1.3$ TeV, $\gamma\gamma + E_T^{\mathrm{miss}}$	$\Lambda \cong 140$ TeV $m_{\tilde{g}/\tilde{q}} \sim 2.0$ TeV, $1l + E_T^{\mathrm{miss}}$ Increase to 2.2 TeV via $Z\gamma + E_T^{\mathrm{miss}}$ if excellent jet-γ rejection is available.
Reach in mSUGRA		$\tilde{m}_{\tilde{g}} \sim 0.35 - 0.4$ TeV E_T^{miss}	~ 1.6 TeV ($m_{\tilde{q}} \gg m_{\tilde{g}}$) $\ell + E_T^{\mathrm{miss}}$ ~ 2.2 TeV ($m_{\tilde{q}} \sim m_{\tilde{g}}$) $\ell + E_T^{\mathrm{miss}}$
Reach in mAMSB			~ 1.4 TeV ($m_{\tilde{q}} \gg m_{\tilde{g}}$) E_T^{miss} ~ 2 TeV ($m_{\tilde{q}} \sim m_{\tilde{g}}$) $\ell^+\ell^- + E_T^{\mathrm{miss}}$

- The first concerns the reach of these machines for the different sparticles. The LHC is a broad band machine, where everything possible will be produced, though cross sections for the production of various sparticles will be very different. At LCs, all sparticles with non-vanishing $SU(2)_L \times U(1)_Y$ quantum numbers will be produced with comparable cross sections, and the reach will essentially be determined by the mass of the lightest visible sparticle. While it would be best to have a program of SUSY searches that is as model-independent as possible, it is also interesting to map out the reach of supercollider experiments for various SUSY models discussed in Chapter 11, and examine this in light of other constraints on the model parameter space.

- The second issue concerns how we would proceed if new physics is indeed discovered at the LHC. As discussed above, we would need to establish that the new physics is indeed softly broken supersymmetry. In this connection, we would embark upon a program of precision measurements of sparticle masses and other properties to unravel the mechanism by which sparticles obtain their masses, and ultimately determine the underlying physics and its associated parameters. We will postpone our discussion of this to the next section, while initially focussing upon the question of the SUSY reach.

15.4.1 Reach of the CERN LHC

We have seen that, in order to obtain an accurate representation of SUSY events for sparticles in the range of masses accessible at the LHC, it is essential to incorporate cascade decays. This is difficult to do within the MSSM because of the large number of free parameters. Instead, we use the various models introduced in Chapter 11 as a guide to our projections for the reach of the LHC. The other advantage of this procedure is that, because a large number of sparticles are expected to be simultaneously produced, contributions from all sparticle reactions to any particular event topology can be included in our exploration of the reach in that topology.

mSUGRA model

As we saw in Chapter 12, $\tilde{g}\tilde{g}$, $\tilde{g}\tilde{q}$, and $\tilde{q}\tilde{q}$ production processes are expected to be the dominant sparticle production mechanisms at the LHC. The cascade decay signatures will generally be very complex and give rise to events with jets, isolated leptons, and possibly isolated photons or Z^0 bosons (re-constructed via their leptonic decays) together with E_T^{miss}. Jets from primary decay of the squark or gluino can be very hard, reflecting the parent sparticle mass. Leptons (as well as other jets) that originate further down the cascade chain are typically softer than the primary jets in these events.

The reach of the CERN LHC in the mSUGRA model has been evaluated by several groups.[10] The event topologies can be classified as before by the number of identified isolated leptons in the events:

1. E_T^{miss} channel: an inclusive channel requiring large E_T^{miss} plus ≥ 2 jets plus any number of identified leptons,
2. 0ℓ channel: a subset of the E_T^{miss} channel which in addition vetoes any isolated leptons,
3. 1ℓ channel: a subset of E_T^{miss} containing a single isolated lepton,
4. OS channel: a subset of E_T^{miss} containing two opposite-sign isolated leptons,
5. SS channel: a subset of E_T^{miss} containing two same-sign isolated leptons,
6. 3ℓ channel: a subset of E_T^{miss} containing three isolated leptons.

Larger lepton multiplicities can also occur, but at lower rates.

The SUSY reach of the LHC within the framework of the mSUGRA model is illustrated in Fig. 15.6 in the $m_0 - m_{1/2}$ plane, with $A_0 = 0$, $\tan \beta = 30$, and $\mu > 0$. As before, the dark (light) shaded regions are excluded by theoretical (experimental) constraints. Also shown are contours where $m_{\tilde{g}}$ or $m_{\tilde{u}_L}$ is 2 TeV. In the figure, many sets of cuts were examined. For each point in the plane, the cuts were chosen to optimize the signal relative to the background. The region below the various curves is where LHC experiments should be able to see a signal at the 5σ level with a minimum of ten signal events in the event topology shown on the contour, assuming an integrated luminosity of 100 fb^{-1}. The cumulative reach in all the channels is shown by the solid contour labelled E_T^{miss}. We see that LHC experiments should be able to explore $m_{1/2}$ values up to 1400 (700) GeV for small (very large) values of m_0, corresponding to $m_{\tilde{g}} = 3(1.8)$ TeV. Moreover, if $m_{\tilde{g}} \lesssim 1.5\text{--}2$ TeV, there should be an observable signal in several channels if the observed signal is to be attributed to SUSY as realized in this framework. The reach results are qualitatively similar for other values of $\tan \beta$ or the opposite sign of μ.

It is also worth mentioning that the trilepton signal from $\widetilde{W}_1 \widetilde{Z}_2$ production may also be observable above backgrounds at the LHC provided $m_{1/2}$ is not too large.[11] For large values of $m_{1/2}$ the two-body decay $\widetilde{Z}_2 \to \widetilde{Z}_1 h$ or $\widetilde{Z}_2 \to \widetilde{Z}_1 Z$ becomes accessible and quickly dominates the $\widetilde{Z}_2$ decay rate unless sleptons are also light so that $\widetilde{Z}_2 \to \tilde{\ell}_{L,R}\ell$ decays are also accessible. Direct production of sleptons leads to an observable signal (above W^+W^- and $t\bar{t}$ backgrounds) in the $\ell^+\ell^- + E_T^{miss}$ channel if sleptons are lighter than 250 GeV (300 GeV if soft jets can be efficiently vetoed).[12]

[10] H. Baer *et al.*, *Phys. Rev.* **D52**, 2746 (1995), *Phys. Rev.* **D53**, 6241 (1996) and *Phys. Rev.* **D59**, 055014 (1999); S. Abdullin and F. Charles, *Nucl. Phys.* **B547**, 60 (1999); S. Abdullin *et al.* (CMS Collaboration), hep-ph/9806366 (1998); B. Allanach *et al.*, *JHEP* **08**, 017 (2000); H. Baer *et al.*, *JHEP* **0306**, 054 (2003).
[11] H. Baer *et al.*, *Phys. Rev.* **D50**, 4508 (1994); I. Iashvili and A. Kharchilava, *Nucl. Phys.* **B526**, 153 (1998).
[12] H. Baer, C. H. Chen, F. Paige and X. Tata, *Phys. Rev.* **D49**, 3283 (1994); D. Denegri, W. Majerotto and L. Rurua, *Phys. Rev.* **D58**, 095010 (1998).

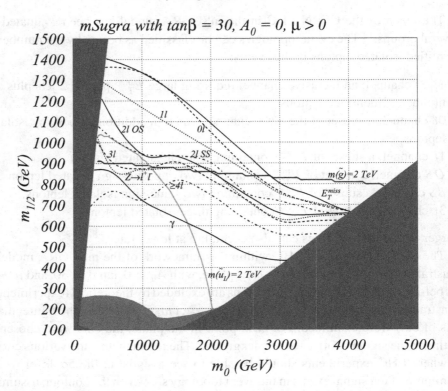

Figure 15.6 The 5σ reach of the CERN LHC in the $m_0 - m_{1/2}$ plane of the mSUGRA model for an integrated luminosity of 100 fb^{-1}. The shaded regions are excluded by theoretical and experimental constraints discussed in the text. Below each of the labelled contours, there should be an observable signal at the LHC in the corresponding channel. Reprinted from H. Baer, C. Balázs, A. Belyaev, T. Krupovnickas and X. Tata, *JHEP* **06**, 054 (2003).

In Fig. 15.7, we illustrate the interplay between various measurements within a constrained framework, using mSUGRA with the same parameters as in the previous figure as an example. The dark shaded regions are excluded by theoretical considerations as shown on the figure, while the light shaded region (labelled LEP2) is excluded by the chargino constraint from LEP2 experiments. Below the unlabeled contour starting around $m_{1/2} = 270$ GeV, $m_h < 114$ GeV. The jagged circular contours labeled 2 and 3 are contours above which $B(B \rightarrow X_s \gamma) > 2(3) \times 10^{-4}$, the region favored by experiment. The slanted lines labeled 1, 2, 5, ... 40 are contours of the SUSY contribution to a_μ, the anomalous magnetic moment of the muon. Between the dotted/dashed contours along the boundaries of the theoretically excluded regions, the neutralino relic density agrees with its determination by the WMAP collaboration, while the corresponding solid line is the contour of $\Omega_{\tilde{Z}_1} h^2 = 1$. The

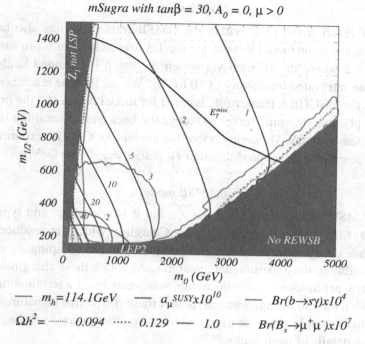

Figure 15.7 The SUSY reach of the CERN LHC within the mSUGRA model, together with contours of $B(B \to X_s \gamma)$, a_μ^{SUSY}, and the neutralino relic density. In the lighter-shaded lower part of the theoretically excluded wedge region on the left, the stau mass squared is negative. Reprinted from H. Baer, C. Balázs, A. Belyaev, T. Krupovnickas and X. Tata, *JHEP* **06**, 054 (2003).

WMAP experiment requires us to lie in the narrow slivers between the theoretically excluded region, and the dashed line, where neutralinos can annihilate efficiently either due to co-annihilation with staus (left side WMAP region) or due to a significant higgsino admixture of the $\widetilde{Z}_1$ in the HB/FP region at large m_0. The contour labeled E_T^{miss} shows the cumulative reach of LHC experiments as we have just discussed. We see that essentially the entire stau co-annihilation region can be probed at the LHC. The HB/FP region, however, continues indefinitely, and new strategies may be needed to extend the reach in this region.[13] An unambiguous observation of a deviation from SM expectation of the muon anomalous magnetic moment or of non-standard flavor-violating decays of B or B_s mesons will preclude nature from being in the part of the HB/FP region that is beyond the reach of the LHC. If such a deviation is to be attributed to the mSUGRA realization of SUSY, then there must be observable signals at the LHC.

---

[13] For very large values of $\tan \beta$ there is another WMAP allowed region where neutralinos can annihilate efficiently via H and A exchange in the s-channel. Again, LHC experiments can probe most, but not all, of this region.

GMSB models

The SUSY reach at the LHC within the GMSB framework has also been computed, using the same model lines as for the Tevatron. The results are summarized in Table 15.2 where the channel via which the reach is obtained is also shown assuming an integrated luminosity of 10 fb^{-1}.[14] We see that the reach is *at least* as good as in the mSUGRA framework, but that for model line **A** (**C**) the presence of additional photons (leptons) serves to reduce the background resulting in a significantly increased reach. We mention that for model line **C**, LHC experiments will be able to search for direct production of $\tilde{\ell}_R$ pairs if $m_{\tilde{\ell}_R} \lesssim 280$ GeV.

mAMSB model

In the mAMSB model, the LSP is the $\widetilde{Z}_1$, but it is wino-like, and typically just $\sim$ 160–200 MeV lighter than the chargino. Charginos which are produced directly or in cascade decays decay to a soft charged pion plus the escaping $\widetilde{Z}_1$ so that it is nearly invisible in the experimental apparatus. Although these charginos typically fly just a few centimeters before decaying, some may leave a terminating track, or a track with a kink in the apparatus. Whether these distinctive signatures of SUSY events (which would have to be triggered by some other means) will be observable depends on details of the detector.[15]

It is interesting to explore the LHC reach using the general search strategies for SUSY. It is expedient to present our results for the reach via various multijet + multilepton + E_T^{miss} channels in the $m_0 - m_{3/2}$ plane. Sample results are shown in Fig. 15.8 for $\tan \beta = 35$ and $\mu > 0$. In this framework, $\tilde{q}_R$ mainly decays to the bino-like $\widetilde{Z}_2$ (if this decay is kinematically allowed); the subsequent $\widetilde{Z}_2$ decays give rise to isolated leptons. In contrast, $\tilde{q}_L$ decays to $\widetilde{Z}_1$ or $\widetilde{W}_1$, and gives jets + E_T^{miss}. The situation with cascades is just the opposite of models with gaugino mass unification where it is $\tilde{q}_L$ that cascade decays while $\tilde{q}_R$ mostly decays directly to the LSP. In the low m_0 large $m_{3/2}$ region, $\tilde{g} \to \tilde{t}_1 t$, which gives rise to leptons from top and stop decay. The best reach is in the OS dilepton channel where values of $m_{\tilde{g}} \gtrsim 2$ TeV can be probed in 10 fb^{-1} of data. At high m_0, $\tilde{g} \to q\bar{q}\widetilde{Z}_1$ or $q\bar{q}'\widetilde{W}_1$, and the best reach

---

[14] H. Baer *et al.*, *Phys. Rev.* **D62**, 095007 (2000).

[15] In a typical collider experiment, it is not possible to record every event because the collision rate is too large for the data acquisition system to handle. Most of these events are small angle elastic or quasi-elastic collisions and not of any interest. In order to ensure that potentially interesting events are all recorded without the data acquisition system being completely swamped, experimentalists set up loose criteria that events must satisfy in order to be recorded. These criteria, referred to as trigger requirements, could for instance require the presence of high E_T jets, isolated hard leptons or photons, or large amounts of E_T^{miss} to reduce event rates to manageable levels. The challenge is to arrive at a decision as to whether or not to record an event in a short time, since collisions are continually occuring in the apparatus. The development of triggers is a complicated but essential issue for all collider experiments, but especially so at the hadron colliders where the total cross section is very large.

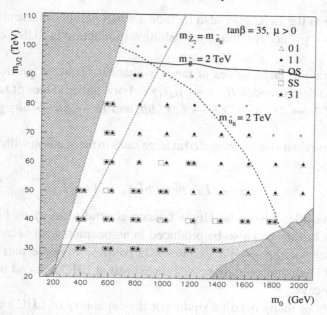

Figure 15.8 The reach of the CERN LHC for mAMSB for $\tan\beta = 35$, $\mu > 0$, and $10\ \mathrm{fb}^{-1}$ of integrated luminosity. Reprinted with permission from H. Baer, J. K. Mizukoshi and X. Tata, *Phys. Lett.* **B488**, 367 (2000).

occurs in the $0\ell + \mathrm{jets} + E_T^{\mathrm{miss}}$ channel, where values of $m_{\tilde{g}} \sim 1350\ \mathrm{GeV}$ may be probed with just $10\ \mathrm{fb}^{-1}$ of data.

LHC reach for SUSY Higgs bosons

The experiments at LEP2 have already placed stringent bounds on Higgs boson masses in the MSSM, and Tevatron experiments may well find evidence for the light scalar h before LHC turns on. Nevertheless, it will be an important task for the CMS and ATLAS experiments to establish the Higgs boson content of the MSSM, and to determine as much as possible about their properties.

The Higgs boson search is complicated and will have to be performed using many channels. For h produced in the s-channel via gg fusion, SM backgrounds preclude the possibility of seeing a signal from its dominant decays $h \to b\bar{b}$ or $h \to \tau^+\tau^-$; the rare decay

$$pp \to hX; \ h \to \gamma\gamma$$

appears to be viable, but will require several years of LHC operation to establish a signal. Excellent electromagnetic calorimetry is essential to see the $h \to \gamma\gamma$ mass bump above the enormous $q\bar{q}, gg \to \gamma\gamma$ continuum background. This will yield an accurate determination of m_h. If squarks and gluinos are not too heavy, the SUSY event sample at the LHC may also include a small number of events with clearly identified $h \to \gamma\gamma$ decays, thus establishing h production in SUSY cascade decay events.

For moderate to large values of $\tan\beta$, s-channel H and A production may be visible via the decay modes $H, A \to \mu\bar{\mu}, \tau\bar{\tau}$. For smaller values of $\tan\beta$, $H, A \to t\bar{t}$, $H \to ZZ^{(*)} \to 4\ell$, $A \to Zh \to \ell^+\ell^- b\bar{b}$, and $H \to hh \to b\bar{b}\gamma\gamma$ may also be observable.

Higgs bosons can also be produced at large rates in association with heavy quarks. The reactions

$$pp \to t\bar{t}h, \; b\bar{b}h, \; b\bar{b}A, \; \text{and} \; b\bar{b}H$$

may all be visible, where the Higgs bosons generally decay to $b\bar{b}$ or $\gamma\gamma$ final states. Higgs bosons can also be produced in association with vector bosons, and their detection via $pp \to Wh \to \ell\nu_\ell\gamma\gamma$ is possible in some part of the plane. The charged Higgs boson may be visible as well at LHC if it can be produced in $t \to bH^+$ decays.

The results of many detailed studies of the capability of LHC experiments are summarized in Fig. 15.9, where it is assumed that Higgs bosons cannot decay to sparticles. It appears that over essentially the entire $m_A - \tan\beta$ parameter space, LHC experiments should be able to discover at least one Higgs boson. The search for SUSY Higgs bosons in many of these channels is difficult, and very large integrated luminosities and excellent detector performance will be necessary. Even so, a small region around $m_A \sim 150$ GeV and $\tan\beta \sim 5-10$ seems difficult, and requires further improvement in the resolution of $b\bar{b}$ dijet invariant masses. Fortunately, Higgs bosons in this "hole" should be easy to study at a 500 GeV e^+e^- collider. It is also gratifying to see that over significant portions of the plane there is an observable signal from more than one Higgs boson: this may serve to distinguish the MSSM Higgs sector from that of the SM.

If SUSY particles are accessible in LHC experiments, it is quite possible that the lightest Higgs scalar h will be discovered first in the SUSY particle event sample as a $h \to b\bar{b}$ mass bump. The parameter space "hole" mentioned above might be explored in this way. Moreover, if some sparticles are light, then Higgs bosons will have significant branching fractions for decays to SUSY particles. Higgs boson decays to SUSY particles will in general diminish the SM decay modes, and may make the search modes listed in Fig. 15.9 more difficult. Decays of neutral Higgs bosons to $\widetilde{Z}_1\widetilde{Z}_1$ states would yield "invisible" Higgs bosons. It is also possible that Higgs boson decays to SUSY particles will open up new, sometimes spectacular, search channels. As an example, H may decay via $H \to \widetilde{Z}_2\widetilde{Z}_2 \to \ell\bar{\ell}\ell'\bar{\ell}' + E_{\mathrm{T}}^{\mathrm{miss}}$. The 4ℓ final state will have an invariant mass $\leq (m_H - 2m_{\widetilde{Z}_1})$, and can be visible over restricted regions of MSSM parameter space.

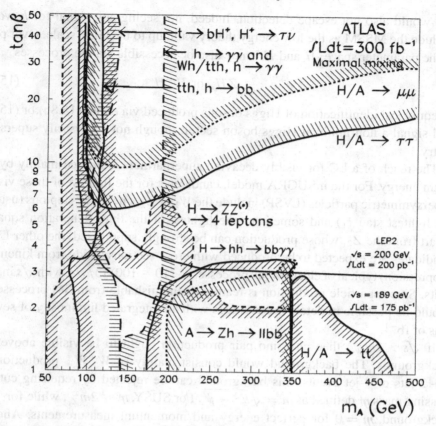

Figure 15.9 The reach of the CERN LHC for SUSY Higgs bosons in the case of heavy sparticles. The signal is detectable on the side of the contour where the shading appears. This figure is reprinted from the ATLAS Technical Design Report.

15.4.2 SUSY reach of e^+e^- colliders

Since $m_h \lesssim 130$–135 GeV in the MSSM, an e^+e^- collider operating at $\sqrt{s} \geq 500$ GeV is sure to access the lightest SUSY Higgs scalar h. If the couplings of h are nearly those of the SM Higgs boson (as it is over much of mSUGRA parameter space) the cross section for the "Higgsstrahlung" process

$$e^+e^- \to Zh \qquad\qquad (15.8a)$$

is large and offers a good channel for h detection above SM backgrounds. The ZZh coupling can become rather small if m_A is light; in this case, the ZhA coupling is necessarily large so that h would be produced via

$$e^+e^- \to Ah, \qquad\qquad (15.8b)$$

and would also not escape detection. Indeed, not seeing any signal for h would exclude the MSSM as the low energy theory valid up to the GUT scale. Over parts of the parameter space, H and $H^{\pm}$ may also be accessible via the processes

$$e^+e^- \to ZH, \text{ or } H^+H^-. \tag{15.8c}$$

Unequivocal identification of Higgs bosons produced via either (15.8b) or (15.8c) will signal a non-minimal Higgs boson sector, though not necessarily supersymmetry.

The reach of a LC for visibly decaying superpartners is limited mainly by the beam energy. For the mSUGRA model, candidates for the lightest of these visible supersymmetric particles (LVSP) include the $\widetilde{W}_1$ or $\widetilde{Z}_2$, one of the sleptons (usually the lightest stau $\tilde{\tau}_1$) and sometimes the lightest of the third generation squarks. Apart from the $\widetilde{Z}_2$, whose production can be strongly suppressed, the other LVSP candidates are expected to be produced with cross sections (aside from kinematic suppression) typical of electroweak processes: $\sim(10 - 100)$ fb/$\sqrt{s}$, with $\sqrt{s}$ in TeV units. Since sparticle production is readily distinguishable from SM processes, it should be possible to detect these at LCs with an integrated luminosity of several tens of fb^{-1}.

If $\sqrt{s} > 2m_{\widetilde{W}_1}$, then chargino pair production ought to be visible above SM backgrounds. The background would consist mainly of W^+W^- production. In $\ell + 2$-jets or 4-jet events, this background can be rejected by requiring cuts on missing mass $m\!\!\!/$ defined as $m\!\!\!/ = \sqrt{E\!\!\!/^2 - p\!\!\!/^2}$. For SUSY, $m\!\!\!/ > 2m_{\widetilde{Z}_1}$, while for WW background, $m\!\!\!/ = 0$ for perfect energy and momentum measurements. Another discriminator in $\ell + 2$-jet events is the distribution in E_{jj}, the energy of all jets: for WW production, $E_{jj} = E_W = \sqrt{s}/2$, while for $\widetilde{W}_1^+\widetilde{W}_1^-$ production with three-body $\widetilde{W}_1^{\pm}$ decays, there is a continuum of values. In the HB/FP region where $|\mu| \lesssim |M_2|$, the chargino and neutralino become close in mass and the visible energy is small. In this case, specialized cuts are needed to select the signal over the various SM backgrounds that, in this case, include $2 \to 3$ and $2 \to 4$ processes.[16]

If instead $\tilde{\tau}_1$ is the LVSP, or several sleptons are co-LVSPs, then the signature is

$$e^+e^- \to \tilde{\ell}^+\tilde{\ell}^- \to \ell^+\ell^- + E\!\!\!/. \tag{15.9}$$

The presence of acoplanar OS dilepton pairs in excess of expectations from WW and ZZ production would signal the production of sleptons. The scalar pair production reactions are suppressed by the usual β^3 factor near threshold. In addition, in mSUGRA it is possible to have nearly degenerate $\tilde{\ell}$ and $\widetilde{Z}_1$, in which case the visible energy from slepton decay will be small, and detection efficiency will be reduced.

---

[16] H. Baer *et al.*, *JHEP* **02**, 007 (2004).

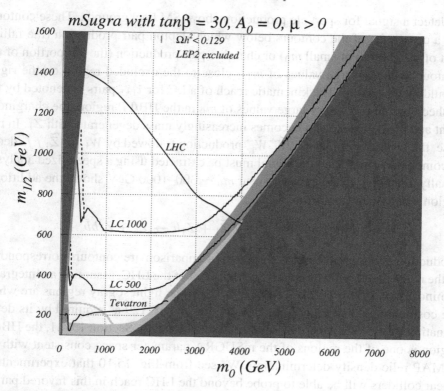

Figure 15.10 The SUSY reach of an e^+e^- LC with $\sqrt{s} = 500$ and 1000 GeV within the mSUGRA model with $A_0 = 0$, $\tan\beta = 30$, and $\mu > 0$, assuming an integrated luminosity of 100 fb^{-1}. The darkest (lightest) shaded regions are excluded by theoretical (experimental) constraints. Below the gray arc in the bottom left corner, $m_h < 114$ GeV. The medium gray shaded regions of the plane run along the boundary of the theoretically excluded wedge at small values of m_0, and in the HB/FP region close to the boundary of the theoretically excluded region on the right: in these regions, the predicted neutralino relic density is consistent with the results of the WMAP collaboration. Finally, contours showing the reaches of Fermilab Tevatron upgrades assuming an integrated luminosity of 25 fb^{-1}, and the CERN LHC with 100 fb^{-1} of integrated luminosity, are also shown for comparison. Reprinted from H. Baer, A. Belyaev, T. Krupovnickas and X. Tata, *JHEP* **02**, 007 (2004).

Our projection for the reach of an e^+e^- LC with $\sqrt{s} = 500$ or 1000 GeV is shown in Fig. 15.10, assuming an integrated luminosity of 100 fb^{-1}. We work within the mSUGRA framework and, as in Fig. 15.6, show the reach in the $m_0 - m_{1/2}$ plane, and fix $A_0 = 0$, $\tan\beta = 30$ and $\mu > 0$. The darkest region is excluded by theoretical constraints that we have already discussed, while the medium gray region at low values of $m_{1/2}$ is excluded by experimental constraints from LEP experiments. The contours labeled "LC 500" and "LC 1000" are the envelope of the regions below which experiments at a LC operating at $\sqrt{s} = 500$ or 1000 GeV should be able to detect a signal for sparticle production above SM backgrounds. These contours are a composite of the contours below which slepton pair production (the falling part of the contour at small m_0) or chargino pair production (the flat portion of the contour, rising to large values of $m_{1/2}$ close to the excluded region on the right) should be detectable. The kinematic reach of a LC for $\tilde{\tau}_1 \bar{\tilde{\tau}}_1$ pairs is denoted by the dashed contours. For very large values of m_0 in the HB/FP region, the chargino is light and higgsino-like and becomes increasingly mass degenerate with $\widetilde{Z}_1$. In this case, the visible energy from $\widetilde{W}_1^+ \widetilde{W}_1^-$ production followed by $\widetilde{W}_1 \to \widetilde{Z}_1 f \bar{f}'$ decay becomes very small, and the signal must be extracted using a specialized analysis. Finally, the bulge in the contours near $m_0 \sim 300$–1000 GeV shows the additional region where the signal from

$$e^+ e^- \to \widetilde{Z}_1 + \widetilde{Z}_2 \to \widetilde{Z}_1 + \widetilde{Z}_1 h \to \widetilde{Z}_1 + \widetilde{Z}_1 b\bar{b}$$

production is observable. Also shown for comparison are contours corresponding to the reach of Tevatron upgrades and the reach of the LHC, assuming an integrated luminosity of 100 fb^{-1}, taken from Fig. 15.6. The lightest gray regions are where the cosmological neutralino relic density $\Omega_{\widetilde{Z}_1} h^2 < 0.129$ as required by its determination by the WMAP collaboration. As mentioned in Section 15.4.1, the HB/FP region is one of the regions of the mSUGRA parameter space consistent with the WMAP relic density determination. We see from Fig. 15.10 that experiments at linear colliders will be able to probe beyond the LHC reach in this favored part of mSUGRA parameter space.

In GMSB models with a low SUSY breaking scale, the gravitino is the LSP. Generally speaking, the presence of additional photons or leptons from NLSP decays should make the detection of any sparticle signal easier. Moreover, if $\widetilde{Z}_1$ is the NLSP, then

$$e^+ e^- \to \widetilde{Z}_1 \widetilde{Z}_1 \to \gamma\gamma + E_T^{\text{miss}} \tag{15.10}$$

should also be observable as long as $\widetilde{Z}_1$ decays inside the detector, though in the case of delayed decays this would require the identification of photons that are very displaced from the primary vertex. As long as this is possible, the reach should be close to the kinematic limit since t-channel neutralino production is not particularly suppressed. If instead a slepton is the NLSP, and decays promptly via $\tilde{\ell} \to \ell \widetilde{G}$, then

$$e^+ e^- \to \tilde{\ell}^+ \tilde{\ell}^- \to \ell^+ \ell^- + E_T^{\text{miss}} \tag{15.11}$$

should have a similar reach as for the case where the slepton is the LVSP in the mSUGRA framework. If the slepton decay is delayed, the reaction can still be detected via searches for tracks with kinks or from searches for quasi-stable slow moving massive exotics that may reveal themselves through highly ionizing tracks.

In the AMSB model, the $SU(2)$ gaugino-like chargino is the LVSP. However, signals from chargino pair production will be difficult to detect because the tiny $\widetilde{W}_1 - \widetilde{Z}_1$ mass gap implies that the visible decay products of the chargino carry very little energy. In this case, the process $e^+e^- \rightarrow \widetilde{W}_1^+ \widetilde{W}_1^- \gamma$ offers the best hope for detection. If $\widetilde{W}_1$ dominantly decays via $\widetilde{W}_1 \rightarrow \widetilde{Z}_1 \pi$, and the pion (whose energy is several hundred MeV) is detectable, its presence serves to reduce background from $e^+e^- \rightarrow \gamma \nu \bar{\nu}$ events in the SM. The background from $e^+e^- \rightarrow e^+e^- \gamma$ events can be controlled as long as there is some instrumentation in the beam direction.

15.5 Beyond SUSY discovery

If new physics is discovered at the LHC in one or more of the several channels that we have discussed above, it will mark the start of the program to establish that it is softly broken SUSY (or something else) and to determine the mechanism by which SUSY is broken. The discovery of several superpartners (with expected spins and gauge quantum numbers), either via their direct production or more likely via a reconstruction of cascade decay chains at the LHC, will make a strong case for SUSY. That the new physics is SUSY can be conclusively established by experiments showing that couplings of superpartners are related to those of their SM partners: this should be possible via precision measurements that are possible at LCs. The determination of the sparticle masses as well as cross sections and branching ratios (these provide information about their couplings) will be the first step to elucidating the mechanism of SUSY breaking, since these will provide information about the underlying SSB parameters. Such measurements, which should be possible at the LHC as well as LCs, will also serve to rule in or rule out various models that we have considered in Chapter 11, and in the former case also provide information about the underlying parameters.

15.5.1 Precision SUSY measurements at the LHC

Once a sufficient number of SUSY scattering events is accumulated, the task will turn to scrutinization of the events to try to make precision measurements of sparticle masses, branching fractions, spin and other quantum numbers, marking the start of sparticle spectroscopy. As discussed at the start of Section 12.2, the environment of hadron collisions poses formidable difficulties for precision measurements. Nevertheless, experience at the CERN $Sp\bar{p}S$ and Fermilab Tevatron, where M_W has been determined very precisely in spite of the undetected neutrino in these events, has taught us that precision measurements are indeed possible. We should, therefore, maintain a positive outlook, and critically examine how well SUSY particle properties can be determined at the LHC.

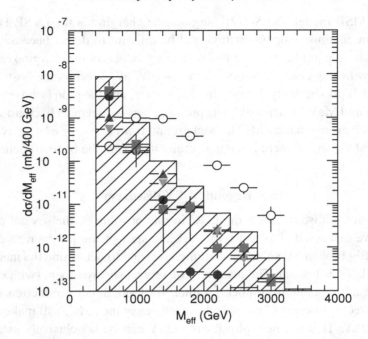

Figure 15.11 Distribution in $M_{\rm eff}$ for SUSY events in the mSUGRA model with $m_0 = 100$ GeV, $m_{1/2} = 300$ GeV, $\tan\beta = 2.1$, $A_0 = 300$ GeV, and $\mu > 0$ (open circles) and for the SM background (histogram) from $t\bar{t}$ production (solid circles), $W + {\rm jets}$ (upright triangles), $Z + {\rm jets}$ (upside down triangles), and QCD jets (squares). Reprinted from the ATLAS Technical Design Report.

Since gluino and squark pair production cross sections are expected to be the dominant SUSY cross sections at the LHC, a first estimate of the SUSY particle mass scale will be obtained from the magnitudes of the momenta of jets and $E_{\rm T}^{\rm miss}$ in these events: heavier sparticles lead to harder jets and $E_{\rm T}^{\rm miss}$. In Fig. 15.11, we show the distribution of the effective mass

$$M_{\rm eff} = E_{\rm T}^{\rm miss} + E_{\rm T}({\rm jet}\ 1) + E_{\rm T}({\rm jet}\ 2) + E_{\rm T}({\rm jet}\ 3) + E_{\rm T}({\rm jet}\ 4) \qquad (15.12)$$

for SUSY events in the mSUGRA model with $m_0 = 100$ GeV, $m_{1/2} = 300$ GeV, $\tan\beta = 2.1$, $A_0 = 300$ GeV, and $\mu > 0$, for which $m_{\tilde{g}} \simeq 767$ GeV and $m_{\tilde{q}} \simeq 680$ GeV.[17] Also shown is the same distribution for SM events. Clearly, for large values of $M_{\rm eff}$, the signal emerges from the falling background distribution. It has been shown that the peak of the SUSY $M_{\rm eff}$ distribution correlates surprisingly well with $M_{\rm SUSY} = \min(m_{\tilde{g}}, m_{\tilde{q}})$, and yields a good first guess as to the SUSY particle mass scale.

---

[17] It is unimportant for the present discussion that this model, which was examined for ATLAS feasibility studies, is now excluded both by the bound on m_h as well as by WMAP constraints.

Detailed determination of sparticle masses is complicated by the fact that every event has *two* undetected particles. Even so, as discussed below, the determination of kinematic "mass edges" constrains particular combinations of masses in SUSY events. If enough such kinematic "end points" can be measured, it may be possible to determine individual sparticle masses. Frequently though, it may be possible to directly determine only mass differences.

The simplest example of a measurable mass edge in SUSY events is the upper limit on the invariant mass of dileptons from $\widetilde{Z}_2 \to \ell\bar{\ell}\widetilde{Z}_1$ decays:

$$m(\ell\bar{\ell}) \leq m_{\widetilde{Z}_2} - m_{\widetilde{Z}_1} \tag{15.13a}$$

regardless of whether the $\widetilde{Z}_2$ is produced directly or in cascade decays. Even allowing for experimental resolution, the end point of this distribution can be well determined as long as the leptonic branching fraction for $\widetilde{Z}_2$ decays is not strongly suppressed. The end point (15.13a) is attained when the two leptons recoil against one another with $\widetilde{Z}_1$ stationary in the rest frame of $\widetilde{Z}_2$. This end point is not kinematically accessible if $\widetilde{Z}_2 \to \tilde{\ell}\ell \to \ell\widetilde{Z}_1\ell$ with the intermediate slepton on its mass shell because kinematic constraints do not allow $\widetilde{Z}_1$ to be at rest. In this case, except for slepton width effects and tiny contributions from off-shell sleptons, the kinematic end point shifts to

$$m(\ell\bar{\ell}) < m_{\widetilde{Z}_2}\sqrt{1 - \frac{m_{\tilde{\ell}}^2}{m_{\widetilde{Z}_2}^2}}\sqrt{1 - \frac{m_{\widetilde{Z}_1}^2}{m_{\tilde{\ell}}^2}} \leq m_{\widetilde{Z}_2} - m_{\widetilde{Z}_1}. \tag{15.13b}$$

Once the overall SUSY mass scale is established using the M_{eff} variable, then attention can be focussed on reconstructing particular decay chains.[18] Although many studies have been performed to examine how this might be done, we use the mSUGRA model with parameters in Fig. 15.11 as an illustration of how one might proceed. The decay $\widetilde{Z}_2 \to \ell\bar{\ell}\widetilde{Z}_1$ just discussed serves as an important starting point. The distribution of opposite sign, same flavor dilepton masses in events with jets plus E_T^{miss} events is shown in Fig. 15.12. Some care must be exercised in extracting information from this measured end point because one does not a priori know the decay pattern of $\widetilde{Z}_2$, though the large number of dileptons may hint at its decay via a real slepton. Indeed, we see a distinct mass edge above SM backgrounds and SUSY contamination close to its expected location, $m(\ell\bar{\ell})^{\text{exp}} = 108.6$ GeV. The large event rate implies that this dilepton mass edge can be measured to a precision of well below a GeV.

---

[18] These studies were pioneered by I. Hinchliffe *et al.*, *Phys. Rev.* **D55**, 5520 (1997) and *Phys. Rev.* **D60**, 095002 (1999); H. Bachacou, I. Hinchliffe, and F. Paige, *Phys. Rev.* **D62**, 015009 (2000); Atlas Collaboration, *Atlas Physics and Detector Performance Technical Design Report*, LHCC 99-14/15.

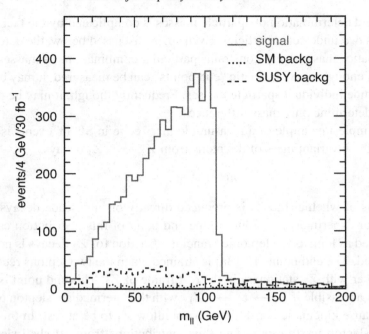

Figure 15.12 Distribution in $m(\ell\bar{\ell})$ for signal (solid) and SM (dots) and SUSY background (dashes) for the mSUGRA point with the same model parameters as in Fig. 15.11. Reprinted from the ATLAS Technical Design Report.

The next step in reconstructing the cascade decay

$$\tilde{q}_{\rm L} \to q\tilde{Z}_2 \to q\tilde{\ell}^\pm \ell^\mp \to q\ell^\pm \ell^\mp \tilde{Z}_1, \qquad (15.14)$$

which has a large branching fraction, is to combine the dilepton invariant mass with one of the high $p_{\rm T}$ jets in the event. Typically there are two or more high $p_{\rm T}$ jets in each SUSY event. One may construct the $m(\ell\bar{\ell}q)$ invariant mass for each of the highest $p_{\rm T}$ jets, and plot the smaller of the two combinations. This distribution is shown in Fig. 15.13, which is plotted for the lepton combinations $e^+e^- + \mu^+\mu^- - e^\pm\mu^\mp$ to statistically remove the contamination from squark decays to chargino pairs. Even for the assumed decay chain, the formula for the kinematic end point depends on the various masses (see exercise below), but an a-posteriori justification of any choice is possible if sparticle masses can be extracted from the data. For our choice of masses, assuming that the combination with the lower mass is the one from the decay of a single squark, we have

$$m(\ell\bar{\ell}q) < m_{\tilde{q}} \sqrt{1 - \frac{m_{\tilde{Z}_2}^2}{m_{\tilde{q}}^2}} \sqrt{1 - \frac{m_{\tilde{Z}_1}^2}{m_{\tilde{Z}_2}^2}} = 552.4 \text{ GeV}. \qquad (15.15)$$

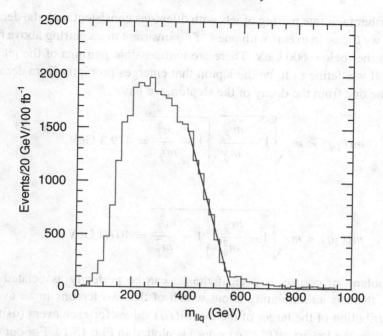

Figure 15.13 Distribution in $m(\ell\bar{\ell}q)$ for the smaller of the two $\ell^+\ell^-q$ invariant mass combinations for the mSUGRA model of the previous figure. The contamination from squark decays to charginos is statistically removed by plotting the distribution for $e^+e^- + \mu^+\mu^-$ pairs minus the same for $e^\pm\mu^\mp$ pairs. Reprinted from the ATLAS Technical Design Report.

Exercise *Consider a chain of two-body decays, $A \rightarrow bB \rightarrow bcC \rightarrow bcdD$, where b, c, d are massless particles. Show that the kinematic end point of the invariant mass $m(bcd)$ is given by,*

$$m(bcd)^2 \leq max \left[\frac{(m_A^2 - m_B^2)(m_B^2 - m_C^2)}{m_B^2}, \frac{(m_A^2 - m_C^2)(m_C^2 - m_D^2)}{m_C^2}, \frac{(m_A^2 m_C^2 - m_B^2 m_D^2)(m_B^2 - m_C^2)}{m_B^2 m_C^2} \right],$$

except for mass ranges where the absolute end point

$$m(bcd) = m_A - m_D$$

can be saturated.

This is in contrast to the case of the three-body decay of $A \rightarrow bB \rightarrow bcC$ where the saturation of the end point $m_A - m_C$ is possible only if $m_B^2 = m_A m_C$.

To further facilitate pairing of jets with dileptons consistent with the decay chain (15.14), we focus on events with one $\ell^+\ell^-q$ invariant mass pairing above 600 GeV and the other below 600 GeV. There are two possible pairings of the jet with the leptons. If we define ℓ_1 to be the lepton that emerges promptly from decay of $\widetilde{Z}_2$, and ℓ_2 the one from the decay of the slepton, we have

$$m(\ell_1 q) < m_{\tilde{q}} \sqrt{1 - \frac{m_{\widetilde{Z}_2}^2}{m_{\tilde{q}}^2}} \sqrt{1 - \frac{m_{\tilde{\ell}}^2}{m_{\widetilde{Z}_2}^2}} = 479.3 \text{ GeV}, \qquad (15.16a)$$

and

$$m(\ell_2 q) < m_{\tilde{q}} \sqrt{1 - \frac{m_{\widetilde{Z}_2}^2}{m_{\tilde{q}}^2}} \sqrt{1 - \frac{m_{\widetilde{Z}_1}^2}{m_{\tilde{\ell}}^2}} = 407.4 \text{ GeV}. \qquad (15.16b)$$

The problem, of course, is even if the jet can be perfectly associated with the leptons, there is an ambiguity about which of the two leptons in an event is ℓ_1. The distribution of the larger of the two $m(\ell q)$ values for each event (using the jet which gives the lowest $m(\ell^+\ell^-q)$ value) is plotted in Fig. 15.14. For our case, this is bounded by (15.16a). The upper edge is not very sharp, but fits to the endpoint come within a few percent of its value. The other mass edge (15.16b) is buried under this distribution.

The three mass edges in the figures constrain, but do not determine, the four masses. To pin these down, we need a fourth mass edge. Unfortunately, except for effects of cuts, the lower edges of these distributions start at $m = 0$ and so provide no information. However, by focussing on events with a minimum value of $m(\ell_1\ell_2)$, we preclude the configuration with $m(q\ell_1\ell_2) = 0$, and the $m(q\ell_1\ell_2)$ distribution starts at a mass value depending on our choice of $m(\ell_1\ell_2)_{\min}$. The corresponding $m(\ell\ell q)$ distribution for events with

$$m(\ell^+\ell^-) > m(\ell^+\ell^-)_{\max}/\sqrt{2}$$

is shown in Fig. 15.15, where the larger of the two $m(\ell^+\ell^-q)$ values is plotted. A lower edge is clearly visible. The expression for this lower edge in terms of the sparticle masses and $m(\ell^+\ell^-)_{\min}$ is complicated and will not be reproduced here. For the present case, the theoretical edge is expected to be at 271.8 GeV, and appears to be smeared to lower values, perhaps because of energy lost to QCD radiation. The main point of this discussion is that at least for the case of a chain of two-body decays considered here, it is possible to extract the four mass values in a *model-independent* manner. Explicit fits to these quantities give sparticle masses to

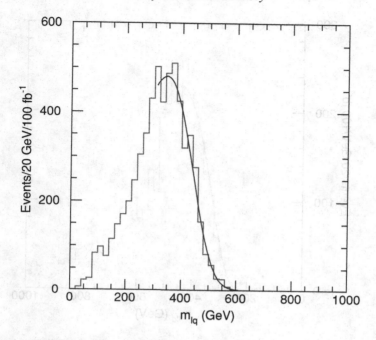

Figure 15.14 Distribution in $m(\ell q)$ for the smaller of the two $\ell^+\ell^- q$ invariant mass combinations for the mSUGRA point under study. Once again, the contamination from squark decays to chargino are removed by using the flavor weighted combination $e^+e^- + \mu^+\mu^- - e^{\pm}\mu^{\mp}$. Reprinted from the ATLAS Technical Design Report.

3–12%.[19] It is not surprising that $m_{\tilde{Z}_1}$ has the largest error, since it is much smaller than the squark mass, and enters only (quadratically) via kinematics.

For the mSUGRA point used in the above example, the decay $\tilde{Z}_2 \to h\tilde{Z}_1$ occurs with a branching fraction of about 50%. We would thus expect that a data sample consisting mainly of SUSY events would contain a significant fraction of events that contain a high p_T Higgs boson h from cascade decays. Since h mostly decays via $h \to b\bar{b}$, such events would contain at least two b-quark jets whose presence is signaled by displaced vertices from B-meson decay, and which have a bump in their invariant mass distribution around the value of m_h; this is illustrated in Fig. 15.16. In general, if h is produced at significant rates in SUSY cascade decay events, it may well first be discovered as a $b\bar{b}$ mass bump in the SUSY event sample! Detection of the $h \to \gamma\gamma$ mode, which may take several years of LHC operation to establish,

[19] If all four sparticle masses can indeed be fit, the ambiguities in the formulae for the end points that we had referred to earlier would automatically be resolved.

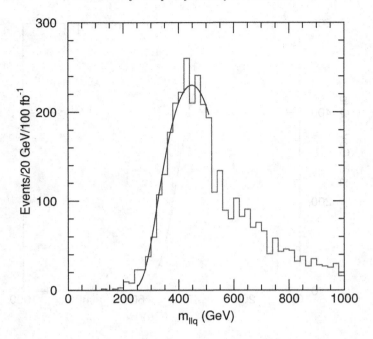

Figure 15.15 Distribution in $m(\ell^+\ell^-q)$ for the larger of the two $\ell^+\ell^-q$ invariant mass combinations for the mSUGRA model parameters in Fig. 15.11, but with the additional requirement that $m(\ell^+\ell^-) > m(\ell^+\ell^-)_{\text{max}}/\sqrt{2}$. Reprinted from the ATLAS Technical Design Report.

is nonetheless very important because the location of the peak in the two-photon distribution yields a very accurate measurement of m_h.

These events may also allow the reconstruction of the decay chain

$$\tilde{q}_L \to q\tilde{Z}_2 \to qh\tilde{Z}_1 \to qb\bar{b}\tilde{Z}_1.$$

Since gluinos are heavier than squarks, $\tilde{q}_L$ comes from either direct production, or from the decay of a gluino. A relatively clean sample may be obtained by focussing on events with just two hard jets (which most likely come from squark decay) and a pair of b-jets. The $m(b\bar{b}j)$ mass distribution from this chain must have both upper and lower end points that can be fixed in terms of $m_{\tilde{q}_L}$, $m_{\tilde{Z}_2}$, $m_{\tilde{Z}_1}$, and m_h.

Exercise *Show that the end points of the $b\bar{b}j$ mass distribution from the cascade decay chain $\tilde{q}_L \to q\tilde{Z}_2 \to qh\tilde{Z}_1 \to qb\bar{b}\tilde{Z}_1$ are given by,*

$$m^2(b\bar{b}j)^{\text{max}}_{\text{min}} = m_{\tilde{q}}^2 + m_{\tilde{Z}_1}^2 - 2E_{\tilde{q}}E_{\tilde{Z}_1} \pm 2p_{\tilde{q}}p_{\tilde{Z}_1},$$

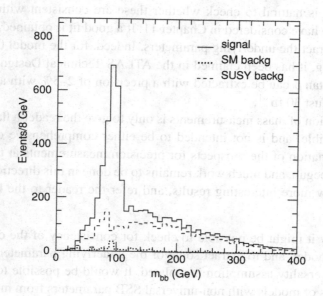

Figure 15.16 Distribution in $m(b\bar{b})$ for signal (solid) and SM (dots) and SUSY background (dashes) for the mSUGRA model with parameters as in Fig. 15.11, assuming an integrated luminosity of 30 fb^{-1}. Reprinted from the ATLAS Technical Design Report.

where

$$E_{\tilde{q}} = \frac{m_{\tilde{q}}^2 + m_{\tilde{Z}_2}^2}{2m_{\tilde{Z}_2}} \quad \text{and} \quad E_{\tilde{Z}_1} = \frac{m_{\tilde{Z}_2}^2 + m_{\tilde{Z}_1}^2 - m_h^2}{2m_{\tilde{Z}_2}}$$

are the energies of the squark and $\tilde{Z}_1$ in the rest frame of $\tilde{Z}_2$, $p_{\tilde{q}} = \sqrt{E_{\tilde{q}}^2 - m_{\tilde{q}}^2}$, and $p_{\tilde{Z}_1} = \sqrt{E_{\tilde{Z}_1}^2 - m_{\tilde{Z}_1}^2}$.

Show that the ideal $m(b\bar{b}j)$ spectrum for the mSUGRA model that we have been examining (where $m_{\tilde{q}} = 688$ GeV, $m_{\tilde{Z}_2} = 233$ GeV, $m_{\tilde{Z}_1} = 122$ GeV, and $m_h = 93$ GeV) extends from 338 GeV to 524 GeV. Compare this with the $m(b\bar{b}j)$ distributions in the ATLAS Technical Design Report, where effects of detector resolution and jet misidentification have been included. Although the distributions are smeared particularly at the lower end point, it may be possible to make corrections to compensate energy losses in b-jets due to escaping neutrinos or losses outside the cone once LHC data are available. Moreover, a more thorough analysis may better isolate events with Higgs bosons.

Once sparticle mass spectra have been extracted from the various reconstructed mass edges, it is natural to check whether these are consistent with any of the models that we have considered in Chapter 11. If a good fit is obtained, it would be possible to extract the underlying parameters. Indeed, for the model that we have been examining, it has been claimed in the ATLAS Technical Design Report that m_0, $m_{1/2}$, and $\tan \beta$ can be extracted with a precision of 2–5% with an integrated luminosity of just 30 fb^{-1}.

Our discussion of mass measurements is only to give the reader a flavor of what might be possible, and is not intended to be either comprehensive or complete. In fact, examination of the prospects for precision measurements at the LHC has only recently begun, and much work remains to be done in this direction. Here, we highlight a few more interesting results, and refer the reader to the literature for more details.

- We saw how it might be possible to check for consistency of the data with the mSUGRA model, and to extract some of the underlying parameters. It may be that the universality assumption is violated. It would be possible to distinguish some classes of models with non-universal SSB parameters from mSUGRA.
- If $\tan \beta$ is large so that decays of charginos and neutralinos to tau leptons become dominant, it may still be possible to reconstruct various mass edges, though with somewhat degraded precision.
- In GMSB models with prompt decay of a bino-like NLSP, the decay chain $\widetilde{Z}_2 \to \tilde{\ell}^\pm \ell^\mp \to \ell^+ \ell^- \widetilde{Z}_1 \to \ell^+ \ell^- \gamma \tilde{G}$ has the same number of steps as the decay chain from $\tilde{q}_L$ decays for the mSUGRA case discussed above, and so can be similarly analyzed. An important difference is that at least for the case study in the ATLAS Technical Design Report, both the $m(\ell_1 \gamma)$ as well as the $m(\ell_2 \gamma)$ edges can be clearly distinguished in the $m(\ell \gamma)$ distribution. The invariant mass edges of $\ell^+ \ell^-$, $\ell_1 \gamma$, $\ell_2 \gamma$, and $\ell^+ \ell^- \gamma$ distributions are sufficient to determine $m_{\widetilde{Z}_2}$, $m_{\tilde{\ell}}$, and $m_{\widetilde{Z}_1}$ to high accuracy. Squark and gluino mass reconstruction is also possible. These measurements allow determination of some of the underlying model parameters: Λ can be determined at the couple of percent level, and, for the case examined, even the messenger scale can be extracted within $\pm 40\%$. If instead, the $\widetilde{Z}_1$ decay is long lived and decays outside the detector, the analysis will be similar to those described above for the mSUGRA model.
- The intermediate possibility that the $\widetilde{Z}_1$ NLSP decays with a decay length of 10 cm to 20 m allows other interesting measurements. If the photon from $\widetilde{Z}_1$ converts to an electron pair, its momentum and point of origin can be well determined, and reconstruction of the entire event appears to be possible.[20] Of course, it is only in a fraction of events that the photon converts. These authors have claimed

---

[20] See K. Kawagoe, T. Kobayashi, M. Nojiri and A. Ochi, *Phys. Rev.* **D69**, 035003 (2004).

that reconstruction is also possible even in events where the photon does not convert: the degradation of the precision is partially compensated by the larger number of these events. Finally, in such scenarios, the lifetime of the NLSP can be determined to within a few percent. This is a very important measurement because the NLSP lifetime is simply related to the fundamental SUSY breaking scale.

- The GMSB case with a slepton co-NLSP has also been examined in the ATLAS Technical Design Report.[21] If $\tilde{\ell}_R$ is quasi-stable and has a distinct track, neutralinos decaying via $\tilde{Z}_i \to \ell\tilde{\ell}_R$ show up as clear mass peaks in appropriate distributions. The decays $\tilde{\ell}_L \to \ell\tilde{Z}_1$ can be used to reconstruct $m_{\tilde{\ell}_L}$. For the case of prompt NLSP decays $\tilde{\ell}_R \to \ell\tilde{G}$, it has been shown that a variety of mass edges involving dileptons and jets can be reconstructed, giving good fits to model parameters. Once again, the underlying model parameters can be extracted. The precision that can be attained is significantly better if the slepton NLSP is quasi-stable. In this case, a determination of the fundamental SUSY breaking scale (via the slepton lifetime) with a precision of tens of percent is possible if the slepton decay length is between $\sim$0.5 m and 1 km.

15.5.2 Precision measurements at an LC

If the discovery of new physics is established, the next step will be to figure out what it is. Taking this new physics to be supersymmetry, this may come about by the discovery of several superpartners. At the LHC, the discovery of several superpartners might occur if signals for new physics in many different channels can be interpreted as different cascade decay chains from superparticle pair production, or via the identification of several "kinematic edges" in appropriate distributions as we have just discussed. Logically, of course, such an observation would only establish the discovery of several new particles. The magnitude of the signal cross sections would tell us whether or not the new particles exhibit strong interactions, and maybe even indicate some of their other gauge quantum numbers.

If superpartners are accessible at linear colliders, the cleanliness of the initial and final states frequently allows their properties to be straightforwardly determined.[22] Since SUSY predicts the existence of superpartners with spins differing by one half, we will first outline how the spin of any new particle may be determined. We will then discuss how sparticle masses may be determined, since these encode the information about the all-important (and as yet completely unknown) mechanism by

[21] See also S. Ambrosanio *et al.*, *JHEP* **01**, 014 (2001) and hep-ph/0012192 (2000).

[22] Studies of the capabilities of linear colliders for SUSY measurements were pioneered by T. Tsukamoto *et al.*, *Phys. Rev.* **D51**, 3153 (1995). H. Baer *et al.*, *Phys. Rev.* **D54**, 6735 (1996) included the effects of cascade decays in the analysis of SUSY mass measurements, and M. Nojiri *et al.*, *Phys. Rev.* **D54**, 6756 (1996) discussed the determination of the properties of third generation sleptons.

which superpartners of SM particles obtain their masses: within specific models, information about the sparticle spectrum may allow us to infer some of the underlying model parameters. If the Higgs bosons A, H or $H^{\pm}$ are also kinematically accessible, we will see that LC experiments will allow further tests of the MSSM framework, and may also yield further information about underlying parameters that may be more difficult to get at otherwise. However, to unambiguously establish (in a model-independent manner) that any new physics is softly broken supersymmetry, we have to show that the dimensionless couplings of the new particles are (aside from radiative corrections) equal to the corresponding SM couplings. We will illustrate the extent to which such a determination is possible in experiments at an $e^{+}e^{-}$ LC.

Spin determination

If sparticle production dominantly occurs via the exchange of vector bosons in the s-channel, it is easy to see from Appendix A.2 that the sparticle angular distribution is given by

$$\sin^2 \theta$$

for spin 0 particles, and by

$$E^2(1 + \cos^2 \theta) + m^2 \sin^2 \theta$$

for equal mass spin $\frac{1}{2}$ particles. If the sparticles are produced with a sufficient boost, the angular distribution of their daughters will be strongly correlated with that of the parent sparticles; the differences between the angular distributions should suffice to readily distinguish between the spin zero and spin $\frac{1}{2}$ cases. An integrated luminosity of several tens of fb^{-1} should suffice to establish the spin 0 nature of smuons at a 500 GeV LC.

We mention in passing that angular distributions may also contain dynamical information. For instance, in $e^{+}e^{-} \rightarrow \tilde{e}_{L(R)}\bar{\tilde{e}}_{L(R)}$ processes, selectrons (antiselectrons) will preferentially be produced along the electron (positron) beam direction if t-channel neutralino exchanges are important, resulting in an angular asymmetry in the distribution of the daughter electron.

Exercise *Consider the reaction $e^{+}e^{-} \rightarrow \tilde{\mu}_{R}\bar{\tilde{\mu}}_{R} \rightarrow \mu^{+}\mu^{-}\widetilde{Z}_{1}\widetilde{Z}_{1}$ at a LC, where $\tilde{\mu}_{R} \rightarrow \mu\widetilde{Z}_{1}$. We will see in the next subsection that it is possible to extract $\tilde{\mu}_{R}$ and $\widetilde{Z}_{1}$ masses from this process. Using the fact that the smuon is a narrow state, show that it is then possible to completely reconstruct (up to a quadratic ambiguity) the smuon momenta from the observable momenta of the final state muons and the missing three-momentum vector, even though each event contains two escaping neutralinos. In this sense, the angular distribution of smuons can be experimentally constructed.*

Mass determination

If sparticles are discovered, determination of their masses will be one of the highest priorities. Measurements at the LHC will, as we have seen, provide some information but at LCs it will be possible to have a systematic program for sparticle spectroscopy. In the approach, initiated by the Japanese Linear Collider group, the idea is to exploit the kinematics of the decays to infer the masses. This is not straightforward since every SUSY event contains two LSPs that escape detection so that a reconstruction of "mass bumps" is not possible.[23] For the production of spinless particles p_1 and p_2 via $e^+e^- \to p_1 + p_2$, followed by the decay $p_2 \to p_3 + p_4$, it is straightforward to check that the energy spectrum of the particle p_3 is flat and kinematically restricted to be between

$$\gamma(E_3^* - \beta p_3^*) \le E_3 \le \gamma(E_3^* + \beta p_3^*), \tag{15.17}$$

where $\quad E_3^* = (m_2^2 + m_3^2 - m_4^2)/2m_2, \quad p_3^* = \sqrt{E_3^{*2} - m_3^2}, \quad \gamma = E_2/m_2,$ $\beta = \sqrt{1 - 1/\gamma^2}$, and $E_2 = (s + m_2^2 - m_1^2)/2\sqrt{s}$, up to corrections from energy mis-measurements, particle losses and bremsstrahlung and beamstrahlung effects.

These considerations can be directly applied to slepton pair production, since sleptons decay via two-body modes. In the case that the sleptons can only decay via $\tilde{\ell} \to \ell \tilde{Z}_1$, the end points of the energy distribution of the final state lepton depend only on the values of $m_{\tilde{\ell}}$ and $m_{\tilde{Z}_1}$ via kinematics. Since sharp end points can be determined rather precisely, it is possible to infer the slepton and neutralino masses.

To illustrate this, we show the muon energy distribution from $e^+e^- \to \tilde{\mu}_R \tilde{\bar{\mu}}_R \to \mu^+\mu^- \tilde{Z}_1 \tilde{Z}_1$ production in Fig. 15.17a, which is taken from the simulation by Tsukamoto *et al.* In this study, the right-handed charged slepton is the NLSP with $m_{\tilde{\ell}_R} = 141.9\,\text{GeV}$, and decays to the neutralino which has a mass $m_{\tilde{Z}_1} = 117.8\,\text{GeV}$. Charginos have a mass $m_{\tilde{W}_1} = 219.3\,\text{GeV}$ and so cannot be produced at the assumed center of mass energy of 350 GeV. By choosing the electron beam to be mainly right-handed, the dominant WW background to the acolinear muon pair signal is greatly diminished, while the right-slepton pair production cross section is enhanced. The data points correspond to a Monte Carlo expectation for an integrated luminosity of just 20 fb^{-1}, while the solid curve is the "best fit" to these data. The corresponding error contours are shown in 15.17b. We see that $m_{\tilde{\mu}_R}$ and $m_{\tilde{Z}_1}$ can both be determined to about 1%. These sparticle masses serve as inputs for determining the smuon spin, as discussed above. In addition, by varying the beam

---

[23] It may be possible to reconstruct mass bumps in R-parity violating scenarios, depending on how the LSP decays.

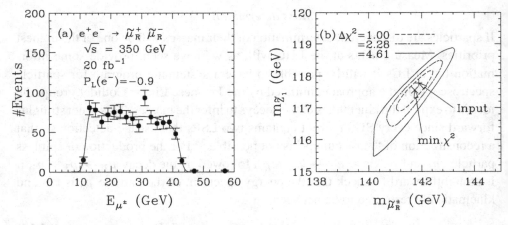

Figure 15.17 (*a*) The energy distribution of final state muons from $e^+e^- \to \tilde{\mu}_R^+\tilde{\mu}_R^- \to \mu^+\mu^- \tilde{Z}_1\tilde{Z}_1$ at $\sqrt{s} = 350$ GeV with $P_L(e^-) = -0.9$, within the mSUGRA framework with $m_0 = 70$ GeV, $M_2 = 250$ GeV, $\mu = 400$ GeV, $A_0 = 0$, and $\tan\beta = 2$. The data points are from Monte Carlo while the smooth curve is from a fit. In (*b*) are shown error contours from a two-parameter fit to $m_{\tilde{\mu}_R}$ and $m_{\tilde{Z}_1}$. Reprinted with permission from T. Tsukamoto, K. Fujii, H. Murayama, M. Yamaguchi and Y. Okada, *Phys. Rev.* **D51**, 3153 (1995), copyright (1995) by the American Physical Society.

polarization and comparing to the cross section, the smuon weak isospin and hypercharge can be extracted, verifying that it is the right-superpartner of the muon. The $\tilde{\mu}_L$ mass and other quantum numbers should be measurable in a similar manner once threshold is passed for $\tilde{\mu}_L\bar{\tilde{\mu}}_L$ production.

The $\tilde{e}_R$ mass can be similarly measured to even better precision since it has a larger cross section because of t-channel neutralino exchange contributions whose presence, as we have noted, will also be reflected in the angular distribution. For selectrons, $\tilde{e}_R\bar{\tilde{e}}_L$, $\bar{\tilde{e}}_R\tilde{e}_L$, and $\tilde{e}_L\bar{\tilde{e}}_L$ may also be accessible, each with unique energy edges in the electron or positron energy distributions. Variable beam polarization will be a key tool in discriminating the different reactions. If we assume that the LSP is dominantly a hypercharge gaugino and that gaugino masses satisfy the unification condition, it should be possible to roughly project the chargino threshold even before charginos are discovered.

Although Tsukamoto *et al.* had confined their analysis to cases where sparticles directly decay to the LSP, it was shown shortly after that cascade decays do not degrade the precision with which sparticle masses can be determined.[24] On the contrary, these decays provide new opportunities: for instance, if the decay $\tilde{\nu}_e \to e\widetilde{W}_1$ has a significant branching ratio, a determination of the end points of the electron

---

[24] H. Baer *et al.*, *Phys. Rev.* **D54**, 6735 (1996).

energy spectrum from $e^+e^- \to \tilde{\nu}_e + \tilde{\nu}_e \to e\widetilde{W}_1 + e\widetilde{W}_1 \to e\mu\nu_\mu\widetilde{Z}_1 + ejj\widetilde{Z}_1$ yields information about electron sneutrino and chargino masses, with a precision at about the percent level. In this case, of course, chargino pair production is also kinematically accessible and, as discussed below, will probably be how the chargino mass will first be determined. Obtaining this same value for $m_{\widetilde{W}_1}$ in $\tilde{\nu}_e\tilde{\nu}_e$ events will be direct evidence for chargino production in SUSY decay cascades. Masses of muon and tau sneutrinos are more difficult to extract since these are produced only via s-channel Z exchange, and so have smaller production cross sections (see Fig. 12.32). We will revisit this later.

The end-point technique that we have just been describing has also been applied to the lighter stau, assuming $\tilde{\tau}_1 \to \tau\widetilde{Z}_1$.[25] In this case, the situation is complicated by the fact that a part of the tau energy is carried off by the tau neutrino, so that the end points of the tau energy spectrum are smeared. Nonetheless, from the spectrum of visible energy of taus decaying via $\tau \to \rho\nu$, it is possible to obtain $m_{\tilde{\tau}_1}$ with a precision of $\sim 2\%$, assuming an integrated luminosity of ~ 100 fb^{-1}. Including tau decays to π and a_1 would improve the precision by about a factor of two.

Tau sleptons differ from other sleptons in that they are expected to have significant mixing between left- and right-states: $\tilde{\tau}_1 = \tilde{\tau}_L \cos\theta_\tau - \tilde{\tau}_R \sin\theta_\tau$. The stau pair production cross section is sensitive to the mixing angle. In Fig. 15.18, we show the result of a simulation to illustrate that the stau mass and mixing angle can be determined to a few percent at a LC. While the fact that taus are unstable was an undesirable complication for stau mass determination, it is now a boon because the energy spectrum of the daughter tau neutrino (and hence of the visible hadronic decay products) is sensitive to the polarization of the tau. Since the tau polarization depends on the stau mixing angle, a study of stau production provides information not accessible in selectron or smuon production (because polarizations of final state electrons and muons are not measured). The tau polarization can be sensitive to the parameter $\tan\beta$, especially in the case where the $\widetilde{Z}_1$ contains a significant higgsino component. In this case, the $\widetilde{Z}_1$ coupling to the tau–stau system also depends on the tau Yukawa coupling. Then, by simultaneously studying selectron pair production (to constrain neutralino mixings) and stau pair production, it may be possible to determine $\tan\beta$.

If charginos are the lightest charged sparticles, it is likely that they will be discovered before sleptons. If the chargino decays via the two-body mode, $\widetilde{W}_1 \to W\widetilde{Z}_1$ and both Ws decay hadronically, it is straightforward to reconstruct each W from the invariant mass of the jets. Aside from spin correlation effects, the chargino and LSP mass can then be obtained via two-body kinematics from the

[25] M. Nojiri *et al.*, *Phys. Rev.* **D54**, 6756 (1996).

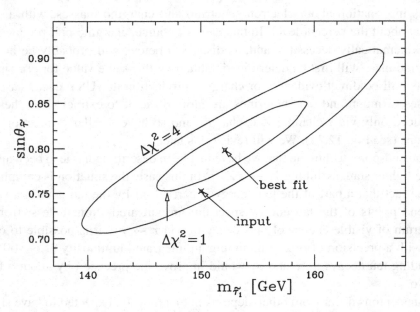

Figure 15.18 Error ellipses from a two-parameter fit to the stau mass and mixing angle. In this illustration, 5000 stau pairs were simulated at $\sqrt{s} = 500$ GeV, assuming that the stau of mass 150 GeV decays exclusively to a 100 GeV neutralino. The stau mixing angle is taken to be given by $\sin\theta_\tau = 0.7526$. A SM background corresponding to an integrated luminosity of 100 fb^{-1} is also included. For more details, we refer the reader to M. Nojiri, K. Fujii and T. Tsukamoto, *Phys. Rev.* **D54**, 6756 (1996), copyright (1996) by the American Society, from which this figure is reprinted with permission.

energy distribution of the W, as in the case of the slepton. An mSUGRA case study by Tsukamoto *et al.* showed that the mass of a chargino as heavy as 220 GeV could be extracted to within a few percent at a 500 GeV LC, assuming an integrated luminosity of 50 fb^{-1}.

What if the chargino decays via three-body decays? In this case, we can force quasi-two-body kinematics by dividing the sample of $e^+e^- \to \widetilde{W}_1^+ \widetilde{W}_1^- \to jj\widetilde{Z}_1 + \ell\nu\widetilde{Z}_1$ events, enriched in signal via suitable cuts, into several narrow bins in m_{jj}.[26] For each m_{jj} bin, the E_{jj} distribution follows the form for $\widetilde{W}_1 \to \widetilde{Z}_1 W^*$ decays, with M_{W^*} close to the central value of the chosen bin. The result of such an analysis is shown in Fig. 15.19 for an mSUGRA model with $m_0 = 300$ GeV, $m_{1/2} = 150$ GeV, $A_0 = -600$ GeV, $\tan\beta = 2$, and $\mu > 0$. The upper frame shows the error ellipse obtained by combining the analyses of the E_{jj} distributions for four different m_{jj}

---

[26] For details, see H. Baer, R. Munroe and X. Tata, *Phys. Rev.* **D54**, 6735 (1996) where this technique is discussed.

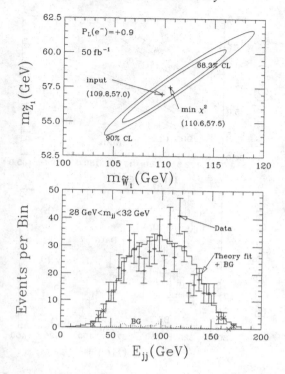

Figure 15.19 The upper frame shows the error ellipses obtained from an analysis of $jj\ell + E_T^{miss}$ events from chargino pair production with forced two-body kinematics, after combining the analysis from four different m_{jj} bins, as discussed in the text. The lower frame shows the $E_{W*} = E_{jj}$ distribution for $M_{W*} = 30 \pm 2$ GeV. Reprinted with permission from H. Baer, R. Munroe and X. Tata, *Phys. Rev.* **D54**, 6735 (1996), copyright (1996) by the American Physical Society.

bins, while the lower frame shows one of these E_{jj} distributions. The result includes SM backgrounds and contamination to the $\ell jj + E_T^{miss}$ signal from other SUSY sources. Once again, we see that a few percent determination of the chargino and LSP mass should be possible at a LC. The precision obtained here is comparable to that obtained by Tsukamoto *et al.* by fitting the shape of the E_{jj} distribution for charginos decaying via three-body decays. It is worth mentioning that for model parameters in the HB/FP region (which yields a favorable value for the neutralino relic density), this technique will be applicable.

In the event that $e^+e^- \to \widetilde{Z}_1\widetilde{Z}_2$ is the only SUSY reaction accessible, mass measurements may still be possible, as illustrated in Fig. 15.20. In this case, $\widetilde{Z}_2 \to \widetilde{Z}_1 h$, $h \to b\bar{b}$, and the missing mass distribution in $b\bar{b} + \not{E}$ events allows $m_{\widetilde{Z}_2}$ and $m_{\widetilde{Z}_1}$ to be determined to a few percent, provided m_h has previously been determined. The missing mass distribution is better suited than the $E_{b\bar{b}}$ distribution for this

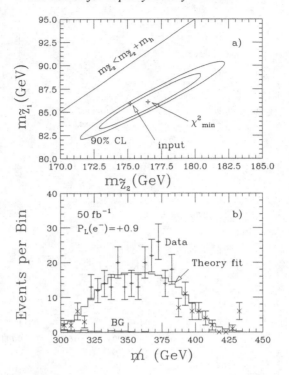

Figure 15.20 Error ellipses and missing mass distributions for $b\bar{b} + \not{E}$ events in a simulation including $\widetilde{Z}_1\widetilde{Z}_2$ production. Reprinted with permission H. Baer, R. Munroe and X. Tata, *Phys. Rev.* **D54**, 6735 (1996), copyright (1996) by the American Physical Society.

measurement because in the determination of missing mass, mismeasurement and losses from undetected neutrinos partially cancel out.

Squark pairs may also be produced in e^+e^- collisions. In many models, the lightest top squark is expected to be the lightest of all squarks, and hence the most likely to be accessible to linear collider searches. Linear collider event generation studies have been performed for an mSUGRA point with $m_{\tilde{t}_1} = 180$ GeV. The signal from $\tilde{t}_1\bar{\tilde{t}}_1$ pair production with $\tilde{t}_1 \to b\widetilde{W}_1$ decay can be almost completely separated from SM backgrounds by requiring ≥ 5 jet events including at least two b-jets. The b-jet energy distribution depends on $m_{\tilde{t}_1}$ and $m_{\widetilde{W}_1}$, and a two-parameter fit gives a measure of these masses to about 5%, as can be seen from Fig. 15.21. By making full use of beam polarization and other capabilities of the LC, it appears that it is possible to also determine the top squark mixing angle to a few percent.[27]

[27] R. Keranen, A. Sopczak, H. Nowak and M. Berggren, *Eur. Phys. J.* **C7**, 1 (2000).

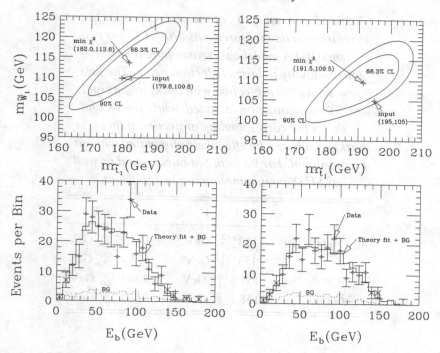

Figure 15.21 Error ellipses for a two-parameter fit to $m_{\tilde{t}_1}$ and $m_{\tilde{W}_1}$ for two nearby parameter space points where $\tilde{t}_1$ pair production is accessible. The corresponding b-jet energy distributions are also shown. Reprinted with permission from H. Baer, R. Munroe and X. Tata, *Phys. Rev.* **D54**, 6735 (1996), copyright (1996) by the American Physical Society.

If pair production of heavier squarks is also kinematically accessible, an LC would be ideal for performing squark spectroscopy. Aside from kinematic determinations of the type that we have been describing, we can see from (A.21a) that by adjusting the polarization of the electron beam, it is also possible to alternate between signals from $\tilde{q}_R\bar{\tilde{q}}_R$ or $\tilde{q}_L\bar{\tilde{q}}_L$ pairs, depending on beam polarization.[28]

It has been suggested that an energy scan of the sparticle production cross section near the production threshold offers a more precise determination of sparticle masses than the "kinematic" measurements described above. The idea is very simple. The shape of the cross section for sparticle pair production close to threshold is a simple function of just the sparticle mass, so that by determining this shape it should be possible to extract the mass very precisely. Indeed, it has been claimed that determining the cross section for ten values of energy each spaced apart by ~ 1 GeV leads to a precision better than a part per mille (a percent) for the masses of

[28] See J. Feng and D. Finnel, *Phys. Rev.* **D49**, 2369 (1994).

Table 15.3 *A summary of the projections for tau sneutrino mass measurements (90% CL) for two mSUGRA model cases, assuming a 95% longitudinally polarized electron beam. The first row shows the projection with backgrounds and SUSY contamination included, while the second shows the corresponding projection if these backgrounds can be effectively eliminated without loss of signal. For $\tilde{\nu}_e$, both SM background as well as SUSY contamination are insignificant.*

	Case I	Case II
$m_{\tilde{\nu}_\tau}$ (500 fb^{-1})	$153^{+12.5}_{-24}$ GeV	$174.9^{+7.1}_{-15.4}$ GeV
	$153^{+11.5}_{-24}$ GeV	$175.4^{+5.6}_{-10.9}$ GeV
$m_{\tilde{\nu}_e}$ (120 fb^{-1})	$157.8^{+0.8}_{-1.2}$ GeV	$178.0^{+0.5}_{-0.8}$ GeV
$m_{\tilde{\nu}_e}$ (500 fb^{-1})	$158.1^{+0.4}_{-0.5}$ GeV	$178.2^{+0.2}_{-0.4}$ GeV

$\tilde{\nu}_e$, $\tilde{e}$, and $\widetilde{W}_1$ ($m_{\tilde{\nu}_\tau}$, $m_{\tilde{\tau}_2}$), assuming an integrated luminosity of just 10 fb^{-1} for each energy scan. The problem is that in order to obtain a relatively background free sample of signal events which is essential for studying the threshold shape, one is forced to focus on particular final states. Not only does this lead to a reduction in the signal but, even more importantly, it also introduces an *unknown* branching fraction on which the cross section depends so that now both the mass as well as the branching fraction have to be extracted from the same counting experiment. This, in turn, leads to a significant degradation in the precision with which sparticle masses may be extracted.

The issue is not simply an academic one because precise determinations of (especially third generation) sparticle masses can provide important information about the underlying physics via which MSSM sparticles obtain their masses. An independent analysis by Mizukoshi *et al.*[29] concludes that the optimal way to make such a mass measurement is to divide the available luminosity between three or four energy points, one of which is chosen at the highest possible energy (this constrains the branching fraction), one close to the threshold and one somewhere in between.[30] The result of their analysis of the precision that is possible for sneutrino mass measurements in two different mSUGRA models is summarized in Table 15.3.

---

[29] J. K. Mizukoshi *et al.*, *Phys. Rev.* **D64**, 115017 (2001).
[30] Since it is not practical to perform a detailed scan of the energy threshold for every sparticle, this is a welcome conclusion. Indeed, running at intermediate values of energy may prove useful for many purposes.

We see that while a precision approaching a part per mille may be possible for $m_{\tilde{\nu}_e}$, and perhaps also for $m_{\tilde{e}}$ and $m_{\widetilde{W}_1}$, an integrated luminosity of 500 fb^{-1} is required. For third generation sneutrinos, the precision is at best several percent.[31] It seems, therefore, that the precision from threshold scans and the kinematic measurements discussed previously is quite comparable.

The Higgs boson sector

The LC is an ideal facility for a study of the Higgs sector, especially if the energy is high enough to access states other than h. The MSSM Higgs boson sector is extremely constrained theoretically, so that precision measurements can serve to experimentally distinguish it from that of the SM, or perhaps exclude it altogether.

Direct observation of the heavier Higgs bosons of the MSSM not only establishes that there is physics beyond the SM, but provides new opportunities. For instance, combining the measurements of $4b$ production from $e^+e^- \rightarrow b\bar{b}A$, $b\bar{b}H$, and HA production processes, together with charged Higgs boson measurements, can lead to a determination of $\tan\beta$ to a high precision.[32] While a study of chargino and neutralino processes may also lead to a determination of $\tan\beta$ if it happens to be small, Higgs boson processes (and to some extent, precise determination of stau properties) offer the best hope for determining $\tan\beta$ when it is large.[33] If $\tan\beta$ is very large, it may also be possible to determine it from the measurements of the widths of the heavy Higgs bosons of the MSSM. Strictly speaking, what is determined are the Yukawa couplings. Although the Yukawa coupling is simply related to $\tan\beta$ at tree level, for large values of $\tan\beta$ one must be careful to include important radiative corrections to reliably extract its value.

If the heavier Higgs bosons are not directly accessible, a precise measurement of the branching ratios of h may still make it possible to exclude the SM, depending on the values of other parameters. For a discussion of these, as well as of many other important measurements possible in the Higgs sector (including a determination of their quantum numbers, couplings to gauge bosons, and their self-couplings), we refer the reader to the literature.

Establishing supersymmetry

The discovery of a few sparticle states will probably convince enthusiasts that nature is supersymmetric. To unambiguously establish that the new physics is indeed (softly broken) supersymmetry, it is necessary to show that the dimensionless couplings of the new particles are equal to the corresponding SM couplings. This

---

[31] We may expect that a determination of $m_{\tilde{\tau}_2}$ will have the same difficulties as that for $m_{\tilde{\nu}_\tau}$.

[32] See J. Gunion, T. Han, J. Jiang and A. Sopczak, *Phys. Lett.* **B565**, 42 (2003); see also V. Barger, T. Han and J. Jiang, *Phys. Rev.* **D63**, 075002 (2001).

[33] Recall that $\tan\beta$ enters via the mass matrices which really depend on $\sin\beta$ and $\cos\beta$, so that its determination becomes difficult if $\tan\beta$ is large.

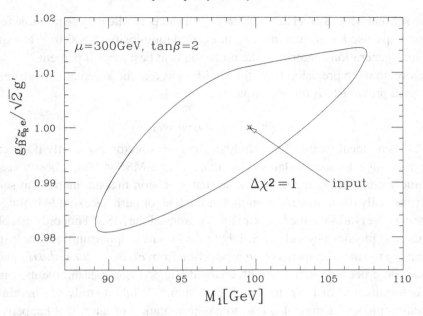

Figure 15.22 The $\Delta\chi^2 = 1$ contour to illustrate the precision with which the bino–selectron–electron coupling may be extracted in experiments at a linear collider from a study of selectron pair production. This study was performed within the framework of a SUSY model with $m_{\tilde{e}_R} = 200\,\text{GeV}$, $M_1 = 99.6\,\text{GeV}$, $\mu = 200\,\text{GeV}$, and $\tan\beta = 2$, and an integrated luminosity of $100\,\text{fb}^{-1}$ was assumed. We see that the ratio of couplings is determined to be unity at about the percent level. This figure is adapted from M. Nojiri, K. Fujii and T. Tsukamoto, *Phys. Rev.* **D54**, 6756 (1996), where more details about this analysis can be found. Reprinted with permission, copyright (1996) by the American Physical Society.

equality is a direct consequence of supersymmetry, and is independent of any underlying model. Radiative corrections from SUSY breaking effects result in small deviations from exact equality which, as we will see, encode information about sparticle masses. The situation is identical to that in spontaneously broken gauge theories in that (tree-level) relationships between dimensionless couplings implied by the symmetry continue to be preserved, while the corresponding relationships between masses may be badly violated.

Experiments at LCs provide a unique opportunity to test such relations if either sleptons or charginos are kinematically accessible.[34] For instance, at tree level, the coupling $g_{\tilde{B}\tilde{e}_R e}$ of the electron to the selectron–bino system is simply the SM hypercharge gauge coupling, aside from a symmetry coefficient of $\sqrt{2}$. Establishing this equality is complicated by the fact that the bino is not a mass eigenstate, so that mixing effects need to be disentangled. Nevertheless, by a careful analysis of $\tilde{e}_R\tilde{e}_R$

[34] J. L. Feng *et al.*, *Phys. Rev.* **D52**, 1418 (1995).

pair production, it is possible to test this relationship in experiments at the LC at the $\sim 1 - 2\%$ level, as illustrated in Fig. 15.22. If squarks are much heavier than sleptons, radiative corrections cause a significant splitting $\delta g' = g_{\tilde{B}\tilde{e}_R e}/\sqrt{2} - g'$ between these couplings: for instance, if squarks are an order of magnitude heavier than sleptons, this difference would be about 2%. A measurement of $\delta g'$ could thus provide an *upper bound* on the squark mass scale, even though the squark production threshold may be far beyond the available center of mass energy.[35]

A similar test of supersymmetry that may be possible if charginos are light instead is the subject of the following exercise. The message of this discussion is that LC experiments offer a unique opportunity for these *direct tests* of SUSY and, further, that these entail a study of just the lightest charged sparticles.

Exercise *Show that supersymmetry implies that the sum of squares of the off-diagonal entries in the MSSM chargino mass matrix is completely determined by* M_W^2. *This follows from the fact that the coupling of the Higgs scalar fields to the charged higgsino–charged gaugino system is determined by the gauge interaction. If charginos are light and have substantial mixing with higgsinos (as is the case, for instance, in the HB/FP region) it is possible to extract the required off-diagonal mixing elements from the chargino mass and production properties.*

If instead charginos are gaugino-like, a test analogous to our discussion in the text may be possible. The point is that it may be possible to extract the gaugino–sneutrino–electron coupling from chargino production data, allowing a test of the SUSY relationship between it and the SU(2) gauge coupling.

Other measurements

Many other measurements are possible at linear colliders, but the details depend on which sparticle states are kinematically accessible, and also on the details of the SUSY model. For instance, if both $\widetilde{W}_1$ and $\widetilde{W}_2$ are kinematically accessible, and their masses as well as production cross sections with longitudinally polarized beams can be measured, a complete reconstruction of the chargino mass matrix would be possible.[36] On a different note, within the GMSB framework, the determination of the lifetime of a charged NLSP with a decay length as small as a millimeter may be possible in experiments at a LC, depending upon capabilities of the detector, though determinations of sparticle decay lengths exceeding ~ 50 m will be difficult.[37] The corresponding determination also appears possible for the case of a neutralino

---

[35] M. Nojiri, D. Pierce and Y. Yamada, *Phys. Rev.* **D57**, 1539 (1998); H-C. Cheng, J. Feng and N. Polonsky, *Phys. Rev.* **D57**, 152 (1998).

[36] S. Y. Choi *et al.*, *Eur. Phys. J.* **C14**, 535 (2000).

[37] P. Mercadante, J. K. Mizukoshi and H. Yamamoto, *Phys. Rev.* **D64**, 015005 (2001).

NLSP, as long as its decay length is in a similar range.[38] We remind the reader that in these models the NLSP lifetime yields a measure of the fundamental SUSY breaking scale.

Finally, we note that in the initial phase of the LC only the lowest lying states will be accessible. Except for the LSP, these will be the easiest to detect and study in detail, since they are free of contamination from other SUSY reactions, have relatively simple decays, and the SM backgrounds to their signals will be well known. In contrast, at the LHC, or when much higher energies are attained at a LC, many SUSY reactions will be occurring simultaneously, and the heavy sparticle decays will be very complex. Knowledge of the lower lying states will prove very useful for disentangling the complicated cascade decay chains expected at the LHC as well as for a study of the more massive sparticles that may be accessible at an energy upgrade of a future LC. For this reason, it would be useful to archive the LHC data in a form suitable for reanalysis once data from a LC becomes available.

15.5.3 Models of sparticle masses: a bottom-up approach

Although the mechanism by which superpartners of SM particles obtain their masses is not known, we saw in Chapter 11 several models of sparticle masses have been proposed. These models differ from one another in that they rely on different assumptions about how the effects of SUSY breaking are communicated from the supersymmetry breaking sector to the MSSM sparticles. Although these models are simple in that sparticle masses and couplings are all determined by just a small set of parameters, it should be remembered that these models are all based on untested assumptions and may turn out to be wrong. Fortunately, if sparticles are discovered, and their properties determined at future colliders, it will be possible to subject these models to experimental tests, and perhaps even determine some of the underlying parameters.

The basic idea is very simple. Any model with a fixed number of adjustable parameters is tested if the number of independent observables exceeds the number of parameters. This is so because the values of parameters that reproduce some of the observables will not automatically also yield the observed values for all of them. In practice, of course, things are more complicated because both experimental measurements as well as theoretical predictions are subject to error, and, further, the sensitivity of observables to the different parameters is not the same. The usual approach for testing any particular framework is to perform a global fit to all relevant experimental data – in addition to sparticle masses, event rates and distributions (possibly, with polarized beams) for various signals, this includes

---

[38] S. Ambrosanio and G. Blair, *Eur. Phys. J.* **C12**, 287 (2000).

low energy measurements such as branching fractions for rare decays, anomalous electric and magnetic moments of leptons or the neutron, as well as cosmological data such as the determination of cold dark matter relic density – and perform statistical tests for the goodness of fit. If a good fit is obtained (some of) the underlying parameters can be extracted; otherwise, a particular framework is excluded.

We have already seen the start of such a program in our discussion of the "allowed" and "excluded" regions of the parameter space of the mSUGRA model. It is straightforward to carry out similar studies for other scenarios. Of course, once direct information about sparticle properties becomes available, such studies will rapidly exclude many scenarios, perhaps even all the simple ones that we discussed in Chapter 11. In this case, we hope that these data will suggest how to proceed, and allow us to synthesize the mechanism by which superpartners acquire their masses.

Several groups have also examined how well experiments at the LHC or the LC will be able to extract the underlying model parameters. These studies have typically been carried out within the mSUGRA as well as GMSB frameworks. What is done is to use Monte Carlo methods to construct a synthetic data sample (within say the mSUGRA model) which is then "analyzed" to see how well the underlying parameters can be reconstructed from the various observables. It is not our purpose to discuss this in detail, and we will refer the reader to the studies in Technical Design Reports (TDRs) of ATLAS and TESLA, as well as to other studies in the literature.

Not surprisingly, the precision with which the underlying parameters can be extracted is sensitive to where one is in parameter space. For the mSUGRA model, m_0 and $m_{1/2}$ set the scale of squark and gluino/gaugino masses and can, in favorable cases, be extracted to better than 5–10% at the LHC, though the errors are somewhat larger if $\tan\beta$ is large. In fortuitous circumstances where isolation of a particular decay chain allows $m_{\widetilde{Z}_2} - m_{\widetilde{Z}_1}$ to be very precisely determined from a dilepton mass edge, $m_{1/2}$ can be determined to within a percent. A more precise determination of m_0 may be possible if the mass edges from $\widetilde{Z}_2 \to \tilde{\ell}_R\ell \to \ell\ell\widetilde{Z}_1$ can be constructed. Determination of $\tan\beta$ and A_0 is more difficult.[39]

If sleptons (charginos) are accessible at a LC, the determination of their masses will yield m_0 and $m_{1/2}$ at the percent level or better depending on the integrated luminosity. The TESLA TDR quotes a precision better than a part per mille on this. Also, for the case study in the TESLA TDR where both $\tilde{\tau}_1$ and $\tilde{\tau}_2$ are kinematically accessible, it is claimed that $\tan\beta = 3 \pm 0.02$, and $A_0 = 0 \pm 6$ GeV. While the sensitivity to these parameters will depend on the precision with which third generation masses are ultimately determined (see our comments in Section 15.5.2),

---

[39] In several of the ATLAS studies, it appears that $\tan\beta$ is determined. Notice, however, that this is because m_h in these studies is relatively light (below the bounds from LEP2); m_h becomes increasingly less sensitive to $\tan\beta$ if it is close to its theoretical upper limit.

experiments at LCs will certainly provide new information. If sleptons are accessible, their masses will pin down m_0 more precisely than experiments at the LHC, and if stau or stop mixing angles can be determined we will obtain information about the other parameters.

At the LHC, the optimal strategy for the extraction of MSSM masses and other weak scale parameters depends sensitively on the model, as well as where we are in parameter space, so that it is not possible to map out how to proceed ahead of time. However, we may say with some confidence that, with some guidance from the data, it will likely be clear how to proceed, and that it is also likely that we will glean more information than is currently thought possible. Experiments at a LC, in contrast, allow a beautiful and systematic program for these measurements that will truly complement the capabilities of the LHC. Here, we have only been able to touch upon some of the exciting capabilities of these machines. Exploration of what might be possible at both these facilities has only just begun, and is an active and fruitful area of research.

15.6 Photon, muon, and very large hadron colliders

Some possibilities for other future colliders include photon–photon and electron–photon colliders operating at a center of mass energy just below that of an available electron–positron collider, muon colliders operating in the TeV region, and also a very large hadron collider (VLHC) which might operate at $\sqrt{s} = 40$–200 TeV to succeed the LHC.

High energy photons can be produced by back scattering laser photons from a high energy electron beam. The maximum photon energy is typically about 80% of the electron beam energy. Moreover, the scattered photons are (partially) polarized if the initial electron and the laser photons are polarized. Since an electron Compton back scatters multiple times as it passes through the laser pulse, a high energy $e^- e^-$ collider can be converted to a $\gamma\gamma$ collider with comparable luminosity, but with a distribution of collision energies and photon polarizations. While there is no particular advantage of this as far as sparticle searches go, the availability of polarized photon beams is especially useful for a study of MSSM Higgs bosons. First, the rate for single Higgs boson production depends on all charged sparticle states that dominantly acquire their mass via a coupling to the Higgs, so that from this rate we can "count" all these new states. For supersymmetry aficionados, it is more interesting that the amplitude for the production of CP-odd and CP-even Higgs scalars by photon–photon collisions depends differently on the polarizations of the initial photons.[40] If CP is not conserved, a study of any Higgs boson resonance for

---

[40] This should not be surprising, since parity arguments would tell us that the leading order matrix element must be proportional to $\epsilon_1 \cdot \epsilon_2$ ($\epsilon_1 \times \epsilon_2 \cdot \hat{\mathbf{p}}_{\text{Higgs}}$) if the CP of the Higgs boson is even (odd), where ϵ_1 and ϵ_2 are the polarization vectors of the two photons.

different photon polarizations would yield information about its CP content. It may also be possible to run the collider in the $e\gamma$ mode, in which case processes such as $e\gamma \to \tilde{e}_{L(R)}\widetilde{Z}_1$ may allow us to access selectrons beyond the kinematic reach of an electron–positron collider with corresponding energy.

A muon beam has the advantage of low energy losses due to synchrotron radiation, so that a circular collider operating in the TeV region and with a much more precisely tuned beam energy relative to an electron–positron collider can be envisioned. The challenge, of course, is that the muons in the beam are unstable, so that storage, acceleration, and collisions must occur before these decay away. In addition, there are significant background problems from decays of muons in the beams. The large muon Yukawa coupling relative to that of the electron provides a unique capability: at a muon collider it is possible to produce neutral Higgs bosons in the s-channel at a large rate, allowing for detailed Higgs boson studies in much the same way that LEP has studied the Z boson. This is especially true for the more massive states such as H and A in the MSSM. Otherwise, capabilities for SUSY particle production are qualitatively similar to those of an e^+e^- collider operating in the same energy regime, except that at a muon collider, smuon pair production would occur at large rates due to t-channel graphs, whereas selectron pair production would only occur via s-channel graphs.

A very large hadron collider (VLHC) is a broad band machine that would search for new physics up to the 10–20 TeV scale, depending on the center of mass energy. While it is reasonable to see what LHC data tell us about new physics, it is worth mentioning that there can be many scenarios where the VLHC may prove essential. These include, for instance, models with additional Z bosons or with (multi-TeV scale) extra spatial dimensions. In the case of weak scale supersymmetry, a VLHC would be useful in the event that SUSY particle masses are in the TeV or multi-TeV region. In the case of GMSB models, it might also be possible at a VLHC to search for the messenger states, along with the superpartners. We note that TeV scale sparticle masses may be realized in the HB/FP region of the mSUGRA model, or in inverted hierarchy models, where just first and second generation squarks and sleptons are in the multi-TeV region. To date, few detailed studies exist for such very high energy hadron colliders.

16

R-parity violation

We have already seen that, unlike the SM, the field content of the MSSM allows gauge-invariant, renormalizable interactions (8.8a) and (8.8b) that violate the conservation of lepton and baryon number, respectively. Within the MSSM these were forbidden by imposing an additional global symmetry that leads to the conservation of a multiplicative quantum number, R-parity, given by,[1]

$$R = (-1)^{3(B-L)-2s}. \tag{16.1}$$

Here B is baryon number, L is lepton number, and s is the spin of the component field. All the SM particles have $R = +1$, while all superpartners have $R = -1$. Imposing the conservation of R-parity has several phenomenological implications: most importantly, superpartners must ultimately decay to the lightest R-odd particle (the LSP), which must be absolutely stable. Since upper limits on the abundance of exotic isotopes exclude stable electrically charged or colored particles at the weak scale, it follows that LSPs produced in SUSY events would escape detection in collider experiments. The resulting E_T^{miss} signals are the hallmark of all models that we have considered up to now. There is, however, no good theoretical argument for excluding renormalizable R-parity-violating operators from the superpotential. However, once excluded, these will not be generated by radiative corrections. If R-parity is not a good quantum number, the arguments that led us to a weakly interacting LSP no longer apply, and the phenomenology may be radically different: except when the effects of R-parity violation are small, even the distinction between

---

[1] Continuous R-symmetries (which are symmetries under which the various components of a superfield do not transform the same way because θ also transforms non-trivially) were introduced by A. Salam and J. Strathdee, *Nucl. Phys.* **B87**, 85 (1975) and P. Fayet, *Nucl. Phys.* **B90**, 104 (1975) to accommodate conservation of lepton number in supersymmetric models. However, these R-symmetries cannot be exact, because they are broken both by gaugino mass terms, as well as by the bilinear μ term in the superpotential. The usually defined R-parity is a linear combination of a discrete parity subgroup of this continuous R-symmetry and other discrete symmetries of the model. To our knowledge, the formula (16.1) was first given by G. Farrar and P. Fayet, *Phys. Lett.* **B76**, 575 (1978).

a particle and a sparticle disappears. An examination of this interesting possibility forms the subject of this chapter.

We begin by rewriting the R-parity-violating superpotential that we introduced in Chapter 8. For later convenience, we reorganize it in terms of trilinear and bilinear terms (rather than baryon- and lepton-number-violating pieces) in the R-parity-violating part of the superpotential, and write it as,

$$\hat{f}_{\not{R}} = \hat{f}_{\text{TRV}} + \hat{f}_{\text{BRV}}, \tag{16.2a}$$

with

$$\hat{f}_{\text{TRV}} = \sum_{i,j,k} \left[\lambda_{ijk} \epsilon_{ab} \hat{L}_i^a \hat{L}_j^b \hat{E}_k^c + \lambda'_{ijk} \epsilon_{ab} \hat{L}_i^a \hat{Q}_j^b \hat{D}_k^c + \lambda''_{ijk} \epsilon_{lmn} \hat{U}_i^{cl} \hat{D}_j^{cm} \hat{D}_k^{cn} \right], \tag{16.2b}$$

and

$$\hat{f}_{\text{BRV}} = \sum_i \mu'_i \epsilon_{ab} \hat{L}_i^a \hat{H}_u^b. \tag{16.2c}$$

Here, i, j, and k are generation indices running from 1–3, a, b are $SU(2)_{\text{L}}$ indices, while l, m, and n are color indices. The first two terms in (16.2b) lead to lepton-number-violating interactions, while the last term leads to baryon-number-violating interactions. Collectively, these terms give rise to explicit *trilinear R-parity viola-tion* (TRV) in the superpotential. Likewise, the operators in (16.2c) violate lepton number conservation and lead to *bilinear R-parity violation* (BRV).[2] We will see later that these provide a parametrization of spontaneous R-parity-violating mod-els. Note that the $SU(2)_{\text{L}}$ and $SU(3)_{\text{C}}$ gauge symmetries require that the couplings λ_{ijk} (λ''_{ijk}) are antisymmetric in the indices i and j (j and k), so that there are $9 + 27 + 9 = 45$ new dimensionless complex parameters and three new dimen-sionful complex parameters in the general R-parity-violating superpotential. In addition, there are also corresponding soft SUSY breaking parameters in the most general parametrization of the model.

The bilinear term in the superpotential can be rotated away by working with the linear combination,

$$\hat{H}'_{da} = \frac{\mu \hat{H}_{da} + \sum_i \epsilon_{ba} \mu'_i \hat{L}_i^b}{\sqrt{\mu^2 + \mu'^2_1 + \mu'^2_2 + \mu'^2_3}},$$

---

[2] It is worth noting that in GUT theories based on higher symmetries (where $U(1)_{B-L}$ is part of the gauge symmetry), e.g. $SO(10)$, some or all of R-parity-violating couplings may not be allowed. As long as the fields that break the gauge symmetry are inert under $(-1)^{3(B-L)}$, R-parity will remain unbroken. Thus, depending on how the larger gauge symmetry is broken, none, some, or all of the R-parity-violating operators in (16.2b) and (16.2c) above would appear in the weak scale SUSY Lagrangian.

together with three other orthogonal combinations $\hat{L}'_i$. Eliminating $\hat{H}_{da}$ in favor of $\hat{H}'_{da}$ and $\hat{L}'_i$ in the R-parity-conserving part of the superpotential results in trilinear R-parity-violating superpotential operators. This field redefinition, which was chosen to eliminate the bilinear $\hat{H}_u \hat{L}_i$ terms from the superpotential, *does not* simultaneously get rid of the corresponding soft SUSY breaking terms,

$$\mathcal{L}_{\text{soft}} \ni \sum_i b_i \epsilon_{ab} \tilde{L}'^a_i H^b_u + \quad \text{h.c.} \tag{16.3}$$

which must be retained in a general analysis. Their existence implies that, in general, the "sneutrinos" will develop VEVs along with the neutral component of H_u.

Our discussion shows that one must be careful when deriving and interpreting limits on R-parity-violating parameters, since these would depend upon the basis that we are working in. We must either carefully and completely specify the basis,[3] or work with "basis-independent" quantities when performing a general analysis.[4] In practice, it is traditional to assume that just one of the many R-parity-violating operators dominates (in a chosen basis), and to examine its effect upon the phenomenology. It is then convenient to consider separately the phenomenological analysis of models with trilinear R-parity violation and bilinear R-parity violation since trilinear and bilinear superpotential terms may well have very different theoretical origins.

Exercise *Consider the MSSM but for a single matter generation. Assume that R-parity conservation is violated only by a bilinear term in the superpotential. Redefine the fields so that R-parity violation in the superpotential appears only as trilinear operators. You will find that the up quark and lepton superpotential Yukawa couplings are basis-independent, while the down quark superpotential Yukawa coupling is altered by the field redefinition. Verify that the down quark mass is basis-independent, as it must be.*

Since there is now no distinction between particles and sparticles, the lepton, the charged gaugino and the charged higgsino can all mix. Work out the charged fermion mass matrix. Check that, though one of the mass eigenvalues is proportional to the lepton Yukawa coupling, the ratio of this eigenvalue to the lepton Yukawa coupling depends on SUSY parameters. In other words, the usual tree-level relation between the fermion mass and its Yukawa coupling is altered.

---

[3] M. Bisset, O. Kong, C. Macesanu and L. Orr, *Phys. Rev.* **D62**, 035001 (2000).
[4] S. Davidson, *Phys. Lett.* **B439**, 63 (1998), and references therein.

16.1 Explicit (trilinear) *R*-parity violation

Here we consider that R-parity is explicitly broken only by dimensionless couplings in the superpotential. We assume, in addition, that there are no soft SUSY breaking bilinears so that we may consistently take all sneutrino VEVs to be zero. The scenario is thus parametrized by 45 additional complex superpotential couplings, together with corresponding trilinear soft SUSY breaking parameters that do not enter our discussion below.

16.1.1 The TRV Lagrangian

Before we can proceed to explore phenomenological implications of the TRV terms in the superpotential, we must first extract the corresponding interactions from $\hat{f}_{\mathrm{TRV}}$. From the master formula (6.44), two sets of terms come from the superpotential:

$$\mathcal{L} \ni -\sum_i \left|\frac{\partial \hat{f}}{\partial \hat{\mathcal{S}}_i}\right|^2_{\hat{\mathcal{S}}=\mathcal{S}} - \frac{1}{2}\sum_{i,j}\left[\left(\frac{\partial^2 \hat{f}}{\partial \hat{\mathcal{S}}_i \partial \hat{\mathcal{S}}_j}\right)_{\hat{\mathcal{S}}=\mathcal{S}} \bar{\psi}_i \frac{1-\gamma_5}{2}\psi_j + \quad \text{h.c.}\right]. \quad (16.4)$$

The first of these leads to new quartic scalar interactions which, while interesting, are not likely to lead to readily observable effects, at least when the scalar fields have no VEVs. We focus, therefore, on the R-parity-violating interactions of matter fermions, starting with the first term of (16.2b):

$$\hat{f} \ni \lambda_{ijk}\left(\hat{\nu}_i \hat{e}_j - \hat{e}_i \hat{\nu}_j\right)\hat{E}^c_k. \quad (16.5)$$

Although the two terms in (16.5) above are identical, for later convenience we will write the contributions from each of these separately. The first of these yields,

$$\mathcal{L} \ni -\frac{1}{2}\cdot 2 \cdot \left[\tilde{e}^\dagger_{Rk}\bar{\psi}_{\nu_i}P_L\psi_{e_j} + \tilde{e}_{Lj}\bar{\psi}_{\nu_i}P_L\psi_{E^c_k} + \tilde{\nu}_i\bar{\psi}_{e_j}P_L\psi_{E^c_k}\right] + \quad \text{h.c.,} \quad (16.6)$$

where we remind the reader that the ψs are all Majorana spinors, whose chiral components make up the Dirac spinor for the massive fermions, as in (8.3). Using this, together with

$$e^c = P_L\psi_{E^c} + P_R\psi_e,$$

and the corresponding equations for the Dirac conjugates, it is straightforward to work out the resulting contributions to the Lagrangian. We find,

$$\mathcal{L}_\lambda = -\lambda_{ijk}\left[\tilde{e}^\dagger_{Rk}\bar{\nu}^c_i P_L e_j + \tilde{e}_{Lj}\bar{e}_k P_L \nu_i + \tilde{\nu}_i\bar{e}_k P_L e_j - \tilde{e}^\dagger_{Rk}\bar{e}^c_i P_L \nu_j \right.$$
$$\left. - \tilde{e}_{Li}\bar{e}_k P_L \nu_j - \tilde{\nu}_j\bar{e}_k P_L e_i\right] + \quad \text{h.c.,} \quad (16.7a)$$

where the last three terms arise from the second term in (16.5). We will leave it as an exercise for the reader to check that the contribution of these last three terms of

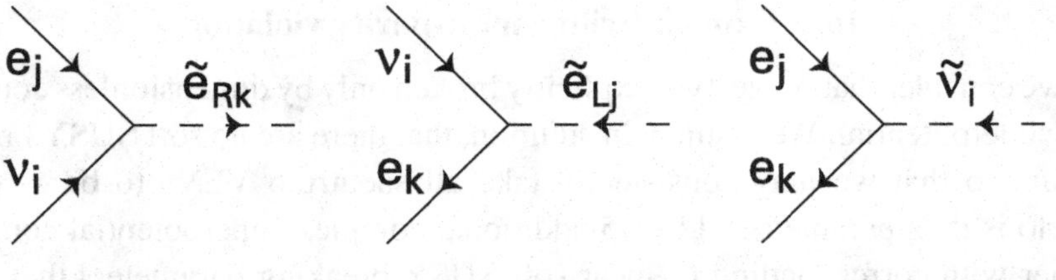

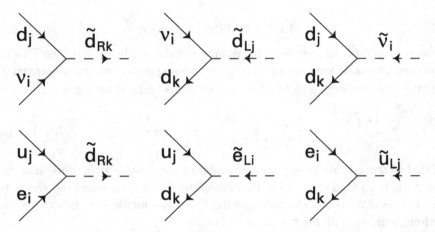

Figure 16.1 *R*-parity-violating interactions arising from the λ_{ijk} coupling in the superpotential. The arrows denote lepton number flow.

Figure 16.2 *R*-parity-violating interactions arising from λ'_{ijk} term in the superpotential. The arrows denote flow of *B* and *L* number.

$\mathcal{L}_\lambda$ is exactly the same as that of the first three. The new *L*-violating vertices are shown in Fig. 16.1.

We see that the *conjugate* fields v_i^c and e_i^c appear in the *R*-parity-violating Lagrangian. We have already encountered this complication before, for instance in our evaluation of the amplitude (12.3b) for the process $d\bar{u} \to \widetilde{W}_i \widetilde{Z}_j$, so that their presence does not pose a new problem. We use the field expansion (3.33) for the conjugate fields in our calculation of any matrix elements that we need for the exploration of the phenomenological implications of these new interactions.

An exactly similar calculation to the one above gives rise to the *R*-parity-violating Lagrangian from the second term of (16.2b). We find that this second set of lepton-number-violating interactions is given by,

$$\mathcal{L}_{\lambda'} = -\lambda'_{ijk}\Big[\tilde{d}_{Rk}^\dagger \bar{v}_i^c P_L d_j + \tilde{d}_{Lj}\bar{d}_k P_L v_i + \tilde{v}_i\bar{d}_k P_L d_j - \tilde{d}_{Rk}^\dagger \bar{e}_i^c P_L u_j$$

$$-\tilde{e}_{Li}\bar{d}_k P_L u_j - \tilde{u}_{Lj}\bar{d}_k P_L e_i\Big] + \quad \text{h.c.} \tag{16.7b}$$

The corresponding vertices are shown in Fig. 16.2.

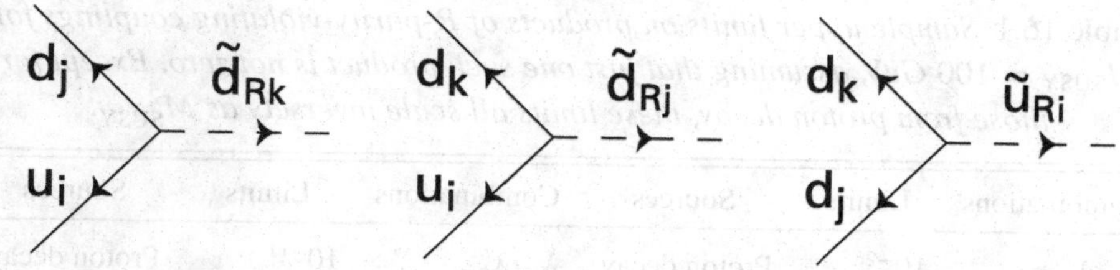

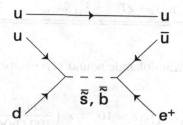

Figure 16.3 *R*-parity-violating interactions arising from λ''_{ijk} term in the superpotential. The arrows denote flow of *B* number.

Figure 16.4 *R*-parity-violating decay of the proton via the λ'_{11k} and λ''_{11k} couplings.

Finally, the *B*-violating superpotential couplings in the third term of (16.2b) give the interactions,

$$\mathcal{L}_{\lambda''} = -\lambda''_{ijk}\left[\tilde{d}^{\dagger}_{Rk}\bar{u}_i P_L d^c_j + \tilde{d}^{\dagger}_{Rj}\bar{u}_i P_L d^c_k + \tilde{u}^{\dagger}_{Ri}\bar{d}_j P_L d^c_k\right] + \quad \text{h.c.} \quad (16.8)$$

The corresponding vertices are shown in Fig. 16.3.

16.1.2 Experimental constraints

Low energy bounds

The new *B*- and *L*-violating interactions can lead to non-standard contributions to a wide variety of physical phenomena. Since the *R*-violation arises from superpotential Yukawa couplings, we expect strong constraints on various couplings from flavor-violating processes. If both λ' as well as λ'' type couplings are present, these interactions can mediate proton decay via the diagrams depicted in Fig. 16.4.

A naive estimate of the proton decay rate gives,

$$\Gamma(p \to \pi^0 e^+) \sim \sigma(ud \to \bar{u}e^+)|\psi(0)|^2 \sim \frac{|\lambda'_{11k}|^2|\lambda''_{11k}|^2}{m^4_{\tilde{d}k}} \frac{m^2_p}{128\pi} \frac{1}{\pi a^3}, \quad (16.9)$$

where $k = 2$ or 3. Here, we have taken the squared wave function factor, which is a measure that the two quarks come together to annihilate by the baryon and lepton number-violating process, to be given by $1/\pi a^3$, where $a \sim 1$ fm is the size

Table 16.1 *Sample upper limits on products of R-parity-violating couplings for* $M_{\rm SUSY} = 100$ *GeV, assuming that just one such product is not zero. Except for those from proton decay, these limits all scale inversely as* $M_{\rm SUSY}^2$.

Combinations	Limits	Sources	Combinations	Limits	Sources
$\lambda'_{11k}\lambda''_{11k}$	10^{-26}	Proton decay	$\lambda'_{ijk}\lambda''_{lmn}$	10^{-11}	Proton decay
$\lambda_{1j1}\lambda_{1j2}$	7×10^{-7}	$\mu \to 3e$	$\lambda_{231}\lambda_{131}$	7×10^{-7}	$\mu \to 3e$
$\lambda'_{i1k}\lambda'_{j2k}$	5×10^{-5}	$K^+ \to \pi^+\nu\nu$	$\lambda'_{i12}\lambda'_{i21}$	1×10^{-9}	Δm_K
$\lambda'_{i13}\lambda'_{i31}$	8×10^{-8}	Δm_B	$\lambda'_{1k1}\lambda'_{2k2}$	8×10^{-7}	$K_L \to \mu e$
$\lambda'_{1k1}\lambda'_{2k1}$	5×10^{-8}	$\mu{\rm Ti} \to e{\rm Ti}$	$\lambda'_{11j}\lambda'_{21j}$	8.5×10^{-8}	$\mu{\rm Ti} \to e{\rm Ti}$

of the proton. The Super-Kamiokande bound $\tau(p \to \pi e^+) > 5 \times 10^{33}$ years then implies that,

$$|\lambda'_{11k}\lambda''_{11k}| \lesssim 8 \times 10^{-27} \times \left(\frac{m_{\tilde{d}k}}{100\,{\rm GeV}}\right)^2. \tag{16.10}$$

We see that unconstrained *R*-parity violation leads to catastrophic *p*-decay rates. This extremely severe bound on the product of couplings strongly suggests that one or the other (or both) of these couplings is zero. It should be remembered that not all combinations of *B*- and *L*-violating interactions are as tightly constrained,[5] and, further, that the limit depends on the basis in which the couplings are written. Nevertheless, it is usually assumed that even if *R*-parity is not a good quantum number, one of *B* or *L* is conserved, which is sufficient to prevent proton decay. Non-observation of $n - \bar{n}$ oscillations or $\Delta B = 2$ "double nucleon decay" of atomic nuclei leads to limits on baryon number violating couplings that do not depend on concomitant lepton number violation.[6] It is clear that if *R*-violating couplings exist, then they must only occur in a restricted set of all the possible new interactions. As we have already noted, it is often assumed that just one of the 45 new couplings is dominant. This allows for tractable phenomenological analyses, and usually leads to the most conservative limits on the couplings. A summary of some of the most important restrictions on products of *R*-violating couplings, along with their sources, is shown in Table 16.1.[7] Here, and in subsequent tables, we have assumed that the couplings are all real. If the couplings are complex, yet new limits may be possible. For instance, the determination of ϵ_K restricts $Im\ \lambda'_{i12}\lambda'^*_{i21} < 8 \times 10^{-12}$ for $M_{\rm SUSY} = 100$ GeV. Upper limits on the electric dipole moments of the electron

[5] See C. Carlson, P. Roy and M. Sher, *Phys. Lett.* **B357**, 99 (1995).

[6] See J. L. Goity and M. Sher, *Phys. Lett.* **B346**, 69 (1995); *ibid* **B385**, 500 (1996) (erratum).

[7] These and the following restrictions on *R*-violating couplings have been adapted from G. Bhattacharyya, hep-ph/9709395 and B. Allanach *et al.*, hep-ph/9906224, where the sources for these limits, as well as others not listed here, can be found.

Table 16.2 *Upper limits (2σ) on the λ_{ijk} couplings of R-violating supersymmetry.*

ijk	λ_{ijk}	Sources
121	$0.049 \times (m_{\tilde{e}_R}/100 \text{ GeV})$	CC universality in μ-decay
122	$0.049 \times (m_{\tilde{\mu}_R}/100 \text{ GeV})$	CC universality in μ-decay
123	$0.049 \times (m_{\tilde{\tau}_R}/100 \text{ GeV})$	CC universality in μ-decay
131	$0.062 \times (m_{\tilde{e}_R}/100 \text{ GeV})$	$\Gamma(\tau \to e\nu\bar{\nu})/\Gamma(\tau \to \mu\nu\bar{\nu})$
132	$0.062 \times (m_{\tilde{\mu}_R}/100 \text{ GeV})$	$\Gamma(\tau \to e\nu\bar{\nu})/\Gamma(\tau \to \mu\nu\bar{\nu})$
133	$0.006 \times \sqrt{m_{\tilde{\tau}}/100 \text{ GeV}}$	ν_e mass
231	$0.070 \times (m_{\tilde{e}_R}/100 \text{ GeV})$	$\Gamma(\tau \to \mu\nu\bar{\nu})/\Gamma(\mu \to e\nu\bar{\nu})$
232	$0.070 \times (m_{\tilde{\mu}_R}/100 \text{ GeV})$	$\Gamma(\tau \to \mu\nu\bar{\nu})/\Gamma(\mu \to e\nu\bar{\nu})$
233	$0.070 \times (m_{\tilde{\tau}_R}/100 \text{ GeV})$	$\Gamma(\tau \to \mu\nu\bar{\nu})/\Gamma(\mu \to e\nu\bar{\nu})$

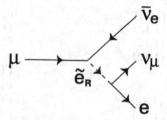

Figure 16.5 An example of an *R*-parity-violating contribution to β decay of the muon.

and the neutron also constrain the imaginary part of some other products at the 10^{-4} level.

In addition to these constraints, there is a variety of limits on individual *R*-parity-violating couplings. For example, the coupling λ_{121} leads to a new contribution to the standard decay of the muon, as shown in Fig. 16.5. Such contributions are strongly constrained by the observed universality of the charged current weak interactions.[8] Comparing muon decay with the β decay of quarks, one finds the limit $\lambda_{121} < 0.049 \times (m_{\tilde{e}_R}/100 \text{ GeV})$. This limit, together with corresponding limits on the λ_{ijk} couplings, along with their sources, is summarized in Table 16.2.

Constraints on the λ'_{ijk} couplings along with their sources are summarized in Table 16.3. While the limits on first generation λ's are rather strict, the corresponding bounds for third generation couplings are generally less severe. Also shown in parentheses are limits that result if we require perturbativity of the *R*-parity-violating couplings up to the GUT scale: if the couplings exceed these bounds at the weak scale, then they will diverge under renormalization group evolution before

[8] See V. Barger, G. F. Giudice and T. Han, *Phys. Rev.* **D40**, 2987 (1989).

Table 16.3 *Upper limits (2σ) on λ'_{ijk} couplings for R-violating SUSY. Bounds from requiring perturbativity up to the GUT scale are shown in parentheses.*

ijk	λ'_{ijk}	Sources
111	$5.2 \times 10^{-4} \times (m_{\tilde{e}}/100 \text{ GeV})^2(m_{\tilde{\chi}_1}/100 \text{ GeV})^{1/2}$	$(\beta\beta)_{0\nu}$
112	$0.021 \times m_{\tilde{s}_R}/100 \text{ GeV}$	CC univ.
113	$0.021 \times m_{\tilde{b}_R}/100 \text{ GeV}$	CC univ.
121	$0.043 \times m_{\tilde{d}_R}/100 \text{ GeV}$	CC univ.
122	$0.043 \times m_{\tilde{s}_R}/100 \text{ GeV}$	CC univ.
123	$0.043 \times m_{\tilde{b}_R}/100 \text{ GeV}$	CC univ.
131	$0.019 \times m_{\tilde{t}_L}/100 \text{ GeV}$	APV
132	$0.28 \times m_{\tilde{t}_L}/100 \text{ GeV}$ (1.04)	A_{FB}
133	$1.4 \times 10^{-3}\sqrt{m_{\tilde{b}}/100 \text{ GeV}}$	ν_e-mass
211	$0.059 \times m_{\tilde{d}_R}/100 \text{ GeV}$	$\Gamma(\pi \to e\nu)/\Gamma(\pi \to \mu\nu)$
212	$0.059 \times m_{\tilde{s}_R}/100 \text{ GeV}$	$\Gamma(\pi \to e\nu)/\Gamma(\pi \to \mu\nu)$
213	$0.021 \times m_{\tilde{b}_R}/100 \text{ GeV}$	$\Gamma(\pi \to e\nu)/\Gamma(\pi \to \mu\nu)$
221	$0.18 \times m_{\tilde{s}_R}/100 \text{ GeV}$ (1.12)	ν_μ DIS
222	$0.21 \times m_{\tilde{s}_R}/100 \text{ GeV}$ (1.12)	$D \to K\ell\nu$
223	$0.21 \times m_{\tilde{b}_R}/100 \text{ GeV}$ (1.12)	$D \to K\ell\nu$
231	$0.18 \times m_{\tilde{b}_L}/100 \text{ GeV}$ (1.12)	ν_μ DIS
232	0.56 (1.04)	$\Gamma(Z \to \text{hadrons})/\Gamma(Z \to \ell\bar{\ell})$
233	$0.15\sqrt{m_{\tilde{b}}/100 \text{ GeV}}$	ν_μ-mass
311	$0.11 \times m_{\tilde{d}_R}/100 \text{ GeV}$ (1.12)	$\Gamma(\tau \to \pi\nu_\tau)/\Gamma(\pi \to \mu\nu)$
312	$0.11 \times m_{\tilde{s}_R}/100 \text{ GeV}$ (1.12)	$\Gamma(\tau \to \pi\nu_\tau)/\Gamma(\pi \to \mu\nu)$
313	$0.11 \times m_{\tilde{b}_R}/100 \text{ GeV}$ (1.12)	$\Gamma(\tau \to \pi\nu_\tau)/\Gamma(\pi \to \mu\nu)$
321	$0.52 \times m_{\tilde{d}_R}/100 \text{ GeV}$ (1.12)	$\Gamma(D_s \to \tau\nu_\tau)/\Gamma(D_s \to \mu\nu_\mu)$
322	$0.52 \times m_{\tilde{s}_R}/100 \text{ GeV}$ (1.12)	$\Gamma(D_s \to \tau\nu_\tau)/\Gamma(D_s \to \mu\nu_\mu)$
323	$0.52 \times m_{\tilde{b}_R}/100 \text{ GeV}$ (1.12)	$\Gamma(D_s \to \tau\nu_\tau)/\Gamma(D_s \to \mu\nu_\mu)$
331	0.45 (1.04)	$\Gamma(Z \to \text{hadrons})/\Gamma(Z \to \ell\bar{\ell})$
332	0.45 (1.04)	$\Gamma(Z \to \text{hadrons})/\Gamma(Z \to \ell\bar{\ell})$
333	0.45 (1.04)	$\Gamma(Z \to \text{hadrons})/\Gamma(Z \to \ell\bar{\ell})$

the GUT scale is reached. If these couplings really become large before $Q = M_{\text{GUT}}$, they would be expected to make a substantial modification to the renormalization group flow, and to the successful prediction of the unification of gauge couplings. Of course, these latter limits are model dependent, since they are obtained assuming a desert between M_{SUSY} and M_{GUT}.

Finally, the limits of the B-violating couplings λ''_{ijk} are summarized in Table 16.4. Note that the bound on the first line is obtained under the assumption that the lifetime for the "double nucleon decay" $^{16}\text{O} \to {}^{14}\text{C} + K^+K^+$ exceeds 10^{30} years.[9] While the bounds on first generation couplings can again be quite severe if

---

[9] This decay could presumably be detected in the Super-Kamiokande experiment which has obtained a limit exceeding 1.9×10^{33} on the decay $p \to K^+\nu$.

Table 16.4 *Upper limits (2σ) on the λ''_{ijk} couplings in R-violating SUSY. The quantity Λ in the first line is some hadronic scale ~ 300 MeV. Most of the direct bounds listed are for $M_{SUSY} = 100$ GeV. Bounds from requiring perturbativity up to M_{GUT} are shown in parentheses.*

ijk	λ''_{ijk}	Sources
112	$10^{-15} \times (M_{SUSY}/\Lambda)^{5/2}$	Double nucleon decay
113	10^{-4}	$n - \bar{n}$ oscillation
123	(1.23)	Perturbativity
212	(1.25)	Perturbativity
213	(1.23)	Perturbativity
223	(1.23)	Perturbativity
312	0.50 (1.00)	$\Gamma(Z \to \text{hadrons})/\Gamma(Z \to \ell\bar{\ell})$
313	0.50 (1.00)	$\Gamma(Z \to \text{hadrons})/\Gamma(Z \to \ell\bar{\ell})$
323	0.50 (1.00)	$\Gamma(Z \to \text{hadrons})/\Gamma(Z \to \ell\bar{\ell})$

squarks are light, most of the second and third generation couplings have no real restriction other than from the requirement of perturbativity up to $Q = M_{GUT}$.

Cosmological bounds

A very interesting bound on *R*-parity-violating couplings follows from considerations of GUT scale baryogenesis in the Big Bang cosmology. This bound arises from the requirement that any GUT scale matter–antimatter asymmetry that can develop in these models not be wiped out by *R*-parity-violating interactions.

It is known that within the SM there are non-perturbative effects from the so-called electroweak sphaleron interactions which violate separate B and L conservation but conserve $B - L$. Sphaleron effects will, therefore, tend to restore the matter–antimatter symmetry as the Universe cools to $T \sim M_{\text{weak}}$. However, any $B - L$ component of the matter–antimatter asymmetry that may have been generated at the high scale cannot be wiped out by these effects, and so will persist to the low scale.

Note that the *R*-parity-violating couplings in (16.2b) do not conserve $B - L$, so that if these remain in thermal equilibrium down to the weak scale, they would wash out any $B - L$ component of the matter–antimatter asymmetry. Together with sphaleron interactions that wash out the $B + L$ component, any matter–antimatter asymmetry that may have been generated at a high scale will be washed away, unless the *R*-parity-violating couplings are small enough so that these interactions fall out of equilibrium before the Universe cools to $T = M_{\text{weak}}$. This leads to a generic upper limit on all TRV couplings:

$$\lambda_{ijk}, \ \lambda'_{ijk}, \ \lambda''_{ijk} < 5 \times 10^{-7} \, (M_{\text{SUSY}}/1 \text{ TeV})^{1/2}. \tag{16.11}$$

We will see in Section 16.1.4 that this limit implies that the LSP will be quasi-stable in that it essentially always decays outside any collider detector. Unless this LSP happens to be charged or colored, it would escape experimental detection exactly as in models where R is a good quantum number. Also, as discussed below, R-violating contributions to sparticle production and decay of heavier sparticles would be negligible, so that R-parity-violating couplings satisfying the bounds (16.11) would be irrelevant to any consideration of SUSY signals at colliders.

The bounds (16.11) clearly do not apply if baryogenesis occurs at the electroweak scale, instead of at the GUT or some intermediate scale. One suggestion (that has not been examined in detail) is that complex λ'' couplings generate the baryon asymmetry below the scale $M_{\text{SUSY}} \sim M_{\text{weak}}$. Electroweak scale baryogenesis is also possible within the MSSM, though this requires that $m_h \lesssim 115\text{–}120$ GeV, and $m_{\tilde{t}_R} < m_t$. It should, therefore, be possible to probe this scenario in collider experiments. If the particle content of the MSSM is extended by a singlet Higgs field and the gauge symmetry by an extra (anomaly free) $U(1)$, it appears possible to accommodate electroweak scale baryogenesis even if the top squark is heavy.

The observation that sphaleron interactions actually conserve $B/3 - L_i$ for each lepton flavor points out another loophole to the general argument that led to the stringent bounds on the TRV couplings, even if the matter–antimatter asymmetry is generated at $T \gg M_{\text{weak}}$.[10] The conserved quantum numbers may equivalently be chosen to be $B - L$ and the two independent combinations of $L_i - L_j$. If a matter–antimatter asymmetry arises asymmetrically between the three lepton flavors, it will clearly be preserved by sphaleron and λ'' interactions even if these are in thermal equilibrium up to the electroweak scale. The surviving lepton number will be converted partially back into a baryon asymmetry at temperatures below the electroweak scale and, as long as L-violating couplings are negligible, the bounds on the λ'' couplings are essentially eliminated. Alternatively, if R-parity-violating couplings conserve baryon number, we can still maintain a GUT scale matter–antimatter asymmetry as long as the set of lepton number violating couplings that violate conservation of one of the lepton flavors falls out of thermal equilibrium sufficiently early – i.e. satisfies the bound (16.11), even if other L-violating couplings are large. In the case of R-parity violation via $\Delta L \neq 0$ couplings, the exact bounds depend on the details of the lepton flavor-violating couplings.

---

[10] See B. Campbell *et al.*, *Phys. Lett.* **B297**, 118 (1992), and H. Dreiner and G. Ross, *Nucl. Phys.* **B410**, 183 (1993) for further details.

The upshot of this discussion is that it is possible to construct scenarios consistent with high scale baryogenesis, and where *R*-parity-violating couplings have an important impact on collider signatures of supersymmetry.

16.1.3 s-channel sparticle production

If *R*-parity-violating couplings exist, then a novel feature of SUSY models is the possibility of resonance production of sparticles.[11] By examining the interactions in Figs. 16.1–16.3, it is easy to see the following processes can occur:

$$e^+e^- \rightarrow \tilde{\nu}_{Lj} \quad \text{(LEP2, NLC)}, \tag{16.12a}$$

$$e^-u_j \rightarrow \tilde{d}_{Rk} \quad \text{(HERA)}, \tag{16.12b}$$

$$e^-\bar{d}_k \rightarrow \bar{\tilde{u}}_{Lj} \quad \text{(HERA)}, \tag{16.12c}$$

$$\bar{u}_jd_k \rightarrow \tilde{e}_{Li} \quad \text{(Tevatron, LHC)}, \tag{16.12d}$$

$$d_j\bar{d}_k \rightarrow \tilde{\nu}_{Li} \quad \text{(Tevatron, LHC)}, \tag{16.12e}$$

$$\bar{u}_i\bar{d}_j \rightarrow \tilde{d}_{Rk} \quad \text{(Tevatron, LHC)}, \tag{16.12f}$$

$$\bar{d}_j\bar{d}_k \rightarrow \tilde{u}_{Ri} \quad \text{(Tevatron, LHC)}. \tag{16.12g}$$

At LEP2 or at an e^+e^- linear collider, it is thus possible to produce the $\tilde{\nu}_\mu$ or $\tilde{\nu}_\tau$ in the *s*-channel via the λ_{121} or λ_{131} couplings, respectively. Neglecting the sneutrino width, the production cross section is given by

$$\sigma(e^+e^- \rightarrow \tilde{\nu}_j) = \frac{\pi|\lambda_{1j1}|^2s\delta(s - m_{\tilde{\nu}_j}^2)}{4m_{\tilde{\nu}_j}^2}, \tag{16.13}$$

where $s = 4E_{\text{beam}}^2$. Although the reaction rate may be suppressed by the magnitude of the *R*-violating Yukawa coupling, it is greatly enhanced compared to sneutrino pair production, provided the energy spread of the beam is smaller than the width of the sneutrino. Once the sneutrino is produced, it may decay via gauge couplings as $\tilde{\nu}_j \rightarrow \ell_j\widetilde{W}_1$ or $\nu_j\widetilde{Z}_i$, or via the *R*-violating coupling back into e^+e^-, if the coupling is large enough. Such reactions have been searched for at LEP2, where limits are usually placed in the $m_{\tilde{\nu}_j}$ vs. λ_{1j1} plane, and depend on the assumed decay modes.

The *R*-violating couplings λ_{122}, λ_{123}, λ_{132}, λ_{133}, and λ_{231} can also be probed at LEP2 and the NLC via the reactions

$$\gamma e^\pm \rightarrow \ell_k^\pm\tilde{\nu}_j, \quad \text{and} \tag{16.14a}$$

$$\gamma e^\pm \rightarrow \tilde{\ell}_j^\pm\nu_k, \tag{16.14b}$$

---

[11] The alert reader will object that the concept of sparticle is ill-defined when *R*-parity is not conserved because odd and even *R* states can now mix to form the mass eigenstates. By "sparticles" we are, in this chapter, referring to those mass eigenstates whose content is dominantly *R*-odd.

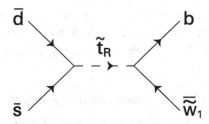

Figure 16.6 An example of resonance production of a top squark via *R*-parity-violating couplings at hadron colliders.

where the photon comes either from initial state radiation or from beamstrahlung. In this case, the production cross section has to be convoluted with a distribution function such as (12.18) that describes the density of photons in the electron or positron. Finally, a sparticle may be produced in association with a SM particle in e^+e^- collisions via t-channel exchange graphs; the resulting cross sections for these $2 \to 2$ processes are quite low because *R*-parity-violating couplings are typically smaller than gauge couplings.

The HERA ep collider at DESY is unique in that it allows for s-channel squark production via the λ'_{1j1} and λ'_{11k} couplings. If the *R*-violating couplings are large enough, and the produced squarks decay back into e and a jet, the analysis becomes very similar to the one for spin-0 leptoquark production. If the produced squarks decay instead into SUSY particles, then the signatures can be very different. Searches have been performed by the H1 and ZEUS collaborations. These searches exclude production of first generation squarks up to 240 GeV assuming that $\lambda' \gtrsim \sqrt{4\pi\alpha_{\rm em}}$, although of course the limit depends strongly on the magnitude of this coupling. Note that for $M_{\rm SUSY} = 240$ GeV, couplings of this size appear to be already excluded by the low energy constraints listed in Table 16.3.

Single squark production is also possible at the Tevatron and LHC colliders, mediated by the λ''_{ijk} couplings: see Fig. 16.6. These couplings are relatively unconstrained for production of second and third generation squarks. An analysis of single top squark production at the Fermilab Tevatron via $\bar{s}\bar{d} \to \tilde{t}_1$ followed by $\tilde{t}_1 \to b\widetilde{W}_1$ decay indicates $m_{\tilde{t}_1} \lesssim 200$–300 GeV can be probed with 2 fb^{-1} of integrated luminosity, if $\lambda''_{3jk} > 0.02$–0.06.[12]

16.1.4 *R* decay of the LSP

If *R*-parity-violating couplings are small compared to gauge couplings, these do not alter sparticle mass patterns in any significant manner and the lightest neutralino

[12] E. Berger, B. W. Harris and Z. Sullivan, *Phys. Rev.* **D83**, 4472 (1999).

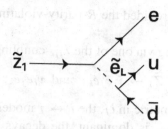

Figure 16.7 *R*-parity-violating decay of the lightest neutralino.

remains as the LSP in many models. However, these couplings render the LSP unstable. In this case, the LSP need not be electrically or color neutral, since if it is unstable, then cosmological bounds on stable relics from the Big Bang no longer apply. Thus, the $\tilde{g}$, $\widetilde{W}_1$, $\tilde{q}$, $\tilde{\ell}$ or $\tilde{v}$ states are viable LSP candidates, as long as these decay quickly enough so as not to disrupt nucleosynthesis in the early Universe.

In models such as mSUGRA, the $\tilde{Z}_1$ is usually the LSP over most of parameter space, just as a consequence of the mSUGRA boundary conditions, and the RGEs. In this case, it is possible that R-violating couplings are so small that they do not affect sparticle production or decay reactions, except for the decay of the LSP, henceforth taken to be $\tilde{Z}_1$. An example of $\tilde{Z}_1 \to eu\bar{d}$ decay via $\tilde{e}_L$ exchange is shown in Fig. 16.7; two other diagrams involving $\tilde{d}_R$ and $\tilde{u}_L$ exchange also contribute. In addition, the λ'_{111} term will also mediate the decay $\tilde{Z}_1 \to v_e d\bar{d}$.

We make an order of magnitude estimate of the decay length of $\tilde{Z}_1$, assuming that it is a pure photino. In this case, the decay rate simplifies to

$$\Gamma(\tilde{Z}_1 \to eu\bar{d}) \sim \frac{3\alpha\lambda'^2_{111}}{128\pi^2} \frac{m^5_{\tilde{Z}_1}}{M^4_{\text{SUSY}}}. \tag{16.15}$$

Roughly speaking, the decay takes place in the detector if $c\gamma\tau(\tilde{Z}_1) \lesssim 1$ m, where γ is the Lorentz boost factor $\gamma = E_{\tilde{Z}_1}/m_{\tilde{Z}_1}$. This implies that

$$\lambda'_{111} > 1.4 \times 10^{-6} \sqrt{\gamma} \left(\frac{M_{\text{SUSY}}}{200\,\text{GeV}}\right)^2 \left(\frac{100\,\text{GeV}}{m_{\tilde{Z}_1}}\right)^{5/2}. \tag{16.16}$$

A similar calculation applies to decays mediated by other λ_{ijk}, λ'_{ijk} or λ''_{ijk} couplings. If the λs are much smaller than this limit, then the $\tilde{Z}_1$ will generally escape the detector, leading to missing energy as in the MSSM with R-parity conservation.

For λ values comparable to the bound in Eq. (16.16), there may exist substantial decay gaps in collider detectors. If the LSP is not $\tilde{Z}_1$ but a charged sparticle, its production will be signalled by highly ionizing tracks in the detector, followed by

B- or *L*-violating decays provided the *R*-parity-violating coupling responsible for the decay is large enough.

For a neutralino decaying via one of the λ_{ijk} couplings, the decay modes are

$$\widetilde{Z}_1 \to \bar{\nu}_i \bar{e}_j e_k \quad \text{and} \quad \nu_i e_j \bar{e}_k. \tag{16.17}$$

Since the λ_{ijk} are antisymmetric in ij, the $i \leftrightarrow j$ modes must be included as well. For instance, assuming λ_{121} is dominant, the decays $\widetilde{Z}_1 \to \bar{\nu}_e \bar{\mu} e$, $\nu_e \mu \bar{e}$, $\bar{\nu}_\mu \bar{e} e$, and $\nu_\mu e \bar{e}$ would each occur with a $\sim 25\%$ branching fraction provided that all the relevant sleptons have the same mass.

If instead the $\widetilde{Z}_1$ decays via the λ'_{ijk} coupling, then the decays are

$$\widetilde{Z}_1 \to e_i u_j \bar{d}_k \quad \text{and} \quad \bar{e}_i \bar{u}_j d_k, \quad \text{as well as} \tag{16.18a}$$
$$\widetilde{Z}_1 \to d_j \nu_i \bar{d}_k \quad \text{and} \quad \bar{d}_j \bar{\nu}_i d_k. \tag{16.18b}$$

The relative branching ratios between the modes containing charged leptons and those containing neutrinos are model dependent. Note that there are several possible *R*-violating $\widetilde{Z}_1$ decay modes for each λ'_{ijk} coupling. For instance, if λ'_{112} is dominant, then $\widetilde{Z}_1 \to eu\bar{s}$, $\bar{e}\bar{u}s$ with a branching fraction B, and $\widetilde{Z}_1 \to d\nu_e \bar{s}$, $\bar{d}\bar{\nu}_e s$ with a branching fraction $1 - B$.

Finally, if $\widetilde{Z}_1$ decays via λ''_{ijk} couplings, the decay modes are

$$\widetilde{Z}_1 \to u_i d_j d_k \quad \text{and} \quad \bar{u}_i \bar{d}_j \bar{d}_k. \tag{16.19}$$

There are nine possibilities, since λ''_{ijk} is antisymmetric on the jk indices. For example, if λ''_{121} is dominant, then $\widetilde{Z}_1 \to uds$ or $\bar{u}d\bar{s}$, each with a branching fraction of 50%.

Exercise *We have focussed on the case that the LSP is a neutralino. Assume instead that the LSP is one of the staus that decays to a pair of SM fermions via one of the λ or λ' couplings. Evaluate its decay rate and estimate the range of the R-violating coupling for which the stau may be detectable as an ionizing track in a collider detector. For what values of this coupling will the stau decay inside the detector? List its possible decay modes and calculate the corresponding branching fractions, assuming that just one of the couplings dominates the decay.*

16.1.5 Collider signatures

If *R*-parity-violating couplings are much smaller than the gauge couplings, the dominant sparticle pair production mechanisms will be the same as those discussed in Chapter 12: i.e. sparticle pair production rates will essentially be the same as in the

MSSM.[13] Moreover, heavier sparticles will dominantly decay to lighter sparticles via their gauge and MSSM superpotential couplings, so that their decay patterns will also be the same as in the MSSM. The difference is that the lightest sparticle will decay as discussed in the last section.

The decay of the LSP inside the experimental apparatus has very important implications for supersymmetric collider signatures.

- The E_T^{miss} signal that we have been considering as the hallmark of sparticle production may be greatly diminished. Except for gravitinos, which are relevant only in some special scenarios, neutrinos from the decay of the LSP or from other stages of SUSY cascades will be the only physics source of E_T^{miss} events. In the case that the LSP dominantly decays via $\widetilde{Z}_1 \rightarrow cds + \bar{c}\bar{d}\bar{s}$, we may expect that the observability of SUSY signals at hadron colliders will be considerably degraded, mostly due to the reduced E_T^{miss}, but also because the excess hadronic activity from LSP decays would also make it more difficult for any leptons produced in SUSY cascade decays to remain isolated.
- If the LSP dominantly decays leptonically into e or μ via λ-type couplings, then the rates for multilepton events from sparticle production would be greatly increased, and the SUSY reach at hadron colliders would be considerably larger than the projections in the last chapter.
- An unstable LSP that decays inside the detector will make it easier to completely reconstruct SUSY events, especially at an e^+e^- collider.
- If the LSP is relatively long-lived, it will decay with a displaced vertex which would serve as an additional handle for selecting the SUSY signal over SM background. Indeed, it may then be possible to determine the lifetime of the LSP, and directly obtain information about R-parity-violating couplings.
- If the LSP is charged and long-lived, it can be searched for by looking for heavily ionizing tracks of relatively slow-moving particles. If it is colored, it would bind with a quark or gluon to make a charged or neutral strongly interacting particle. A number of handles, some of which are quite spectacular, may be possible, but the signals are somewhat dependent on how this particle loses energy in traversing the material of the detector.[14]

In the difficult case where the LSP decays hadronically and without any displaced vertex, simulations within the mSUGRA framework extended by R-parity violation have shown that experiments at the Fermilab Tevatron may have no observable signal if gluinos are heavier than about 200 GeV and $m_{\tilde{q}} \gg m_{\tilde{g}}$.[15] On the other hand,

---

[13] We will, of course, have the "resonant $2 \rightarrow 1$" s-channel production mechanisms occuring with a rate that is directly dependent on the corresponding R-parity-violating coupling.

[14] See e.g. M. Drees and X. Tata, *Phys. Lett.* **B252**, 695 (1990).

[15] H. Baer, C. Kao and X. Tata, *Phys. Rev.* **D51**, 2180 (1995).

if the LSP decays dominantly via $\widetilde{Z}_1 \to \ell\ell\nu$ ($\ell = e, \mu$), there will be observable signals in the $\geq 4\ell$ channel even if $m_{\tilde{g}}$ exceeds 800 GeV with 10 fb^{-1} of integrated luminosity.

It is interesting to ask whether sparticles can remain hidden at the LHC in the case that $\widetilde{Z}_1 \to cds + c\bar{d}\bar{s}$. A detailed study, again within the mSUGRA model extended to include the λ''_{212} coupling, has shown that the reach in the E_T^{miss} channel is indeed greatly degraded relative to that in the mSUGRA model.[16] Fortunately, the reach via multijet plus various $n_\ell \geq 1$ lepton channels introduced in the last chapter, where the leptons come from cascade decays, remains robust for squarks or gluinos up to just over 1 TeV.

At electron–positron colliders, we do not expect the decays of a neutralino LSP to significantly alter the mass reach for charged sparticles since this frequently extends most of the way to the kinematic limit. For the case of an unstable LSP it may in fact be easier to reconstruct SUSY events as we have already noted. *R*-parity-violating couplings may, however, greatly expand the model parameter space for which there is an observable signal at an e^+e^- collider because then $e^+e^- \to \widetilde{Z}_1\widetilde{Z}_1$ also leads to detectable signals.[17]

16.2 Spontaneous (bilinear) *R*-parity violation

Instead of adding TRV couplings to the superpotential, some authors have suggested that *R*-parity may be a symmetry of the Lagrangian, but not a symmetry of the ground state: i.e. *R*-parity conservation is broken spontaneously.

A model to exhibit the spontaneous violation of *R*-parity can be constructed by adding several new gauge singlet superfields ($\hat{\Phi}$, $\hat{\nu}_i^c$, $\hat{S}_i$) which carry lepton number $(0, -1, 1)$, respectively, but no baryon number (i is a generation index) to the MSSM.[18] The superpotential of the model is given by,

$$\hat{f} = \sum_{i,j=1,2,3} \left[(\mathbf{f}_u)_{ij}\epsilon_{ab}\hat{Q}_i^a\hat{H}_u^b\hat{U}_j^c + (\mathbf{f}_d)_{ij}\hat{Q}_i^a\hat{H}_{da}\hat{D}_j^c + (\mathbf{f}_e)_{ij}\hat{L}_i^a\hat{H}_{da}\hat{E}_j^c \right.$$
$$\left. + (\mathbf{f}_\nu)_{ij}\epsilon_{ab}\hat{L}_i^a\hat{H}_u^b\hat{\nu}_j^c + (\mathbf{f})_{ij}\hat{\Phi}\hat{S}_i\hat{\nu}_j^c \right] + (f_0\hat{H}_u\hat{H}_d - \epsilon^2)\hat{\Phi}. \tag{16.20}$$

This superpotential which trivially conserves B, also conserves L, and hence also *R*-parity. Upon minimization, the corresponding scalar potential develops VEVs in the directions $\phi = \tilde{\nu}_{iR}$, $\tilde{S}_i$, $\hat{\Phi}$, h_u^0, h_d^0 and $\tilde{\nu}_{iL}$. To illustrate the general idea, it is sufficient to only consider just one generation. The resulting Lagrangian, written

[16] H. Baer, C. Chen and X. Tata, *Phys. Rev.* **D55**, 1466 (1997).

[17] For a discussion of branching fractions and relative rates into various event topologies at an e^+e^- collider, see R. Godbole, P. Roy and X. Tata, *Nucl. Phys.* **B401**, 67 (1992).

[18] A. Masiero and J. W. F. Valle, *Phys. Lett.* **B251**, 273 (1990); for a review, see J. W. F. Valle, hep-ph/9603307 (1996).

in terms of the shifted fields, appears to violate lepton number, and hence R-parity conservation, but preserves B so that the proton is safe from decay. Since $U(1)_L$ is spontaneously broken, there is a dominantly gauge singlet, massless Goldstone boson J, the Majoron, with very weak couplings to the Z boson. The Majoron may be eliminated by the Higgs mechanism if this model is embedded into one with a higher gauge symmetry.

Many of the phenomenological effects of spontaneous R-parity violation can be incorporated into the MSSM by adding just the bilinear terms,

$$\hat{f} \ni \sum_i \mu_i' \epsilon_{ab} \hat{L}_i^a \hat{H}_u^b \tag{16.21a}$$

to the superpotential along with the corresponding soft SUSY breaking terms,

$$\mathcal{L}_{\text{soft}} \ni \sum_i b_i \epsilon_{ab} \tilde{L}_i^a \tilde{H}_u^b + \text{ h.c.,} \tag{16.21b}$$

but no TRV couplings. As we have already discussed, it is possible to go to a basis where the superpotential bilinear R-violating (BRV) interactions are rotated away, resulting in trilinear couplings in the superpotential, with "sneutrinos" of this basis developing VEVs.

The BRV model as defined by (16.21a) and (16.21b) leads to several interesting consequences.[19] In the basis where lepton number is violated only by bilinear terms, there are many new sources of mixing that need to be included to deduce the phenomenological implications. This happens because in the absence of conservation of the lepton numbers L_i, there is no distinction between the three matter doublet superfields $\hat{L}_i$ and the doublet superfields $\hat{H}_d$ and $\hat{H}_u$.

- The neutralino fields now mix with the neutrino fields, leading to a 7×7 neutralino/neutrino mass matrix. One linear combination of *neutrino* fields develops a Majorana mass via tree-level mixing with Higgsinos, while the other combinations acquire masses upon including one-loop corrections if the corresponding lepton number is also not conserved. While it is possible to accommodate small neutrino masses, this evidently requires that the parameters be carefully adjusted to ensure that the tree-level neutrino masses are at the sub-eV level or smaller as required by phenomenology. It is, perhaps, worth emphasizing that neutrinos generically acquire Majorana masses in *all* R-parity-violating models where the corresponding lepton number is not conserved because there is no symmetry that precludes these masses from being radiatively generated. Indeed, it is exactly this

---

[19] Although we used the model with spontaneous R-violation as motivation for the BRV model, the two models are different. In the BRV model, L and R-parity are explicitly broken so that there is no Majoron. Many aspects of the phenomenology are similar because the Majoron is weakly coupled, and so is mainly relevant for neutrino physics since it allows for neutrino decays.

that leads to the bound on λ_{133} and λ'_{133} in Tables 16.2 and 16.3, respectively. What is unique to the BRV model is that one of the neutrinos acquires an (albeit too large) mass even at the tree level.

- Likewise, charged gauginos and charged Higgsinos can now mix with the charged leptons, leading to a 5×5 mass matrix for charged fermions. As noted in the exercise at the start of this chapter, the SM relation between the Yukawa coupling and the corresponding "fermion mass" is modified.
- In the bosonic sector, the CP-even Higgs fields mix with the real components of the sneutrino fields, leading to a 5×5 mass matrix. The lightest Higgs scalar always has mass less than the corresponding lightest scalar in the MSSM.
- The imaginary components of the sneutrino fields mix with the CP-odd scalars, leading to a 5×5 mass matrix that includes the massless (would-be) neutral Goldstone boson which is subsequently eaten up by the Higgs mechanism.
- The fields $h_u^\pm$ and $h_d^\pm$ mix with the charged sleptons; including $\tilde{\ell}_L - \tilde{\ell}_R$ mixing effects, an 8×8 mass matrix is obtained that includes the (would-be) charged Goldstone boson.

In addition to low energy effects, e.g. in neutrinoless $\beta\beta$ decay, a variety of R-violating signals are possible at colliders. In particular, the LSP is unstable, and can decay for example into $\tilde{Z}_1 \to \tau W^{(*)}$ or $\nu_\tau Z^{(*)}$, where the W and Z can be real or virtual. Decay gaps in $\tilde{Z}_1$ decay are likely, and the LSP may appear to be quasi-stable in a collider detector. In models with a Majoron J, the lightest neutralino may also decay via $\tilde{Z}_1 \to \nu J$.

The BRV model may also be embedded in the mSUGRA model framework. In this case, one assumes the MSSM augmented by the bilinear terms (16.21a) and (16.21b) is valid up to $Q = M_{\text{GUT}}$ where, in addition to the usual mSUGRA boundary conditions (9.18b)–(9.18d), we assume that the three μ'_i unify to μ'_0, and the $B'_i \equiv b_i/\mu'_i$ unify with the usual Higgs sector B parameter. Compared to the mSUGRA model, there is then just one more GUT scale parameter in the theory. The RGEs of the MSSM must be supplemented with corresponding RGEs for the B'_is and the μ'_is.[20] The weak scale scalar potential of the model can now be minimized, exactly as we did in the mSUGRA model, although because there are now three additional field directions $\text{Re}(\tilde{\nu}_{iL})$ that can potentially acquire VEVs the details are more complicated. The five minimization conditions for the potential fix these VEVs in terms of the potential parameters. Then, exactly as in the mSUGRA framework, one of the GUT scale parameters is fixed by the experimental value of M_Z, so that it would appear that the unified BRV model contains one more parameter than the mSUGRA framework. We should remember

---

[20] Renormalization group evolution does not generate trilinear superpotential couplings or soft parameters if these have been set to zero.

however, that the mass of the heaviest neutrino (the other neutrinos are massless at tree level) must be in accord with atmospheric neutrino data, given its successful interpretation in terms of neutrino oscillations.[21] This would then mean that the model does not contain any additional free parameters. It has been argued that if the neutrino mass is constrained to be $\lesssim 1$ eV, many R-parity-violating effects are also suppressed within this constrained framework.[22] Even so, the "TRV" couplings in the superpotential induced upon rotating away the bilinear terms may be as big as $\sim 10^{-4}$ (depending on other parameters) of the original (R-conserving) superpotential coupling, in which case neutralino LSPs would still decay inside the detector.

---

[21] The charged fermion masses are all fixed by the corresponding Yukawa couplings exactly as in mSUGRA, and so do not enter our parameter counting. The neutrino which acquires a Majorana mass via mixing with the Higgsino, however, has no Yukawa couplings, so that its mass serves as a constraint on the other parameters, which must be fine-tuned to m_ν in the sub-eV range.

[22] See J. Ferrandis, *Phys. Rev.* **D60**, 095012 (1999).

17
Epilogue

Almost no one doubts that the Standard Model is only an effective theory that has to be incorporated into a larger framework. What this framework will ultimately look like, we do not know. Empirical facts that we cannot account for in the Standard Model, such as neutrino masses, dark matter, and dark energy, provide some guidance. Aesthetic considerations such as the desire for unification of interactions and for an understanding of the patterns of matter fermion masses and mixing angles also guide our thinking. Although this seems rather removed from particle physics today, we also hope that one day we will have a framework that consistently incorporates gravity.

It was, however, efforts to resolve the fine-tuning problem of the Standard Model that led us to arrive at the exciting conclusion that there must be new physics at the TeV scale that can be probed at high energy colliders such as the LHC or a TeV electron–positron linear collider. Weak scale supersymmetry provides an attractive resolution of this problem, and continues to hold promise also for several other reasons, detailed at the end of Chapter 2. Indeed, many of these positive aspects of supersymmetric models have become evident only in the last 10–15 years – many years after the discovery of supersymmetry, and well after the effort to explore its phenomenological implications had begun in earnest. We believe that the motivations for seriously examining supersymmetry remain as strong as ever.

These promising features notwithstanding, SUSY is not a panacea. By itself, it has nothing to say about the choice of gauge group or particle multiplets, the replication of generations, or the patterns of matter fermion masses and mixing angles (though specific SUSY models that incorporate these patterns have been constructed). In fact, *generic* SUSY models lead to new problems *not present* in the Standard Model.

1. Why do baryon and lepton numbers appear to be conserved when we can write down renormalizable $SU(3)_C \times SU(2)_L \times U(1)_Y$ invariant interactions that violate their conservation?
2. What is the origin of SUSY breaking, and what makes the SUSY breaking scale required to avoid fine tuning so much smaller than the Planck or GUT scales?
3. Why is the supersymmetric parameter μ so much smaller than the Planck scale?
4. What makes the flavor-violating interactions of scalar quarks and leptons so small when we can write gauge invariant renormalizable flavor-violating couplings for these?
5. What makes the potentially large CP-violating effects in supersymmetry so small?

We stress that these are problems only of a generic SUSY theory, that can with suitable (but seemingly ad hoc) assumptions be overcome in specific models. Indeed, we have studied such models in the text. The point, however, is that while none of these were issues in the Standard Model, they appear to be so in the supersymmetric context. We speculate that once the mechanism of supersymmetry breaking is understood, the answers to these questions will appear evident; in the meantime, these should serve to guide our thinking about how supersymmetry is broken.

If nature turns out to be supersymmetric, it will change the physicist's view of the Universe. Indeed, the wide range of issues that might be addressed by the inclusion of supersymmetry in particle physics has led many physicists to expect that supersymmetry is realized in nature. While we know that supersymmetry – if it exists – must be broken, the scale of supersymmetry breaking is not known. However, if supersymmetry is the new physics that stabilizes the scalar electroweak symmetry breaking sector, supersymmetric matter will ultimately be revealed at or near the weak scale. With the LHC set to begin operation in 2007, and with the high energy physics community seriously considering the possibility of a TeV scale e^+e^- linear collider, this is an exciting prospect.

Only experiments can tell whether weak scale supersymmetry is realized in nature. The important thing is that the idea of weak scale supersymmetry can be directly tested in experiments at various collider and non-accelerator facilities. The fact that supersymmetric theories can sensibly be extrapolated to much higher energy scales suggests that if superpartners are discovered and their properties measured, we may be able to learn about physics at energy scales not directly accessible to experiment. Whether or not supersymmetric particles are discovered soon, it is clear that the exploration of the TeV scale will provide clues for unravelling the nature of electroweak symmetry breaking interactions. We must look to see what we find.

Appendix A
Sparticle production cross sections

In this appendix, we list all $2 \to 2$ sparticle production subprocess cross sections, first for hadron colliders, and then for e^+e^- colliders.

A.1 Sparticle production at hadron colliders
A.1.1 Chargino and neutralino production

The production subprocess cross section for $d\bar{u} \to \widetilde{W}_i \widetilde{Z}_j$ has already been presented in Chapter 12, so will not be repeated here.

For $d\bar{d} \to \widetilde{W}_1 \overline{\widetilde{W}}_1$, we find (here, $\widetilde{W}_1 \equiv \widetilde{W}_1^-$ and $\overline{\widetilde{W}}_1 \equiv \widetilde{W}_1^+$)

$$\frac{d\sigma}{dz}(d\bar{d} \to \widetilde{W}_1 \overline{\widetilde{W}}_1) = \frac{p_{\widetilde{W}_1}}{192\pi \hat{s}^{3/2}} \left[M_\gamma + M_Z + M_{\tilde{u}} + M_{\gamma Z} + M_{\gamma \tilde{u}} + M_{Z\tilde{u}} \right],$$

(A.1)

where

$$M_\gamma = \frac{16e^4 Q_d^2}{\hat{s}} \left[E^2(1+z^2) + m_{\widetilde{W}_1}^2(1-z^2) \right],$$

$$M_Z = 16e^4 \cot^2 \theta_W \hat{s} |D_Z(\hat{s})|^2 \left\{ (x_c^2 + y_c^2)(\alpha_d^2 + \beta_d^2)[E^2(1+z^2) + m_{\widetilde{W}_1}^2(1-z^2)] \right.$$
$$\left. - 2y_c^2(\alpha_d^2 + \beta_d^2)m_{\widetilde{W}_1}^2 - 8x_c y_c \alpha_d \beta_d Epz \right\},$$

$$M_{\tilde{u}} = \frac{e^4 \sin^4 \gamma_R \hat{s}}{\sin^4 \theta_W [E^2 + p^2 - 2Epz + m_{\tilde{u}_L}^2]^2} (E - pz)^2,$$

$$M_{\gamma Z} = 32e^4 Q_d \cot \theta_W (\hat{s} - M_Z^2) |D_Z(\hat{s})|^2$$
$$\times \left\{ \alpha_d x_c [E^2(1+z^2) + m_{\widetilde{W}_1}^2(1-z^2)] - 2\beta_d y_c Epz \right\},$$

$$M_{\gamma \tilde{u}} = \frac{4e^4 Q_d \sin^2 \gamma_R}{\sin^2 \theta_W} \frac{[(E - pz)^2 + m_{\widetilde{W}_1}^2]}{[E^2 + p^2 - 2Epz + m_{\tilde{u}_L}^2]}, \quad \text{and}$$

$$M_{Z\bar{u}} = \frac{4e^4 \cot\theta_W \sin^2\gamma_R}{\sin^2\theta_W}(\hat{s} - M_Z^2)(\alpha_d - \beta_d)\hat{s}|D_Z(\hat{s})|^2$$

$$\times \left\{ \frac{(x_c - y_c)[(E - pz)^2 + m_{\widetilde{W}_1}^2] + 2y_c m_{\widetilde{W}_1}^2}{E^2 + p^2 - 2Epz + m_{\tilde{u}_L}^2} \right\}.$$

The corresponding expression for $u\bar{u} \rightarrow \widetilde{W}_1\overline{\widetilde{W}}_1$ is obtained from the one above by making several changes. Of course, the contribution from $\tilde{u}$ exchange is replaced by that from $\tilde{d}$ exchange. To obtain the chargino production cross section from $u\bar{u}$ collisions, replace, (i) $\alpha_d \rightarrow \alpha_u$, $\beta_d \rightarrow \beta_u$, $Q_d \rightarrow Q_u$, $\tilde{u}_L \rightarrow \tilde{d}_L$, and $\gamma_R \rightarrow \gamma_L$ everywhere, (ii) $z \rightarrow -z$ in just $M_{\tilde{d}}$, $M_{\gamma\tilde{d}}$, and $M_{Z\tilde{d}}$, (iii) change the sign of $M_{\gamma\tilde{d}}$ and $M_{Z\tilde{d}}$, and finally, (iv) change $y_c \rightarrow -y_c$, in just $M_{Z\tilde{d}}$. The corresponding cross sections for $\widetilde{W}_2\overline{\widetilde{W}}_2$ production can be obtained from those for $\widetilde{W}_1\overline{\widetilde{W}}_1$ production by replacing $m_{\widetilde{W}_1} \rightarrow m_{\widetilde{W}_2}$, $\sin\gamma_R \rightarrow \cos\gamma_R$, $\sin\gamma_L \rightarrow \cos\gamma_L$, and $(x_c, y_c) \rightarrow (x_s, y_s)$.

The cross section for $d\bar{d} \rightarrow \widetilde{W}_1\overline{\widetilde{W}}_2$ is given by

$$\frac{d\hat{\sigma}}{dz}(d\bar{d} \rightarrow \widetilde{W}_1\overline{\widetilde{W}}_2) = \frac{e^4 p}{192\pi\hat{s}^{1/2}}(M_Z + M_{\tilde{u}} + M_{Z\tilde{u}}), \tag{A.2}$$

where

$$M_Z = 4(\cot\theta_W + \tan\theta_W)^2|D_Z(\hat{s})|^2 \times \left[(x^2 + y^2)(\alpha_d^2 + \beta_d^2)(E^2 + p^2z^2 - \Delta^2 - \xi m_{\widetilde{W}_1}m_{\widetilde{W}_2}) + 2x^2\xi(\alpha_d^2 + \beta_d^2)m_{\widetilde{W}_1}m_{\widetilde{W}_2} - 8xy\alpha_d\beta_d Epz\right],$$

$$M_{\tilde{u}} = \frac{\sin^2\gamma_R\cos^2\gamma_R}{\sin^4\theta_W} \frac{[(E - pz)^2 - \Delta^2]}{[2E(E - \Delta) - 2Epz + m_{\tilde{u}_L}^2 - m_{\widetilde{W}_1}^2]^2},$$

$$M_{Z\tilde{u}} = -\frac{2\theta_y(\cot\theta_W + \tan\theta_W)\sin\gamma_R\cos\gamma_R(\hat{s} - M_Z^2)(\alpha_d - \beta_d)|D_Z(\hat{s})|^2}{\sin^2\theta_W}$$

$$\times \frac{(x - y)[(E - pz)^2 - \Delta^2 - \xi m_{\widetilde{W}_1}m_{\widetilde{W}_2}] + 2x\xi m_{\widetilde{W}_1}m_{\widetilde{W}_2}}{[2E(E - \Delta) - 2Epz + m_{\tilde{u}_L}^2 - m_{\widetilde{W}_1}^2]},$$

where $\Delta = \frac{m_{\widetilde{W}_2}^2 - m_{\widetilde{W}_1}^2}{4E}$ and $\xi = (-1)^{\theta_{\widetilde{W}_1} + \theta_{\widetilde{W}_2} + 1}$. The corresponding expression for $u\bar{u} \rightarrow \widetilde{W}_1\overline{\widetilde{W}}_2$ is obtained from the one above by replacing, (i) $\alpha_d \rightarrow \alpha_u$, $\beta_d \rightarrow \beta_u$, $\tilde{u}_L \rightarrow \tilde{d}_L$, and $\gamma_R \rightarrow \gamma_L$, everywhere, (ii) $z \rightarrow -z$ in $M_{\tilde{d}}$ and $M_{Z\tilde{d}}$, and finally, (iii) just in $M_{Z\tilde{d}}$, $\theta_y \rightarrow \theta_x$, $\xi \rightarrow -\xi$, $(x - y) \rightarrow (x + y)$ (in the first term) and $x \rightarrow y$ (in the second term containing $m_{\widetilde{W}_1}m_{\widetilde{W}_2}$).

The cross section for neutralino pair production is given by,

$$\frac{d\sigma}{dz}(q\bar{q} \rightarrow \widetilde{Z}_i\widetilde{Z}_j) = \frac{p}{48\pi\hat{s}^{3/2}}\left\{|A_{\widetilde{Z}_i}^q|^2|A_{\widetilde{Z}_j}^q|^2 G_t(m_{\widetilde{Z}_i}, m_{\widetilde{Z}_j}, m_{\tilde{q}_L}, z)\right.$$

$$+|B^q_{\widetilde{Z}_i}|^2|B^q_{\widetilde{Z}_j}|^2 G_t(m_{\widetilde{Z}_i}, m_{\widetilde{Z}_j}, m_{\tilde{q}_R}, z) + 4e^2|W_{ij}|^2(\alpha_q^2 + \beta_q^2)|D_Z(\hat{s})|^2$$

$$\times \left[\hat{s}^2 - (m_{\widetilde{Z}_i}^2 - m_{\widetilde{Z}_j}^2)^2 + 4(-1)^{\theta_i+\theta_j+1}\hat{s}m_{\widetilde{Z}_i}m_{\widetilde{Z}_j} + 4\hat{s}p^2z^2 \right]$$

$$-\frac{1}{2}e(\alpha_q - \beta_q)(\hat{s} - M_Z^2)|D_Z(\hat{s})|^2 \left[\mathrm{Re}(W_{ij}A^{q*}_{\widetilde{Z}_i}A^q_{\widetilde{Z}_j})G_{st}(m_{\widetilde{Z}_i}, m_{\widetilde{Z}_j}, m_{\tilde{q}_L}, z) \right.$$

$$\left. + (-1)^{\theta_i+\theta_j}\mathrm{Re}(W_{ij}A^q_{\widetilde{Z}_i}A^{q*}_{\widetilde{Z}_j})G_{st}(m_{\widetilde{Z}_i}, m_{\widetilde{Z}_j}, m_{\tilde{q}_L}, -z) \right] - \frac{1}{2}e(-1)^{\theta_i+\theta_j+1}$$

$$\times (\alpha_q + \beta_q)(\hat{s} - M_Z^2)|D_Z(\hat{s})|^2 \left[\mathrm{Re}(W_{ij}B^{q*}_{\widetilde{Z}_i}B^q_{\widetilde{Z}_j})G_{st}(m_{\widetilde{Z}_i}, m_{\widetilde{Z}_j}, m_{\tilde{q}_R}, z) \right.$$

$$\left. \left. + (-1)^{\theta_i+\theta_j}\mathrm{Re}(W_{ij}B^q_{\widetilde{Z}_i}B^{q*}_{\widetilde{Z}_j})G_{st}(m_{\widetilde{Z}_i}, m_{\widetilde{Z}_j}, m_{\tilde{q}_R}, -z) \right] \right\}, \tag{A.3}$$

where

$$p = \frac{\lambda^{1/2}(\hat{s}, m_{\widetilde{Z}_i}^2, m_{\widetilde{Z}_j}^2)}{2\sqrt{\hat{s}}}, \tag{A.4a}$$

$$G_t(m_{\widetilde{Z}_i}, m_{\widetilde{Z}_j}, m_{\tilde{q}}, z)$$

$$= \frac{1}{16}\left\{ \left[\frac{\hat{s}^2 - (m_{\widetilde{Z}_i}^2 - m_{\widetilde{Z}_j}^2)^2 - 4p\hat{s}^{3/2}z + 4p^2\hat{s}z^2}{[\frac{1}{2}(\hat{s} - m_{\widetilde{Z}_i}^2 - m_{\widetilde{Z}_j}^2) - \sqrt{\hat{s}}pz + m_{\tilde{q}}^2]^2} \right. \right.$$

$$\left. + \frac{\hat{s}^2 - (m_{\widetilde{Z}_i}^2 - m_{\widetilde{Z}_j}^2)^2 + 4p\hat{s}^{3/2}z + 4p^2\hat{s}z^2}{[\frac{1}{2}(\hat{s} - m_{\widetilde{Z}_i}^2 - m_{\widetilde{Z}_j}^2) + \sqrt{\hat{s}}pz + m_{\tilde{q}}^2]^2} \right]$$

$$\left. - \frac{8(-1)^{\theta_i+\theta_j}m_{\widetilde{Z}_i}m_{\widetilde{Z}_j}\hat{s}}{[\frac{1}{2}(\hat{s} - m_{\widetilde{Z}_i}^2 - m_{\widetilde{Z}_j}^2) + \sqrt{\hat{s}}pz + m_{\tilde{q}}^2][\frac{1}{2}(\hat{s} - m_{\widetilde{Z}_i}^2 - m_{\widetilde{Z}_j}^2) - \sqrt{\hat{s}}pz + m_{\tilde{q}}^2]} \right\},$$

$$\tag{A.4b}$$

and

$$G_{st}(m_{\widetilde{Z}_i}, m_{\widetilde{Z}_j}, m_{\tilde{q}}, z)$$

$$= \frac{\hat{s}^2 - (m_{\widetilde{Z}_i}^2 - m_{\widetilde{Z}_j}^2)^2 - 4\hat{s}^{3/2}pz + 4\hat{s}p^2z^2 + 4(-1)^{\theta_i+\theta_j+1}\hat{s}m_{\widetilde{Z}_i}m_{\widetilde{Z}_j}}{\frac{1}{2}(\hat{s} - m_{\widetilde{Z}_i}^2 - m_{\widetilde{Z}_j}^2) - \sqrt{\hat{s}}pz + m_{\tilde{q}}^2}.$$

$$\tag{A.4c}$$

A.1.2 Gluino and squark production

Next, we turn to production of strongly interacting sparticles. Gluino pair production takes place via either gluon–gluon annihilation, or via quark–antiquark annihilation. The subprocess cross sections are usually presented as differential distributions in the Mandelstam variable $\hat{t}$:

$$\frac{d\sigma}{d\hat{t}}(gg \rightarrow \tilde{g}\tilde{g})$$

$$= \frac{9\pi\alpha_s^2}{4\hat{s}^2}\left\{ \frac{2(m_{\tilde{g}}^2 - \hat{t})(m_{\tilde{g}}^2 - \hat{u})}{\hat{s}^2} + \frac{(m_{\tilde{g}}^2 - \hat{t})(m_{\tilde{g}}^2 - \hat{u}) - 2m_{\tilde{g}}^2(m_{\tilde{g}}^2 + \hat{t})}{(m_{\tilde{g}}^2 - \hat{t})^2} \right.$$

$$+ \frac{(m_{\tilde{g}}^2 - \hat{t})(m_{\tilde{g}}^2 - \hat{u}) - 2m_{\tilde{g}}^2(m_{\tilde{g}}^2 + \hat{u})}{(m_{\tilde{g}}^2 - \hat{u})^2} + \frac{m_{\tilde{g}}^2(\hat{s} - 4m_{\tilde{g}}^2)}{(m_{\tilde{g}}^2 - \hat{t})(m_{\tilde{g}}^2 - \hat{u})}$$

$$\left. - \frac{(m_{\tilde{g}}^2 - \hat{t})(m_{\tilde{g}}^2 - \hat{u}) + m_{\tilde{g}}^2(\hat{u} - \hat{t})}{\hat{s}(m_{\tilde{g}}^2 - \hat{t})} - \frac{(m_{\tilde{g}}^2 - \hat{t})(m_{\tilde{g}}^2 - \hat{u}) + m_{\tilde{g}}^2(\hat{t} - \hat{u})}{\hat{s}(m_{\tilde{g}}^2 - \hat{u})} \right\},$$

$$(A.5a)$$

and

$$\frac{d\sigma}{d\hat{t}}(q\bar{q} \rightarrow \tilde{g}\tilde{g})$$

$$= \frac{8\pi\alpha_s^2}{9\hat{s}^2}\left\{ \frac{4}{3}\left(\frac{m_{\tilde{g}}^2 - \hat{t}}{m_{\tilde{q}}^2 - \hat{t}} \right)^2 + \frac{4}{3}\left(\frac{m_{\tilde{g}}^2 - \hat{u}}{m_{\tilde{q}}^2 - \hat{u}} \right)^2 \right.$$

$$+ \frac{3}{\hat{s}^2}\left[(m_{\tilde{g}}^2 - \hat{t})^2 + (m_{\tilde{g}}^2 - \hat{u})^2 + 2m_{\tilde{g}}^2\hat{s} \right] - 3\frac{[(m_{\tilde{g}}^2 - \hat{t})^2 + m_{\tilde{g}}^2\hat{s}]}{\hat{s}(m_{\tilde{q}}^2 - \hat{t})}$$

$$\left. - 3\frac{[(m_{\tilde{g}}^2 - \hat{u})^2 + m_{\tilde{g}}^2\hat{s}]}{\hat{s}(m_{\tilde{q}}^2 - \hat{u})} + \frac{1}{3}\frac{m_{\tilde{g}}^2\hat{s}}{(m_{\tilde{q}}^2 - \hat{t})(m_{\tilde{q}}^2 - \hat{u})} \right\}.$$

$$(A.5b)$$

Gluinos can also be produced in association with squarks. The subprocess cross section is independent of whether the squark is the right- or the left-type:

$$\frac{d\sigma}{d\hat{t}}(gq \rightarrow \tilde{g}\tilde{q}) = \frac{\pi\alpha_s^2}{24\hat{s}^2}\frac{\left[\frac{16}{3}(\hat{s}^2 + (m_{\tilde{q}}^2 - \hat{u})^2) + \frac{4}{3}\hat{s}(m_{\tilde{q}}^2 - \hat{u}) \right]}{\hat{s}(m_{\tilde{g}}^2 - \hat{t})(m_{\tilde{q}}^2 - \hat{u})^2}$$

$$\times \left((m_{\tilde{g}}^2 - \hat{u})^2 + (m_{\tilde{q}}^2 - m_{\tilde{g}}^2)^2 + \frac{2\hat{s}m_{\tilde{g}}^2(m_{\tilde{q}}^2 - m_{\tilde{g}}^2)}{(m_{\tilde{g}}^2 - \hat{t})} \right). \quad (A.6)$$

There are many different subprocesses for production of squark pairs. Since left- and right-squarks generally have different masses and different decay patterns, we present the differential cross section for each subprocess separately. (In early literature, these subprocesses were usually summed over squark types.) The results are:

$$
\frac{d\sigma}{d\hat{t}}(gg \to \tilde{q}_i \bar{\tilde{q}}_i) = \frac{\pi \alpha_s^2}{4\hat{s}^2} \left\{ \frac{1}{3} \left(\frac{m_{\tilde{q}}^2 + \hat{t}}{m_{\tilde{q}}^2 - \hat{t}} \right)^2 + \frac{1}{3} \left(\frac{m_{\tilde{q}}^2 + \hat{u}}{m_{\tilde{q}}^2 - \hat{u}} \right)^2 \right.
$$

$$
+ \frac{3}{32\hat{s}^2} \left(8\hat{s}(4m_{\tilde{q}}^2 - \hat{s}) + 4(\hat{u} - \hat{t})^2 \right) + \frac{7}{12}
$$

$$
- \frac{1}{48} \frac{(4m_{\tilde{q}}^2 - \hat{s})^2}{(m_{\tilde{q}}^2 - \hat{t})(m_{\tilde{q}}^2 - \hat{u})}
$$

$$
+ \frac{3}{32} \frac{\left[(\hat{t} - \hat{u})(4m_{\tilde{q}}^2 + 4\hat{t} - \hat{s}) - 2(m_{\tilde{q}}^2 - \hat{u})(6m_{\tilde{q}}^2 + 2\hat{t} - \hat{s}) \right]}{\hat{s}(m_{\tilde{q}}^2 - \hat{t})}
$$

$$
+ \frac{3}{32} \frac{\left[(\hat{u} - \hat{t})(4m_{\tilde{q}}^2 + 4\hat{u} - \hat{s}) - 2(m_{\tilde{q}}^2 - \hat{t})(6m_{\tilde{q}}^2 + 2\hat{u} - \hat{s}) \right]}{\hat{s}(m_{\tilde{q}}^2 - \hat{u})}
$$

$$
\left. + \frac{7}{96} \frac{\left[4m_{\tilde{q}}^2 + 4\hat{t} - \hat{s} \right]}{m_{\tilde{q}}^2 - \hat{t}} + \frac{7}{96} \frac{\left[4m_{\tilde{q}}^2 + 4\hat{u} - \hat{s} \right]}{m_{\tilde{q}}^2 - \hat{u}} \right\}, \qquad \text{(A.7a)}
$$

$$
\frac{d\sigma}{d\hat{t}}(q_1 \bar{q}_2 \to \tilde{q}_{1L} \bar{\tilde{q}}_{2R}) = \frac{d\sigma}{dt}(q_1 \bar{q}_2 \to \tilde{q}_{1R} \bar{\tilde{q}}_{2L})
$$

$$
= \frac{2\pi \alpha_s^2}{9\hat{s}^2} \frac{m_{\tilde{g}}^2 \hat{s}}{(\hat{t} - m_{\tilde{g}}^2)^2}, \qquad \text{(A.7b)}
$$

$$
\frac{d\sigma}{d\hat{t}}(q_1 \bar{q}_2 \to \tilde{q}_{1L(R)} \bar{\tilde{q}}_{2L(R)}) = \frac{2\pi \alpha_s^2}{9\hat{s}^2} \frac{-\hat{s}\hat{t} - (\hat{t} - m_{\tilde{q}_{L(R)}}^2)^2}{(\hat{t} - m_{\tilde{g}}^2)^2}, \qquad \text{(A.7c)}
$$

$$
\frac{d\sigma}{d\hat{t}}(q_1 q_2 \to \tilde{q}_{1L} \tilde{q}_{2R}) = \frac{d\sigma}{d\hat{t}}(q_1 q_2 \to \tilde{q}_{1R} \tilde{q}_{2L})
$$

$$
= \frac{2\pi \alpha_s^2}{9\hat{s}^2} \frac{-\hat{s}\hat{t} - (\hat{t} - m_{\tilde{q}_L}^2)(\hat{t} - m_{\tilde{q}_R}^2)}{(\hat{t} - m_{\tilde{g}}^2)^2}, \qquad \text{(A.7d)}
$$

$$
\frac{d\sigma}{d\hat{t}}(q_1 q_2 \to \tilde{q}_{1L(R)} \tilde{q}_{2L(R)}) = \frac{2\pi \alpha_s^2}{9\hat{s}^2} \frac{m_{\tilde{g}}^2 \hat{s}}{(\hat{t} - m_{\tilde{g}}^2)^2}, \qquad \text{(A.7e)}
$$

$$
\frac{d\sigma}{d\hat{t}}(q\bar{q} \to \tilde{q}_L \bar{\tilde{q}}_R) = \frac{d\sigma}{d\hat{t}}(q\bar{q} \to \tilde{q}_R \bar{\tilde{q}}_L) = \frac{2\pi \alpha_s^2}{9\hat{s}^2} \frac{m_{\tilde{g}}^2 \hat{s}}{(\hat{t} - m_{\tilde{g}}^2)^2}, \qquad \text{(A.7f)}
$$

$$
\frac{d\sigma}{d\hat{t}}(q\bar{q} \to \tilde{q}_{L(R)} \bar{\tilde{q}}_{L(R)}) = \frac{2\pi \alpha_s^2}{9\hat{s}^2} \left\{ \frac{1}{(\hat{t} - m_{\tilde{g}}^2)^2} + \frac{2}{\hat{s}^2} - \frac{2/3}{\hat{s}(\hat{t} - m_{\tilde{g}}^2)} \right\}
$$

$$
\times \left[-\hat{s}\hat{t} - (\hat{t} - m_{\tilde{q}_{L(R)}}^2)^2 \right], \qquad \text{(A.7g)}
$$

$$\frac{d\sigma}{d\hat{t}}(q\bar{q} \to \tilde{q}'_{\text{L(R)}}\bar{\tilde{q}}'_{\text{L(R)}}) = \frac{4\pi\alpha_s^2}{9\hat{s}^4}\left[-\hat{s}\hat{t} - (\hat{t} - m_{\tilde{q}'_{\text{L(R)}}}^2)^2\right], \tag{A.7h}$$

$$\frac{d\sigma}{d\hat{t}}(qq \to \tilde{q}_{\text{L}}\tilde{q}_{\text{L}}) = \frac{d\sigma}{d\hat{t}}(qq \to \tilde{q}_{\text{R}}\tilde{q}_{\text{R}})$$

$$= \frac{\pi\alpha_s^2}{9\hat{s}^2}m_{\tilde{g}}^2\hat{s}\left\{\frac{1}{(\hat{t} - m_{\tilde{g}}^2)^2} + \frac{1}{(\hat{u} - m_{\tilde{g}}^2)^2} - \frac{2/3}{(\hat{t} - m_{\tilde{g}}^2)(\hat{u} - m_{\tilde{g}}^2)}\right\}, \tag{A.7i}$$

and

$$\frac{d\sigma}{d\hat{t}}(qq \to \tilde{q}_{\text{L}}\tilde{q}_{\text{R}})$$

$$= \frac{2\pi\alpha_s^2}{9\hat{s}^2}\left\{\frac{[-\hat{s}\hat{t} - (\hat{t} - m_{\tilde{q}_{\text{L}}}^2)(\hat{t} - m_{\tilde{q}_{\text{R}}}^2)]}{(\hat{t} - m_{\tilde{g}}^2)^2} + \frac{[-\hat{s}\hat{u} - (\hat{u} - m_{\tilde{q}_{\text{L}}}^2)(\hat{u} - m_{\tilde{q}_{\text{R}}}^2)]}{(\hat{u} - m_{\tilde{g}}^2)^2}\right\}. \tag{A.7j}$$

Because there are essentially no third generation quarks in the proton, these cross sections for gluino and squark production are fixed by SUSY QCD, and depend only on the masses of the squarks and gluinos: in particular, cross sections for producing third generation squarks do not depend on the intra-generational mixing. In this case, the squark type should be read as 1 or 2 instead of L or R. Refer back to the exercise in Section 8.4.1 in this connection.

A.1.3 Gluino and squark associated production

Gluinos and squarks may also be produced in association with charginos and neutralinos. Here, the subprocess cross sections are given by:

$$\frac{d\sigma}{d\hat{t}}(\bar{u}g \to \widetilde{W}_i\bar{\tilde{d}}_{\text{L}}) = \frac{\alpha_s}{24\hat{s}^2}|A_{\widetilde{W}_i}^u|^2\psi(m_{\tilde{d}_{\text{L}}}, m_{\widetilde{W}_i}, \hat{t}), \tag{A.8}$$

$$\frac{d\sigma}{d\hat{t}}(dg \to \widetilde{W}_i\tilde{u}_{\text{L}}) = \frac{\alpha_s}{24\hat{s}^2}|A_{\widetilde{W}_i}^d|^2\psi(m_{\tilde{u}_{\text{L}}}, m_{\widetilde{W}_i}, \hat{t}), \tag{A.9}$$

$$\frac{d\sigma}{d\hat{t}}(qg \to \widetilde{Z}_i\tilde{q}) = \frac{\alpha_s}{24\hat{s}^2}\left(|A_{\widetilde{Z}_i}^q|^2 + |B_{\widetilde{Z}_i}^q|^2\right)\psi(m_{\tilde{q}}, m_{\widetilde{Z}_i}, \hat{t}), \tag{A.10}$$

$$\frac{d\sigma}{d\hat{t}}(q\bar{q} \to \widetilde{Z}_i\tilde{g}) = \frac{\alpha_s}{18\hat{s}^2}\left(|A_{\widetilde{Z}_i}^q|^2 + |B_{\widetilde{Z}_i}^q|^2\right)\left[\frac{(m_{\widetilde{Z}_i}^2 - \hat{t})(m_{\tilde{g}}^2 - \hat{t})}{(m_{\tilde{q}}^2 - \hat{t})^2}\right.$$

$$\left. + \frac{(m_{\widetilde{Z}_i}^2 - \hat{u})(m_{\tilde{g}}^2 - \hat{u})}{(m_{\tilde{q}}^2 - \hat{u})^2} - \frac{2(-1)^{\theta_i + \theta_{\tilde{g}}}m_{\tilde{g}}m_{\widetilde{Z}_i}\hat{s}}{(m_{\tilde{q}}^2 - \hat{t})(m_{\tilde{q}}^2 - \hat{u})}\right], \tag{A.11}$$

and

$$\frac{d\sigma}{d\hat{t}}(\bar{u}d \to \widetilde{W}_i \tilde{g}) = \frac{\alpha_s}{18\hat{s}^2} \left[|A^u_{\widetilde{W}_i}|^2 \frac{(m^2_{\widetilde{W}_i} - \hat{t})(m^2_{\tilde{g}} - \hat{t})}{(m^2_{\tilde{d}_L} - \hat{t})^2} \right.$$

$$+ |A^d_{\widetilde{W}_i}|^2 \frac{(m^2_{\widetilde{W}_i} - \hat{u})(m^2_{\tilde{g}} - \hat{u})}{(m^2_{\tilde{u}_L} - \hat{u})^2}$$

$$\left. + \frac{2(-1)^{\theta_{\tilde{g}}}\text{Re}(A^u_{\widetilde{W}_i} A^d_{\widetilde{W}_i}) m_{\tilde{g}} m_{\widetilde{W}_i} \hat{s}}{(m^2_{\tilde{d}_L} - \hat{t})(m^2_{\tilde{u}_L} - \hat{u})} \right], \tag{A.12}$$

where

$$\psi(m_1, m_2, t) = \frac{s + t - m^2_1}{2s} - \frac{m^2_1(m^2_2 - t)}{(m^2_1 - t)^2} + \frac{t(m^2_2 - m^2_1) + m^2_2(s - m^2_2 + m^2_1)}{s(m^2_1 - t)}. \tag{A.13}$$

A.1.4 Slepton and sneutrino production

The subprocess cross section for $\tilde{\ell}_L \bar{\tilde{\nu}}_L$ production is given by

$$\frac{d\sigma}{d\hat{t}}(d\bar{u} \to \tilde{\ell}_L \bar{\tilde{\nu}}_L) = \frac{g^4 |D_W(\hat{s})|^2}{192\pi \hat{s}^2} \left(\hat{t}\hat{u} - m^2_{\tilde{\ell}_L} m^2_{\tilde{\nu}_L} \right). \tag{A.14}$$

The production of $\tilde{\tau}_1 \bar{\tilde{\nu}}_\tau$ is given as above, replacing $m_{\tilde{\ell}_L} \to m_{\tilde{\tau}_1}$, $m_{\tilde{\nu}_L} \to m_{\tilde{\nu}_\tau}$, and multiplying by an overall factor of $\cos^2 \theta_\tau$. Similar substitutions hold for $\tilde{\tau}_2 \bar{\tilde{\nu}}_\tau$ production, except the overall factor is $\sin^2 \theta_\tau$.

The subprocess cross section for $\tilde{\ell}_L \bar{\tilde{\ell}}_L$ production is given by

$$\frac{d\sigma}{d\hat{t}}(q\bar{q} \to \tilde{\ell}_L \bar{\tilde{\ell}}_L)$$

$$= \frac{e^4}{24\pi \hat{s}^2} \left(\hat{t}\hat{u} - m^4_{\tilde{\ell}_L} \right) \left\{ \frac{q^2_\ell q^2_q}{\hat{s}^2} + (\alpha_\ell - \beta_\ell)^2 (\alpha^2_q + \beta^2_q) |D_Z(\hat{s})|^2 \right.$$

$$\left. + \frac{2q_\ell q_q \alpha_q (\alpha_\ell - \beta_\ell)(\hat{s} - M^2_Z)}{\hat{s}} |D_Z(\hat{s})|^2 \right\}. \tag{A.15a}$$

The cross section for sneutrino production is given by the same formula, but with α_ℓ, β_ℓ, q_ℓ, and $m_{\tilde{\ell}_L}$ replaced by α_ν, β_ν, 0, and $m_{\tilde{\nu}_L}$, respectively. The cross section for $\tilde{\tau}_1 \bar{\tilde{\tau}}_1$ production is obtained by replacing $m_{\tilde{\ell}_L} \to m_{\tilde{\tau}_1}$ and $\beta_\ell \to \beta_\ell \cos 2\theta_\tau$. The cross section for $\tilde{\ell}_R \bar{\tilde{\ell}}_R$ production is given by substituting $\alpha_\ell - \beta_\ell \to \alpha_\ell + \beta_\ell$ and $m_{\tilde{\ell}_L} \to m_{\tilde{\ell}_R}$ in (A.15a). The cross section for $\tilde{\tau}_2 \bar{\tilde{\tau}}_2$ production is obtained from the formula for $\tilde{\ell}_R \bar{\tilde{\ell}}_R$ production by replacing $m_{\tilde{\ell}_R} \to m_{\tilde{\tau}_2}$ and $\beta_\ell \to \beta_\ell \cos 2\theta_\tau$. Finally, the cross section for $\tilde{\tau}_1\bar{\tilde{\tau}}_2$ production is given by

$$\frac{d\sigma}{d\hat{t}}(q\bar{q} \to \tilde{\tau}_1\bar{\tilde{\tau}}_2) = \frac{d\sigma}{d\hat{t}}(q\bar{q} \to \bar{\tilde{\tau}}_1\tilde{\tau}_2)$$

$$= \frac{e^4}{24\pi\hat{s}^2}(\alpha_q^2 + \beta_q^2)\beta_\ell^2 \sin^2 2\theta_\tau |D_Z(\hat{s})|^2(\hat{u}\hat{t} - m_{\tilde{\tau}_1}^2 m_{\tilde{\tau}_2}^2).$$

$$(A.15b)$$

A.2 Sparticle production at e^+e^- colliders

The degree of longitudinal beam polarization has been parametrized as

$$P_L(e^-) = f_L - f_R, \quad \text{where} \qquad (A.16a)$$

$$f_L = \frac{n_L}{n_L + n_R} = \frac{1 + P_L}{2} \quad \text{and} \qquad (A.16b)$$

$$f_R = \frac{n_R}{n_L + n_R} = \frac{1 - P_L}{2}. \qquad (A.16c)$$

Here, $n_{L,R}$ are the number of left-(right-)polarized electrons in the beam, and $f_{L,R}$ is the corresponding fraction. Thus, a 90% right-polarized beam would correspond to $P_L(e^-) = -0.8$, and a completely unpolarized beam corresponds to $P_L(e^-) = 0$.

We present in this section relevant $2 \to 2$ SM and SUSY cross sections, retaining information on the polarization of the incoming beams. The calculations can be performed using usual techniques, but in addition inserting projection operators $P_L = \frac{1-\gamma_5}{2}$ and $P_R = \frac{1+\gamma_5}{2}$ to select the desired left- or right-polarized initial state particles.[1]

We begin by listing lowest order Standard Model cross sections for right- or left-polarized incoming electrons and positrons. For SM fermion pair production, we have:

$$\frac{d\sigma}{dz}(e_{R/L}\bar{e}_{L/R} \to f\bar{f}) = \frac{N_f}{4\pi}\frac{p}{E}\Phi_{fR/L}(z) \qquad (A.17a)$$

where $z = \cos\theta$ (θ is the angle between the final state particle and the electron), E is the beam energy, and p is the magnitude of the momentum of the final state SM fermions, $f = \mu, \tau, \nu_\mu, \nu_\tau$, and q, and:

$$\Phi_{fR/L}(z) = e^4 \left[\frac{q_f^2}{s^2}\left(E^2(1+z^2) + m_f^2(1-z^2)\right) + \frac{(\alpha_e \pm \beta_e)^2}{(s-M_Z^2)^2 + M_Z^2\Gamma_Z^2} \right.$$
$$\left. \times \left([(\alpha_f^2 + \beta_f^2)(E^2 + p^2z^2) \pm 4\alpha_f\beta_f Epz + (\alpha_f^2 - \beta_f^2)m_f^2]\right) \right.$$

[1] We assume that we are at high energy $E \gg m_e$ so that the difference between chirality and helicity can be safely ignored.

$$- \frac{2(\alpha_e \pm \beta_e)(s - M_Z^2)q_f}{s[(s - M_Z^2)^2 + M_Z^2\Gamma_Z^2]} \left(\alpha_f \left[E^2(1 + z^2) + m_f^2(1 - z^2)\right]\right.$$

$$\left.\pm 2\beta_f Epz)\right]. \tag{A.17b}$$

The upper (lower) signs are for Φ_{fR} (Φ_{fL}). For electron pair production t-channel photon exchange contributions must be included. For Z pair production, we have

$$\frac{d\sigma}{dz}(e_{R/L}\bar{e}_{L/R} \to ZZ)$$

$$= \frac{e^4(\alpha_e \pm \beta_e)^4 p}{4\pi s\sqrt{s}} \left[\frac{u(z)}{t(z)} + \frac{t(z)}{u(z)} + \frac{4M_Z^2 s}{u(z)t(z)} - M_Z^4\left(\frac{1}{t^2(z)} + \frac{1}{u^2(z)}\right)\right], \tag{A.18}$$

where s, $t(z)$, and $u(z)$ are the Mandelstam variables. For W^+W^- production we have,

$$\frac{d\sigma}{dz}(e_{R/L}\bar{e}_{L/R} \to W^+W^-) = \frac{e^4 p}{16\pi s\sqrt{s}\sin^4\theta_W}\Phi_{WWR/L}(z), \tag{A.19a}$$

where

$$\Phi_{WWR}(z) = \frac{4(\alpha_e + \beta_e)^2\tan^2\theta_W|D_Z|^2}{s^2}\left[U_T(z)(p^2 s + 3M_W^4) + 4M_W^2 p^2 s^2\right], \tag{A.19b}$$

and

$$\Phi_{WWL}(z) = \frac{U_T(z)}{s^2}\left[3 + 2(\alpha_e - \beta_e)\tan\theta_W(s - 6M_W^2)\text{Re}(D_Z)\right.$$

$$+ 4(\alpha_e - \beta_e)^2\tan^2\theta_W(p^2 s + 3M_W^4)|D_Z|^2\right] + \frac{U_T(z)}{t^2(z)}$$

$$+ 8(\alpha_e - \beta_e)\tan\theta_W M_W^2\text{Re}(D_Z) + 16(\alpha_e - \beta_e)^2\tan^2\theta_W M_W^2 p^2|D_Z|^2$$

$$+ 2\left[1 - 2(\alpha_e - \beta_e)\tan\theta_W M_W^2\text{Re}(D_Z)\right]\left[\frac{U_T(z)}{st(z)} - \frac{2M_W^2}{t(z)}\right], \tag{A.19c}$$

with $U_T(z) = u(z)t(z) - M_W^4$, and $D_Z = (s - M_Z^2 + iM_Z\Gamma_Z)^{-1}$.

The reader will have noticed that we have not listed cross sections from the $e_{L/R}\bar{e}_{L/R}$ initial states. This is because these cross sections vanish in the chiral limit, i.e. when the electron mass is neglected. The reason is that gauge interactions couple members of the electroweak doublets (singlets) to one another, but never a doublet to a singlet. The same reasoning, therefore, applies to SUSY processes that involve only s-channel photon and Z exchange. The reader should understand that these unlisted cross sections also vanish.

For lowest order MSSM Higgs boson production, we have

$$\frac{d\sigma}{dz}(e_{R/L}\bar{e}_{L/R} \to Zh) = \frac{p}{16\pi\sqrt{s}} \frac{e^4 \sin^2(\alpha + \beta)}{\sin^2\theta_W \cos^2\theta_W}$$
$$\times \frac{(\alpha_e \pm \beta_e)^2}{(s - M_Z^2)^2 + M_Z^2\Gamma_Z^2}(M_Z^2 + E_Z^2 - p^2z^2). \quad (A.20a)$$

To obtain the corresponding cross section for ZH production, replace $\sin^2(\alpha + \beta)$ with $\cos^2(\alpha + \beta)$. The angular distribution for the production of h (or H) bosons in association with A is given by,

$$\frac{d\sigma}{dz}(e_{R/L}\bar{e}_{L/R} \to hA)$$
$$= \frac{p^3}{16\pi\sqrt{s}} \frac{e^4 \cos^2(\alpha + \beta)}{\sin^2\theta_W \cos^2\theta_W} \frac{(\alpha_e \pm \beta_e)^2}{(s - M_Z^2)^2 + M_Z^2\Gamma_Z^2}(1 - z^2); \quad (A.20b)$$

for HA production, replace $\cos^2(\alpha + \beta)$ with $\sin^2(\alpha + \beta)$. Finally, the differential cross section for H^+H^- production is given by,

$$\frac{d\sigma}{dz}(e_{R/L}\bar{e}_{L/R} \to H^+H^-) = \frac{e^4}{4\pi} \frac{p^3}{\sqrt{s}}(1 - z^2)$$
$$\times \left[\frac{1}{s^2} + \left(\frac{2\sin^2\theta_W - 1}{2\cos\theta_W \sin\theta_W}\right)^2 \frac{(\alpha_e \pm \beta_e)^2}{(s - M_Z^2)^2 + M_Z^2\Gamma_Z^2}\right.$$
$$\left. + \frac{1}{s}\left(\frac{2\sin^2\theta_W - 1}{\cos\theta_W \sin\theta_W}\right) \frac{(\alpha_e \pm \beta_e)(s - M_Z^2)}{(s - M_Z^2)^2 + M_Z^2\Gamma_Z^2}\right].$$
$$(A.20c)$$

For sfermion pair production ($\tilde{f}_i\bar{\tilde{f}}_i$, with $f = \mu$, ν_μ, ν_τ, u, d, c, s, and $i = L$ or R), we find:

$$\frac{d\sigma}{dz}(e_{R/L}\bar{e}_{L/R} \to \tilde{f}_i\bar{\tilde{f}}_i) = \frac{N_f}{256\pi} \frac{p^3}{E^3} \Phi_{\tilde{f}_iR/L}(z), \quad (A.21a)$$

where

$$\Phi_{\tilde{f}_iR/L}(z) = e^4(1 - z^2)\times$$
$$\left[\frac{8q_f^2}{s} + \frac{2A_{\tilde{f}_i}^2(\alpha_e \pm \beta_e)^2 s - 8(\alpha_e \pm \beta_e)q_f A_{\tilde{f}_i}(s - M_Z^2)}{(s - M_Z^2)^2 + M_Z^2\Gamma_Z^2}\right], \quad (A.21b)$$

and $A_{f_{L,R}} = 2(\alpha_f \mp \beta_f)$. The cross sections for producing $\tilde{f}_L\tilde{f}_R$ pairs are zero since both the photon and the Z boson only couple to pairs of like type (LL or RR) sfermions. In the corresponding expressions for third generation sfermions, we need to include intragenerational mixing. For the case of $\tilde{t}_1\bar{\tilde{t}}_1$ production, we have $A_{t_1} = 2(\alpha_t - \beta_t)\cos^2\theta_t + 2(\alpha_t + \beta_t)\sin^2\theta_t$; for $\tilde{t}_2\bar{\tilde{t}}_2$ production, simply switch $\cos^2\theta_t$ with $\sin^2\theta_t$. Since Z couples to $\tilde{t}_1\tilde{t}_2$ pairs we also have,

$$\frac{d\sigma}{dz}(e_{R/L}\bar{e}_{L/R} \to \tilde{t}_1\bar{\tilde{t}}_2) = \frac{48\pi\alpha^2}{\sqrt{s}}\frac{(\alpha_e \pm \beta_e)^2\beta_t^2\cos^2\theta_t\sin^2\theta_t}{[(s - M_Z^2)^2 + M_Z^2\Gamma_Z^2]}p^3(1 - z^2).$$

(A.21c)

The differential cross sections for stau and sbottom pair production are given by analogous formulae.

For selectron pair production, we find

$$\frac{d\sigma}{dz}(e_{R/L}\bar{e}_{L/R} \to \tilde{e}_L\bar{\tilde{e}}_L) = \frac{1}{256\pi}\frac{p^3}{E^3}\Phi_{\tilde{e}_L R/L}(z),$$

(A.22a)

where

$$\Phi_{\tilde{e}_L R}(z) = \Phi_{\tilde{\mu}_L R}(z),$$

(A.22b)

and

$$\Phi_{\tilde{e}_L L}(z) = \Phi_{\tilde{\mu}_L L}(z) + \sum_{i=1}^{4}\frac{2|A_{\tilde{Z}_i}^e|^4 s(1 - z^2)}{[2E(E - pz) - m_{\tilde{e}_L}^2 + m_{\tilde{Z}_i}^2]^2}$$

$$- 8e^2(1 - z^2)\sum_{i=1}^{4}\frac{|A_{\tilde{Z}_i}^e|^2}{[2E(E - pz) - m_{\tilde{e}_L}^2 + m_{\tilde{Z}_i}^2]}$$

$$\times\left[1 + \frac{(\alpha_e - \beta_e)^2 s(s - M_Z^2)}{(s - M_Z^2)^2 + M_Z^2\Gamma_Z^2}\right]$$

$$+ \sum_{i<j=1}^{4}\frac{4|A_{\tilde{Z}_i}^e|^2|A_{\tilde{Z}_j}^e|^2 s(1 - z^2)}{[2E(E - pz) - m_{\tilde{e}_L}^2 + m_{\tilde{Z}_i}^2][2E(E - pz) - m_{\tilde{e}_L}^2 + m_{\tilde{Z}_j}^2]}.$$

(A.22c)

Similarly,

$$\frac{d\sigma}{dz}(e_{R/L}\bar{e}_{L/R} \to \tilde{e}_R\bar{\tilde{e}}_R) = \frac{1}{256\pi}\frac{p^3}{E^3}\Phi_{\tilde{e}_R R/L}(z),$$

(A.23a)

where

$$\Phi_{\tilde{e}_R L}(z) = \Phi_{\tilde{\mu}_R L}(z),$$

(A.23b)

and

$$\Phi_{\tilde{e}_R R}(z) = \Phi_{\tilde{e}_L L}(z) \quad \text{but with the substitutions,}$$

$$A_{\tilde{Z}_i}^e \to B_{\tilde{Z}_i}^e, \quad m_{\tilde{e}_L} \to m_{\tilde{e}_R}, \quad \text{and} \quad (\alpha_e - \beta_e) \to (\alpha_e + \beta_e). \tag{A.23c}$$

For $\tilde{e}_L \tilde{e}_R$ production, we have,[2]

$$\frac{d\sigma}{dz}(e_R \bar{e}_L \to \tilde{e}_L \tilde{e}_R) = \frac{d\sigma}{dz}(e_L \bar{e}_R \to \tilde{e}_R \bar{\tilde{e}}_L) = 0, \tag{A.24a}$$

while

$$\frac{d\sigma}{dz}(e_L \bar{e}_L \to \tilde{e}_L \bar{\tilde{e}}_R) = \frac{1}{32\pi s} \frac{p}{E} \left[\sum_{i=1}^{4} \frac{|A_{\tilde{Z}_i}^e|^2 |B_{\tilde{Z}_i}^e|^2 m_{\tilde{Z}_i}^2}{[E_{\tilde{e}_L} - pz + a_{\tilde{Z}_i}]^2} \right.$$
$$\left. + \sum_{i<j=1}^{4} \frac{2 m_{\tilde{Z}_i} m_{\tilde{Z}_j} \mathrm{Re}(A_{\tilde{Z}_i}^e A_{\tilde{Z}_j}^{e*} B_{\tilde{Z}_i}^{e*} B_{\tilde{Z}_j}^e)}{[E_{\tilde{e}_L} - pz + a_{\tilde{Z}_i}][E_{\tilde{e}_L} - pz + a_{\tilde{Z}_j}]} \right], \tag{A.24b}$$

where $a_{\tilde{Z}_i} = \frac{m_{\tilde{Z}_i}^2 - m_{\tilde{e}_L}^2}{2E}$, and

$$\frac{d\sigma}{dz}(e_R \bar{e}_R \to \tilde{e}_R \bar{\tilde{e}}_L) = \frac{1}{32\pi s} \frac{p}{E} \left[\sum_{i=1}^{4} \frac{|A_{\tilde{Z}_i}^e|^2 |B_{\tilde{Z}_i}^e|^2 m_{\tilde{Z}_i}^2}{[E_{\tilde{e}_R} - pz + a_{\tilde{Z}_i}]^2} \right.$$
$$\left. + \sum_{i<j=1}^{4} \frac{2 m_{\tilde{Z}_i} m_{\tilde{Z}_j} \mathrm{Re}(A_{\tilde{Z}_i}^e A_{\tilde{Z}_j}^{e*} B_{\tilde{Z}_i}^{e*} B_{\tilde{Z}_j}^e)}{[E_{\tilde{e}_R} - pz + a_{\tilde{Z}_i}][E_{\tilde{e}_R} - pz + a_{\tilde{Z}_j}]} \right], \tag{A.24c}$$

where now $a_{\tilde{Z}_i} = \frac{m_{\tilde{Z}_i}^2 - m_{\tilde{e}_R}^2}{2E}$.

For $\tilde{\nu}_e$ pair production, we find

$$\frac{d\sigma}{dz}(e_R \bar{e}_L \to \tilde{\nu}_e \bar{\tilde{\nu}}_e) = \frac{d\sigma}{dz}(e_R \bar{e}_L \to \tilde{\nu}_\mu \bar{\tilde{\nu}}_\mu), \tag{A.25a}$$

while

$$\frac{d\sigma}{dz}(e_L \bar{e}_R \to \tilde{\nu}_e \bar{\tilde{\nu}}_e)$$
$$= \frac{p^3 E}{8\pi}(1 - z^2) \left[\frac{4e^4 (\alpha_\nu - \beta_\nu)^2 (\alpha_e - \beta_e)^2}{(s - M_Z^2)^2 + M_Z^2 \Gamma_Z^2} \right.$$
$$+ \frac{g^4 \sin^4 \gamma_R}{[2E(E - pz) + m_{\tilde{W}_1}^2 - m_{\tilde{\nu}_e}^2]^2} + \frac{g^4 \cos^4 \gamma_R}{[2E(E - pz) + m_{\tilde{W}_2}^2 - m_{\tilde{\nu}_e}^2]^2}$$
$$\left. - \frac{4e^2 g^2 (\alpha_\nu - \beta_\nu)(\alpha_e - \beta_e)(s - M_Z^2) \sin^2 \gamma_R}{[(s - M_Z^2)^2 + M_Z^2 \Gamma_Z^2][2E(E - pz) + m_{\tilde{W}_1}^2 - m_{\tilde{\nu}_e}^2]} \right.$$

[2] As long as we neglect electron Yukawa couplings, neutralinos couple doublets (singlets) to doublets (singlets). Thus $\tilde{e}_L \tilde{e}_R$ production occurs only from the $e_{L/R} \bar{e}_{L/R}$ initial state via t-channel neutralino exchange, in sharp contrast to other SUSY processes that we have seen. We will leave it to the reader to work out in advance those polarizations of the initial electron and positron beams that contribute to chargino and neutralino pair production processes which, we recall, have both s- and t-channel contributions.

$$-\frac{4e^2 g^2 (\alpha_v - \beta_v)(\alpha_e - \beta_e)(s - M_Z^2)\cos^2 \gamma_R}{[(s - M_Z^2)^2 + M_Z^2 \Gamma_Z^2][2E(E - pz) + m_{\widetilde{W}_2}^2 - m_{\bar{\nu}_e}^2]}$$

$$+ \frac{2g^4 \sin^2 \gamma_R \cos^2 \gamma_R}{[2E(E - pz) + m_{\widetilde{W}_1}^2 - m_{\bar{\nu}_e}^2][2E(E - pz) + m_{\widetilde{W}_2}^2 - m_{\bar{\nu}_e}^2]} \Bigg]. \quad \text{(A.25b)}$$

The differential cross sections for neutralino pair production are given by,

$$\frac{d\sigma}{dz}(e_{R/L}\bar{e}_{L/R} \rightarrow \widetilde{Z}_i \widetilde{Z}_j) = \frac{p}{8\pi s \sqrt{s}} \left(M_{\bar{e}eR/L} + M_{ZZR/L} + M_{Z\bar{e}R/L} \right), \quad \text{(A.26)}$$

with

$$M_{\bar{e}eR} = 2|B^e_{\widetilde{Z}_i}|^2 |B^e_{\widetilde{Z}_j}|^2 G_t(m_{\widetilde{Z}_i}, m_{\widetilde{Z}_j}, m_{\bar{e}_R}, z),$$

$$M_{\bar{e}eL} = 2|A^e_{\widetilde{Z}_i}|^2 |A^e_{\widetilde{Z}_j}|^2 G_t(m_{\widetilde{Z}_i}, m_{\widetilde{Z}_j}, m_{\bar{e}_L}, z),$$

$$M_{ZZR/L} =$$

$$\frac{4e^2 |W_{ij}|^2 (\alpha_e \pm \beta_e)^2}{(s - M_Z^2)^2 + M_Z^2 \Gamma_Z^2} \left[s^2 - (m_{\widetilde{Z}_i}^2 - m_{\widetilde{Z}_j}^2)^2 - 4(-1)^{\theta_i + \theta_j} s m_{\widetilde{Z}_i} m_{\widetilde{Z}_j} + 4sp^2 z^2 \right],$$

$$M_{Z\bar{e}R} = \frac{-e(-1)^{(\theta_i + \theta_j + 1)}(\alpha_e + \beta_e)(s - M_Z^2)}{2[(s - M_Z^2)^2 + M_Z^2 \Gamma_Z^2]}$$

$$\times \left[\text{Re}(W_{ij} B^{e*}_{\widetilde{Z}_i} B^e_{\widetilde{Z}_j}) G_{st}(m_{\widetilde{Z}_i}, m_{\widetilde{Z}_j}, m_{\bar{e}_R}, z) \right.$$

$$\left. + (-1)^{(\theta_i + \theta_j)} \text{Re}(W_{ij} B^e_{\widetilde{Z}_i} B^{e*}_{\widetilde{Z}_j}) G_{st}(m_{\widetilde{Z}_i}, m_{\widetilde{Z}_j}, m_{\bar{e}_R}, -z) \right],$$

and

$$M_{Z\bar{e}L} = \frac{-e(\alpha_e - \beta_e)(s - M_Z^2)}{2[(s - M_Z^2)^2 + M_Z^2 \Gamma_Z^2]}$$

$$\times \left[\text{Re}(W_{ij} A^{e*}_{\widetilde{Z}_i} A^e_{\widetilde{Z}_j}) G_{st}(m_{\widetilde{Z}_i}, m_{\widetilde{Z}_j}, m_{\bar{e}_L}, z) \right.$$

$$\left. + (-1)^{\theta_i + \theta_j} \text{Re}(W_{ij} A^e_{\widetilde{Z}_i} A^{e*}_{\widetilde{Z}_j}) G_{st}(m_{\widetilde{Z}_i}, m_{\widetilde{Z}_j}, m_{\bar{e}_L}, -z) \right].$$

The functions G_t and G_{st} are defined in (A.4b) and (A.4c), respectively.

For chargino pair production we have,

$$\frac{d\sigma}{dz}(e_L \bar{e}_R \rightarrow \widetilde{W}_1 \overline{\widetilde{W}}_1) = \frac{1}{64\pi s}\frac{p}{E} \left(M_{\gamma\gamma L} + M_{ZZL} + M_{\gamma ZL} \right.$$

$$\left. + M_{\bar{\nu}\bar{\nu}L} + M_{\gamma\bar{\nu}L} + M_{Z\bar{\nu}L} \right), \quad \text{(A.27a)}$$

and

$$\frac{d\sigma}{dz}(e_R \bar{e}_L \to \widetilde{W}_1 \overline{\widetilde{W}}_1) = \frac{1}{64\pi s} \frac{p}{E} \left(M_{\gamma\gamma R} + M_{ZZR} + M_{\gamma ZR} \right), \quad \text{(A.27b)}$$

with

$$M_{\gamma\gamma L} = M_{\gamma\gamma R} = \frac{16e^4}{s} \left[E^2(1+z^2) + m_{\widetilde{W}_1}^2(1-z^2) \right],$$

$$M_{ZZR/L} = \frac{16e^4 \cot^2 \theta_W s}{(s-M_Z^2)^2 + M_Z^2\Gamma_Z^2} \left[(x_c^2 + y_c^2)(\alpha_e \pm \beta_e)^2 \times \right.$$

$$\left[E^2(1+z^2) + m_{\widetilde{W}_1}^2(1-z^2) \right] - 2y_c^2(\alpha_e \pm \beta_e)^2 m_{\widetilde{W}_1}^2 \mp 4x_c y_c(\alpha_e \pm \beta_e)^2 Epz \right],$$

$$M_{\gamma ZR/L} =$$

$$\frac{-32e^4(\alpha_e \pm \beta_e)\cot\theta_W(s-M_Z^2)}{(s-M_Z^2)^2 + M_Z^2\Gamma_Z^2} \left\{ x_c[E^2(1+z^2) + m_{\widetilde{W}_1}^2(1-z^2)] \mp 2y_c Epz \right\},$$

$$M_{\tilde{\nu}\tilde{\nu}L} = \frac{2e^4 \sin^4 \gamma_R}{\sin^4 \theta_W} \frac{s(E-pz)^2}{[E^2 + p^2 - 2Epz + m_{\tilde{\nu}}^2]^2},$$

$$M_{\gamma\tilde{\nu}L} = \frac{-8e^4 \sin^2 \gamma_R}{\sin^2 \theta_W} \frac{[(E-pz)^2 + m_{\widetilde{W}_1}^2]}{[E^2 + p^2 - 2Epz + m_{\tilde{\nu}}^2]},$$

and

$$M_{Z\tilde{\nu}L} = \frac{8e^4(\alpha_e - \beta_e)\cot\theta_W \sin^2 \gamma_R}{\sin^2 \theta_W} \frac{s(s-M_Z^2)}{(s-M_Z^2)^2 + M_Z^2\Gamma_Z^2}$$

$$\times \left[\frac{(x_c - y_c)[(E-pz)^2 + m_{\widetilde{W}_1}^2] + 2y_c m_{\widetilde{W}_1}^2}{E^2 + p^2 - 2Epz + m_{\tilde{\nu}}^2} \right].$$

To obtain the differential cross section for $\widetilde{W}_2 \overline{\widetilde{W}}_2$ production, we simply replace x_c with x_s, y_c with y_s, $\sin \gamma_R$ with $\cos \gamma_R$, and $m_{\widetilde{W}_1}$ with $m_{\widetilde{W}_2}$ in the corresponding expression for $\widetilde{W}_1 \overline{\widetilde{W}}_1$ production. Finally for $\widetilde{W}_1 \overline{\widetilde{W}}_2$ production we have,

$$\frac{d\sigma}{dz}(e_L \bar{e}_R \to \widetilde{W}_1 \overline{\widetilde{W}}_2) = \frac{e^4}{64\pi} \frac{p}{E} [M_{ZZL} + M_{\tilde{\nu}\tilde{\nu}L} + M_{Z\tilde{\nu}L}], \quad \text{(A.27c)}$$

and

$$\frac{d\sigma}{dz}(e_R \bar{e}_L \to \widetilde{W}_1 \overline{\widetilde{W}}_2) = \frac{e^4}{64\pi} \frac{p}{E} M_{ZZR}, \quad \text{(A.27d)}$$

where

$$M_{ZZR/L} = \frac{4(\alpha_e \pm \beta_e)^2(\cot\theta_W + \tan\theta_W)^2}{(s - M_Z^2)^2 + M_Z^2\Gamma_Z^2}\left[(x^2 + y^2)(E^2 + p^2z^2\right.$$
$$\left. - \Delta^2 - \xi m_{\widetilde{W}_1}m_{\widetilde{W}_2}) + 2x^2\xi m_{\widetilde{W}_1}m_{\widetilde{W}_2} \mp 4xyEpz\right]$$

$$M_{\bar{\nu}\bar{\nu}L} = \frac{2\sin^2\gamma_R\cos^2\gamma_R}{\sin^4\theta_W}\frac{[(E - pz)^2 - \Delta^2]}{[2E(E - \Delta) - 2Epz + m_{\bar{\nu}}^2 - m_{\widetilde{W}_1}^2]^2},$$

and

$$M_{Z\bar{\nu}L} = \frac{-4\theta_y(\alpha_e - \beta_e)(\cot\theta_W + \tan\theta_W)\sin\gamma_R\cos\gamma_R(s - M_Z^2)}{\sin^2\theta_W[(s - M_Z^2)^2 + M_Z^2\Gamma_Z^2]}$$
$$\times\frac{(x - y)[(E - pz)^2 - \Delta^2 - \xi m_{\widetilde{W}_1}m_{\widetilde{W}_2}] + 2x\xi m_{\widetilde{W}_1}m_{\widetilde{W}_2}}{[2E(E - \Delta) - 2Epz + m_{\bar{\nu}}^2 - m_{\widetilde{W}_1}^2]},$$

with

$$\Delta = \frac{(m_{\widetilde{W}_2}^2 - m_{\widetilde{W}_1}^2)}{4E} \text{ and } \xi = (-1)^{\theta_{\widetilde{W}_1} + \theta_{\widetilde{W}_2} + 1}.$$

The cross sections for unpolarized or partially polarized beams can be obtained using,

$$\sigma = f_L(e^-)f_L(e^+)\sigma_{LL} + f_L(e^-)f_R(e^+)\sigma_{LR}$$
$$+ f_R(e^-)f_L(e^+)\sigma_{RL} + f_R(e^-)f_R(e^+)\sigma_{RR}, \quad\quad (A.28)$$

where f_L and f_R have been defined at the start of this subsection, and σ_{ij} ($i, j =$ L, R) refers to the cross section from $e_i^- e_j^+$ annihilation.

Appendix B
Sparticle decay widths

In this appendix, we list formulae for the partial widths for the $1 \to 2$ and $1 \to 3$ tree-level decays of sparticles that are relevant to SUSY searches at colliders.

B.1 Gluino decay widths

B.1.1 Two-body decays

The decay width for $\tilde{g} \to \bar{q}\tilde{q}_i$ ($i = $ L or R), is given by

$$\Gamma(\tilde{g} \to \bar{q}\tilde{q}_i) = \frac{\alpha_s}{8} m_{\tilde{g}} \lambda^{1/2}(1, \frac{m_q^2}{m_{\tilde{g}}^2}, \frac{m_{\tilde{q}_i}^2}{m_{\tilde{g}}^2}) \left(1 + \frac{m_q^2}{m_{\tilde{g}}^2} - \frac{m_{\tilde{q}_i}^2}{m_{\tilde{g}}^2}\right). \tag{B.1a}$$

If intra-generation squark mixing and quark mass effects are included, the formula is slightly more complicated:

$$\Gamma(\tilde{g} \to \bar{q}\tilde{q}_{1,2}) = \frac{\alpha_s}{8} m_{\tilde{g}} \lambda^{1/2}(1, \frac{m_q^2}{m_{\tilde{g}}^2}, \frac{m_{\tilde{q}_{1,2}}^2}{m_{\tilde{g}}^2})$$
$$\times \left(1 + \frac{m_q^2}{m_{\tilde{g}}^2} - \frac{m_{\tilde{q}_{1,2}}^2}{m_{\tilde{g}}^2} \pm 2(-1)^{\theta_{\tilde{g}}} \sin 2\theta_q \frac{m_q}{m_{\tilde{g}}}\right), \tag{B.1b}$$

where the upper (lower) sign is for the decay to $\tilde{q}_1$ ($\tilde{q}_2$).

In models with a very light gravitino, the gluino decay to gluon and a gravitino may be relevant:

$$\Gamma(\tilde{g} \to \tilde{G}g) = \frac{m_{\tilde{g}}^5}{48\pi(m_{3/2}M_{\rm P})^2}, \tag{B.2}$$

where $M_{\rm P}$ is the reduced Planck mass, $M_{\rm P} \simeq 2.4 \times 10^{18}$ GeV, and we have ignored the gravitino mass in the phase space.

B.1.2 Three-body decays to light quarks

If two-body gluino decays are kinematically suppressed, then three-body decays can be important. The three-body decay $\tilde{g} \to \widetilde{W}_i u \bar{d}$ width is given by

$$
\Gamma(\tilde{g} \to \widetilde{W}_i u \bar{d}) = \frac{\alpha_s}{16\pi^2} \Big[|A^d_{\widetilde{W}_i}|^2 \psi(m_{\tilde{g}}, m_{\tilde{u}_L}, m_{\widetilde{W}_i}) + |A^u_{\widetilde{W}_i}|^2 \psi(m_{\tilde{g}}, m_{\tilde{d}_L}, m_{\widetilde{W}_i})
$$
$$
- 2(-1)^{\theta_{\tilde{g}}} \mathrm{Re}(A^u_{\widetilde{W}_i} A^d_{\widetilde{W}_i}) \phi(m_{\tilde{g}}, m_{\tilde{u}_L}, m_{\tilde{d}_L}, m_{\widetilde{W}_i}) \Big], \tag{B.3}
$$

where

$$
\psi(m_{\tilde{g}}, m_{\tilde{q}}, m) = \int dq \frac{q^2 (m_{\tilde{g}}^2 - 2m_{\tilde{g}}q - m^2)^2}{(m_{\tilde{g}}^2 - 2m_{\tilde{g}}q - m_{\tilde{q}}^2)^2 (m_{\tilde{g}}^2 - 2m_{\tilde{g}}q)}
$$

and

$$
\phi(m_{\tilde{g}}, m_{\tilde{q}_1}, m_{\tilde{q}_2}, m) = \frac{m}{2} \int \frac{dq}{m_{\tilde{g}}^2 - m_{\tilde{q}_1}^2 - 2m_{\tilde{g}}q} \left[\frac{-q(m_{\tilde{g}}^2 - m^2 - 2m_{\tilde{g}}q)}{m_{\tilde{g}}(m_{\tilde{g}} - 2q)} \right.
$$
$$
\left. - \frac{2m_{\tilde{g}}q - m_{\tilde{q}_2}^2 + m^2}{2m_{\tilde{g}}} \log \frac{m_{\tilde{q}_2}^2 (m_{\tilde{g}} - 2q) - m_{\tilde{g}}m^2}{(m_{\tilde{g}} - 2q)(m_{\tilde{q}_2}^2 - 2m_{\tilde{g}}q - m^2)} \right],
$$

and where the range of integration on ψ and ϕ ranges from 0 to $(m_{\tilde{g}}^2 - m^2)/2m_{\tilde{g}}$.

The decay rate for $\tilde{g} \to \widetilde{Z}_i q \bar{q}$ is given by

$$
\Gamma(\tilde{g} \to \widetilde{Z}_i q \bar{q}) = \frac{\alpha_s}{8\pi^2} \Big[|A^q_{\widetilde{Z}_i}|^2 \left(\psi(m_{\tilde{g}}, m_{\tilde{q}_L}, m_{\widetilde{Z}_i}) \right.
$$
$$
\left. - (-1)^{\theta_i + \theta_{\tilde{g}} - 1} \phi(m_{\tilde{g}}, m_{\tilde{q}_L}, m_{\tilde{q}_L}, m_{\widetilde{Z}_i}) \right) + |B^q_{\widetilde{Z}_i}|^2 \left(\psi(m_{\tilde{g}}, m_{\tilde{q}_R}, m_{\widetilde{Z}_i}) \right.
$$
$$
\left. - (-1)^{\theta_i + \theta_{\tilde{g}} - 1} \phi(m_{\tilde{g}}, m_{\tilde{q}_R}, m_{\tilde{q}_R}, m_{\widetilde{Z}_i}) \right) \Big]. \tag{B.4}
$$

The formulae for gluino decay to third generation particles are more complicated. They involve Yukawa coupling contributions, squark mixing effects and all final state fermion masses are non-negligible.

B.1.3 $\tilde{g} \to \widetilde{Z}_i t \bar{t}$ and $\tilde{g} \to \widetilde{Z}_i b \bar{b}$

The partial width for $\tilde{g} \to t \bar{t} \widetilde{Z}_i$ can be written as

$$
\Gamma(\tilde{g} \to t \bar{t} \widetilde{Z}_i) = \frac{\alpha_s}{8\pi^4 m_{\tilde{g}}} \left[\Gamma_{\tilde{t}_1} + \Gamma_{\tilde{t}_2} + \Gamma_{\tilde{t}_1 \tilde{t}_2} \right], \tag{B.5}
$$

with,

$$
\Gamma_{\tilde{t}_1} = \Gamma_{LL}(\tilde{t}_1) \cos^2 \theta_t + \Gamma_{RR}(\tilde{t}_1) \sin^2 \theta_t
$$
$$
- \sin \theta_t \cos \theta_t \left\{ \Gamma_{L_1 R_1} + \Gamma_{L_1 R_2} + \Gamma_{L_2 R_1} + \Gamma_{L_2 R_2} \right\} (\tilde{t}_1), \tag{B.6a}
$$

$$\Gamma_{\tilde{t}_2} = \Gamma_{LL}(\tilde{t}_2)\sin^2\theta_t + \Gamma_{RR}(\tilde{t}_2)\cos^2\theta_t$$
$$+ \sin\theta_t\cos\theta_t\left\{\Gamma_{L_1R_1} + \Gamma_{L_1R_2} + \Gamma_{L_2R_1} + \Gamma_{L_2R_2}\right\}(\tilde{t}_2), \qquad \text{(B.6b)}$$

and

$$\Gamma_{\tilde{t}_1\tilde{t}_2} = (\Gamma_{LL}(\tilde{t}_1, \tilde{t}_2) + \Gamma_{RR}(\tilde{t}_1, \tilde{t}_2))\sin\theta_t\cos\theta_t$$
$$+ \Gamma_{LR}(\tilde{t}_1, \tilde{t}_2)\cos^2\theta_t + \Gamma_{RL}(\tilde{t}_1, \tilde{t}_2)\sin^2\theta_t. \qquad \text{(B.6c)}$$

The Γ_{ij} contributions to the partial width are all written in terms of one-dimensional integrals. The various $\Gamma_{ij}(\tilde{t}_1)$ that enter the expression (B.6a) for $\Gamma_{\tilde{t}_1}$ are:

$$\Gamma_{LL}(\tilde{t}_1)$$
$$= (\alpha_1^2 + \beta_1^2)\psi(m_{\tilde{g}}, m_{\tilde{t}_1}, m_{\tilde{Z}_i}) - 4m_t m_{\tilde{Z}_i}(-1)^{\theta_i}\alpha_1\beta_1\chi(m_{\tilde{g}}, m_{\tilde{t}_1}, m_{\tilde{Z}_i})$$
$$+ (-1)^{\theta_{\tilde{g}}}m_{\tilde{g}}\left[(-1)^{\theta_i}m_{\tilde{Z}_i}\left(\frac{\alpha_1^2}{m_{\tilde{g}}m_{\tilde{Z}_i}}\phi(m_{\tilde{g}}, m_{\tilde{t}_1}, m_{\tilde{Z}_i}) + \beta_1^2 m_t^2\rho(m_{\tilde{g}}, m_{\tilde{t}_1}, m_{\tilde{Z}_i})\right)\right.$$
$$\left. - \alpha_1\beta_1 m_t\left(\xi(m_{\tilde{g}}, m_{\tilde{t}_1}, m_{\tilde{t}_1}, m_{\tilde{Z}_i}) - m_{\tilde{Z}_i}^2\rho(m_{\tilde{g}}, m_{\tilde{t}_1}, m_{\tilde{Z}_i})\right)\right], \qquad \text{(B.7)}$$

where

$$\alpha_1 = \tilde{A}_{\tilde{Z}_i}^t\cos\theta_t - f_t v_1^{(i)}\sin\theta_t \quad \text{and} \qquad \text{(B.8a)}$$
$$\beta_1 = f_t v_1^{(i)}\cos\theta_t + \tilde{B}_{\tilde{Z}_i}^t\sin\theta_t, \qquad \text{(B.8b)}$$

and

$$\tilde{A}_{\tilde{Z}_i}^t = \frac{g}{\sqrt{2}}v_3^{(i)} + \frac{g'}{3\sqrt{2}}v_4^{(i)}, \qquad \text{(B.9a)}$$

$$\tilde{A}_{\tilde{Z}_i}^b = -\frac{g}{\sqrt{2}}v_3^{(i)} + \frac{g'}{3\sqrt{2}}v_4^{(i)}, \qquad \text{(B.9b)}$$

$$\tilde{B}_{\tilde{Z}_i}^t = \frac{4}{3}\frac{g'}{\sqrt{2}}v_4^{(i)}, \quad \text{and} \qquad \text{(B.9c)}$$

$$\tilde{B}_{\tilde{Z}_i}^b = -\frac{2}{3}\frac{g'}{\sqrt{2}}v_4^{(i)}. \qquad \text{(B.9d)}$$

Also,

$$\Gamma_{RR}(\tilde{t}_1) = \Gamma_{LL}(\tilde{t}_1), \qquad \text{(B.10)}$$

but with $\alpha_1 \to \alpha_2$ and $\beta_1 \to \beta_2$, where

$$\alpha_2 = \tilde{B}_{\tilde{Z}_i}^t \sin\theta_t + f_t v_1^{(i)} \cos\theta_t \quad \text{and} \tag{B.11a}$$

$$\beta_2 = -f_t v_1^{(i)} \sin\theta_t + \tilde{A}_{\tilde{Z}_i}^t \cos\theta_t. \tag{B.11b}$$

Furthermore,

$$\Gamma_{\mathrm{L}_1\mathrm{R}_1}(\tilde{t}_1) = 2m_{\tilde{g}} m_t (-1)^{\theta_{\tilde{g}}} \left[(\alpha_1\alpha_2 + \beta_1\beta_2)(-1)^{\theta_i} m_t m_{\tilde{Z}_i} \zeta(m_{\tilde{g}}, m_{\tilde{t}_1}, m_{\tilde{t}_1}, m_{\tilde{Z}_i}) \right.$$
$$\left. - (\alpha_2\beta_1 + \alpha_1\beta_2)X(m_{\tilde{g}}, m_{\tilde{t}_1}, m_{\tilde{t}_1}, m_{\tilde{Z}_i}) \right], \tag{B.12}$$

with

$$\Gamma_{\mathrm{L}_2\mathrm{R}_2}(\tilde{t}_1) = \Gamma_{\mathrm{L}_1\mathrm{R}_1}(\tilde{t}_1). \tag{B.13}$$

Finally,

$$\Gamma_{\mathrm{L}_1\mathrm{R}_2}(\tilde{t}_1) = \beta_1\beta_2 Y(m_{\tilde{g}}, m_{\tilde{t}_1}, m_{\tilde{t}_1}, m_{\tilde{Z}_i}) + \alpha_1\alpha_2 m_t^2 \xi(m_{\tilde{g}}, m_{\tilde{t}_1}, m_{\tilde{t}_1}, m_{\tilde{Z}_i})$$
$$- m_t m_{\tilde{Z}_i} (-1)^{\theta_i} (\alpha_1\beta_2 + \alpha_2\beta_1)\chi'(m_{\tilde{g}}, m_{\tilde{t}_1}, m_{\tilde{t}_1}, m_{\tilde{Z}_i}) \tag{B.14}$$

with

$$\Gamma_{\mathrm{L}_2\mathrm{R}_1}(\tilde{t}_1) = \Gamma_{\mathrm{L}_1\mathrm{R}_2}(\tilde{t}_1). \tag{B.15}$$

Turning to $\Gamma_{\tilde{t}_2}$, we have

$$\Gamma_{\mathrm{LL}}(\tilde{t}_2) = \Gamma_{\mathrm{LL}}(\tilde{t}_1), \tag{B.16}$$

but with the replacements

$$m_{\tilde{t}_1} \to m_{\tilde{t}_2}, \tag{B.17a}$$

$$\alpha_1 \to \tilde{A}_{\tilde{Z}_i}^t \sin\theta_t + f_t v_1^{(i)} \cos\theta_t, \tag{B.17b}$$

$$\beta_1 \to f_t v_1^{(i)} \sin\theta_t - \tilde{B}_{\tilde{Z}_i}^t \cos\theta_t, \tag{B.17c}$$

and

$$\Gamma_{\mathrm{RR}}(\tilde{t}_2) = \Gamma_{\mathrm{RR}}(\tilde{t}_1), \tag{B.18}$$

with

$$m_{\tilde{t}_1} \to m_{\tilde{t}_2}, \tag{B.19a}$$

$$\alpha_2 \to -\tilde{B}_{\tilde{Z}_i}^t \cos\theta_t + f_t v_1^{(i)} \sin\theta_t, \quad \text{and} \tag{B.19b}$$

$$\beta_2 \to f_t v_1^{(i)} \cos\theta_t + \tilde{A}_{\tilde{Z}_i}^t \sin\theta_t. \tag{B.19c}$$

Also,

$$\Gamma_{L_1R_1}(\tilde{t}_2) = \Gamma_{L_2R_2}(\tilde{t}_2), \quad \text{and} \tag{B.20a}$$

$$\Gamma_{L_1R_2}(\tilde{t}_2) = \Gamma_{L_2R_1}(\tilde{t}_2), \tag{B.20b}$$

where these expressions can be obtained from the previous $\Gamma_{L_iR_j}$ formulae by replacing $m_{\tilde{t}_1} \to m_{\tilde{t}_2}$ and using the revised α_1, α_2, β_1 and β_2 values.

The expression for $\Gamma_{\tilde{t}_1\tilde{t}_2}$ contains,

$$
\begin{aligned}
\Gamma_{LL}(\tilde{t}_1, \tilde{t}_2) = {}& 2(\alpha_1\alpha_2 + \beta_1\beta_2)\tilde{\psi}(m_{\tilde{g}}, m_{\tilde{t}_1}, m_{\tilde{t}_2}, m_{\tilde{Z}_i}) \\
& - (-1)^{\theta_i} 4 m_t m_{\tilde{Z}_i}(\alpha_1\beta_2 + \alpha_2\beta_1)\tilde{\chi}(m_{\tilde{g}}, m_{\tilde{t}_1}, m_{\tilde{t}_2}, m_{\tilde{Z}_i}) \\
& - (-1)^{\theta_{\tilde{g}}} m_{\tilde{g}}\Big\{ 2m_{\tilde{Z}_i}(-1)^{\theta_i-1}\big[\frac{\alpha_1\alpha_2}{m_{\tilde{g}}m_{\tilde{Z}_i}}\tilde{\phi}(m_{\tilde{g}}, m_{\tilde{t}_1}, m_{\tilde{t}_2}, m_{\tilde{Z}_i}) \\
& + \beta_1\beta_2 m_t^2 \tilde{\rho}(m_{\tilde{g}}, m_{\tilde{t}_1}, m_{\tilde{t}_2}, m_{\tilde{Z}_i})\big] \\
& + (\alpha_1\beta_2 + \alpha_2\beta_1)m_t[\tilde{\xi}(m_{\tilde{g}}, m_{\tilde{t}_1}, m_{\tilde{t}_2}, m_{\tilde{Z}_i}) \\
& - m_{\tilde{Z}_i}^2 \tilde{\rho}(m_{\tilde{g}}, m_{\tilde{t}_1}, m_{\tilde{t}_2}, m_{\tilde{Z}_i})]\Big\},
\end{aligned}
\tag{B.21}
$$

where

$$\alpha_1 = \tilde{A}_{\tilde{Z}_i}^t \cos\theta_t - f_t v_1^{(i)} \sin\theta_t, \tag{B.22a}$$

$$\alpha_2 = \tilde{A}_{\tilde{Z}_i}^t \sin\theta_t + f_t v_1^{(i)} \cos\theta_t, \tag{B.22b}$$

$$\beta_1 = f_t v_1^{(i)} \cos\theta_t + \tilde{B}_{\tilde{Z}_i}^t \sin\theta_t, \quad \text{and} \tag{B.22c}$$

$$\beta_2 = f_t v_1^{(i)} \sin\theta_t - \tilde{B}_{\tilde{Z}_i}^t \cos\theta_t. \tag{B.22d}$$

Also,

$$
\begin{aligned}
\Gamma_{RR}(\tilde{t}_1, \tilde{t}_2) = {}& -2(\alpha_1\alpha_2 + \beta_1\beta_2)\tilde{\psi}(m_{\tilde{g}}, m_{\tilde{t}_1}, m_{\tilde{t}_2}, m_{\tilde{Z}_i}) \\
& - (-1)^{\theta_i} 4 m_t m_{\tilde{Z}_i}(\alpha_1\beta_2 + \alpha_2\beta_1)\tilde{\chi}(m_{\tilde{g}}, m_{\tilde{t}_1}, m_{\tilde{t}_2}, m_{\tilde{Z}_i}) \\
& + (-1)^{\theta_{\tilde{g}}} m_{\tilde{g}}\Big\{ 2m_{\tilde{Z}_i}(-1)^{\theta_i-1}\big[\frac{\alpha_1\alpha_2}{m_{\tilde{g}}m_{\tilde{Z}_i}}\tilde{\phi}(m_{\tilde{g}}, m_{\tilde{t}_1}, m_{\tilde{t}_2}, m_{\tilde{Z}_i}) \\
& + \beta_1\beta_2 m_t^2 \tilde{\rho}(m_{\tilde{g}}, m_{\tilde{t}_1}, m_{\tilde{t}_2}, m_{\tilde{Z}_i})\big] \\
& - (\alpha_1\beta_2 + \alpha_2\beta_1)m_t[\tilde{\xi}(m_{\tilde{g}}, m_{\tilde{t}_1}, m_{\tilde{t}_2}, m_{\tilde{Z}_i}) \\
& - m_{\tilde{Z}_i}^2 \tilde{\rho}(m_{\tilde{g}}, m_{\tilde{t}_1}, m_{\tilde{t}_2}, m_{\tilde{Z}_i})]\Big\},
\end{aligned}
\tag{B.23}
$$

where

$$\alpha_1 = -\tilde{B}^t_{\tilde{Z}_i} \sin\theta_t - f_t v_1^{(i)} \cos\theta_t, \tag{B.24a}$$

$$\alpha_2 = \tilde{B}^t_{\tilde{Z}_i} \cos\theta_t - f_t v_1^{(i)} \sin\theta_t, \tag{B.24b}$$

$$\beta_1 = -f_t v_1^{(i)} \sin\theta_t + \tilde{A}^t_{\tilde{Z}_i} \cos\theta_t, \quad \text{and} \tag{B.24c}$$

$$\beta_2 = f_t v_1^{(i)} \cos\theta_t + \tilde{A}^t_{\tilde{Z}_i} \sin\theta_t. \tag{B.24d}$$

Next,

$$\begin{aligned}
\Gamma_{\text{LR}}&(\tilde{t}_1, \tilde{t}_2) \\
&= 4 m_{\tilde{g}} m_t (-1)^{\theta_g} \{ (-1)^{\theta_i} (-\alpha_1\alpha_2 + \beta_1\beta_2) m_t m_{\tilde{Z}_i} \zeta(m_{\tilde{g}}, m_{\tilde{t}_1}, m_{\tilde{t}_2}, m_{\tilde{Z}_i}) \\
&\quad + (\alpha_2\beta_1 - \alpha_1\beta_2) X(m_{\tilde{g}}, m_{\tilde{t}_1}, m_{\tilde{t}_2}, m_{\tilde{Z}_i}) \} + 2\beta_1\beta_2 Y(m_{\tilde{g}}, m_{\tilde{t}_1}, m_{\tilde{t}_2}, m_{\tilde{Z}_i}) \\
&\quad + 2 m_t m_{\tilde{Z}_i} (-1)^{\theta_i} (\beta_1\alpha_2 - \alpha_1\beta_2) \chi'(m_{\tilde{g}}, m_{\tilde{t}_1}, m_{\tilde{t}_2}, m_{\tilde{Z}_i}) \\
&\quad - 2\alpha_1\alpha_2 m_t^2 \xi(m_{\tilde{g}}, m_{\tilde{t}_1}, m_{\tilde{t}_2}, m_{\tilde{Z}_i}),
\end{aligned} \tag{B.25}$$

where

$$\alpha_1 = \tilde{A}^t_{\tilde{Z}_i} \cos\theta_t - f_t v_1^{(i)} \sin\theta_t, \tag{B.26a}$$

$$\alpha_2 = \tilde{B}^t_{\tilde{Z}_i} \cos\theta_t - f_t v_1^{(i)} \sin\theta_t, \tag{B.26b}$$

$$\beta_1 = f_t v_1^{(i)} \cos\theta_t + \tilde{B}^t_{\tilde{Z}_i} \sin\theta_t, \quad \text{and} \tag{B.26c}$$

$$\beta_2 = f_t v_1^{(i)} \cos\theta_t + \tilde{A}^t_{\tilde{Z}_i} \sin\theta_t. \tag{B.26d}$$

Finally,

$$\Gamma_{\text{RL}}(\tilde{t}_1, \tilde{t}_2) = -\Gamma_{\text{LR}}(\tilde{t}_1, \tilde{t}_2) \tag{B.27}$$

but using

$$\alpha_1 = \tilde{A}^t_{\tilde{Z}_i} \sin\theta_t + f_t v_1^{(i)} \cos\theta_t, \tag{B.28a}$$

$$\alpha_2 = -\tilde{B}^t_{\tilde{Z}_i} \sin\theta_t - f_t v_1^{(i)} \cos\theta_t, \tag{B.28b}$$

$$\beta_1 = f_t v_1^{(i)} \sin\theta_t - \tilde{B}^t_{\tilde{Z}_i} \cos\theta_t, \quad \text{and} \tag{B.28c}$$

$$\beta_2 = -f_t v_1^{(i)} \sin\theta_t + \tilde{A}^t_{\tilde{Z}_i} \cos\theta_t, \tag{B.28d}$$

and interchanging $m_{\tilde{t}_1} \leftrightarrow m_{\tilde{t}_2}$ in the arguments of the functions ζ, X, Y; and χ' (the first three of which are automatically symmetric).

The functions appearing above are defined as,

$$\tilde{\psi}(m_{\tilde{g}}, m_{\tilde{t}_1}, m_{\tilde{t}_2}, m_{\tilde{Z}}) = \pi^2 m_{\tilde{g}} \int dE_t\, p_t E_t \frac{\lambda^{1/2}(m_{\tilde{g}}^2 + m_t^2 - 2m_{\tilde{g}}E_t, m_{\tilde{Z}}^2, m_t^2)}{m_{\tilde{g}}^2 + m_t^2 - 2m_{\tilde{g}}E_t}$$

$$\times \frac{m_{\tilde{g}}^2 - m_{\tilde{Z}}^2 - 2m_{\tilde{g}}E_t}{(m_{\tilde{g}}^2 + m_t^2 - 2m_{\tilde{g}}E_t - m_{\tilde{t}_1}^2)(m_{\tilde{g}}^2 + m_t^2 - 2m_{\tilde{g}}E_t - m_{\tilde{t}_2}^2)}, \quad \text{(B.29a)}$$

$$\tilde{\phi}(m_{\tilde{g}}, m_{\tilde{t}_1}, m_{\tilde{t}_2}, m_{\tilde{Z}}) = \frac{1}{2}\pi^2 m_{\tilde{g}} m_{\tilde{Z}} \int \frac{dE_t}{m_{\tilde{g}}^2 + m_t^2 - 2m_{\tilde{g}}E_t - m_{\tilde{t}_1}^2}$$

$$\times \left[-[E_{\tilde{t}}(\max) - E_{\tilde{t}}(\min)] - \frac{m_{\tilde{Z}}^2 - m_t^2 + 2m_{\tilde{g}}E_t - m_{\tilde{t}_2}^2}{2m_{\tilde{g}}} \log Z(m_{\tilde{t}_2}) \right],$$

$$\text{(B.29b)}$$

$$\tilde{\chi}(m_{\tilde{g}}, m_{\tilde{t}_1}, m_{\tilde{t}_2}, m_{\tilde{Z}}) = \pi^2 m_{\tilde{g}} \int dE_t\, p_t E_t \frac{\lambda^{1/2}(m_{\tilde{g}}^2 + m_t^2 - 2m_{\tilde{g}}E_t, m_{\tilde{Z}}^2, m_t^2)}{m_{\tilde{g}}^2 + m_t^2 - 2m_{\tilde{g}}E_t}$$

$$\times \frac{1}{(m_{\tilde{g}}^2 + m_t^2 - 2m_{\tilde{g}}E_t - m_{\tilde{t}_1}^2)(m_{\tilde{g}}^2 + m_t^2 - 2m_{\tilde{g}}E_t - m_{\tilde{t}_2}^2)}, \quad \text{(B.29c)}$$

$$\xi(m_{\tilde{g}}, m_{\tilde{t}_1}, m_{\tilde{t}_2}, m_{\tilde{Z}}) = \frac{1}{2}\pi^2 \int \frac{dE_t}{m_{\tilde{g}}^2 + m_t^2 - 2m_{\tilde{g}}E_t - m_{\tilde{t}_1}^2}$$

$$\times \left[[E_{\tilde{t}}(\max) - E_{\tilde{t}}(\min)] - \frac{m_{\tilde{g}}^2 - m_t^2 - 2m_{\tilde{g}}E_t + m_{\tilde{t}_2}^2}{2m_{\tilde{g}}} \log Z(m_{\tilde{t}_2}) \right],$$

$$\text{(B.29d)}$$

$$\tilde{\rho}(m_{\tilde{g}}, m_{\tilde{t}_1}, m_{\tilde{t}_2}, m_{\tilde{Z}}) = -\frac{\pi^2}{2m_{\tilde{g}}} \int \frac{dE_t}{m_{\tilde{g}}^2 + m_t^2 - 2m_{\tilde{g}}E_t - m_{\tilde{t}_1}^2} \log Z(m_{\tilde{t}_2}),$$

$$\text{(B.29e)}$$

$$\zeta(m_{\tilde{g}}, m_{\tilde{t}_1}, m_{\tilde{t}_2}, m_{\tilde{Z}})$$
$$= \pi^2 \int \frac{dE_t [E_{\tilde{t}}(\max) - E_{\tilde{t}}(\min)]}{(m_{\tilde{g}}^2 + m_t^2 - 2m_{\tilde{g}}E_t - m_{\tilde{t}_1}^2)(m_{\tilde{g}}^2 + m_t^2 - 2m_{\tilde{g}}E_t - m_{\tilde{t}_2}^2)}, \quad \text{(B.29f)}$$

$$X(m_{\tilde{g}}, m_{\tilde{t}_1}, m_{\tilde{t}_2}, m_{\tilde{Z}}) = \frac{\pi^2}{2} \int dE_t\, p_t \frac{m_{\tilde{g}}^2 - m_{\tilde{Z}}^2 - 2m_{\tilde{g}}E_t}{m_{\tilde{g}}^2 + m_t^2 - 2m_{\tilde{g}}E_t}$$

$$\times \frac{\lambda^{1/2}(m_{\tilde{g}}^2 + m_t^2 - 2m_{\tilde{g}}E_t, m_{\tilde{Z}}^2, m_t^2)}{(m_{\tilde{g}}^2 + m_t^2 - 2m_{\tilde{g}}E_t - m_{\tilde{t}_1}^2)(m_{\tilde{g}}^2 + m_t^2 - 2m_{\tilde{g}}E_t - m_{\tilde{t}_2}^2)}, \quad \text{(B.29g)}$$

$$Y(m_{\tilde{g}}, m_{\tilde{t}_1}, m_{\tilde{t}_2}, m_{\tilde{Z}}) = \frac{\pi^2}{2} \int \frac{dE_t}{m_{\tilde{g}}^2 + m_t^2 - 2m_{\tilde{g}} E_t - m_{\tilde{t}_1}^2}$$

$$\times \left[[E_{\tilde{t}}(\max) - E_{\tilde{t}}(\min)](m_{\tilde{g}}^2 + m_t^2 - 2m_{\tilde{g}} E_t) \right.$$

$$\left. + \frac{1}{2m_{\tilde{g}}}(m_{\tilde{g}}^2 m_{\tilde{Z}}^2 - m_{\tilde{g}}^2 m_{\tilde{t}_2}^2 + m_t^4 + 2m_{\tilde{g}} E_t m_{\tilde{t}_2}^2 - m_{\tilde{t}_2}^2 m_t^2) \log Z(m_{\tilde{t}_2}) \right], \quad \text{(B.29h)}$$

$$\chi'(m_{\tilde{g}}, m_{\tilde{t}_1}, m_{\tilde{t}_2}, m_{\tilde{Z}}) = -\frac{\pi^2}{2} \int \frac{dE_t E_t}{m_{\tilde{g}}^2 + m_t^2 - 2m_{\tilde{g}} E_t - m_{\tilde{t}_2}^2} \log Z(m_{\tilde{t}_1}). \quad \text{(B.29i)}$$

The functions with three arguments ψ, χ, ϕ and ρ that appear in various expressions for $\Gamma_{ij}(\tilde{t}_1)$ and $\Gamma_{ij}(\tilde{t}_2)$ are simply the corresponding functions $\tilde{\psi}$, $\tilde{\chi}$, $\tilde{\phi}$ and $\tilde{\rho}$, but with the two top squark mass arguments being the same, i.e.

$$\psi(m_{\tilde{g}}, m_{\tilde{t}_1}, m_{\tilde{Z}_i}) = \tilde{\psi}(m_{\tilde{g}}, m_{\tilde{t}_1}, m_{\tilde{t}_1}, m_{\tilde{Z}_i}), \quad \text{etc.}$$

The limits of integration on E_t range from m_t to $(m_{\tilde{g}}^2 - 2m_t m_{\tilde{Z}} - m_{\tilde{Z}}^2)/2m_{\tilde{g}}$, and

$$Z(m) = \frac{m_{\tilde{g}}^2 + m_t^2 - 2m_{\tilde{g}} E_{\tilde{t}}(\max) - m^2}{m_{\tilde{g}}^2 + m_t^2 - 2m_{\tilde{g}} E_{\tilde{t}}(\min) - m^2} \quad \text{(B.30)}$$

and

$$E_{\tilde{t}}\binom{\max}{\min} = \frac{\zeta(m_{\tilde{g}} - E_t) \pm [p_t^2 \zeta^2 - 4p_t^2 m_t^2(m_{\tilde{g}}^2 + m_t^2 - 2m_{\tilde{g}} E_t)]^{1/2}}{2(m_{\tilde{g}}^2 + m_t^2 - 2m_{\tilde{g}} E_t)}, \quad \text{(B.31)}$$

where $\zeta = 2m_t^2 + m_{\tilde{g}}^2 - m_{\tilde{Z}}^2 - 2m_{\tilde{g}} E_t$.

The partial width $\Gamma(\tilde{g} \rightarrow b\bar{b}\tilde{Z}_i)$ can be obtained from the formula for $\Gamma(\tilde{g} \rightarrow t\bar{t}\tilde{Z}_i)$ by making the following substitutions:

$$m_{\tilde{t}_i} \rightarrow m_{\tilde{b}_i}, \quad \text{(B.32a)}$$

$$\tilde{A}_{\tilde{Z}_i}^t \rightarrow \tilde{A}_{\tilde{Z}_i}^b, \quad \text{(B.32b)}$$

$$\tilde{B}_{\tilde{Z}_i}^t \rightarrow \tilde{B}_{\tilde{Z}_i}^b, \quad \text{(B.32c)}$$

$$f_t \rightarrow f_b, \quad \text{(B.32d)}$$

$$v_1^{(i)} \rightarrow v_2^{(i)}, \quad \text{(B.32e)}$$

$$\theta_t \rightarrow \theta_b, \quad \text{(B.32f)}$$

$$m_t \rightarrow m_b, \quad \text{(B.32g)}$$

where,

$$\tilde{A}_{\tilde{Z}_i}^b = -\frac{g}{\sqrt{2}} v_3^{(i)} + \frac{g'}{3\sqrt{2}} v_4^{(i)}, \quad \text{(B.33a)}$$

$$\tilde{B}_{\tilde{Z}_i}^b = -\frac{2}{3} \frac{g'}{\sqrt{2}} v_4^{(i)}. \quad \text{(B.33b)}$$

B.1.4 $\tilde{g} \to \widetilde{W}_i t \bar{b}$ *decays*

These decays proceed through the exchange of each of the four top and bottom squark mass eigenstates. The formula given below includes effects from t and b Yukawa couplings as well as from intra-generation squark mixing, but with m_b ignored in the squared matrix element (though not in the phase space).

The partial width for the decay $\tilde{g} \to t\bar{b}\widetilde{W}_i^-$ can be written as

$$\Gamma(\tilde{g} \to t\bar{b}\widetilde{W}_i^-) = \frac{\alpha_s}{16\pi^2 m_{\tilde{g}}} \left(\Gamma_{\tilde{t}_1} + \Gamma_{\tilde{t}_2} + \Gamma_{\tilde{t}_1\tilde{t}_2} + \Gamma_{\tilde{b}_1} + \Gamma_{\tilde{b}_2} + \sum_{k,l=1}^{2} \Gamma_{\tilde{t}_k\tilde{b}_l} \right).$$

(B.34)

Note that in the limit $m_b \to 0$ the two sbottom exchange diagrams do not interfere with each other. The individual terms in Eq. (B.34) are given by:

$$\Gamma_{\tilde{t}_k} = \left[\left(\alpha_{\widetilde{W}_i}^{\tilde{t}_k}\right)^2 + \left(\beta_{\widetilde{W}_i}^{\tilde{t}_k}\right)^2 \right] \left[G_1(m_{\tilde{g}}, m_{\tilde{t}_k}, m_{\widetilde{W}_i}) \right.$$
$$\left. - (-1)^k \sin(2\theta_t) G_8(m_{\tilde{g}}, m_{\tilde{t}_k}, m_{\tilde{t}_k}, m_{\widetilde{W}_i}) \right],$$

(B.35a)

$$\Gamma_{\tilde{t}_1\tilde{t}_2} = -2 \left(\alpha_{\widetilde{W}_i}^{\tilde{t}_1} \alpha_{\widetilde{W}_i}^{\tilde{t}_2} + \beta_{\widetilde{W}_i}^{\tilde{t}_1} \beta_{\widetilde{W}_i}^{\tilde{t}_2} \right) \cos(2\theta_t) G_8(m_{\tilde{g}}, m_{\tilde{t}_1}, m_{\tilde{t}_2}, m_{\widetilde{W}_i}),$$

(B.35b)

$$\Gamma_{\tilde{b}_k} = \left[\left(\alpha_{\widetilde{W}_i}^{\tilde{b}_k}\right)^2 + \left(\beta_{\widetilde{W}_i}^{\tilde{b}_k}\right)^2 \right] G_2(m_{\tilde{g}}, m_{\tilde{b}_k}, m_{\widetilde{W}_i})$$
$$- \alpha_{\widetilde{W}_i}^{\tilde{b}_k} \beta_{\widetilde{W}_i}^{\tilde{b}_k} G_3(m_{\tilde{g}}, m_{\tilde{b}_k}, m_{\widetilde{W}_i}),$$

(B.35c)

$$\Gamma_{\tilde{t}_1\tilde{b}_1} = \left(\cos\theta_t \sin\theta_b \alpha_{\widetilde{W}_i}^{\tilde{b}_1} \beta_{\widetilde{W}_i}^{\tilde{t}_1} + \sin\theta_t \cos\theta_b \beta_{\widetilde{W}_i}^{\tilde{b}_1} \alpha_{\widetilde{W}_i}^{\tilde{t}_1} \right) G_6(m_{\tilde{g}}, m_{\tilde{t}_1}, m_{\tilde{b}_1}, m_{\widetilde{W}_i})$$
$$- \left(\cos\theta_t \cos\theta_b \alpha_{\widetilde{W}_i}^{\tilde{b}_1} \alpha_{\widetilde{W}_i}^{\tilde{t}_1} + \sin\theta_t \sin\theta_b \beta_{\widetilde{W}_i}^{\tilde{b}_1} \beta_{\widetilde{W}_i}^{\tilde{t}_1} \right) G_4(m_{\tilde{g}}, m_{\tilde{t}_1}, m_{\tilde{b}_1}, m_{\widetilde{W}_i})$$
$$+ \left(\cos\theta_t \cos\theta_b \beta_{\widetilde{W}_i}^{\tilde{b}_1} \alpha_{\widetilde{W}_i}^{\tilde{t}_1} + \sin\theta_t \sin\theta_b \alpha_{\widetilde{W}_i}^{\tilde{b}_1} \beta_{\widetilde{W}_i}^{\tilde{t}_1} \right) G_5(m_{\tilde{g}}, m_{\tilde{t}_1}, m_{\tilde{b}_1}, m_{\widetilde{W}_i})$$
$$- \left(\cos\theta_t \sin\theta_b \beta_{\widetilde{W}_i}^{\tilde{b}_1} \beta_{\widetilde{W}_i}^{\tilde{t}_1} + \sin\theta_t \cos\theta_b \alpha_{\widetilde{W}_i}^{\tilde{b}_1} \alpha_{\widetilde{W}_i}^{\tilde{t}_1} \right) G_7(m_{\tilde{g}}, m_{\tilde{t}_1}, m_{\tilde{b}_1}, m_{\widetilde{W}_i}).$$

(B.35d)

The couplings $\alpha_{\widetilde{W}_i}^{\tilde{t}_j}$ and $\beta_{\widetilde{W}_i}^{\tilde{t}_j}$ are given by,

$$\alpha_{\widetilde{W}_1}^{\tilde{t}_1} = -g \sin\gamma_R \cos\theta_t + f_t \cos\gamma_R \sin\theta_t,$$

(B.36a)

$$\beta_{\widetilde{W}_1}^{\tilde{t}_1} = -f_b \cos\gamma_L \cos\theta_t,$$

(B.36b)

$$\alpha_{\widetilde{W}_1}^{\tilde{b}_1} = -g \sin\gamma_L \cos\theta_b + f_b \cos\gamma_L \sin\theta_b,$$

(B.36c)

$$\beta_{\widetilde{W}_1}^{\tilde{b}_1} = -f_t \cos\gamma_R \cos\theta_b.$$

(B.36d)

The corresponding couplings for heavy sfermions $\tilde{f}_2$ ($f = t$, b) can be obtained from those above by replacing $\cos\theta_f \to \sin\theta_f$ and $\sin\theta_f \to -\cos\theta_f$. The couplings for heavy charginos $\widetilde{W}_2$ can be obtained from those above by replacing $\cos\gamma_{\mathrm{L,R}} \to -\theta_{x,y}\sin\gamma_{\mathrm{L,R}}$ and $\sin\gamma_{\mathrm{L,R}} \to \theta_{x,y}\cos\gamma_{\mathrm{L,R}}$. Finally, the other stop–sbottom interference terms can be obtained from (B.35d) by substituting appropriate couplings, squark masses, and squark mixing angle factors.

The eight functions that enter Eq. (B.35a)–(B.35d) are given by,

$$
G_1(m_{\tilde{g}}, m_{\tilde{t}}, m_{\widetilde{W}}) = m_{\tilde{g}} \int \frac{\mathrm{d}E_t\, p_t\, E_t \left(m_{\tilde{g}}^2 + m_t^2 - 2E_t m_{\tilde{g}} - m_{\widetilde{W}}^2\right)^2}{\left(m_{\tilde{g}}^2 + m_t^2 - 2E_t m_{\tilde{g}} - m_{\tilde{t}}^2\right)^2 \left(m_{\tilde{g}}^2 + m_t^2 - 2E_t m_{\tilde{g}}\right)},
\tag{B.37a}
$$

$$
G_2(m_{\tilde{g}}, m_{\tilde{b}}, m_{\widetilde{W}}) = m_{\tilde{g}} \int \mathrm{d}E_{\tilde{b}}\, E_{\tilde{b}}^2 \lambda^{1/2}(m_{\tilde{g}}^2 + m_b^2 - 2E_{\tilde{b}} m_{\tilde{g}}, m_{\widetilde{W}}^2, m_t^2)
$$
$$
\times \frac{m_{\tilde{g}}^2 + m_b^2 - m_t^2 - 2E_{\tilde{b}} m_{\tilde{g}} - m_{\widetilde{W}}^2}{\left(m_{\tilde{g}}^2 + m_b^2 - 2E_{\tilde{b}} m_{\tilde{g}} - m_{\tilde{b}}^2\right)^2 \left(m_{\tilde{g}}^2 + m_b^2 - 2E_{\tilde{b}} m_{\tilde{g}}\right)},
\tag{B.37b}
$$

$$
G_3(m_{\tilde{g}}, m_{\tilde{b}}, m_{\widetilde{W}}) = (-1)^{\theta_{\widetilde{W}}} \int \mathrm{d}E_{\tilde{b}}\, E_{\tilde{b}}^2 \lambda^{1/2}(m_{\tilde{g}}^2 + m_b^2 - 2E_{\tilde{b}} m_{\tilde{g}}, m_{\widetilde{W}}^2, m_t^2)
$$
$$
\times \frac{4 m_{\tilde{g}} m_{\widetilde{W}} m_t}{\left(m_{\tilde{g}}^2 + m_b^2 - 2E_{\tilde{b}} m_{\tilde{g}} - m_{\tilde{b}}^2\right)^2 \left(m_{\tilde{g}}^2 + m_b^2 - 2E_{\tilde{b}} m_{\tilde{g}}\right)},
\tag{B.37c}
$$

$$
G_4(m_{\tilde{g}}, m_{\tilde{t}}, m_{\tilde{b}}, m_{\widetilde{W}}) = (-1)^{\theta_{\tilde{g}} + \theta_{\widetilde{W}}} m_{\tilde{g}} m_{\widetilde{W}} \int \frac{\mathrm{d}E_t}{m_{\tilde{g}}^2 + m_t^2 - 2E_t m_{\tilde{g}} - m_{\tilde{t}}^2}
$$
$$
\times \left[E_{\tilde{b}}(\mathrm{max}) - E_{\tilde{b}}(\mathrm{min}) - \frac{m_{\tilde{b}}^2 + m_t^2 - 2E_t m_{\tilde{g}} - m_{\widetilde{W}}^2}{2 m_{\tilde{g}}} \log X \right],
\tag{B.37d}
$$

$$
G_5(m_{\tilde{g}}, m_{\tilde{t}}, m_{\tilde{b}}, m_{\widetilde{W}}) = (-1)^{\theta_{\tilde{g}}} \frac{m_t}{2} \int \mathrm{d}E_t\, \frac{m_{\tilde{g}}^2 + m_t^2 - 2E_t m_{\tilde{g}} - m_{\widetilde{W}}^2}{m_{\tilde{g}}^2 + m_t^2 - 2E_t m_{\tilde{g}} - m_{\tilde{t}}^2} \log X,
\tag{B.37e}
$$

$$
G_6(m_{\tilde{g}}, m_{\tilde{t}}, m_{\tilde{b}}, m_{\widetilde{W}})
$$
$$
= \frac{1}{2} \int \frac{\mathrm{d}E_t}{m_{\tilde{g}}^2 + m_t^2 - 2E_t m_{\tilde{g}} - m_{\tilde{t}}^2} \left\{ \left[m_{\tilde{g}} \left(m_{\tilde{g}}^2 + m_t^2 - 2E_t m_{\tilde{g}} - m_{\widetilde{W}}^2\right) \right. \right.
$$
$$
\left. \left. - \frac{m_{\tilde{b}}^2 - m_{\tilde{g}}^2}{m_{\tilde{g}}} \left(2E_t m_{\tilde{g}} - m_t^2 - m_{\tilde{g}}^2\right) \right] \log X \right.
$$
$$
\left. + 2 \left(2E_t m_{\tilde{g}} - m_t^2 - m_{\tilde{g}}^2\right) [E_{\tilde{b}}(\mathrm{max}) - E_{\tilde{b}}(\mathrm{min})] \right\},
\tag{B.37f}
$$

$$G_7(m_{\tilde{g}}, m_{\tilde{t}}, m_{\tilde{b}}, m_{\tilde{W}}) = (-1)^{\theta_{\tilde{W}}} \frac{1}{2} m_{\tilde{W}} m_t \int \frac{\mathrm{d}E_t}{m_{\tilde{g}}^2 + m_t^2 - 2E_t m_{\tilde{g}} - m_{\tilde{t}}^2}$$

$$\times \left\{ 2 \left[E_{\tilde{b}}(\max) - E_{\tilde{b}}(\min) \right] - \frac{m_{\tilde{b}}^2 - m_{\tilde{g}}^2}{m_{\tilde{g}}} \log X \right\}, \tag{B.37g}$$

$$G_8(m_{\tilde{g}}, m_{\tilde{t}_1}, m_{\tilde{t}_2}, m_{\tilde{W}})$$

$$= (-1)^{\theta_{\tilde{g}}} m_t m_{\tilde{g}} \int \mathrm{d}E_t \frac{\left(m_{\tilde{g}}^2 + m_t^2 - 2E_t m_{\tilde{g}} - m_{\tilde{W}}^2 \right) \left[E_{\tilde{b}}(\max) - E_{\tilde{b}}(\min) \right]}{\left(m_{\tilde{g}}^2 + m_t^2 - 2E_t m_{\tilde{g}} - m_{\tilde{t}_1}^2 \right) \left(m_{\tilde{g}}^2 + m_t^2 - 2E_t m_{\tilde{g}} - m_{\tilde{t}_2}^2 \right)}. \tag{B.37h}$$

The quantities $E_{\tilde{b}}(\min, \max)$, p_t and X in the functions for G_i are given by,

$$\frac{(m_{\tilde{g}}^2 + m_t^2 - 2m_{\tilde{g}}E_t + m_b^2 - m_{\tilde{W}}^2)(m_{\tilde{g}} - E_t) \mp p_t \lambda^{1/2}(m_{\tilde{g}}^2 + m_t^2 - 2m_{\tilde{g}}E_t, m_b^2, m_{\tilde{W}}^2)}{2 \left(m_{\tilde{g}}^2 + m_t^2 - 2E_t m_{\tilde{g}} \right)},$$

$$p_t = \sqrt{E_t^2 - m_t^2}, \quad \text{and}$$

$$X = \frac{m_{\tilde{b}}^2 + 2E_{\tilde{b}}(\max)m_{\tilde{g}} - m_{\tilde{g}}^2}{m_{\tilde{b}}^2 + 2E_{\tilde{b}}(\min)m_{\tilde{g}} - m_{\tilde{g}}^2}.$$

Finally, the limits of integration over E_t in Eq. (B.37a)–(B.37h) range from m_t to $\left(m_{\tilde{g}}^2 + m_t^2 - (m_{\tilde{W}} + m_b)^2 \right) / 2m_{\tilde{g}}$, while the integration over $E_{\tilde{b}}$ ranges from m_b to $[m_{\tilde{g}}^2 - (m_t + m_{\tilde{W}})^2]/2m_{\tilde{g}}$.

B.2 Squark decay widths

The general expression for the rate for squarks to decay to gluinos, including quark masses and intra-generation mixing is given by,

$$\Gamma(\tilde{q}_{1,2} \to q\tilde{g}) = \frac{2\alpha_s}{3} m_{\tilde{q}_{1,2}} \lambda^{1/2}(1, \frac{m_{\tilde{g}}^2}{m_{\tilde{q}_{1,2}}^2}, \frac{m_q^2}{m_{\tilde{q}_{1,2}}^2})$$

$$\times \left(1 - \frac{m_{\tilde{g}}^2}{m_{\tilde{q}_{1,2}}^2} - \frac{m_q^2}{m_{\tilde{q}_{1,2}}^2} \mp 2(-1)^{\theta_{\tilde{g}}} \sin(2\theta_q) \frac{m_q m_{\tilde{g}}}{m_{\tilde{q}_{1,2}}^2} \right). \tag{B.38a}$$

For the first two generations, the quark Yukawa coupling and the concomitant intra-generation mixing can be neglected, and this reduces to

$$\Gamma(\tilde{q}_i \to q\tilde{g}) = \frac{2\alpha_s}{3} m_{\tilde{q}_i} \left(1 - \frac{m_{\tilde{g}}^2}{m_{\tilde{q}_i}^2} - \frac{m_q^2}{m_{\tilde{q}_i}^2} \right) \lambda^{1/2}(1, \frac{m_{\tilde{g}}^2}{m_{\tilde{q}_i}^2}, \frac{m_q^2}{m_{\tilde{q}_i}^2}), \tag{B.38b}$$

where $i = L, R$.

The partial width for up-type squarks to decay to neutralinos including effects of Yukawa couplings and intra-generational mixing is given by,

$$\Gamma(\tilde{t}_1 \to t\widetilde{Z}_i) = \frac{m_{\tilde{t}_1}}{8\pi} \lambda^{1/2} \left(1, \frac{m_{\widetilde{Z}_i}^2}{m_{\tilde{t}_1}^2}, \frac{m_t^2}{m_{\tilde{t}_1}^2}\right)$$

$$\times \left\{ |a|^2 \left[1 - (\frac{m_t}{m_{\tilde{t}_1}} + \frac{m_{\widetilde{Z}_i}}{m_{\tilde{t}_1}})^2\right] + |b|^2 \left[1 - \left(\frac{m_t}{m_{\tilde{t}_1}} - \frac{m_{\widetilde{Z}_i}}{m_{\tilde{t}_1}}\right)^2\right] \right\},$$

$$\text{(B.39)}$$

where

$$a = \frac{1}{2} \{ [iA_{\widetilde{Z}_i}^t - (i)^{\theta_i} f_t v_1^{(i)}] \cos\theta_t - [iB_{\widetilde{Z}_i}^t - (-i)^{\theta_i} f_t v_1^{(i)}] \sin\theta_t \}, \quad \text{(B.40a)}$$

$$b = \frac{1}{2} \{ [-iA_{\widetilde{Z}_i}^t - (i)^{\theta_i} f_t v_1^{(i)}] \cos\theta_t - [iB_{\widetilde{Z}_i}^t + (-i)^{\theta_i} f_t v_1^{(i)}] \sin\theta_t \}. \quad \text{(B.40b)}$$

The formula for $\Gamma(\tilde{t}_2 \to t\widetilde{Z}_i)$ is the same, except that we must replace $m_{\tilde{t}_1} \to m_{\tilde{t}_2}$, and $\cos\theta_t \to \sin\theta_t$ and $\sin\theta_t \to -\cos\theta_t$ in the corresponding expressions for a and b.

The widths for the decays $\tilde{b}_i \to b\widetilde{Z}_i$ can be obtained from these by the substitutions,

$$m_{\tilde{t}_i} \to m_{\tilde{b}_i}, \quad \text{(B.41a)}$$

$$\tilde{A}_{\widetilde{Z}_i}^t \to \tilde{A}_{\widetilde{Z}_i}^b, \quad \text{(B.41b)}$$

$$\tilde{B}_{\widetilde{Z}_i}^t \to \tilde{B}_{\widetilde{Z}_i}^b, \quad \text{(B.41c)}$$

$$f_t \to f_b, \quad \text{(B.41d)}$$

$$v_1^{(i)} \to v_2^{(i)}, \quad \text{(B.41e)}$$

$$\theta_t \to \theta_b, \quad \text{(B.41f)}$$

$$m_t \to m_b. \quad \text{(B.41g)}$$

If quark Yukawa coupling effects are neglected (but quark masses retained), the partial widths for squark decays to neutralinos simplify to,

$$\Gamma(\tilde{u}_L \to u\widetilde{Z}_i) = \frac{|A_{\widetilde{Z}_i}^u|^2}{16\pi} m_{\tilde{u}_L} \left(1 - \frac{m_{\widetilde{Z}_i}^2}{m_{\tilde{u}_L}^2} - \frac{m_u^2}{m_{\tilde{u}_L}^2}\right) \lambda^{1/2}(1, \frac{m_{\widetilde{Z}_i}^2}{m_{\tilde{u}_L}^2}, \frac{m_u^2}{m_{\tilde{u}_L}^2}),$$

$$\text{(B.42a)}$$

$$\Gamma(\tilde{u}_R \to u\widetilde{Z}_i) = \frac{|B_{\widetilde{Z}_i}^u|^2}{16\pi} m_{\tilde{u}_R} \left(1 - \frac{m_{\widetilde{Z}_i}^2}{m_{\tilde{u}_R}^2} - \frac{m_u^2}{m_{\tilde{u}_R}^2}\right) \lambda^{1/2}(1, \frac{m_{\widetilde{Z}_i}^2}{m_{\tilde{u}_R}^2}, \frac{m_u^2}{m_{\tilde{u}_R}^2}),$$

$$\text{(B.42b)}$$

$$\Gamma(\tilde{d}_L \to d\tilde{Z}_i) = \frac{|A^d_{\tilde{Z}_i}|^2}{16\pi} m_{\tilde{d}_L} \left(1 - \frac{m^2_{\tilde{Z}_i}}{m^2_{\tilde{d}_L}} - \frac{m^2_d}{m^2_{\tilde{d}_L}}\right) \lambda^{1/2}(1, \frac{m^2_{\tilde{Z}_i}}{m^2_{\tilde{d}_L}}, \frac{m^2_d}{m^2_{\tilde{d}_L}}),$$

(B.42c)

$$\Gamma(\tilde{d}_R \to d\tilde{Z}_i) = \frac{|B^d_{\tilde{Z}_i}|^2}{16\pi} m_{\tilde{d}_R} \left(1 - \frac{m^2_{\tilde{Z}_i}}{m^2_{\tilde{d}_R}} - \frac{m^2_d}{m^2_{\tilde{d}_R}}\right) \lambda^{1/2}(1, \frac{m^2_{\tilde{Z}_i}}{m^2_{\tilde{d}_R}}, \frac{m^2_d}{m^2_{\tilde{d}_R}}).$$

(B.42d)

The rate for third generation squarks to decay to charginos, including Yukawa coupling effects is given by,

$$\Gamma(\tilde{t}_1 \to b\tilde{W}^+_i) = \frac{m_{\tilde{t}_1}}{16\pi} \lambda^{1/2}(1, \frac{m^2_{\tilde{W}_i}}{m^2_{\tilde{t}_1}}, \frac{m^2_b}{m^2_{\tilde{t}_1}})$$

$$\times \left[[(iA^d_{\tilde{W}_i}\cos\theta_t - B_{\tilde{W}_i}\sin\theta_t)^2 + B'^2_{\tilde{W}_i}\cos^2\theta_t] \left(1 - \frac{m^2_{\tilde{W}_i}}{m^2_{\tilde{t}_1}} - \frac{m^2_b}{m^2_{\tilde{t}_1}}\right) \right.$$

$$\left. - 4\frac{m_{\tilde{W}_i} m_b}{m^2_{\tilde{t}_1}} (iA^d_{\tilde{W}_i}\cos\theta_t - B_{\tilde{W}_i}\sin\theta_t)B'_{\tilde{W}_i}\cos\theta_t \right]$$

(B.43a)

for the lighter top squark, and

$$\Gamma(\tilde{b}_1 \to t\tilde{W}^-_i) = \frac{m_{\tilde{b}_1}}{16\pi} \lambda^{1/2}(1, \frac{m^2_{\tilde{W}_i}}{m^2_{\tilde{b}_1}}, \frac{m^2_t}{m^2_{\tilde{b}_1}})$$

$$\times \left[[(iA^u_{\tilde{W}_1}\cos\theta_b - B'_{\tilde{W}_i}\sin\theta_b)^2 + B^2_{\tilde{W}_i}\cos^2\theta_b](1 - \frac{m^2_{\tilde{W}_i}}{m^2_{\tilde{b}_1}} - \frac{m^2_t}{m^2_{\tilde{b}_1}}) \right.$$

$$\left. - 4\frac{m_{\tilde{W}_i} m_t}{m^2_{\tilde{b}_1}} (iA^u_{\tilde{W}_i}\cos\theta_b - B'_{\tilde{W}_i}\sin\theta_b)B_{\tilde{W}_i}\cos\theta_b \right],$$

(B.43b)

for bottom type squarks. The widths for the corresponding decays of $\tilde{t}_2$ and $\tilde{b}_2$ can be obtained from those for the lighter squarks by simply replacing,

$$m_{\tilde{q}_1} \to m_{\tilde{q}_2}, \quad \cos\theta_q \to \sin\theta_q, \quad \text{and} \quad \sin\theta_q \to -\cos\theta_q.$$

(B.43c)

The various couplings $A^f_{\tilde{W}_i}$, $B^f_{\tilde{W}_i}$, $B_{\tilde{W}_i}$, and $B'_{\tilde{W}_i}$ as well as those involving neutralino decays are as defined in Section 8.4.2.

Neglecting the couplings of the higgsino components as well as intra-generational mixing, we obtain simplified formulae for the widths:

$$\Gamma(\tilde{u}_L \to d\widetilde{W}_1^+) = \frac{g^2 \sin^2 \gamma_R}{16\pi} m_{\tilde{u}_L} \left(1 - \frac{m_{\widetilde{W}_1}^2}{m_{\tilde{u}_L}^2} - \frac{m_d^2}{m_{\tilde{u}_L}^2}\right) \lambda^{\frac{1}{2}}(1, \frac{m_{\widetilde{W}_1}^2}{m_{\tilde{u}_L}^2}, \frac{m_d^2}{m_{\tilde{u}_L}^2}),$$

(B.44a)

and

$$\Gamma(\tilde{d}_L \to u\widetilde{W}_1^-) = \frac{g^2 \sin^2 \gamma_L}{16\pi} m_{\tilde{d}_L} \left(1 - \frac{m_{\widetilde{W}_1}^2}{m_{\tilde{d}_L}^2} - \frac{m_d^2}{m_{\tilde{d}_L}^2}\right) \lambda^{\frac{1}{2}}(1, \frac{m_{\widetilde{W}_1}^2}{m_{\tilde{d}_L}^2}, \frac{m_u^2}{m_{\tilde{d}_L}^2}).$$

(B.44b)

The rates for $\tilde{u}_L \to d\widetilde{W}_2^+$ and $\tilde{d}_L \to u\widetilde{W}_2^-$ decays are the same, except that we must replace $m_{\widetilde{W}_1} \to m_{\widetilde{W}_2}$, $\sin^2 \gamma_R \to \cos^2 \gamma_R$, and $\sin^2 \gamma_L \to \cos^2 \gamma_L$.

For third generation squarks (for which we have included intragenerational mixing effects) additional two-body decays to $W^\pm$, Z or the various charged and neutral Higgs bosons may be possible. We write the partial widths for decays $\tilde{t}_i \to \tilde{b}_j W^+$ or $\tilde{b}_i \to \tilde{t}_j W^-$ as,

$$\Gamma(\tilde{q}_i \to \tilde{q}_f W) = \frac{g^2}{32\pi} \frac{1}{m_{\tilde{q}_i}^3} \frac{1}{M_W^2} \lambda^{\frac{3}{2}}(m_{\tilde{q}_i}^2, m_{\tilde{q}_f}^2, M_W^2)\Theta_i \Theta_f,$$

(B.45)

where Θ_i and Θ_f take into account intra-generational mixing for the initial and final squarks, $\tilde{q}_i$ and $\tilde{q}_f$, respectively. These factors are given by $\Theta_i = \cos^2 \theta_{t/b}$ ($\Theta_i = \sin^2 \theta_{t/b}$) if the parent squark is a lighter (heavier) stop/sbottom, and likewise for Θ_f. For instance, for the decay $\tilde{t}_2 \to \tilde{b}_1 W^+$, $\Theta_i = \sin^2 \theta_t$ and $\Theta_f = \cos^2 \theta_b$, etc.

The partial width for the decays $\tilde{q}_2 \to \tilde{q}_1 Z$ can be written as,

$$\Gamma(\tilde{q}_2 \to \tilde{q}_1 Z) = \frac{g^2}{64\pi} \frac{1}{\cos^2 \theta_W} \frac{1}{m_{\tilde{q}_2}^3 M_Z^2} \lambda^{\frac{3}{2}}(m_{\tilde{q}_2}^2, m_{\tilde{q}_1}^2, M_Z^2) \cos^2 \theta_q \sin^2 \theta_q. \quad (B.46)$$

Turning to the rates for squarks to decay to charged Higgs bosons we find,

$$\Gamma(\tilde{t}_i \to \tilde{b}_j H^+) = \frac{1}{16\pi} \frac{|A_{ij}|^2}{m_{\tilde{t}_i}^3} \lambda^{\frac{1}{2}}(m_{\tilde{t}_i}^2, m_{\tilde{b}_j}^2, m_{H^\pm}^2), \quad (B.47a)$$

with

$$A_{11} = \frac{g}{\sqrt{2}M_W} \{m_t m_b (\cot \beta + \tan \beta) \sin \theta_t \sin \theta_b$$

$$+ m_t(\mu + A_t \cot \beta) \sin \theta_t \cos \theta_b + m_b(\mu + A_b \tan \beta) \sin \theta_b \cos \theta_t$$

$$+ \left[(m_b^2 \tan \beta + m_t^2 \cot \beta) - M_W^2 \sin 2\beta\right] \cos \theta_t \cos \theta_b\}, \quad (B.47b)$$

and

$$A_{12} = A_{11}(\cos\theta_b \to \sin\theta_b, \sin\theta_b \to -\cos\theta_b). \tag{B.47c}$$

The couplings A_{2j} that enter the partial widths for $\tilde{t}_2 \to H^+\tilde{b}_j$ decays can be obtained by replacing $\cos\theta_t \to \sin\theta_t$ and $\sin\theta_t \to -\cos\theta_t$ in the coefficients A_{1j} listed above. Also,

$$\Gamma(\tilde{b}_i \to H^-\tilde{t}_j) = \frac{1}{16\pi} \frac{|A_{ji}|^2}{m_{\tilde{b}_i}^3} \lambda^{\frac{1}{2}}(m_{\tilde{b}_i}^2, m_{\tilde{t}_j}^2, m_{H^\pm}^2). \tag{B.48}$$

The partial widths for the decays $\tilde{q}_2 \to \tilde{q}_1\phi$ ($\phi = h$, H, or A) are given by,[1]

$$\Gamma(\tilde{t}_2 \to \tilde{t}_1\phi) = \frac{1}{16\pi} \frac{|A_\phi|^2}{m_{\tilde{t}_2}^3} \lambda^{\frac{1}{2}}(m_{\tilde{t}_2}^2, m_{\tilde{t}_1}^2, m_\phi^2), \tag{B.49a}$$

$$\Gamma(\tilde{b}_2 \to \tilde{b}_1\phi) = \frac{1}{16\pi} \frac{|B_\phi|^2}{m_{\tilde{b}_2}^3} \lambda^{\frac{1}{2}}(m_{\tilde{b}_2}^2, m_{\tilde{b}_1}^2, m_\phi^2), \tag{B.49b}$$

where

$$\begin{aligned} A_h &= \frac{gM_W}{4}\sin(\beta-\alpha)\left(1 - \frac{5}{3}\tan^2\theta_W\right)\sin 2\theta_t \\ &\quad + \frac{gm_t}{2M_W\sin\beta}\cos 2\theta_t\,(A_t\cos\alpha - \mu\sin\alpha), \end{aligned} \tag{B.50a}$$

$$\begin{aligned} A_H &= -\frac{gM_W}{4}\cos(\beta-\alpha)\left(1 - \frac{5}{3}\tan^2\theta_W\right)\sin 2\theta_t \\ &\quad + \frac{gm_t}{2M_W\sin\beta}\cos 2\theta_t\,(A_t\sin\alpha + \mu\cos\alpha), \end{aligned} \tag{B.50b}$$

$$A_A = -\mathrm{i}\frac{gm_t}{2M_W}(A_t\cot\beta + \mu), \tag{B.50c}$$

and

$$\begin{aligned} B_h &= \frac{gM_W}{4}\sin(\beta-\alpha)\left(-1 + \frac{1}{3}\tan^2\theta_W\right)\sin 2\theta_b \\ &\quad + \frac{gm_b}{2M_W\cos\beta}\cos 2\theta_b\,(A_b\sin\alpha - \mu\cos\alpha), \end{aligned} \tag{B.51a}$$

$$\begin{aligned} B_H &= -\frac{gM_W}{4}\cos(\beta-\alpha)\left(-1 + \frac{1}{3}\tan^2\theta_W\right)\sin 2\theta_b \\ &\quad + \frac{gm_b}{2M_W\cos\beta}\cos 2\theta_b\,(A_b\cos\alpha + \mu\sin\alpha), \end{aligned} \tag{B.51b}$$

---

[1] Although we write these for the third generation, it should be clear that these formulae also apply (with obvious changes) to the first two generations. Notice that in the absence of intra-generation mixing, these decays occur only via superpotential Yukawa interactions as pointed out in the exercise at the end of Section 13.2.

and

$$B_A = -\mathrm{i}\frac{gm_b}{2M_W}\left(A_b\tan\beta + \mu\right).$$ (B.51c)

Finally, if the decay to a gravitino is allowed, we find that (assuming that the gravitino is much lighter than the squark so that the decay rate can be well approximated by that to goldstinos, as discussed in the last section of Chapter 13)

$$\Gamma(\tilde{q}_i \to q\tilde{G}) = \frac{(m_{\tilde{q}}^2 - m_q^2)^4}{48\pi m_{\tilde{q}_i}^3 (M_{\mathrm{P}}m_{3/2})^2},$$ (B.52)

where M_{P} is the reduced Planck mass. Notice that there is no dependence on the squark mixing angle.

B.3 Slepton decay widths

We begin by listing the partial widths for various two-body decays of the first two generations of sleptons and sneutrinos for which intrageneration mixing is negligible. For left-slepton decay to neutralinos, we have ($\ell = e$ or μ),

$$\Gamma(\tilde{\ell}_{\mathrm{L}} \to \ell\widetilde{Z}_i) = \frac{|A^{\ell}_{\widetilde{Z}_i}|^2}{16\pi}m_{\tilde{\ell}_{\mathrm{L}}}\left(1 - \frac{m_{\widetilde{Z}_i}^2}{m_{\tilde{\ell}_{\mathrm{L}}}^2} - \frac{m_{\ell}^2}{m_{\tilde{\ell}_{\mathrm{L}}}^2}\right)\lambda^{1/2}(1, \frac{m_{\widetilde{Z}_i}^2}{m_{\tilde{\ell}_{\mathrm{L}}}^2}, \frac{m_{\ell}^2}{m_{\tilde{\ell}_{\mathrm{L}}}^2}),$$ (B.53a)

while for right-slepton decay to neutralinos, we have

$$\Gamma(\tilde{\ell}_{\mathrm{R}} \to \ell\widetilde{Z}_i) = \frac{|B^{\ell}_{\widetilde{Z}_i}|^2}{16\pi}m_{\tilde{\ell}_{\mathrm{R}}}\left(1 - \frac{m_{\widetilde{Z}_i}^2}{m_{\tilde{\ell}_{\mathrm{R}}}^2} - \frac{m_{\ell}^2}{m_{\tilde{\ell}_{\mathrm{R}}}^2}\right)\lambda^{1/2}(1, \frac{m_{\widetilde{Z}_i}^2}{m_{\tilde{\ell}_{\mathrm{R}}}^2}, \frac{m_{\ell}^2}{m_{\tilde{\ell}_{\mathrm{R}}}^2}).$$ (B.53b)

The partial width for sneutrino decay to a neutralino is,

$$\Gamma(\tilde{\nu}_{\ell} \to \nu_{\ell}\widetilde{Z}_i) = \frac{|A^{\nu}_{\widetilde{Z}_i}|^2}{16\pi}m_{\tilde{\nu}_{\ell}}\left(1 - \frac{m_{\widetilde{Z}_i}^2}{m_{\tilde{\nu}_{\ell}}^2}\right)^2.$$ (B.53c)

For slepton decay to charginos, we have

$$\Gamma(\tilde{\ell}_{\mathrm{L}} \to \nu_{\ell}\widetilde{W}_i^-) = \frac{g^2\sin^2\gamma_{\mathrm{L}}}{16\pi}m_{\tilde{\ell}_{\mathrm{L}}}\left(1 - \frac{m_{\widetilde{W}_i}^2}{m_{\tilde{\ell}_{\mathrm{L}}}^2}\right)^2,$$ (B.54a)

while for sneutrino decay, we have

$$\Gamma(\tilde{\nu}_{\ell} \to \ell\widetilde{W}_i^+) = \frac{g^2\sin^2\gamma_{\mathrm{R}}}{16\pi}m_{\tilde{\nu}_{\ell}}\left(1 - \frac{m_{\widetilde{W}_i}^2}{m_{\tilde{\nu}_{\ell}}^2} - \frac{m_{\ell}^2}{m_{\tilde{\nu}_{\ell}}^2}\right)\lambda^{1/2}(1, \frac{m_{\widetilde{W}_i}^2}{m_{\tilde{\nu}_{\ell}}^2}, \frac{m_{\ell}^2}{m_{\tilde{\nu}_{\ell}}^2}).$$

(B.54b)

Yukawa coupling effects can be important for decays of third generation sleptons and sneutrinos. For $\tilde{\tau}_1 \to \tau \widetilde{Z}_i$, we have

$$
\Gamma(\tilde{\tau}_1 \to \tau \widetilde{Z}_i) = \frac{m_{\tilde{\tau}_1}}{8\pi} \lambda^{1/2}(1, \frac{m_{\widetilde{Z}_i}^2}{m_{\tilde{\tau}_1}^2}, \frac{m_\tau^2}{m_{\tilde{\tau}_1}^2})
$$

$$
\times \left\{ |a|^2 \left[1 - (\frac{m_\tau}{m_{\tilde{\tau}_1}} + \frac{m_{\widetilde{Z}_i}}{m_{\tilde{\tau}_1}})^2 \right] + |b|^2 \left[1 - (\frac{m_\tau}{m_{\tilde{\tau}_1}} - \frac{m_{\widetilde{Z}_i}}{m_{\tilde{\tau}_1}})^2 \right] \right\}, \tag{B.55a}
$$

with

$$
a = \frac{1}{2} \left\{ [iA_{\widetilde{Z}_i}^\tau - (i)^{\theta_i} f_\tau v_2^{(i)}] \cos\theta_\tau - [iB_{\widetilde{Z}_i}^\tau - (-i)^{\theta_i} f_\tau v_2^{(i)}] \sin\theta_\tau \right\},
$$

$$
b = \frac{1}{2} \left\{ [-iA_{\widetilde{Z}_i}^\tau - (i)^{\theta_i} f_\tau v_2^{(i)}] \cos\theta_\tau - [iB_{\widetilde{Z}_i}^\tau + (-i)^{\theta_i} f_\tau v_2^{(i)}] \sin\theta_\tau \right\}.
$$

The formula for $\tilde{\tau}_2 \to \tau \widetilde{Z}_i$ is the same, except that we must replace $m_{\tilde{\tau}_1} \to m_{\tilde{\tau}_2}$, $\cos\theta_\tau \to \sin\theta_\tau$ and $\sin\theta_\tau \to -\cos\theta_\tau$.

For stau decays to charginos, we have

$$
\Gamma(\tilde{\tau}_1 \to \nu_\tau \widetilde{W}_i^-) = \frac{|iA_{\widetilde{W}_i}^\nu \cos\theta_\tau - B_{\widetilde{W}_i}'' \sin\theta_\tau|^2}{16\pi} m_{\tilde{\tau}_1} \left(1 - \frac{m_{\widetilde{W}_i}^2}{m_{\tilde{\tau}_1}^2} \right)^2, \tag{B.55b}
$$

where $B_{\widetilde{W}_1}'' = -f_\tau \cos\gamma_L$ and $B_{\widetilde{W}_2}'' = f_\tau \theta_x \sin\gamma_L$. Also,

$$
\Gamma(\tilde{\tau}_2 \to \nu_\tau \widetilde{W}_i^-) = \frac{|iA_{\widetilde{W}_i}^\nu \sin\theta_\tau + B_{\widetilde{W}_i}'' \cos\theta_\tau|^2}{16\pi} m_{\tilde{\tau}_2} \left(1 - \frac{m_{\widetilde{W}_i}^2}{m_{\tilde{\tau}_2}^2} \right)^2. \tag{B.55c}
$$

Finally,

$$
\Gamma(\tilde{\nu}_\tau \to \tau \widetilde{W}_i^+) = \frac{m_{\tilde{\nu}_\tau}}{16\pi} \lambda^{1/2}(1, \frac{m_{\widetilde{W}_i}^2}{m_{\tilde{\nu}_\tau}^2}, \frac{m_\tau^2}{m_{\tilde{\nu}_\tau}^2}) \left\{ \left[|A_{\widetilde{W}_i}^\tau|^2 + B_{\widetilde{W}_i}''^2 \right] \right.
$$

$$
\left. \times (1 - \frac{m_{\widetilde{W}_i}^2}{m_{\tilde{\nu}_\tau}^2} - \frac{m_\tau^2}{m_{\tilde{\nu}_\tau}^2}) - 4\frac{m_{\widetilde{W}_i} m_\tau}{m_{\tilde{\nu}_\tau}^2} B_{\widetilde{W}_i}''(iA_{\widetilde{W}_i}^\tau) \right\}. \tag{B.55d}
$$

Turning to the decays to gauge bosons, we have

$$
\Gamma(\tilde{\tau}_2 \to \tilde{\nu}_\tau W) = \frac{g^2 \sin^2\theta_\tau}{32\pi m_{\tilde{\tau}_2}^3 M_W^2} \lambda^{3/2}(m_{\tilde{\tau}_2}^2, m_{\tilde{\nu}_\tau}^2, M_W^2), \tag{B.56a}
$$

$$\Gamma(\tilde{\nu}_\tau \to \tilde{\tau}_1 W) = \frac{g^2 \cos^2 \theta_\tau}{32\pi m_{\tilde{\nu}_\tau}^3 M_W^2} \lambda^{3/2}(m_{\tilde{\nu}_\tau}^2, m_{\tilde{\tau}_1}^2, M_W^2), \tag{B.56b}$$

$$\Gamma(\tilde{\nu}_\tau \to \tilde{\tau}_2 W) = \frac{g^2 \sin^2 \theta_\tau}{32\pi m_{\tilde{\nu}_\tau}^3 M_W^2} \lambda^{3/2}(m_{\tilde{\nu}_\tau}^2, m_{\tilde{\tau}_2}^2, M_W^2), \tag{B.56c}$$

and

$$\Gamma(\tilde{\tau}_2 \to \tilde{\tau}_1 Z) = \frac{g^2 \cos^2 \theta_\tau \sin^2 \theta_\tau}{64\pi \cos^2 \theta_W m_{\tilde{\tau}_2}^3 M_Z^2} \lambda^{3/2}(m_{\tilde{\tau}_2}^2, m_{\tilde{\tau}_1}^2, M_Z^2). \tag{B.56d}$$

Third generation sleptons may also decay with significant rates to Higgs bosons. The partial widths for decays to charged Higgs bosons are given by,

$$\Gamma(\tilde{\nu}_\tau \to \tilde{\tau}_i H^+) = \frac{|A|^2}{16\pi m_{\tilde{\nu}_\tau}^3} \lambda^{1/2}(m_{\tilde{\nu}_\tau}^2, m_{\tilde{\tau}_i}^2, m_{H^+}^2), \tag{B.57a}$$

with

$$A(\tilde{\nu}_\tau \to \tilde{\tau}_1 H^+) = \frac{g}{\sqrt{2}M_W} \left\{ \left[m_\tau^2 \tan\beta - M_W^2 \sin 2\beta \right] \cos\theta_\tau \right.$$
$$\left. + m_\tau \left[\mu + A_\tau \tan\beta \right] \sin\theta_\tau \right\} \tag{B.57b}$$

and

$$A(\tilde{\nu}_\tau \to \tilde{\tau}_2 H^+) = \frac{g}{\sqrt{2}M_W} \left\{ \left[m_\tau^2 \tan\beta - M_W^2 \sin 2\beta \right] \sin\theta_\tau \right.$$
$$\left. - m_\tau \left[\mu + A_\tau \tan\beta \right] \cos\theta_\tau \right\}. \tag{B.57c}$$

Finally,

$$\Gamma(\tilde{\tau}_i \to \tilde{\nu}_\tau H^-) = \frac{|A|^2}{16\pi m_{\tilde{\tau}_i}^3} \lambda^{1/2}(m_{\tilde{\tau}_i}^2, m_{\tilde{\nu}_\tau}^2, m_{H^-}^2), \tag{B.58a}$$

with

$$A(\tilde{\tau}_i \to \tilde{\nu}_\tau H^-) = A(\tilde{\nu}_\tau \to \tilde{\tau}_i H^+). \tag{B.58b}$$

For stau decays to neutral Higgs bosons $\phi = h, H, A$, we find

$$\Gamma(\tilde{\tau}_2 \to \tilde{\tau}_1 \phi) = \frac{|A_\phi|^2}{16\pi m_{\tilde{\tau}_2}^3} \lambda^{1/2}(m_{\tilde{\tau}_2}^2, m_{\tilde{\tau}_1}^2, m_\phi^2), \tag{B.59a}$$

with

$$A_h = \frac{gM_W}{4} \sin(\beta - \alpha) \sin 2\theta_\tau [-1 + 3\tan^2 \theta_W]$$
$$+ \frac{gm_\tau}{2M_W \cos\beta} \cos 2\theta_\tau [-\mu \cos\alpha + A_\tau \sin\alpha], \tag{B.59b}$$

$$A_H = \frac{-g M_W}{4} \cos(\beta - \alpha) \sin 2\theta_\tau [-1 + 3 \tan^2 \theta_W]$$

$$+ \frac{g m_\tau}{2 M_W \cos \beta} \cos 2\theta_\tau [\mu \sin \alpha + A_\tau \cos \alpha], \tag{B.59c}$$

and

$$A_A = \frac{-i g m_\tau}{2 M_W}(\mu + A_\tau \tan \beta). \tag{B.59d}$$

Finally, if the decay to a gravitino is allowed, we find that

$$\Gamma(\tilde{\ell} \to \ell \tilde{G}) = \frac{(m_{\tilde{\ell}}^2 - m_\ell^2)^4}{48\pi m_{\tilde{\ell}}^3 (M_P m_{3/2})^2}, \tag{B.60}$$

again assuming that the goldstino approximation used for $\tilde{q}_i \to q \tilde{G}$ decays is valid. Here, $\tilde{\ell}$ denotes any of the sleptons or sneutrinos. Notice that there is no dependence on the slepton mixing angle.

B.4 Neutralino decay widths

B.4.1 Two-body decays

We list the partial widths for two-body decays of the neutralino, beginning with their decays to gauge bosons.

$$\Gamma(\tilde{Z}_i \to \tilde{W}_j^- W^+) = \frac{g^2}{16\pi m_{\tilde{Z}_i}^3} \lambda^{1/2}(m_{\tilde{Z}_i}^2, m_{\tilde{W}_j}^2, M_W^2)$$

$$\times \left[(X_j^{i2} + Y_j^{i2}) \left(m_{\tilde{Z}_i}^2 + m_{\tilde{W}_j}^2 - M_W^2 + \frac{(m_{\tilde{Z}_i}^2 - m_{\tilde{W}_j}^2)^2 - M_W^4}{M_W^2} \right) \right.$$

$$\left. - 6 m_{\tilde{Z}_i} m_{\tilde{W}_j} (X_j^{i2} - Y_j^{i2}) \right], \tag{B.61a}$$

where the couplings X_j^i and Y_j^i are given in Eq. (8.103a) and (8.103b). Also,

$$\Gamma(\tilde{Z}_i \to \tilde{Z}_j Z) = \frac{|W_{ij}|^2}{4\pi m_{\tilde{Z}_i}^3} \lambda^{\frac{1}{2}}(m_{\tilde{Z}_i}^2, m_{\tilde{Z}_j}^2, M_Z^2)$$

$$\times \left[(m_{\tilde{Z}_i}^2 + m_{\tilde{Z}_j}^2 - M_Z^2) + \frac{(m_{\tilde{Z}_i}^2 - m_{\tilde{Z}_j}^2)^2 - M_Z^4}{M_Z^2} + 6(-1)^{\theta_i}(-1)^{\theta_j} m_{\tilde{Z}_i} m_{\tilde{Z}_j} \right], \tag{B.61b}$$

with W_{ij} as defined in (8.101).

Turning to neutralino decays to Higgs bosons, we have

$$\Gamma(\tilde{Z}_i \to \tilde{W}_j^- H^+) = \Gamma(\tilde{Z}_i \to \tilde{W}_j^+ H^-) = \frac{\lambda^{1/2}(m_{\tilde{Z}_i}^2, m_{\tilde{W}_j}^2, m_{H^+}^2)}{16\pi m_{\tilde{Z}_i}^3}$$

$$\times \left[(a_j^2 + b_j^2)(m_{\tilde{Z}_i}^2 + m_{\tilde{W}_j}^2 - m_{H^+}^2) + 2(a_j^2 - b_j^2)m_{\tilde{Z}_i}m_{\tilde{W}_j} \right], \quad \text{(B.62)}$$

where

$$a_1 = \frac{1}{2} \left((-1)^{\theta_{\tilde{W}_1}} \cos\beta A_2^{(i)} - (-1)^{\theta_i} \sin\beta A_4^{(i)} \right), \quad \text{(B.63a)}$$

and

$$b_1 = \frac{1}{2} \left((-1)^{\theta_{\tilde{W}_1}} \cos\beta A_2^{(i)} + (-1)^{\theta_i} \sin\beta A_4^{(i)} \right). \quad \text{(B.63b)}$$

To obtain a_2 and b_2, replace $A_2^{(i)} \to \theta_y A_1^{(i)}$ and $A_4^{(i)} \to \theta_x A_3^{(i)}$ in the expressions for a_1 and b_1. The coefficients $A_j^{(i)}$ are given in Eq. (8.122a)–(8.122d).

For the partial width of the decay $\tilde{Z}_i \to \tilde{Z}_j h$, we have

$$\Gamma(\tilde{Z}_i \to \tilde{Z}_j h) = \frac{(X_{ij}^h + X_{ji}^h)^2}{16\pi m_{\tilde{Z}_i}^3} \lambda^{\frac{1}{2}}(m_{\tilde{Z}_i}^2, m_{\tilde{Z}_j}^2, m_h^2)$$

$$\times \left[(m_{\tilde{Z}_i}^2 + m_{\tilde{Z}_j}^2 - m_h^2) + 2(-1)^{\theta_i + \theta_j} m_{\tilde{Z}_i} m_{\tilde{Z}_j} \right], \quad \text{(B.64)}$$

with the couplings X_{ij}^h as given in Eq. (8.117). The same formula with the replacements $m_h \to m_H$ and $X_{ij}^h \to X_{ij}^H$ yields $\Gamma(\tilde{Z}_i \to \tilde{Z}_j H)$. This formula also applies to $\tilde{Z}_i \to \tilde{Z}_j A$ if $m_h \to m_A$, $X_{ij}^h \to X_{ij}^A$ and the sign of the second term (proportional to the product of the neutralino masses) in the square brackets is flipped.

For $\tilde{Z}_i$ decays to fermion–sfermion pairs, we have, including effects of Yukawa couplings and intra-generational mixing,

$$\Gamma(\tilde{Z}_i \to f\bar{\tilde{f}}_k) = N_c \frac{\lambda^{1/2}(m_{\tilde{Z}_i}^2, m_{\tilde{f}_k}^2, m_f^2)}{16\pi m_{\tilde{Z}_i}^3} \left(|a|^2 \left[(m_{\tilde{Z}_i} + m_f)^2 - m_{\tilde{f}_k}^2 \right] \right.$$

$$\left. + |b|^2 \left[(m_{\tilde{Z}_i} - m_f)^2 - m_{\tilde{f}_k}^2 \right] \right), \quad \text{(B.65)}$$

where $k = 1, 2$ and the color factor $N_c = 3$ if $f = q$, and $N_c = 1$ if $f = \ell$ or ν. The coefficients a and b are exactly the same as those that enter the decays $\tilde{f}_k \to f\tilde{Z}_i$, and may be found in Eq. (B.40a)–(B.41g) for squarks, or directly below (B.55a) for sleptons. If intra-generational mixing can be neglected, these reduce to,

$$\Gamma(\tilde{Z}_i \to \tilde{f}_L) = \Gamma(\tilde{Z}_i \to \bar{f}\tilde{f}_L)$$

$$= \frac{N_c |A_{\tilde{Z}_i}^f|^2}{32\pi m_{\tilde{Z}_i}^3} \lambda^{\frac{1}{2}}(m_{\tilde{Z}_i}^2, m_{\tilde{f}_L}^2, m_f^2)\left(m_{\tilde{Z}_i}^2 + m_f^2 - m_{\tilde{f}_L}^2\right). \quad (B.66)$$

The corresponding widths for decays to $\tilde{f}_R$ may be obtained from this by replacing $A_{\tilde{Z}_i}^f \to B_{\tilde{Z}_i}^f$, and $m_{\tilde{f}_L} \to m_{\tilde{f}_R}$.

The widths for two-body decay to longitudinal gravitinos, which are essentially goldstinos in the limit that $m_{3/2}$ is much smaller than the mass of the decaying sparticle, are given by

$$\Gamma(\tilde{Z}_i \to \tilde{G}\gamma) = \frac{(v_4^{(i)}\cos\theta_W + v_3^{(i)}\sin\theta_W)^2}{48\pi m_{3/2}^2 M_P^2} m_{\tilde{Z}_i}^5, \quad (B.67)$$

$$\Gamma(\tilde{Z}_i \to \tilde{G}Z)$$
$$= \frac{2(v_4^{(i)}\sin\theta_W - v_3^{(i)}\cos\theta_W)^2 + (v_1^{(i)}\sin\beta - v_2^{(i)}\cos\beta)^2}{96\pi m_{3/2}^2 M_P^2 m_{\tilde{Z}_i}^3}(m_{\tilde{Z}_i}^2 - M_Z^2)^4, \quad (B.68)$$

and

$$\Gamma(\tilde{Z}_i \to \tilde{G}\phi) = \frac{|\kappa_\phi|^2}{16\pi m_{\tilde{Z}_i}^3}\left(m_{\tilde{Z}_i}^2 - m_\phi^2\right)^4, \quad (B.69a)$$

where $\phi = h, H$ or A, and

$$\kappa_h = -\frac{(i)^{\theta_i+1}}{\sqrt{6}M_P m_{3/2}}[v_1^{(i)}\cos\alpha + v_2^{(i)}\sin\alpha], \quad (B.69b)$$

$$\kappa_H = -\frac{(i)^{\theta_i+1}}{\sqrt{6}M_P m_{3/2}}[-v_1^{(i)}\sin\alpha + v_2^{(i)}\cos\alpha], \quad \text{and} \quad (B.69c)$$

$$\kappa_A = -\frac{(i)^{\theta_i+2}}{\sqrt{6}M_P m_{3/2}}[v_1^{(i)}\cos\beta + v_2^{(i)}\sin\beta], \quad (B.69d)$$

as in Chapter 13 of the text. In deriving the decay rates to gravitinos, we have neglected the gravitino mass, except of course in the coupling of the goldstino.

B.4.2 $\tilde{Z}_i \to \tilde{Z}_j f\bar{f}$ decays

Here we present the partial width for neutralino three-body decays $\tilde{Z}_i \to \tilde{Z}_j f\bar{f}$, where f is a SM fermion. We neglect SM fermion masses in the evaluation of final state spin sums that have to be performed after squaring the matrix element, but retain these in the kinematics. This would be a poor approximation for $\widetilde{Z}_i \to t\bar{t}\widetilde{Z}_j$. However, when this decay is kinematically accessible, so are the two-body decay modes $\widetilde{Z}_i \to \widetilde{Z}_j Z$ and $\widetilde{Z}_i \to \widetilde{Z}_j h$: these two-body decays dominate the branching fraction of the neutralino, and the inapplicability of our approximation becomes essentially irrelevant.

This decay proceeds via the exchange of the two sfermion mass eigenstates $\tilde{f}_{1,2}$ or their antiparticles, via the exchange of a Z boson, or via the exchange of one of the three neutral Higgs bosons of the MSSM. The partial width can therefore be written as

$$\Gamma(\widetilde{Z}_i \to \widetilde{Z}_j f\bar{f}) = \frac{1}{2} N_{\mathrm{c}}(f) \frac{1}{(2\pi)^5} \frac{1}{2m_{\widetilde{Z}_j}}$$
$$\times \left(\Gamma_{\tilde{f}} + \Gamma_Z + \Gamma_{h,H} + \Gamma_A + \Gamma_{Z\tilde{f}} + \Gamma_{\phi\tilde{f}} \right), \quad \text{(B.70)}$$

where the color factor $N_{\mathrm{c}}(f) = 3\,(1)$ for $f = b\,(\tau)$. The Higgs and Z exchange diagrams do not interfere with each other in the approximation that the spin sums are evaluated with $m_f = 0$. For decays to the first two generations of fermions, the Higgs exchange contributions are also negligible.

The pure sfermion exchange contribution is given by

$$\Gamma_{\tilde{f}} = \Gamma_{\tilde{f}_1} + \Gamma_{\tilde{f}_2} + \Gamma_{\tilde{f}_{1,2}}, \quad \text{(B.71)}$$

where

$$\Gamma_{\tilde{f}_k} = \Gamma_{\mathrm{LL}}^{\tilde{f}_k} + \Gamma_{\mathrm{RR}}^{\tilde{f}_k} + \Gamma_{\mathrm{LR}}^{\tilde{f}_k} \quad (k = 1, 2), \quad \text{(B.72a)}$$
$$\Gamma_{\tilde{f}_{1,2}} = \Gamma_{\mathrm{L}}^{\tilde{f}_1} \Gamma_{\mathrm{L}}^{\tilde{f}_2} + \Gamma_{\mathrm{L}}^{\tilde{f}_1} \Gamma_{\mathrm{R}}^{\tilde{f}_2} + \Gamma_{\mathrm{R}}^{\tilde{f}_1} \Gamma_{\mathrm{L}}^{\tilde{f}_2} + \Gamma_{\mathrm{R}}^{\tilde{f}_1} \Gamma_{\mathrm{R}}^{\tilde{f}_2}. \quad \text{(B.72b)}$$

Here, the subscripts L and R refer to the chirality of the SM fermion coupling to the heavier neutralino $\widetilde{Z}_j$. The quantities appearing in Eq. (B.72a) and (B.72b) are:

$$\Gamma_{\mathrm{LL}}^{\tilde{f}_k} = 4 \left(\alpha_{\widetilde{Z}_i}^{\tilde{f}_k} \right)^2 \left\{ \left[\left(\alpha_{\widetilde{Z}_j}^{\tilde{f}_k} \right)^2 + \left(\beta_{\widetilde{Z}_j}^{\tilde{f}_k} \right)^2 \right] \psi(m_{\widetilde{Z}_i}, m_{\tilde{f}_k}, m_{\widetilde{Z}_j}) \right.$$
$$\left. + (-1)^{\theta_i + \theta_j} \left(\alpha_{\widetilde{Z}_j}^{\tilde{f}_k} \right)^2 \phi(m_{\widetilde{Z}_i}, m_{\tilde{f}_k}, m_{\widetilde{Z}_j}) \right\}; \quad \text{(B.73a)}$$

$$\Gamma_{\mathrm{RR}}^{\tilde{f}_k} = 4 \left(\beta_{\widetilde{Z}_i}^{\tilde{f}_k} \right)^2 \left\{ \left[\left(\alpha_{\widetilde{Z}_j}^{\tilde{f}_k} \right)^2 + \left(\beta_{\widetilde{Z}_j}^{\tilde{f}_k} \right)^2 \right] \psi(m_{\widetilde{Z}_i}, m_{\tilde{f}_k}, m_{\widetilde{Z}_j}) \right.$$
$$\left. + (-1)^{\theta_i + \theta_j} \left(\beta_{\widetilde{Z}_j}^{\tilde{f}_k} \right)^2 \phi(m_{\widetilde{Z}_i}, m_{\tilde{f}_k}, m_{\widetilde{Z}_j}) \right\}; \quad \text{(B.73b)}$$

$$\Gamma_{LR}^{\tilde{f}_k} = -8\alpha_{\tilde{Z}_j}^{\tilde{f}_k}\beta_{\tilde{Z}_j}^{\tilde{f}_k}\alpha_{\tilde{Z}_i}^{\tilde{f}_k}\beta_{\tilde{Z}_i}^{\tilde{f}_k}Y(m_{\tilde{Z}_i}, m_{\tilde{f}_k}, m_{\tilde{f}_k}, m_{\tilde{Z}_j}); \tag{B.73c}$$

$$\Gamma_L^{\tilde{f}_1}\Gamma_L^{\tilde{f}_2} = 8\alpha_{\tilde{Z}_i}^{\tilde{f}_1}\alpha_{\tilde{Z}_i}^{\tilde{f}_2}\left\{\left[\alpha_{\tilde{Z}_j}^{\tilde{f}_1}\alpha_{\tilde{Z}_j}^{\tilde{f}_2} + \beta_{\tilde{Z}_j}^{\tilde{f}_1}\beta_{\tilde{Z}_j}^{\tilde{f}_2}\right]\tilde{\psi}(m_{\tilde{Z}_i}, m_{\tilde{f}_1}, m_{\tilde{f}_2}, m_{\tilde{Z}_j})\right.$$
$$\left. + (-1)^{\theta_i+\theta_j}\alpha_{\tilde{Z}_j}^{\tilde{f}_1}\alpha_{\tilde{Z}_j}^{\tilde{f}_2}\tilde{\phi}(m_{\tilde{Z}_i}, m_{\tilde{f}_1}, m_{\tilde{f}_2}, m_{\tilde{Z}_j})\right\}; \tag{B.73d}$$

$$\Gamma_R^{\tilde{f}_1}\Gamma_R^{\tilde{f}_2} = 8\beta_{\tilde{Z}_i}^{\tilde{f}_1}\beta_{\tilde{Z}_i}^{\tilde{f}_2}\left\{\left[\alpha_{\tilde{Z}_j}^{\tilde{f}_1}\alpha_{\tilde{Z}_j}^{\tilde{f}_2} + \beta_{\tilde{Z}_j}^{\tilde{f}_1}\beta_{\tilde{Z}_j}^{\tilde{f}_2}\right]\tilde{\psi}(m_{\tilde{Z}_i}, m_{\tilde{f}_1}, m_{\tilde{f}_2}, m_{\tilde{Z}_j})\right.$$
$$\left. + (-1)^{\theta_i+\theta_j}\beta_{\tilde{Z}_j}^{\tilde{f}_1}\beta_{\tilde{Z}_j}^{\tilde{f}_2}\tilde{\phi}(m_{\tilde{Z}_i}, m_{\tilde{f}_1}, m_{\tilde{f}_2}, m_{\tilde{Z}_j})\right\}; \tag{B.73e}$$

$$\Gamma_L^{\tilde{f}_1}\Gamma_R^{\tilde{f}_2} = -8\alpha_{\tilde{Z}_i}^{\tilde{f}_1}\beta_{\tilde{Z}_i}^{\tilde{f}_2}\alpha_{\tilde{Z}_j}^{\tilde{f}_2}\beta_{\tilde{Z}_j}^{\tilde{f}_1}Y(m_{\tilde{Z}_i}, m_{\tilde{f}_1}, m_{\tilde{f}_2}, m_{\tilde{Z}_j}); \tag{B.73f}$$

$$\Gamma_L^{\tilde{f}_2}\Gamma_R^{\tilde{f}_1} = -8\alpha_{\tilde{Z}_i}^{\tilde{f}_2}\beta_{\tilde{Z}_i}^{\tilde{f}_1}\alpha_{\tilde{Z}_j}^{\tilde{f}_1}\beta_{\tilde{Z}_j}^{\tilde{f}_2}Y(m_{\tilde{Z}_i}, m_{\tilde{f}_1}, m_{\tilde{f}_2}, m_{\tilde{Z}_j}). \tag{B.73g}$$

We have already encountered the kinematic functions $\tilde{\psi}$, $\tilde{\phi}$, Y, ψ, and ϕ, that appear in the expressions above, in our discussion of the decay $\tilde{g} \to f\bar{f}\tilde{Z}_i$; see Eq. (B.29a)–(B.29i), and the discussion immediately following these. The *real* couplings $\alpha_{\tilde{Z}_i}^{\tilde{f}_k}$ and $\beta_{\tilde{Z}_i}^{\tilde{f}_k}$ that enter (B.73a)–(B.73g) are given by,[2]

$$\alpha_{\tilde{Z}_i}^{\tilde{f}_1} = \tilde{A}_{\tilde{Z}_i}^f \cos\theta_f - f_f v_a^{(i)}\sin\theta_f, \tag{B.74a}$$

$$\beta_{\tilde{Z}_i}^{\tilde{f}_1} = \tilde{B}_{\tilde{Z}_i}^f \sin\theta_f + f_f v_a^{(i)}\cos\theta_f, \tag{B.74b}$$

where $a = 1$ if $T_{3f} = 1/2$ and $a = 2$ if $T_{3f} = -1/2$. The corresponding couplings for the heavy sfermions ($k = 2$) are obtained via the replacements,

$$\cos\theta_f \to \sin\theta_f, \quad \sin\theta_f \to -\cos\theta_f.$$

The couplings $\tilde{A}_{\tilde{Z}_i}^f$ and $\tilde{B}_{\tilde{Z}_i}^f$ are listed in (B.9a)–(B.9d) for $f = q$. For leptons, these are given by,

$$\tilde{A}_{\tilde{Z}_i}^\ell = -\frac{gv_3^{(i)}}{\sqrt{2}} - \frac{g'v_4^{(i)}}{\sqrt{2}}, \tag{B.75a}$$

$$\tilde{B}_{\tilde{Z}_i}^\ell = -\sqrt{2}g'v_4^{(i)}, \tag{B.75b}$$

$$\tilde{A}_{\tilde{Z}_i}^\nu = \frac{gv_3^{(i)}}{\sqrt{2}} - \frac{g'v_4^{(i)}}{\sqrt{2}}, \tag{B.75c}$$

$$\tilde{B}_{\tilde{Z}_i}^\nu = 0. \tag{B.75d}$$

---

[2] We caution the reader that these differ from these same couplings defined in Eq. (8.91a)–(8.91d) of Chapter 8 by phases that we have removed, purely for convenience. We trust that our abuse of notation in using the same symbol to denote different, though closely related, quantities will not cause a problem. These real couplings are only used in the formulae in this Appendix.

The squared Z exchange contribution, which is not affected by sfermion mixing, is given by

$$\Gamma_Z = 64e^2 |W_{ij}|^2 \left(\alpha_f^2 + \beta_f^2\right) m_{\tilde{Z}_i} \pi^2$$
$$\times \int_{m_{\tilde{Z}_j}}^{E_{\max}} dE \frac{B_f \sqrt{E^2 - m_{\tilde{Z}_j}^2}}{\left(m_{\tilde{Z}_i}^2 + m_{\tilde{Z}_j}^2 - M_Z^2 - 2Em_{\tilde{Z}_i}\right)^2}$$
$$\times \left\{ E \left[m_{\tilde{Z}_i}^2 + m_{\tilde{Z}_j}^2 - (-1)^{\theta_i + \theta_j} 2m_{\tilde{Z}_i} m_{\tilde{Z}_j} \right] \right.$$
$$- m_{\tilde{Z}_i} \left(E^2 + m_{\tilde{Z}_j}^2 + \frac{B_f}{3}(E^2 - m_{\tilde{Z}_j}^2) \right)$$
$$\left. + (-1)^{\theta_i + \theta_j} m_{\tilde{Z}_j} \left(m_{\tilde{Z}_i}^2 + m_{\tilde{Z}_j}^2 - 2m_f^2 \right) \right\}. \tag{B.76}$$

Here, α_f and β_f are the vector and axial vector couplings of Z to SM fermions, W_{ij} is the $Z\tilde{Z}_i\tilde{Z}_j$ coupling given by (8.101), with

$$B_f = \sqrt{1 - \frac{4m_f^2}{m_{\tilde{Z}_i}^2 + m_{\tilde{Z}_j}^2 - 2Em_{\tilde{Z}_j}}}, \tag{B.77a}$$

and the upper integration limit

$$E_{\max} = \frac{m_{\tilde{Z}_i}^2 + m_{\tilde{Z}_j}^2 - 4m_f^2}{2m_{\tilde{Z}_i}^2}. \tag{B.77b}$$

The squared scalar Higgs exchange contributions can also be written as a single integral:

$$\Gamma_{h,H} = 2\pi^2 \left(\frac{gm_f}{M_W \cos\beta} \right)^2 m_{\tilde{Z}_i} \int_{m_{\tilde{Z}_j}}^{E_{\max}} dE\, B_f \sqrt{E^2 - m_{\tilde{Z}_j}^2}$$
$$\times \left(m_{\tilde{Z}_i}^2 + m_{\tilde{Z}_j}^2 - 2m_{\tilde{Z}_i} E - 2m_f^2 \right) \left[E + (-1)^{\theta_i + \theta_j} m_{\tilde{Z}_j} \right]$$
$$\times \left[\frac{\sin\alpha \left(X_{ij}^h + X_{ji}^h \right)}{m_{\tilde{Z}_i}^2 + m_{\tilde{Z}_j}^2 - 2m_{\tilde{Z}_i} E - m_h^2} + \frac{\cos\alpha \left(X_{ij}^H + X_{ji}^H \right)}{m_{\tilde{Z}_i}^2 + m_{\tilde{Z}_j}^2 - 2m_{\tilde{Z}_i} E - m_H^2} \right]^2. \tag{B.78a}$$

The couplings $X_{ij}^{h,H}$ are given by (8.117), and the upper limit of integration by (B.77b).

The squared pseudoscalar Higgs exchange contribution can be cast in a similar form:

$$\Gamma_A = 2\pi^2 \left[\frac{g m_f \tan\beta}{M_W} \left(X_{ij}^A + X_{ji}^A \right) \right]^2 m_{\tilde{Z}_i} \int_{m_{\tilde{Z}_j}}^{E_{\max}} dE\, B_f \sqrt{E^2 - m_{\tilde{Z}_j}^2}$$

$$\times \frac{\left(m_{\tilde{Z}_i}^2 + m_{\tilde{Z}_j}^2 - 2m_{\tilde{Z}_i} E - 2m_f^2 \right) \left[E - (-1)^{\theta_i + \theta_j} m_{\tilde{Z}_j} \right]}{\left(m_{\tilde{Z}_i}^2 + m_{\tilde{Z}_j}^2 - 2m_{\tilde{Z}_i} E - m_A^2 \right)^2},$$

$$\tag{B.78b}$$

where the coupling X_{ij}^A is given in (8.120).

We now turn to the various interference terms. The Z–sfermion interference contributions can be written as

$$\Gamma_{Z\tilde{f}} = \Gamma_{Z\tilde{f}_1} + \Gamma_{Z\tilde{f}_2}, \tag{B.79a}$$

with

$$\Gamma_{Z\tilde{f}_k} = 32 e \widetilde{W}_{ij} \left[\alpha_{\tilde{Z}_i}^{\tilde{f}_k} \alpha_{\tilde{Z}_j}^{\tilde{f}_k} (\alpha_f - \beta_f) - \beta_{\tilde{Z}_i}^{\tilde{f}_k} \beta_{\tilde{Z}_j}^{\tilde{f}_k} (\alpha_f + \beta_f) \right] \frac{\pi^2}{2 m_{\tilde{Z}_i}}$$

$$\times \int_{4m_f^2}^{(m_{\tilde{Z}_i} - m_{\tilde{Z}_j})^2} \frac{ds}{s - M_Z^2} \left\{ -\frac{1}{2} Q' \left(m_{\tilde{Z}_i} E_Q + m_{\tilde{f}_k}^2 - m_{\tilde{Z}_i}^2 - s - m_f^2 \right) \right.$$

$$- \frac{1}{4 m_{\tilde{Z}_i}} \left[\left(m_{\tilde{f}_k}^2 - m_{\tilde{Z}_j}^2 - m_f^2 \right) \left(m_{\tilde{f}_k}^2 - m_{\tilde{Z}_i}^2 - m_f^2 \right) \right.$$

$$\left. + (-1)^{\theta_i + \theta_j} m_{\tilde{Z}_i} m_{\tilde{Z}_j} (s - 2m_f^2) \right] \log \frac{m_{\tilde{Z}_i} (E_Q + Q') - \mu^2}{m_{\tilde{Z}_i} (E_Q - Q') - \mu^2} \right\}. \tag{B.79b}$$

Here we have introduced the quantities

$$\mu^2 = s + m_{\tilde{f}_k}^2 - m_{\tilde{Z}_j}^2 - m_f^2, \quad E_Q = \frac{s + m_{\tilde{Z}_i}^2 - m_{\tilde{Z}_j}^2}{2 m_{\tilde{Z}_i}}, \tag{B.80a}$$

and

$$Q = \sqrt{E_Q^2 - s}, \quad Q' = Q \sqrt{1 - \frac{4m_f^2}{s}}. \tag{B.80b}$$

The real coupling $\widetilde{W}_{ij}$ is defined to be,

$$\widetilde{W}_{ij} = (-i)^{\theta_i + \theta_j} (-1)^{\theta_j} W_{ij}. \tag{B.81}$$

Finally, the Higgs boson–sfermion interference contributions can be written as,

$$\Gamma_{\phi\tilde{f}} = \Gamma_{h\tilde{f}_1} + \Gamma_{h\tilde{f}_2} + \Gamma_{H\tilde{f}_1} + \Gamma_{H\tilde{f}_2} + \Gamma_{A\tilde{f}_1} + \Gamma_{A\tilde{f}_2}, \tag{B.82a}$$

with

$$\Gamma_{h\tilde{f}_k} = \frac{2\pi^2}{m_{\tilde{Z}_i}} \frac{gm_f \sin\alpha}{M_W \cos\beta} (X^h_{ji} + X^h_{ij}) \left[\alpha^{\tilde{f}_k}_{\tilde{Z}_i}\beta^{\tilde{f}_k}_{\tilde{Z}_j} + \alpha^{\tilde{f}_k}_{\tilde{Z}_j}\beta^{\tilde{f}_k}_{\tilde{Z}_i}\right]$$
$$\times (-1)^{\theta_i + \theta_j} J(m_{\tilde{Z}_i}, m_{\tilde{f}_k}, m_h, m_{\tilde{Z}_j}, \theta_i + \theta_j), \qquad (B.82b)$$

$$\Gamma_{H\tilde{f}_k} = \frac{2\pi^2}{m_{\tilde{Z}_i}} \frac{gm_f \cos\alpha}{M_W \cos\beta} (X^H_{ji} + X^H_{ij}) \left[\alpha^{\tilde{f}_k}_{\tilde{Z}_i}\beta^{\tilde{f}_k}_{\tilde{Z}_j} + \alpha^{\tilde{f}_k}_{\tilde{Z}_j}\beta^{\tilde{f}_k}_{\tilde{Z}_i}\right]$$
$$\times (-1)^{\theta_i + \theta_j} J(m_{\tilde{Z}_i}, m_{\tilde{f}_k}, m_H, m_{\tilde{Z}_j}, \theta_i + \theta_j), \qquad (B.82c)$$

$$\Gamma_{A\tilde{f}_k} = \frac{2\pi^2}{m_{\tilde{Z}_i}} \frac{gm_f \tan\beta}{M_W} (X^A_{ji} + X^A_{ij}) \left[\alpha^{\tilde{f}_k}_{\tilde{Z}_i}\beta^{\tilde{f}_k}_{\tilde{Z}_j} + \alpha^{\tilde{f}_k}_{\tilde{Z}_j}\beta^{\tilde{f}_k}_{\tilde{Z}_i}\right]$$
$$\times (-1)^{1+\theta_i + \theta_j} J(m_{\tilde{Z}_i}, m_{\tilde{f}_k}, m_A, m_{\tilde{Z}_j}, 1+\theta_i + \theta_j). \qquad (B.82d)$$

The function J is defined as

$$J(m_{\tilde{Z}_i}, m_{\tilde{f}}, m_H, m_{\tilde{Z}_j}, \theta) = \int_{4m_f^2}^{(m_{\tilde{Z}_i} - m_{\tilde{Z}_j})^2} \frac{ds}{s - m_H^2}$$
$$\times \left[\frac{1}{2}sQ' + \frac{sm_{\tilde{f}}^2 - m_f^2(m_{\tilde{Z}_i}^2 + m_{\tilde{Z}_j}^2) + (-1)^\theta m_{\tilde{Z}_i} m_{\tilde{Z}_j}(s - 2m_f^2)}{4m_{\tilde{Z}_j}}\right.$$
$$\times \left. \log \frac{m_{\tilde{Z}_i}(E_Q + Q') - \mu^2}{m_{\tilde{Z}_i}(E_Q - Q') - \mu^2}\right], \qquad (B.83)$$

where μ^2, E_Q, Q, and Q' have been defined previously.

B.4.3 $\tilde{Z}_j \rightarrow \tilde{W}_i^+ \tau^- \nu_\tau$ decays

The partial width for the decay $\tilde{Z}_j \rightarrow \tilde{W}_i^+ \tau^- \nu_\tau$ is related to that for the decay $\tilde{W}_i^- \rightarrow \tilde{Z}_j \tau^- \nu_\tau$, as in (B.106) of the next section. These neutralino decays are usually not very important because they are either phase space suppressed, or are dwarfed by other two-body decays of the parent neutralino.

B.5 Chargino decay widths
B.5.1 Two-body decays

We list the tree-level partial widths for all two-body decays of the charginos. The partial width for a mode and its charge conjugate are the same. Also, in the following, whether $\tilde{W}_1$ refers to positive or negative chargino should be clear from the context.

Starting with decays to gauge bosons, we find that

$$\Gamma(\widetilde{W}_i \to \widetilde{Z}_j W) = \frac{g^2}{16\pi m_{\widetilde{W}_i}^3} \lambda^{\frac{1}{2}}(m_{\widetilde{W}_i}^2, m_{\widetilde{Z}_j}^2, M_W^2)$$

$$\times \left[(X_i^{j2} + Y_i^{j2})\left(m_{\widetilde{W}_i}^2 + m_{\widetilde{Z}_j}^2 - M_W^2 + \frac{(m_{\widetilde{W}_i}^2 - m_{\widetilde{Z}_j}^2)^2 - M_W^4}{M_W^2} \right) \right.$$

$$\left. - 6(X_i^{j2} - Y_i^{j2})m_{\widetilde{W}_i} m_{\widetilde{Z}_j} \right], \tag{B.84}$$

where X_i^j and Y_i^j are given in Eq. (8.103a) and (8.103b), and

$$\Gamma(\widetilde{W}_2 \to \widetilde{W}_1 Z) = \frac{e^2}{64\pi m_{\widetilde{W}_2}^3}(\cot\theta_{\mathrm{W}} + \tan\theta_{\mathrm{W}})^2 \lambda^{\frac{1}{2}}(m_{\widetilde{W}_2}^2, m_{\widetilde{W}_1}^2, M_Z^2)$$

$$\times \left[(x^2 + y^2)\left(m_{\widetilde{W}_2}^2 + m_{\widetilde{W}_1}^2 - M_Z^2 + \frac{(m_{\widetilde{W}_2} - m_{\widetilde{W}_1})^2 - M_Z^4}{M_Z^2} \right) \right.$$

$$\left. + 6(x^2 - y^2)(-1)^{\theta_{\widetilde{W}_1}}(-1)^{\theta_{\widetilde{W}_2}} m_{\widetilde{W}_1} m_{\widetilde{W}_2} \right], \tag{B.85}$$

where x and y are given in Eq. (8.100e) and (8.100f).

Turning to decays to various Higgs bosons, we find,

$$\Gamma(\widetilde{W}_i \to \widetilde{Z}_j H^-) = \frac{1}{16\pi m_{\widetilde{W}_i}^3} \lambda^{\frac{1}{2}}(m_{\widetilde{W}_i}^2, m_{\widetilde{Z}_j}^2, m_{H^-}^2)$$

$$\times \left[(a^2 + b^2)(m_{\widetilde{W}_i}^2 + m_{\widetilde{Z}_j}^2 - m_{H^-}^2) + 2(a^2 - b^2)m_{\widetilde{W}_i} m_{\widetilde{Z}_j} \right], \tag{B.86}$$

where the coefficients a and b are exactly the same as those that enter the decay $\widetilde{Z}_j \to \widetilde{W}_i^- H^+$; these are given in (B.63a) and (B.63b), and in the discussion following for the decay $\widetilde{Z}_i \to \widetilde{W}_j^- H^+$ (so that the reader must remember to interchange i and j). Charginos may also decay to neutral Higgs bosons $\phi = h, H$ or A with partial widths given by,

$$\Gamma(\widetilde{W}_2 \to \widetilde{W}_1 \phi) = \frac{g^2}{32\pi m_{\widetilde{W}_2}^3} \lambda^{\frac{1}{2}}(m_{\widetilde{W}_2}^2, m_{\widetilde{W}_1}^2, m_\phi^2)$$

$$\times \left[(S^{\phi 2} + P^{\phi 2})(m_{\widetilde{W}_2}^2 + m_{\widetilde{W}_1}^2 - m_\phi^2) + 2(S^{\phi 2} - P^{\phi 2})m_{\widetilde{W}_1} m_{\widetilde{W}_2} \right], \tag{B.87}$$

where $S^{h(H)}$ and $P^{h(H)}$ are given in (8.116c), and the corresponding couplings to A given in (8.119c).

Charginos may also decay via $\widetilde{W}_i \to \tilde{f}_j f'$ if these decays are kinematically accessible. We write the partial widths for decays to the third generation, but these can be used with obvious modifications for the first two generations. For $\widetilde{W}_i$ decay to squark plus quark, we find

$$\Gamma(\widetilde{W}_i^+ \to \tilde{t}_1 \bar{b}) = \frac{3m_{\widetilde{W}_i}}{32\pi} \lambda^{1/2}(1, \frac{m_b^2}{m_{\widetilde{W}_i}^2}, \frac{m_{\tilde{t}_1}^2}{m_{\widetilde{W}_i}^2}) \left\{ [|\mathcal{A}|^2 + B_{\widetilde{W}_i}'^2 \cos^2 \theta_t] \right.$$

$$\left. \times (1 + \frac{m_b^2}{m_{\widetilde{W}_i}^2} - \frac{m_{\tilde{t}_1}^2}{m_{\widetilde{W}_i}^2}) + 4\mathcal{A} B_{\widetilde{W}_i}' \cos \theta_t \frac{m_b}{m_{\widetilde{W}_i}} \right\}, \qquad \text{(B.88a)}$$

where $\mathcal{A} \equiv iA_{\widetilde{W}_i}^d \cos \theta_t - B_{\widetilde{W}_i} \sin \theta_t$ is real. The width for $\widetilde{W}_i \to \tilde{t}_2 \bar{b}$ can be obtained from this by replacing $m_{\tilde{t}_1} \to m_{\tilde{t}_2}$, $\cos \theta_t \to \sin \theta_t$, and $\sin \theta_t \to -\cos \theta_t$. For $\widetilde{W}_i \to \bar{\tilde{b}}_1 t$ decay, we find

$$\Gamma(\widetilde{W}_i \to \tilde{b}_1 \bar{t}) = \frac{3m_{\widetilde{W}_i}}{32\pi} \lambda^{1/2}(1, \frac{m_t^2}{m_{\widetilde{W}_i}^2}, \frac{m_{\tilde{b}_1}^2}{m_{\widetilde{W}_i}^2}) \left\{ [|\mathcal{A}|^2 + B_{\widetilde{W}_i}^2 \cos^2 \theta_b] \right.$$

$$\left. \times (1 + \frac{m_t^2}{m_{\widetilde{W}_i}^2} - \frac{m_{\tilde{b}_1}^2}{m_{\widetilde{W}_i}^2}) + 4\mathcal{A} B_{\widetilde{W}_i} \cos \theta_b \frac{m_t}{m_{\widetilde{W}_i}} \right\}, \qquad \text{(B.88b)}$$

where this time $\mathcal{A} = iA_{\widetilde{W}_i}^u \cos \theta_b - B_{\widetilde{W}_i}' \sin \theta_b$. Again, the replacements $m_{\tilde{b}_1} \to m_{\tilde{b}_2}$, $\cos \theta_b \to \sin \theta_b$, and $\sin \theta_b \to -\cos \theta_b$, yield the width for the decay $\widetilde{W}_i \to \tilde{b}_2 \bar{t}$.

These formulae simplify considerably if we ignore couplings to the higgsino components and intra-generation mixing. The decay rate for $\widetilde{W}_i^+ \to \tilde{u}_L \bar{d}$ can then be obtained from $\Gamma(\widetilde{W}_i^+ \to \tilde{t}_1 \bar{b})$ by replacing $m_b \to m_d$, $m_{\tilde{t}_1} \to m_{\tilde{u}_L}$ and setting $\cos \theta_t \to 1$, $\sin \theta_t \to 0$ and setting the Yukawa couplings in $B_{\widetilde{W}_i}$ and $B_{\widetilde{W}_i}'$ to zero. Similarly, the decay $\widetilde{W}_i \to \tilde{d}_L \bar{u}$ can be obtained from the formula for $\widetilde{W}_i \to \tilde{b}_1 \bar{t}$ by replacing $m_t \to m_u$, $m_{\tilde{b}_1} \to m_{\tilde{d}_L}$ and setting $\cos \theta_b \to 1$, $\sin \theta_b \to 0$ and again setting the Yukawa couplings in $B_{\widetilde{W}_i}$ and $B_{\widetilde{W}_i}'$ to zero. Of course, charginos do not decay into $\tilde{q}_R$ in this limit.

Finally, the partial widths to leptons and sleptons are given by

$$\Gamma(\widetilde{W}_i \to \tilde{\nu}_\tau \bar{\tau}) = \frac{m_{\widetilde{W}_i}}{32\pi} \lambda^{1/2}(1, \frac{m_\tau^2}{m_{\widetilde{W}_i}^2}, \frac{m_{\tilde{\nu}_\tau}^2}{m_{\widetilde{W}_i}^2})$$

$$\times \left\{ [|A_{\widetilde{W}_i}^\tau|^2 + B_{\widetilde{W}_i}''^2] \left(1 + \frac{m_\tau^2}{m_{\widetilde{W}_i}^2} - \frac{m_{\tilde{\nu}_\tau}^2}{m_{\widetilde{W}_i}^2}\right) + 4(iA_{\widetilde{W}_i}^\tau) B_{\widetilde{W}_i}'' \frac{m_\tau}{m_{\widetilde{W}_i}} \right\},$$

$$\text{(B.89a)}$$

(notice that $iA^\tau_{\widetilde{W}_i}$ is real) and

$$\Gamma(\widetilde{W}_i \rightarrow \widetilde{\tau}_1 \bar{\nu}_\tau) = \frac{|\mathcal{A}|^2}{32\pi} m_{\widetilde{W}_i} \left(1 - \frac{m^2_{\widetilde{\tau}_1}}{m^2_{\widetilde{W}_i}}\right)^2 , \qquad (B.89b)$$

where the real coefficient $\mathcal{A} = iA^\nu_{\widetilde{W}_i} \cos\theta_\tau - B''_{\widetilde{W}_i} \sin\theta_\tau$. The couplings $A^\tau_{\widetilde{W}_i}$, $A^\nu_{\widetilde{W}_i}$, and $B''_{\widetilde{W}_i}$ are given in (8.98a)–(8.98d). The decay width to $\widetilde{\tau}_2$ is obtained via the replacements, $\cos\theta_\tau \rightarrow \sin\theta_\tau$, $\sin\theta_\tau \rightarrow -\cos\theta_\tau$, and $m_{\widetilde{\tau}_1} \rightarrow m_{\widetilde{\tau}_2}$ in the formula for $\Gamma(\widetilde{W}_i \rightarrow \widetilde{\tau}_1 \nu_\tau)$ and the corresponding coefficient $\mathcal{A}$. How these formulae simplify if coupling via higgsino components and $\widetilde{\ell}_L$–$\widetilde{\ell}_R$ mixing are neglected should be evident.

B.5.2 Three-body decay: $\widetilde{W}_i \rightarrow \widetilde{Z}_j \tau \bar{\nu}_\tau$

We present a formula for the partial width for the decay $\widetilde{W}_i \rightarrow \widetilde{Z}_j \tau \nu_\tau$. We will see later that partial widths for other relevant three-body decays of the chargino can be readily obtained from this. This decay proceeds via the exchange of a W boson, a charged or neutral third generation slepton, and a charged Higgs boson. The partial width can be written as

$$\Gamma(\widetilde{W}_i \rightarrow \widetilde{Z}_j \tau^- \bar{\nu}_\tau)$$
$$= \frac{1}{2} \frac{1}{(2\pi)^5} \frac{1}{2m_{\widetilde{W}_i}} (\Gamma_W + \Gamma_{\widetilde{\nu}} + \Gamma_{\widetilde{\tau}} + \Gamma_H + \Gamma_{W\widetilde{\nu}} + \Gamma_{W\widetilde{\tau}} + \Gamma_{\widetilde{\nu}\widetilde{\tau}} + \Gamma_{H\widetilde{\nu}} + \Gamma_{H\widetilde{\tau}}).$$
$$(B.90)$$

The Higgs and W exchange contributions do not interfere, since we neglect terms $\propto m_\tau$ when doing the Dirac algebra.

The squared W exchange contribution is given by

$$\Gamma_W = 4g^4 \frac{\pi^2}{3} \int_{m_{\widetilde{Z}_j}}^{E_{max}} dE \frac{\sqrt{E^2 - m^2_{\widetilde{Z}_j}}}{\left(m^2_{\widetilde{W}_i} + m^2_{\widetilde{Z}_j} - 2m_{\widetilde{W}_i} E - M^2_W\right)^2}$$
$$\times \left\{\left(\left|X^j_i\right|^2 + \left|Y^j_i\right|^2\right) \left[3\left(m^2_{\widetilde{W}_i} + m^2_{\widetilde{Z}_j}\right) m_{\widetilde{W}_i} E - 2m^2_{\widetilde{W}_i} \left(2E^2 + m^2_{\widetilde{Z}_j}\right)\right]\right.$$
$$\left. - 3\left(\left|X^j_i\right|^2 - \left|Y^j_i\right|^2\right) m_{\widetilde{W}_i} m_{\widetilde{Z}_j} \left(m^2_{\widetilde{W}_i} + m^2_{\widetilde{Z}_j} - 2Em_{\widetilde{W}_i}\right)\right\}. \qquad (B.91)$$

Here X^j_i and Y^j_i are the $W\widetilde{W}_i\widetilde{Z}_j$ couplings given in (8.103a) and (8.103b), and the upper integration limit $E_{max} = (m^2_{\widetilde{W}_i} + m^2_{\widetilde{Z}_j})/2m_{\widetilde{W}_i}$.

The squared sneutrino exchange contribution is given by

$$\Gamma_{\tilde{\nu}} = 2\left(\tilde{A}_{\tilde{Z}_j}^\nu\right)^2 \left[\left(\tilde{A}_{\tilde{W}_i}^\tau\right)^2 + \left(B_{\tilde{W}_i}''\right)^2\right]^2 \psi(m_{\tilde{W}_i}, m_{\tilde{\nu}_\tau}, m_{\tilde{Z}_j}),$$ (B.92)

where $\tilde{A}_{\tilde{Z}_j}^\nu$ has been defined in (B.75d) and

$$\tilde{A}_{\tilde{W}_1}^\tau = -g\sin\gamma_R.$$ (B.93a)

The pure scalar tau exchange terms can be written as

$$\Gamma_{\tilde{\tau}} = \Gamma_{\tilde{\tau}_1} + \Gamma_{\tilde{\tau}_2} + \Gamma_{\tilde{\tau}_1\tilde{\tau}_2},$$ (B.94)

where

$$\Gamma_{\tilde{\tau}_k} = 2\left(\alpha_{\tilde{W}_i}^{\tilde{\tau}_k}\right)^2 \left[\left(\alpha_{\tilde{Z}_j}^{\tilde{\tau}_k}\right)^2 + \left(\beta_{\tilde{Z}_j}^{\tilde{\tau}_k}\right)^2\right] \psi(m_{\tilde{W}_i}, m_{\tilde{\tau}_k}, m_{\tilde{Z}_j}),$$ (B.95a)

$$\Gamma_{\tilde{\tau}_1\tilde{\tau}_2} = 4\alpha_{\tilde{W}_i}^{\tilde{\tau}_1}\alpha_{\tilde{W}_i}^{\tilde{\tau}_2}\left[\alpha_{\tilde{Z}_j}^{\tilde{\tau}_1}\alpha_{\tilde{Z}_j}^{\tilde{\tau}_2} + \beta_{\tilde{Z}_j}^{\tilde{\tau}_1}\beta_{\tilde{Z}_j}^{\tilde{\tau}_2}\right] \tilde{\psi}(m_{\tilde{W}_i}, m_{\tilde{\tau}_1}, m_{\tilde{\tau}_2}, m_{\tilde{Z}_j}),$$ (B.95b)

and where

$$\alpha_{\tilde{W}_1}^{\tilde{\tau}_1} = -g\sin\gamma_L\cos\theta_\tau + f_\tau\cos\gamma_L\sin\theta_\tau,$$ (B.96a)

$$\alpha_{\tilde{W}_1}^{\tilde{\tau}_2} = -g\sin\gamma_L\sin\theta_\tau - f_\tau\cos\gamma_L\cos\theta_\tau,$$ (B.96b)

$$\alpha_{\tilde{W}_2}^{\tilde{\tau}_1} = (-g\cos\gamma_L\cos\theta_\tau - f_\tau\sin\gamma_L\sin\theta_\tau)\theta_x,$$ (B.96c)

$$\alpha_{\tilde{W}_2}^{\tilde{\tau}_2} = (-g\cos\gamma_L\sin\theta_\tau + f_\tau\sin\gamma_L\cos\theta_\tau)\theta_x.$$ (B.96d)

The squared charged Higgs boson exchange contribution is

$$\Gamma_H = \pi^2 m_{\tilde{W}_i}\left(\frac{gm_\tau\tan\beta}{M_W}\right)^2 \int_{m_{\tilde{Z}_j}}^{E_{max}} dE\sqrt{E^2 - m_{\tilde{Z}_j}^2}\left(m_{\tilde{W}_i}^2 + m_{\tilde{Z}_j}^2 - 2Em_{\tilde{W}_i}\right)$$

$$\times \frac{\left\{E\left[\left(\alpha_{\tilde{W}_i}^{(j)}\right)^2 + \left(\beta_{\tilde{W}_i}^{(j)}\right)^2\right] + 2(-1)^{\theta_{\tilde{W}_i}+\theta_j}m_{\tilde{Z}_j}\alpha_{\tilde{W}_i}^{(j)}\beta_{\tilde{W}_i}^{(j)}\right\}}{\left(m_{\tilde{W}_i}^2 + m_{\tilde{Z}_j}^2 - 2Em_{\tilde{W}_i} - m_{H^+}^2\right)^2}.$$ (B.97)

Here, E_{max} is as defined just after (B.91), and

$$\alpha_{\tilde{W}_1}^{(j)} = \cos\beta A_2^{(j)},$$ (B.98a)

$$\beta_{\tilde{W}_1}^{(j)} = -\sin\beta A_4^{(j)},$$ (B.98b)

$$\alpha_{\tilde{W}_2}^{(j)} = \cos\beta A_1^{(j)}\theta_y,$$ (B.98c)

$$\beta_{\tilde{W}_2}^{(j)} = -\sin\beta A_3^{(j)}\theta_x,$$ (B.98d)

with the coefficients $A_i^{(j)}$ as defined in (8.122a)–(8.122d).

The W–sneutrino interference contribution is not affected by $\tilde{\tau}_L$–$\tilde{\tau}_R$ mixing and contributions $\propto f_\tau$; it can be written as

$$\Gamma_{W\tilde{\nu}} = -4\sqrt{2}g^2(-1)^{\theta_{\tilde{W}_i}+\theta_j}\tilde{A}^\tau_{\tilde{W}_i}\tilde{A}^\nu_{\tilde{Z}_j}$$

$$\times\left[\left(X_i^j - Y_i^j\right)I_1(m_{\tilde{W}_i}, m_{\tilde{\nu}_\tau}, m_{\tilde{Z}_j}) - \left(X_i^j + Y_i^j\right)I_2(m_{\tilde{W}_i}, m_{\tilde{\nu}_\tau}, m_{\tilde{Z}_j})\right],$$

$$(B.99)$$

where we have introduced the functions

$$I_1(m_{\tilde{W}}, m_{\tilde{f}}, m_{\tilde{Z}}) = \frac{\pi^2}{2m_{\tilde{W}}}\int\frac{ds}{s - M_W^2}\left[-\frac{1}{2}Q\left(m_{\tilde{W}}E_Q + m_{\tilde{f}}^2 - m_{\tilde{W}}^2 - s\right)\right.$$

$$\left. -\frac{\left(m_{\tilde{f}}^2 - m_{\tilde{Z}}^2\right)\left(m_{\tilde{f}}^2 - m_{\tilde{W}}^2\right)}{4m_{\tilde{W}}}\log\frac{m_{\tilde{W}}(E_Q + Q) - \mu^2}{m_{\tilde{W}}(E_Q - Q) - \mu^2}\right],$$

$$(B.100a)$$

$$I_2(m_{\tilde{W}}, m_{\tilde{f}}, m_{\tilde{Z}}) = \frac{\pi^2}{8m_{\tilde{W}}}\int\frac{ds}{s - M_W^2}m_{\tilde{Z}}s\log\frac{m_{\tilde{W}}(E_Q + Q) - \mu^2}{m_{\tilde{W}}(E_Q - Q) - \mu^2}, \quad (B.100b)$$

and the limits of integration on I_1 and I_2 run from zero to $(m_{\tilde{W}} - m_{\tilde{Z}})^2$. The quantities μ^2, E_Q, and Q are defined in (B.80a) and (B.80b) but with $m_{\tilde{Z}_i} \to m_{\tilde{W}}$, $m_{\tilde{Z}_j} \to m_{\tilde{Z}}$, and $m_{\tilde{f}_k} \to m_{\tilde{f}}$.

The same functions also appear in the W–scalar tau interference contributions:

$$\Gamma_{W\tilde{\tau}} = \Gamma_{W\tilde{\tau}_1} + \Gamma_{W\tilde{\tau}_2}, \quad (B.101a)$$

where

$$\Gamma_{W\tilde{\tau}_k} = 4\sqrt{2}g^2\alpha^{\tilde{\tau}_k}_{\tilde{W}_i}\alpha^{\tilde{\tau}_k}_{\tilde{Z}_j}\left[\left(X_i^j + Y_i^j\right)I_1(m_{\tilde{W}_i}, m_{\tilde{\tau}_k}, m_{\tilde{Z}_j})\right.$$

$$\left. -\left(X_i^j - Y_i^j\right)I_2(m_{\tilde{W}_i}, m_{\tilde{\tau}_k}, m_{\tilde{Z}_j})\right].$$

$$(B.101b)$$

The sneutrino–scalar tau interference terms can be written as

$$\Gamma_{\tilde{\nu}\tilde{\tau}} = \Gamma_{\tilde{\nu}\tilde{\tau}_1} + \Gamma_{\tilde{\nu}\tilde{\tau}_2}, \quad (B.102a)$$

where

$$\Gamma_{\tilde{\nu}\tilde{\tau}_k} = -4\tilde{A}^\nu_{\tilde{Z}_j}\alpha^{\tilde{\tau}_k}_{\tilde{W}_i}\left[B''_{\tilde{W}_i}\beta^{\tilde{\tau}_k}_{\tilde{Z}_j}Y(m_{\tilde{W}_i}, m_{\tilde{\nu}_\tau}, m_{\tilde{\tau}_k}, m_{\tilde{Z}_j})\right.$$

$$\left. -(-1)^{\theta_i+\theta_j}\tilde{A}^\tau_{\tilde{W}_i}\alpha^{\tilde{\tau}_k}_{\tilde{Z}_j}\tilde{\phi}(m_{\tilde{W}_i}, m_{\tilde{\nu}_\tau}, m_{\tilde{\tau}_k}, m_{\tilde{Z}_j})\right].$$

$$(B.102b)$$

The functions Y and $\tilde{\phi}$ have already been defined in (B.29a) and (B.29i), respectively.

The charged Higgs–sneutrino interference term is given by

$$\Gamma_{H\tilde{\nu}} = 2\sqrt{2}\tilde{A}^{\nu}_{\tilde{Z}_j}B''_{\tilde{W}_i}\frac{gm_\tau \tan\beta}{m_W}I_H(m_{\tilde{W}_i}, m_{H^+}, m_{\tilde{\nu}_\tau}, m_{\tilde{Z}_j}), \tag{B.103}$$

where we have introduced the function

$$I_H(m_{\tilde{W}_i}, m_H, m_{\tilde{f}}, m_{\tilde{Z}_j}) = \frac{\pi^2}{2m_{\tilde{W}_i}}\int_0^{(m_{\tilde{W}_i}-m_{\tilde{Z}_j})^2}\frac{ds}{s-m_H^2} \tag{B.104}$$

$$\times \left\{ \frac{1}{2}sQ\beta^{(j)}_{\tilde{W}_i} + \frac{1}{4m_{\tilde{W}_i}}\left[\beta^{(j)}_{\tilde{W}_i}sm_{\tilde{f}}^2 + (-1)^{\theta_{\tilde{W}_i}+\theta_j}\alpha^{(j)}_{\tilde{W}_i}m_{\tilde{W}_i}m_{\tilde{Z}_j}s\right] \right.$$

$$\left. \times \log\frac{m_{\tilde{W}_i}(E_Q+Q)-\mu^2}{m_{\tilde{W}_i}(E_Q-Q)-\mu^2} \right\}. \tag{B.104}$$

The coupling $\tilde{A}^{\nu}_{\tilde{Z}_j}$ is as defined in (B.75d), and the quantities μ^2, E_Q, and Q are as defined below (B.100b).

The same function also appears in the charged Higgs–scalar tau interference contributions:

$$\Gamma_{H\tilde{\tau}} = \Gamma_{H\tilde{\tau}_1} + \Gamma_{H\tilde{\tau}_2}, \tag{B.105a}$$

where

$$\Gamma_{H\tilde{\tau}_k} = 2\sqrt{2}\alpha^{\tilde{\tau}_k}_{\tilde{W}_i}\beta^{\tilde{\tau}_k}_{\tilde{Z}_j}\frac{gm_\tau \tan\beta}{M_W}I_H(m_{\tilde{W}_i}, m_{H^+}, m_{\tilde{\tau}_k}, m_{\tilde{Z}_j}). \tag{B.105b}$$

The partial widths for the analogous neutralino to chargino decays are given by crossing. The partial width for the neutralino to chargino decay can be obtained from the formula for the corresponding width for the chargino decay by simply interchanging the masses. In other words,

$$\Gamma(\tilde{Z}_j \to \tilde{W}_i^+\tau^-\bar{\nu}_\tau) = \Gamma(\tilde{W}_i^- \to \tilde{Z}_j\tau^-\bar{\nu}_\tau)(m_{\tilde{W}_i} \leftrightarrow m_{\tilde{Z}_j}). \tag{B.106}$$

Note that $\tilde{Z}_j$ can also decay into $\tilde{W}_i^-\tau^+\nu_\tau$ final states, with equal probability. However, these neutralino decays are usually not very important, since they are either phase space suppressed, or have to compete with two-body decays of the heavy neutralinos.

Our formula for $\Gamma(\tilde{W}_i \to \tilde{Z}_j\tau^-\bar{\nu}_\tau)$ decay can be readily adapted to three-body chargino decays into fermion–antifermion pairs of the first two generations. Ignoring Yukawa couplings and intra-generation mixing, we have just three contributions to this amplitude: W exchange, and the exchanges of the "left-handed" up and down type sfermions. We retain only the W, $\tilde{\tau}_1$, and sneutrino exchange contributions and set $\cos\theta_\tau = 1$ to obtain the partial width for the decays $\tilde{W}_i \to \ell\bar{\nu}_\ell\tilde{Z}_j$. The replacements $\tilde{\nu}_\ell \to \tilde{u}_L$, $\tilde{\ell}_L \to \tilde{d}_L$ in the formula for $\Gamma(\tilde{W}_i \to \tilde{Z}_j\ell\bar{\nu}_\ell)$ will yield $\Gamma(\widetilde{W}_i \to \widetilde{Z}_j d\bar{u})$ if we remember to include the color factor of 3. The decay $\widetilde{W}_i \to b\bar{t}\widetilde{Z}_j$ always has a small branching fraction, since two-body decays $\widetilde{W}_i \to W\widetilde{Z}_j$ are also accessible whenever the three-body decay to top is.

B.6 Top quark decay to SUSY particles

If charged Higgs bosons or sparticles are light enough, new two-body decays of the top quark may be allowed. These include, $t \to bH^+$, $t \to \tilde{t}_1\widetilde{Z}_i$, and $t \to \tilde{b}_1\widetilde{W}_i$. The partial width for the decay $t \to fS$, where f is a spin $\frac{1}{2}$ fermion and S a spin zero particle, is given by,

$$\Gamma(t \to fS) = \frac{m_t}{16\pi}\lambda^{\frac{1}{2}}(1, \frac{m_f^2}{m_t^2}, \frac{m_S^2}{m_t^2})$$

$$\times \left[(|\alpha|^2 + |\beta|^2)(1 + \frac{m_f^2}{m_t^2} - \frac{m_S^2}{m_t^2}) + 2(|\alpha|^2 - |\beta|^2)\frac{m_f}{m_t}\right],$$

$$(B.107)$$

where α and β are the scalar and pseudoscalar couplings of S to the t–f system.

For the decay $t \to bH^+$, $f = b$ and $S = H^+$ and we have,

$$\alpha = \frac{g}{2\sqrt{2}M_W}(m_b \tan\beta + m_t \cot\beta),$$

$$\beta = \frac{g}{2\sqrt{2}M_W}(m_b \tan\beta - m_t \cot\beta). \qquad (B.108a)$$

For the decay $t \to \tilde{t}_1\widetilde{Z}_i$, $f = \widetilde{Z}_i$, $S = \tilde{t}_1$, and

$$\alpha = \frac{1}{2}\left\{\left[iA_{\widetilde{Z}_i}^t - (i)^{\theta_i} f_t v_1^{(i)}\right]\cos\theta_t - \left[iB_{\widetilde{Z}_i}^t - (-i)^{\theta_i} f_t v_1^{(i)}\right]\sin\theta_t\right\}$$

and

$$\beta = \frac{1}{2}\left\{\left[-iA_{\widetilde{Z}_i}^t - (i)^{\theta_i} f_t v_1^{(i)}\right]\cos\theta_t - \left[iB_{\widetilde{Z}_i}^t + (-i)^{\theta_i} f_t v_1^{(i)}\right]\sin\theta_t\right\}.$$

$$(B.108b)$$

Finally, for the decay $t \to \tilde{b}_1\widetilde{W}_i$, $f = \widetilde{W}_i$, $S = \tilde{b}_1$, and

$$\alpha = \frac{1}{2}\left[iA_{\widetilde{W}_i}^t \cos\theta_b - B_{\widetilde{W}_i}' \sin\theta_b + B_{\widetilde{W}_i} \cos\theta_b\right],$$

$$\beta = \frac{1}{2}\left[-iA_{\widetilde{W}_i}^t \cos\theta_b + B_{\widetilde{W}_i}' \sin\theta_b + B_{\widetilde{W}_i} \cos\theta_b\right]. \qquad (B.108c)$$

Appendix C
Higgs boson decay Widths

Here, we list the partial widths of all possible tree-level two-body decays of the various Higgs bosons of the MSSM.

C.1 Decays to SM fermions

The partial widths for the decays of MSSM Higgs bosons to SM fermions are given by:

$$\Gamma(h \to u\bar{u}) = \frac{g^2}{32\pi} N_c \frac{\cos^2\alpha}{\sin^2\beta} \left(\frac{m_u}{M_W}\right)^2 m_h \left(1 - \frac{4m_u^2}{m_h^2}\right)^{\frac{3}{2}}, \tag{C.1a}$$

$$\Gamma(h \to d\bar{d}) = \frac{g^2}{32\pi} N_c \frac{\sin^2\alpha}{\cos^2\beta} \left(\frac{m_d}{M_W}\right)^2 m_h \left(1 - \frac{4m_d^2}{m_h^2}\right)^{\frac{3}{2}}, \tag{C.1b}$$

$$\Gamma(h \to \ell^+\ell^-) = \frac{g^2}{32\pi} \frac{\sin^2\alpha}{\cos^2\beta} \left(\frac{m_\ell}{M_W}\right)^2 m_h \left(1 - \frac{4m_\ell^2}{m_h^2}\right)^{\frac{3}{2}}, \tag{C.1c}$$

$$\Gamma(H \to u\bar{u}) = \frac{g^2}{32\pi} N_c \frac{\sin^2\alpha}{\sin^2\beta} \left(\frac{m_u}{M_W}\right)^2 m_H \left(1 - \frac{4m_u^2}{m_H^2}\right)^{\frac{3}{2}}, \tag{C.2a}$$

$$\Gamma(H \to d\bar{d}) = \frac{g^2}{32\pi} N_c \frac{\cos^2\alpha}{\cos^2\beta} \left(\frac{m_d}{M_W}\right)^2 m_H \left(1 - \frac{4m_d^2}{m_H^2}\right)^{\frac{3}{2}}, \tag{C.2b}$$

$$\Gamma(H \to \ell^+\ell^-) = \frac{g^2}{32\pi} \frac{\cos^2\alpha}{\cos^2\beta} \left(\frac{m_\ell}{M_W}\right)^2 m_H \left(1 - \frac{4m_\ell^2}{m_H^2}\right)^{\frac{3}{2}}, \tag{C.2c}$$

and

$$\Gamma(A \to u\bar{u}) = \frac{g^2}{32\pi} N_c \cot^2\beta \left(\frac{m_u}{M_W}\right)^2 m_A \left(1 - \frac{4m_u^2}{m_A^2}\right)^{\frac{1}{2}}, \tag{C.3a}$$

$$\Gamma(A \to d\bar{d}) = \frac{g^2}{32\pi} N_c \tan^2 \beta \left(\frac{m_d}{M_W}\right)^2 m_A \left(1 - \frac{4m_d^2}{m_A^2}\right)^{\frac{1}{2}}, \qquad \text{(C.3b)}$$

$$\Gamma(A \to \ell^+\ell^-) = \frac{g^2}{32\pi} \tan^2 \beta \left(\frac{m_\ell}{M_W}\right)^2 m_A \left(1 - \frac{4m_\ell^2}{m_A^2}\right)^{\frac{1}{2}}, \qquad \text{(C.3c)}$$

where the color factor $N_c = 3$ for decays to quarks.

For charged Higgs boson decays, we find

$$\Gamma(H^+ \to u\bar{d}) = \Gamma(H^- \to d\bar{u}) = \frac{g^2}{32\pi M_W^2 m_{H^+}} N_c \lambda^{\frac{1}{2}} \left(1, \frac{m_u^2}{m_{H^+}^2}, \frac{m_d^2}{m_{H^+}^2}\right)$$
$$\times \left[(m_d^2 \tan^2 \beta + m_u^2 \cot^2 \beta)(m_{H^+}^2 - m_u^2 - m_d^2) - 4m_u^2 m_d^2\right].$$
$$\text{(C.4)}$$

To get $\Gamma(H^+ \to \nu_\ell \bar{\ell})$ simply replace $m_d \to m_\ell$, $m_u \to 0$, and $N_c = 1$ in (C.4).

The dominant radiative corrections can be included by replacing the fermion masses that enter the prefactors of these formulae via the corresponding Yukawa couplings by running masses evaluated at the scale $Q = m_{h,H,A}$.

C.2 Decays to gauge bosons

The heavy scalar H may decay to ZZ or WW with partial widths given by

$$\Gamma(H \to Z^0 Z^0) = \frac{g^2 \cos^2(\alpha + \beta) M_W^2}{32\pi \cos^4 \theta_W m_H} \left[3 - \frac{m_H^2}{M_Z^2} + \frac{m_H^4}{4M_Z^4}\right] \lambda^{\frac{1}{2}} \left(1, \frac{M_Z^2}{m_H^2}, \frac{M_Z^2}{m_H^2}\right)$$
$$\text{(C.5a)}$$

and

$$\Gamma(H \to W^+ W^-)$$
$$= \frac{g^2 \cos^2(\alpha + \beta) M_W^2}{16\pi m_H} \left[3 - \frac{m_H^2}{M_W^2} + \frac{m_H^4}{4M_W^4}\right] \lambda^{\frac{1}{2}} \left(1, \frac{M_W^2}{m_H^2}, \frac{M_W^2}{m_H^2}\right). \quad \text{(C.5b)}$$

The A has no tree-level couplings to vector boson pairs, but a coupling can be induced at the one-loop level. The h is too light to decay to electroweak vector boson pairs. Note, however, that the branching fractions for the three-body decays of h or H to WW^* or ZZ^* may be large, since these have only to compete with two-body decays mediated by bottom Yukawa couplings; formulae for these partial widths in the SM are given by Keung and Marciano.[1] It is simple to modify these by inserting the appropriate factor that arises in the hVV or HVV ($V = W, Z$) coupling in the MSSM.

---

[1] W. Y. Keung and W. Marciano, *Phys. Rev.* **D30**, 248 (1984).

In the MSSM, charged Higgs bosons cannot decay via $H^\pm \to W^\pm Z^0$ at the tree level.

C.3 Decays to sfermions

The partial widths for *scalar* neutral Higgs boson decays to a pair of squarks or sleptons are given by,

$$\Gamma(h,\ H \to \tilde{f}_i \bar{\tilde{f}}_j) = \frac{|\mathcal{A}^{h,H}_{\tilde{f}_i \bar{\tilde{f}}_j}|^2}{16\pi m_{h,H}} N_c(f) \lambda^{\frac{1}{2}}\left(1, \frac{m^2_{\tilde{f}_1}}{m^2_{h,H}}, \frac{m^2_{\tilde{f}_2}}{m^2_{h,H}}\right) \qquad \text{(C.6)}$$

where $i, j = 1, 2$ and $N_c(f) = 3\ (1)$ for squarks (sleptons). The relevant couplings are given by

$$\mathcal{A}^{h,H}_{\tilde{f}_1 \bar{\tilde{f}}_1} = \mathcal{A}_{\tilde{f}_L \bar{\tilde{f}}_L} \cos^2\theta_f + \mathcal{A}_{\tilde{f}_R \bar{\tilde{f}}_R} \sin^2\theta_f - 2\mathcal{A}_{\tilde{f}_L \bar{\tilde{f}}_R} \cos\theta_f \sin\theta_f, \qquad \text{(C.7a)}$$

$$\mathcal{A}^{h,H}_{\tilde{f}_2 \bar{\tilde{f}}_2} = \mathcal{A}_{\tilde{f}_L \bar{\tilde{f}}_L} \sin^2\theta_f + \mathcal{A}_{\tilde{f}_R \bar{\tilde{f}}_R} \cos^2\theta_f + 2\mathcal{A}_{\tilde{f}_L \bar{\tilde{f}}_R} \cos\theta_f \sin\theta_f, \qquad \text{(C.7b)}$$

$$\mathcal{A}^{h,H}_{\tilde{f}_1 \bar{\tilde{f}}_2} = \mathcal{A}_{\tilde{f}_L \bar{\tilde{f}}_L} \cos\theta_f \sin\theta_f - \mathcal{A}_{\tilde{f}_R \bar{\tilde{f}}_R} \cos\theta_f \sin\theta_f + \mathcal{A}_{\tilde{f}_L \bar{\tilde{f}}_R} \cos 2\theta_f, \qquad \text{(C.7c)}$$

and $\mathcal{A}^{h,H}_{\tilde{f}_2 \bar{\tilde{f}}_1} = \mathcal{A}^{h,H}_{\tilde{f}_1 \bar{\tilde{f}}_2}$.

The couplings $\mathcal{A}^{h,H}_{\tilde{q}_L \bar{\tilde{q}}_L}$, $\mathcal{A}^{h,H}_{\tilde{q}_R \bar{\tilde{q}}_R}$ and $\mathcal{A}^{h,H}_{\tilde{q}_L \bar{\tilde{q}}_R}$ are given by,

$$\mathcal{A}^h_{\tilde{u}_L \bar{\tilde{u}}_L} = g\left[M_W(\frac{1}{2} - \frac{1}{6}\tan^2\theta_W)\sin(\beta - \alpha) - \frac{m^2_u \cos\alpha}{M_W \sin\beta}\right], \qquad \text{(C.8a)}$$

$$\mathcal{A}^h_{\tilde{d}_L \bar{\tilde{d}}_L} = g\left[M_W(-\frac{1}{2} - \frac{1}{6}\tan^2\theta_W)\sin(\beta - \alpha) - \frac{m^2_d \sin\alpha}{M_W \cos\beta}\right], \qquad \text{(C.8b)}$$

$$\mathcal{A}^h_{\tilde{u}_R \bar{\tilde{u}}_R} = g\left[\frac{2}{3}M_W \tan^2\theta_W \sin(\beta - \alpha) - \frac{m^2_u \cos\alpha}{M_W \sin\beta}\right], \qquad \text{(C.8c)}$$

$$\mathcal{A}^h_{\tilde{d}_R \bar{\tilde{d}}_R} = g\left[-\frac{1}{3}M_W \tan^2\theta_W \sin(\beta - \alpha) - \frac{m^2_d \sin\alpha}{M_W \cos\beta}\right], \qquad \text{(C.8d)}$$

and

$$\mathcal{A}^H_{\tilde{u}_L \bar{\tilde{u}}_L} = g\left[-M_W(\frac{1}{2} - \frac{1}{6}\tan^2\theta_W)\cos(\beta - \alpha) + \frac{m^2_u \sin\alpha}{M_W \sin\beta}\right], \qquad \text{(C.9a)}$$

$$\mathcal{A}^H_{\tilde{d}_L \bar{\tilde{d}}_L} = g\left[M_W(\frac{1}{2} + \frac{1}{6}\tan^2\theta_W)\cos(\beta - \alpha) - \frac{m^2_d \cos\alpha}{M_W \cos\beta}\right], \qquad \text{(C.9b)}$$

$$\mathcal{A}^H_{\tilde{u}_R \bar{\tilde{u}}_R} = g\left[-\frac{2}{3}M_W \tan^2\theta_W \cos(\beta - \alpha) + \frac{m^2_u \sin\alpha}{M_W \sin\beta}\right], \qquad \text{(C.9c)}$$

$$\mathcal{A}^H_{\tilde{d}_R \bar{\tilde{d}}_R} = g\left[\frac{1}{3}M_W \tan^2\theta_W \cos(\beta - \alpha) - \frac{m^2_d \cos\alpha}{M_W \cos\beta}\right]. \qquad \text{(C.9d)}$$

Furthermore,

$$A^h_{\tilde{u}_L \bar{\tilde{u}}_R} = \frac{g m_u}{2 M_W \sin \beta}(-\mu \sin \alpha + A_u \cos \alpha), \tag{C.10a}$$

$$A^h_{\tilde{d}_L \bar{\tilde{d}}_R} = \frac{g m_d}{2 M_W \cos \beta}(-\mu \cos \alpha + A_d \sin \alpha), \tag{C.10b}$$

and

$$A^H_{\tilde{u}_L \bar{\tilde{u}}_R} = \frac{g m_u}{2 M_W \cos \beta}(-\mu \cos \alpha - A_u \sin \alpha), \tag{C.11a}$$

$$A^H_{\tilde{d}_L \bar{\tilde{d}}_R} = \frac{g m_d}{2 M_W \cos \beta}(\mu \sin \alpha + A_d \cos \alpha). \tag{C.11b}$$

The pseudoscalar A cannot decay into $\tilde{f}_i \bar{\tilde{f}}_i$ pairs because of CP conservation. It may, however, decay into unlike sfermion–antisfermion pairs with a width,

$$\Gamma(A \to \tilde{f}_1 \bar{\tilde{f}}_2) = \Gamma(A \to \tilde{f}_2 \bar{\tilde{f}}_1) = \frac{|A^A_{\tilde{f}_L \bar{\tilde{f}}_R}|^2}{16 \pi m_A} N_c(f) \lambda^{\frac{1}{2}} \left(1, \frac{m^2_{\tilde{f}_1}}{m^2_A}, \frac{m^2_{\tilde{f}_2}}{m^2_A} \right), \tag{C.12}$$

where the relevant couplings for decays to squarks are given by,

$$A^A_{\tilde{u}_L \bar{\tilde{u}}_R} = \frac{g m_u}{2 M_W}(\mu + A_u \cot \beta), \tag{C.13a}$$

$$A^A_{\tilde{d}_L \bar{\tilde{d}}_R} = \frac{g m_d}{2 M_W}(\mu + A_d \tan \beta). \tag{C.13b}$$

Notice that this decay rate does not depend on the sfermion mixing angle.

The couplings for decays of h, H or A decays to sleptons can be obtained from those for their decays to squarks via the substitutions listed below Eq. (8.125d).

The partial width for the decay of the charged Higgs boson to squarks is given by,

$$\Gamma(H^+ \to \tilde{q}_i \bar{\tilde{q}}'_j) = \Gamma(H^- \to \tilde{q}'_j \bar{\tilde{q}}_i) = \frac{C^2_{\tilde{q}_i \bar{\tilde{q}}'_j}}{16 \pi m_{H^+}} N_c \lambda^{\frac{1}{2}} \left(1, \frac{m^2_{\tilde{q}_i}}{m^2_{H^+}}, \frac{m^2_{\tilde{q}'_j}}{m^2_{H^+}} \right), \tag{C.14}$$

where

$$C_{\tilde{u}_1 \tilde{d}_1} = C_{\tilde{u}_L \tilde{d}_L} \cos \theta_u \cos \theta_d + C_{\tilde{u}_R \tilde{d}_R} \sin \theta_u \sin \theta_d$$
$$- C_{\tilde{u}_L \tilde{d}_R} \cos \theta_u \sin \theta_d - C_{\tilde{u}_R \tilde{d}_L} \sin \theta_u \cos \theta_d, \tag{C.15a}$$

$$C_{\tilde{u}_2 \tilde{d}_2} = C_{\tilde{u}_L \tilde{d}_L} \sin \theta_u \sin \theta_d + C_{\tilde{u}_R \tilde{d}_R} \cos \theta_u \cos \theta_d$$
$$+ C_{\tilde{u}_L \tilde{d}_R} \sin \theta_u \cos \theta_d + C_{\tilde{u}_R \tilde{d}_L} \cos \theta_u \sin \theta_d, \tag{C.15b}$$

$$C_{\tilde{u}_1 \tilde{d}_2} = C_{\tilde{u}_L \tilde{d}_L} \cos \theta_u \sin \theta_d - C_{\tilde{u}_R \tilde{d}_R} \sin \theta_u \cos \theta_d$$
$$+ C_{\tilde{u}_L \tilde{d}_R} \cos \theta_u \cos \theta_d - C_{\tilde{u}_R \tilde{d}_L} \sin \theta_u \sin \theta_d, \tag{C.15c}$$

$$C_{\tilde{u}_2 \bar{d}_1} = C_{\tilde{u}_L \bar{d}_L} \sin \theta_u \cos \theta_d - C_{\tilde{u}_R \bar{d}_R} \cos \theta_u \sin \theta_d$$
$$- C_{\tilde{u}_L \bar{d}_R} \sin \theta_u \sin \theta_d + C_{\tilde{u}_R \bar{d}_L} \cos \theta_u \cos \theta_d, \tag{C.15d}$$

with

$$C_{\tilde{u}_L \bar{d}_L} = \frac{g}{\sqrt{2}} \left[-M_W \sin 2\beta + \frac{m_d^2 \tan \beta + m_u^2 \cot \beta}{M_W} \right], \tag{C.16a}$$

$$C_{\tilde{u}_R \bar{d}_R} = \left[\frac{g m_u m_d (\cot \beta + \tan \beta)}{\sqrt{2} M_W} \right], \tag{C.16b}$$

$$C_{\tilde{u}_L \bar{d}_R} = \left[\frac{-g m_d}{\sqrt{2} M_W} (A_d \tan \beta + \mu) \right], \tag{C.16c}$$

$$C_{\tilde{u}_R \bar{d}_L} = \left[\frac{-g m_u}{\sqrt{2} M_W} (A_u \cot \beta + \mu) \right]. \tag{C.16d}$$

For $H^+ \to \tilde{\nu}_L \bar{\ell}_{1,2}$ decay, replace $C_{\tilde{q}_i \bar{q}'_j} \to C_{\tilde{\nu}_L \bar{\ell}_{1,2}}, m_{\tilde{q}_i} \to m_{\tilde{\nu}_L}, m_{\tilde{q}_j} \to m_{\tilde{\ell}_{1,2}}, N_c = 1$ and use

$$C_{\tilde{\nu}_L \bar{\ell}_1} = C_{\tilde{\nu}_L \bar{\ell}_L} \cos \theta_\ell - C_{\tilde{\nu}_L \bar{\ell}_R} \sin \theta_\ell, \tag{C.17a}$$

$$C_{\tilde{\nu}_L \bar{\ell}_2} = C_{\tilde{\nu}_L \bar{\ell}_L} \sin \theta_\ell + C_{\tilde{\nu}_L \bar{\ell}_R} \cos \theta_\ell, \tag{C.17b}$$

with

$$C_{\tilde{\nu}_L \bar{\ell}_L} = \frac{g}{\sqrt{2}} \left[-M_W \sin 2\beta + \frac{m_\ell^2 \tan \beta}{M_W} \right], \quad \text{and} \tag{C.18a}$$

$$C_{\tilde{\nu}_L \bar{\ell}_R} = \left[\frac{-g m_\ell}{\sqrt{2} M_W} (A_\ell \tan \beta + \mu) \right]. \tag{C.18b}$$

C.4 Decays to charginos and neutralinos

The partial width for the decays of neutral Higgs bosons, $\phi = h, H$ or A, to chargino pairs is given by,

$$\Gamma(\phi \to \widetilde{W}_i^+ \widetilde{W}_i^-) = \frac{g^2}{4\pi} |S_i^\phi|^2 m_\phi \left(1 - 4 \frac{m_{\widetilde{W}_i}^2}{m_\phi^2} \right)^{\delta_\phi}, \tag{C.19}$$

where $\delta_{h,H} = 3/2$ and $\delta_A = 1/2$, and

$$\Gamma(\phi \to \widetilde{W}_1^+ \widetilde{W}_2^-) = \lambda^{\frac{1}{2}} \left(1, \frac{m_{\widetilde{W}_1}^2}{m_\phi^2}, \frac{m_{\widetilde{W}_2}^2}{m_\phi^2} \right) \frac{g^2}{16\pi m_\phi} \left\{ |S^\phi|^2 \left[m_\phi^2 - (m_{\widetilde{W}_2} + m_{\widetilde{W}_1})^2 \right] \right.$$
$$\left. + |P^\phi|^2 \left[m_\phi^2 - (m_{\widetilde{W}_2} - m_{\widetilde{W}_1})^2 \right] \right\}, \tag{C.20}$$

with $\Gamma(h, H, A \to \widetilde{W}_1^+ \widetilde{W}_2^-) = \Gamma(h, H, A \to \widetilde{W}_2^+ \widetilde{W}_1^-)$ by CP invariance. The various couplings $S_i^\phi, P_i^\phi, S^\phi$, and P^ϕ have been listed in Eq. (8.116a)–(8.116c) and in the accompanying discussion for $\phi = h, H$, and in (8.119a)–(8.119c) for $\phi = A$.

The neutral Higgs bosons can also decay into neutralino pairs with partial widths given by,

$$\Gamma(\phi \to \widetilde{Z}_i \widetilde{Z}_j) = \frac{\Delta_{ij}}{8\pi m_\phi} \left(X_{ij}^\phi + X_{ji}^\phi \right)^2 \left[m_\phi^2 - (m_{\widetilde{Z}_i} + (-1)^{\theta_i + \theta_j + \theta_\phi} m_{\widetilde{Z}_j})^2 \right]$$

$$\times \lambda^{\frac{1}{2}} \left(1, \frac{m_{\widetilde{Z}_i}^2}{m_\phi^2}, \frac{m_{\widetilde{Z}_j}^2}{m_\phi^2} \right), \tag{C.21}$$

where $\theta_\phi = 0$ if $\phi = h, H$ and $\theta_\phi = 1$ if $\phi = A$, $\Delta_{ij} = \frac{1}{2} (1)$ for $i = j$ $(i \neq j)$, and where $X_{ij}^{h,H,A}$ are given in Eq. (8.117) and (8.120).

Finally, the partial width for a charged Higgs boson to decay into a chargino and a neutralino is given by

$$\Gamma(H^+ \to \widetilde{W}_i^+ \widetilde{Z}_j) = \Gamma(H^- \to \widetilde{W}_i^- \widetilde{Z}_j) = \frac{1}{8\pi m_{H^+}} \lambda^{\frac{1}{2}} \left(1, \frac{m_{\widetilde{W}_i}^2}{m_{H^+}^2}, \frac{m_{\widetilde{Z}_j}^2}{m_{H^+}^2} \right)$$

$$\times \left[(R_{ij}^2 + S_{ij}^2)(m_{H^+}^2 - m_{\widetilde{W}_i}^2 - m_{\widetilde{Z}_j}^2) - 2(R_{ij}^2 - S_{ij}^2) m_{\widetilde{W}_i} m_{\widetilde{Z}_j} \right], \tag{C.22}$$

where

$$R_{1j} = \frac{1}{2} \left[(-1)^{\theta_{\widetilde{w}_1}} A_2^j \cos\beta - (-1)^{\theta_j} A_4^j \sin\beta \right], \tag{C.23a}$$

$$R_{2j} = \frac{1}{2} \left[(-1)^{\theta_{\widetilde{w}_2}} \theta_y A_1^j \cos\beta - (-1)^{\theta_j} \theta_x A_3^j \sin\beta \right], \tag{C.23b}$$

and

$$S_{1j} = \frac{1}{2} \left[(-1)^{\theta_{\widetilde{w}_1}} A_2^j \cos\beta + (-1)^{\theta_j} A_4^j \sin\beta \right], \tag{C.24a}$$

$$S_{2j} = \frac{1}{2} \left[(-1)^{\theta_{\widetilde{w}_2}} \theta_y A_1^j \cos\beta + (-1)^{\theta_j} \theta_x A_3^j \sin\beta \right], \tag{C.24b}$$

with $A_1^j - A_4^j$ as given in (8.122a)–(8.122d).

C.5 Decays to Higgs bosons

Finally, we list Higgs boson decay widths to other Higgs bosons, including decays to Higgs boson–gauge boson final states:

$$\Gamma(H \to hh) = \frac{\xi_{Hhh}^2}{8\pi m_H} \left(1 - \frac{4m_h^2}{m_H^2} \right)^{\frac{1}{2}}, \tag{C.25a}$$

where

$$\xi_{Hhh} = \frac{gM_Z}{4\cos\theta_W} [\cos 2\alpha \cos(\beta - \alpha) + 2\sin 2\alpha \sin(\beta - \alpha)] ; \qquad \text{(C.25b)}$$

$$\Gamma(H \to AA) = \frac{\xi_{HAA}^2}{8\pi m_H} \left(1 - \frac{4m_A^2}{m_H^2}\right)^{\frac{1}{2}} , \qquad \text{(C.26a)}$$

where

$$\xi_{HAA} = -\frac{gM_Z}{4\cos\theta_W} \cos(\beta - \alpha) \cos 2\beta ; \qquad \text{(C.26b)}$$

$$\Gamma(H \to H^+H^-) = \frac{\xi_{H+-}^2}{16\pi m_H} \left(1 - \frac{4m_{H^+}^2}{m_H^2}\right)^{\frac{1}{2}} , \qquad \text{(C.27a)}$$

where

$$\xi_{H+-} = gM_W \left[\cos(\beta + \alpha) - \frac{\cos(\beta - \alpha)\cos 2\beta}{2\cos^2\theta_W}\right] ; \qquad \text{(C.27b)}$$

$$\Gamma(h \to AA) = \frac{\xi_{hAA}^2}{8\pi m_h} \left(1 - \frac{4m_A^2}{m_h^2}\right)^{\frac{1}{2}} , \qquad \text{(C.28a)}$$

where

$$\xi_{hAA} = \left(\frac{gM_Z}{4\cos\theta_W}\right) \sin(\beta - \alpha) \cos 2\beta . \qquad \text{(C.28b)}$$

Higgs bosons may also decay into gauge bosons and a lighter Higgs boson. The partial widths for these decays are given by,

$$\Gamma(H \to Z^0 A) = \frac{(g\cos\theta_W + g'\sin\theta_W)^2 \sin^2(\alpha + \beta)m_H^3}{64\pi m_Z^2} \lambda^{\frac{3}{2}}\left(1, \frac{m_A^2}{m_H^2}, \frac{M_Z^2}{m_H^2}\right),$$

$$\text{(C.29a)}$$

$$\Gamma(A \to Z^0 h) = \frac{(g\cos\theta_W + g'\sin\theta_W)^2 \cos^2(\alpha + \beta)m_A^3}{64\pi m_Z^2} \lambda^{\frac{3}{2}}\left(1, \frac{m_h^2}{m_A^2}, \frac{M_Z^2}{m_A^2}\right),$$

$$\text{(C.29b)}$$

and

$$\Gamma(H^\pm \to W^\pm h) = \frac{g^2 \cos^2(\alpha + \beta)m_{H^+}^3}{64\pi M_W^2} \lambda^{\frac{3}{2}}\left(1, \frac{M_W^2}{m_{H^+}^2}, \frac{m_h^2}{m_{H^+}^2}\right) . \qquad \text{(C.29c)}$$

The decays $H^\pm \to W^\pm A$ and $W^\pm H$ are kinematically forbidden in the MSSM, assuming tree-level formulae for their masses.

References

1. L. Okun, *Leptons and Quarks*, North-Holland (1982). A conceptual introduction to the Standard Model of electroweak interactions and its phenomenology.
2. C. Quigg, *Gauge Theories for the Strong, Weak and Electromagnetic Interactions*, Addison-Wesley (1986). An exposition of the Standard Model and phenomenology.
3. V. Barger and R. J. N. Phillips, *Collider Physics*, Addison-Wesley (1987). QCD and electroweak phenomenology for collider experiments.
4. F. Halzen and A. D. Martin, *Quarks and Leptons*, J. Wiley and Sons, Inc. (1984). An introduction to the Standard Model and its phenomenology.
5. R. K. Ellis, W. J. Stirling and B. R. Webber, *QCD and Collider Physics*, Cambridge University Press (1996). SM collider physics phenomenology with an emphasis on QCD.
6. J. F. Gunion, H. E. Haber, G. Kane and S. Dawson, *The Higgs Hunter's Guide*, Perseus Press (1990). An overview of phenomenology associated with SM and SUSY Higgs bosons.
7. M. Peskin and D. V. Schroeder, *Introduction to Quantum Field Theory*, Perseus Press (1995). A modern text on quantum field theory and the Standard Model.
8. R. N. Mohapatra, *Unification and Supersymmetry*, Springer-Verlag (1992). An introduction to grand unification, supersymmetry and supergravity.
9. G. G. Ross, *Grand Unified Theories*, Frontiers in Physics series, Benjamin/Cummings (1985). An introduction to grand unification and supersymmetry in particle physics.
10. E. Kolb and M. Turner, *The Early Universe*, Frontiers in Physics series, Addison-Wesley (1990). An account of Big Bang cosmology and its connection to the modern ideas of particle physics.
11. P. J. E. Peebles, *Principles of Physical Cosmology*, Princeton University Press (1993). An introduction to modern cosmology from the perspective of one of the pioneers of the subject.
12. S. Dodelson, *Modern Cosmology*, Academic Press (2003). An account of cosmology, cold dark matter and the cosmic microwave background from a more modern perspective.
13. G. L. Kane and M. A. Shifman, editors, *The Supersymmetric World: The Beginnings of the Theory*, World Scientific (2001). An historical account of the early development of supersymmetry.
14. P. D. B. Collins, A. D. Martin and E. J. Squires, *Particle Physics and Cosmology*, Wiley (1989). Introductory chapters on many advanced topics, including supersymmetry, superstrings and cosmology.

15. P. Ramond, *Journeys Beyond the Standard Model*, Perseus Press (1999). Physics beyond the Standard Model, including massive neutrinos, axions and supersymmetry.

16. D. Bailin and A. Love, *Supersymmetric Gauge Field Theory and String Theory*, Institute of Physics Publishing (1996). A concise overview of supersymmetry, supergravity and string theory.

17. G. L. Kane, editor, *Perspectives on Supersymmetry*, World Scientific (1998). Chapters by experts on diverse topics in supersymmetry theory and phenomenology, including a nice introductory chapter by S. Martin.

18. M. Drees, R. Godbole and P. Roy, *Theory and Phenomenology of Sparticles*, World Scientific (2004). Theory and phenomenology of supersymmetry, similar in scope to this text.

19. S. Weinberg, *The Quantum Theory of Fields: Vol. III, Supersymmetry*, Cambridge University Press (2000). Final volume of a three volume set addressing supersymmetry and supergravity by one of the twentieth century's great physicists.

20. H. J. W. Müller-Kirsten and A. Wiedemann, *Supersymmetry: An Introduction with Conceptual and Calculational Details*, World Scientific (1987). Out-of-print, hard to find, but very detailed development of the representations of super-Poincaré algebra and Lagrangians for gauge theories with global supersymmetry.

21. J. Wess and J. Bagger, *Supersymmetry and Supergravity*, Princeton (1992). A concise exposition of supersymmetry and supergravity.

22. P. Nath, R. Arnowitt and A. Chamseddine, *Applied N=1 Supergravity*, ICTP Series in Theoretical Physics, V1, World Scientific (1984). An early introduction to the theory and phenomenology of supergravity grand unified theories by pioneers of the subject.

23. N. Polonsky, *Supersymmetry: Structure and Phenomena: Extensions of the Standard Model*, Lecture Notes in Physics, Monograph M68, Springer-Verlag (2001). An informal introduction to supersymmetry and supersymmetry phenomenology.

24. P. West, *Introduction to Supersymmetry and Supergravity*, World Scientific (1989). A somewhat formal overview of supersymmetry, supergravity and some material on superstrings.

25. P. P. Srivastava, *Supersymmetry, Superfields and Supergravity: an Introduction*, Adam Hilger (1986). An exposition of superfield formalism with a short chapter on supergravity.

Index

$B_s \rightarrow \mu^+\mu^-$ decay, 217–220
D-term, 69
F-term, 69
$K_L - K_S$ mass difference, 191
$\gamma\gamma$ collider, 452
γ_5-dependent fermion mass matrices, 124
$\mu \rightarrow e\gamma$, 194
$\mu \rightarrow e\gamma$, 103
μ problem, 269
$\tan\beta$ determination, 447
θ-identities, 56
  bilinear, 57
  quartic, 58
  trilinear, 57
$b \rightarrow s\gamma$ decay, 103, 214–217
$e\gamma$ collider, 452

aesthetics, 19
AMSB models, 278–285
  D-term improved AMSB model, 284
  minimal, 280
anomalous dimension matrix, 216
anomalous isotopes, 222
anomalous magnetic moment, 103, 220
anomaly-mediated SUSY breaking, 278–285
  D-term improved AMSB model, 284
  mAMSB model, 280
auxiliary field, 72, 95, 99

Baker–Campbell–Hausdorff formula, 86
baryogenesis constraints on R-violating couplings,
   463
beam polarization, 298, 331–335
beamstrahlung, 335–337
beta function, 199
  MSSM gauge couplings, 201
  MSSM Yukawa couplings, 203
  SM gauge couplings, 200
Big Bang model, 221
Big Bang nucleosynthesis, 223
Boltzmann equation, 223
branes, 278
bremsstrahlung, 335–337

Callan–Symanzik equation, 199
canonical quantization, 32
cascade decays, 379
Casimir operator, 41
CERN LHC, 414
  precision measurements, 427
  reach, 416–422
  reach within mSUGRA, 420
charge conjugation matrix, 24
chargino pair production, 308
chargino–neutralino production, 301
charginos
  couplings to $W^\pm$, 173
  couplings to Z^0, γ, 173
  couplings to matter, 170
  decays, 357–361, 516–523
  mass, 149
chiral superfield, 29, 61–63, 82
co-annihilation, 227
cold dark matter, 21
Coleman–Mandula theorem, 45
compactification scale, 295
connection field, 239
cosmic microwave background, 223
cosmological constant, 223, 243, 258
cosmological implications, 221–231
cosmology, 221
covariant derivative, 65, 100, 239, 244, 245
CP problem, 195–198
CP violation, 195–198
curl superfield, 64, 71

D-term, 69
D-term scalar mass splitting, 275
dark matter, 222, 228
dark matter detection, 222
dimensional reduction, 201
dimensional regularization, 200
Dirac field, 32
Dirac spinor, 44, 128
direct detection
  seasonal modulation, 229
direct test of SUSY, 447

direct WIMP detection, 228
DRED, 201
DREG, 200

E821 experiment, 220
effective mass, 428
Einstein Lagrangian, 242
electroweak model, 6
endpoints, 439
equivalence principle, 236, 240
event generation, 374–393
    beam remnants, 383
    cascade decays, 379
    hadronization, 382
    hard scattering, 377
    parton showers, 377
event generators, 383
extra dimensions, 17, 278, 294

F-term, 69
Fayet–Iliopoulos D-term, 98, 284
FCNC, 191–194
Fermilab Tevatron, 322, 402
Fierz re-arrangement, 30, 57
Fierz transformation, 30
fine tuning, 211
    cosmological constant, 258
fine-tuning parameter, 213
fine-tuning problem, 16, 104
flat direction, 108, 140
flavor-changing neutral currents, 190
flavor problem, 191–194
    alignment solution, 193
    decoupling solution, 194
    degeneracy solution, 193
focus point, 214, 227, 409
Friedmann–Robertson–Walker universe,
    223

gauge-coupling unification, 199–203
gauge invariance, 1
gauge kinetic function, 116, 251
gauge-mediated SUSY breaking, 285–293
    minimal GMSB model, 287
    non-minimal GMSB models, 293
gauge transformation, 80, 84
gauge transformations
    Abelian, 84
    non-Abelian, 86–89
gaugino, 95
gaugino condensation, 116
gaugino-mediated SUSY breaking,
    294
Gauss' theorem, 25
general co-ordinate transformations,
    236
general relativity, 236–245
    field equations, 242
    Lagrangian, 242
    spinors, 243
    tensors, 237

gluino decay, 342–350, 491–501
gluino mass, 148
gluino QCD interaction, 163
GMSB models, 285–293
    determining the SUSY breaking scale, 437,
        450
    LHC searches, 420
    Tevatron searches, 407
goldstino, 111, 115, 189
    composite, 115
    coupling of, 117
    decays to, 368–373, 491, 506, 509,
        511
    interactions, 368–371
Goldstone's theorem, 111
graded Lie algebra, 31, 46
Grassmann numbers, 50
    differentiation, 55
    integration, 76
gravitinos, 189, 246
    as LSPs, 286, 291
    decays to, 291, 368–373, 491, 506, 509,
        511
    interactions, 368
    mass, 259
gravitons, 189, 246
gravity-mediated SUSY breaking,
    264–278
    gaugino mass, 267
    scalar interactions, 266
    scalar mass, 266
GUTs, 270

Haag–Lopuszanski–Sohnius theorem, 47
hadronization
    cluster, 382
    independent, 382
    string, 382
HB/FP region, 409, 419, 424, 426
HERWIG, 383
hidden sector, 264
Higgs boson, SM, 8
    mass limits, 21
Higgs bosons, MSSM, 144–148
    couplings, 174
    decays, 364–367, 524–530
    production, 485
    searches, 411, 421, 423
Higgs mechanism, 4, 141
Hubble parameter, 223
hyperbolic branch, 213, 227, 409

IMH model, 276
initial state radiation in e^+e^- collisions,
    335
inoMSB, 294
inverted mass hierarchy, 276
ISAJET, 383
    event generation, 388
    set-up, 384
    SUSY models in, 385

Jacobi identity, 30, 53, 55

Kähler function, 252
Kähler metric, 252
Kähler potential, 69, 70
  gauge theory, 82
  non-renormalizable theory, 251
  renormalizable theory, 70
Kaluza–Klein, 295
Kobayashi–Maskawa, 9
Kuraev–Fadin distribution, 335

LEP, 398
LEP2, 322
  SUSY searches, 399–402
leptogenesis, 234
lepton flavor violation, 233
Lie algebra, 41, 53, 80
  adjoint representation, 91
  conjugate representation, 97
  representations of, 53
lifetime, 339
lifetimes of sparticles, 338
linear e^+e^- collider, 323, 437–450
Little Higgs models, 18
local gauge transformations, 79
local Lorentz transformations, 243
local supersymmetry, 245–260
  sum rule, 259
  supersymmetry breaking, 257
  transformations, 250, 255
Lorentz group, 42
LSP lifetime in RPV models, 467

Majorana field, 32
  two-point function, 33, 317
Majorana mass
  CP-violating, 134
Majorana spinor identities, 26
Majorana spinors, 24, 44, 128
  bilinears, 26
Majoron, 471
mAMSB model
  LHC searches, 420
mass insertions, 195
mass sum rule, 118, 120
Master Lagrangian
  gauge theories, 99
  non-gauge theories, 75
matter parity, 132
metric tensor, 241
minimal coupling, 3
minimal supergravity model, 207, 269–270
Minimal Supersymmetric Standard Model, *see*
  MSSM 127
motivations for supersymmetry, 19
MSSM, 127–189
  beta functions, 201
  construction, 127
  EWSB, 138–141
  field content, 128

gauge symmetry, 127
  interactions, 161–183
    electroweak, 164
    Higgs bosons, 174–183
    QCD, 161
  masses
    charginos, 149–154
    gauge bosons, 141
    gluinos, 148
    Higgs bosons, 144–148
    matter fermions, 142
    neutralinos, 149–152
    sleptons, 155–160
    squarks, 155–160
  parameter space, 134–138
  radiative corrections, 184–188
  soft SUSY breaking terms, 134
  superpotential, 130
mSUGRA model, 207, 211, 269
mu problem, 269
muon anomalous magnetic moment, 220–221
muon collider, 452

naturalness, 211–214
neutralino relic density, 223–228
neutralinos
  couplings to $W^\pm$, 173
  couplings to Z^0, 173
  couplings to matter, 168, 169
  dark matter, 223
    direct detection, 228
    indirect detection, 230
  decays, 361–364, 509–516
  mass, 149
neutrino mass, 231–234, 471
new physics
  motivations for, 11
next-to-lightest SUSY particle (NLSP), 292
  decay length, 372
NMSSM, 161, 177
Noether procedure, 247
Noether's theorem, 27
non-renormalization theorem, 104
non-universal
  gaugino masses, 273
  scalar masses, 267, 268

O'Raifeartaigh model, 107

Palatini formalism, 242
partial decay rate, 339
parton distribution function, 299, 307
parton model, 299
Planck mass, 243
Planck slop, 265
Poincaré group, 45
Polonyi model, 259
proton decay, 21, 271, 459
PYTHIA, 383

QCD, 6, 161
quadratic divergences
  cancellation of, 36, 104

R-parity, 132
R-parity violation
  bilinear, 470
  collider signatures, 468
  constraints, 459–466
  explicit, 457
  interactions, 457
  LSP decay, 467
  muon decay, 461
  s-channel production, 465
  spontaneous, 470
  trilinear, 457
radiative decays, 214
  supersymmetry limit, 103
radiative EWSB, 21, 209–211
Rarita–Schwinger field, 246
  Lagrangian, 247
reduced Planck mass, 243
relic density, 222, 223–228
renormalization group equations, 199–209
renormalized perturbation theory, 199
resonance annihilation, 228
REWSB, 209–211
RGEs, 199–209
  gauge couplings, 200
  soft parameters, 204–206
  Yukawa couplings, 203
Ricci scalar, 242
Ricci tensor, 242
Riemann curvature tensor, 240, 245
rotations, 41
Runge–Kutta integration, 206

scalar potential, 258
scattering events, 390
see-saw mechanism, 232
sleptons, 129
  decays, 353–357, 506–509
  electroweak gauge couplings, 165
  masses, 155
SO(10) SUSY GUT model, 274
sparticle decays, 338–373
  a sample calculation, 342
sparticle lifetime, 340, 341
sparticle production, 298–337, 476–490
  e^+e^- colliders
    chargino/neutralino, 328–331, 488–490
    polarized beams, 332, 490
    sleptons/sneutrinos/squarks, 325–328, 485–488
    two-photon background, 337
  a sample calculation, 301
  at the LHC, 322
  at the Tevatron, 322
  hadron colliders
    associated production, 319–320, 481
    chargino pair, 308–309, 476
    chargino/neutralino, 306–308

gluino/squark, 314–319, 479–481
    neutralino pair, 310–311, 478
    NLO corrections, 321
    slepton/sneutrino, 312–313, 482
sphaleron, 234, 463
spin connection, 245
spontaneous symmetry breaking, 3–6, 138
spurion, 123
squarks, 129
  decays, 350–353, 501–506
  electroweak gauge couplings, 165
  masses, 155
  QCD interactions, 162
Standard Model, 1, 6–10
SU(2), 41
SU(5) SUSY GUT model, 270
sum rule, 118, 120, 259, 262
super-algebra
  irreducible representations, 47
supercurrent, 27
super-GIM mechanism, 196
super-Higgs mechanism, 258
super-Poincaré group, 19
supercovariant derivative, 66–68
superfields, 49–102
  SUSY transformation, 60
  chiral, 61–63
  curl, 64, 89
    in Wess–Zumino gauge, 91
  expansion of, 51
  gauge, 79–102
  gauge potential, 81, 87
  gauge transformation, 84
  left-chiral scalar, 61
  left-chiral spinor, 82
  products, 64
  real, 52
  right-chiral scalar, 63
  scalar, 51
supergravity, 20, 245–260
  breaking, 257
  Lagrangian, 251
  models, 264–278
  mSUGRA model, 269
supernovae, 223
superpotential, 69, 72
  non-renormalizable theory, 251
  renormalizable theory, 72
superspace, 50, 76
superstring theories, 20
supersymmetric gauge theories
  action, 102
  recipe, 101
supersymmetric Lagrangian
  as integral over superspace, 78, 102
  gauge theory, 98
    gauge kinetic terms, 92
    gauge-matter interactions, 95
  non-gauge theory, 68–76
    D-term contribution, 72
    F-term contribution, 74

supersymmetry
  algebra, 30, 41–48
  collider simulation, 374–393
  defined, 18
  new problems with, 474
  representations of, 47, 60
supersymmetry breaking, 105–126
  F-type, 108
  dynamical, 116
  explicit, 40, 121
  mediation of, 263
  soft, 40, 122, 134
    non-analytic terms, 123
  spontaneous, 106
    D-type, 106, 113
    F-type, 106
supersymmetry generators
  represented as derivatives,
    54
supertrace, 110, 120
SUSY searches, 395–427
  Z pole, 398
  early experiments, 395
  LC
    mass measurements, 439–447
    reach, 423
    spin determination, 438
  LEP2, 399–402
  LHC
    precision measurements, 427–437
    reach, 416–422
  Tevatron
    reach, 407–414
SUSYGEN, 383

technicolor model, 16
top quark, 210
  SUSY decays of, 367, 523
torsion tensor, 239
total decay rate, 339
trilepton signal at the Tevatron, 406
trileptons, 409
two spinors, 44

universality, 196
universality hypothesis, 206

vierbein, 244
VLHC, 453

weak hypercharge, 130
Wess–Zumino gauge, 85, 87, 89
Wess–Zumino model, 23–40, 76, 247
  algebra, 30
  auxiliary fields, 24
  chiral multiplet, 28
  interactions, 35
  Lagrangian, 23
  local supersymmetry, 248
  quadratic divergences, 36
  quantization, 32
  superpotential for, 76
  SUSY breaking, 39
  SUSY transformations, 25
Wilson coefficients, 216
WIMPs, 222, 228
WMAP, 223

Yang–Mills gauge theories, 3

Printed in the United States
By Bookmasters